PLANT INVADERS

'PEOPLE AND PLANTS' CONSERVATION MANUALS

Manual Series Editor:
Martin Walters

Manual Series Originator:
Alan Hamilton
Plants Conservation Officer
WWF International

'People and Plants' is a joint initiative of the World Wide Fund for Nature (WWF), the United Nations Educational, Scientific, and Cultural Organization (UNESCO) and the Royal Botanic Gardens, Kew, UK. Partial funding has been provided by the Darwin Initiative for the Survival of Species (Department of the Environment, UK).

Titles in this series

Ethnobotany: A methods manual
Gary J. Martin

Plant Invaders: The threat to natural ecosystems
Quentin C.B. Cronk and Janice L. Fuller

People and Wild Plant Use
Anthony B. Cunningham

Botanical Surveys for Conservation and Land Management
Peggy Stern and Peter Ashton

Botanical Databases for Conservation and Development

PLANT INVADERS

The threat to natural ecosystems

Quentin C.B. Cronk and Janice L. Fuller

WWF International
(World Wide Fund for Nature)

UNESCO
(United Nations Educational, Scientific, and Cultural Organization)

Royal Botanic Gardens, Kew, UK

CHAPMAN & HALL

London · Weinheim · NewYork · Tokyo · Melbourne · Madras

Published by Chapman & Hall, 2–6 Boundary Row, London SE1 8HN, UK

Chapman & Hall, 2–6 Boundary Row, London SE1 8HN, UK

Chapman & Hall GmbH, Pappelallee 3, 69469 Weinheim, Germany

Chapman & Hall USA, One Penn Plaza, 41st Floor, New York NY 10119, USA

Chapman & Hall Japan, ITP-Japan, Kyowa Building 3F, 2-2-1 Hirakawacho, Chiyoda-ku, Tokyo 102, Japan

Chapman & Hall Australia, Thomas Nelson Australia, 102 Dodds Street, South Melbourne, Victoria 3205, Australia

Chapman & Hall India, R. Seshadri, 32 Second Main Road, CIT East, Madras 600 035, India

First edition 1995
Reprinted 1996

© 1995 World Wide Fund for Nature (WWF International)

Typeset in Goudy 10/13 by ROM-Data Corporation Ltd., Falmouth, Cornwall
Printed in Great Britain at the University Press, Cambridge

ISBN 0 412 48380 7

A catalogue record for this book is available from the British Library

Library of Congress Catalog Card Number: 94-69375

∞ Printed on permanent acid-free text paper, manufactured in accordance with ANSI/NISO Z39.48-1992 and ANSI/NISO Z39.48-1984 (Permanence of Paper).

WWF

The World Wide Fund for Nature (WWF), founded in 1961, is the world's largest private nature conservation organization. It consists of 29 national organizations and associates, and works in more than 100 countries. The coordinating headquarters are in Gland, Switzerland. The WWF mission is to conserve biodiversity, to ensure that the use of renewable natural resources is sustainable and to promote actions to reduce pollution and wasteful consumption.

UNESCO

The United Nations Educational, Scientific, and Cultural Organization is the only UN agency with a mandate spanning the fields of science (including social sciences), education, culture and communication. UNESCO has over 40 years of experience in testing interdisciplinary approaches to solving environment and development problems, in programs such as that on Man and the Biosphere (MAB). An international network of biosphere reserves provides sites for conservation of biological diversity, long-term ecological research, and testing and demonstrating approaches to the sustainable use of natural resources.

The Royal Botanic Gardens, Kew

The Royal Botanic Gardens, Kew has 150 professional staff and associated researchers and works with partners in over 42 countries. Research focuses on taxonomy, preparation of floras, economic botany, plant biochemistry and many other specialized fields. The Royal Botanic Gardens has one of the largest herbaria in the world and an excellent botanic library.

Darwin Initiative for the Survival of Species

At the Earth Summit in June 1992, the Prime Minister of the United Kingdom announced the Darwin Initiative as a demonstration of the UK's commitment to the aims of the Biodiversity Convention. The Initiative will build on Britain's scientific, educational and commercial strengths in the field of biodiversity to assist in the conservation of the world's biodiversity and natural habitats, particularly in those countries rich in biodiversity but poor in resources.

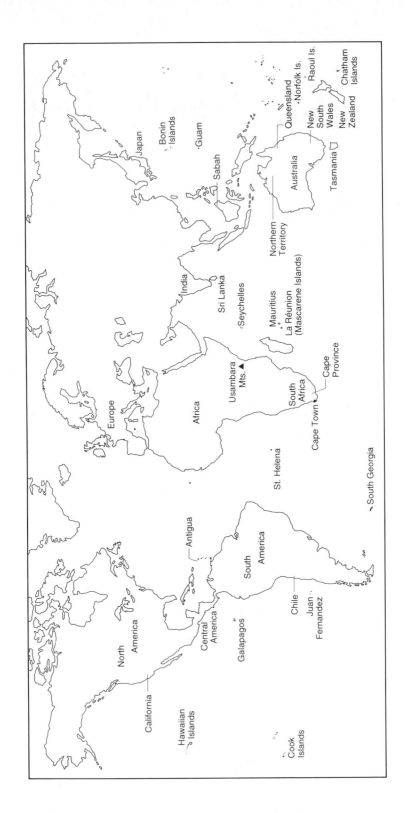

Map 1 The world, showing some of the places and regions mentioned in this book.

Contents

The 'People and Plants' Initiative

This manual is one of a series, forming a contribution to the People and Plants Initiative of the World Wide Fund for Nature (WWF), the United Nation Educational, Scientific, and Cultural Organization (UNESCO) and the Royal Botanic Gardens, Kew (UK). The Initiative has received financial support from the Darwin Initiative for the Survival of Species (Department of the Environment, UK) and the Tropical Forestry Program (US Department of Agriculture). The main objective of the People and Plants Initiative is to build up the capacity for work with local communities on botanical aspects of conservation, especially in countries with tropical forests. The principal intention in the manual series is to provide information which will assist botanists and others to undertake such practical conservation work. Other components of the People and Plants Initiative include demonstration projects in Bolivia, Kenya, Mozambique, Malaysia, Mexico, and Uganda, as well as support for workshops and publication of working papers.

Our aim in producing the present manual is to draw attention to the growing problem of invasive plant species – plants introduced by people to new areas in which they then have become established and spread into natural or seminatural ecosystems. Invasive plants are already a very serious threat to the conservation of biodiversity in many parts of the world, for example on some oceanic islands, in some areas of Mediterranean-type ecosystems, such as the fynbos of South Africa, and in some wetlands. Certain types of ecosystem, like continental tropical forest, are commonly regarded as typically resistant to plant invaders, but this is not true of all tropical forests and the present-day low levels of plant invaders in most tropical forests may, at least in part, be due to history as much as ecology.

We know little of the factors which lead to biological invasions, making it difficult to predict which species will become invasive and where invasion will occur. It does however seem certain that the threat of invasions is growing, as more and more species of plants are moved around the world, planted in gardens or used in agriculture or forestry. Often very little thought is given to the risks of plant invasion, for example by those promoting widespread use of fast-growing legum-inous trees in agroforestry schemes in the tropics.

A number of groups of people can contribute to the control of invasive plants, including legislators and customs agents (to prevent initial introductions), scientists (for instance, to devise much needed effective and environmentally friendly methods of control) and land managers (to put control measures into effect). The subject of invasive plants is not so obviously central to the basic theme of the People and Plants Initiative as some of the other titles in the manual series, but nevertheless there certainly are occasions when local communities play a role in the control of invasive plants, as with clearance of invasive *Rhododendron* in woodlands and *Hippophae* on sand-dunes in Britain and Ireland and clearance of *Leucaena* and *Flacourtia* by local conservation groups on Ile aux Aigrettes, Mauritius.

Alan Hamilton
Plants Conservation Officer
WWF International

Panel of advisers

Pierre Binggeli
c/o School of Biological and Chemical Science
University of Ulster
Coleraine
Co. Londonderry BT52 1SA
United Kingdom

David Given
PO Box 84
Lincoln University
Canterbury
New Zealand

Vernon Heywood
School of Plant Sciences
University of Reading
Reading RG6 2AS
United Kingdom

Ian Macdonald
Southern African Nature Foundation
456 Stellenbosch 7600
Republic of South Africa

Clifford Smith
Botany Department
University of Hawaii at Manoa
3190 Maile Way
Honolulu HI 06822
United States of America

Charles Stirton
Deputy Director
Royal Botanic Gardens
Kew, Richmond
Surrey TW9 3AB
United Kingdom

Preface

Cases could be given of introduced plants which have become common throughout whole islands in a period of less than ten years. Several of the plants now most numerous over the wide plains of La Plata, clothing square leagues of surface almost to the exclusion of all other plants, have been introduced from Europe; and there are plants which now range in India, as I hear from Dr Falconer, from Cape Comorin to the Himalaya, which have been imported from America since its discovery. In such cases, and endless instances could be given, no one supposes that the fertility of these animals and plants has been suddenly and temporarily increased in any sensible degree. The obvious explanation is that the conditions of life have been very favourable, and that there has consequently been less destruction of the old and young, and that nearly all the young have been enabled to breed. In such cases the geometrical ratio of increase, the result of which never fails to be surprising, simply explains the extraordinarily rapid increase and wide diffusion of naturalised productions in their new homes.

The Origin of Species: Charles Darwin, 1859

This book is intended as a review of the problem of invasive plants worldwide. Invasion by animals or other organisms is of similar importance, but this study is confined to plant invasions. Its aim is to address the following needs:

1. To increase public awareness of the threat of invasive plants to natural and seminatural ecosystems. Increased awareness will help prevent the translocation of invasive species to new areas. Numerous instances of invasion are given to indicate the nature of the problem.

2. To alert research and governmental organizations to the threat to biodiversity posed by invasive plants, in order to focus and strengthen research and management strategies.

3. To give practical information and advice to managers, conservationists and researchers on how to deal with invasive plants. The use of case-studies of important invasive species is designed to illustrate the wide variety of invasions and control methods, and to make available the results of studies in one region

to managers working in another. The book discusses the importance of recording, evaluation of invasive potential and development of appropriate control strategies.

This work approaches the subject in the context of conservation and concentrates on plant invasion as a threat to wild biodiversity. The definition of invasive plant used here (see section 1.1) reflects this conservation bias and deliberately avoids discussion of agricultural weeds. Our intention is to attempt to change current thinking on plant invaders: from obscure problem to major hazard to biodiversity. The book is intended to be as free from jargon as possible to reach a wide constituency, including amateur naturalists, land managers and decision-makers at all levels, for instance those who can tackle issues such as legislation and prevention. The case-studies of specific invaders are intended to give evidence of the scale and diversity of invasion.

Special mention must be made here of the important contribution of the SCOPE Program. In 1982, SCOPE (Scientific Committee on Problems of the Environment), a subsidiary body of the International Council of Scientific Unions, initiated a project on the 'Ecology of Biological Invasions'. This project has resulted in a number of symposia and associated publications [145, 192, 213, 252, 285]. The main questions that SCOPE has attempted to answer are [6]:

1. What factors determine whether a species will become an invader or not?
2. What site properties determine whether a species will become an invader or not?
3. How should management systems be developed to best advantage, given the knowledge gained from attempting to answer questions 1 and 2?

In studying biological invasions SCOPE has examined a variety of organisms including plants, mammals, microorganisms, insects and fish. There seems to be no consistent use of the term 'invasive plant' in the SCOPE program. Some authors deal with species invading agricultural habitats and threatening crop production, others with adventives and comparatively few consider those that invade natural or seminatural habitats. Many of the authors lump all these categories under the term 'invasive', and this should be borne in mind in interpreting the ideas and conclusions generated by the SCOPE program. Many of the ideas emerging from the SCOPE meetings are discussed in Chapters 1 and 2.

Quentin C.B. Cronk
Janice L. Fuller

Acknowledgements

We wish to acknowledge the many people who generously gave their help in various stages of the compilation of data for this book, and provided valuable information or comments relating to invasive plants, especially the following: P. Adam, P. Binggeli, T. Goodland, A. Hamilton, C.E. Hughes, W. Joenje, D.J. MacQueen, C.H. Stirton, B.T. Styles, C.J. West, T.C. Whitmore, P.A. Williams.

The following contributed comments on the list of invasive plants in Chapter 5: J. Akeroyd, P. Adam, B.A. Auld, P. Binggeli, M. Crawley, L. Cuddihy, N. de Zoysa, C. den Hartog, M. Diaz, I. Edwards, S.V. Fowler, T. Goodland, R.H. Groves, A. Hamilton, R. Hengeveld, V.H. Heywood, G. Howard, R. Jeffrey, W. Joenje, M.H. Julien, T. Lasseigne, W. Lonsdale, D.H. Lorence, I.A.W. Macdonald, R.N. Mack, J.L. Mayall, E. Medina, P.S. Mitchell, P.S. Motooka, M. Numata, A. Rabinovich, P.S. Ramakrishnan, L. Raulerson, M. Rejmanek, C. Ryall, D. Schroeder, C.W. Smith, J.M.B. Smith, C.H. Stirton, W. Strahm, H. Sukkopp, J. Swarbrick, W.R. Sykes, M.B. Usher, P.M. Wade, J. Waage, C. West, H.L.K. Whitehouse, T.C. Whitmore, S. Wiejusundara, P.A. Williams, G. Williams, M.H. Williamson, R. Wise.

The portraits of invasive plants were drawn by Rosemary Wise.

The authors take full responsibility for any remaining errors, and would be pleased to have corrections drawn to their attention. Finally, we would like to record special thanks to Alan Hamilton of WWF and to the series editor, Martin Walters, for their guidance throughout.

The preparation of this work was assisted by a grant to WWF by the Darwin Initiative, Department of the Environment, UK.

1

The nature of plant
invasion

1.1 What are invasive plants?

One of the main threats to biodiversity in the world is the direct destruction of habitats by people (through inappropriate resource use or pollution). Another serious but underestimated problem is the threat to natural and seminatural habitats by invasion of alien organisms, which potentially is a lasting and pervasive threat [73]. It is a lasting threat because when exploitation or pollution stops, ecosystems often begin to recover. However, when the introduction of alien organisms stops, the existing aliens do not disappear; in contrast they sometimes continue to spread and consolidate, and so may be called a more pervasive threat. Conservationists are becoming increasingly aware of the problem of invasive plants as they observe the native vegetation in many areas of interest succumbing to the 'revolutionary' [365] effect of vigorous aliens. The problem of biological invasions has been recognized by SCOPE (see Preface) as a central problem in the conservation of biological communities. Heywood [166] remarks that 'Invasion of natural communities, in many parts of the world, by introduced plants, especially woody species, constitutes one of the most serious threats to their survival, although it is one that is not fully acknowledged by conservationists'.

We define an invasive plant as:

an alien plant spreading naturally (without the direct assistance of people) in natural or seminatural habitats, to produce a significant change in terms of composition, structure or ecosystem processes.

This definition is intended to draw a clear line between the 'invasive plants' considered here and plants invading highly disturbed man-made or agricultural habitats (ruderals and weeds) [212]. It is necessary to make this distinction not only for practical conservation purposes, but also in order to reach any ecologically meaningful conclusions about the nature of invasion. In the literature this distinction is only rarely made [166, 259]. A difficulty arises when one tries to draw the line between 'natural/seminatural' and 'unnatural' habitats. An agricultural habitat is generally a highly managed system that is extremely artificial, differing from

1

most natural habitats in terms of competition, nutrient levels, diversity, disturbance and composition. As well as agricultural habitats, there are areas which are
highly disturbed by humans and vegetated by opportunistic species.

Our definition of natural or seminatural habitats, for the purposes of this book,
is as follows:

> *communities of plants and animals with some conservation significance, either*
> *where direct human disturbance is minimal or where human disturbance serves to*
> *encourage communities of wild species of interest to conservation.*

Several other definitions of invasive plants have been used. An invasive species
has been described as one that 'enters a territory in which it has never before
occurred regardless of circumstances' [259]. This catch-all description includes
species introduced into a new region, without considering whether they become
established, spread or have any effect on the native organisms. 'Adventive' species
is a better term for this kind of plant. Alternatively, 'invasive species are introduced
species that expand their population (and distributional range) in the new geographical location, without further human intervention' [406]. However, this
definition makes no reference to the impact on the habitat or ecosystem invaded
or to the type of habitat invaded. The definition: 'invasive species are introduced
species that have become pests' [404] merely shifts the question to that of defining
a 'pest'.

A much better definition of an invasive species is as an 'introduced species
which must be capable of establishing self-sustaining populations in areas of
natural or seminatural vegetation (i.e. untransformed ecosystems)' [253]. This
definition distinguishes between invaders of natural habitats and weeds ('ruderals')
of agricultural or highly disturbed habitats, and also between transient plants
('casuals') and well-established plants with self-sustaining populations. However,
this definition does not consider the impact on the native flora and fauna, although
aggressiveness of spread is a feature of importance to conservationists. Sukopp
[382] has proposed the term 'Verdrangung' (pushing out) to distinguish alien plants
which displace native species, significantly altering the habitat or ecosystem, from
plants which are introduced and become naturalized as constituents of the native
plant community ('Einpassung' or fitting in) without modifying it greatly. Similarly, Stirton [377] coined the term 'plant invader' to emphasize their capacity to
spread aggressively and cause rapid, often irreversible, changes in the landscape.

1.2 Plant invasion and conservation

1.2.1 General issues

Invasive plants are undesirable in conservation areas not simply because they are
aliens. There should be no serious objection to the introduction of a plant which
spreads naturally in a way that does not decrease the diversity of the native flora

2

and fauna, or alter ecosystem processes, although some would disagree with this view. *Impatiens parviflora*, for example, has been introduced to Britain and now occurs naturally in some ancient woodlands but does not pose any threat to the native vegetation. Invasive plants, on the other hand, modify natural and semi-natural habitats, for example, by replacing a diverse system with single species stands, introducing a new life form to the habitat, altering the water or fire regime, changing the nutrient status of the soil and humus, removing a food source or introducing a food source where none existed before, or altering sedimentation processes [253, 291, 321, 417]. Such alterations may have profound effects on the composition of both the flora and fauna of the region and on the landscape as a whole.

Plants which are now considered as invasive may have once appeared to be non-invasive when their populations were small or they were only found in habitats influenced by people. For conservation purposes, in environmentally sensitive areas, all aliens should perhaps be considered as potential threats and therefore monitored carefully. There has been a tendency not to record the spread of alien plants (e.g. *Rhododendron ponticum* in Britain [404]), at least when their populations seemed insignificant. Botanists and ecologists generally prefer to study the native flora and may often disregard the aliens. In many regions the number of introduced species that have had no detrimental effect on the native flora and fauna is far greater than the number of 'invasive species' (in Hawaii, for example, it has been estimated that there are 900 native plant species, 4000 introduced plants, of which 870 species which have naturalized and 91 'invasive plants' present significant management problems [361, 416]). As so few introduced species become 'invasive', it has been argued that there is no need for caution when introducing species to new areas. However, this argument does not take into account that those that do become invasive often have effects far out of proportion to their numbers.

1.2.2 Invasion and succession

There is a blurred distinction between invasion by a native species, as part of a succession, and invasion by an alien species. In some nature reserves, the conservation objective is to protect a species assemblage that may be invaded by native species as part of the natural succession. However, invasion by alien species may alter the native habitat more dramatically, in such a way that all native species are pushed out; in extreme cases, species may become extinct. An example of a single species attaining complete dominance is *Psidium cattleianum* in Mauritius and Hawaii. However, as many invasions are relatively recent, it is too early to say what the 'final' stage of their succession will be, particularly for long-lived invaders whose stands have not yet started to degenerate. In New Zealand, the invaders *Cytisus scoparium* (broom) and *Sambucus nigra* (elder) may come to dominate disturbed sites, but in other parts of New Zealand they behave only as early successional species, which are in turn invaded by the native *Melicytus ramiflorus* (mahoe). Broom and elder can therefore be desirable invasives in the re-establishment of native forest [442, 443]. Another example of the same is *Rosa rubiginosa* (sweet-briar) which is an efficient and readily bird-dispersed exotic colonizer of forest degraded by cattle-ranching in Argentina. It aids the recovery of the forest by acting as a thorny protective nurse for the regeneration of native trees such as *Lomatia hirsuta* [100]. However, the *Rosa* invasion lowers species richness [94]. Change is a feature of all ecosystems and what is at issue is not change itself but the rate and direction of change, both of which may be dramatically altered by invasion.

1.2.3 The threats

The examples of invasive species in Chapter 5, although not exhaustive or entirely representative of the global situation, do give an idea of the magnitude of the problem of alien plants spreading into and disrupting native plant communities and ecosystems. This account differs from many previous studies in that it is restricted to plants invading natural and seminatural habitats and having a significant impact on the native organisms. The threat posed by invasive plants to natural ecosystems is illustrated in the following selected examples:

1. *Replacement of diverse systems with single species stands of aliens* In Mauritius and Hawaii, *Psidium cattleianum* has spread to such an extent that it dominates large tracts of wet evergreen forest [239, 361] and has replaced much of the native vegetation with foreign but vigorously reproducing vegetation. Another oceanic island example is *Miconia calvescens* which is now estimated to cover 25% of Tahiti, primarily invading upland cloud forest. In the British Isles, *Rhododendron ponticum* has invaded a wide range of habitats [90]. In suitable conditions, it forms dense, impenetrable stands that prevent regeneration of the native species and removes a habitat for birds and other animals. *Acacia* species have spread over large areas of lowland and montane fynbos in South Africa, often forming stands with few other species present, leading to a reduction in biodiversity [377]. *Salvinia molesta* is an aquatic waterweed which has spread to dominate many freshwater systems in tropical and subtropical regions [279, 391]. The dense, floating mats formed can spread to cover the whole waterbody under sheltered conditions. These mats reduce light and oxygen levels, resulting in the displacement of the native animals and plants.

2. *Invasion that poses a direct threat to the native fauna* Invasion by *Lantana camara*, forming dense impenetrable stands in the Galápagos, threatens to remove the breeding site of an endangered bird, the dark-rumped petrel (*Pterodroma phaeopygia*) [91]. *Casuarina equisetifolia* has spread to such an extent on coastal areas in Florida (as well as other areas) that it is interfering with nesting sea-turtles (*Caretta*

Figures 1.1 & 1.2 Ecosystem effects of invasion. These photographs show the difference after tropical cyclone floods have cleared all the water hyacinth, *Eichhornia crassipes*, from the Enseleni River in Natal (South Africa). This species, originally from South America, can now be found smothering aquatic ecosystems throughout the tropics and subtropics on every continent. (Photos: I.A.W. Macdonald)

caretta) and American crocodiles (*Crocodylus acutus*)[21, 253]. However, while some animals decline in response to invasive plants other members of the native fauna may benefit. For example, there has been considerable range extension of the pied barbet (*Lybius leucomelas*) in South Africa by increased provision of nesting sites and fruit in introduced *Acacia* thicket [244].

3. *Alteration of soil chemistry* *Myrica faya* enriches the nutrient status of young volcanic soils in Hawaii by fixing nitrogen in its root nodules. This may have serious long-term consequences for this naturally nutrient-poor ecosystem, and may yet help ('facilitate') the invasion of other alien species which otherwise would have been unable to invade due to the low nutrient levels [293, 414, 417]. *Mesembryanthemum crystallinum* accumulates large quantities of salt, which is released after it dies. In this way it salinizes invaded areas and may prevent the native vegetation from establishing [204, 418].

4. *Alteration of geomorphological processes* *Spartina anglica* has invaded estuarine mud flats in many areas, including several localities in New Zealand, and as a result has altered sedimentation processes. Invasion by *Casuarina equisetifolia* in some areas is contributing to dune erosion. *Ammophila arenaria* is an aggressive invader in many areas where it was introduced for dune stabilization [253], and has significantly altered the dunes in places. In New Zealand it is said to be increasing the instability of some dune systems [445].

5. *Invasion leading to plant extinction* There are few documented examples of invasion leading to extinction of another species as invasives are usually just one of a number of contributory factors. However, on oceanic islands some extinctions can be attributed almost entirely to plant invasion. An example is the extinct endemic genus *Astiria* of Mauritius, represented by the single species *Astiria rosea* (Sterculiaceae). The last known site for this species is now a dense thicket of exotics and this is the most likely cause of its extinction. Many species are invading areas with endemic, rare or endangered species, and might contribute to their extinction. The spread of *Acacia saligna* in South Africa is directly threatening several species listed as endangered by IUCN [241]; *Passiflora mollissima* is seriously threatening Hawaiian rainforests, home to many endemic species [222]; *Rhododendron* is threatening the survival of rare Atlantic bryophytes in woodlands in the west of Ireland [199]; in Mauritius, *Ligustrum robustum* has invaded wet evergreen forests with many endemic species [239, 240]; *Lantana* invasion threatens several rare composite species in the Galápagos [91]; the invasion of *Pennisetum setaceum* in Hawaii is threatening several species listed as endangered by the US Wildlife Service [422]; *Ageratina adenophorum* is threatening the survival of two endemic shrubs, *Acomis acoma* and *Euphrasia bella*, in New South Wales, Australia [119].

6. *Alteration of fire regime* Invasive plants can alter both fire frequency and intensity. Fire frequency is a function of both the frequency of dry periods, ignition events and the flammability of the vegetation which determines the success rate of the ignition events. Many fire-adapted species have adaptations, such as the production of volatile oil or standing dead matter, to increase the frequency of successful ignition, and thus fire is important in their competitive relations with other plants. In Florida, the introduction of *Melaleuca quinquenervia*, which is almost perfectly adapted to fire, has increased the frequency of fires, damaging the native vegetation [115, 296]. Dense infestations of *Hakea* species in South Africa have increased the intensity of fires due to their rapid accumulation of large quantities of inflammable matter [301–303]. Like *Melaleuca*, and many other fire-adapted species such as *Pinus radiata*, species of *Hakea* have seeds with delayed dispersal ('serotiny') and fire promotes their release. By encouraging regeneration and the release of seed by fire-adapted invasives, regular burning in previously non-inflammable vegetation can have dramatic effects on species composition. In this way, fire-adapted bunchgrasses (*Schizachyrium*) in Hawaii are preventing regeneration of native species that are not adapted to regular burning [361].

7. *Alteration of hydrology* *Andropogon virginicus* now dominates areas in Hawaii which otherwise would have supported tropical rainforest vegetation. Its seasonal pattern of growth and formation of dense mats of dead vegetation are increasing water run-off in the area and leading to accelerated erosion [291]. Conversely, taller growing alien plants can invade an open native vegetation and cause a reduction in run-off. *Hakea sericea*, by producing a larger and denser canopy than the native fynbos shrubs, substantially increases evapotranspiration, and thus decreases the amount of water draining out of the area and available to people [303].

These examples show the greater variety of plant invasion in natural, as opposed to agricultural or highly disturbed ecosystems. This is a reflection of the variety of natural ecosystems and invaders. This variety is expressed in the taxonomy, morphology, biology and the ecology of the invading species, as well as in the habitats invaded and the effects (short and long term) on the native animals and plants.

1.3 Where is invasion happening?

Information on the geographical distribution of the problem of plant invasion is difficult to evaluate. In some areas, such as Hawaii, South Africa and Australia, there is an acute awareness of invasive plants, which have been recorded in detail. However, in parts of several large regions, such as South America, north and central Asia and India, there are few, if any, records of the spread of alien plants. This may be due to a lack of resources, local awareness or active recorders, rather than a lack of actual cases. It is therefore difficult to compare numbers of invasive

Figure 1.3 shows the dense invasive colonization of logged land in the East Usambara Mountains, Tanzania by *Maesopsis eminii* (not native to the region). Selective logging has created highly suitable conditions for the spread of this and other invasive species, including *Clidemia hirta* and *Lantana camara*. (Photo: Alan Hamilton)

plants between regions. The regions that contain most species from the sample of invasives in Chapter 5 are Australasia, Africa (mainly South Africa) and the Pacific islands (Oceania).

1.3.1 Invasion on oceanic islands

Islands are often reported to be more susceptible to invasion than continental ecosystems [413]. This susceptibility is illustrated by the high numbers of invasive species found on oceanic islands in the sample of invasive species in Chapter 5. We suggest the following explanations for the greater susceptibility of oceanic islands to invasion.

1. *Species poverty* This is usually attributed to their remoteness from continental sources of animals and plants, making them relatively species-poor. This poverty precludes vigorous competition between species as there is less chance of two species requiring identical or very similar conditions [113, 236, 361]. The lack of competition is demonstrated by the phenomenon, characteristic of oceanic islands, of 'adaptive radiation' or the evolution of plants into new habitats and life-forms in the absence of competition from existing plants in these life-forms.

8

Figure 1.4 shows the tall canopy of a mature individual of *Maesopsis eminii*, which had seeded into a natural gap in the Amani-Sigi Forest Reserve, East Usambara Mountains. *Maesopsis* is able to invade even apparently undisturbed forest and it is estimated that, unless it is controlled, it will come to form a substantial proportion of the canopy over the next century, seriously threatening the survival of the many endemic species in this forest. (Photo: Alan Hamilton)

By this reasoning alien plants are more likely to establish and spread on oceanic islands than in regions with a rich flora.

2. *Evolution in isolation* The native flora and fauna of remote islands have evolved in isolation, often without adaptation to high levels of competition, grazing and trampling by mammalian herbivores or to regular burning. The absence of mammals (with the exception of bats) is a characteristic of oceanic islands, and oceanic island plants seldom have the typical grazing adaptations, shown, for instance, by Mediterranean plants, of spines, thorns, pungent leaf chemicals, dense thicket-forming growth habit and vigorous resprouting from dormant buds on trunks and shoots. In the presence of grazing or browsing animals, adapted aliens will have a

competitive advantage over natives. For example, the introduction of goats to St Helena, shortly after the island's discovery in 1502, led to the destruction of many endemic species not adapted to grazing by vertebrates, such as the St Helena Ebony [84]. In contrast, the successful alien plants in St Helena, such as *Nicotiana glauca*, are often grazing resistant. In Hawaii, *Sophora chrysophylla* is one example of a native, highly palatable or 'goat ice cream' plant. In some Hawaiian communities, release from grazing produces an increase of native cover, indicating that differences in grazing tolerance are primary. However, in other communities an increase of alien cover follows goat eradication, indicating a greater importance of differences in competitive ability between natives and aliens [379].

3. *Early colonization* Islands were the first European colonies and they were settled to protect trade routes at a time when climate, disease and indigenous populations made permanent European settlement on continents difficult. Islands that lie in the trade routes have borne the brunt of this colonial activity, particularly those in the Atlantic and Indian Oceans. They were often the first landfalls of explorers and have been subject to disturbance by Europeans for far longer than many continental regions (St Helena was discovered in 1502 and alien plants and animals were introduced immediately, although the island was not finally settled until 1659). They also have a long history of intentionally introduced crops and ornamental plants. However, the situation is different in the Pacific. Although the Hawaiian Islands were settled by Polynesians in around AD 1000, regular contact with European vessels dates from the start of the sandalwood trade (1815) and the first visits by whalers (1819). It was only after the 'Greate Mahele' of 1848 that non-Polynesians were allowed to own land.

4. *Small scale* Their small geographical size means that history is 'concentrated' in a small area and no physical features are large enough to prevent exploitation, disturbance and introduction throughout the island. Often the effect of the small scale can be seen in the relatively high population densities. For example, Mauritius has a population of roughly 530 people per km^2 whereas India, a country with which Mauritius is culturally and historically linked, has only about 240 people per km^2. Islands are small in relation to dispersal distances, so introduced plants can spread quite rapidly throughout an island, giving a worse impression of invasion than an apparently localized invasion on a continent that in fact covers a larger area. Finally, distances between different communities are often small on islands relative to dispersal distances and consequently an introduction can more easily disperse to the best habitat. For instance, in St Helena the zonation between semidesert and cloud forest occurs in 3–4 km and some 600 m altitude, whereas on continents greater distances often separate such divergent communities.

5. *Crossroads of intercontinental trade*   Often oceanic islands were colonized to promote trade routes, and therefore they have been at the crossroads of intercontinental trade and used quite deliberately to transfer living material for propagation ('germplasm') between continents. Islands were often key staging posts for trade and plant transport (the earliest British colonial botanical garden in the tropics was at St Vincent in the Caribbean, established for just this purpose). Often, too, they have had an eventful history. Mauritius changed hands between the Dutch (1598), French (1715) and British (1810) at various points in its history, increasing its chance of receiving plants from different parts of the tropics. The garden at Pamplemousses, to which many introductions came, was established by Governor Labourdonnais as early as 1735. This deluge of disturbance and alien plants goes some way towards explaining the apparent susceptibility of islands to invasion, at least in the case of Atlantic, Caribbean and Indian Ocean islands.

6. *Exaggeration of ecological release*   Alien species come to islands free from their natural array of pests and diseases, thus having an advantage over the native vegetation. This is particularly noticeable on islands, which usually have a species-poor fauna and therefore few generalist insect or other plant herbivores to prey on the alien species. If generalist plant-feeders are introduced, the alien species are more likely than native plants to be adapted to resist them.

1.3.2 Climatic patterns of invasion

Taking the sample of 70 invasive species in Chapters 4 and 5 for which climatic data are given, 18 are invaders of the cool temperate zone, 50 of the warm temperate zone, 47 of the subtropical zone and 16 of the tropical zone (see beginning of Chapter 5 for a definition of these zones). The striking preponderance of the warm temperate and subtropical zones may reflect a sample bias, with South Africa, Australia and Hawaii, where invasive plants have been the focus of intensive studies, weighing the balance in favor of these climate zones. More information is needed for other areas, especially South America and East Asia, to assess the significance of the climatic and geographic distribution of invasives. However, Australia, New Zealand, South Africa, the Mascarene Islands and Oceania (Pacific Islands), which are all regions that include warm temperate and subtropical climate zones, do genuinely seem to have significantly higher numbers of invasive species than other areas.

Few invasions seem to occur in areas with extreme climatic or environmental conditions, such as those which are extremely dry (desert, thorn steppe) or cold (polar regions), and it has been suggested that non-extreme ('mesic') habitats are more susceptible to invasion [324]. Certainly Antarctica is free of invasive species [405], although as continental Antarctica has only two indigenous species of vascular plants the absence of invaders is hardly surprising. The main invasive threat in Antarctica comes from fungi, protozoans and bacteria. However, *Poa*

11

annua is a serious invader on the subantarctic island of South Georgia [226] and aliens have had an impact on the subantarctic Campbell Island [271].

The cool temperate zone appears to have a low number of invasive species relative to the warm temperate zone. Three reasons may be suggested:

1. Some cool temperate regions, such as northern and central Europe, have long been influenced by people and much of the natural vegetation has been destroyed and replaced with vigorous invasion-resistant secondary types of vegetation. Where this vegetation is invaded, it is usually by Northern Hemisphere species, such as *Rhododendron ponticum*. Although Europe has a long list of Northern Hemisphere aliens associated with its disturbed secondary vegetation, it may be that these weedy communities are too vigorous for most Southern Hemisphere species to join, except as casuals. The difference is that there are few areas of natural vegetation left and therefore, under our definition of invasive plant, only a few species qualify.

2. Most regions of the world were affected in one way or another by the climatic changes associated with the ice age cycles of the last 1.6 M yr (the Quaternary). Much of the cool temperate regions was glaciated during the ice ages, and repeated disturbance by glaciation may have encouraged the native flora to adopt characteristics associated with invasive species, making the vegetation inherently resistant to other invasive plants. At the end of the last ice age in northern Europe it is known that a diverse flora, which included many weeds and vigorous herbs, such as *Plantago major* and *Urtica dioica*, grew around the ice sheet, on soils disturbed by frost heave [40, 434]. Furthermore, the trees of formerly glaciated cool temperate regions underwent repeated long-distance migrations from refugia at the end of each cold stage. Most native trees of cool temperate regions therefore have many characteristics of invasive species. The rate of distribution change ('migration') has differed from species to species, but in some cases has been extremely rapid [35].

3. Climatic extremes in cool temperate regions may limit the spread of vigorous invaders. Marked seasonality reduces the length of the growing season and consequently may reduce the fecundity and the potential rates of increase of invasives. Environmental extremes such as frost increase density-independent mortality, while suboptimal growing conditions, for instance when temperatures are low, reduce growth and the aggressive competitiveness of invaders.

Species tend to invade, or grow well, in regions of the same, or similar, climate [23, 144], a characteristic which agriculturalists and foresters use in computerized climatic matching systems to predict the potential ranges of organisms [42, 384]. However, some species are apparently able to invade a wide range of climate zones or habitats outside their native limits. Examples are: *Lantana camara*, which is

12

invading tropical, subtropical and warm temperate habitats; *Pueraria lobata*, which is native to cool temperate regions, but is invading both warm temperate and subtropical areas; and *Ailanthus altissima*, which is native in subtropical and warm temperate China, but is invading areas ranging from cool temperate to tropical. However, the climate zones used for analysis (cool temperate, warm temperate, subtropical and tropical) are broad categories, and it is possible that a closer examination of seasonality and extremes may be more useful in predicting invasion.

1.3.3 Invasions in particular biomes

1. *Tropical forest* Invasive species in large areas of the tropics are under-recorded, but undisturbed tropical rainforest does seem, at least in the Old World, to be fairly resistant to invasion [440]. However, there are some exceptions, such as *Pittosporum undulatum* and *Maesopsis eminii*(see Chapters 4 and 5). Large, intact areas of continental tropical forest, such as central Amazonia, are comparatively isolated and have not been exposed to the same numbers of aliens as more populated and long-colonized areas, and this may contribute to the apparent lack of invasive plants in most continental tropical rainforests. Secondary rainforest and island rainforest does appear to be prone to invasion as seen by the high numbers of alien species in Hawaii [359], Mauritius [239, 240] and Australia [119]. One study, of a sample of tropical nature reserves (Hawaii – Maui, the Galápagos and Australia – Kakadu National Park), against a similar study of Britain [446], suggested that no tropical ecosystems are free from invasive species and that the proportion of introduced plants that have become invasive is far higher in the tropics than in temperate regions [406]. However, nature reserves may not be fully representative of natural vegetation and some nature reserves are significantly disturbed by visitor pressure. A different picture might be gained from a wider survey.

2. *Mediterranean-climate shrublands and grasslands* In areas with a Mediterranean climate, the reduction of woodland cover, a common phenomenon in these regions, has apparently favored invasion; in turn, these invasions have often affected the native ecosystem by altering the natural fire regime. In California, it appears that accelerated soil erosion may be caused by plant invasion, which in the long term may seriously destabilize the native ecosystem [250]. In some Mediterranean-type ecosystems the native flora has been almost entirely displaced, but in others a large decline in native species diversity does not appear to have resulted. In California, Chile and many Australian habitats, the alien flora is mostly composed of annual herbs [300], although annual herbs have not displaced the native bunchgrasses in very similar vegetation in western Guatemala. This may be due to the less severe dry season and the presence of sheep rather than cattle as grazers [411]. In contrast, the South African fynbos vegetation has been invaded mainly by trees and shrubs. This diversity of situations involving invasive

plants illustrates the problem of understanding why certain species become invaders [216].

3. *Arid lands* Plant invasions in tropical savannas and dry woodlands are less frequent in the drier parts of these vegetation types. Here, fire and herbivory by large animals, both accentuated by drought, appear to limit invasion. In moister areas, particularly where the savannas appear to be man-made, shrubs and scramblers such as *Lantana camara*, *Chromolaena odorata*, *Mikania scandens* and *Parthenium hysterophorus* can be a serious problem [248]. Areas with more extreme aridity (deserts and semideserts) have not been invaded to a great extent by plants, except along perennial or intermittent water courses [237].

4. *Wetlands* Wetlands have been strongly invaded in many parts of the world and five of the 17 case studies in Chapter 4 are aquatic or wetland species. There are three main types of wetland invader: floating aquatics, submerged aquatics (e.g. *Myriophyllum aquaticum* and *Lagarosiphon major*) and plants of seasonally-flooded low-lying areas (e.g. *Mimosa pigra*, *Sesbania punicea* and *Melaleuca quinquenervia*). Floods and flowing water are efficient seed dispersal agents for invasives and most species have water dispersed propagules, which may be either vegetative fragments, fruits or seeds. Mineral nutrients brought down as silt during floods, provide growing conditions favorable to the growth of vigorous invasives. Many aquatic species (e.g. *Salvinia molesta*, *Eichhornia crassipes* and *Pistia stratiotes*) are unusual in that they have invaded a large number of regions. This may be due to freshwater habitats being so frequent worldwide and to easy accidental transport by boat traffic. Many of the invasive aquatic species have broad environmental tolerances, which also contribute to their success.

2

How invasion occurs

2.1 Process of invasion

2.1.1 Stages of invasion

There are several stages in the process of invasion. These may conveniently be divided into the following: introduction, naturalization, facilitation, spread, interaction with other animals and plants, and stabilization.

1. *Introduction* The introduction of invasive plants involves translocation of living material by people from one region to another, either accidentally or deliberately.

2. *Naturalization* Once introduced, a plant to be considered invasive, must become established beyond the site of initial introduction to form large self-sustaining populations in natural or seminatural vegetation. While the population remains small, the plant is at risk genetically and ecologically and may be unable to increase. This stage is known as naturalization and its success depends on the species being planted or otherwise introduced sufficiently close to natural vegetation for 'escape' to occur, and also on certain biological features of the plant (e.g. breeding system, successful reproduction) and on the environment (climate, seasonality, soil conditions).

3. *Facilitation* A naturalized plant may remain a rarity unless it is 'facilitated' in some way in order to spread, for example by the introduction of a suitable dispersal agent or pollinator, provision of disturbance in the ecosystem or lack of pests and diseases. Genetic adaptation to the new environment by the selection of the fittest individuals ('microevolutionary adaptation') is a form of facilitation. Another type of facilitation neglected by scientists is the 'fostering' of the plant in the wild by people, for instance *Psidium cattleianum* in Mauritius, where the collection of fruits for Guava wine makes contemplation of the introduction of biological control for this species difficult or impossible. In some estates in Britain *Rhododendron ponticum* has, in the past, been protected from clearance during standard forestry procedures by virtue of its attractive flowers, which have also encouraged people

15

to spread the plant (in some cases by scattering the seed at beauty spots!). Features of invasive plants which promote fostering, such as attractive flowers, edible fruit or nectar for honey, are important adaptations for invasion which can hinder control by promoting management conflicts (Chapter 3).

4. *Spread* If the spread of an alien plant has been facilitated, the rate of its spread depends on the intrinsic growth and reproduction rates of the plant as well as the nature of the invaded habitat including the presence of suitable places ('safe sites' [154]) for reproduction. Efficiency of seed dispersal is an important determinant of rate of spread and knowledge of both average and maximum dispersal distances is important for understanding population expansion. A successful invader often has adaptations for short- and long-distance dispersal, the short-distance bulking up existing populations and the long-distance establishing new foci for further spread, remote from the original site of invasion.

It is often observed that spread may be delayed and that an introduced plant initially occurs at low numbers in its new environment and later suddenly undergoes a population explosion [284]. This poorly understood phenomenon may occur for one or more of the following reasons: the population is actually expanding exponentially, but this is not noticed when numbers are very low; the plant was under-recorded when at low population levels due to its alien status; the population was unable to increase until facilitated in some way, for example through the arrival of a pollinator or the disturbance of habitat; genetic changes allow adaptation to new habitat (this may take a very long time); or the species spreads to an optimal habitat, the first habitat colonized being suboptimal. Once spread starts and if conditions remain suitable, the population of the alien species may increase to such levels that the native animals and plants are endangered and a control program is desirable.

5. *Interaction with animals and other plants* Sooner or later (as when originally planted), the alien species will encounter plants and animals already present in the area. The outcome of this interaction will determine whether the alien will be a 'fitting in' invasive or a 'pushing out' invasive (Chapter 1) with respect to the native animals and plants, and therefore whether it will have a significant effect on ecosystem processes, composition and structure. In some cases competition may restrict the alien species to disturbed sites on which the native vegetation has been or is unable to survive due to extreme environmental conditions.

6. *Stabilization* Some invasions appear to stabilize as single or near single species stands (e.g. *Psidium cattleianum* and *Rhododendron ponticum* – see the case studies in Chapter 4). However, it is often uncertain whether this stabilization is actually illusory. Many invasions are relatively recent and it is possible that the populations

Figure 2.1 shows the wild ginger, *Hedychium gardnerianum*, from India invading a wet gully in Madeira. (Photo: Quentin Cronk)

Figure 2.2 Here on the Indian Ocean island of Réunion *Hedychium gardnerianum* totally dominates the understorey of a stand of native rainforest. (Photo: I.A.W. Macdonald)

Figure 2.3 Seedlings of sycamore, *Acer pseudoplatanus*, growing vigorously alongside the rare endemic orchid, *Dactylorhiza foliosa*, in Madeira. Base-rich sites suitable for this orchid are ideal for sycamore seedlings, whose presence will, in turn, alter the ecosystem. (Photo: Quentin Cronk)

will eventually undergo senescence. It may be more appropriate to regard single species dominance of communities by invasives as analogous to a successional stage. An alternative possibility is that these massive infestations represent an 'ecological overshoot', as seems to have happened in the case of the nineteenth-century invasion of the waterweed *Elodea canadensis* in Britain, which rapidly became a major waterway nuisance before subsiding to its present relatively infrequent level. A similar example is provided by the invasion of *Lagarosiphon major* in Lake Rotorua, New Zealand which, from the massive weed-banks at the height of the invasion in 1958, has now subsided to an innocuous level (Chapter 4).

2.1.2 Modelling invasion

A 'model' is a simplified description of a process or system, sometimes given in mathematical terms, which aids understanding and may even allow predictions to be made given specified initial conditions. Most models of plant invasion do not distinguish between natural or seminatural habitats and disturbed or people-modified habitats [446]. Defining the characteristics and factors for the 'typical plant invasion' is difficult considering the ecological variety of plant invaders and natural habitats. Many models of invasion deal with disease epidemics [283] and there appears to have been little success in modelling plant invasion, particularly of relatively undisturbed natural habitats. Models of community dynamics often assume that the principal barriers to invasion include [83]: (1) competition with established native species; (2) losses caused by generalist natural enemies (including disease); (3) lack of necessary coadapted animals (mutualists) to pollinate, disperse or otherwise help ('facilitate') the invader; and (4) deleterious low density effects operating on the invader itself. Environmental factors, such as climate, particularly seasonality of climate, fire regime, natural disturbance (such as hurricanes, grazing and trampling by native herbivores and flooding) and soil chemistry are rarely considered. However, analysis of the locations of apparently suitable climatic conditions, based on the climatic tolerances of a species in its home range ('homocline analysis'), is potentially a useful method for predicting the future distribution of an invasive plant. Models attempting to simulate invasion (usually for predictive purposes) often break down the process of invasion into stages: arrival, establishment and spread [283, 446], which can be further elaborated. However, even very simple models are useful in generating hypotheses, for instance of the relative effects of altering certain parameters, which can be tested in the field. For example, on the basis of one simple model, the rate of spread appears to differ widely from exponential to linear, depending on changes in simple population parameters, particularly dispersal pattern [15].

2.2 Historical aspects of plant translocation

2.2.1 Beginnings

From the start of agriculture, whenever people have moved, plants have moved too, both through the deliberate spread of domesticated crops and the accidental associated spread of weeds and ruderals [191]. Crop plants (particularly cereals) and associated arable weeds were introduced to Britain from SW Asia and S. Europe by Neolithic cultures some 6000 years ago [133]. Another example is the Japanese honeysuckle, Lonicera japonica. This species is native to E. Asia and is a serious problem in the US, where it flourishes as an agricultural weed in fields, thickets and bordering woodland, from Florida and Texas to New York State and Massachusetts. However, it is only since the seventeenth century and especially more recently (with the increase in ease of long-distance transport) that there has been large scale intercontinental changes in plant distributions. The activities of people have introduced a new order of magnitude into distances of dispersal [154].

2.2.2 Colonialism

The expansion of the European colonial powers (notably Britain, France, Germany, the Netherlands, Portugal and Spain) increased dramatically the transport of living material. One major aim was to discover and exploit new economic crops for the empires [166]. In particular, it was the opening up of the tropics and the discovery of the New World that led to a great wave of plant exchange or, as it has been called, 'ecological imperialism' [87]. The European colonization of Australia, for example, has been described as an 'apocalyptic' event for Australian ecosystems. In the tropics of Asia and Australasia, and on islands, the effects have been magnified by the extent and duration of the colonial period. In tropical Africa the impact on native ecosystems, although significant, has been less severe as, in the main, these countries were not colonized by European powers until the late nineteenth century during the 'scramble for Africa'. Most botanic gardens in tropical Africa date from the late nineteenth or early twentieth centuries, whereas, in contrast, the Calcutta Botanic Garden in India was established in 1787 and the Royal Botanic Garden at Peradeniya, Sri Lanka in 1821. In Latin America, the early withdrawal of Portugal and Spain and the minor involvement of Britain (by far the most active power in plant transport) has meant that the vegetation has suffered very much less than elsewhere in the tropics. The introduction and planting of useful or potentially useful crops and other plants was usually a characteristic and essential part of colonialism, although the dates and purposes of introduction may differ widely. In South Africa, for instance, *Pinus pinaster* was introduced (possibly by French Huguenots in 1688) as a timber tree for the largely treeless Western Cape and was certainly being grown in the Dutch East India Company's Garden in Cape Town around 1690; *Acacia cyclops* and *A. saligna* were both probably introduced to Baron von Ludwig's private garden in the 1830s and were used to plant the Cape Flats from *c*. 1847 onwards; in contrast, *Nasella trichotoma* was probably introduced from Argentina during the second South African War (1899–1902) in the 138 000 tons of fodder for horses imported by the British army [353, 431].

Islands that lie in the trade routes have borne the brunt of this colonial activity. They were often the first landfall for explorers and have been subject to disturbance far longer than most continental regions (St Helena was discovered in 1502 and alien plants and animals were introduced immediately). Islands were often key staging posts for trade and plant transport (the earliest British colonial botanical garden in the tropics was at St Vincent in the Caribbean, established for just this purpose). Often, too, they had an eventful history. Mauritius changed hands between Dutch (1598), French (1715) and British (1810) at various points in its history, and the garden at Pamplemousses, to which many introductions came, was established by Governor Labourdonnais as early as 1735. This deluge of disturbance and alien plants goes some way towards explaining the apparent susceptibility of islands to invasion, at least in the case of Atlantic, Caribbean and Indian Ocean islands.

2.2.3 Botanic gardens

Although many European botanic gardens founded from the sixteenth century onwards were responsible for a considerable amount of plant introduction (for medicinal, ornamental and amenity uses, as well as for scientific study), it was the

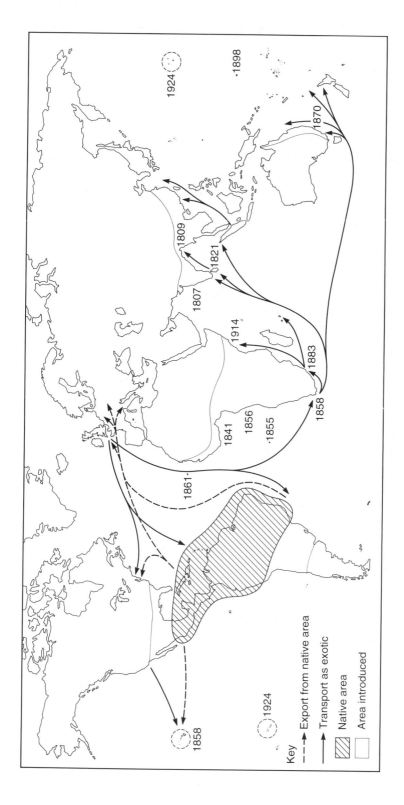

Map 2 Transport of *Lantana camara* around the world. This species has become a major pest in many areas. (After Stirton (1978))

tropical botanic gardens that served as the main staging posts or centres of exchange [165]. The Royal Botanic Gardens, Kew, as a colonial institution, introduced Pará rubber, *Hevea brasiliensis*, to the Singapore Botanic Garden (which had been founded in 1859) from Brazil [56]; rubber later significantly altered the Malayan economy, landscape and ecosystems. Although it is a well-known example of plant introduction, *Hevea* is not invasive, but other similar introductions arranged by Kew for economic reasons have invaded important habitats. Examples include *Cinchona succirubra* and *Phormium tenax*, introduced for quinine and fiber production respectively and now established in the endemic-rich tree fern thicket on the central ridge of St Helena [86]. Britain was responsible for a considerable amount of plant introduction as a result of its policy of forming 'networks' of botanic gardens which exchanged vast quantities of living material under the direction of Kew. 'Over-introduction' is a common feature of these activities. Baron von Ludwig set out to introduce useful and ornamental exotic plants to his garden in South Africa, and with the help of his botanical contacts worldwide managed to introduce 1600 species [377].

2.2.4 This century

Modern agriculture and forestry have taken many crops, fruitplants and trees to new regions, and often they have escaped from cultivation and spread into the native vegetation. Examples include *Passiflora mollissima*, *Psidium cattleianum*, *Pennisetum clandestinum*, *Pinus radiata* and *Acacia melanoxylon*. Ever since the seventeenth century, enthusiastic gardeners have, in increasing quantities, introduced exotic species to countries for their ornamental value. Examples are *Pittosporum undulatum*, *Passiflora mollissima* (both an ornamental and a fruit crop), *Sesbania punicea*, *Acacia longifolia*, *Eichhornia crassipes* and *Rhododendron ponticum*. Many species have been introduced to areas for other qualities such as binding sand dunes (*Ammophila arenaria*, *Acacia saligna* in South Africa and Libya, *Casuarina equisetifolia*), erosion control (*Pueraria lobata*), game cover (*Rhododendron ponticum*), aquarium plants (*Crassula helmsii*, *Myriophyllum aquaticum*) and stabilizing mud-flats (*Spartina anglica*). Reforestation and agroforestry projects in developing countries often use fast-growing, alien 'weedy' trees or shrubs instead of suitable native species. These projects have often backfired because the plants have invaded the native vegetation [184]. This century has seen the rise of tourism, which is proving a considerable force for translocating ornamental plants, as well as seeds and spores unintentionally on shoes or clothing. Some countries have recognized this route for the introduction of undesirable weeds and pests and have taken measures to try and prevent it. This is discussed further in Chapter 3.

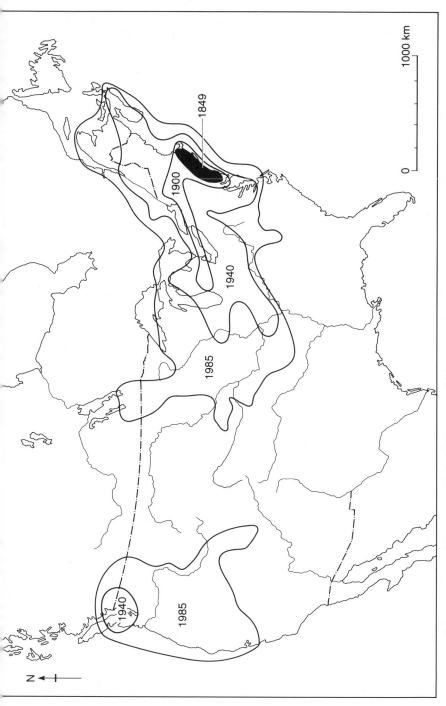

Map 3 The spread of *Lythrum salicaria* in the period 1849–1985, from a New England focus (probably introduced in ballast) followed by initial spread along the Erie Canal. A western focus was subsequently established, probably by long distance dispersal. (After Thompson (1987))

BOX 2.1 Dates of establishment of some selected botanic gardens, which have been major conduits of plant transfer

1543	Pisa, Italy
1621	Oxford, England
1652	founding of the Dutch East India Company's Garden at Cape Town by Van Riebeck to grow grain and vegetables for the new colony
1735	Pamplemousses, Ile de France (Mauritius)
1759	Royal Gardens at Kew, England becomes a botanic garden with the appointment of W. Aiton as Superintendent
1764	St Vincent, Caribbean
1770	Pierre Poivre introduces cloves, nutmeg and pepper to Pamplemousses
1774	Jamaica Botanic Garden
1787	Calcutta Botanical Garden, India established by the English East India Company
1796	Penang, Malaysia
1808	site prepared for the Real Horto, later the Jardim Botanico, Rio de Janeiro, Brazil
1816	inauguration of the Sydney Botanic Garden, Australia, on the Governor's Farm site
1817	Bogor, Java
1821	Peradeniya, Sri Lanka
1822	Singapore (closed 1846)
1841	role of the Royal Botanic Gardens, Kew expands with the appointment of Sir William Hooker as Director
1845	Royal Botanic Garden, Melbourne
1848	refounding of the Cape Town Botanic Gardens (now the Public Gardens), South Africa
1859	Missouri Botanical Garden, USA established by Henry Shaw
1859	Singapore Botanic Garden refounded
1863	Botanic Garden, Christchurch, New Zealand
1892	Botanical Garden of Buenos Aires, Argentina
1898	Entebbe Botanical Garden, Uganda
1913	South African National Botanic Garden at Kirstenbosch near Cape Town

2.3 Characteristics of invasive species

Can we predict which species will invade? Attempts have been made to describe the characteristics of the 'ideal weed' of agricultural habitats [22]. To define an 'ideal invasive plant' of diverse natural habitats is more problematic, but several authors have suggested factors and plant characteristics that appear to be associated with invasion [23, 144, 320]. However, the properties of the habitat seem to be as important in determining invasion as the properties of the invasive plant [306, 330]. Invasive plants are not confined to any particular growth form, although the most successful species in natural or seminatural habitats tend to be trees [122] and by far the majority are perennials (Chapter 5). In New Zealand, for example, over 50% of the invasives in protected natural areas are trees or shrubs over 3m high [395]. The following features seem to be important in characteristics of invasive species.

2.3.1 Seed dispersal mechanism

The 'ideal weed' [23] has adaptations for both long-distance and short-distance dispersal. The most commonly found mechanism of seed dispersal among invasive species of conservation importance appears to be by birds. Over half the woody invaders of New Zealand have bird-dispersed fleshy fruits [443] and this mechanism allows for both short- and long-distance transport of the seeds. Seed dispersal by animals appears to be the most effective method for the widest variety of species [371]. Fruit collected by birds or mammals may be eaten where it is found and the seed ejected, or it may be transported externally or in the crop or gut (for up to 100 hours [319]). Efficient dispersal of seeds by wind works best in open habitats and is thus suited to a more restricted range of habitats. Long-distance dispersal by wind is erratic and seeds need to be light, thus limiting the nutrient reserves they contain. In contrast, animal-dispersed seeds may be large, with additional nutrients to sustain the growing seedling in shade, which may be critical in determining the success of an alien species. The advantage of long- distance dispersal for an invasive species is the potential to establish new points ('foci') of invasion. Each new colony increases the size of the expanding margin of the invasion more than mere expansion of the original colony. This increases the rate of spread and makes control more difficult [284]. Examples of successful invasive species dispersed by birds are:

1. In Australia, many of the indigenous *Acacia* species are associated with ant dispersal, others with bird dispersal or both ant and bird dispersal. In South Africa, where several Australian *Acacia* species have become invasive, it is the species with bird-dispersed seeds (*Acacia cyclops*, *A. melanoxylon*, *A. saligna* [47, 48, 282]) that are the most successful. *Acacia pycnantha* is ant-dispersed and has not spread far in South Africa. The bird-dispersed species (except *A. saligna* which is also dispersed by ants and mammals [47]) have swollen, bright red funicles to attract birds. Exceptions to the association between bird dispersal and success are *A. longifolia* and *A. mearnsii* which are both highly invasive, particularly along watercourses and river valleys. Both are water-dispersed, though *A. longifolia* attracts some bird dispersal which is important in establishing new points of infestation.

2. The dispersal of the large fleshy fruits of *Passiflora mollissima* by birds and feral pigs [222] has confounded all attempts to control it. Many isolated infestations, some in inaccessible areas, have presumably been established originally by long-distance bird dispersal, while local infestation from these foci is promoted by pigs.

3. *Lantana camara* is a ubiquitous tropical weed with bird-dispersed seeds. In Australia it invades in sclerophyllous habitats which have few native bird-dispersed species. *Lantana*'s success is helped by the lack of competition for dispersal agents. In St Helena it is dispersed widely and effectively by introduced mynah birds (*Acridotheres tristis*).

4. *Pittosporum undulatum* produces sticky seeds which are transported both in the gut of birds and by adhering externally. It is successfully invading undisturbed tropical rainforest [136].

2.3.2 Breeding system

For the 'ideal weed' the most suitable breeding system for colonizing new areas is 'self-compatible, but not obligatorily self-pollinated or apomictic' [22]. Self-compatibility allows seed set at low population levels, in the absence of suitable pollinators. Obligate self-pollination, on the other hand, could lead to inbreeding and associated lack of variation and vigour in the population.

This rule does not appear to hold for invasive plants of natural habitats. Taking the plants for which information on breeding system is given in this book, 13% are dioecious and 11% monoecious (and the rest hermaphrodite). This is a similar level of dioecy to that found in the native flora of New Zealand (dioecy 12–13%), which is a flora usually considered to have a high level of dioecy, higher than those of many other places such as the British Isles (dioecy 3%) [30]. A rather high percentage of invasive plants thus appears to be dioecious or monoecious – two mechanisms which usually promote cross-pollination and therefore outbreeding. However, many of the successful dioecious invasive plants have either developed ways of overcoming the disadvantages of dioecy or have been introduced in large numbers and, as such, had no problems in setting seed initially. Several dioecious invaders have effective means of vegetative reproduction. This is particularly relevant for the aquatic species, such as *Lagarosiphon major*, *Myriophyllum aquaticum* and *Hydrilla verticillata*, which reproduce predominantly by vegetative means and are effectively dispersed by water. *Dioscorea bulbifera*, a woody climber, has also spread vegetatively from the botanic gardens in Singapore, where it was introduced, into neighboring rainforest, but its spread has been limited (only 4 km) due to lack of efficient dispersal. *Casuarina equisetifolia*, a dioecious tree and a serious invader, has been widely planted for dune stabilization [21], thus overcoming initial problems of adequate seed set.

Monoecy is associated with similar problems as dioecy. Some monoecious invasive species are self-compatible, allowing them to produce seed when they are

Figure 2.4 A small clump of water hyacinth, *Eichhornia crassipes* (originally from S. America) dispersing down the River Nile (Africa). This species is adapted to dispersing downstream as mat-fragments. (Doug Sheil)

isolated or at low population levels. An example is *Pinus radiata*, in which outbreeding is favored by the way the cones are distributed on the trees, but, in the absence of neighboring individuals of the same species, self-pollination may occur. This ensures seed set in all situations without promoting inbreeding and associated problems [25].

The majority of invasive species are, of course, hermaphrodite, but information on compatibility is unknown for many invasive species. Obligate self-fertilization is a rare condition in plants generally [325], although some opportunistic annuals show a high degree of self-fertilization. Most plants are outbreeders with some capacity for self-fertilization and this seems true of invasive plants. Species with a greater frequency of self-fertilization may be at an advantage when colonizing a new site. However, many successful hermaphrodite invasive species have strong outbreeding mechanisms. Invasive species have often been introduced intentionally in large numbers for horticultural or agricultural purposes, e.g. *Casuarina*

27

equisetifolia, Rhododendron ponticum, Ammophila arenaria, Pinus radiata and *Passiflora mollissima*. They may not, therefore, be disadvantaged by strong outbreeding mechanisms, although it is possible that, if they start from a single location, many small foci may not be successful in establishing viable populations, possibly limiting the rate of spread.

The following examples illustrate the main patterns:

1. *Acacia* spp., invasive in South Africa, are apparently 'obligate outbreeders' [274]. Several of these species, e.g. *Acacia saligna*, have become widespread in South Africa and pose enormous control problems due to the extent of infestation. Their extensive spread has been aided by planting for dune stabilization and timber, but it is mainly due to copious seed production (despite lack of self-fertilization) and efficient dispersal by birds.

2. *Passiflora mollissima* is a self-compatible woody climber which takes advantage of both self-fertilization and cross-fertilization. This characteristic, along with its adaptation for long-distance dispersal, has allowed it to spread into isolated areas, away from the main focus of invasion, where it is invading evergreen rainforest.

3. *Eichhornia crassipes* is also self-compatible, although it mainly reproduces vegetatively. This allows isolated individuals to form new infestations rapidly from seed. *Eichhornia crassipes* shows remnants of a self-incompatibility mechanism known as 'tristyly', which has broken down [27]. Tristylous plants have flowers of one of three lengths, which must be pollinated by a plant with styles of a different length.

4. *Pennisetum clandestinum*, a grass which is causing problems in Hawaii, can reproduce sexually or asexually ('facultative apomixis'). This may help rapid spread in new areas. Partial apomixis allows for some outbreeding and keeps variation in the population. Facultative apomixis 'appears the ideal' breeding system for an invasive species [23].

5. *Ageratina adenophora* is an obligate apomict, producing seed without fertilization ('agamospermy'). It cannot produce seed by sexual means because of its irregular (triploid) number of chromosomes. Apomixis has been suggested to be characteristic of invasive species [144], but, more accurately, it is a characteristic of weediness. *Ageratina* grows best in grasslands and agricultural habitats and is not always successful at invading intact natural or seminatural habitats. However, many habitats of conservation importance are also disturbed and in Australia *Ageratina* threatens the survival of two endemic shrubs (*Acomis acoma* and *Euphrasia bella*) [119]. *Chromolaena odorata* is also an apomictic species and, like *Ageratina*, a member of the Compositae. It is an extremely serious weed in disturbed places, but is not always successful in natural habitats as it is intolerant of shade. However, it regenerates well after burning and has become the most successful invader of savannas in South Africa.

2.3.3 Seed ecology

Longevity of seed is an advantage for weeds [22] and there is some evidence that many invasive species have seeds that can remain dormant for some time, as in *Mimosa pigra, Acacia melanoxylon, Acacia longifolia, Passiflora mollissima, Ulex europaeus* and *Acacia saligna*. It is a feature which often frustrates efforts at control. Several invasives that are successful in frequently burnt habitats have serotinous seeds (i.e. with delayed dispersal). Typically, such species have seeds which are only released after fire, as in *Hakea suaveolens, H. sericea, Melaleuca quinquenervia* and *Pinus radiata*.

Many invasive species, even those that grow tall when mature, reach reproductive maturity relatively early (e.g. *Mimosa pigra, Clematis vitalba, Melaleuca quinquenervia* and *Acacia saligna*), as well as having copious seed production. Those species which are successful in invading forested habitats often display characters of both the early successional (high seed production, rapid growth) and late successional species (high competitive ability, shade-tolerance – ability to persist under the canopy of a dominant species). *Rhododendron ponticum, Psidium cattleianum* and *Ligustrum robustum* are good examples of shrubs combining weediness with high competitive ability.

2.4 Taxonomic patterns of invasion

2.4.1 The taxonomic distribution of weediness

The taxonomic distribution of weediness is very different from the taxonomic distribution of invasiveness. Heywood [166] used *A Geographical Atlas of World Weeds* [177], including both introduced and native weeds, to show that the families containing the most weeds are the Compositae and the Gramineae, a conclusion similar to that of Groves who predicted that most 'invasive' species (i.e. weeds) will belong to families such as the Compositae, Cruciferae or Amaranthaceae [144]. The Compositae are viewed as one of the most advanced families in evolutionary terms, containing a high number of the widespread 'weedy' species, with features ensuring both survival under adverse conditions and a high reproductive rate. This family has many herbs and shrubs with relatively few tree species. Other important families are the Leguminosae (subfamily Papilionoideae), Euphorbiaceae, Labiatae, Cruciferae, Convolvulaceae, Cyperaceae, Solanaceae, Umbelliferae, Rosaceae, Scrophulariaceae, Polygonaceae and Malvaceae. Thus, weediness seems to occur very selectively amongst the angiosperms. The ratio of weed species between monocotyledons and dicotyledons is 16% : 84% (as opposed to a global ratio of 28% : 72%). The smaller than expected number of invaders among the monocotyledons may be partly explained by the almost total absence of weeds in the Orchidaceae (only a few species, such as *Arundina graminifolia* in Hawaii, are weedy), the largest monocotyledon family with about 20 000 species. The Pinaceae and the Cupressaceae are the weedy families in the gymnosperms.

2.4.2 The taxonomic distribution of invasiveness

Invasive species follow different patterns of taxonomic distribution and are found in a wider range of families than those in which weeds of cultivation occur. Data for agricultural weeds cannot therefore be used to draw taxonomic conclusions about invasive plants. Many of the conclusions on weeds above are quite different from those of an analysis of the list of invasives in this book. The sample of 80 *World's Worst Weeds* [178] are distributed among 28 families. However, a random sample of 80 of the invasive species treated in this book are distributed among 40 families. Weeds are therefore found in a smaller number of families than invasive plants, probably reflecting a greater variety of natural as opposed to weedy, habitats. The Leguminosae appear have have the highest number of invasive species and include serious invaders such as: *Acacia* spp., *Mimosa pigra*, *Sesbania punicea* and *Leucaena leucocephala*. The nitrogen-fixing ability of leguminous species may help them to invade nutrient-poor habitats. Many of the leguminous invasive species are rapidly-growing shrubs or trees, with copious production of seeds which often have a high ability to remain dormant and which are efficiently dispersed by birds or water. The majority of the invasive legumes are in the subfamily Mimosoideae, whereas most of the weedy legumes are in the Papilionoideae. In the protected areas of New Zealand the commonest family for invasives is the Leguminosae (14%) and the only other significant families are the Compositae, Rosaceae, Pinaceae and Gramineae (8% each); 27 other families are represented [395].

Examination of the plant list in Chapter 5 reveals that, within the dicotyledons, plants with predominantly unjoined (free) petals (such as the Leguminosae) are often the worst invaders of natural or seminatural habitats. These species often have characters such as bird or mammal dispersal, high seed production and the tree habit, all of which are associated with highly invasive species [122]. In contrast, more specialized plants with joined petals (such as the Compositae) tend to be more weedy, invading highly disturbed habitats, and are often associated with characters such as wind dispersal and the herbaceous habit.

2.5 Predicting invasion

2.5.1 The ecological theory of invasion

Can the invasiveness of a plant, or the susceptibility of a habitat to invasion, be predicted from theory? Two types of hypothesis can be recognized, general theories of invasion, which have little predictive value, and hypotheses relating to specific ecological situations [194] which may give more reliable predictions of invasion in certain more specified instances. Examples of these two types of hypothesis (which may not be mutually exclusive) are given below:

1. *Absence of predators hypothesis* Invasion occurs in the absence of a species' natural pests and diseases, which are usually not introduced along with the invasive plant. *Pinus radiata*, *Myrica faya* and *Clidemia hirta* have been reported to grow and reproduce more vigorously in the absence of specialist pests and diseases [170, 276, 435]. In addition, some species are unpalatable to many generalist herbivores (e.g. *Sesbania punicea* and *Rhododendron ponticum*). Predator-free invasives may have an advantage over the native flora, but predators do not always limit the growth of a plant. In cases where a major contributor to death is crowding of plants (density dependent mortality), pests may have little effect on the plant population [175].

2. *Greater reproductive potential hypothesis* Rapid invasion is made possible by the greater reproductive potential of the invading species than those in the community being invaded. Many invasive plants produce prodigious quantities of seed, which are often retained in the soil as seed-banks. Such seed-banks may be much larger than those of native species [180, 233].

3. *Poorly adapted native species hypothesis* Invasions occur when the native species are not 'well adapted', and when invasive species are more tolerant of suboptimal levels of resources, thus having a competitive advantage. In only a few cases has this been studied experimentally in native and alien species. However, in South Africa different patterns of resource utilization between the native *Protea repens* and the invasive *Acacia saligna* seem to account, at least in part, for the invasiveness of the Australian *Acacia* [451].

4. *Chemical change hypothesis* Plant invasions occur after the chemical characteristics of a habitat have changed, for instance by fertilizer or sewage pollution (eutrophication). This is commonly used as an explanation of aquatic plant invasions, by species better able to take advantage of a superabundant nutrient supply.

5. *Balance of nature hypothesis* Elton, in his classic work on biological invasions [113], suggested that higher 'complexity' of a community leads to greater stability. The more complex a community, the more resistant it is to invasion. Complexity is defined as the total number of interactions between all organisms in the community. However, this assumes that ecosystems are in 'balance', for which there is little evidence, rather than constantly reacting to spatial and temporal environmental variation (the individualistic hypothesis) [163, 164].

6. *Empty-niche hypothesis* The presence of an 'empty niche' allows invasion, as has been argued for the invasion of *Pinus lutchuensis* in the Bonin Islands [355]. Invasions of this type should be of the 'fitting in' (Einpassung) rather than 'pushing out' (Verdrängung) type. If immigration and evolution fill all a community's

niches, then that community is said to be 'species-saturated' and so immune to invasion. However, there is little evidence for this [449].

7. *Disturbance produced gaps hypothesis* Disturbance is commonly supposed to be important or even essential for invasion by plants, especially where disturbance results in reduced competition [169, 217, 321]. Increased disturbance, for instance due to greater fire frequency [54], may increase the invasibility of a habitat by lowering competition for suitable establishment sites and then, once an alien plant is established, by reducing competition with the native flora. However, there are some examples of alien plants invading relatively undisturbed vegetation such as *Pittosporum undulatum* in Jamaica, *Pinus radiata* in Australia and *Maesopsis eminii* in the Usambara Mountains, Tanzania. However, most invasions do seem to be dependent on the scale and type of disturbance, together with the numbers and timing of invasive propagules deposited in the community per year. Pines, for instance, tend to invade disturbed habitats at all latitudes, particularly grasslands which are often maintained by disturbance such as grazing or burning [327].

The average proportion of bare ground has even been suggested as a workable predictor of the invasibility of ecosystems [83], reflecting the frequency and intensity of soil surface disturbance. However, the amount of bare ground in a community is only one measure of disturbance. Although disturbance is usually defined as the removal of plant biomass [142], other activities, such as alteration of soil nutrient levels or water regime, are often referred to as disturbance (in a broad sense) and may be more subtle and difficult to measure. The concept of 'safe sites' [154] has been applied to invasion [194]. Invasions are said to occur at a rate determined by the intrinsic population growth rate of the invader and the availability of 'safe sites', free of specific hazards such as competition or shade – as in a forest gap. The corollary of this view is that invasions take place under conditions of zero environmental resistance (through safe sites) and the rate and occurrence of invasion can theoretically be predicted from the intrinsic properties of the organism and the habitat. This contrasts with hypotheses of invasion that emphasize the competitive or reproductive superiority of the invader.

2.5.2 Examples of characters associated with invaders in particular habitats

In considering theoretical aspects of invasion in general it is helpful to classify invasive species into the following four main 'ecological groups' with associated collections of characters ('syndromes'), with respect to the habitats they invade. Note, however, that there are many exceptions; the following are merely some of the more common types.

1. *Aquatic habitats* Syndrome: aquatic invaders are generally tolerant of a wide variety of aquatic conditions, have very effective means of vegetative reproduction, grow rapidly, have a free-floating habit, are perennial and herbaceous. They

are often dioecious, but are capable of vigorous asexual reproduction if one sex is not present. Where reproduction is sexual, seeds are efficiently dispersed by water. Examples: *Crassula helmsii*, *Eichhornia crassipes*, *Hydrilla verticillata*, *Lagarosiphon major*, *Myriophyllum aquaticum* and *Salvinia molesta*.

2. *Forest and open woodland habitats* Syndrome: the most 'important' invasive species (those having the most serious impact on natural or seminatural habitats) appear often to be shrubs or small trees, with high seed production, bird-dispersed seeds, rapid rates of growth, early reproductive maturity and which are shade-tolerant. Examples: *Acacia melanoxylon*, *Myrica faya*, *Clidemia hirta*, *Passiflora mollissima* (climber), *Pittosporum undulatum*, *Psidium cattleianum*, *P. guajava*. Exceptions include *Pinus radiata*, which invades undisturbed *Eucalyptus* woodland, and *Rhododendron ponticum*, which invades woodland and other habitats: both have wind-dispersed seeds.

3. *Open habitats* Syndrome: generally herbaceous (perennial herbs or small shrubs), have high seed production, early reproductive maturity, wind-dispersed seeds, often vegetative means of reproduction and rapid rates of growth. They are most commonly dicotyledons from those families characterized by joined petals (these are families that have many weedy members, like the Compositae) or, more rarely, monocotyledons. Examples: *Ageratina adenophorum*, *Erica lusitanica*, *Eupatorium odoratum*, *Mikania micrantha*, *Nassella trichtoma*, *Nicotiana glauca*.

4. *Fire vulnerable habitats* Syndrome: these are fire-adapted invasive species which often promote increases in the frequency of burning or its intensity by producing burnable growth, such as leaves and stems which may contain flammable oils. These are often grasses or shrubs, with vigorous vegetative reproduction (grasses) or high seed production (shrubs or small trees). The seeds usually have mechanisms to survive fire and to delay dispersal (serotiny) or germination, until stimulated by fire. The seeds are usually numerous, light and wind-dispersed. Other characteristics include vigorous resprouting after disturbance, rapid rate of growth and early reproductive maturity. Examples: *Andropogon virginicus*, *Pennisetum setaceum* (grasses), *Hakea sericea*, *Melaleuca quinquenervia*, *Pinus radiata* (shrubs/trees).

2.5.3 Practical prediction

These ecological groups and syndromes of characters are of course extremes, but they represent a series in which there is an increasing amount of disturbance in the ecosystem (disturbance being defined as removal of biomass, whether this is natural or human-induced). The combinations of characters mentioned above may be pointers to invasive potential. Indeed, in attempting to predict invasion, the attributes of the habitat and the species should be considered together. Are the ecological requirements of the species provided by the habitat? An examination of the invader in its native habitat should consider the following: the position of the species in succession, e.g. pioneer or late successional species; its environmental requirements – water or fire regime, nutrients, natural disturbance regime (e.g. frequent hurricanes, wind throw, flooding); its interactions with other animals and plants such as pests and diseases and competitive ability. Dominance or rarity in its native habitat is not necessarily a good predictor on its own, as species apparently of modest vigour have become invasive when introduced elsewhere.

It may not be possible to develop general models or theories which will successfully predict invasion for all species and in all habitats. However, studies restricted to particular species and specific habitats can achieve a high degree of success. As discussed in the section on dispersal, *Acacia* species with a high reproductive output and bird-dispersed seeds should come under immediate suspicion in suitable habitats. Pioneering work on the mountain fynbos has created a flow chart solution for assessing species risk, although it may be easier to predict species that are of low risk rather than to identify the major invasives of the future [330]. An example of ease of prediction within a single genus is provided by *Pinus*. The most invasive pines conform to a remarkably uniform syndrome of characters: early reproductive maturity, fire-adapted serotiny and small seeds with relatively large wings. Extreme caution should be exercised with any pine with this combination of characters. These empirical data on pine invasion have been used to predict risk species in the Australian genus *Banksia* which has analogous ecological behavior to *Pinus*. On this basis, several taxa, such as B. *burdettii*, B. *hookeriana* and B. *leptophylla*, appear to be high risk species if introduced to habitats at risk from invasive pines. There are complicating factors however, such as the susceptibility of B. *burdettii* and B. *hookeriana* to infection by the root-infecting pathogenic fungus *Phytophthora cinnamoni*, which may limit invasiveness [331].

Studies that consider the biological and ecological characteristics of the alien species, and the attributes of both the native habitat and the habitat to which it will be or is introduced, may in time come to achieve a high degree of predictive success. However, the huge volume of species introductions (many accidental and unrecorded) to so many places precludes the possibility of detailed studies for all aliens. Nevertheless, in environmentally sensitive areas the stakes are high enough to investigate thoroughly those species which pose potential threats.

3

Action against invasive
plants

The main categories of action to be taken concerning invasive plants are the following: education and awareness, legislation, prevention of introduction, information, and control. All are essential for effective measures against invasives.

3.1 Education and awareness

The movement of plants by people from one region to another ('plant translocation') is as old as agriculture and is a characteristic feature of human behavior. Not only have colonists carried with them plants as a means of livelihood and accidentally as weeds, but colonists commonly have a desire to 'transport landscapes', as shown by the Europeanization of the landscape of New Zealand by the mass introduction of homely plants. Furthermore, there is a tendency to believe that 'the grass is greener on the other side of the fence' and that the importation of new plants will improve the beauty or utility of a place more readily than proper or improved exploitation of the native flora. In reality, it may lead to devastating landscape change if introduced species become intractable. General awareness of the problem is essential. This involves awareness of the following: (a) the difference between native and alien (exotic) flora; (b) the importance of native plants over and above alien plants (a pride in uniqueness [360]); (c) that alien plants can, in some instances, threaten native ones; (d) that apparently harmless activities, such as gardening or forestry, can result in the naturalization of plants.

Education is essential in order to minimize accidental introductions of species. Education should foster awareness of the potential threat that alien species pose to natural ecosystems and awareness that plant introductions should be accompanied by assessment of their potential to spread and cause damage. Those who are mainly responsible for the introduction of alien plants (such as agriculturalists, horticulturalists, foresters and tourists) need to be made aware of the dangers. Ideally, environmental studies should be a required element in schools, adult education should be conducted through the mass media and tourist education should be introduced by means of visitor literature. Tourist education can also take the form of information video-films for showing on aeroplanes prior to landing [360]. The general public contribute to the spread of weeds, but they could equally well act as watchdogs against the problem if properly informed [187].

Environmental education receives little or no attention in many countries. In 1989, New Zealand, with a population of 3.3 million, had only 30 full-time specialists in environmental education assisting classroom teachers and Hawaii had an even lower ratio of three such professionals at State level for a population of 1.2 million [128]. However, curriculum requirements are changing and, in New Zealand, the environmental component taught by general teachers is increasing.

Staff members who work in nature reserves also need training and this applies to both managers and operations staff. Managers have to allocate resources, often from dwindling budgets, to competing worthy management needs; they must be fully aware of the importance of stemming plant invasions before they get out of hand. Invasive plant control is often the management task most likely to become impossible or costly if neglected, and is best done before the plants become a problem. Operations staff should be aware of the importance and purpose of what are often painstaking and tedious control exercises. They should also be aware of the potential effect of their activities on plant invasion, for instance allowing the introduction of plants by poor equipment and vehicle hygiene, and creating sites for invasion by disturbing native vegetation.

Educational booklets are important, and these should be colorful and easy to use, such as the booklet *Making Your Garden Bush-Friendly*, aimed at informing Sydney suburban gardeners about the dangers to the bush from dumping garden waste in the forest and from seeds washed into the forest by stormwater drains [270]. *The Banana Poka Caper* introduces *Passiflora mollissima* in Hawaii as 'public enemy number 1' with its 'gang' (dispersal agents) [370]. However, it should always be remembered that active involvement with projects in the field is by far the most vivid educational experience.

In several countries conservation groups organize events where laypeople volunteer to work in nature reserves removing problem plant invaders, such as *Rhododendron ponticum* in Ireland. These activities serve to raise awareness about conservation and the problem of invasive plants, as well as achieving some control of invaders in reserves, and are usually thoroughly enjoyed by the participants.

3.2 Legislation

Importation and spread of invasive plants can be prevented by either voluntary restraint or legislation, building on educational awareness. Databases or weed-lists may alert those involved to risk species, based on experience in other regions. The IUCN position statement on translocations [189] advises the adoption of legislation to curtail introductions; it proposes that deliberate introductions should be subject to a permit system. Governments should be aware of international agreements relevant to translocation. These include: the ICES Revised Code of Practice to Reduce the Risks from Introduction of Marine Species (1982); ASEAN Agreement on the Conservation of Nature and Natural Resources; and the Protocol for Protected Areas and Wild Fauna and Flora in Eastern African Region. Special care should be taken to prevent invasive species spreading across international boundaries: the Stockholm Declaration on the Human Environment (Principle 210) enjoins states to 'ensure that activities within their jurisdiction or control do not cause damage to the environment of other states'. Similarly, Article 196.1 of the convention on the Law of the Sea requires states to 'take all measures necessary to prevent, reduce or control . . . the intentional or accidental introduction of species, alien or new, to a particular part of the marine environment which may cause significant and harmful changes thereto'. There are binding obligations (section 11.2.b) in the Council of Europe's Berne Convention on the Conservation of European Wildlife and Natural Habitats calling for action to 'strictly control the introduction of non-native species' [294].

Laws at the level of individual countries are required too. Most import control legislation, such as the Australian Federal Quarantine Act of 1908, controls weed entry by restricting the importation of certain weeds. However, an alternative approach is to prohibit all non-native plants unless the importer can show that the plant will not become weedy: 'guilty until proven innocent'. Unfortunately, plants are more likely to be placed on the schedules after they have become serious problems, not before. Most plant quarantine legislation is anyway aimed at agricultural pests and is rarely implemented seriously in the case of 'environmental weeds' (invasive plants). In the United States the importation of plants is controlled under the Plant Quarantine Act of 1912 and importations need United States Department of Agriculture (USDA) approval, although again this is mainly aimed at agricultural pests.

In cases where invaders are distributed by the horticultural trade, such as *Lythrum salicaria* in North America [392], these laws could be effective at shutting down movement of invasive wild-type horticultural stock. Enforcement should be visible, both for the deterrent effect and to increase public awareness of the problem. Airlines should distribute, and require passengers to fill out, the necessary declaration forms that are required by most countries for living plant material. In Hawaii for instance the import of plants from other states of the US is controlled under the Hawaii Revised Statutes (HRS) and regulated under Department of Agriculture Administrative Rules (DOAAR). If plants are introduced and do become invasive, an 'introducer pays principle' has been suggested [360], whereby the introducer is liable for the costs of clearance operations. Although these measures are excessive for countries with little or no invasive plant problem, in highly invaded subtropical countries (for instance) they may be not only acceptable (especially in conjunction with public education programs) but also essential.

Transport of seed is regulated usually by separate legislation, such as the United States Federal Seed Act of (1939) 1980. However, the ease of seed transport by mail and the difficulty of extensive inspection of mail make this legislation rather ineffective.

Laws requiring the clearance of invasives may be enshrined in protected areas legislation. In New Zealand the National Parks Act 1980 requires that 'native plants and animals of the parks shall as far as possible be preserved and the introduced plants and animals shall as far as possible be exterminated' – an onerous task given that there are so many introduced species [445]. This act is now seldom used to exterminate species and some species, such as Russell lupins (*Lupinus* hybrids), are tolerated in the parks because of their tourist value. Many countries have legislation allowing certain plants to be given 'noxious weed status'. This empowers the state, often in the form of a local weed board, to force landowners to control weeds on their land. Although this can be useful, it can also be a mixed blessing for reserve managers faced with bills for spraying plants which may not even be those prioritized by the manager for control in the reserve. Noxious weed status is usually conferred on plants which already exist as large infestations, calling its utility into question. In the USA only weeds in their invasive phase are treated as noxious and, once completely established, they are no longer listed, which again limits the usefulness of this legal device to conservationists. However, declaring a plant a noxious weed does draw it to the attention of the public [187].

3.3 Preventing introduction and spread

Introduction may be defined as 'the intentional or accidental dispersal by human agency of a living organism outside its historically known native range' [189]. The overwhelming majority of invasive plants have been introduced deliberately and the largest number of these can be traced to gardeners. Some 42% of South African aliens were originally introduced as ornamentals [184] and some 50% of the naturalized plants of Victoria, Australia are readily available in nurseries. In Hawaii, botanic gardens and private gardeners have both contributed, as have agricultural stations [360]. Agriculture and forestry follow as the next most significant sources. Usually little thought is given to the impact these alien species may have on native animals and plants or their ecosystems. Very large numbers of plants have been translocated in this way, but most introduced species do not naturalize and most of those that do naturalize do not become important invasives. However, if plants are translocated in sufficiently large numbers, serious invasives are likely to arise sooner or later.

Foresters continue to introduce rapid-growing woody legumes into many developing countries for reforestation and 30 of these species are known to have become major pests [184]. The following agencies have programs for the systematic exploration, documentation and collection of tropical tree genetic resources for distribution: ACIAR/CSIRO in Australia, CAMCORE/North Carolina State University (USA), the Oxford Forestry Institute and International Plant Genetic Resources Institute (IPGRI) (Italy). These programs are in response to very serious problems of environmental degradation, poverty and fuelwood shortage. However, native alternatives are not usually properly investigated and unwise introductions have been made. A rational approach to introductions is needed [184,189] and the national administration concerned should take the lead. Firstly, governments should formulate national policies on translocation, prevention and control of invasives, based on adequate scientific consultation. Regional development plans that may increase landscape invasibility should take invasives into account. Secondly, pre-existing governmental structures for agriculture and conservation should be used to control introductions, collect information on introductions and conduct scientific research. Government sponsored landscaping (for instance in Hawaii [361]) should use species known not to naturalize. Here it is best to err on the side of caution as the situation can change with the introduction of new seed dispersers or pollinators. *Citharexylum spinosum* has been used extensively as an innocuous landscaping tree in Hawaii, but since the escape of the red-whiskered bulbul (*Pycnonotus jocosus*) on Oahu it has spread and is rapidly becoming a problem.

3.3.1 Planned beneficial introduction

For planned beneficial introductions, the following procedures could be adopted to ensure safe introductions.

1. *Assessment phase* In this phase, basic biology and potential invasive hazard should be examined. The simplest indicator of invasive potential is a thorough review of the literature to see if the species has become invasive elsewhere, while bearing in mind that many invasions have not been adequately reported. The assessment should be carried out in the light of the ecology of the relevant natural habitats, the likelihood of the plant hybridizing with native species to produce new aggressive species or biotypes [1] and prospects for eventual control. Characters, such as breeding system, dispersal mechanism, growth rate, seed production and life cycle, may indicate invasive potential. Habitat information may also be assessed. At the very least this should include assessment of climate and frequency of fire in both the native habitat of the plant and the proposed area of introduction.

2. *Experimental trial* If initially carried out on a small scale this permits complete eradication if there is cause for concern. It is important to ensure that the same genotype is tested as the one eventually introduced.

3. *Extensive introduction* This should be carried out preferably only in areas with little conservation or economic value, for example areas highly disturbed by human activity and away from natural vegetation. Suitable sites might be those resulting from reclamation work in areas denuded of native vegetation by mismanagement, overpopulation, pollution or erosion. Any such introductions should be monitored carefully.

4. *Recording of information* This needs to be in a readily available form, such as a database of invasive plants or potential problems. Records of the introduction and any trials should be kept and made public.

5. *Agreement of liability* It should be clearly agreed at the outset if organizations introducing plants are required to bear the cost of control, should control become necessary.

3.3.2 Accidental introductions

Accidental introductions are much more difficult to foresee and control. However, particular care should be taken with islands and other isolated habitats, for instance by insisting that visitors observe strict hygiene – removing seeds or fruits from shoes, clothing and tents (mud on shoes and trouser turn-ups are traditional routes for plant introduction). The contamination of agricultural seed and other materials should also be controlled. Military bases on islands present particular difficulties, as the large amount of freight transported is uncheckable due to military secrecy [360]. The effect of major engineering projects transporting vast quantities of material and creating disturbed sites potentially suitable for invasion should also be considered [4]. At sensitive sites vigilant surveillance should be maintained as much as possible. In 1984 a scientist visiting Gough Island was able to pull out the South African weed *Senecio burchellii* which had become established around the magnetometer hut, probably introduced with construction materials [420].

The prevention of introductions must be built on existing educational awareness and legislation. These have been discussed above, but explanatory leaflets, in-flight videos shown on passenger aircraft, increasing public awareness of legislation (often people don't know that the relevant laws exist) are obvious suggestions. Introductions can come from unlikely sources: in 1983 some 800 athletes agreed to have the soles of their running shoes scrubbed by volunteers before a marathon run through the Volcanoes National Park in Hawaii. From the washings, 16 species of plants germinated, including one serious invader – *Melinis minutiflora* (although this plant was present in parts of the Park already) [167]. It is not just important to lobby the public, but politicians may be lobbied for vigilant enforcement of legislation. Likewise horticultural firms, garden societies, forestry organizations, developers and tourist operators may be more influential than individual members of the public, and thus may be more suitable targets for lobbying.

Most ecosystems would probably be less invasible if they were not subject to some form of disturbance by people, and habitats should be managed to minimize disturbance and optimize resistance to invasion. Where invasive species are dependent on alien animals for seed dispersal (e.g. *Passiflora mollissima*, dispersed by feral pigs in Hawaii) as much care should be taken concerning the introduction of the animal as with the plant. Containment of an invasion is an important prevention measure. For instance, *Mimosa pigra* has the potential to spread very widely in northern Australia outside its present invasive range. Even though it is expensive, the patrolling of Arnhem Land by the Northern Lands Council to identify incipient invasions has been recognized as a priority. Similarly, careful hygiene should be observed where an invasive has the potential to spread. In the Kakadu National Park vehicles are inspected for propagules of *Salvinia*, an ecologically damaging invader [187].

3.4 Information and recording

The species accounts which follow in Chapter 4 give some indication of the research and information needs of a program to prevent or control invasion. Useful information for the prevention and control of plant invasion is categorized below.

3.4.1 Information about invasive species

Databases and weed lists published in various countries give information on which species are likely to become invasive (see also Chapter 5 and Appendix 2). Knowledge of risk species and patterns of invasion is the basis on which rational policies of import control and reserve surveillance can be built. A single database, global in scope, maintained by international collaboration would be ideal, although expensive, but no such database exists. A more practical solution would be a network of local databases capable of exchanging information in a coordinated way.

3.4.2 Taxonomy and biogeography

Correct identification of the species is essential, but further taxonomic study may be needed (e.g. if the species is thought to be a hybrid). The importance of taxonomic work is evident from the case of the *Lantana camara* complex, in which there are numerous biotypes differing in invasiveness, and from the examples of *Salvinia molesta* and the *Euphorbia esula/waldsteinii (virgata)* complex, in which biological control proved difficult or impossible in the absence of detailed taxonomic studies [157, 196]. Taxonomic studies will often reveal biotypes that are non-invasive or even sterile, which may be safely used in the horticultural trade.

3.4.3 Pattern and rate of spread

Accurate recording of alien species is necessary, particularly their patterns of spread, from the original sites of introduction into seminatural or natural vegetation. Many species are introduced accidentally, and prediction of potential to invade is often not possible or very precise. For these reasons, accurate recording of alien species and their rates of spread when they first appear are important pieces of information for attempts to evaluate population dynamics and invasive potential. Valuable information about invasive behavior can be gained from historical studies, for instance using old herbarium specimens and other evidence. The

41

number of plant species invading native bushland in Australia has been estimated at 'thousands' [187] and in this case a uniform geographical database has been suggested for analysis and planning. Information for such a geographical database can come from printed records, herbarium specimens, aerial photos, questionnaires, remote sensing and field survey.

3.4.4 Impact assessment

When an alien species is spreading, its likely impact should be assessed. Information should be collected to evaluate its likely further spread and likely effects on native organisms and ecosystem processes. This impact assessment is important in deciding which species should be given priority for control, as ecosystem effects have important consequences for conservation and control [51, 141, 429]. Impact assessment is also an important stage in the cost/benefit analysis involved in planning control measures, particularly biological control.

3.4.5 Ecological information on plant invaders: growth, reproduction and survival in relation to the environment

Biological and ecological characteristics such as the following, in both native and introduced habitats, should be noted to aid the prediction of spread and control:

1. Seed dispersal mechanism – adaptations for both short- and long-distance dispersal.

2. Seed ecology – high or low seed production; continuous or seasonal; longevity of seeds; dormancy mechanisms.

3. Breeding system – adaptations for self-fertilization and cross-fertilization; ability to reproduce vegetatively.

4. Rate of growth under favorable conditions.

5. Ability to resprout after cutting.

6. Requirements for germination and establishment.

7. Environmental factors – tolerance of frost, fire and shade; nutrient and water requirements, etc.

8. Susceptibility to pests and disease – generalist or specialist pests.

9. Comparative ecology of the invasive species with that of those species, native or alien, which are most likely to replace it after control measures have been taken [429]. There is no point in eradicating an invader to have it replaced by a worse one.

It is important to establish the native range of an alien, along with the biological and environmental characteristics of its natural habitat. This information invariably gives useful insights into the biology and ecology of the plant, and thus into its invasive potential.

3.4.6 Information on control: the effects of herbicides and potential for biological control

The information detailed above will help managers of nature reserves evaluate the threat posed by an invasive species and to draw up a suitable control program, if it is necessary. However, whatever control method is chosen further information is required in implementing it. If it is decided that chemical control is a necessary option, special care is needed. Plant responses to chemicals differ widely, and it may be necessary to conduct trials of different chemicals at different application rates to different parts of the plant, and at different times of year, to optimize effectiveness and minimize cost. With physical control too, the plant and ecosystem responses vary according to method of cutting and time of the year. This sort of information is helpful to maximize cost-effectiveness. Biological control requires detailed taxonomic and ecological information about the plant and biological control species, both in the native and invaded range, together with information about all other species in the ecosystem that may be affected. Nevertheless, biological control may offer the best possibility for many species in terms of minimizing the impact on the native flora and fauna.

3.5 Planning a control program

As a last resort, where prevention has failed, there are four main options for control (although they are often used in combination). These are: **physical, chemical, biological,** and **environment management** control. Control measures should have careful **planning** – strategic and tactical decisions must be made at the outset when the **coordination of control methods (integrated pest management (IPM))** must be considered. This section considers the issues involved in planning. As Coblentz [73] has suggested: 'Programs to eradicate exotic organisms provide an opportunity to combine good science and good conservation into functioning conservation biology'.

3.5.1 Prioritizing species

The choice of species for control priority should take into account that prevention is the best method of control and that early intervention is desirable. Control should therefore be aimed at newcomer invaders as well as established ones. Here, delays can be critical as most successful invaders reproduce readily – *Salvinia* (an extreme case) can double its population in 10 days. Unfortunately, funding is much easier to obtain if an invasion is already very serious. On the other hand, ecological research and research on biological control have high capital outlay and work on widespread invaders can maximize the return from research input. For instance, research on the biological control of *Salvinia* has been applied all over the tropics, with transfer of technology between regions at comparatively low cost. Although widespread invaders are often very difficult to eradicate, such methods can reduce their impact.

Box 3.1 Control and conservation

Almost any program of control of invasive plants raises conservation questions which are impossible to resolve fully. If the use of pesticides is contemplated, then attention should be given to the possible effects on non-targeted organisms both at the site of application and more distantly. If biological control is envisaged, then consideration must be given to possible adverse effects on the native fauna and flora of organisms introduced as control agents.

Most human action has some environmentally damaging consequences. In the case of invasive species it is up to managers and their scientific advisers to evaluate which control measures, if any, are most appropriate in particular cases. There is also a need to be acquainted with relevant laws and regulations and to take all recommended safety precautions. The intention of this manual is not to make decisions for managers; but to provide information which will assist them in this task.

Decisions on whether and how to control invasive species can bring into sharp focus the difficulties of making wise environmental decisions. WWF insists that invasive plants sometimes represent serious threats to the conservation of biodiversity, and that their control is often highly desirable and in some cases urgent. On the other hand, it is also WWF policy to advocate reduction in herbicide use. As an ideal, WWF advocates integrated pest control measures (in which a systems model is used to to optimize control by different measures, while seeking to minimize ecosystem disturbance). However, WWF is also fully aware that this ideal has hardly been applied in practice to the control of invasive plants, and may be considered very theoretical by a manager faced with controlling a serious invasion on a minimal budget.

WWF advocates the precautionary principle in attempts to engineer the environment. Ecosystems are full of non-linear relationships and feedback loops, making it impossible to predict fully the consequences of human activities. It thus makes sense to take a careful step-wise experimentally-based (or adaptive management) approach in tackling plant invasions, reviewing progress and assessing side-effects on a continuing basis.

Further information on some known adverse environmental effects of some major pesticides is given in Appendix 1.

The most important criterion must be conservation impact and, where invaders are threatening other species with extinction, their eradication should always take precedence. Invasions in areas which are poorly known scientifically should also take precedence because there is a possibility that undescribed species may be threatened. Control should, however, be seen as the 'art of the possible' and no control measures, except perhaps containment, should be taken against species whose control is inherently impossible. Species with a realistic control potential should therefore be prioritized, particularly where there is a possibility of total eradication. Similarly, control measures should start with those species whose biology is well known so that the causal mechanisms of the invasion can be addressed, not just the symptoms. When the ecology and biology of a species are well known, there is a better chance of preventing re-infestation based on informed habitat management.

Careful consideration of the invaders on a case by case basis is often revealing, as has been done for the Kruger National Park in South Africa [247]. This reserve of some 19 400 km^2 of tropical and subtropical savanna is relatively little affected by invasive plants. The native flora is well adapted to regular fires and heavy grazing by large hoofed mammals, which keeps many alien plants at bay. However, it has 113 invasive plants [249], of which seven have serious ecological impacts. A long-term strategy of control has succeeded in eliminating 10 non-native plants from the park and control is feasible for another 14. The other plants are either 'out-of-control' (and thus control is too expensive to contemplate), not serious enough in their ecological impacts to merit control or regularly renewed from outside the park (so control would be pointless).

3.5.2 Prioritizing areas

In some situations a geographically planned approach will be more suitable than a species based one, especially where a variety of invaders are threatening localized reserves. In Hawaii, success has been achieved by concentrating control on small areas with high conservation value – special ecological areas (SEAs) [400]. In all planning a clear statement of what the control is intended to achieve is important; usually this is the protection of the most pristine and biologically important reserve areas. More generally, the aim of control should be to further wider conservation aims, usually those of the government or regional organization concerned or that of the World Conservation Strategy 'to conserve species diversity, genetic diversity and the ecological processes that sustain them' [187]. In the early stages of invasion a containment strategy may be adopted: early detection of the invasion followed by complete local eradication is used to contain spread, perhaps leaving the core-area of the invasion for later attention. Continuous monitoring is required, especially of non-reserve areas around key reserves. If an invasion is contained, it is not only very much easier to deal with but total eradication remains a possibility. Some nature reserves are intrinsically more invasible than others and this should be considered not only in reserve design, but also in planning control strategies. Reserves that are less likely to be re-invaded should be given priority. In a survey of 95 New Zealand reserves, the most invasible were found to be small, narrow (high boundary to area ratio) disturbed remnants with fertile soil close to towns, with road or railway lines nearby [394]. Similarly, ecosystems which have been severely damaged by invasive plants, such that rehabilitation with native plants is unlikely, should not command high priority.

3.5.3 Choosing control methods

Indiscriminate herbicidal or mechanical clearance is counter-productive in nature reserves, because this is likely to have adverse effects on the ecosystem as a whole and may endanger diversity and rare species. Instead, a control program of 'spot' herbicide use (minimum herbicide use, carefully targeted), careful physical removal or biological control must be tailored to the particular demands of the ecosystem. Systems of invasive plant control have occasionally been devised for particular vegetation types, such as the 'Bradley method' for Australian bushland near Mosman [50]. However, sometimes even the total destruction of all vegetation in a small area can be justified if this is the only means to save the reserve in the long term, as has been the case with *Pereskia aculeata* in coastal for est reserves of Natal, South Africa.

What is apparently the cheapest control in the short term is not always cost-effective in the long term. In agricultural systems, cost-benefit models, together with knowledge of patterns of weed spread, have been used to optimize cost-effectiveness [18, 19, 412]. In the case of *Nasella* in the Australian Tablelands, the control measure cheapest for the government (containment herbicide spraying of light infestations) does not give the greatest public economic benefit in the long term, mainly due to lost revenues ('opportunity costs') from highly degraded pastures. The expensive option of wholesale *Nasella* eradication and pasture rehabilitation makes economic sense in the long term [16, 20]. These analyses are difficult to apply to nature conservation because of the difficulty of quantifying the 'opportunity costs', not only of possible lost revenues from tourism and recreation caused by invasive plants, but also the intrinsic aesthetic, scientific and cultural 'costs' of landscape degradation by invasion. These may, however, be roughly quantified as the costs the state or the public are prepared to bear out of pride and concern for natural ecosystems (a function of environmental education).

Similarly, biological control may appear to be an unattractive option because of the very high outlay costs, typically requiring government support. The development costs of chemical control are typically less and borne entirely by the chemical industry and passed on to users. On the face of it, chemical methods of control may appear to be financially attractive, but may involve ongoing and repeated costs, so that in the long run biological control may thus be more cost-effective [396]. However, there is usually a problem in raising the required initial investment.

3.5.4 Setting up a voluntary action group

Informal groups ('hack groups') to work on their own or alongside reserve management teams have been formed very successfully in Australia and South Africa [299, 377]. Such groups are well advised to establish clear aims before starting, based on a careful assessment of the site. An action plan with a tightly defined, manageable objective is a good spur. Liaison with local authorities and local conservation organizations will prevent duplication of work. Provision of tools and instruction in plant identification must precede action and any special skills of group members should be used to the full. Safety and first-aid equipment should also be provided and appropriate safety procedures established. Written and photographic records of work done are valuable. Not only can 'before and after' comparisons be made, but good records can provide useful research data and help in the planning of further work. This information might include lists of invasive plants and natives, perhaps backed by a collection of pressed plants for reference. Identification problems should be solvable with the help of an herbarium or museum if adequate field-guides are not available.

3.5.5 Coordinating control measures

Integration of the various control and prevention measures is possible on two levels. Firstly, the different control, regulatory, research and conservation management bodies, both statutory and voluntary, should work in harmony. Secondly the control and management techniques used against invasives should operate in concert. Different control techniques used together can often strikingly reinforce each other. In one example, biological control of *Chondrilla juncea* by rust fungus gave 55% control and control by improved pasture competition gave 35%. However, the two used together gave 95% control [146]. Many approaches may be needed. Research to identify critical points in the life cycle will be ineffective unless this is translated into management action. Effective communication is important. Recently, attention has focused on the concept of 'Integrated Pest Management' (IPM) as a formalized, concerted, many-sided control strategy [44, 348]. A systems model is used to optimize control by natural processes, application of herbicides, biological control and other measures, while seeking to minimize ecosystem disturbance. Where adequate research data exist to support it, this approach holds out great promise.

3.6 Physical control

This form of control includes hand-pulling (for annual herbs and tree seedlings), cutting and slashing (for lianes, tree saplings and trees), digging/levering (with mattock or crowbar, for plants regenerating from underground parts, or tree seedlings/saplings which will regenerate from cut shoots) or mowing/discing (for herbaceous plants) [270, 299]. It may be very effective for controlling some species,

Figure 3.1 Physical control by hand-weeding, Mondrain, Mauritius. In this species-rich ecosystem chemical control cannot be used, so careful removal must be carried out by hand. The invasive privet, *Ligustrum robustum*, is a serious problem on the island, preventing regeneration of native species. It has now been removed from this plot, which is fenced to keep out introduced pigs and deer. (Photo: Alan Hamilton)

Figure 3.2 The boundary between a weeded and an unweeded plot at Mondrain, Mauritius. In the unweeded plot the mortality of native tree seedlings is much higher. No biological control system has been devised for *Ligustrum robustum*. (Photo: Quentin Cronk)

for example pine trees, which do not resprout or regenerate from underground shoots. However, many species do resprout after cutting and are only killed if chemicals are applied to the cut surface. Physical control of shrubs or trees is very labor intensive, especially if they cover a large area. Other forms of physical control, such as burning, which grades into environment management control (below), may be effective but can only be used where such techniques do not damage the native flora or fauna. Some aquatic weeds may be removed by hand or using large mechanical harvesters and the organic matter used as compost for neighboring agricultural land. This method is reasonably effective and produces a useful by-product for local farmers. However, aquatic weeds are never eradicated completely, and physical disturbance and removal may in fact encourage their spread. Thus, while for some species physical control is adequate and effective it does not work for many others, including most of the more serious invasive species, in which physical control is frustrated by features such as resprouting after cutting, a long-lived seed-bank, regeneration from fragments and copiously produced, efficiently dispersed seeds which re-infest from neighbouring areas. However, the tedious and painstaking work involved is often worthwhile in isolated ecosystems, such as small islets, where complete eradication is possible. Here extreme caution must be taken to prevent accidental re-infestation, or introduction of new species by the weeding parties. A supply of tents and bedding should preferably be left on the islet, to reduce the need for rigorous cleaning on each visit.

3.6.1 The situation in Mauritius

One of the most interesting and important studies of physical control comes from Strahm's work on the WWF project on rare plants in Mauritius [381]. This island is so badly invaded that only a very small part of the natural vegetation can realistically be preserved. Lowland vegetation has entirely disappeared except on two off-shore islands on which physical clearance projects have been initiated. These islands are Round Island (where rabbits were finally completely eradicated in 1986) and Ile aux Aigrettes. Much more upland vegetation remains, but almost all is badly invaded. As a holding measure the Mauritius Forestry Service, in association with WWF, is weeding eight small plots, covering a total area of 3 ha. As these are of great scientific importance, often as the last examples of particular vegetation types, and are rich in endemic species growing amongst the aliens, special care has to be taken with the weeding, which is consequently slow and painstaking work. In 1990 the Forestry Service in Mauritius spent 1969 work-days weeding the 3 ha of plots at a cost of Rs 275 730. Initially, at two sites (Macabé and Brise Fer) some plants of the main invasive species, *Psidium cattleianum*, were left to provide partial shade. Without this, the weedy herb *Laurentia* invades, but now the last *Psidium* plants have been weeded out. Hasty eradication of invasives can have deleterious ecosystem effects, the lesson being that, with severe infestation, clearance should proceed with caution [436].

There is evidence that repeated weeding of plots on Mauritius eventually results in a reduction in the effort needed to maintain invasive-free plots. In the case of the 5 ha reserve of Mondrain [380], the first year of intensive weeding (1989) required 594 work-days (although some weeding had been carried out previously), while in 1990 only 460 work-days were needed, representing a considerable decrease [381]. This suggests that given regular weeding the 'maintenance load' will eventually diminish, even in the most intractably invaded areas. However, the main invasive species at Mondrain, *Ligustrum robustum* var. *walkeri*, is very widespread and abundant on Mauritius and is bird-dispersed, so re-infestation will constantly occur and there are no prospects of total eradication.

The effect on regeneration of indigenous trees after weeding has been studied in a 1.3 ha plot at Brise Fer. The response is quite dramatic. In comparison with an adjacent control (unweeded) plot, over a four-year period there was more seedling recruitment (4×), better seedling survival (3×) and higher seedling growth rates (6×) [381].

On the off-shore islands of Mauritius, the invasive plant covering the largest area on Ile aux Aigrettes is *Flacourtia indica*, which became established after widespread military disturbance on the island during the Second World War. Unfortunately the fairly large stumps left after cutting resprout well, so that they have to be treated with 10% Tordon, which usually prevents regeneration. On Round Island the problem is quite different. The introduced rabbit had previously kept both the native and the introduced vegetation under severe check. However, after total eradication of rabbits by repeated capture methods in 1986 (with the purpose of allowing regeneration of the native vegetation), the introduced small weedy shrubs *Desmanthus virgatus* and *Desmodium incanum* expanded, requiring hand pulling. Fortunately, these species were only present as localized populations which can be eradicated, with the eventual aim of total elimination from the island. In this instance the ideal weeding party consists of six volunteer weeders, with the turnover kept down to two novice weeders per trip. There are a number of reasons why experienced weeders are at a premium: there is less chance of them pulling out important plants, and items of natural history interest will be 'old hat', so they will spend less time on what have been dubbed 'snake breaks' [381].

3.7 Chemical control

There are several problems with the use of chemicals in areas of conservation importance. Herbicides may reduce the numbers of an invasive species, but, unless repeatedly applied, they will not limit its spread or prevent re-invasion. Many herbicides are unspecific and may damage the non-target flora and fauna. Many are also very persistent and may accumulate in soil or leaf tissue. Herbicides are expensive and an unwelcome large recurrent cost in the budgets of reserve managers (although physical control may be more expensive). When infestations are large, the herbicides have to be mixed and transported in bulk, which is not easy in difficult terrain. If the species are fairly tolerant of herbicides, as are many woody invaders, and regenerate fast, the herbicides will have to be applied almost continually. However, herbicides are often the only option available to managers and are a front-line defence.

3.7.1 Methods of application

There are many different methods of application [270, 290]. Adequate protective gear should be worn when working with all herbicides. **Careful compliance with the latest directions on safety and use is essential.** Appendix 2 gives notes on some herbicides commonly referred to in the literature about invasive plants.

1. *Woody species*

Frilling (notching or 'hack and squirt'). In this method the base of the stem is notched with an axe or a machete at about 10 cm intervals all round, into a frill of bark slivers, behind which the herbicide (usually 2,4-D or glyphosate) is applied to the moist sapwood. Care must be taken not to bark the tree completely, as this will prevent transport of herbicide within the plant. The herbicide may be applied with a brush or squirted from a plastic bottle. Each stem of a multistemmed plant must be treated, as translocation is not very efficient between major portions of the tree.

Stem injection. A drill ('brace and bit') is generally used to penetrate below the bark (although not into the heartwood). An injection gun is then used to deliver a 2 ml herbicide dose into each hole. This herbicide is then taken up by the sapwood and translocated around the tree. Again, if the plants are multistemmed, each stem must be injected.

Side branch reservoir application. Herbicide is applied from reservoirs on lateral shoots. This is a technique that has been developed for the control of *Myrica faya* in the Volcanoes National Park in Hawaii [124]. A lower side branch is cut near the main stem and a length of surgical tubing is stretched over the cut end to form a small reservoir into which 1–3 ml of undiluted 'roundup' (glyphosate) is inserted; this is absorbed in 30–40 minutes. The technique is often easier than stem-injection and translocation around the plant may be more efficient. It is a potentially useful method when stem injection or frilling is time-consuming or costly because plants have multiple stems, or the main stems are too small to drill or made inaccessible by lateral branches.

Basal bark application. Herbicide may be painted or sprayed onto the bark at the base of trees. An oil-based formulation is used (often of 2,4-D), using a light oil (e.g. diesel) capable of penetrating the waterproof bark of the tree and being translocated (mainly upwards). The bark should be treated all round and as close to the soil as possible [290].

Cut stump application. If cut stumps are likely to sprout, they should be treated (usually with undiluted glyphosate or a 2,4-D formulation) within 30 seconds of being cut, when translocation from the damaged sapwood is still occurring.

2. Vines

Painting of stem bunches. Treatment of vines is often difficult as (with the exception of some large woody vines (lianes)) there may be no obvious main stem. Bunches of stems may be gathered up, cut and the cut ends painted with herbicide.

Stem scraping. Vines with aerial tubers need special care since, when the plants are cut, the tubers may fall from the dead stems and regenerate massively on the forest floor below. In such cases the plants should be left intact, with herbicide applied to a scrape on the side of the vine stem, allowing translocation from the exposed sapwood throughout the plant.

3. Herbaceous species

Foliar spray. This is the only economically feasible method for large scale application. It can be used at any scale from knapsack sprayers to aircraft. Application techniques can affect efficacy of application greatly, and modern electrostatic spraying techniques can reduce the volumes of herbicide needed and go some way to preventing inaccurate delivery of herbicides and drift [168]. Foliar herbicides should be sprayed on still, cool, dry days during the growing season. However, woody species are almost never killed by a single application (rapid defoliation is not necessarily an indication of kill). It is very important that the optimum timing, dose and choice of herbicide are determined before use. An alternative to foliar spraying is the use of a 'weed wand', which allows more accurate application to small plants by touching them with a wick.

Soil application. This is usually in the form of picloram granules scattered around the plant. Application by broadcasting the pellets is easy and avoids the need to transport bulky oil- or water-based preparations. Many plants are highly susceptible, even when other herbicides have proved ineffective. However, the persistence and mobility of picloram in the soil are serious problems and this method should not be used in watershed areas. It is not normally used in conservation work.

3.7.2 Choice of treatment

Plants vary greatly in their susceptibility to different herbicides, methods of application, formulations (carriers or surfactants), concentrations and times of application. Often very extensive trials are needed to work out the optimum treatment. This information is usually readily available for agricultural weeds but not for plants invasive of natural ecosystems. However, some work has been carried out. In Hawaii a trial of different herbicides on a range of introduced plants suggested (for instance) that the optimum control methods for Hedychium gardnerianum, Rubus argutus and Tibouchina urvilleana differed widely. The chosen methods were: broadcast picloram pellets (Tordon 10K), although problems with persistence were evident (Hedychium), 2% foliar spray of roundup in water (Rubus) and 20% Garlon 4 in diesel oil on cut stumps (Tibouchina) [344].

Different countries have their own regulations for the use of herbicides. In particular it is difficult to find herbicides that are safe to spray near water [304, 305]. Early control of *Lythrum salicaria* in the USA used Dicamba with modest success, but in 1982 a new formulation of glyphosate (Rodeo-EPA) with a new surfactant Ortho X-77 was approved for spraying over water, which aided control significantly. However, spray efficacy in this case is much greater if the herbicide is sprayed in the late flowering season (August). Not only is the kill rate better, but there is less re-infestation with *Lythrum* seedlings [392]. Glyphosate is a particularly useful herbicide, as it is inactivated almost at once in soil and is not thought to have any residual effects in terrestrial ecosystems [368].

The Kruger National Park in South Africa is an example of a natural area where herbicides have been used extensively [247]. In fact the chemical control measures are said to take up a large part of the reserve budget (58 000 Rand in 1983 for the control of *Lantana* and *Salvinia*). Very extensive whole plant spraying of *Lantana* and *Opuntia* has been carried out, initially using 2,4,5-T (which was discontinued after concern about dioxin contamination emerged), now with glyphosate. On the whole this program has been successful and *Opuntia* has been effectively eradicated from the park. However, some concern has been expressed about effects on non-target species. Likewise *Salvinia* appears to have been eradicated by aerial spraying of clarason, but re-infestation by this water-borne plant is almost certain to occur eventually. Control has been least successful where circumstances promote regular re-infestation. *Eichhornia* is not controlled as it is abundant and uncontrolled in headwaters outside the park; *Melia*, which responds to ringbarking, spreads along the Crocodile River and complete control is impossible as it is not controlled on the south bank. In the case of *Xanthium*, the infestation is so massive that no attempts at control could possibly be realistic, although some cosmetic control is possible by mowing road verges. These examples highlight the weaknesses of chemical control and the need for integrated approaches.

3.8 Biological control (biocontrol)

This method uses natural enemies to regulate the numbers of invasive plants. When successful, the utilization of natural enemies is an inexpensive, non-hazardous means of reducing pest populations and maintaining them, often permanently, well below economic or conservation injury levels. The critical aspect of biological control of plants is the selection of organisms that are highly host-specific and will not themselves become pests on other plant species. Fortunately many invertebrates are highly host specific, but to ensure that host switching will not occur requires lengthy trials of the potential biological control agent with the plants it may encounter in the habitat to which it is introduced, as well as with the intended host. In the past, there have been notable disasters where organisms were introduced to control weeds with little regard to non-target organisms, but now extensive codes of practice and legislation are in place to set out required

Figure 3.3   The long-leaved wattle, *Acacia longifolia*, native to Australia, is highly invasive in moist sites in the Mediterranean-type climate zone of the southern Cape Province. (Photo: I.A.W. Macdonald)

Figure 3.4   The gall-wasp, *Trichilogaster acaciae-longifoliae*, was introduced deliberately from Australia in the mid 1980s, reducing the reproductive success of the wattle almost to nil. This has enabled managers of invaded nature reserves to concentrate on clearing out established stands without having to worry about further spread of this species. Biological control is less labour-intensive than physical control but the initial cost of development is often higher. (Photo: I.A.W. Macdonald)

procedures for testing and quarantine. Nevertheless, procedures can break down. For exaple, agriculturalists in Florida have introduced *Cactoblastis* near to sites for native *Opuntia* species, with the result that one species has been very badly affected. Despite these very real concerns about mistakes, there have been many well-regulated successes in regard to the use of biological control for weed species and, increasingly, biological control is being used for invasive species too.

3.8.1 Advantages

Properly conceived biological control is highly host-specific, with little or no ecosystem damage. However, there is still considerable resistance to biological control use, resulting from examples of disastrous ecosystem effects after some early introductions. Macdonald compared attitudes of North American and South African reserve managers to biological control and found the latter more sympathetic to it [246]. This he attributed to the fact that South African problems were more intractable. As more plant invaders get 'out-of-control' he predicted that biological control will grow in importance. For some invasives there is no alternative method. Biocontrol also has the advantage of permanence and, despite high initial development costs, may be the cheapest in the long term.

3.8.2 Disadvantages

Damage of native plants by biological control agents is a considerable worry but seems rarely to happen. This is partly because biological control organisms are now thoroughly screened against an extensive range of native plants. However, even when native plants are attacked to a degree in the laboratory this is rarely carried over into the wild, probably due to competitive exclusion by more closely adapted native insects [196]. Certainly there are cases of one biological control organism displacing another, such as the moth, *Cactoblastis cactorum*, replacing the previously effective coreid (squash bug) *Chelinidea vittiger*.

One problem with biological control is the high failure rate, with some 60–75% of control attempts proving ineffective [196]. This may be due to failure to establish, to the established population densities being too low or to ineffective control even after successful establishment. A biological control agent must still be able to survive, even after it has brought the invasive plant down to a low level. However, the rate of success is likely to increase with better procedures for selecting control agents, based on better understanding of the ecological mechanisms behind population regulation [341, 419] and more accurate taxonomy at the species and biotype level (the better to match invasive plant and control organism). The taxonomic problems inherent in the biological control of *Salvinia molesta* and the *Euphorbia esula/virgata* complex have already been mentioned. Looking for control organisms in inappropriate biotypes, because of misidentification, can be an expensive mistake.

Another problem with biological control is that its development is slow and expensive. Programs for a single plant may take 5–10 years from conception to release and the cost can be in excess of £500 000 (US$ 815 000). A Canadian study in 1979 estimated biological control costs of plants by insect-agents at $1.2–1.5 million per species [155]. Once released, the beneficial effect can be further delayed by the length of time taken for the biological control agent to build up its numbers [263]. However, perhaps the most important problem is due to often sharply conflicting interests surrounding biological control [5]. These conflicts may be economic (the invasive plant may be a honey source), ecological (in places, invasive plants may bind the soil and prevent erosion) and aesthetic (invasives can sometimes be very attractive and can be stoutly defended by local people). Grasses are used as food and fodder crops throughout the world and any proposed biological control measure for a grass species would immediately provoke suspicion, and hence conflict. A conflict arises too when the native and non-native range of a species lie close together. There is thus a conflict over the possible introduction of the South African noctuid, *Conservula cinisigna*, to control *Pteridium aquilinum* in the UK where it has become a weed of pastures [225], since this introduction would also affect the perfectly natural populations of *Pteridium* which are characteristic of certain woods and heaths.

3.8.3 Cost-benefit analysis

Biological control of *Lythrum salicaria* in North America is highly promising, as it is attacked by 120 species of plant-feeding insect in Europe. A cost-benefit analysis has been carried out, which estimates costs of *Lythrum* control at $1.7 million per year (mainly the cost of lost honey production after *Lythrum* eradication, since the direct annual cost of the biological control program is only $100 000) [392]. The benefit is estimated at $45.9 million per year (which includes increased land values, wild hay, muskrat fur, duck hunting and wildlife tourism values). The cost benefit ratio is therefore 1 : 27, amply sufficient to justify a biological control program on economic grounds alone. Such exercises may be valuable in helping to commit resources, but may be a hindrance in natural ecosystems where the main benefit may be the prevention of the extinction of rare endemic species. Economic valuation of endemic species is difficult, although not impossible to compute in financial terms. Often the scientific and heritage value of endemic species may be much greater than their wildlife tourism or natural product value.

3.8.4 Biocontrol procedure

The procedure for classical biological control follows a number of steps.

1. Find region of origin of invasive plants. This will involve taxonomic and biogeographical research.
2. Examine the ecology of the target plant in its native area and identify candidate

biological control agents (such as invertebrates (principally insects) and diseases).

3. Organize with host country and program country arrangements and authorization for collecting and shipping candidates.

4. Construct certified quarantine facilities in program country for testing and screening candidates, with no danger of accidental release.

5. Determine biology of candidate biological control agents and methods for raising them in large numbers.

6. Evaluate candidates. Which ones are most likely to achieve control? Up to 20 candidates may have to be evaluated.

7. Test host specificity of likely candidates. This involves the intensive study of candidates over one to two years. It may involve investigating the feeding habits of up to 10 candidates on some 100 test species of plants.

8. Approve agent for release. Results of specificity testing are reviewed by expert panels of scientists, agriculturalists and government officers.

9. Release agent. A carefully planned release program should be conducted, with timing and locations planned to maximize impact and establishment.

10. Monitor the agent's populations for 5–10 years to determine the impact on the invasive plant or the reasons for failure, as well as effects on non-target species.

3.8.5 Prospects for biological control

In Hawaii, biological control of most of the introduced weeds is the long term goal of conservation managers. At present only 21 Hawaiian weeds are controlled biologically, following 70 insect introductions with a 50% success rate [263]. Biocontrol of invasive plants of natural areas poses special problems not applying to weeds. For instance, the insect quarantine facilities at Honolulu are at sea-level, whereas the invasive Passiflora mollissima will only grow at the altitude of the montane forests; an insect quarantine facility therefore had to be built in the comparative isolation of the Volcanoes National Park. This problem was solved by sealing and caulking the joints of a greenhouse to make it insect proof and then ventilating it with insect-proof fans. Entry is through two darkened anterooms with light traps and fully sealing doors, only one of which is open at any one time [125]. Taking the facilities to the reserve may be better than taking the reserve to the facilities, although prevention of escapes becomes doubly important. There is much scope for research and new developments, not only in classical biological control but also in other biological control fields. These include augmentative biological control (in which repeated releases of a non-persistent agent supplement other control [156]) and mycoherbicides (inundative biological control), in which a fungus is used in the same way as a herbicide [19, 159].

Mycoherbicides are a particularly promising development, although so far success has been achieved against only a small number of plant species. In some cases, common fungi have been used, such as *Colletotrichum gloeosporoides* and *Fusarium solani*, with specificity achieved by using a host-specific type ('forma *specialis*') [49]. Rust fungi have also been used for highly specific biological control [427]. Good control of *Morrenia odorata* (introduced from South America to the Florida citrus groves) has been obtained using *Phytophthora citrophthora* [334], by means of a fungal suspension which is stable for up to six months and is manufactured commercially [201]. After the initial kill there is some residual control from increased levels of pathogen in the soil. Similarly, *Eichhornea crassipes* has been successfully treated with the fungus *Cercospora rodmannii* used in conjunction with arthropod biological control for a reinforced (synergistic) effect [64, 67]. There are prospects of important further developments in this field.

3.9 Environment management control

Measures to reduce disturbance are measures that affect the whole ecosystem and are not just targeted at the invasives. However, this sort of 'environment management' is important in reducing invasion. Reduction of disturbance complements other control measures. Gap size is often crucial. In the Cape Forests of South Africa, natural gaps are usually very small, caused by single standing dead trees. Artificial gaps larger than 0.1 ha led to a deteriorated microclimate, dry soil and the establishment of herbaceous species, and there is no regeneration of native trees [126]. In these circumstances, as in disturbance at the forest fringe by fynbos fires, the invasive *Acacia melanoxylon* can invade.

The prevention of disturbance may require the resolution of conflicts. In Hawaii, introduced goats and pigs are major causes of disturbance but in some areas the State maintains populations of these animals for hunting [361]. Another conflict is the planting of alien grasses to 'improve' native rangeland [187]. Even if these grasses do not prove to be invasive, the disturbance of planting may allow other invasions. The control of invasives, particularly when these form large single species stands, is in itself a form of disturbance and some attention should be paid to the effect of clearance, in particular to its timing with regard to the regeneration of other species. In Australia, the clearance of invasive *Tradescantia* has been found to be followed only by massive regeneration of *Ligustrum*, another invader [378]. In these situations, it may be necessary to sow or plant non-invasive temporary replacement species. In the USA, removal of *Lythrum salicaria* from wetlands which experience lowering of the water-table in summer, merely results in massive germination of *Lythrum salicaria* seeds on the exposed mud. To pre-empt this, the non-invasive replacement species, *Echinochloa frumentacea* (a grass), has been sown [392]. This species is more favorable to waterfowl than *Lythrum salicaria*. Ideally, replacement species should gradually yield to a more diverse range of species (preferably native). Other management techniques include the maintenance of grazing pressure above the optimal level for invasives (where the native vegetation is adapted to grazing or browsing by hoofed mammals), maintaining the frequency of fires and adjusting water regimes, for instance by the judicious control of sluices in wetlands, so that the ecosystem remains resistant to invasives. Such techniques usually require a close knowledge of the ecology of invaders.

3.10 Prospects for the future

Species introductions will continue to take place, either accidentally or intentionally. As most do not cause any significant problems in relation to conservation there is no cause for concern for the majority – although the pressure of accelerating environmental and climatic change may increase the instability of natural ecosystems and allow more species, which are not at present a problem, to invade. Kangaroo Island near Adelaide, Australia has gained over 200 alien plants at a rate of one to two per year and the rate appears to be accelerating [205]. However, the relatively small invasive fraction poses huge conservation problems, with the risk of diverse natural vegetation being replaced by monocultures of aliens. Dominance is the opposite of diversity, and in many parts of the world there is no hope of conserving native biodiversity unless we can improve our ability to regulate ecosystems for diversity, as opposed to dominance imposed by plant invaders [85]. Invasion may lead to species extinction, either due to direct replacement by aliens or the indirect effects of alien species on the ecosystem. The problem is one worthy of further research and resources for practical management as it poses a very serious threat to the conservation of biodiversity, perhaps second only to that of direct habitat destruction.

Foresters, gardeners and agriculturalists should accept their responsibility as principal agents in species introductions and be cautious when introducing species. A global database network listing the invasive species worldwide, and relevant ecological details, would be an extremely useful source of information when deciding whether or not a species should be introduced. In the absence of a worldwide database of invasive plants, species that are a serious problem in one area may be unwittingly introduced to another by those unaware of the potential threat. For example, *Passiflora mollissima*, one of the most serious invasive plants of the Hawaiian forests, was noticed recently spreading in native forest in South Africa [245]. Fortunately, the forestry board in South Africa was quickly alerted to the danger. It had apparently been introduced by an enthusiastic gardener and was being sold in plant nurseries. This is a good example of how conservationists need to be vigilant about threatening species as well as about threatened species.

4

Case studies of some important invasive species

IMPORTANT NOTE

These accounts show measures of control which have been tried. This information should be considered not so much as recommendations, but rather it is to be used in the making of informed decisions.

4.1 Introduction

The following case studies are intended to show the range of invasive species, covering different habitats, life forms and plant families, to demonstrate the importance of the phenomenon of invasion, causes of invasive behavior and the range of control strategies which have been employed to combat these species. The species below do not include all the 'worst' invaders but are chosen as examples of particular types. The system of climatic zones used is that of Holdridge, which is based on mean annual temperature and rainfall (see Table 5.1, p. 129). Although crude, its use does give some indication of the climates of regions of natural occurrence and introduced range of invasive species. The 'invasive categories' are as follows:

0   Not weedy or invasive
1   Minor weed of highly disturbed or cultivated land (man-made artificial landscapes)
1.5 Serious or widespread weeds of 1
2   Weeds of pastures managed for livestock, forestry plantations or artificial waterways
2.5 Serious or widespread weeds of 2
3   Invading seminatural or natural habitats (some conservation interest)
3.5 Serious or widespread invaders of 3
4   Invading important natural or seminatural habitats (i.e. species-rich vegetation, nature reserves, areas containing rare or endemic species)
4.5 Serious or widespread invaders of 4
5   Invasion threatening other species of plants or animals with extinction

As Table 4.1 illustrates, the examples include a range of serious invaders with a variety of life forms. Choice of species is also conditioned partly by the available information; only well-studied species are included.

Table 4.1 Some examples of serious invaders and their characteristic life forms and habitats

Species name	Life form	Invasive category	Habitat invaded
Acacia saligna	tree	5	Open
Andropogon virginicus	herb	3.5	Fire-prone
Clematis vitalba	climber	4.5	Forest
Clidemia hirta	shrub	3.5	Forest
Hakea sericea	tree	5	Fire-prone
Lagarosiphon major	herb	4	Aquatic
Lantana camara	shrub	4.5	Open (dry)
Melaleuca quinquenervia	tree	4.5	Fire-prone (wetland)
Mimosa pigra	shrub	4.5	Open (wetland)
Myrica faya	tree	4.5	Forest (open)
Passiflora mollissima	climber	5	Forest
Pinus radiata	tree	4	Forest (open)
Pittosporum undulatum	tree	5	Forest
Psidium cattleianum	shrub	5	Forest
Rhododendron ponticum	shrub	4	Forest
Salvinia molesta	herb	4.5	Aquatic

4.2 Species accounts

Acacia saligna (Labill.) Wendl. {*Acacia cyanophylla* Lindl.} (Leguminosae)

Port Jackson willow (South Africa); golden wreath wattle, blue-leaved wattle, orange wattle (Australia)

Description and Distribution

Habit Dense shrub or small tree, 2–6 m tall (in South Africa reaches 9 m [47]); bark smooth, gray to red-brown becoming dark gray and fissured with age.

Phyllodes (flattened leaf-like stalks) variable, linear to lanceolate, 8–25 × 0.4–2 cm (often much larger towards the base of the plant), straight or sickle-shaped, often pendulous, hairless, green to glaucous, midrib conspicuous. Solitary gland situated on upper margin of phyllode, oblong to circular, 1–2 mm diameter.

Figure 4.1 *Acacia saligna*

Inflorescence racemose (occasionally reduced to a single flower head), usually axillary but sometimes terminal; heads stalked, globular, 5–10 mm in diameter, with 25–55(–78) flowers.

Flowers bright yellow, calyx ½–⅓ length of corolla, shortly 5-lobed; petals 5, (1.5–) 2–3 mm long, joined for ⅔–¾ their length.

Fruit a legume, linear, (3) 8–12 × 0.4–0.6 cm, slightly contracted between seeds, surface slightly undulate.

Seed oblong (4) 5–6 × (2.5) 3–3.5 mm, dark brown to black, shiny, hard and long-lived. A large seed-bank develops under the mature trees [47, 265].

Invasive category 5.

Region of origin Australasia – southwest Australia [265].

Native climate subtrop. arid. *Acacia saligna* is commonly found on sandy soil in areas with yearly rainfall greater than 380 mm [47] and a dry period during the summer months (3–7 months) with low to high humidity during the remainder of the year. In more arid areas, *Acacia saligna* is restricted to creeks and rivers.

Regions where introduced Acacia saligna has been extensively cultivated outside its native range, both for its horticultural value and as a source of tannin. It has become naturalized or invasive in the following areas: Africa – South Africa (invasive) [127, 180, 265]; North America – California (naturalized) [143].

Climate where invading wmtemp. dry, wmtemp. moist. In South Africa *Acacia saligna* is confined mainly to the coastal plain in areas with yearly rainfall of more than 250 mm. The climate is similar to that in its native range with a dry period in the summer lasting for three to five months but fairly humid for the rest of the year.

Acacia saligna is included here as an example of one of the worst woody invaders, a plant that has run amuck in a threatened biome, rich in endemic plant species. In its native Australia it grows best in the deep sands and loams associated with water courses. Where *Acacia saligna* occurs on the coastal dune system, it often forms dense thickets in the hollows between the sandhills [265]. From Australia, *Acacia saligna* was introduced to the Cape in about 1833 and planted to bind drifting sand dunes. From there it has spread to mountain fynbos, lowland fynbos, eastern Cape forest, southern forest, succulent karoo, grassveld and to the southern margins of the karoo[47]. The spread of *Acacia saligna* is threatening several species cited as threatened by IUCN [241], such as *Restio acockii, Chondropetalum acockii, Serruria ciliata, Leucadendron verticillatum* and *Gladiolus aureus.* In areas it is replacing native vegetation. It occurs on all substrates where adequate water is available. It is ranked as the most serious alien plant invader in the fynbos biome on the grounds of both the extent of its current infestation and its potential to spread [179].

Acacia saligna is a variable species [265], which matures early and has a relatively fast growth rate [274]. It is tolerant of drought and sprouts readily after cutting or burning, regenerating easily from the large seed-bank that develops under the canopy of mature trees. The seeds germinate rapidly after fire. They have a water impermeable seed-coat so they remain dormant until heat ruptures the lens (a specialized area of the seed-coat in legumes) allowing water uptake and breaking of dormancy [180]. In common with other *Acacia* species, *Acacia saligna* produces large numbers of hard dry seeds and one square metre of canopy can produce 10 500 seeds per year. The seedlings are robust and have an extensive root system. The seeds may be bird-dispersed (starlings and doves in South Africa [47]) but most fall directly to the ground [180] and may be transported further by water or people. Despite poor long-distance dispersal, *Acacia saligna* has become widespread in the west, south and eastern coastal zones of the Cape [127].

The above-ground biomass of dense *Acacia saligna* infestations is much greater than that of uninvaded native vegetation in South Africa [409]. Falling leaves have a higher nitrogen content than those of native species and inputs of nitrogen from the litter are consequently higher, leading to greater levels of nitrogen in the litter and top layer of the soil. The invasion of the coastal lowlands of the Cape by *Acacia saligna* results in an increase in the nitrogen status of the fynbos during the early stages of invasion. This may have important consequences for adjacent native vegetation, possibly leading to changes in species composition and structure [450].

The following ecological characteristics of *Acacia saligna* may be suggested as reasons for its success as an invasive species:

1. widespread planting of *Acacia saligna* for dune stabilization;
2. comparatively rapid growth rate on soils low in nutrients;
3. early reproductive maturity;
4. copious production of seeds that may remain dormant, producing a large seed-bank in the soil;
5. ability to survive fires as seed;
6. ability to sprout after cutting or burning;
7. tolerance of a wide variety of substrates;
8. nitrogen fixation;
9. preadaptation to the Mediterranean climate and nutrient poor soils [450, 451] (this may be said for all the invasive *Acacia* species invading the fynbos);
10. extensive root system, with abundant root nodules and mycorrhizal association (the root systems of invasive *Acacia* spp. have more extensive laterals than those legumes that are native to the Cape [171], and *Acacia saligna* has very rapid depth penetration of the soil by the seedling taproot [451]);
11. higher stature (more than 3 m) than fynbos plants (less than 2 m), allowing it to overtop and shade them.

Control and Management

Physical control Any program to clear *Acacia* scrub manually must always include treatment to kill the seedlings, as large amounts of viable seed remain in the soil even after burning, and also treatment of the cut stumps, since these readily produce new shoots.

Chemical control If trees are cut, the stumps must be painted with herbicide to avoid regrowth [47].

Management program The large seed-bank of *Acacia saligna* is a major obstacle to successful control [179], which is possible only if every individual is removed (it is a prolific seed producer) and the seed-bank is reduced to zero. The optimal program

Figure 4.2 *Andropogon virginicus*

[256] is to clear the mature trees, treat with chemicals to prevent regrowth and burn the site (burning stimulates germination), repeating the burning treatment and applying herbicide once the seeds have germinated (or the young seedlings may be pulled by hand). This may need to be repeated. The aim is to remove all existing and potential seed-producing individuals, which may act as foci for reinvasion; one surviving seed-producer may re-infest the whole area. Follow-up treatment must involve repeated physical control (mattocking) in subsequent years. This treatment is expensive but repeated burning and application of herbi-cides may damage the native vegetation, which can be slow to recolonize due to lack of propagules of native species. Costs of the first five years of such a program for 1 ha have been estimated to exceed the cost of one person-year of labour [256]. There are some 425 600 ha of densely infested fynbos.

Biological control There are no fully effective biological control measures against *Acacia saligna* yet available, although some are under development. Biological control is the only possible cost-effective method. It may not decrease dramatically the area invaded by *Acacia saligna*, but it could slow the rate of spread and prevent the formation of such dense stands as now occur without causing unacceptable environmental damage.

Androctopon virginicus L. (Gramineae)

Broom sedge

Description and Distribution

Habit Perennial tall bunchgrass with tufted stems, 50–100 cm tall, branches 1–3 at node.

Leaves leaf-sheaths, more or less tuberculate-hirsute on the margins with long usually lax hairs; ligule yellow-brown, membranous, truncate, white-fringed at edge; blades 40 cm long or less, 2–5 mm wide, rough or roughish, hirsute on the upper surface near the base; spathes 3–5 cm long, extending beyond the racemes.

Racemes 2 (–3–4), 2–3 cm long.

Spikelets sessile spikelet 3–4 mm long, twice to half again as long as the internode, the awn straight, 10–15 mm long; pedicellate spikelet wanting or rarely present as a minute scale, pedicel exceeding the sessile spikelet [357].

Flowers either sessile and hermaphrodite, or stalked and staminate, sterile or not developed [374].

Invasive category 3.5.

Region of origin North America – Florida to Texas and Mexico, north to Massachusetts, New York, Ohio, Indiana, Illinois, Missouri, Kansas and Oklahoma; Central America – West Indies and Central America [374]. Occurs in prairies, fallow or abandoned fields, along railroad tracks; rarely in wet open and swampy places. The occurrence of this grass usually indicates an acid soil. *Androctopon virginicus* is found on dry or moist soil [357] and is an occasional weed in waste places [295]. It forms part of the primary succession in abandoned pasture lands, followed by *Pinus taeda* L. and later by southern mixed hardwood forest [134].

Native climate wmtemp. moist, subtrop. moist.

Regions where introduced Oceania – Hawaii (invasive) [291]; Australasia – Australia (invasive).

Climate where invading subtrop. dry, subtrop. moist. In Hawaii there is a summer dry period lasting four to five months with the remainder of the year humid, though in places humidity may be high all year round.

In Hawaii, *Andropogon virginicus* is considered one of the most threatening aliens. It occurs in disturbed grassland and scrub on Oahu from about 50–250 m, on red clay soils in places where the native forest vegetation has been replaced by introduced woody and herbaceous plants. It was introduced to Oahu inadvertently in 1932 and probably to the other Hawaiian islands at about the same time. In Australia, *Andropogon virginicus* has invaded communities which are extremely deficient in nutrients and were thought for this reason to be uninvasible. As *Andropogon virginicus* is highly flammable (due to accumulated standing-dead material), it alters the fire regime in areas where the native flora is not adapted to frequent fires, which has serious consequences for these ecosystems.

Andropogon virginicus rapidly invades burnt or bare areas [359]. It grows vigorously, forming an extremely dense cover with dry shoots which remain standing, together with the active, green shoots. Stands of *Andropogon virginicus* appear yellow throughout the year because of these accumulated dead shoots. They are shed annually during the season of higher rainfall (October to April), at which time the grass goes into partial dormancy, indicated by the drying up of most of the current-year crop of photosynthetic shoots. Only a central core of shoots remains active during the winter rains, so there is little transpiration of excess soil water. This drying up of the shoots mulches the soil, which also prevents direct soil evaporation at a time when rainfall is excessive [291]. The mulch effect lasts all year so there is continual excess water under *Andropogon virginicus* communities (evaporation and transpiration through the grass cover is insufficient) not only during its period of partial dormancy but even during the more productive summer season.

The monthly run-off of water from the areas covered with *Andropogon virginicus* is much greater than would be expected, because water does not penetrate readily into a soil that is already water-saturated. The result in lowland areas is accelerated erosion, as seen in the form of slumps on steep slopes and deeply cut erosion channels in the grassland. In addition to being a poor utilizer of the productive capacity of the rainforest habitat, introduced *Andropogon* is causing damage to the landscape and probably contributing to siltation in the Knaeohe Bay area on Oahu. Reforestation with climatically adapted evergreen species, and fire protection, may be the answer to the problem [291]. Unless a solution can be found, *Andropogon virginicus* may have serious long-term consequences in Hawaii by altering the hydrology and fire regime, with resulting soil erosion. Infestations of *Andropogon* are preventing the natural re-establishment of rainforest vegetation in these areas.

In Australia, *Andropogon virginicus* disrupts native communities, which were once thought to be resistant to invasion due to the low nutrient status of the soil, by increasing the frequency of fires. *Andropogon* is tolerant of very low nutrient levels, but it is outcompeted by other species if nutrients are added to the area in which it occurs [140]. The following ecological characteristics of *Andropogon virginicus* may be suggested as reasons for its success as an invasive species:

1. vigorous growth, producing a continuous, dense cover;

2. production of allelopathic substances [359];

3. well adapted to burning. The dead material provides an excellent fuel, while *Andropogon virginicus* spreads rapidly onto burnt ground.

Control and Management

Biological control Any attempts to control *Andropogon* by introducing a natural pest or disease will probably be resisted by the sugar industry [359].

Chemical control As *Andropogon virginicus* is also a weed of agricultural habitats several chemical control methods have been devised [140]. Effective control can be achieved by application of bromacil, hexazinone, tebuthiuron, bromacil & diuron and buthidazole at 4.5 kg ha^{-1}. Use of these herbicides with addition of fertilizers can accelerate the removal of *Andropogon virginicus* from infested pastures.

Clematis vitalba L. (Ranunculaceae)
Old man's beard, traveller's joy, mile-a-minute, hedge feathers, graybeard (New Zealand), herb aux gueux (France)

Description and Distribution

Habit Deciduous, perennial woody petiole-climber [433] with stems up to 30 m long, 6-angled, strongly ribbed.

Leaves pinnate with usually 3–5 leaflets; leaflets 3–10 cm long, ovate, acute to acuminate, rounded or subcordate at base, coarsely toothed or entire, hairless or slightly pubescent.

Flowers greenish-white, *c*.2 cm in diameter, in terminal axillary cymes, fragrant, hermaphrodite. Perianth segments obtuse, pubescent. Anthers 1–2 mm. Slightly protogynous. Nectar is secreted from the filaments and the flowers are visited by various bees and flies for both pollen and nectar [71]. Generally thought to be wind-pollinated, but insect pollination has been reported [433].

Achenes in large heads on the pubescent receptacle, scarcely compressed, with long white plumose styles. Copious seed production – it has been estimated that 17 000

Figure 4.3 *Clematis vitalba*

viable seeds are produced for every 0.5 m^2 of *Clematis vitalba* canopy [433, 445]. Seeds dispersed as achenes by wind, water, people and other vertebrates [445].

Invasive category 4.5

Region of origin Europe – south, west and central Europe. Its distribution extends north to the Netherlands, south to the Mediterranean and east to the Caucasus mountains [433].

Native climate Temp. moist. Annual rainfall of less than 800 mm and low summer temperatures found at high altitude appear to be limiting.

Regions where introduced Australasia – New Zealand (invasive), South Australia (naturalized) [433]; Europe – Ireland, Scotland, Poland, Denmark (naturalized) [433]; North America – Oregon, USA (naturalized) [433].

Climate where invading wmtemp. moist–wet (New Zealand).

Clematis vitalba appears to have different ecological behavior in its native and alien ranges. An innocuous climber in Europe, in New Zealand it can have a devastating impact on various ecosystems. In Britain, *Clematis vitalba* is native to Wales and southern England, associated with chalk and limestone, although in central Europe it is found on a wider range of soils from weakly acid to weakly basic [112, 433]. It requires a substrate with moderate to high fertility and medium to good drainage [200]. Low calcium levels in the soil appear to retard the growth of *Clematis vitalba* in Britain [62], but on base-rich soils it is common in hedgerows, thickets and on wood margins [71].

In New Zealand *Clematis vitalba* is invading tall and low forest, scrub and forest margins, shrubland, waste land, willow vegetation along river courses and hedgerows in coastal and lowland areas [428, 445]. *Clematis vitalba* was first recorded as a naturalized plant in New Zealand in the mid-1930s in both southern North Island and northern South Island [318]. *Clematis* invasion has a dramatic impact on the native vegetation. The vine can easily smother native trees over 20 m high, including podocarps [428]. After the native vegetation has been killed by *Clematis vitalba* it gradually collapses, and the *Clematis* then continues to grow in thick layers (which can become metres deep) along the ground, preventing regeneration of native trees [445].

Clematis vitalba can regrow from fragments after cutting and this is an important feature in its spread. Regeneration of fragments is related to age, since older stem sections have better water retention and larger nutrient resources than softer young tissue [200]. *Clematis vitalba* has a high growth rate (young plants and new shoots can grow up to 2 m per year) and it reaches reproductive maturity relatively early if exposed to full sun (it can reproduce sexually after one to three years, and asexually after one year). Vegetative reproduction involves rooting from stem fragments and attached stems. It appears to be tolerant of only moderate shade [112, 200, 433] and requires high light for growth and reproduction. *Clematis vitalba* usually produces many climbing shoots per plant which cover the host, often resulting in its death. It is tolerant of frost partly because it is deciduous. It spreads wherever land is not intensively managed or grazed, especially along river margins [445].

The following ecological characteristics of *Clematis vitalba* may be suggested as reasons for its success as an invasive species:

1. rapid growth rate;

2. early reproductive maturity;

3. ability to spread easily vegetatively and from fragments;

4. rapid recovery from physical damage by resprouting;

5. high seed production and effective wind dispersal of seeds.

Control and Management

Physical control Small seedlings can be pulled, but larger stems have to be cut and removed from the area and the roots grubbed out, otherwise they will resprout [363].

Chemical control A variety of herbicides have been shown to be effective in controlling *Clematis vitalba* [317]. In New Zealand, the application of picloram granules to soil near the base of the plant has shown to be effective. However, the picloram affects the surrounding vegetation and has residual effects in the soil [363]. The method recommended by the Department of Lands and Survey (now the Department of Conservation) in New Zealand is to cut the vines at ground level and waist high and treat both cut ends with 2,4,5-T in diesel [317]. Roundup has also been used, with varying levels of success: when applied at a suitable time of year, it results in a complete kill of the vine, and it does not appear to affect the surrounding bush [363]. Tordon Brushkiller and Garlon 520 applied to vine bases in fall (autumn) have been found to be very effective in killing *Clematis vitalba* [106].

Biological control A survey of insect predators of *Clematis vitalba* in England has produced a list of some possible candidates for biological control in New Zealand. Further research on host specificity will be required before these insects can be considered for introduction to New Zealand as there are several native species of *Clematis*. Possible biological control candidates [62] are: *Horisme vitalbata* (Lepidoptera: Geometridae), which attacks the leaves of *Clematis*; *Melanthia procellata* (Lepidoptera); *Eupithica haworthiata* (Lepidoptera), which damages the flowers; and *Xylocleptes bispinus* (Coleoptera: Scolytidae), which attacks the structural and vascular tissue of the stems.

Prospects The inherent difficulties with physical and chemical control, due to the vigorous resprouting response of *Clematis vitalba* when cut, mean that research into biological control is essential for practical and economic reasons. As the biology and ecology of *Clematis vitalba* have been fairly well studied in its native region (unusual for an invasive plant), it may present a good case study to determine the basis for its different ecological behavior in New Zealand. Fortunately, this species does not appear to have been introduced elsewhere where it might become a pest. It is important that the problems associated with it are made known so that future introductions to potentially 'invasible' areas are prevented.

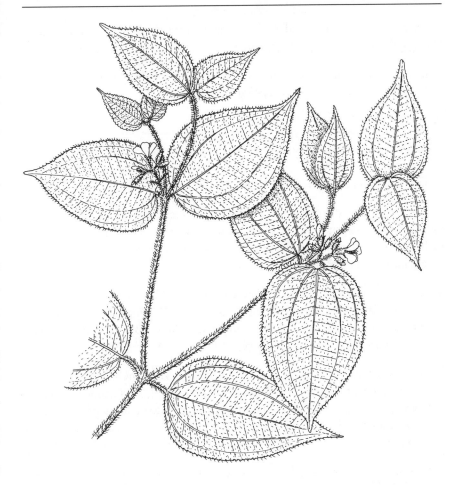

Figure 4.4   *Clidemia hirta*

Clidemia hirta (L.) D.Don. {*Melastoma hirta* L.} (Melastomataceae)

Koster's curse (Fiji, Hawaii), soap bush (Jamaica)

Description and Distribution

Habit Densely branching shrub rarely more than 1.5 m high in native region [2]; reaches a height of 4 m in Hawaii [435]; all parts densely hairy.

Leaves papery, petiolate, mostly 4–15 cm long, acute or short-acuminate, usually 5-nerved, margins often crenulate and denticulate.

Inflorescence 3–5 cm long, axillary.

Flowers white or pinkish, 8–10 mm.

73

Fruit a berry, sweet, variously described as red-purple to blackish (Jamaica) or dark blue (Hawaii), containing 100 or more seeds.

Seeds c. 0.5mm in diameter, most likely dispersed by birds. It fruits continuously with prolific seed production [132, 435].

Invasive category 3.5.

Region of origin Central and South America. *Clidemia hirta* has a wide natural range in humid tropical America, extending from southern Mexico to Argentina and including the islands of the West Indies [2, 435].

Native climate trop. dry, trop. moist, usually characterized by very high humidity for most of the year with a dry period of about two months (February–March) prior to the summer months; however, in some areas, the dry period is during the winter, lasting for three months.

Regions where introduced Oceania – Hawaii (invasive [362]), Fiji (invasive/ruderal), Java, Samoa, British Solomon Is, Tonga and Palau [435]; Malagassia – Madagascar (naturalized) [186, 435]; Africa – Tanzania (naturalized) [435]; S., E. and SE Asia – Sri Lanka, India, Singapore (invasive) and Sabah [78, 335, 435].

Climate where invading subtrop. wet–dry. In Oahu, *Clidemia hirta* is widespread in areas that receive more than 1270 mm of rain annually [435] (although in Fiji it apparently occurs in drier habitats). It does not appear to be limited by elevation as it is found from almost sea-level up to the highest ridges at 900 m. It can also be found in areas of annual rainfall in excess of 7600 mm, with no dry season.

Clidemia hirta is an example of a pantropical weed which, in places, has been able to spread out of managed habitats to become a conservation nuisance in native forest. In its native range, *Clidemia hirta*, like many of the other members of its genus, is a plant of secondary succession [435]. In Trinidad and Jamaica it is characteristically found in moist, shaded localities, on the edges of clearings and stream-banks, in ditches, along paths and roadways and in moist pastures and thickets; the altitude range is 30–900 m^2.

It has now become very widespread, being introduced to Fiji from Guyana before 1890 [309] and to Java in the late nineteenth century [435]. By 1934, *Clidemia hirta* had become established in Samoa, British Solomon Islands and Tonga. It was noticed in Hawaii 1941 and Palau in 1971. The routes and methods of its introduction are unknown, but it was probably introduced to most places accidentally. In both Hawaii and Fiji, approximately 30 years elapsed between its first sighting and the time when it was perceived as a serious pest [309].

In Hawaii it is established in a variety of habitats. On Oahu, more than 40 000 ha are infested [398]. It is spreading into undisturbed vegetation such as rainforest, as well as into disturbed vegetation. It can be found under many of the common introduced species such as *Psidium cattleianum*, *Melaleuca quinquenervia* and *Eucalyptus robusta*, as well as with the native *Metrosideros collina* and *Acacia koa* [435]. Under the forest canopy, *Clidemia hirta* may be replacing the native fern *Dicranopteris linearis* as well as other native species. More recently, it has also spread into forests on the islands of Hawaii, Maui, Molokai and Kauai, with the greatest infestations occurring in the Puna and Waiakea forests of Hawaii.

In Fiji it has, in the past, invaded rubber and cocoa plantations, grasslands and forested areas, but is now controlled by an introduced insect in many areas, excluding the native forests and shaded places where the biological control agent (see biological control section) has not been very successful due to intolerance of shade [309, 435]. In Singapore it invades older secondary and primary forests [78].

Clidemia hirta is found in light conditions ranging from full sunlight to 100% canopy cover, demonstrating broad tolerance [435]. Forest fire seems to give *Clidemia* an advantage over the native and non-native flora as several months after burning it has often become the dominant species. *Clidemia* has a rapid growth rate and, where it is established outside its native range, it appears to grow larger, producing more fruit and forming dense, almost impenetrable thickets over such extensive areas that it is evidently able to compete successfully in a greater range of environmental conditions than where it is native. This is probably due to the absence of the burden of native pests and diseases.

The following ecological characteristics of *Clidemia hirta* may be suggested as reasons for its success:

1. tolerance of a wide range of environmental conditions;
2. prolific seed production;
3. rapid growth rate;
4. long-distance dispersal by birds;
5. formation of dense thickets which prevent regeneration of native tree species.

Control and Management

Biological control Biocontrol agents were sought in Trinidad and feeding experiments indicated that a thrips, *Liothrips urichi* (Thysanoptera: Phlaeothripidae), was highly host-specific and would not attack plants of economic importance. The thrips were imported to Fiji in 1930 and field releases were made immediately. Within a few months they multiplied rapidly, spread to other islands and caused a widespread collapse of the *Clidemia hirta* population [435]. The thrips provide excellent control in most areas by reducing the plant's competitive ability [196], the exception being in moist and shady habitats.

Liothrips urichi is a sucking insect which attacks terminal shoots, causing leaf drop and reduced vitality, which results in the smothering of *Clidemia hirta* by other plants [435]. The insect shows a strong preference for sunlight and, although it has been found in deeper shade than in its native range, it has not significantly suppressed *Clidemia hirta* in Fijian forests. *Liothrips* was also introduced to Hawaii in 1953 but was not as successful as in Fiji, possibly due to predation by *Montadoniola moraguesi* (Hemiptera: Anthocoridae) and *Pheidde megacephala* (Hymenoptera: Formicidae) [323] and also because it cannot tolerate the low light levels found under a forest canopy. *Liothrips* provides moderate control in open pastureland but is ineffective in conservation and watershed areas where the greatest problem exists. *Liothrips* was introduced to the Solomon Islands on several occasions but failed to establish [196].

Other organisms are being tested as potential biological control agents in Hawaii [297]. For example, the fungal pathogen *Colletotrichum gloeosporioides* has been isolated from *Clidemia hirta* plants growing in Panama [398]. Tests have shown it to be a highly aggressive pathogen of cultivars from Hawaii. The disease symptoms are severe premature defoliation and tip dieback of the plants. This pathogen shows promise as an agent for biological control and specificity tests are underway. Another organism, a pyralid moth, *Bleparomastix ebulealis* (Lepidoptera: Pyralidae), was introduced to Hawaii from Puerto Rico. It has become established but so far has been ineffective in controlling *Clidemia hirta* [196, 435].

More research into a suitable biological control agent for forest habitats is needed. Physical or chemical control is not a viable option considering the extensive areas covered by *Clidemia hirta*.

Hakea sericea Schrad. {*H. acicularis* R. Br., *H. tenuifolia* (Salisb.) Domin} (Proteaceae)

Syerige hakea, silky wattle (South Africa); needlebush, prickly hakea (New Zealand)

Description and Distribution

Habit An erect single stemmed, much branched shrub or small tree, 2–5 m high. Shoots densely hairy, somewhat angular.

Leaves dark green, simple, terete, 2–6 × 1 cm, hairless except when very young, rigid and spiny.

Flowers few, in axillary clusters, perianth white.

Fruit a woody follicle, 2–3 × 1.4–2.5 cm, mostly very corrugated, with 2 small horns at the beaked apex; purplish-brown changing to gray with age; serotinous.

Seeds 2 per follicle, winged, wind-dispersed [332, 452].

Invasive category 5.

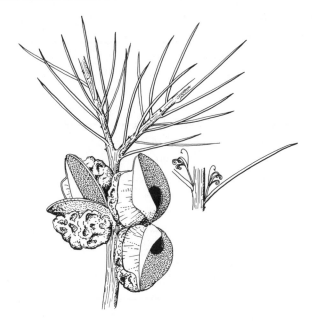

Figure 4.5   *Hakea sericea*

Region of origin Australasia – SE Australia (Victoria, New South Wales and some Bass Strait islands). *Hakea sericea* is found on coastal heath, in open sclerophyll forests on the highlands and on coastal dividing ranges [303].

Native climate wmtemp. moist; usually in areas that are humid all year round with very high rainfall in places.

Regions where introduced Africa – South Africa (invasive). Nearly all the coastal mountain ranges in the western, southern, and southeastern Cape [303]. Its potential range corresponds roughly to the areas of mountain fynbos and to sandstone and granite areas on mountain ranges, with rainfall in winter or throughout the year. Australasia – New Zealand (ruderal). Common to abundant on North Island in north Auckland, the Auckland region, Great Barrier Island, Coromandel Peninsula, Bay of Plenty and the Wellington area; on South Island it is found in Waikakaho valley and is locally common in northwest Nelson [428].

Climate where invading wmtemp. dry–moist; dry in summer with winter rain or else humid all year round.

Hakea sericea is a fire-adapted species, an important factor in its achievement of remarkable dominance in species-rich vegetation of great nature conservation importance in South Africa. It occurs up to 100 km inland in mountain fynbos vegetation and, by forming in places dense impenetrable thickets, threatens the indigenous vegetation. *Hakea sericea* and *Pinus pinaster*, another alien invader, are considered the 'greatest threat to the mountain fynbos vegetation' [303]. It was probably introduced into South Africa in the nineteenth century and has been used to bind sand dunes and also as a hedge plant [303]. In New Zealand it occurs in gumlands and scrub and on open hillsides. It was introduced into New Zealand as a hedge plant but now is rarely used [428].

Hakea sericea is well adapted to survive the periodic fires which are a feature of the fynbos in South Africa. It reaches reproductive maturity relatively early, after about two years, and is strongly outbreeding [109]. Lack of specialized pre-dispersal seed predators and prolific seed production ensure a large seed load (seed production appears to be higher in South Africa than in its native Australia). The winged seed is retained in woody, heat resistant follicles and is released only on death of the parent plant or branch. Standing plants killed by fire liberate seeds within a few days. Germination is equally good on burnt and unburnt substrates [332]. The seeds may then be dispersed by wind over great distances. Infestations of *Hakea sericea* may become so dense that they smother the surrounding vegetation. The invasion of fynbos vegetation by *Hakea sericea* results in a 60% increase in fuel load for the area and the low moisture content of the leaves ensures that they burn vigorously [409] upsetting the natural fire regime and threatening the native flora and fauna [303].

The following ecological characteristics of *Hakea sericea* may be suggested as reasons for its success as an invasive species:

1. serotinous seed dispersal which leads to massive and simultaneous release of canopy-stored seed after fire or felling;
2. winged seeds which aid in long-distance dispersal;
3. formation of dense stands which smother the native vegetation;
4. prolific seed production [333];
5. germination requirements not specific;
6. early reproductive maturity;
7. adaptation to fire.

Control and Management

Biological control Several biological control agents have been introduced to South Africa in the past but have had limited success. *Carposina autologa* (Lepidoptera: Carposinidae) was introduced from Australia in 1972 but did not survive long after introduction [196]. *Cydmaea binotata* (Coleoptera: Curculionidae) was introduced in 1980 but only a few colonies have survived and the damage they caused is probably insignificant. *Erytenna consputa* Pascoe (Coleoptera: Curculionidae), introduced in 1972 with poor results, was reintroduced in 1974 causing significant seed destruction in dry inland localities [208]. Recently [206, 288, 332] *Hakea sericea* has suffered from a gummosis and die-back disease in many areas of the Cape. The symptoms are stem and branch cankers exuding quantities of gum. These cankers may eventually girdle the stem, killing the host plant. Shoot tips of mature plants may also be affected and die back progressively. The growing points of young shoots are attacked, necrosis extending down the stems to kill the whole plant [289]. The origin of this disease is uncertain [206], but the causal organism, a form of the fungus *Colletotrichum gloesporioides* [289], is endemic to South Africa and appears to have a limited host range. The disease has been spread artificially but this has been found to be unnecessary as natural spread is rapid [332]. It is now widespread in the Cape and is causing extensive mortality [196]. It is thought that this fungus combined with the other biological control agent, *Erytenna consputa*, may cause a significant reduction in the extent and spread of *Hakea sericea*.

Physical control Areas invaded by *Hakea sericea* are systematically cleared and then burnt 9–12 months later [206]. This form of control is not very satisfactory considering the expense [332], time taken and need for continual vigilance to maintain an area free of *Hakea sericea*, which is a prolific seed producer. The intense fires produced by burning *Hakea sericea* may damage the native flora and fauna [303, 332].

Lagarosiphon major (Ridl.) Moss {*L. muscoides* Harvey var. *major* Ridl.} (Hydrocharitaceae)

Description and Distribution

Habit Submerged aquatic herb, with elongated, branched, brittle stems, about 3 mm in diameter.

Leaves alternate to subwhorled, *c.* 1.6 cm × 0.2 cm, tapering, stiff, recurved, with short blunt teeth.

Spathes axillary, shortly and bluntly toothed, solitary and unisexual (dioecious). Male spathe is ovate and many flowered (*c.* 50 flowers per spathe); female spathes one-flowered.

Flowers with 6 perianth segments; male: stamens 3, staminoids 3; female: staminoids 3, styles 3, stigmatic surface red.

Invasive category 4.

Region of origin Africa – South Africa. An endemic species found in high mountain freshwater streams and ponds [421].

79

Figure 4.6 *Lagarosiphon major*

Regions where introduced Europe – Britain (naturalized), Italy (naturalized); Australasia – New Zealand (invasive) [182, 266]; Malagassia – Mascarenes (naturalized).

Lagarosiphon major provides an extraordinary example of a restricted endemic, able to spread explosively when introduced to freshwater lakes and streams in New Zealand, to the extent of coming near to eliminating the indigenous aquatic vegetation in Lake Rotorua [66]. The first report of *Lagarosiphon major* naturalized in New Zealand was in 1950 in the Hutt Valley and, by 1957, it was established in nuisance proportions in Lake Rotorua [182]. *Lagarosiphon major* was introduced to Lake Rotorua with the intention of improving the oxygenation levels. It is thought to have been introduced to Lake Taupo around 1966 by recreational boat traffic. It has since spread to many other freshwater lakes in New Zealand.

Lagarosiphon major has the ability to form tall closed-canopy stands in sheltered freshwater habitats. In Lake Rotorua, New Zealand, it once formed thick, dense mats of weed, up to 0.4 km wide and to depths of 2–3 m, although it has declined since the early 1980s. The greatest height and biomass are produced in sites with fine sediment and steep slopes, and it is not tolerant of exposure to large wind-runs. *Lagarosiphon major* has pushed out *Elodea canadensis*, previously the dominant alien species in Lake Taupo. *Elodea*, which did not form monospecific stands in the oligotrophic waters of Lake Taupo, had been coexisting with the native vegetation rather than replacing it. The barrier to further invasion by *Lagarosiphon major* is likely to be wave action. *Lagarosiphon major* directly displaces the native flora over large areas of the lake littoral zone. It attracts large herbivores such as swans and detritivores such as crayfish, which adversely affect the native flora. As in Rotorua, the native flora of Lake Taupo has been directly replaced or reduced in many areas as a result of its spread [182]. *Lagarosiphon major* is intolerant of low water transparency associated with accelerated eutrophication and declined in abundance in some lakes in New Zealand after they became more eutrophic [74]. In Lake Rotorua the population collapse, associated with turbidity and storms in the early 1980s, has now reduced *Lagarosiphon major* to approximately the same abundance as *Elodea canadensis* (common but not dominant). However, another alien water plant, *Egeria densa*, promises to form similar weed beds to those that used to be formed by *Lagarosiphon major* [432].

Lagarosiphon major is dioecious and in New Zealand spreads entirely by vegetative means (only the female plant has been introduced). Efficient vegetative reproduction and water dispersal allows it to invade New Zealand lakes despite being unable to set seed. Before its decline in Lake Rotorua, *Lagarosiphon major* formed dense growths up to 5 m high reaching to the surface between 2–6 m depth. It now grows sparsely, with growths most commonly around 1 m long, in sheltered sites growing to within 50 cm of the surface [432].

The following ecological characteristics may be suggested as reasons for the success of *Lagarosiphon major*.

1. formation of dense stands which crowd out the native flora;

2. rapid growth rate;

3. vegetative spread.

Control and Management

Biological control There have been proposals to introduce grass carp to control adventive plants such as *Lagarosiphon major*, but it is feared that this will damage the native flora, since an examination of the food preference of the fish shows that some of the native species are highly preferred [182].

Chemical control In places *Lagarosiphon major* has been controlled by spraying with diquat [66].

Prospects Lagarosiphon major has directly displaced native species in New Zealand [66, 182] and therefore the development of some control method is essential. The use of chemicals in a freshwater system is not desirable, as residues may enter the human food chain as well as having deleterious effects on the native flora and fauna. A biological control agent would therefore be the most desirable and economic method, but host specificity of any introduced control organism will be difficult to achieve.

Lantana camara L. {*L.aculeata* L., *L.brittonii* Moldenke, *L.scabrida* Ait.} (Verbenaceae)

White sage, wild sage (Caribbean), tickberry (South Africa)

Description and Distribution

Habit Perennial, erect or (in shady places) straggling, aromatic shrub, 1–2 (–6) m tall. Stems four-angled, often armed with recurved prickles, more or less pubescent.

Leaves opposite, ovate to ovate-lanceolate, with a strong odour when crushed, 2.5–10 cm × 1.75–7.5 cm, margins crenate to dentate, rough above, hairy below.

Inflorescences axillary or terminal in dense corymbs, 2.5 cm in diameter, flowers maturing from the outside of the head inwards.

Flowers yellow or pink on opening, changing to orange or red, rarely blue or purple (in wild type *Lantana camara* the flowers change from yellow to red [218]), corolla a slender tube, shortly but densely pubescent outside, with four unequal lobes. Stamens 4, in 2 pairs, epipetalous.

Fruit a small drupe, globular, dark purple to black, *c.* 6 mm across at maturity.

Seed c. 1.5 mm [364].

Taxonomy Lantana camara is not homogeneous, but consists of a number of forms of mixed hybrid origin [375]. These plants are collectively referred to as the *Lantana*

Figure 4.7   *Lantana camara*

camara complex or *Lantana camara* sensu lato. The most aggressive taxon, *Lantana camara* L. var. *aculeata* (L.) Moldenke [387] is characterized by its vigor, long weak branches and numerous recurved prickles.

Invasive category 4.5.

Region of origin Central and South America. *Lantana camara* L. sensu stricto occurs as a wild plant in dry thickets in the West Indies and widely in the Neotropics [2], but the distribution is complicated by the introduction of non-wild forms in these countries.

Regions where introduced Oceania – Hawaiian Is [359], Society Is, Cook Is, Samoa, Fiji, New Caledonia, New Hebrides, New Britain, New Guinea, Micronesia (e.g. Ponape, Guam [103]), Tonga, Niue and Makatea; Isolated Oceanic islands – Galápagos [91], St Helena; Australasia – Australia (New South Wales) [211, 364],

New Zealand (North Island) [387], Norfolk I. [387]; Africa – South Africa, where infestations are said to cover some 400 000 ha in Natal alone [70, 247, 375]; Malagassia – Madagascar [211]; S. Asia [211] – Indo-China, Indonesia, SE Asia, India; North America – USA [91].

Climate where invading wmtemp.-trop. dry–moist. In South Africa it does not invade very dry regions, such as the Orange Free State, and it is abundant only where the temperature does not frequently drop below 5°C [70]. Generally, however, the complex grows throughout a considerable climatic range.

Lantana camara is probably familiar to anyone who has been in warm-temperate, subtropical or tropical regions. The main adaptation responsible for its rapid spread has been its brightly colored flowers, which have induced gardeners to spread it widely. The early hybridizations, carried out by European gardeners during the eighteenth and nineteenth centuries, probably involved two or more species [375]. The height of popularity of *Lantana camara* in Europe happened to coincide with the expansion of colonial powers into the tropics and many imported varieties escaped from gardens to become serious weeds. Diploid, triploid, tetraploid, pentaploid and hexaploid forms of *Lantana camara* have been reported [375], resulting from auto- and allo-polyploidy [298]. The breeding system is both sexual and asexual [202]; most weedy types appear to be tetraploid.

Lantana camara is mainly a weed of highly disturbed habitats and rather rarely invades natural and seminatural habitats to threaten native flora and fauna. It is included here mainly as an example of a man-made hybrid becoming a serious pest. In South Africa, since its introduction in about 1880, *Lantana camara* has come to invade natural forests, plantations, forest margins, overgrazed or burnt veld, waste ground, orchards, relatively undisturbed rocky hillsides and fields [70, 108, 376]. In Australia, it occurs at the margins of rainforest or in rainforest clearings [364]. In the Galápagos *Lantana camara* is found from the arid zone to the *Scalesia* forest regions [151] and has replaced, on Floreana, *Scalesia penduculata* forest and a dry vegetation of *Croton, Macraea* and *Darwiniothamnus*. Some small populations of rare native plants in these areas are in danger of being eliminated. This is true of two of the three populations of *Lecocarpus pinnatifidus* and of a population of *Scalesia villosa*, both of which are endemic to Floreana Island. The impending spread of *Lantana camara* to the crater area of Cerro Pajas, where the dark-rumped petrel (*Pterodroma phaeopygia*) nests, is of grave concern. The resultant dense thicket of *Lantana camara* will keep the petrels, which nest in burrows, from occupying their breeding site [91], the last remaining nesting colony on the Galápagos. Likewise, rare endemic plant species may also become extinct [91]. In Ponape (Caroline Is) it 'has replaced the native vegetation in many areas and formed closed communities' [387]. In New Caledonia it 'has replaced the native vegetation and infested the coastal districts on both the windward and leeward coasts and penetrated into the mountain passes of the central range. It has transformed the native "niaouli" (*Melaleuca leucadendron*) savannas into impenetrable thickets, colonized both pasture and cultivated land, and transformed previously native forests into impenetrable thickets as well as suffocating both mature trees and seedlings which served to support it' [387]. In Hawaii, *Lantana camara* is found up to 600 m on all islands, principally in dry areas, though it also grows in mesic and wet habitats [359].

Lantana camara is capable of growing on poor soils in full sunlight or semi-shade. It can survive fire by regenerating from basal shoots [359]. *Lantana camara* only occurs at temperatures above 5°C and is susceptible to frost which kills the leaves and stems; it cannot tolerate permanently wet soils [387]. The branches are brittle and easily broken. It can reproduce vegetatively from branch fragments, aiding rapid spread [387]. It has naturalized in a range of habitats, from coastal areas to elevations of over 300 m on both the windward and drier leeward sides of most major Pacific islands.

Lantana camara is unable to compete with taller native forest tree species, although it is commonly found scrambling through the branches of taller vegetation. Once established, it quickly covers open areas forming dense thorny thickets [178]. The leaves and seeds cause photosensitivity and gastrointestinal disorders in sheep and calves [376]. Allelopathic substances are produced by shoots and roots [359]. Hybridity and polyploidy may contribute to its success [298, 356]. *Lantana camara* is apparently adapted to butterfly pollination [28], but in India it appears to be pollinated by thrips [268]. The flowers of *Lantana camara* undergo colour changes triggered by pollination [281]. It is a facultatively apomictic outbreeder [202].

The following ecological characteristics of *Lantana camara* may be suggested as reasons for its success as an invasive species:

1. bird dispersal of the fleshy drupes may result in long-distance dispersal providing new foci for invasion; *Lantana camara* is particularly well dispersed by the mynah bird (*Acridotheres tristis*) which has been widely introduced in the tropics;

2. toxicity to many mammals means that it is not prone to herbivory by large herbivores;

3. ability to reproduce easily by vegetative means;

4. ability to invade a wide range of environments;

5. introduction of alien animals to isolated islands has disturbed the native vegetation and allowed *Lantana camara* to spread, as it thrives on disturbed or mildly eroded soils and is resistant to grazing;

6. production of allelopathic substances increases competitive ability against native species;

7. flowers profusely for much of the year under favorable conditions, setting copious seed.

Control and Management

Biological control A large number of insects (at least 33 species [196]) have been tried in many of the countries where there is a serious *Lantana camara* invasion, often because it is an agricultural pest. One insect, *Calcomyza lantanae* Frick (Diptera), was introduced from Trinidad to Australia and South Africa in 1974 and 1982 respectively. It spread rapidly in Australia causing considerable damage initially, but since then has been declining except in low rainfall areas. In South Africa this insect is still expanding its range and its effects are still being evaluated. *Cremastobombycia lantanella* Busk (Lepidoptera) was introduced to Hawaii from Mexico in 1902. It is now on all the Hawaiian islands providing partial control in dry areas. *Teleonemia scrupulosa* Stal (Hemiptera) has been introduced to Australia, Hawaii and Fiji from Mexico or Brazil [196]. In Australia it caused extensive defoliation but appears to be declining. In Hawaii it causes considerable damage and provides control in dry areas. In Fiji it once prevented flower production and seed set over considerable areas but its influence has been adversely affected by wet weather and predation.

Physical control Mechanical eradication can be an effective but labor intensive and expensive method of control. The bushes need to be cleared and the roots grubbed out [376].

Prospects This case study highlights the importance of bibliography, taxonomy and autecology in the control of widespread invasive plants [375]. The need for a taxonomic review of the *Lantana camara* complex cannot be overstated, since the exact type of *Lantana camara* seems to be significant in determining its susceptibility to pests and diseases. Control experience in agricultural ecosystems needs to be applied to *Lantana camara* in natural ecosystems.

Melaleuca quinquenervia (Cav.) Blake {*M.leucadendron* (L.) L., *Metrosideros quinquenervia* Cav.} (Myrtaceae)

Melaleuca, cajeput (Australia), punktree (Florida)

Description and Distribution

Habit Slender upright tree with papery bark reaching 12–25 m in height.

Leaves 70 × 5–18 mm, lanceolate to oblanceolate with 5 longitudinal veins, 3 of which are usually quite prominent.

Flowers white, in cylindrical spikes up to 5 cm long, borne in the axils on the current season's growth with further growth and leaves appearing beyond the flowers; stamens conspicuous.

Figure 4.8 *Melaleuca quinquenervia*

Fruit woody, capsule 3-celled, containing approximately 250 seeds, serotinous.

Seeds minute and unwinged (34 000 seeds per gram); dispersed by wind or water [116, 296].

Invasive category 4.5.

Region of origin Australasia – eastern Australia, Papua New Guinea and New Caledonia [115]. In its native habitat *Melaleuca quinquenervia* forms single species stands which burn regularly [115]. It is a common tree around the margins of lagoons and swamps along the coast northwards from New South Wales to Queensland [116].

Native climate subtrop. moist–wet, trop. dry, ranging from high humidity all year round to seasonally dry for six months of the year.

Regions where introduced North America – Florida [21, 115, 296] (invasive). It occurs in the following major natural areas of southern Florida [296]: (1) the Atlantic coastal ridge and eastern flatlands; (2) the Everglades; (3) the big Cypress swamp; and (4) the western flatlands.

Climate where invading subtrop. moist, characterized by high humidity for about six months of the year (including summer), with the rest of the year wet or dry.

Melaleuca quinquenervia has a unique combination of adaptations to disturbance, fire and flooding which has allowed a dramatic invasion of disturbed wetlands in Florida, where the native species are not adapted to such high levels of disturbance. It was first introduced into Florida in 1906 by a forester [21] and shortly afterwards began to invade wet prairies and marshes. Seeds of *Melaleuca quinquenervia* were scattered from a plane in 1936 with the aim of reforesting the Everglades.

Melaleuca quinquenervia readily invades depressions in drier pine flatwoods [296], wet pine flatwoods, the herbaceous perimeter frequently found around pond cypress swamps, the transition zone between pine forest and cypress swamps and shallow pond cypress swamps that have been burned. *Melaleuca quinquenervia* appears to be displacing native species such as pond cypress, *Taxodium ascendens*, in the zone between pine and cypress forests [115]. When *Melaleuca quinquenervia* invades wet prairies or marsh systems the species diversity decreases by 60–80% [21]. It is capable of invading relatively undisturbed areas, although favored by fire and disturbance. When colonizing new areas it increases rapidly, from 5% to 95% infestation in about 25 years [220].

Several factors, including soil moisture, light, fire, soil type and competition interact to influence seed germination and survival of *Melaleuca quinquenervia*, determining which areas are most susceptible to colonization by this species. Huge numbers of *Melaleuca quinquenervia* seeds are released, usually following a late dry season fire. Initial establishment, although greatly influenced by fire, is ultimately controlled by the water regime [296], that is when and for how long critical moisture requirements are met. Timing of seed release is crucial. Seedlings that germinate either before a site becomes flooded or which germinate immediately after flood water recedes, have the best chance of surviving the next dry season. Successful establishment depends on adequate growth during the wet season to ensure survival during the dry season. Soils that remain moist to saturated, but rarely submerged during the four to six months wet season, are ideal for establishment. Throughout its native range, *Melaleuca quinquenervia* is reported to grow on sandy soils and, in Florida, acid sandy soils appear to be favored for its germination and growth.

Melaleuca quinquenervia is almost perfectly adapted to fire. The thick, spongy bark insulates the cambium and the outer layers are flaky and burn vigorously. This conducts flame into the canopy, igniting the oil-laden foliage. The leaves and branches are killed, but dormant lateral buds on the trunk germinate within weeks. This prolific resprouting greatly increases the surface area of small branches and therefore the tree's reproductive potential. Furthermore, it can flower within weeks after a fire. Each serotinous capsule on a *Melaleuca quinquenervia* tree contains about 250 minute seeds which are released after fire or frost or any event which severs the vascular connections to the fruit. After fire the tree may release millions of seeds, which are dispersed short distances by wind and water [115].

Natural communities into which *Melaleuca quinquenervia* invades, such as pine and cypress forests, are maintained by fire. In some cases, *Melaleuca quinquenervia* appears to pre-empt sites where the native vegetation would normally regenerate following fire. It accomplishes this by producing seed rapidly after fire. Once established it forms a dense canopy, shading out or preventing the establishment of seedlings of other species, resulting in a *Melaleuca*-dominated community maintained by fire. It tolerates a broad range of site conditions, growing in places subject to extended flooding, under conditions of moderate drought and also under conditions of moderate salinity [115]. Under flooded conditions, *Melaleuca quinquenervia* produces a fibrous sheath of 'water roots' that clothe the base of the trunk up to the high water level. An allelopathic influence of this species on pine seedlings has also been demonstrated [296]. *Melaleuca quinquenervia* is intolerant of frost [452] and this may limit its spread northwards in Florida.

The following ecological features of *Melaleuca quinquenervia* may be suggested as reasons for its success as an invasive species:

1. widespread planting and dispersal of seed of *Melaleuca quinquenervia* for reforestation of the Everglades;
2. fire adaptation;
3. prolific seed production;
4. production of allelopathic substances;
5. toleration of extended flooding, moderate drought and some salinity;
6. ability to flower and produce seed soon after fire.

Control and Management

Physical and chemical Melaleuca quinquenervia control is complicated by the fact that death of the aerial portion of the tree results in both seed release and sprouting. Present control efforts use mechanical cutting, followed by application of herbicides [296], and take advantage of present knowledge of its ecology and reproductive biology. Fire and herbicides may be applied at critical stages in its life cycle. For example, treatment of seed trees with herbicide following frost reduces resprouting. A late wet season or early dry season prescribed burn puts the seed on the ground at an unsuitable time.

Prospects Melaleuca quinquenervia is expanding rapidly in southern Florida and threatening to alter the structure and composition of the Everglades. Wetlands are a diminishing habitat worldwide and extra care should be taken in introducing this species to other tropical wetland areas. *Melaleuca quinquenervia* may be considered in the future by foresters for commercial introduction in various parts of the world [43], but should not be introduced without a thorough environmental impact assessment.

Mimosa pigra L. {M. *pellita* Humb. & Bonpl. ex Willd., M. *asperata* sensu Bentham} (Leguminosae)

Giant sensitive plant, zaraz, dormilona

Description and Distribution

Habit Shrub, height to 6 m, stems armed with broad based prickles up to 7 mm long.

Leaves bipinnate, sensitive to touch. Straight, erect or forward pointing, prickle at the junction of each of the 6–14 pairs of pinnae and sometimes with stouter spreading or deflexed prickles between the pairs. Leaflets 20–42 pairs per pinna, linear – oblong, 3–8 (–12.5) mm long, 0.5–1.25 (–2) mm wide, venation nearly parallel with midrib, margins often bearing minute bristles.

Inflorescence of tight, subglobose pendunculate heads 1 cm in diameter, each head containing c. 100 flowers, produced 1–2(–3) together in the upper axils.

Flowers mauve or pink, calyx minute, lacinate, 0.75–1 mm long; corolla about 2.25–3 mm long; stamens 8.

Pods clustered, brown, densely bristly all over, breaking transversely into about 21 (14–26) partially dehiscent segments, each containing a seed, the pod sutures persisting as an empty frame.

Seeds light brown to brown or olive green, oblong, light, dispersed by water and floating for an indefinite period [51, 53, 234]. The *Mimosa pigra* group comprises a complex of six tropical American species [26]; *Mimosa pigra* being the conserved name for the widespread weed.

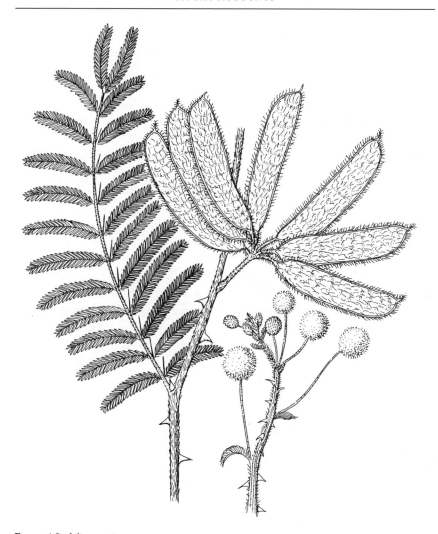

Figure 4.9   *Mimosa pigra*

Invasive Category 4.5.

Region of Origin South and Central America. *Mimosa pigra* is thought to have originated in tropical America and is now widespread in the tropics [234]. Bentham noted in 1875 [37] that is was widely found by early explorers in tropical Africa and the Mascarene Islands, yet was unlikely to represent a relict of some primitive flora as it had not diverged into specifically African varieties and was present in the form most commonly found in America [234]. The early widespread distribution in Africa has not yet been adequately explained, and indeed some people consider *Mimosa pigra* to be native to Africa as well as tropical America [230].

Native climate trop. dry, characterized by a dry period for four months and very high humidity for the rest of the year.

Regions where introduced Australasia – Australia (Northern Territory) (invasive) [51, 232, 234]; S. Asia – Thailand (invasive/ruderal) [234]; Central America – Costa Rica (naturalized) [190]; Africa – Zambia (Kafue flats) (invasive).

Climate where invading trop. dry. *Mimosa pigra* favors a wet/dry tropical climate with a high summer rainfall and relatively pronounced drought during the winter months [234]. Except in permanently wet areas, it is not likely to be a problem in regions with less than 750 mm annual rainfall [272].

Mimosa pigra is a pantropical weed which poses the most serious of all invasive threats to tropical wetlands. Where native, *Mimosa pigra* occurs in cultivated clearings, cocoa plantations, swampland and heavily disturbed land by rivers [229], and when introduced it also favors moist situations such as floodplains and river banks, on soils ranging from cracking clays through sandy soils to coarse siliceous river sand. In northern Australia, *Mimosa pigra* invades sedgeland and grassland in open floodplains particularly where water buffalo (*Bubalus arnee*) or fires have removed the vegetation [234]. It invades the paperbark (*Melaleuca*) swamp forests fringing floodplains, forming a dense understorey and shading out native tree seedlings. *Mimosa pigra* also invades billabongs (seasonally dry river beds), leaving only small remnants of open water at the centre.

Mimosa pigra was introduced to the Darwin Botanic Gardens prior to 1891, probably as a botanical curiosity because of its sensitive leaves [273]. At first it remained at low densities in the Darwin area, occasionally causing a problem, and was noticed upstream from Adelaide River township in 1952. By 1968, it had spread downstream on the Adelaide River to the Marrakai crossing and by 1975 had reached the Arnhem Highway bridge. The population subsequently increased dramatically until by 1981 much of the Adelaide River floodplain was covered by practically single-species stands [234]. It now poses a huge conservation problem in Australia where a largely intact natural landscape is being completely altered [51]. *Mimosa pigra* is invading these fertile wetlands with little human help. It has been estimated that 450 km^2 of floodplain and swamp forest have been covered by dense monospecific stands of this species.

Mimosa pigra thickets contain fewer birds and lizards, less herbaceous vegetation, and fewer tree seedlings than the native vegetation [51]. It is probable that the magpie goose (*Anseranas semipalmata*), which once had a range extending as far south as New South Wales, but now depends on the wetlands of northern Australia for its survival, is endangered by the spread of this invasive plant, since it requires dense stands of native sedges for nesting and food [234]. Kakadu National Park, a World Heritage site in northern Australia, is under serious threat from *Mimosa pigra*. Thousands of visitors come to the park every year to see the abundant bird life of the wetlands, thus the area is important in economic as well as conservation terms. The paperbark forest, which is being invaded by *Mimosa pigra* [51], provides the main rookery sites for the sacred ibis (*Threskiornis aethiopica*), royal spoonbill (*Platalea regia*) and little pied cormorant (*Phalacrocorax melanoleucos*), as well as the main nesting sites for most of the raptors in the area. Large groups of flying-foxes (*Pteropus alecto* and *P. scapulatus*) roost and feed in these forests for much of the year, which is important as these animals are major pollinators and seed dispersers for trees throughout northern Australia.

In Thailand *Mimosa pigra* is considered to be a serious weed, particularly in irrigation systems. It was introduced to Thailand in 1943 as a green manure and cover crop [234]. In Zambia *Mimosa pigra* is invading the wetland area of Kafue flats, which is of considerable conservation importance.

Beneath the impenetrable thickets of *Mimosa pigra* a dense carpet of seedlings emerges at the start of the dry season [232]. There is considerable mortality of seedlings, probably from drought rather than shade or competition from other seedlings. Under ideal conditions plants begin to flower six to eight months after germinating. Although pollinated by bees where native, the prolific seed production observed in isolated plants indicates that it is self-compatible [234]. Seed production is copious, estimated at 9103 seeds per m^2 per year in a typical mature stand [231]. The seed pods are covered with bristles that help them float and assist in dispersal by rivers. *Mimosa pigra* has a rapid growth rate and, under favorable conditions, it can grow at about 1 cm per day [232]. Development from flower bud to seed takes about five weeks [234].

In seasonally moist sites, mature plants survive the dry season (which may last from May to December in the Australian Northern Territory savanna zone) by steadily losing leaves until around August, when 40–50% have fallen [234]. In permanently moist sites, such as river banks, growth and flowering continues almost all year round. *Mimosa pigra* also survives the dry season as seeds buried in the soil [233]. The seed-bank densities at some sites in northern Australia average over 12 000 m^2. Seedlings of *Mimosa pigra* are susceptible to competition from grasses and its success was probably facilitated initially by the overgrazing of floodplains by massive herds of feral water buffalo in the mid to late 1970s. However, *Mimosa pigra* is also capable of invading relatively undisturbed sites [51].

The following ecological characteristics of *Mimosa pigra* may be suggested as reasons for its success as an invasive species:

1. prolific seed production;
2. autogamy, if it occurs, allowing an isolated individual in a new site to reproduce without being dependent on the presence of other plants of the same species for outbreeding and also eliminating any possible reliance on the presence of a suitable pollinator;

3. large seed-bank which makes control difficult and allows the species to survive adverse weather conditions as seed;

4. growth into dense monospecific stands, suppressing the growth of the native vegetation which is not able to survive the low light conditions and reducing intraspecific competition;

5. early age of reproductive maturity in favorable conditions;

6. dispersal by water resulting in long-distance dispersal downstream;

7. rapid growth rate.

Control and Management

Chemical control Mimosa pigra is susceptible to a number of herbicides [234].

Biological control For long-term control, biological methods are probably the most cost-effective considering the extent and ecology of this species. Palatibility to higher animals is low but, in its native range, it is attacked by more than 200 species of insect herbivores and fungal pathogens [234]. The first insects introduced to Australia as controlling agents were the seed-feeding beetles *Acanthoscelides quadridentatus* and *A. puniceus* (Bruchidae) from Mexico. They were released in Australia and Thailand in 1984 and 1985 respectively, but have not attained high population densities and have had little impact on seed production of *Mimosa pigra* [448]. Two stem-boring moths, *Neurostrota gunniella* (Gracillariidae) and *Carmenta mimosa* (Sesiidae), were released in Australia in 1989; of these, *N. gunniella* established readily. The young larvae mine leaf pinnules and the older larvae tunnel in the stems, causing them to die [98]. *Carmenta mimosa* complements the action of *N. gunniella* by tunnelling stems of larger diameter. Other important insects currently being tested for their host specificities in Mexico and Australia are the seed- and flower-feeding weevils *Apion* sp., *Chalcodermus serripes*, *Sibinia fastigiata*, *S. ochreosa*, *S. pervana* and *S. seminicola*.

Two fungal pathogens, *Phloeosporella* sp. (Coelomycetes), and a rust, *Diabole cubensis* (Uredenales), severely debilitate *Mimosa pigra* in Mexico [234]. *Phloeosporella* sp. attack leaves, branches, main stems and seed pods, causing leaf fall and cankers of the stems and leading to ring barking and die-back. *Diabole cubensis* causes chlorosis in stems and leaves resulting in premature leaf fall. Both fungi are attacked by hyperparasitic fungi in their native range and it seems likely that their effects on *Mimosa pigra* could be even more damaging in Australia if they were to be introduced without their natural enemies. These fungi are under investigation in Mexico and Britain. The number of insects which feed on leaves and flowers (generally) in Northern Territory, Australia, compared with Central America (the place of origin of *Mimosa pigra*) is low. It has been suggested that leaf- and flower-feeding insects be sought in Central America as future biological control agents [118]. It is possible too that some control may be assisted by insects native to Australia. Some of these have been found on *Mimosa pigra* and two, *Mictis profana* (Hemiptera: Coreidae) and *Platymopsis humeralis* (Coleptera: Cerambycidae), have been observed to cause considerable damage.

Management and prospects Prevention of overgrazing by water buffalo and other large herbivores may limit the success of *Mimosa pigra*, as it spreads only slowly into undisturbed vegetation. While it is already a serious weed in the wetlands of northern Australia, southeast Asia and parts of Africa, many of which are of great conservation importance, it has the potential to become a serious pest in other areas. Further research into the prospects of biological control is needed, to halt its spread.

Myrica faya Ait. (Myricaceae)

Fire tree, faya (Hawaii)

Figure 4.10 *Myrica faya*

Description and Distribution

Habit Evergreen shrub or small tree, to 4–16 m, with shoots covered with peltate hairs [52].

Leaves oblanceolate, alternate, aromatic, 4–12 cm, smooth, shiny, dark green, cuneate at base, margins somewhat revolute.

Catkins usually branched, borne amongst the leaves of the current year's growth. Dioecious, but sometimes pistillate plants may have a few staminate flowers and vice versa.

Flowers male with four stamens, borne on small catkins near the branch tips; female flowers in threes, accompanied by a bract, grouped in small catkins further back from the branch tip.

Fruits small edible drupes in dense clusters, red to purple when ripe, prolific.

Invasive category 4.5.

Region of origin Oceanic Islands (Macaronesia) – The Azores: mountainous volcanic areas associated with *Erica azorica*, *Laurus azorica*, *Picconia azorica* and *Juniperus oxycedrus* [107]; Madeira: an important component of the understorey along with *Erica arborea*, *Clethra arborea*, *Ilex perado*, *I. canariensis* and *Juniperus oxycedrus*; Canary Islands: in forests and degraded forest scrub on the western islands, often associated with *Laurus azorica*, *Persea indica*, *Apollonias barbujana* and *Ocotea foetens* [52, 170, 224, 358].

Native climate wmtemp. dry, wmtemp. moist.

Regions where introduced Oceania – Hawaii (invasive) [224, 358, 437, 438].

Climate where invading wmtemp. moist, wmtemp. wet. Within Volcanoes National Park *Myrica faya* occurs at elevations of 666–1210 m, normally only in places with mean annual rainfall greater than 875 mm.

Myrica faya is a remarkable case of an oceanic island endemic being introduced to oceanic islands on the other side of the globe with very similar environmental conditions. However, while it is a valued constituent of protected native forest in Macaronesia, it is a pernicious forest weed in Hawaii. In the Azores *Myrica faya* is a colonist of young volcanic soils and often forms extensive stands on old lava flows [170]. It is locally common in the Canary Islands, particularly the western islands, and in Madeira is abundant in both undisturbed and secondary native forests up to 900 m elevation. *Myrica faya* grows in a type of relict warm temperate rainforest known as 'laurisilva' or laurel forest.

It was introduced to Hawaii in the late 1800s by Portuguese immigrants, presumably as an ornamental. Later, the Hawaiian Sugar Planters Association obtained seeds from a Portuguese farmer in Hawaii for reforestation [437]. *Myrica faya* has been planted in reforestation projects on the islands of Kauai, Oahu and Hawaii, but by 1937 the invasive character of this species had been recognized. It now occurs on nearly all the major Hawaiian islands and covers a total area of approximately 34 365 hectares [437]. Populations are densest in open seasonal montane forest and pastures, but it is also found in the understory of closed *Metrosideros polymorpha* forests, in forest stands damaged by volcanic ash and in open volcanic ash deposits as young as 10–12 years [401, 415]. Although *Myrica faya* often invades disturbed areas, it can compete successfully with the native vegetation in such areas [170]. It seems to be able to adapt to a wide range of habitats with soils ranging from recent thin ash over lava to deep well-developed silty clay loam soil [437]. Dense stands of *Myrica faya* prevent regeneration of native species.

Myrica faya is relatively shade-intolerant and has bird-dispersed seeds. Consequently, it is a ready colonizer of open sites with some perch trees for birds [224]. It forms a symbiotic association with a nitrogen-fixing actinomycete, *Frankia* [414]. No native nitrogen-fixing plants occur in primary rainforest in Hawaii in the early stages of succession, despite the occurrence of *Acacia koa* in the later stages. Invasion by *Myrica faya* is altering primary successional ecosystems in Hawaii by increasing the amount and availability of fixed nitrogen [415, 417]. *Myrica faya* may enhance its own survival, perhaps to the exclusion of native species, by enriching the soil with nitrogen [359]. Young volcanic soils have characteristically low levels of nitrates and the native plants are adapted to survive in these conditions. As susceptibility to invasion in Hawaii is apparently dependent in part on soil fertility, other alien species which have previously been confined to the richer soils may now be able to invade areas where *Myrica faya* has enriched the soil with nitrogen [401], although this has not yet occurred [293].

Myrica faya forms a dense, interlocking canopy with little or no understory due to heavy shading and possibly the production of allelopathic substances. The microhabitat under *Metrosideros collina* appears to favor germination and seedling development of *Myrica faya* [358]. In dry sites it seems to grow poorly, especially when mature.

Low germination rates of *Myrica faya* seeds collected in the field, contrasted with copious seed production and rapid dispersal of the species, has led to the hypothesis that scarification from bird ingestion greatly improves germination rates [437]. Many species of birds are probably involved in the seed dispersal of *Myrica faya*, including the Japanese white-eye (*Zosterops japonica*) from Asia [32], house finches (*Carpodacus mexicanus*) and 'oma'o (*Phaeornis obscurus*), the latter being only an occasional visitor to *Myrica*, though, when it does so, it eats large amounts of fruit [224]. Feral pigs also disperse the seeds.

The following ecological characteristics of *Myrica faya* may be suggested as reasons for its success as an invasive species:

1. prolific seed production;
2. long-distance dispersal of seeds by birds;
3. symbiotic association with nitrogen-fixing organism;
4. possibly production of allelopathic substances which may inhibit potential competitors;
5. formation of a dense canopy under which the native species are unable to regenerate.

Control and Management

Chemical control To date, herbicides have been the main method of control.

Biological control In 1955 an expedition visited the native habitat of *Myrica faya* in the Azores, Maderia and the Canary Islands to study its natural pests and diseases. A number of insects and diseases that appeared initially promising were later rejected either because of lack of host specificity or because they did not thrive in Hawaii. Another more recent search for control agents in the native habitat concluded that pests and diseases are not limiting growth, reproduction and distribution of *Myrica faya* [170]. Research is still underway with several organisms being investigated, including the fungus *Ramularia destructiva*, in the hope of finding a suitable biological control agent.

Management Since feral alien birds and other animals are implicated as dispersal agents for *Myrica faya* seeds, they should be controlled as far as possible to limit further spread.

Prospects Myrica faya may have serious long-term consequences for the nutrient-poor ecosystems in Hawaii in which it is invading. With the failure of biological control, it would be desirable to develop an integrated control program using physical control, herbicides and ecosystem management.

Passiflora mollissima (H B K.) Bailey {*P.tomentosa* Lam., *Murucuju mollissima* Spreng.} (Passifloraceae)

Banana poka (Hawaii), banana passion fruit (Australia and New Zealand), curuba, tintin, tumbo, trompos (South America), granadilla cimarrona (Mexico).

Figure 4.11 *Passiflora mollissima*

Description and Distribution

Habit Woody climber to 20 m or more, shoots densely covered with white or yellowish hairs.

Leaves three-lobed, lobes lanceolate to ovate, minutely glandular-toothed, leathery, lower surface usually densely hairy, upper surface usually hairless; stipules not persistent, somewhat kidney-shaped; petioles channeled, 1–7 cm long, with 4–9 (rarely 0) obscure, sessile to subsessile, glands.

Flowers solitary in leaf axils, pendant; epicalyx of 3 ovate bracts, 3–5 cm long, margins entire; hypanthium tubular, 5–8 × 1cm, dilated at the base, olive green and occasionally red-tinged on outer surface, white within.

Fruit elliptic to ovate, yellow at maturity with a soft downy (rarely hairless) leathery pericarp.

Seeds broadly obovate, dark brown at maturity, 50–200 per fruit; aril fleshy, translucent, orange.

Taxonomy Passiflora mollissima (as now recognized) is a morphologically variable species and may be a hybrid between certain original, less variable species, (such as *P. mollissima* in a restricted taxonomic sense and another, unknown species of *Passiflora*) [221, 222].

Invasive category 5.

Region of origin South America – eastern Cordillera of the Andes of Colombia, southeastern Andean slopes of Peru and western slopes of the Bolivian and Venezuelan Andes [203].

Native climate temp. moist–wet, wmtemp. moist–wet.

Regions where introduced Oceania – Hawaii (invasive), in a variety of habitats including both open and closed forests of *Acacia koa* and *Metrosideros collina*, mixed native species associations and tree fern (*Cibotium*) forests [221–223]; Africa – South Africa (naturalized) [245]; Australasia – New Zealand (naturalized) [428, 445].

Climate where invading Hawaii: temp. moist–wet, wmtemp. moist–wet [222]; South Africa: wmtemp. dry. Knysna Forest, where it is found in South Africa, has a climate which is fairly humid all year round with little in the way of a dry period and no frost.

Passiflora mollissima is at first sight an innocuous gardener's plant, spread for ornament and fruit and apparently modified by hybridization. However, in Hawaii it has become a destroyer of forests. It grows wild in the Andean upper montane forest (known as 'ceja de la montaña') above 2000 m, a forest type composed of evergreen woody vegetation with abundant epiphytic ferns, orchids, mosses and bromeliads, and with a cool, moist, foggy climate [222]. *Passiflora mollissima* is extensively cultivated in the Andean highlands from Venezuela to Bolivia and has been introduced to many subtropical and tropical mountainous areas both as an ornamental and for its fruit. It was first reported in Hawaii in 1921, probably planted in gardens from which it then spread. It is now proliferating in mid to high elevation forests, both disturbed and undisturbed, on the islands of Hawaii and Kauai. The total area covered by a continuous distribution of *Passiflora mollissima* was over 190 km^2 in 1983 [425] and it has successfully invaded areas of diverse climate and vegetation. On the island of Hawaii it can be found from 300 to 2500 m elevation, in habitats ranging from dry lava flows with sparse open scrub to montane rainforests and pastures [222]. Dense curtains of the vine extend to the ground from canopy branches, sometimes causing branches to break and toppling trees during storms. Where the canopy has been opened, dense mats of vines also mantle the understory trees and shrubs and inhibit regeneration of the native trees [292, 351]. Endangered endemic forest birds are affected by the increase of *Passiflora mollissima*, which alters the structure and composition of the forest [425]. On Kauai, the populations are centered in Koke'e and are found in both open and closed *Acacia* forests from 850 to 1300 m elevation.

In South Africa it was only noticed in the wild as recently as 1987; it is available for sale in nurseries [245]. It appears to have naturalized in some forests in South Africa but is not yet widespread. It has been seen in Knysna Forest of the southeast Cape Province as well as in other areas. In New Zealand it occurs mainly in forest plantations, margins and on isolated trees and is sometimes a serious weed [428, 445].

In its native habitat of the moist Andes from 2000 to 3600 m [264], populations of *Passiflora mollissima* are sparse, with only about two to three plants per hectare; its flower and fruits are heavily predated by numerous insects [425]. In Hawaii, it is found at densities far in excess of this. *Passiflora mollissima* grows best in cool regions and can tolerate occasional frosts to −2°C [222], although it occurs in Hawaii under a broad range of environmental conditions and on several types of soils, from ash to weathered basalt [425]. The relatively shade-tolerant seedlings (there is usually a large seedling bank resulting from the continuous and prolific seed rain) grow rapidly in full sun. *Passiflora mollissima* can invade closed forests through gap-phase replacement involving its rapid growth in gaps caused by fallen trees; it forms a dense tangle of vegetation which smothers the undergrowth. Individuals reach reproductive maturity at an early age and mortality is low after establishment: the life span may exceed 20 years [222].

Flowers can be found during all months of the year and fruit is copiously produced. The abundant fruit set observed in Hawaii seems to be due to a mixture of spontaneous self-pollination and pollination by alien insects. The newly opened flowers have exposed stamens, favorable to cross-pollination by insects; if cross-pollination does not occur, each flower later pollinates itself through movement of the stigmas to touch the stamens. Where native, it is thought to be pollinated

Figure 4.12   In Hawaii, the weight of the South America climber banana poka, *Passiflora mollissima*, tears branches from native forest trees which are poorly adapted to climbers. Alien plants often do most damage to native communities where they invade an apparently vacant niche. (Photo: I.A.W. Macdonald)

by hummingbirds and large bees. *Passiflora mollissima* exhibits continuous growth and reproduction, but peak flowering occurs in the dry season in both Hawaii and South America [222]. The seeds are dispersed by frugivorous animals, in Hawaii principally by feral pigs (*Sus scrofa* L.). Birds aid in long-distance dispersal to uninfested areas, providing new foci for invasion. Pigs provide a fertile medium for seedling growth in the early stages of establishment [222] and their rooting activities create an environment with low competition, favorable for *Passiflora*.

The following ecological characteristics of *Passiflora mollissima* may be suggested as reasons for its success as an invasive species:

1. prolific, continuous seed production;

2. effective dispersal by feral pigs in Hawaii; long-range dispersal by birds;

3. facilitation by the 'rooting' activities of pigs, disturbing the soil and providing suitable areas for seedlings to establish and a fertile medium of pig dung in which the seedlings initially grow;

4. ability to tolerate low light levels and exploit gaps;

5. combination of auto- and allogamy;

6. relatively fast growth rates leading to early reproductive maturity.

Control and Management

Physical and chemical control Since the 1970s, several attempts by the State of Hawaii and the National Park Service at control by physical and chemical means have met with little success [425]. The extent and density of infestations make these methods uneconomical as well as ineffective. Long-distance dispersal of *Passiflora mollissima* initiates new populations in isolated or inaccessible areas and provides new foci for invasion. However, one exception is where forests of the endemic *Acacia koa* are being re-established on degraded montane forest land infested by *Passiflora mollissima*. The application of a high dose of glyphosate (6 kg/ha) prior to planting *Acacia* significantly reduced the mortality of *Acacia* by *Passiflora mollissima* after 10 years; in contrast, all *Acacia* trees were killed on untreated plots [350].

Biological control The only realistic hope for control of *Passiflora mollissima* in Hawaii is biological control [425]. *Passiflora mollissima* is attacked by many pests and diseases in its native range [222] but at present only one candidate, a moth, *Cyanotricha necyrina*, has been cleared by officials for release [263]. Studies on the potential of *Fusarium oxysporum* f. sp. *passiflorae* are in progress [361]. High host specificity is needed due to the large commercial passion fruit (*Passiflora edulis* Sims) industry on Hawaii and the potential damage which might be caused to this by more generalist control agents.

Pinus radiata D.Don {*P.insignis* Dougl. ex Loud., *P.californica* H.&A.} (Pinaceae)

Monterey pine

Description and Distribution

Habit Symmetrical tree, flat-topped at maturity, 15–25 m high, branchlets orange to dark red-brown.

Bark dark brown with narrow ridges.

Leaves in threes, deep glossy green, 8–15 cm long, slender.

Figure 4.13 *Pinus radiata*

Cones male cones yellow, about 12 mm long; female cones asymmetrical, ovoid, 7–15 cm long, brown, shiny, bluntly pointed, lower scales on outer side thickened, armed with slender prickles that usually wear off, the cones remaining closed and persisting for years.

Seeds winged, dark, rough, 6–7 mm long, wings 12–18 mm long, dispersed by wind [276, 295].

Invasive category 4.

Region of origin North America – California: limited natural range – 4000 ha in three small areas on the Californian coast [59] and few little-known populations in Mexico (var. *binata*).

Native climate wmtemp. dry, characterized by a dry period of three to five months during the summer months and high humidity for the rest of the year.

Regions where introduced Pinus radiata has been introduced to warm temperate regions all over the world and cultivated as a timber crop. In the following regions it has spread from cultivation and is invading natural and seminatural vegetation: Australasia – Australia [68], New Zealand [445]. Africa – South Africa (southwestern Cape Province) [329].

Climate where invading wmtemp. dry (Australia, South Africa), wmtemp. moist– wet (New Zealand). In Australia it occurs in areas with no marked dry season but with maximum humidity during the winter months. In South Africa it occurs in areas characterized by a dry period lasting for two to three months in the summer and very high humidity during the rest of the year.

Pinus radiata has been the subject of an intensive selection program by foresters, who have converted the poor wild form into a straight-trunked, fast-growing tree crop which is highly invasive. In the wild it occurs on dry bluffs and slopes below 1200 m in coastal areas of California [295]. As an exotic *Pinus radiata* is remarkably successful, having been planted to cover nearly one million hectares of land in different parts of the world [68], which contrasts strongly with its limited natural range estimated to cover only 4000 ha [276]. *Pinus radiata* has been widely planted in Australia and New Zealand where it now forms the major softwood source [68, 276]. In Australia *Pinus radiata* is invading an open and relatively undisturbed type of dry sclerophyll *Eucalyptus* forest. In New Zealand it is reported to be a 'significant problem' in scrub and forest margins, shrubland, sand dunes, open land, and short and tall tussockland [445]; while in South Africa it is a major weed of the mountain fynbos [331]. *Pinus radiata* has been present in South Africa since 1865 but has only recently (since about 1930) been planted on a large scale [410].

In natural forests invaded by *Pinus radiata* in Australia, the seedlings of *Pinus* grow faster than the native *Eucalyptus* [59, 68], but the rate of *Pinus radiata* recruitment appears to be slow due to high seedling mortality, apparently caused by drought. The small, light, winged seeds of *Pinus radiata* allow for easy long-distance dispersal by wind [410]. *Pinus radiata* is self-compatible (although there is considerable variation in self-fertility) and thus isolated individuals can produce viable seed. Colonization is likely to be accompanied at first by an increase in inbreeding, but outbreeding will tend to be restored as population density increases. This flexible breeding system may help it to cope well with both the colonizing and the sedentary phases [25].

The invasion of fynbos by *Pinus radiata* is characterized by a rapid but sparse influx of initial colonizers throughout an area. This is followed, in the absence of fire, by slow and erratic establishment of seedlings around these trees, and, with fire, the recruitment of dense daughter stands. *Pinus radiata* is reported to produce viable seed after seven years in California, after about 10 years in Australia and after 10–12 years in New Zealand. Cones are first produced after six to seven years in South Africa but ripe cones require three to four years to dry out sufficiently to allow opening [329]. Seeds may be held in the cones for five years or longer with no loss of viability. Seeds are released from the cones in hot dry summers in South Africa when the strongest winds blow from the southeast [329].

Fire stimulates simultaneous release of large quantities of seed from the cones and creates favorable conditions for germination and establishment [269, 329]. However, *Pinus radiata* is often killed by serious fires, whereas native *Eucalyptus* species have dormant buds under the bark which resprout after fire. Regeneration of *Pinus radiata* after fire (particularly surface fires in which parent trees survive) results in dense stands which radically alter the structure and composition of natural plant communities [329]. However, as the cones vary from semiserotinous to serotinous, *Pinus radiata* can establish on sites burnt only infrequently. Native animals and birds, as well as rabbits and introduced animals, are known to damage alien pines in Australia, but the rapid growth of *Pinus radiata* soon puts it out of reach of mammalian herbivores. In one study of invasion of *Eucalyptus* forest in Australia, invasion did not begin until 20–21 years after the establishment of local plantations [68], possibly because of the elimination of rabbits by an epidemic of myxomatosis about four years before invasion commenced.

In general, the densest invasions in Australia occur on open dry sites with poor shallow soils. Wet sclerophyll forests in Australia seem to be resistant to invasion. Those dry sclerophyll forests prone to invasion are more open and this feature, coupled with the pendent leaves of some species of *Eucalyptus*, permits ample light through the canopy. Once established, *Pinus radiata* intercepts much more light than *Eucalyptus* [68] and it has been estimated that trees may persist up to 150 years [276].

The following ecological characteristics may be suggested as reasons for the success of *Pinus radiata* as an invasive species:

1. large and widespread planting as a timber crop;
2. vigorous growth when released from the burden of pests and diseases [59];
3. reproductively mature at a young age and prolific seed production;
4. degree of serotiny insuring that regeneration by seed is rapid after fire;
5. small, light, winged seeds easily dispersed long-distances by wind;
6. self-compatibility allowing isolated individuals to produce seed.

Control and Management

Biological control As *Pinus radiata* is such an economically important species and *Pinus* such an important genus, biological control methods have never been sought.

Physical control A combination of cutting followed, a few months later after the seeds have germinated, by fire, may be a method of control in habitats which are naturally adapted to fire.

Prospects Pinus radiata poses a threat to 'Mediterranean' ecosystems worldwide and foresters should be aware of the potential dangers before including it in forestry schemes. There are good prospects for experiments to determine appropriate ecological management regimes in order to minimize invasion.

Pittosporum undulatum Vent. {*Pittosporum phillyraeoides* Haw., non DC.} (Pittosporaceae)

Sweet pittosporum, mock orange, Victorian box, Victorian laurel

Figure 4.14 *Pittosporum undulatum*

Description and Distribution

Habit Slender branched shrub or tree, with smooth gray bark, 5–13 m tall.

Leaves alternate, elliptic-oblong to oblanceolate, 6–16 cm long, entire, green above, paler beneath with undulate margins.

Inflorescence terminal, 4–5 flowered, about 2 cm long, borne on the youngest branches.

Flowers white, fragrant; petals 5, recurved above; stamens 5–11 mm long, sometimes reduced to sterile rudiments; stigma slightly longer or shorter than the stamens. The flowers are usually hermaphrodite, but dimorphic flowers occur in the native area and unisexual individuals have even been found [138].

Fruit a globular dehiscent capsule, 2-valved, yellow to brown [77].

Seeds 12–22 per capsule, sticky, black or red; birds feed on the seeds after the ripe fruits have burst open [328].

Invasive category 5.

Region of origin Australasia – Australia. *Pittosporum undulatum* occurs naturally seawards of the Great Dividing Range from Brisbane to Western Port in Victoria [130], also in Tasmania. It extends inland for about 280 km in New South Wales but only 120 km inland in Victoria, where it reaches altitudes of around 400 m [130].

Native climate subtrop. moist (subtropical rainforest and wet sclerophyll forest of New South Wales), wmtemp. dry (dry sclerophyll forest of E. Victoria). Although found in a wide range of habitats, the climate throughout its range is characterized by the absence of a pronounced dry season [8,130].

Regions where introduced Australasia – Australia (south central Victoria) (invasive) [130], Norfolk I. (invasive) [328], Lord Howe I. (invasive) [314], New Zealand (naturalized) [328]; Central America – Jamaica (invasive) [136]; Africa – South Africa (invasive) [328]; Oceania – Hawaii (ruderal) [158]; Oceanic Islands – Macaronesia (Azores, Pico Island (invasive) [150], Canary Islands (naturalized) [130]), Bermuda (naturalized) [77], St Helena (naturalized).

Climate where invading trop. wet (Jamaica), with mean annual rainfall 2690 mm at 1500 m altitude [147], wettest between August and December [136]. Wmtemp.

dry (South Africa). In Jonkershoek Valley, where it is invading, there is a summer dry period, but otherwise a moist climate. In Australia it is extending westwards into areas with slightly less rainfall than in its native habitat; its spread may perhaps have been encouraged by a reduction in the frequency of fires, following human control.

Pittosporum undulatum is an example of an invasive ornamental spread initially by the network of British colonial botanic gardens. It is found sporadically in subtropical rainforest in New South Wales and in 'tall open' *Eucalyptus* forests [130] (wet sclerophyll forest, height greater than 30 m [366]) and dry sclerophyll forest in Victoria. In Australia it is invading up to 200 km outside its natural range into coastal or subcoastal areas at low elevations covered by relatively undisturbed *Eucalyptus* forest or scrub. Invasion of forests in south and west Australia may be restricted by the very dry summers, although the wet karri (*Eucalyptus diversicolor*) forests in southwest Australia may be prime candidates for invasion when the right vectors are present [130]. It is also invading Norfolk and Lord Howe Islands, which have a continually wet climate.

Pittosporum undulatum is often planted as an ornamental or hedge plant. In Jamaica, it was introduced in 1870 to the Cinchona Botanic Garden and is now invading fairly undisturbed montane sclerophyll rainforest, containing many endemic species [147]. In South Africa it was introduced in 1901 to the Tokai Arboretum [328] and is now invading mountain fynbos with vegetation dominated by *Olea europaea* ssp. *africana*, *Kiggelaria africana*, *Rhus angustifolia* and *Maytenus oleoides*, as well as undisturbed patches of riparian forest [328]. Introduced to Hawaii in 1875, probably as a timber crop, it now occurs in moist to wet regions along roadsides, in forests and as a weed of pasture and rangelands [158]. In the Azores it has spread widely in lowland cloud forest, which contains many endemic species [150].

Pittosporum undulatum flowers when only four or five years old. In Jamaica it flowers and sets fruit earlier in the year than most native trees [136] and the lack of competitors for pollinators may be an advantage. Early emergence may also give seedlings a competitive advantage. The flowers appear to be suited to nonspecialized insect pollinators, as are the flowers of most tree species in the Blue Mountains of Jamaica [136]. Prolific quantities of sticky seeds fall directly below the crown or are eaten by, or adhere to, animals and birds. In Australia and South Africa invading *Pittosporum undulatum* initially tends to be clustered around the base of established trees [129, 328], and seedlings seem to have greater resistance to drought under shade than in the open. It can exploit high light levels but appears to have the capacity to endure shade, and is a successful gap colonizer. It often forms dense growth of tree seedlings in forest gaps, in which it can out-compete native tree species. Seedlings of *Pittosporum undulatum* in Jamaica and Australia seem to emerge after the winter rains [131, 136], which usually last for three months. Seedlings establish more easily on humus rather than on bare soil and appear to prefer sites which are less heavily shaded. The mortality rate of recently germinated seedlings appears to be considerably less in gaps [136]. Throughout the areas of undisturbed forest invaded in Jamaica, all size classes of *Pittosporum undulatum* individuals may be observed.

The shade cast by the low, very dense, canopy of *Pittosporum* dramatically reduces the amount of light that reaches the forest floor. Apical dominance is weak, hence a bushy habit develops and the crown diameter may not be much less than the height, although it is capable of relatively rapid growth. The foliage of *Pittosporum undulatum* contains many oils, resins and saponins which may be allelopathic. When tested in the laboratory, leachates from the foliage significantly inhibited the germination of several species of *Eucalyptus* such as *E. obliqua*, *E. melliodora* and *E. goniocalyx* [130]. However, no such effects, other than those expected from deep shade, have been demonstrated under canopies in the field. Litter production of *Pittosporum undulatum* is high. Soil from beneath *Pittosporum undulatum* clumps is often more fertile than that under native *Eucalyptus* in Australia, the leaf litter containing higher levels of nutrients [130]. The capacity of *Pittosporum undulatum* to withstand fire is not great, as the bark is thin and resinous. If the fire burns through downwards to the mineral soil, the basal buds in the trunk are killed. However, light fires do little permanent damage and saplings (above 2 m high) will sprout vigorously.

The following ecological features of *Pittosporum undulatum* may be suggested as reasons for its success as an invasive species:

1. prolific seed production;
2. bird-dispersal results in long-distance dispersal of seeds, which may become foci of new invasions;
3. toleration of shade and exploitation of gaps;
4. dense canopy shades out ground vegetation and inhibits regeneration of native tree species, thereby reducing competition;
5. not adapted to a specific pollinator;
6. artificial suppression of fire by people in Australia, outside the native region of *Pittosporum undulatum*, and the infrequency of fire in Jamaica;
7. the litter enriches soils which may not favor plants such as sclerophyll shrubs adapted to poor soils [130];
8. the adaptable root system allows it survive in a wide range of soil types [130];
9. early age of reproductive maturity [136].

Control and Management

Physical control Seedlings can easily be hand-pulled but stumps sprout vigorously [328].

Chemical control Application of 2,4,5 T and diesel mixture to stumps cut just above ground level prevents coppicing [328].

Biological control No biological control program has been initiated to date.

Prospects Pittosporum undulatum is unusual as it appears to be invading relatively intact species-rich rainforest vegetation in Jamaica. It may become a serious pest in other areas, for example in South Africa, as it continues to be spread by gardeners. Studies of its ecology are needed to devise appropriate control strategies.

Psidium cattleianum Sabine {*P.littorale* Raddi} (Myrtaceae)
Strawberry, Chinese, purple, or pineapple guava

Description and Distribution
Habit Large shrub or small tree to 6 m; trunk smooth, pale brown; shoots terete or subterete, covered with fine, short dense hairs when young.

Leaves 4–8 × 2.4–4.5 cm, obovate, leathery, hairless, shiny, dark green above, dotted with glands beneath.

Flowers white, solitary in axils; bracteoles very small, caducous, ovate; calyx with 4–5 rounded lobes, 3–4 mm; petals *c*. 5 mm, elliptic.

Fruit 2.5–3.5 cm long, globose, reddish-purple, occasionally yellow; flesh usually purple, edible.

Seeds numerous [2, 65].

Invasive category 5.

Region of origin South America – Brazil [240]. Little is known about its ecology in its natural habitat.

Native climate trop. dry–moist (tropical moist forest).

Regions where introduced Oceania – Hawaii (invasive) [183], Tropical Polynesia (invasive); Australasia – Norfolk I. (invasive) [428]; Malagassia – Mauritius (invasive) [239, 240, 253].

Climate where invading Hawaii: wmtemp. wet, subtrop. wet, in areas at 100m–1300m in elevation with a wet aseasonal climate; Mauritius: wmtemp. moist–wet,

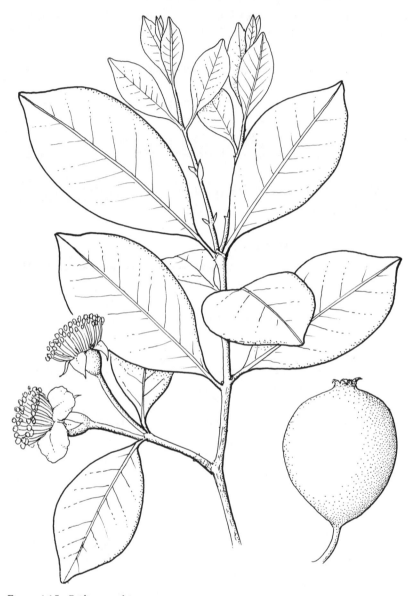

Figure 4.15 *Psidium cattleianum*

subtrop. moist–wet, relatively wet all year with maximum rainfall during the summer months and a brief (*c.* 1–2 months) dry period prior to the summer; Norfolk I.: subtrop. moist, relatively wet all year, with maximum rainfall during the winter months.

Psidium cattleianum is one of the most serious invaders of montane tropical

Figure 4.16   Relict native trees stand above a forest badly invaded by *Psidium cattleianum* on Mauritius. (Photo: Quentin Cronk)

rainforest on islands, and its effect on the native vegetation of Mauritius has been catastrophic. Surprisingly there have been no studies of its ecology in its native habitat. It is now a familiar cultivated plant throughout the tropics. *Psidium cattleianum* was introduced to Hawaii in 1821 for its fruit, but it soon escaped [361]. It is now invading a variety of native ecosystems, principally rainforest, where it is considered the worst pest [359]. It apparently invades intact and undisturbed rainforest, but also occurs along roadsides and in other disturbed habitats [183]. In Mauritius it was introduced around 1822 and is now invading upland evergreen wet forest, lower montane moist and wet forest, including areas with low disturbance [239, 240]. It has spread to nearly all areas on Norfolk Island.

   Psidium cattleianum is a thicket-forming tree with rapid growth which shades out all other plants. It may occur in large numbers within dense forest, indicating that it is shade-tolerant. It can produce dense populations of suckers and seedlings, with dense mats of surface feeder roots, even in thickets with low light levels. It has broad environmental tolerances.

   It flowers nearly all year round, although the peak of reproductive activity in Hawaii occurs from June to October. The prolifically produced seeds germinate rapidly under a wide range of environmental conditions. Feral pigs and non-native birds disperse the seeds. *Psidium cattleianum* seedlings occur on the same substrates as native woody species, such as bryophyte clumps, usually in undisturbed sites. Passage of the seeds through the guts of feral pigs has been found to have little effect on germination success, but does appear to shorten the time required for germination. Seedling establishment in the field seems to be independent of soil disturbance, such as that resulting from the rooting activity of pigs. However, dung piles of feral pigs may provide the same sort of protection from desiccation as bryophyte covered substrates. High densities of *Psidium cattleianum* seeds have been found in pig droppings [183].

Psidium cattleianum reproduces by seed and suckers, and the clonal offspring appear to have higher rates of growth and survival than the seedlings [183]. Native tree seedlings are often suppressed by litter from the canopy trees, but *Psidium cattleianum* can send up vigorous vegetative shoots. Both in Hawaii and Mauritius [183, 240] there are red and yellow fruited forms.

The following ecological characteristics of *Psidium cattleianum* may be suggested as reasons for its success as an invasive species:

1. effective dispersal of seeds over moderate distances by animals;
2. sprouting response of *Psidium cattleianum* to falling leaf litter, which is an important difference between this and many of the native species in Hawaii;
3. prolific seed production all year round;
4. broad environmental tolerances, including shade tolerance;
5. rapid growth, forming dense shade-casting thickets with dense mats of surface feeder roots, preventing the regeneration of native species.

Control and Management

Chemical control Chemical treatments are being tested in Hawaii Volcanoes and Haleakala National Parks [359].

Biological control No biological control program has been established to date in Hawaii. The prospects for biological control are relatively slim as it and the common guava, *Psidium guajava*, are cultivated widely for fruit. Any biological control agent would need to be very species-specific and not damage native endemic Myrtaceous trees [361]. However, the extent and success of *Psidium cattleianum* would suggest that biological control is the only viable method.

Prospects Psidium cattleianum is threatening the unique endemic floras of Hawaii, Mauritius and Norfolk Island, as well as other islands in tropical Polynesia. Control methods devised in one area would be of great interest elsewhere.

Rhododendron ponticum L. {*R. lancifolium* Moench, *R. speciosum* Salisb.} (Ericaceae)

Rhododendron

Figure 4.17 *Rhododendron ponticum*

Description and Distribution

Habit Evergreen shrub, 2–8 m high, when mature consisting of several major axes arising from a large irregular base, shoots hairless.

Leaves spirally arranged in lax clusters, oblong – elliptic, up to 22 cm long, leathery,

115

hairless, entire, dark green above, paler below.

Inflorescence a compact raceme, protected by leafy scales in bud.

Flowers purple magenta, spotted with brown and orange; calyx small; corolla campanulate, 5-lobed, *c.* 5 cm across; stamens 10.

Fruit a woody capsule, persisting up to three years.

Seeds 1.5 × 0.5 mm, weighing 0.066 mg, dispersed by wind.

Invasive category 4.

Taxonomy Two subspecies are described from Europe [403]: ssp. *ponticum,* which has leaves 12–18 (–25) cm long, 2.5–3.5 times as long as wide and an inflorescence axes which are more or less hairless; and ssp. *baeticum* (Boiss. & Reuter) Hand.-Mazz., which has elliptic-oblong leaves 6–12 (–16) cm long and 3–5 times as long as wide, with tomentose inflorescence axes. The plants in the British Isles resemble ssp. *ponticum* [89] but Cox and Hutchinson [82] consider that they are mostly hybrids between *R. ponticum* and *R. catawbiense* Michaux; they have been considered to be nearer to ssp. *baeticum* by some authors [354]. *Rhododendron ponticum* in Britain has certainly been altered by hybridization [31].

Region of Origin Europe – a small area in the Stranja Mountains of northwest European Turkey and bordering parts of southeast Bulgaria (ssp. *ponticum*); parts of Portugal and southern Spain (ssp. *baeticum*) [89]; N. Asia – Turkey. *Rhododendron ponticum* has its main and most continuous distribution in the region of the Black sea (ssp. *ponticum*) [31]. It also occurs in the mountains of Lebanon and Syria (sometimes distinguished as ssp. *brachycarpum*).

Native climate Turkey: wmtemp. dry, temp. moist, seasonal, with a dry or drought summer period lasting from three to four months. In the colder months temperatures often drop below −10° C, and humidity is high [424]. Spain and Portugal: wmtemp. dry, a seasonal climate with a summer dry period lasting three to five months but with warmer winters than in the eastern part of the range.

Regions where introduced Europe – British Isles and Ireland (invasive) [88, 123, 339, 354, 386, 404], Belgium (naturalized) and France (naturalized) [89, 403].

Climate where invading temp. moist.

Rhododendron ponticum eliminates the ground flora of temperate deciduous woodlands and threatens rare species of Atlantic bryophytes in Ireland. Ironically there is fossil evidence that it occurred naturally in Ireland before the last ice age. In the present postglacial period it has, however, remained localized in refuges in southern Europe and Turkey. South of the Black Sea it grows in mixed deciduous forests of lime, oak and chestnut, and also in forests of beech, Fagus orientalis, [89] associated with species such as Ilex aquifolium, Prunus laurocerasus, Vaccinium arctostaphylos, Buxus sempervirens and Daphne pontica [423]. Similar vegetation types are found in SE Bulgaria [402]. In SW Iberia it grows in evergreen Mediterranean forest associated with other members of the family Ericaceae, such as Erica arborea and Arbutus unedo. East of the Black Sea Rhododendron ponticum is an important ingredient of the rich forests of western Transcaucasia (Colchis), growing under a canopy of Fagus orientalis, Picea orientalis or Abies nordmanniana. It ascends above the tree line as a dwarfed form well above 1800 m altitude [31] and grows right down to sea level in the neighborhood of Batum, forming thickets up to 6 m high in association with the cherry laurel (Prunus laurocerasus).

In Britain, Rhododendron ponticum was introduced as an ornamental. It became valued as game cover in large estates, from where it has invaded many seminatural woodlands on acid soils. It has been used as a rootstock for other Rhododendrons, which, when neglected, can then revert to Rhododendron ponticum [354, 386]. It came to Kew in 1793 [31] and was first distributed by the nurseryman Conrad Lodiges. Early introductions were probably of ssp. baeticum [31], but typical Rhododendron ponticum soon followed. Early hybridization with Rhododendron maximum has given rise to many garden forms. Rhododendron ponticum was probably introduced to Ireland in the late eighteenth century [90], and it now can be found invading western Quercus petraea woodlands dominated by Quercus petraea and Ilex aquifolium on acid soils with a rich bryophyte flora [199]. It also occurs in mixed oak woods, heaths, upland acid Nardus grassland and occasionally on dune heaths and bogs [89]. Invasion replaces the woodland understory with single species stands of Rhododendron. Reduction of light, along with the accumulating litter, eliminates bryophyte, herbaceous and dwarf shrub layers. Native trees and shrubs are unable to regenerate under the dense shade cast by Rhododendron ponticum. It can readily invade woods in which regeneration of native trees is inhibited by overgrazing of sheep and deer.

Rhododendron ponticum reproduces mainly by seed, although there is limited vegetative spread by layering, and cut stems sprout vigorously. It flowers after about 12 years and is predominantly an outbreeder, with the flowers mainly pollinated by members of the Hymenoptera and Syrphidae (this mining-bee group are specialized pollinators). Rhododendron ponticum flowers most profusely in the open or under a light canopy. It is a prolific seed producer and viable seeds, which are dispersed by wind, are produced every year. Light is essential for germination, the required temperature range for which is 10–15° C [89]. Shade cast by Rhododendron ponticum thickets reduces both the rate and quantity of germination of the ground flora compared to shade cast by an oak or holly canopy. Germination of *Rhododendron ponticum* seeds occurs on many substrates, including peat, mor humus and brown earth, but successful establishment of seedlings does not appear to occur where there is dense ground cover of litter or vegetation. The most favored sites are patches of bare mineral soil or humus, or on rotting wood.

Large mature *Rhododendron ponticum* bushes transmit only about 2% total daylight and growth of other species underneath is precluded [89]. Root competition is also likely to be important. In waterlogged conditions, growth is very slow and accompanied by signs of mineral deficiency. Like other ericaceous plants, *Rhododendron ponticum* has endotrophic mycorrhizae. Despite its presence in the British Isles for over 200 years very few insects are yet associated with this species. The most important disease is the fungus *Pycnostysanus azaleae*, which causes bud blast. This fungus occasionally attacks young stems and leaves. In Berkshire up to 50% of buds have been observed killed but in Ireland it is of minor importance only. *Rhododendron* produces an andromedo-toxin and the leaves are unpalatable to herbivores [121]. It has few associated animals or epiphytes and invasion causes a dramatic decline in ecosystem diversity at the species level [90].

The following ecological characteristics may be suggested as reasons for its success as an invasive species:

1. prolific annual production of small light seeds of high viability, which are dispersed efficiently by the wind;
2. toleration of a wide range of temperatures, including frost;
3. unpalatibility to herbivores, both vertebrate and invertebrate;
4. high competitive ability – its dense canopy shades out native trees preventing regeneration, while thick accumulation of litter which decomposes only slowly prevents the establishment of seedlings of native species;
5. ability to sprout vigorously after being cut or to send up shoots from fallen branches or stems;
6. rootstock rarely damaged by fire since it sprouts readily from underground buds;
7. disturbance of many native woodlands by people and grazing animals, providing sites for regeneration.

Control and Management

Physical and chemical control Cut stems of *Rhododendron ponticum* resprout vigorously, so all cut stems must be treated with herbicide and seedlings must be removed to prevent rerooting adventitiously [33, 386]. Control is labor-intensive and expensive. Spray treatment of cut stems with Amcide has been shown to be effective, although the surrounding vegetation may be affected [195]. Glyphosate is commonly used in Ireland and in English nature reserves.

Management Senescence of older *Rhododendron ponticum* bushes may allow native trees to regenerate and re-establish themselves, although this is unlikely to occur if the native vegetation is disturbed in any way. Management of woodlands for a rich ground flora (*Rhododendron ponticum* prefers bare patches for seedlings) helps, but the prevalent overgrazing by sheep and deer in many woods encourages invasion.

Biological control No biological control program has been initiated, and is anyway hampered by the low number of pests of *Rhododendron ponticum*, even in its natural range.

Salvinia molesta D.Mitch. {*Salvinia auriculata* auct.} (Pteridophyta: Salviniaceae)
Water fern, kariba weed, African payal

Description and Distribution
Habit Perennial, free-floating, aquatic fern, forming dense mats with plagiotropic shoots and tightly overlapping leaves [428].

Leaves floating leaves of different sizes, elliptic, entire, folded, light or brownish-green, becoming somewhat darker near the entire margins, densely covered on upper surface by hydrophobic papillae bearing groups of 2 or 4 uniseriate hairs united at their distal ends; papillae to 3 mm long; submerged leaf greatly dissected, hanging into the water, functioning as a root [336].

Sporocarps in long straight secund chains, hairy, about 1 mm in diameter, containing mostly empty sporangia [428].

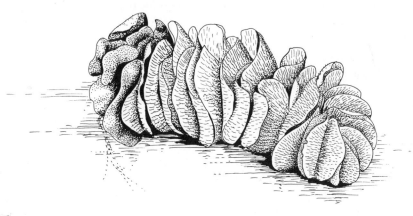

Figure 4.18   *Salvinia molesta*

Taxonomy morphologically plastic [278], it is distinguished from *S. auriculata, S. biloba* and *S. herzogii* by its small sporocarps bearing empty sporangia. It is sterile (pentaploid [391]) and probably of hybrid origin [277].

Invasive category 4.5.

Region of origin South America: *Salvinia molesta* has recently been discovered occurring naturally in southern Brazil [120]. It is found up to 200 km inland and at altitudes up to 500 m. *Salvinia molesta* occurs in natural lagoons, artificial dams, swamps, drains and along margins of rivers.

Native climate subtrop. dry-moist, trop. moist.

Regions where introduced Africa – Africa (invasive) [69, 247, 278]; S., E. and SE Asia – India (invasive) [389, 390], Sri Lanka (invasive) [352], Ceylon (invasive) [388]; Australasia – New Zealand (naturalized) [428], Australia (invasive) [337]; Malesia – Papua New Guinea (invasive) [278], Indonesia (invasive) [388], Singapore (invasive).

Climate where invading It is a serious pest in most tropical, subtropical and warm temperate regions of the Southern Hemisphere [389]. In the laboratory it is killed if the temperature drops below −3° C or rises above 43° C for 2–3 hours [439].

Salvinia molesta provides a good example of a case in which the control of an invasion sweeping the tropics was initially hampered by misidentification, as the species was not correctly named until 1972. It arrived in Asia via European botanical gardens [391], introduced as a botanical curiosity and aquarium plant. During the last 50 years it has spread widely through the tropics. In South Africa it has been widely distributed for aquaria and fish ponds and, although now proscribed as a noxious weed, it is still illegally grown in fish ponds and aquaria [148].

Salvinia molesta was first recorded in Australia in 1952, and first noticed as a pest in Kerala in 1956. By 1976 it had spread into many tropical and subtropical rivers and lakes in Australia and was more widespread than the other aquatic pest, *Eichhornia crassipes*. An infestation of *Salvinia molesta* leads to a decline in native plant species. In Kerala, for example, emergent hydrophytes with floating leaves (such as *Nymphaea stellata, Nelumbium speciosum* and *Limnanthemum cristatum*), 'suspended' hydrophytes (such as *Hydrilla verticillata*) and free-floating hydrophytes (such as *Pistia stratiotes, Azolla pinnata* and *Lemna minor*) are now rarely seen due to the invasion and spread of *Salvinia molesta* and *Eichhornia* [390].

In the early 1970s a few plants of *Salvinia molesta* were accidentally or intentionally introduced to a water body in the Sepik River floodplain, Papua New Guinea. By 1977 it had invaded many of the oxbow lakes associated with the river and it was continuing to spread [278]. By 1980 it covered 250 km^2 of water surface [391]. *Salvinia molesta* impedes river transport, prevents fishing and the collection of sago palm in canoes (an essential source of carbohydrate in the diet of the people living near the floodplain of the Sepik River), blocks access to drinking water for people, domestic stock and wildlife, clogs irrigation and drainage canals during floods, invades rice fields and creates a habitat for hosts of human diseases such as schistosomiasis and bilharzia [148, 278, 390]. The detritus of *Salvinia molesta* infills ponds and leads to eutrophication of small water bodies.

Where it is native in Brazil its population density is generally lower than in Australia, where it is invading, presumably due to attack by natural enemies [120]. However, little is known of the ecology of *Salvinia molesta* in its native habitat. It rapidly colonizes tropical and subtropical waters which are sheltered from strong winds and currents [278], forming extensive and relatively stable mats. These mats are often colonized by sedges and other plants. By cutting off light to submerged plants, mats of *Salvinia molesta* depress oxygen concentrations and increase those of carbon dioxide and hydrogen sulphide in the waters below, which can lead to extinction of most of the benthic flora and fauna [391]. Not only does *Salvinia molesta* invasion affect the native flora, it dramatically alters the habitat for wildlife, including some types of invertebrates, birds and fish, which depend on open water. By depleting water of nutrients, it can cause a decline in fish production [389].

The doubling time in the biomass of *Salvinia molesta* may be as short as 2.2 days in summer and 40–60 days in winter [117]. It can cover lakes and slow moving rivers with mats up to 1 m thick [391], although its highest rate in terms of individual plants growth is in uncrowded conditions [342]. The rapid spread of *Salvinia molesta* depends not on any intrinsic photosynthetic advantage, compared with other plants, but on its free-floating habit and many-branched growth pattern, which enables it to remain in an active vegetative form until the water surface is covered. Intrinsic growth and net assimilation rates are generally significantly positively correlated with air temperature and NPK (nitrogen, phosphorus, potassium) contents of the plant [338]. The upper surface of the leaves is unwettable, thus helping flotation. *Salvinia molesta* is thus dispersed easily by wind and water. This species is capable of surviving severe winters and also tolerating saline water for some time [148].

Salvinia molesta exhibits three distinct growth forms (morphs) [9, 280]: (1) a delicate, fragile 'primary invading form' with long internodes and small leaves which are up to 1.5 cm in width and which float on the water surface; (2) an open water colonizing form with leaves more than about 2 cm in width, becoming deeply keeled and assuming a boat-shaped form, found on the margins of weed mats; (3) a 'mat form' with normally large floating leaves (up to 6 cm in width), found when the plants become compressed together.

The following ecological characteristics of *Salvinia molesta* may be suggested as reasons for its success as an invasive species:

1. morphological plasticity, shown by its ability to modify the shape of vegetative structures in response to environmental conditions;

2. rapid vegetative propagation;

3. potentially very high growth rates;

4. readily dispersed through pieces of the parent plant being broken off by the action of wind, waves or people (boats); long-distance dispersal can result if plant material is transported downstream and new colonies are initiated.

Control and Management

Chemical control Several chemicals have been tried that have remarkably toxic effects on *Salvinia molesta* [391]. However, the use of herbicides creates further problems of water pollution and threatens water supplies of the local people. Chemical control should therefore only be used in extreme circumstances in certain small and isolated water bodies.

Biological control Research into biological control was initiated following the appearance of the weed on Lake Kariba [391]. At the time, the plant was misidentified as *Salvinia auriculata* sensu stricto. The Commonwealth Institute for Biological Control was commissioned to find a suitable control agent and they made their collections for possible controlling agents in South America. Three insects were recommended, a moth, *Samea multiplicalis*, a grasshopper, *Paulinia acuminata*, and a beetle, apparently *Cyrtobagous singularis*. During the 1960s and 1970s, all three insects were released in several countries with varying levels of success and failure. It was not until 1970 that it was realized that the weedy *Salvinia* species was not *Salvinia auriculata* [277]. In 1978, apparently native populations of *Salvinia molesta* in southern Brazil were discovered [120] and collections of insects were made [337], including a new species, *Cyrtobagous salviniae* Calder and Sands. This beetle has successfully controlled *Salvinia molesta* in parts of Australia and Papua New Guinea, after earlier successes in Namibia [391]. It is now achieving control throughout southern Africa [69]. The snail, *Pila globosa*, has been found to feed voraciously on *Salvinia molesta* in Kerala [388]. Control has also been attempted using Khaki Campbell ducks (*Anas platyrhynchos*) imported into Kerala from Britain [389], but without much success.

Prospects *Salvinia molesta* is the only invasive representative of a complex of four closely related species [7, 9], distinguished by differences in the branching structure of sporocarp chains and chromosome number: *S. molesta* – $2n = 45$; *S. auriculata* – $2n = 54$; *S. herzogii* – $2n = 63$; and *S. biloba* – unknown. Studies of the comparative ecology of these species might help in the formulation of control strategies.

Sesbania punicea (Cav.) Benth. (Leguminosae)
Coffeeweed, sesbania

Description and distribution

Habit Semideciduous shrub or small tree, up to 6 m tall [173].

Leaves 10–20 cm long, pendulous, pinnate, 10–40 pinnae per leaf; pinnae opposite, oblong and ending in pointed tips.

Inflorescence up to 25 cm long, pendulous, dense, a raceme of *c.* 14 flowers.

Flowers 2–3 cm long.

Fruit a legume, longitudinally 4-winged, oblong, 6–8 × 1 cm, short stalked, water dispersed, dehiscent, prolific.

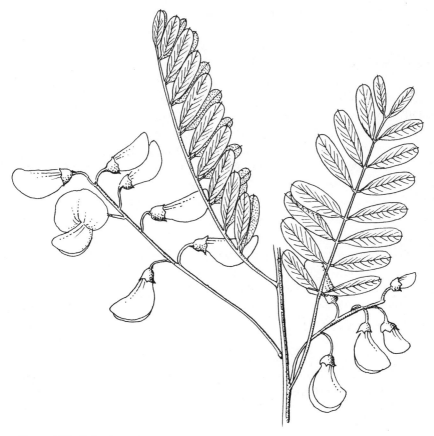

Figure 4.19 *Sesbania punicea*

Seeds 4–10 per pod [286, 315].

Invasive category 4.5.

Region of origin South America – northeast Argentina, Uruguay and southern Brazil [114, 315].

Native climate wmtemp. moist, subtrop. dry – moist.

Regions where introduced Africa – South Africa (invasive) [173, 315, 316]; North America – USA (Florida to Texas) (ruderal) [315].

Climate where invading wmtemp. dry – moist.

Sesbania punicea is another successful leguminous species invading South Africa. The nitrogen-fixing ability of these species may give them a significant advantage over the native vegetation. A native of riverside gallery forest vegetation in central South America [114], it is most common in areas frequently flooded. Initially introduced to South Africa as a garden ornamental, probably early in the present century, it now forms dense thickets along river banks, in wetlands and in damp lowland areas, posing a considerable conservation threat to native vegetation [173]. It invades various natural and seminatural vegetation types, including southern forest, eastern Cape forest, lowland fynbos and grassveld. In winter rainfall areas *Sesbania punicea* is mainly confined to major rivers, unable to survive the hot dry summers without perennial groundwater, but this is not the case in summer rainfall regions. The first mention of *Sesbania punicea* as a weed was in 1966. In the USA, it has spread from cultivation and invaded river banks, ditches and waste places [315].

Sesbania punicea seeds are readily water-dispersed, so the species poses a serious threat to riverside vegetation. *Sesbania punicea* infestations in the southwestern Cape are sometimes interspersed with other woody trees, in particular the Australian *Acacia* species and/or the native willow *Salix capensis*. Usually, however, the infestations take the form of almost impenetrable single-species stands, with dense interlocking canopies and lacking any grazing by native animals. The leaves and seeds of *Sesbania punicea* are highly poisonous to many mammals, birds and other animals [315]. The replacement of native vegetation by monotypic stands of *Sesbania punicea* affects native animals both by poisoning and by removing other food sources. The presence of *Sesbania punicea* in rivers obstructs the flow of water and this sometimes leads to erosion of water courses and widening of stream-beds, creating a perfect substrate for further lateral expansion of the infestations [173].

Sesbania punicea has a high growth rate [174], begins to set seed in its second growing season and does not usually survive for longer than 10 years [172]. Seeds germinate in the soil within about two years, but may persist longer if conditions are unfavorable for germination. Dormancy of seeds may pose a problem in weed control programs, as a source of re-infestation after clearance [139]. Crowding does not increase mortality in *Sesbania punicea* shrubs over two years of age but does decrease the number of seeds set, as well as the survival of seedlings and juveniles. This regulates plant density in herbivore-free populations [172].

The following ecological characteristics of *Sesbania punicea* may be suggested as reasons for its success as an invasive species:

1. prolific seed production;
2. adaptations for effective dispersal by water;
3. toxicity to many herbivores;
4. early reproductive maturity;
5. high growth rate;
6. seeds may remain dormant in the soil.

Control and Management

Biological control In 1982, four species of weevils, whose adults feed on the leaflets and buds of *Sesbania punicea*, were quarantined for screening tests in South Africa [173]. The smallest of the four weevil species, the bud-feeding apionid *Trichapion lativentre*, has so far proved to be the most successful biological control agent tested. Later it was discovered that *T. lativentre* was already present in Natal, probably inadvertently introduced by tourists from South America. It feeds on and damages the leaflets and growing points of the host [286] and has now been established over most of the range of *Sesbania punicea* in South Africa. Its presence also increases premature leaf abcission, reduces the growth of branches and greatly reduces the numbers of flowers, pods and seeds produced by the plants. Larval development of *T. lativentre* in the buds and flowers of *Sesbania punicea* results in a dramatic decrease in seed set, by more than 98%. Remarkably, *Sesbania punicea* populations are scarcely thinned by this, because of density dependent effects, although the ability of the plant to colonize new areas is severely reduced [175].

Trichapion lativentre, when used in combination with other weevil species which have been shown to be relatively effective such as *Rhyssomatus marginatus* and *Neodiplogrammus quadrivittatus*, may well result in complete control of *Sesbania punicea* [173]. These species and others are presently under investigation in South Africa. Models based on the population dynamics of *Sesbania punicea* show that a combination of a stem-boring weevil with two other weevils, which destroy the flower-buds and seeds respectively, would provide effective control. Ideally the stem-borer should be released around five years after the other two species [172].

5

Representative
invasive species

The following list is not intended to be exhaustive, but to be a representative sample of the world's invasive species to demonstrate the wide range of plant families, geographical areas, life forms and habitats associated with invasive plants. The accounts are of two types: (1) brief accounts, in which the salient facts of the plant's invasive behavior are given, together with references, and (2) listings, used in cases where detailed information is not known or not recorded, and only the basic geographical information is therefore given.

5.1 Notes on brief accounts of invasive species
The headings used at the beginning of each account are: species name, {synonyms}, (family) [242], common names.

Information given under the heading description
§Life form: Life form characteristics: herb, shrub, climber, tree, aquatic, moss; annual, biennial, perennial; deciduous, evergreen, semideciduous.
§Seed disp.: Seed dispersal mechanism: bird, mammal, wind, water, human, other.
§Seed prod.: Seed production: high, medium, low.
§Breeding syst.: Breeding system: hermaphrodite, monoecious, dioecious, seed apomixis, vegetative apomixis.
§Invasive category (scale of 0 to 5; 5 is most serious):

Category	Description
0	Not weedy or invasive
1	Minor weed of highly disturbed or cultivated land (man-made artificial landscapes)
1.5	Serious or widespread weeds of 1
2	Weeds of pastures managed for livestock, forestry plantations, or artificial waterways
2.5	Serious or widespread weeds of 2
3	Invading seminatural or natural habitats (of some conservation interest)

3.5	Serious or widespread invaders of 3
4	Invading important natural or seminatural habitats (i.e. species-rich vegetation, nature reserves, areas containing rare or endemic species)
4.5	Serious or widespread invaders of 4
5	Invasion threatening other species of plants or animals with extinction.

Regions of origin and invasion are recorded according to the following recognized areas on a practical rather than floristic basis:

1. **N. America** (United States, including Alaska and the Aleutian Is, Canada, Greenland)

2. **C. America** (= Central America and Caribbean) (Mexico including Baja California, Guatamala, Honduras, Costa Rica, Panama etc., Bahamas, Greater Antilles including Jamaica and Cuba etc., Lesser Antilles, Trinidad, Tobago)

3. **S. America** (Venezuela, the Guianas, Colombia and Ecuador south to Chile and Argentina)

4. **Australasia** (Australia, Tasmania, New Zealand, Lord Howe I., Norfolk I., Kermadecs, Chatham Is, Auckland Is and Campbell I.)

5. **Malagassia** (Madagascar, Comores, Mauritius, Réunion, Rodrigues, Seychelles, Aldabra)

6. **Africa** (North Africa and Egypt, south to the Cape of Good Hope, including Socotra and the Gulf of Guinea Is)

7. **Europe** (includes Iceland, Scandinavia, Russia to the Urals, Turkey-in-Europe and Greece to Portugal)

8. **N. Asia** (= north, central and west Asia) (Arabia, Levant including Turkey and Cyprus, Iranian Highlands including Afghanistan and Pakistan (part), Caucasus, Siberia, former Soviet Middle Asia, Chinese Central Asia including Mongolia and Tibet, former Soviet Far East including Kamchatka and Sakhalin)

9. **S. Asia** (= south, east and southeast Asia) (Pakistan (part), India, Sri Lanka, Bangladesh, Himalaya, China excluding Chinese Central Asia, Japan, Korea, Indochina including Vietnam, Burma and Thailand etc., Andaman Is, Nicobar Is)

10. **Malesia** (Peninsula Malaysia, Singapore, Sumatra, Borneo, Java, Sulawesi, Philippines, New Guinea, Solomon Is, Bismarcks etc.)

11. **Pacific** (= Oceania) (New Caledonia, south Central Pacific including Fiji and Samoa etc., Micronesia including Carolines and Marianas, Central Pacific Is including Tuvalu and Line Is, southeast Polynesia including Tahiti, Pitcairn and Marquesas etc., Hawaiian Is)

12. **Oceanic Is** (= most isolated oceanic islands) (Eastern Pacific Is: Clipperton, Cocos, Galápagos, Desventuradas, Juan Fernandez archipelago; Atlantic Is: Macaronesia, Bermuda, Fernando de Noronha and Trindade, Ascension, St Helena, Tristan da Cunha group, Gough I.; Indian Ocean Is: Laccadives, Maldives, Chagos, New Amsterdam and St Paul, Cocos (Keeling) Is, Christmas I.; Subantarctic Is: Macquarie, Kerguelen, Crozet, Marion, South Georgia)

Climatic zones. These are based only on mean annual precipitation and mean annual temperature [176] and do not take seasonality into account. They are intended as a rough guide to the climate of the areas in which the plants listed are found naturally or invade. In Table 5.1 the basic climatic information is given, including potential evapotranspiration (PET), which is a measure of the dryness of a climate taking both temperature and rainfall into account (higher values indicate greater dryness).

Table 5.1 Potential evapotranspiration values of the main climatic zones, with respect to temperature and rainfall

Climatic zone		PET	Precipitation (mm)	Temp (°C)
Temp. arid	Cool temperate arid	>2	<250	6–12
Temp. dry	Cool temperate dry	1–2	250–500	6–12
Temp. moist	Cool temperate moist	0.5–1	500–1000	6–12
Temp. wet	Cool temperate wet	<0.25	>1000	6–12
Wmtemp. arid	Warm temperate arid	>2	<500	12–18
Wmtemp. dry	Warm temperate dry	1–2	500–1000	12–18
Wmtemp. moist	Warm temperate moist	0.5–1	1000–2000	12–18
Wmtemp. wet	Warm temperate wet	<0.25	>2000	12–18
Subtrop. arid	Subtropical arid	>2	<500	18–24
Subtrop. dry	Subtropical dry	1–2	500–1000	18–24
Subtrop. moist	Subtropical moist	0.5–1	1000–2000	18–24
Subtrop. wet	Subtropical wet	<0.25	>2000	18–24
Trop. arid	Tropical arid	>2	<1000	>24
Trop. dry	Tropical dry	1–2	1000–2000	>24
Trop. moist	Tropical moist	0.5–1	2000–4000	>24
Trop. wet	Tropical wet	<0.25	>4000	>24

5.2 Species list

General References 9, 55, 57, 58, 81, 104, 110, 113, 119, 127, 143, 162, 166, 227, 239, 243, 254, 282, 314, 346, 347, 359, 361, 377, 445

Acacia cyclops A. Cunn. ex G. Don (Leguminosae)

Rooikrans

Description §Life form: shrub or shrubby tree (to 6 m), evergreen §Seed disp.: mammal (mouse, baboon), bird §Seed prod.: medium §Breeding syst.: hermaphrodite §Invasive category **4.5**.

Region of origin SW Australia. Climatic zone: wmtemp. arid, subtrop. arid. Grows scattered, rarely forming dense stands, typically on calcareous sands, with a rainfall of at least 250 mm.

Region invaded Africa (Cape Province, South Africa), N. America (California). Climatic zone: subtrop. dry–moist, wmtemp. arid–dry. It is the most widespread Australian wattle invading the lowland fynbos of Cape Province. It is also now well established in mountain fynbos and beginning to invade southern forest, eastern Cape forest and succulent karroo.

Notes Spread into natural vegetation by seed is particularly rapid after fire. It forms dense impenetrable stands with interlocking crowns, causing complete destruction of native vegetation.

Control Rooikrans rarely sprouts after intense burning or effective felling. Mechanical clearance is thus possible as long as the stems are cut low enough, preferably below the ground [48]. Biological control using seed-eating insects is under investigation in South Africa.

References 37, 48, 143, 209, 450.

Acacia dealbata Link (Leguminosae)

Blue wattle, silver wattle

Description shrub/tree

Region of origin Australasia (Australia)

Region invaded Africa (South Africa) [162]

Acacia longifolia (Andr.) Willd. (Leguminosae)

Long-leaved wattle, Sydney golden wattle

Description §Life form: shrub or tree (to 10 m), evergreen §Seed disp.: wind, water §Seed prod.: high §Breeding syst.: hermaphrodite §Invasive category **4.5**.

Region of origin Australasia (coastal areas of eastern Australia, New South Wales, Victoria and South Australia). Climatic zone: wmtemp. dry–moist, subtrop. moist. Characteristic of coastal areas, particularly sandy soils near creeks etc.

Region invaded Australasia (New Zealand), Africa (southwestern and eastern parts of Cape Province), N./W. Asia (Israel – especially the sand dunes of the coastal plain). Climatic zone: wmtemp. dry, subtrop. dry–moist. In South Africa it was originally planted on sand dunes but it has spread onto wet clay and dry sandy soils. It is one of the worst threats to mountain fynbos vegetation. It is established in lowland fynbos, southern forest, eastern cape forest and grassland vegetation.

Notes It produces large quantities of long-lived seed and regenerates rapidly after fire.

Control Mechanical clearance possible as long-leaved wattle does not sprout after cutting. The pteromalid gall wasp, *Trichilogaster acaciae longifoliae*, has been successfully used to reduce reproductive potential and vegetative growth [102] in South Africa by galling the reproductive buds.

References 46, 102, 196, 251.

Acacia mangium Willd. (Leguminosae)

Description shrub/tree

Region of origin Moluccas

Region invaded Sabah, Africa

Acacia mearnsii De Wild. {*Racosperma mearnsii* (De Wild.) Pedley} (Leguminosae)

Black wattle

Description §Life form: tree (to 20 m), evergreen §Seed disp.: mammal, water §Seed prod.: high §Breeding syst.: hermaphrodite §Invasive category **3.5**.

Region of origin Australasia (southern Queensland, southern New South Wales, Victoria, Tasmania and southeastern areas of South Australia). Climatic zone: wmtemp. dry–moist, subtrop. moist. It forms the understorey to *Eucalyptus* forest in areas with rainfall above 500 mm, typically on soils derived from sandstone, shale, granite or dolerite.

Region invaded Africa (Cape), Australasia (New Zealand), Pacific (Hawaii). Climatic zone: temp. moist, wmtemp. dry–moist, subtrop. moist. In South Africa it grows in fynbos vegetation, especially along rivers, streams and ditches, where it destroys the native vegetation and impedes water flow. In Hawaii it invades disturbed mesic habitats between 600 and 1700 m. It has been grown in the Cape since it was planted in the Cape Town Botanic Garden in the 1850s.

Notes It has a very high seed production, with a seed-bank of up to 20 000 per m^2 developing under a mature canopy. Fire stimulates germination and basal sprouting. It forms dense impenetrable thickets.

Control Mechanical control is only effective if the roots are removed or the stem is cut below the junction of the roots and stems. In South Africa, stems cut above the ground sprout and are treated with herbicide, such as 2,4,5-T in diesel oil [45]. Glyphosate controls seedlings and saplings.

References 45, 119, 161, 312, 359.

Acacia melanoxylon R.Br. (Leguminosae)

Australian blackwood, Australian ysterhout

Description §Life form: tree, evergreen §Seed disp.: bird, water §Seed prod.: high §Breeding syst.: hermaphrodite §Invasive category **3.5**.

Region of origin Australasia (South Australia, Victoria, New South Wales, south-eastern Queensland). Climatic zone: subtrop. moist, wmtemp. dry–moist. Pioneer species in rainforest succession, usually on deep humic soils but tolerant.

Region invaded Africa (South Africa), S. America (Argentina), N. America (California). Climatic zone: wmtemp. dry. It is a timber tree in South Africa producing valuable wood, but it invades disturbed native forest and also fynbos.

Notes It produces abundant seed which remains dormant until fire or disturbance. It also suckers.

Control The roots must be grubbed up or herbicides applied when clearing an area to prevent resprouting from suckers [282].

Acacia nilotica (L.) Del. ssp. *indica* (Benth.) Brenan (Leguminosae)

Babul

Description shrub/tree

Region of origin Arabian peninsula, Pakistan, India, Burma

Region invaded Antigua, Barbuda, Anguilla, Ecuador, Australia (Queensland, Northern Territory), Indonesia (Baluran National Park).

Control The seed-feeding bruchid, *Bruchidius sahlbergi*, has been released in Australia [447].

Acacia saligna (Labill.) Wendl. (Leguminosae)

Description tree

Region of origin Australia

Region invaded South Africa, California [47]. See main account in Chapter 4.

Acer pseudoplatanus L. (Aceraceae)

Sycamore, great maple

Description §Life form: tree, deciduous §Seed disp.: wind §Seed prod.: high §Breeding syst.: monoecious §Invasive category **4**.

Region of origin Europe (south and central). Climatic zone: temp. moist. Common in woods, hedges, scree slopes and streamsides.

Region invading Australasia (New Zealand), Europe (Britain and Ireland), Oceanic Is (Madeira), S. America (Chile). Climatic zone: temp. moist, wmtemp. dry. It was introduced into Britain in Tudor times and invades strongly into seminatural woods and nutrient-rich waste land.

Notes It is one of the fastest growing of the European hardwood trees, tolerant of shade and exposure but preferring moist rich soils. It is mainly outbreeding, with nectar-bearing flowers; trees regularly produce *c.* 10 000 seeds annually.

Control Stems resprout unless treated with herbicides.

Reference 445.

Acorus calamus L. (Araceae)

Sweet-flag

Description herb

Region of origin S. Asia

Region invaded Europe

Ageratina adenophora (Spreng.) King & Robinson {Eupatorium adenophorum Spreng.} (Compositae)

Crofton weed, pamakani, white thoroughwort

Description §Life form: herb, perennial §Seed disp.: wind §Seed prod.: high §Breeding syst.: hermaphrodite §Invasive category **2.5**.

Region of origin C. America (Mexico). Climatic zone: subtrop. dry.

Region invaded S. Asia (N. India, N. Thailand, SE Asia), Australasia (subtropical E. Australia, New Zealand), Pacific Is including Hawaii, Africa (S. and W. Africa), N. America (California). Climatic zone: trop. arid–dry, subtrop. dry–moist, wmtemp. arid–moist. It invades rangelands, grasslands and woodland margins. It is unpalatable to cattle and tolerant of salinity.

Notes Agamospermous (triploid) with a high seed production, although low germination rates have been reported. It was introduced from Mexico to Britain in 1826 as an ornamental and from there was distributed around the world, escaping from cultivation in Australia around 1900 [13].

Control Easily removed by mechanical methods such as slashing. *Ageratina adenophora* is susceptible to a number of herbicides such as 2,4-D amine applied in a 0.6–0.8% solution [11]. Successful biological control has been achieved in Hawaii by the introduction of the Trypetid gallfly, *Procecidochares utilis* [11], a species which has also caused considerable mortality in parts of Australia, South Africa, New Zealand and India.

References 10–12, 14, 196.

Ageratina riparia King & Robinson {*Eupatorium riparium* Regel} (Compositae)

Description herb

Region of origin C. America

Region invaded Australia, New Zealand, Hawaii

Notes Biocontrol by a combination of the plume moth, *Oidaemaophorus beneficus*, a stem-galling fly, *Procecidochares alani*, and a pathogenic fungus, *Entyloma* sp., has been achieved in Hawaii [97].

Ailanthus altissima (Miller) Swingle {*Rhus cacodendron* Ehrh., *Albonia peregrina* Buc'hoz} (Simaroubaceae)
Tree of heaven

Description §Life form: tree, deciduous §Seed disp.: wind, water, bird §Seed prod.: high §Breeding syst.: dioecious §Invasive category **2.5**.

Region of origin E. Asia (temperate and subtropical China). Climatic zone: wmtemp. moist, subtrop. moist.

Region invaded Australia (Victoria), N. America (SE USA), C. America, S. America, Europe (France, Hungary, Greece). Climatic zone: temp. moist, wmtemp. arid–moist, subtrop. moist, trop. arid. Spreads rapidly on undisturbed grazing land, roadsides and waste places, on both clay and sandy soils.

Control Chemical control by spraying (using, for example, an 80% solution of 2,4,5-T) has proved successful in Australia [310]. Cut stumps are treated with herbicide to prevent reprouting, and for this Picloram is effective but expensive (as Tordon 50-D dilution 1 : 100).

References 310, 373.

Albizia lebbeck (L.) Benth. (Leguminosae)

East Indian walnut, kokko, siris

Description tree

135

Region of origin Tropical Asia

Region invaded Venezuela, Caribbean

Albizia lophantha (Willd.) Benth. (Leguminosae)

Description shrub

Region of origin W. Australia

Region invaded South Africa, California, New Zealand

Albizia procera (Roxb.) Benth. (Leguminosae)

Description tree

Region of origin NE Africa

Region invaded Venezuela

Alocasia macrorrhiza (L.) G.Don (Araceae)

Giant taro

Description herb

Region of origin Tropical Asia

Region invaded Kermadecs (Raoul)

Alternanthera philoxeroides (Mart.) Griseb. {*Telanthera philoxeroides* Moquin-Tandon} (Amaranthaceae)

Alligator weed

Description §Life form: herb, perennial, aquatic §Seed disp.: water, wind §Seed prod.: low §Breeding syst.: hermaphrodite §Invasive category **3.5.**

Region of origin S. America. Climatic zone: trop. arid. Widespread in freshwater habitats.

Region invaded S. Asia (India), N. America (southern USA – S. Carolina to Florida and Louisiana). Climatic zone: trop. arid, subtrop. moist. Forms dense mats in stagnant or slow-moving water.

Notes Seed set is rare; propagation is by vigorous production of below-surface shoots.

References 261, 343, 373.

Amaranthus spinosus L. (Amaranthaceae)

Description herb

Region of origin Tropics (including Vietnam)

Region invaded N. America

Ammophila arenaria (L.) Link {A. *arundinacea* Host, *Psamma arenaria* (L.) Roem. & Schult.} (Gramineae)

Marram grass

Description §Life form: herb, perennial, evergreen §Seed disp.: wind §Seed prod.: low §Breeding syst.: hermaphrodite §Invasive category **3.5**.

Region of origin Europe (coastal sand dunes of western Europe). Climatic zone: temp. moist, wmtemp. dry–moist. Dominant on foredunes, where it is often planted for stabilization, it colonizes bare sand with creeping underground stems.

Region invaded Australasia (New Zealand), N. America (California). Climatic zone: wmtemp. dry–moist, temp. wet.

Notes Vegetative reproduction predominant. It replaces native plants, altering the landscape significantly.

Control May be removed manually but care must be taken to remove all the rhizomes. Economic value for dune stabilization makes biological control unlikely.

Reference 445.

Andropogon glomeratus (Walt.) B.S.P. {*Schizachyrium condensatum* Nees} (Gramineae)

Bush beardgrass

Description herb

Region of origin USA, Mexico

Region invaded Hawaii

Andropogon virginicus L. (Gramineae)

Description herb

Region of origin SE USA

Region invaded Hawaii, Australia [140]. See main account in Chapter 4.

Anemone hupehensis (Lem. & Lem.f.) Lem. & Lem.f (Ranunculaceae)

Hupeh anemone

Description herb

Region of origin China

Region invaded Hawaii

Anredera cordifolia (Ten.) Steenis (Basellaceae)

Madeira vine, mignonette vine

Description climber

Region of origin Subtropical South America

Region invaded Australia (New South Wales)

Notes Tree-smotherer in rainforests with large underground tuber and clusters of aerial tubers.

Anthoxanthum odoratum L. (Gramineae)

Sweet vernal grass

Description herb

Region of origin Eurasia

Region invaded Chile, Hawaii

Antigonon leptopus H. & A. (Polygonaceae)

Coral vine, corallita

Description climber

Region of origin Mexico

Region invaded Guam

Ardisia crenata Sims (Myrsinaceae)

Description shrub

Region of origin NE India – Japan

Region invaded Mauritius, Réunion

Ardisia elliptica Thunb. {A. solanacea Roxb.} (Myrsinaceae)

Description tree

Region of origin S. Asia

Region invaded Hawaii

Ardisia humilis Vahl (Myrsinaceae)

Description climber

Region of origin Tropical Himalayas

Region invaded Cook Is

Aronia x *prunifolia* (Marshall) Rehder (Rosaceae)

Chokeberry

Description shrub

Region of origin hybrid

Region invaded Netherlands

Aster subulatus Michx (Compositae)

Description herb

Region of origin N. America

Region invaded Israel, Saudi Arabia

Baccharis halimifolia L. (Compositae)

Tree groundsel

Description dioecious shrub

Region of origin Eastern N. America

Region invaded Australia

Berberis darwinii Hook. (Berberidaceae)
Darwin's barberry

Description §Life form: shrub (to 4 m), evergreen §Seed disp.: mammal, bird. §Seed prod.: low §Breeding syst.: hermaphrodite §Invasive category 3.

Region of origin S. America (Southern Chile, Patagonia). Climatic zone: temp. arid–wet. Temperate forests dominated by *Saxegothaea* and *Nothofagus* at 0–500 m altitude.

Region invaded Australasia (New Zealand – Canterbury to Dunedin and the

Foveaux Strait). Climatic zone: temp. dry–wet. Remnant forest stands, scrub, along forest and plantation margins, roadsides. Locally common in Wellington and Wairarapa.

Notes It can invade undisturbed mature secondary woodland dominated by native species. Many tree species appear to be relatively unaffected, but *Kunzea ericoides* (with light-demanding seedlings) suffers, apparently from competition for seedling regeneration sites.

References [3, 322, 428, 445].

Boehmeria macrophylla D.Don (Urticaceae)

False nettle

Description herb

Region of origin subtrop. Himalayas and W. China

Region invaded Réunion

Brachiaria mutica (Forssk.) Stapf (Gramineae)

Description herb

Region of origin Tropical Africa and Brazil (wetlands)

Region invaded Hawaii, Australia (Northern Territory)

Bromus tectorum L. (Gramineae)
Downy brome, downy chess, cheatgrass

Description §Life form: herb, annual §Seed disp.: wind §Breeding syst.: hermaphrodite §Invasive category **2.5**.

Region of origin Europe, N. Asia. Herbivore adapted grass of the arid Eurasian steppes. Climatic zone: wmtemp. arid–dry.

Region invaded N. America., E. Asia (Japan), Oceanic Is (Tenerife). Climatic zone: wmtemp. arid–dry. In N. America it has replaced native colonizers such as *Festuca octoflora*, *F. microstachys* and *Bromus carinatus* in a variety of grassland types and along roadsides.

Figure 5.1 Invasive paper mulberry, *Broussonetia papyrifera*, a native of East Asia, growing over abandoned agricultural machinery, Budongo Forest, Uganda. (Photo: Doug Shiel)

Notes Cleistogamous and self-fertile, it is adapted to disturbance (trampling and grazing by large mammals). It behaves as a winter annual, surviving summer drought as seed.

References [105, 185, 257, 258].

Broussonetia papyrifera (L.) Vent. (Moraceae)
Paper mulberry

Description tree

Region of origin Tahiti

Region invaded N. America (SE USA), S. America (Peru), Africa (Ghana, Uganda – especially Budongo Forest Reserve), India

Caesalpinia decapetala (Roth.) Alston (Leguminosae)

Mysore thorn

Description shrub

142

Region of origin India

Region invaded Kermadecs (Raoul), South Africa (Transvaal) [162]

Calluna vulgaris L. (Ericaceae)

Heather, ling

Description shrub

Region of origin Europe

Region invaded New Zealand

Calotropis procera (Aiton) Aiton f. (Asclepiadaceae)

Apple of Sodom

Description shrub

Region of origin Old World tropics

Region invaded Pantropical Australia

Campylopus introflexus (Hedw.) Brid. (Dicranales – Bryophyta)

Description §Life form: moss §Spore disp.: wind, water §Breeding syst.: dioecious §Invasive category **3**.

Region of origin S. America, Africa, Australia, Pacific. Climatic zone: temp. moist, wmtemp. moist.

Region invaded Europe. Climatic zone: temp. moist, wmtemp. moist. Invades and dominates dune bryophyte communities.

References [326].

Carduus nutans L. (Compositae)

Musk thistle

Description herb

Region of origin Europe

Region invaded Canadian and New Zealand pastures (agricultural ecosystems only, characteristic of overgrazed sward)

Notes Biocontrol by receptacle weevil, *Rhinocyllus conicus*, has proved moderately successful in Canada but unsuccessful in New Zealand.

Carpobrotus edulis (L.) N.E. Br. (Aizoaceae)
Hottentot fig

Description succulent herb

Region of origin South Africa (Cape)

Region invaded N. America (California) [92, 93], Europe (Portugal, SW England)

Casuarina equisetifolia L. (Casuarinaceae)
Common ironwood, common ru, Australian pine, horsetail tree (South Africa)

Description §Life form: tree (to 50 m), evergreen §Seed disp.: wind §Breeding syst.: dioecious (sometimes monoecious) §Invasive category **3.5**.

Region of origin Malesia (Melanesia, Polynesia), Australasia (Australia – N. and NE coastlines). Climatic zone: subtrop. dry–wet. Pioneer tree of tropical and subtropical coasts, forming dense stands on the foreshore, sometimes suckering from the roots, which possess nodules containing nitrogen fixing bacteria.

Region invaded Pacific (Hawaii), N. America (Florida), C. America (Bahamas), Africa (South Africa), Malagassia (Réunion), S. Asia (Japan). Climatic zone: wmtemp. arid–dry, subtrop. dry–moist, trop. arid. Where planted for shelter on sandy shores, it spreads to form monotypic stands with little understorey. In Hawaii it is now widely distributed on all the islands up to 500 m altitude, in all but the wettest and driest places. In Florida it has become so abundant on some sandy shores that it interferes with the nesting of sea turtles and American crocodiles. It regenerates from basal shoots after fire.

References [21, 115, 135, 359].

Figure 5.2 A South American scrambling shrub of the compositae (daisy family), *Chromolaena odorata*, is one of the most important alien invaders of the tropics and subtropics. Here it is shown dominating a woodland fringe in the South African savanna region. (Photo: I.A.W. Macdonald)

Cedrela odorata A.Juss (Meliaceae)

West Indian cedar

Description tree

Region of origin Central Mexico to Brazil

Region invaded Galápagos

Cenchrus ciliaris L. (Gramineae)

Buffel grass (Australia)

Description herb

Region of origin SW Asia

Region invaded Hawaii, Australia

Cestrum nocturnum L. (Solanaceae)

Lady of the night

Description shrub

Region of origin West Indies

Region invaded Rarotonga

Chromolaena odorata (L.) King & Robinson {*Eupatorium odoratum* L.} (Compositae)

Christmas bush, archangel (Jamaica), sam-solokh (India), awolowo weed, Siam weed (Nigeria)

Description Life form: woody herb or shrub scrambling to 20 m §Seed disp.: wind §Seed prod.: high §Breeding syst.: hermaphrodite §Invasive category 2.5.

Region of origin N. America (Florida), C. America (West Indies), S. America (south to Paraguay). Climatic zone: subtrop. arid–dry, trop. arid. Common and variable species of roadside banks, fields, hillsides, clearings on limestone and wastelands.

Notes Apomictic. It does not grow in heavy shade but thrives on poor, rocky soils. It grows fast and regenerates from the roots after cutting or burning. It contains turpentines and is highly flammable, increasing the fire frequency often with disastrous results.

Control Chromolaena odorata rapidly recolonizes an area after physical clearing or application of herbicides [34]. However, in South Africa, Tordon 101 (0.75% in water) and Roundup (3%) have been used with some success. Biological control has been attempted in several countries but has not proved very successful. In Sri Lanka, partial control has been achieved by the introduction of *Pareuchaetes pseudoinsulara* Rego Barros (Lepidoptera), which causes defoliation. Management to reduce ecosystem disturbance reduces the spread of *C. odorata*.

References [34, 135, 219, 243, 340].

Chrysanthemoides monilifera (DC.) Norl. (ssp. *rotundata* and ssp. *monilifera*) (Compositae)

Bitou bush, bone-seed, salt bush, bietou

Description §Life form: shrub/tree (to 3 m) §Seed disp.: bird §Seed prod.: high, over an extended period §Breeding syst.: monoecious, outer florets female, inner florets male §Invasive category **3.5**.

Region of origin Africa (South Africa). Climatic zone: subtrop. dry–moist, wmtemp. dry. Widespread and occurs on coastal dunes.

Region invaded Australasia (Australia – southern Queensland, New South Wales, Victoria; New Zealand), Oceanic Is (St Helena), Europe (Mediterranean). Climatic zone: wmtemp. arid–moist, subtrop. moist. In New Zealand it invades coastal cliffs, waste places and scrubland. In Australia it invades littoral areas with native communities of *Acacia longifolia* and *Banksia integrifolia*, which it ousts.

Notes Mature individuals resprout after fire. Long decumbent branches can root at the nodes giving it sandbinding properties.

References [287, 429, 430].

Cinchona succirubra Pav. ex Klotsch (Rubiaceae)
Red quinine tree

Description §Life form: tree, evergreen §Seed disp.: wind §Seed prod.: high §Breeding syst.: hermaphrodite §Invasive category **4**.

Region of origin S. America (Ecuador). Climatic zone: trop. dry. It is not a dominant in its native forest habitats.

Region invaded Oceanic Is (Galápagos, St Helena). Climatic zone: wmtemp. dry, subtrop. moist. In the Galápagos it occurs in *Miconia*, *Robinsonia* and fern–sedge vegetation zones, with infestations covering some 4000 ha.

Notes Germination and seedling growth occur under a wide range of conditions, including deep shade. Regrowth after cutting is rapid and the plants have a wide ecological tolerance, mature early (flowering after 1–2 years) and can compete with and shade out native species.

Control Cinchona succirubra may be removed manually as has been done in 1000 ha of the Galápagos National Park [255]. It appears that it can regrow from root remnants [255]. Care must therefore be taken when removing roots that no fragments are left behind. In the Galápagos, painting the stump with herbicides such as Tordon 101, D.M.A.6., Certox 3,34 and Esterpac (concentration 1045 g/ha) was found to be effective in killing cut stumps [255].

References [151, 255, 347].

Cinnamomum camphora (L.) J. Presl (Lauraceae)

Camphor

Description tree

Region of origin China, Taiwan, Japan

Region invaded Australia, Japan, South Africa

Cinnamomum zeylanicum Nees (Lauraceae)

Cinnamon

Description tree

Region of origin East Indies

Region invaded Seychelles

Notes Introduced in the eighteenth century for spice cultivation and spread by the endemic blue pigeons (*Alectroenas pulcherrima*) into native vegetation.

Citrus limetta Risso (Rutaceae)

Sweet lime

Description tree

Region of origin Eurasia, North Africa

Region invaded Galápagos

Clematis vitalba L. (Ranunculaceae)

Description climber

Region of origin Europe

Region invaded New Zealand [200]. See main account in Chapter 4.

Clidemia hirta (L.) D. Don (Melastomataceae)

Description shrub

Region of origin Tropical America

Region invaded Hawaii, Guam, Fiji, Solomon Is, American Samoa, Vanuatu, Wallis and Futuna Is, Mascarene Is, Seychelles, Comores [435]. See main account in Chapter 4.

Coccinia grandis (L.) Voigt (Curcurbitaceae)

Description climber

Region of origin Tropical Africa

Region invaded Guam

Cortaderia selloana (Schultes & Schultes f.) Asch. & Graebner (Gramineae)

Pampas grass

Description herb, subdioecious

Region of origin S. America

Region invaded Australia, New Zealand

Crassula helmsii (T.Kirk) Cockayne {Tillaea recurva Hooker f.} (Crassulaceae)

Swamp stonecrop

Description §Life form: herb, perennial, aquatic §Seed disp.: water §Breeding syst.: hermaphrodite (probably self-pollinated) §Invasive category **4**.

Region of origin Australasia (SE Australia, Tasmania). Climatic zone: wmtemp. dry–moist. At altitudes up to 900 m in still shallow fresh water, swamps, edging streams, lagoons and channels up to 50 cm in depth. Occasionally occurs to 1m depth or terrestrially as dense mats on damp soil.

Region invaded Europe (southern England). Climatic zone: temp. moist. Wide range of habitats from shallow temporary pools to permanent water to depth of 3 m. Tolerant of a wide range of pH and nutrient status. Fast growing, readily dispersed by vegetative fragments, it now dominates many pools in the New Forest (England), excluding rare native plants such as *Ludwigia palustris* and *Galium debile*.

Notes Introduced as an aquarium plant. It is dispersed by people, birds (such as herons and geese) and ponies that drink from the pools.

References [71, 99, 369].

Crataegus monogyna Jacq. (Rosaceae)
Hawthorn

Description §Life form: shrub/tree (to 10 m), deciduous §Seed disp.: bird, mammal (possums in Australia) §Seed prod.: medium, hard seed coat promotes dormancy §Breeding syst.: hermaphrodite, the flowers are self-sterile and insect pollinated §Invasive category **3**.

Region of origin Europe. Climatic zone: temp. moist–wet. Abundant in hedgerows and woods of northern Europe and invading abandoned pasture. It does not regenerate under shade but is an effective colonizing shrub of base-rich soils.

Region invaded Australasia (Australia – New South Wales; New Zealand – South Island). Climatic zone: temp. moist, wmtemp. dry. Invades wasteland, riverbeds, forest remnants and hill country farmland.

Notes In Britain colonization by *Crataegus monogyna* enriches the nutrient status of nutrient-poor grasslands, making restoration of the original grassland difficult.

Control Spread of *Crataegus monogyna* may be controlled by cutting, although it sprouts from cut stumps.

References [29, 142, 444, 445].

Cryptostegia grandiflora R. Br. (Asclepiadaceae (Periplocaceae))
Rubber vine (Australia)

Description climber

Region of origin Africa

Region invaded Australia (Queensland rainforest)

Cupressus lusitanica Miller (Cupressaceae)

Mexican cypress

Description tree, monoecious

Region of origin Mexico

Region invaded Malawi

Cytisus scoparius (L.) Link (Leguminosae)

Broom

Description §Life form: shrub (to 2.5 m), perennial, deciduous §Seed disp.: explosive dehiscence of the legume scatters seed, mammals (livestock and humans), water §Seed prod.: high §Breeding syst.: hermaphrodite §Invasive category **3.5.**

Region of origin Europe (southern Europe), N. Asia (Asia Minor, Russia). Climatic zone: wmtemp. arid–dry. Distribution limited by drought in the south and winter cold in the north. It occurs on heaths, rocky places and open places in woods.

Region invaded Australasia (New Zealand, Australia), Africa (South Africa), S. Asia (India). Climatic zone: wmtemp. dry–moist, subtrop. moist. In Australia it can invade eucalypt forest, suppressing herbaceous vegetation and the regeneration of trees. In New Zealand it invades river beds, native grasslands and previously forested hillsides.

Notes Tolerant of a wide range of soils, it fixes nitrogen and can grow all year in an equable climate. Seed set is little affected by altitude or drought. Drought adaptations include the xeromorphic photosynthetic stems.

Control Frustrated by long-lived soil seed-bank.

References [166, 426, 441].

Dactylis glomerata L. (Gramineae)

Cocks-foot

Description herb

Region of origin Eurasia

Region invaded Hawaii

Datura inoxia Mill. (Solanaceae)

Description herb

Region of origin Southern N. America

Region invaded Namibia

Digitaria decumbens Stent. (Gramineae)

Pangola grass

Description herb

Region of origin South Africa

Region invaded Galápagos

Dioscorea bulbifera L. {*D. sativa* Thunb.} (Dioscoreaceae)
Air potato, abobo (Malaysia)

Description §Life form: climber, perennial, herbaceous with a woody tuber §Seed disp.: water, wind (winged seeds) §Seed prod.: high §Breeding syst.: dioecious §Invasive category **4**.

Region of origin S. Asia (Japan), Pacific (Polynesia). Climatic zone: subtrop. wet, trop. arid–dry. Mountains (to 2000 m), disturbed woodland, coastal woodland where the rainfall is more than 700 mm during the wetter six months of the year.

Region invaded Malesia (Singapore). Climatic zone: trop. dry. Introduced to the Singapore Botanic Garden from where it has spread into the edges of the relatively undisturbed primary rainforest. It has not spread far from the original point of introduction, although where it occurs it has a serious effect on the vegetation.

Notes It is cultivated for its edible bulbils and tubers. In Singapore it reproduces solely by means of vegetative axillary bulbils, borne aerially. Where both sexes occur it is probably pollinated by insects.

References [78, 152, 307, 372, 440].

Egeria densa Planchon (Hydrocharitaceae)

Description aquatic herb

Region of origin South America

Region invaded New Zealand (Lakes Rotorua, Rotoiti, Tarawera) [432]

Ehrharta calycina Sm. (Gramineae)

Description herb

Region of origin South Africa

Region invaded Australia

Ehrharta stipoides Lab. {*Microlaena stipoides* (Lab.) R. Br.} (Gramineae)

Description herb

Region of origin Australia

Region invaded Hawaii

Eichhornia crassipes (Mart.) Solms-Laub. {*E.speciosa* Kunth., *Piaropus mesomelas* Raf.} (Pontederiaceae)

Water hyacinth, aquape (Brazil), falkumbhi (India)

Description §Life form: herb, perennial, aquatic (free floating) §Seed disp.: water, wind §Seed prod.: high §Breeding syst.: hermaphrodite (tristyly) §Invasive category **4.5**.

Region of origin S. America (freshwater habitats in NE Brazil). Climatic zone: trop. arid–dry.

Region invaded S. Asia (India, Thailand, Malaysia), Africa (tropical and southern Africa), Australasia (Australia), N. America (Florida). Climatic zone: wmtemp. dry, subtrop. moist, trop. arid–dry. Widespread in the tropics and subtropics.

Notes Potential to produce high numbers of seeds (self-pollinated) but usually reproduces vegetatively. Tristylous and self-compatible populations are found and the flowers are insect pollinated. It grows rapidly in tropical freshwaters to form large floating mats of vegetation which cannot tolerate water temperatures above 34° C. The leaves are killed by frost but the plant is killed only if the rhizome tip is frozen.

Control Biological control, using the beetle *Neochetina eichhorniae* Warner (Coleoptera: Curculionidae), has been successful in some areas but not in others [196]. Inoculation of the plants with the fungus, *Cercospora rodmanii*, may improve success of the insect pest. Large areas of freshwater have been treated with 2,4-D in an attempt to control this pest but reinvasion from seed usually occurs.

References [9, 95, 137, 178, 181, 196, 313].

Elaeagnus angustifolia L. (Elaeagnaceae)

Russian olive, Trebizond date

Description shrub

Region of origin SE Europe, W. Asia

Region invaded N. America

Elettaria cardamomum (L.) Maton (Zingiberaceae)

Cardamom

Description herb

Region of origin India

Region invaded Sri Lanka, S. India

Elodea canadensis Michx. (Hydrocharitaceae)

Canadian pondweed

Description aquatic herb

Region of origin N. America

Region invaded C. Europe, Britain, New Zealand, Australia (New South Wales, Victoria)

Notes Introduced to Britain in 1840s initially as a female clone only, populations have now declined from the early massive infestations. Introduced to Australia sometime before 1960 [345].

Elodea nutallii (Planchon) St John (Hydrocharitaceae)

Nutall's pondweed

Description aquatic herb

Region of origin North America

Region invaded Britain

Erica lusitanica Rud. in Schrad. (Ericaceae)

Spanish heath

Description §Life form: shrub, evergreen §Seed disp.: wind §Seed prod.: high §Breeding syst.: hermaphrodite §Invasive category 2.

Region of origin Europe (south Portugal to SW France). Climatic zone: wmtemp. arid–moist. Damp heaths and woodland margins.

Region invaded Australasia (New Zealand). Climatic zone: wmtemp. dry–moist. Invades open disturbed areas and poor hill country pasture, replacing the native species *Leptospermum scoparium*.

Notes Has a high seed-bank (up to 480 000 per m^2) with germination stimulated by fluctuating temperatures. Fire creates ideal conditions for seed germination. It withstands grazing and trampling by producing abundant epicormic shoots. It is strongly mycorrhizal.

References [267, 403].

Eucalyptus camaldulensis Dehnh. (Myrtaceae)
Red gum

Description tree

Region of origin Australia

Region invaded South Africa (northern Cape savanna) [160]

Eucalyptus globulus Lab. (Myrtaceae)

Blue gum, fever tree

Description tree

Region of origin Australia

Region invaded California, USA

Eugenia jambos L. {Syzygium jambos (L.) Alston} (Myrtaceae)

Rose apple

Description shrub

Region of origin SE Asia

Region invaded Galápagos, Réunion, Cook Is, Hawaii

Flacourtia jangomas (Lour.) Rauschel (Flacourtiaceae)

Description shrub/tree

Region of origin Assam, Burma

Region invaded Cook Is

Fuchsia boliviana Carr. (Onagraceae)

Description shrub

Region of origin Bolivia

Region invaded Réunion

Fuchsia magellanica Lam. (Onagraceae)

Description shrub

Region of origin Chile, Argentina

Region invaded Réunion

Furcraea cubensis (Jacq.) Vent. {*F. hexapetala* (Jacq.) Urb.} (Agavaceae)

Cuban hemp, hemp, cabuya (Galápagos)

Description §Life form: shrub, evergreen §Breeding syst.: hermaphrodite §Invasive category **3**.

Region of origin C. America (Greater Antilles), S. America (NW S. America). Climatic zone: trop. arid, subtrop. moist. Common along roadsides and in waste places.

Region invaded Oceanic Is (Galápagos). Climatic zone: trop. arid. Habitats on Galápagos include the arid zone, fern–sedge vegetation (120–360 m), openings in forests, along trails and in abandoned settlements. Forms extensive thickets that exclude all other species.

Notes Reproduces vegetatively by means of bulbils and suckering.

References [2, 151, 347].

Furcraea foetida (L.) Haw. (Agavaceae)

Mauritius hemp

Description shrubby rosette

Region of origin Cuba S. America

Region invaded Kermadecs (Raoul), Hawaii, Mascarenes

Glyceria maxima (Hartm.) Holmb. (Gramineae)
Reed sweetgrass

Description water grass

Region of origin Europe

Region invaded Waterbodies up to 1m deep in southern Australia

Grevillea robusta Cunn. (Proteaceae)

Silky oak

Description tree

Region of origin E. Australia

Region invaded Hawaii

Hakea gibbosa (Sm.) Cav. (Proteaceae)
Rock hakea, downy hakea

Description §Life form: shrub/tree (up to 2 m where native, up to 4 m in South Africa) §Seed disp.: wind §Seed prod.: medium §Breeding syst.: hermaphrodite §Invasive category **4**.

Region of origin Australasia (Australia – New South Wales). Climatic zone: wmtemp. moist. Coastal heaths and scrub on sandy soil.

Region invaded Africa (South Africa – mountain fynbos), Australasia (New Zealand – gumlands and roadside scrub). Climatic zone: wmtemp. dry–moist. Forms dense thickets in which the native vegetation is suppressed. It was introduced into New Zealand as a hedge plant.

Control Hakea gibbosa can be removed from an area by cutting. Biological control using *Erytenna consputa* Pascoe (Coleoptera) has been tried since 1972 but consistent large scale fruit destruction has not been achieved [196].

Notes It is relatively long-lived with early reproductive maturity and shade toler-ance.

References [116, 196, 301, 428].

Hakea sericea Schrad. (Proteaceae)

Description shrub

Region of origin Australia

Region invaded South Africa [303], Mediterranean. See main account in Chapter 4.

Hakea suaveolens R.Br. (Proteaceae)
Sweet hakea, fork-leaved hakea

Description §Life form: shrub/tree §Seed disp.: wind, water §Seed prod.: medium §Breeding syst.: hermaphrodite §Invasive category **4**.

Region of origin Australasia (endemic to Western Australia). Climatic zone: wmtemp. arid–dry. Shallow soils on rock outcrops.

Region invaded Africa (South Africa – mountain and lowland fynbos). Climatic zone: wmtemp. dry–moist. Forms fire adapted dense stands on granitic soils.

Notes Serotinous winged seeds which are not released from their capsules until after the death of the tree or branch. Fire usually leads to large-scale release of seeds.

Control Mechanical control is effective provided that all the plants are removed [302]. Potential biological control agents are under investigation.

Hedychium flavescens Carey ex Roscoe (Zingiberaceae)

Description large rhizomatous herb

Region of origin Himalayas

Region invaded New Zealand

Hedychium flavescens Carey ex Roscoe (Zingiberaceae)

Description large rhizomatous herb

Region of origin Himalayas

Region invaded New Zealand

Helianthus tuberosus L. (Compositae)

Jerusalem artichoke

Description herb

Region of origin USA

Region invaded C. Europe

Heracleum mantegazzianum Sommier & Levier (Umbelliferae)

Giant hogweed

Description herb

Region of origin Former USSR

Region invaded British Isles

Hieracium praealtum Gochnat (Compositae)

Description herb

Region of origin Europe

Region invaded New Zealand

Hieracium pilosella L. (Compositae)
Mouse-ear hawkweed

Description §Life form: herb, perennial §Seed disp.: wind §Seed prod.: low §Breeding syst.: hermaphrodite §Invasive category 2.

Region of origin Europe. Climatic zone: temp. moist–wet, wmtemp. arid–moist. Pasture, heaths, banks, rocks, walls.

Region invaded Australasia (New Zealand). Climatic zone: temp. moist, wmtemp. moist. Tussock grasslands, lawns, roadsides and pastures. Can displace tussock and fescue-tussock vegetation.

Notes Relies heavily on vegetative propagation by stolons. Variable, polyploid complex of sexually reproducing and partially apomictic forms. Produces allelopathic substances.

Control The herbicide 'Versatil' with 2,4-D esters gives relatively good control [349].

References [262, 349, 445].

Hiptage benghalensis (L.) Kurz (Malpighiaceae)

Description woody climber

Region of origin Indomalaysia

Region invaded Mauritius, Réunion

Holcus lanatus L. (Gramineae)

Yorkshire-fog

Description herb

Region of origin Eurasia

Region invaded Hawaii, New Zealand

Hydrilla verticillata (L.f.) Royle (Hydrocharitaceae)

Hydrilla, Florida elodea, water thyme

Description §Life form: herb, perennial, aquatic §Seed disp.: wind, water §Seed prod.: low §Breeding syst.: dioecious §Invasive category **3**.

Region of origin Australasia (NE Australia), E. and SE Asia, Africa (E.

Africa). Climatic zone: subtrop. moist–wet, trop. arid. Still or slow moving freshwater habitats.

Region invaded N. America (Florida – male plants, California – introduced by 1976, first to Sacramento Valley, now widespread), C. America (Panama), Pacific (Fiji), Australasia (New Zealand). Climatic zone: temp. moist, subtrop. moist, trop. arid. Forms dense submerged stands displacing native species.

Notes Mainly reproduces vegetatively. It is tolerant of a wide range of nutrient levels from oligotrophic to eutrophic and can overwinter as dormant shoots.

Control In the USA, *Paraponyx diminutalis* (Lepidoptera: Pyralidae) was introduced accidentally; it causes some damage but does not result in full control [196]. Other insects are under investigation as suitable biological control agents [24], as are fish such as the Chinese grass carp (a triploid strain of the fish has been used which lives to five years and dies without reproducing). Several herbicides such as diquat at 5 ppm or paraquat at 1–2 ppm, endothal and xylene have been used to control *Hydrilla*. Lowering the water level to expose *Hydrilla* to the sun for a number of days has been reported to control the weed. Manual weeding is used in the Philippines.

References [24, 196, 345, 385].

Hyparrhenia rufa (Nees) Stapf (Gramineae)

Jaragua grass

Description herb

Region of origin Tropical Africa

Region invaded Venezuela, Hawaii

Hypochoeris radicata L. (Compositae)

Cat's-ear

Description herb

Region of origin Europe

Region invaded Hawaii

Impatiens glandulifera Royle (Balsaminaceae)

Indian balsam, policeman's helmet

Description §Life form: herb, annual §Seed disp.: explosive dehiscence, water §Seed prod.: medium §Breeding syst.: hermaphrodite §Invasive category **3**.

Region of origin S. Asia (Himalayas). Climatic zone: subtrop. moist–wet. Riverine rainforest.

Region invaded Europe (Britain). Climatic zone: temp. moist. Along riverbanks and in wet disturbed habitats.

Notes Long cultivated in gardens and dispersed by people. The flowers are relatively large and pollinated by bumble-bees. The seeds sink in water.

References [96, 397].

Imperata conferta (J.S.Presl) Ohwi (Gramineae)

Description herb

Region of origin SE Asia

Region invaded Guam

Jacaranda mimosifolia D.Don (Bignoniaceae)

Description shrub

Region of origin NW Argentina

Region invaded Subtropical Africa

Kalanchoe pinnata (Lam.) Pers. {*Bryophyllum pinnatum* (Lam.) Kurz} (Crassulaceae)

Description succulent herb/shrub

Region of origin Madagascar

Region invaded Galápagos, Hawaii, Raoul

Lagarosiphon major (Ridl.) Moss (Hydrocharitaceae)

Description herb

Region of origin South Africa

Region invaded New Zealand, Mascarenes, Italy, Britain. See main account in Chapter 4.

Lantana camara L. (Verbenaceae)

Description shrub

Region of origin New World Tropics

Region invaded Pantropical, pansubtropical. See main account in Chapter 4.

Leptospermum laevigatum (Gaertn.) Muell. (Myrtaceae)
Australian myrtle, coastal tea-tree

Description §Life form: shrub/tree (to 12 m) §Seed disp.: wind, water §Seed prod.: high, but low viability reported §Breeding syst.: hermaphrodite §Invasive category **4**.

Region of origin Australasia (Australia – Queensland, New South Wales, Victoria, South Australia to Tasmania). Climatic zone: wmtemp. dry–moist, subtrop. moist. Heath communities on calcareous or coastal sands. Characteristic of the closed-shrub stage of dune succession.

Region invaded Africa (SE Cape). Climatic zone: wmtemp. arid–moist. Sandy flats, lowland and mountain fynbos, southern forest. Introduced via the Cape Town Botanic Garden.

Notes Plants are killed by fire but the fruits open after burning to give prolific simultaneous germination. *Leptospermum laevigatum* has an efficient and extensive root system which competes for water with native plants.

Control In South Africa, the mature trees are cut and the area burnt about four years later to kill any seedlings that may have germinated from seed before they mature and produce fruit [193].

References [60, 193, 251, 428].

Leucaena leucocephala (Lam.) de Wit {*Leucaena glauca* Benth.} (Leguminosae)

Wild tamarind, lead tree, ko haole (Hawaii), guaje (Mexico)

Description §Life form: shrub/tree (2–7 m), evergreen §Seed disp.: gravity, insect activity on the ground §Seed prod.: high §Breeding syst.: hermaphrodite and largely self pollinated §Invasive category **3.5**.

Region of origin C. America (southern Mexico to Guatamala). Climatic zone: subtrop. moist, trop. arid–dry. Roadsides thickets and sandy scrub.

Region invaded Africa (Kenya, Tanzania), N. America (Florida), C. America (West Indies), S. America (south to Brazil), Pacific (Hawaii), S. Asia (Japan, Bonin Is), Malagassia (Mascarene Is), Australasia (N. Australia). Climatic zone: subtrop. dry–wet, trop. arid. It spread from Mexico to the Philippines as early as the seventeenth century and was widely introduced throughout the tropics for fodder in the nineteenth century. In the Hawaiian Is it is replacing native *Metrosideros–Diospyros* open forest and is possibly threatening *Erythrina sanwichensis* in parts of its range. It is found in dry to mesic habitats on all the Hawaiian Is to 700 m.

Notes Forms dense monospecific thickets. The stands are not very flammable but, if burning occurs, *Leucaena leucocephala* regenerates rapidly from basal shoots. Flowers and fruits continuously. Two subspecies are recognized, of which ssp. *leucocephala* is the shrubby form and the main invasive.

Control Biological control is frustrated by its economic importance. A psyllid insect pest, *Heteropsylla cubana*, which causes defoliation, spread by chance from Central America to Hawaii in 1984 and has recently spread through Asia into East Africa.

References [2, 79, 80, 237, 311, 422].

Ligustrum lucidum Aiton (Oleaceae)

Glossy privet

Description shrub or small tree

Region of origin China, Korea

Region invaded Australia, New Zealand, N. Argentina

Ligustrum robustum Blume ssp. walkeri (Decne.) P.S. Green {L.walkeri Decne., L. ceylanicum Decne.} (Oleaceae)

Privet

Description §Life form: shrub, evergreen §Seed disp.: bird §Seed prod.: medium §Breeding syst.: hermaphrodite §Invasive category **4.5**.

Region of origin S. Asia (India, Sri Lanka). Climatic zone: wmtemp. moist, subtrop. dry–wet. Characteristic of disturbed montane forest, often along streams, at 450–2000 m

Region invaded Malagassia (Mauritius, Réunion and Rodrigues). Climatic zone: subtrop. moist. Strongly invades the lower montane evergreen forest, with *Psidium cattleianum*. It is the most invasive species of the Mauritius uplands.

Notes In Mauritius it is widely dispersed by the introduced red-whiskered bulbul (*Pycnonotus jocosus*). Germination and growth are rapid and it forms dense monospecific thickets.

Control It is laboriously cleared by hand from special patches in nature reserves, an operation that has to be repeated yearly.

References [96, 239, 240, 381].

Ligustrum sinense Lour. (Oleaceae)

Chinese privet

Description shrub

Region of origin China

Region invaded Australia

Linaria genistifolia (L.) Mill. ssp. dalmatica (L.) Maire & Petitmengin (Scrophulariaceae)

Description herb

Region of origin Balkans

Region invaded USA

166

Lonicera japonica Thunb. (Caprifoliaceae)

Description climber

Region of origin E. Asia

Region invaded USA, Hawaii, Australia, New Zealand

Ludwigia peploides (Kunth) Raven (Onagraceae)

Description herb

Region of origin Australia

Region invaded Subtropical and Tropical America

Lupinus arboreus Sims (Leguminosae)

Tree lupin

Description shrub

Region of origin California

Region invaded New Zealand

Lupinus polyphyllus Lindley (Leguminosae)

Description herb

Region of origin Western N. America

Region invaded New Zealand

Lygodium japonicum (Thunb.) Sw. (Schizaeaceae – Pteridophyta)

Description fern

Region of origin Japan

Region invaded USA (Florida–Texas)

Lythrum salicaria L. (Lythraceae)

Purple loosestrife

Description herb

Region of origin Old World

Region invaded N. America

Maesopsis eminii Engl. (Rhamnaceae)

Musizi

Description §Life form: tree (to 40 m) §Seed disp.: bird (hornbills in the E. Usambaras) §Seed prod.: high §Breeding syst.: hermaphrodite, but protogynous and probably mainly outbreeding §Invasive category **4**.

Region of origin Africa (Uganda, Zaire, NW Tanzania, Zambia, Kenya, Angola, central Africa west to Liberia). In places it is a dominant canopy species in colonizing forest growing with an understorey of *Caloncoba schweinfurthii*. Climatic zone: subtrop. moist, trop.

Region invaded Africa (East Usambaras, E. Tanzania, Rwanda). It invades submontane (800–1200 m) evergreen forest of the East Usambaras, rich in endemics, as a gap-replacement species. It has spread from plantations into natural forest. It was first introduced to the East Usambaras by German foresters around 1913, with more planting being carried out in the 1960s and 70s. Climatic zone: subtrop. moist.

Notes Seeds are early and copiously produced and may remain dormant for some months in soil or damp litter. Seeds can germinate in shade but require a canopy gap within a few months of germination to survive.

References [38, 39].

Mangifera indica L. (Anacardiaceae)

Mango

Description tree

Region of origin Asia

Region invaded Antigua, Mauritius

Melaleuca quinquenervia (Cav.) Black (Myrtaceae)

Description tree

Region of origin Australia

Region invaded Florida [21], Hawaii, South Africa. See main account in Chapter 4.

Melia azedarach L. {M. dubia Cav.} (Meliaceae)
White cedar, Cape lilac, China berry, tulip cedar, syringa

Description §Life form: tree (10–20 m), deciduous §Seed disp.: bird §Seed prod.: high §Breeding syst.: hermaphrodite/monoecious §Invasive category **4.5**.

Region of origin Asia and Australasia (Australia). Climatic zone: wmtemp. dry–moist. On fertile soils in coastal rainforests

Region invaded Africa, Pacific. Climatic zone: subtrop. dry–wet. In South Africa it colonizes riparian habitats in subtropical veld and is one of the most widespread invaders in the Transvaal. It competes well with native plants.

Notes It flowers throughout the year and comes to reproductive maturity early. It is widely planted as an ornamental, growing rapidly. Its timber is used in cabinet making.

References [162, 247, 369].

Melinis minutiflora Beauv. {Panicum minutiflora (Beauv.) Rasp., P. melinis Trin.} (Gramineae)
Molasses grass, wynne grass

Description §Life form: herb, perennial §Seed disp.: wind §Breeding syst.: hermaphrodite §Invasive category **3.5**.

Region of origin Africa (Tropical Africa). Climatic zone: trop. arid–dry. Open grasslands.

Region invaded Pacific (Hawaii), C. America (Jamaica), S. America (Venezuela), Oceanic Is (Ascension). Climatic zone: subtrop. dry–wet. In Hawaii it occurs in dry habitats on all islands to 1500 m.

Notes Spreads mainly by means of runners. Apomictic. It forms dense monotypic stands, smothering surrounding vegetation. It is adapted to fire. Although relatively unpalatable to stock, its importance as a pasture grass prevents biological control.

References [359, 361].

Memecylon floribundum Blume (Melastomataceae)

Description shrub

Region of origin Indonesia

Region invaded Seychelles (Mahé)

Mesembryanthemum crystallinum L. {*Gasoul crystallinum* (L.) Rothm.} (Aizoaceae)

Ice-plant

Description §Life form: herb (succulent), annual §Seed disp.: wind §Breeding syst.: hermaphrodite §Invasive category **2.5**.

Region of origin Africa (probably native along west coast) trop. arid–dry. Locally frequent on saline soils in coastal areas.

Region invaded N. America. (California), Australasia (Australia, on samphire flats and saline ground), Europe – Africa – N./W. Asia (Mediterranean, including Israel). Climatic zone: wmtemp. arid wmtemp. dry, subtrop. arid, trop. arid. Invades degraded coastal pastures; it is encouraged by overgrazing. It is said to accumulate salt in the surface soil horizons, making grassland re-establishment more difficult after it has been removed.

References [204, 418].

Miconia calvescens (Schr. & Mart.) DC. (Melastomataceae)

Bush currant

Description shrub

Region of origin Tropical America

Region invaded Pacific Islands (Tahiti, Moorea)

Mikania micrantha H.B.K. (Compositae)

Description §Life form: climber, perennial §Seed disp.: wind §Seed prod.: medium §Breeding syst.: hermaphrodite §Invasive category **2.5**.

Region of origin S. America, C. America (including Caribbean). Climatic zone: subtrop. dry–moist, trop. arid. Damp clearings in forest from lowlands to 2000 m, streamsides and roadsides.

Region invaded S. Asia (India), Malesia (Malaysia, Philippines, Solomon Is), Pacific (Rarotonga). Climatic zone: subtrop. moist, trop. arid. It is mainly a weed of pasturelands but it invades forest margins.

Notes It is intolerant of deep shade; produces allelopathic substances. It combines effective seed production with vigorous vegetative reproduction.

References [178].

Mikania scandens (L.) Willd. (Compositae)

Description climber

Region of origin USA

Region invaded Guam, Rota, Saipan, Sri Lanka

Mimosa invisa Mart. (Leguminosae)

Description shrub

Region of origin Tropical America

Region invaded Pacific, Mariana Is, including Rota, Tinian, Saipan

Mimosa pigra L. *{M. pellita* Humb. & Bonpl. ex Willd., nom. rej.} (Leguminosae)

Description shrub

Region of origin Tropical America

Region invaded SE Asia, Tropical Africa, N. Australia [51, 234]. See main account in Chapter 4.

Myrica faya Ait. (Myricaceae)

Description shrub

Region of origin Azores

Region invaded Hawaii. See main account in Chapter 4.

Myriophyllum aquaticum (Vell. Conc.) Verdc. *{M. brasiliense* Cambess.} (Haloragidaceae)

Parrot's feather, water milfoil

Description §Life form: herb, perennial, aquatic §Seed disp.: wind, water §Breeding syst.: dioecious §Invasive category **3.5.**

Region of origin S. America (Brazil, Peru, Uruguay, Chile, Argentina). Climatic zone: subtrop. moist, trop. dry. Lakes, rivers and streams.

Region invaded N. America (E. Texas, Edwards Plateau), Australasia (Australia – Western Australia, Queensland, New South Wales, Victoria, Tasmania; New Zealand), Africa (South Africa). Climatic zone: temp. moist, wmtemp. dry–moist, subtrop. dry. Invades flowing and still water. Tolerant of a wide variety of environmental conditions and will persist in brackish or polluted water.

Notes In South Africa propagation is entirely vegetative.

References [149].

Myriophyllum spicatum L. (Haloragidaceae)

Description aquatic herb

172

Region of origin Eurasia

Region invaded USA (e.g. Lake George, NY [260]), Canada

Myroxylon toluiferum Humb. (Leguminosae)

Description tree

Region of origin Tropical America

Region invaded Sri Lanka

Myrsiphyllum asparagoides Willd. (Liliaceae)

Description climber to 3 m

Region of origin South Africa

Region invaded Australia (South Australia, Victoria)

Nassella trichotoma (Nees) Arech. {Stipa trichotoma Nees} (Gramineae)

Serrated tussock grass

Description §Life form: herb, perennial §Seed disp.: wind §Seed prod.: hermaphrodite §Invasive category **2.5**.

Region of origin S. America (Peru, Chile, Argentina, Uruguay). Climatic zone: temp. arid–wet, wmtemp. moist. Pampas where vegetation is sparse and in disturbed or cultivated areas. It is not abundant in its native region.

Region invaded Australasia (SE Australia), Africa (South Africa), Europe (Italy). Climatic zone: temp. moist, wmtemp. dry–moist, subtrop. dry. In Australia it invades the Tablelands grasslands and, in South Africa, grassveld. It is capable of replacing native grasslands once they have been disturbed.

Notes It is a drought resistant tussock-forming grass with a deep root system and seed which may remain viable in the soil for up to 20 years.

Control Burning at appropriate times to prevent seed production and release, and repeated plowing of stands. Improving pastures (addition of fertilizers) results in *Nassella trichotoma* diminishing with increasing competition. Herbicides (2,2-DPA

173

or tetrapion) have been used in Australia, but they are uneconomical for use on large stands due to rapid re-infestation from buried seed. Minimizing disturbance in adjacent areas slows spread of *Nassella trichotoma*.

References [16, 20, 63, 431].

Nerium oleander L. (Apocynaceae)

Oleander

Description shrub

Region of origin Eurasia

Region invaded S. Africa

Nicotiana glauca Grah. (Solanaceae)

Wild tobacco

Description §Life form: shrub, perennial, evergreen §Seed disp.: wind, water §Seed prod.: high §Breeding syst.: hermaphrodite §Invasive category 3.

Region of origin S. America (northwestern and central Argentina, Paraguay, Bolivia). Climatic zone: wmtemp. arid, subtrop. dry–moist. Roadsides and along river banks, to 3000 m.

Region invaded C. America (Mexico), Africa (South Africa – Kruger National Park and the lower reaches of the Orange River; Namibia – lower reaches of the Ugab River), N./W. Asia (Israel), Australasia (Australia), Oceanic Is (St Helena). Climatic zone: wmtemp. arid–moist, subtrop. dry–moist. Waste places, dry river beds, roadsides and along river banks.

Notes Drought resistant, tolerant of a wide range of environmental conditions and poisonous to most stock. Where it grows vigorously, it forms dense monospecific stands.

Control In South Africa the plants are cut and the stumps treated with 2,4,5-T. Successful control of *Nicotiana glauca* has been achieved where the plants were sprayed with herbicide and then exposed to the beetle, *Malabris aculeata*.

References [58, 237, 254, 377].

Olea europaea L. ssp. africana Mill. (Oleaceae)

Olive

Description shrub

Region of origin Tropical and South Africa

Region invaded Norfolk I., Hawaii

Operculina ventricosa (Bert.) Peter (Convolvulaceae)

Description climber

Region of origin Tropical America

Region invaded Rota, Tinian, Saipan

Opuntia aurantiaca Lindley (Cactaceae)

Tiger pear

Description shrub

Region of origin South America, West Indies

Region invaded Australia [17], S. Africa

Opuntia ficus-indica (L.) Mill. (Cactaceae)

Common prickly pear, spiny pest pear

Description §Life form: shrub or small tree (to 5 m) §Seed disp.: mammals (including humans) §Seed prod.: medium §Breeding syst.: hermaphrodite §Invasive category 3.

Region of origin C. America (Mexico) Climatic zone: trop. arid–dry. Typically occurs in valley floodplains and at the mouths of small canyons.

Region invaded Africa (Red Sea coasts and South Africa), Pacific (Hawaii), N. America (California), Europe (Mediterranean), N. Asia (Arabia) Climatic zone: wmtemp. arid–dry. In South Africa it invades pastureland and native vegetation (karoo). It forms dense infestations and the seeds are dispersed by baboons.

Notes Widely introduced for its edible fruit. Pollinated by bees and beetles. It is intolerant of fire.

Control This cactus has been successfully controlled in most areas in Hawaii and South Africa by two introduced insects, *Dactylopius opuntiae* Cockerell (Hemiptera: Dactylopiidae) and *Cactoblastis cactorum* Bergroth (Lepidoptera: Pyralidae). Physical control is very difficult as it readily regenerates from spiny leaf pads.

References [36, 161, 453].

Opuntia imbricata (Haw.) DC. (Cactaceae)

Description shrub

Region of origin C. America

Region invaded S. Africa

Opuntia rosea DC. (Cactaceae)

Description shrub

Region of origin C. America

Region invaded S. Africa

Opuntia stricta (Haw.) Haw. {Opuntia inermis DC.} (Cactaceae)

Description §Life form: shrub (sprawling or weakly ascending, 0.5–2 m), perennial §Seed disp.: mammals (including humans), water §Seed prod.: medium §Breeding syst.: hermaphrodite §Invasive category 3.

Region of origin N. America (Florida, Texas), C. America (Cuba). Climatic zone: wmtemp. arid, subtrop. dry–moist, trop. arid. On sandy soils, typically of coastal woodlands or stabilized dunes.

Region invaded Australasia (Australia – Queensland). Climatic zone: subtrop. moist, trop. arid. Invades *Acacia–Casuarina* scrubland and other habitats, converting open scrub into impenetrable *Opuntia stricta* thickets.

Control Opuntia stricta has been successfully controlled in Australia by the introduction of the cactus-consuming moth, *Cactoblastis cactorum* Bergroth (Lepidoptera: Pyralidae).

References [36, 308].

Opuntia vulgaris Mill. (Cactaceae)

Description shrub to 3–4 m

Region of origin C. America

Region invaded E. Africa (Uganda – Queen Elizabeth National Park), S. Africa, India

Orthodontium lineare Schwaegr. (Bryales – Bryophyta)

Description apocarpous moss

Region of origin Southern Hemisphere

Region invaded NW Europe

Ossaea marginata (Desr.) Triana (Melastomataceae)

Description shrub

Region of origin Brazil

Region invaded Mauritius

Panicum maximum Jacq. (Gramineae)

Guinea grass

Description herb

Region of origin Africa

Region invaded Antigua, Barbuda, Anguilla, S. America (Venezuela)

Parkinsonia aculeata L. (Leguminosae)

Jerusalem thorn (South Africa), parkinsonia (Australia)

Description shrub or small tree

Region of origin N. and S. America

Region invaded Australia (Northern Territory, Queensland)

Paspalum conjugatum L. (Gramineae)

Description herb

Region of origin West Indies

Region invaded Hawaii, Raoul

Paspalum digitatum (L.) Poiret (Gramineae)

Description herb

Region of origin S. America

Region invaded Hawaii

Passiflora mollissima (H.B.K.) Bailey (Passifloraceae)

Description climber

Region of origin S. America

Region invaded Hawaii [222], South Africa, New Zealand. See main account in Chapter 4.

Passiflora rubra L. (Passifloraceae)

Passion flower

Description climber

Region of origin Tropical America, West Indies

Region invaded Rarotonga

Passiflora suberosa L. (Passifloraceae)

Passion flower

Description climber

Region of origin C. and S. America, West Indies

Region invaded Hawaii

Pennisetum clandestinum Chiov. {*P.longistylum* Hochst. var. *clandestinum* (Chiov.) Leeke} (Gramineae)

Kikuyu grass

Description §Life form: herb, perennial (sward-forming) §Seed disp.: wind §Breeding syst.: hermaphrodite/monoecious §Invasive category **3.5.**

Region of origin Africa (tropical eastern Africa). Climatic zone: subtrop. dry–moist, wmtemp. dry–moist. Humid tropical highlands, 1400–3300 m, tolerant of some frost. An important fodder and pasture grass.

Region invaded Pacific (Hawaii), Africa (South Africa), Australasia (Australia, New Zealand). Climatic zone: wmtemp. moist. subtrop. moist. It prevents the regeneration of native trees and shrubs in degraded Hawaiian forests, 500–2000 m.

Notes Facultative apomict, reproducing vegetatively and rarely setting seed except at high elevations. It has been widely introduced around the world as a pasture and lawn grass. It is said to release allelopathic substances.

Control Two insect pests, *Sphenophorus ventus vestitus* and *Herpetogramma licarsicalis*, have caused severe injury to Kikuyu grass in Hawaii. Its economic importance prevents release of further biological control agents. The application of 0.5% glyphosate has proved effective.

References [72, 251, 359, 361, 445].

Pennisetum polystachion (L.) Schult. (Gramineae)

Mission grass (Australia)

Description herb

Figure 5.3 *Pinus patula*, a native of Mexico, invading herb-rich grassland in the Mazeka valley, Mount Mulanje, Malawi. Efforts have been made to control its spread out of forestry plantations, but control will depend not only on removal of young trees but also of the source seed trees in the plantations. (Photo: Alan Hamilton)

Region of origin Tropical Africa

Region invaded Guam, Sri Lanka, Australia (eucalypt forest)

Pennisetum purpureum Schumach. {*P. macrostachyum* Benth., *P. blepharideum* Gilli} (Gramineae)

Elephant grass, Napier grass

Description §Life form: herb, perennial (1–6 m) §Seed disp.: wind §Seed prod.: low §Breeding syst.: hermaphrodite/monoecious §Invasive category **4**.

Region of origin Africa (Tropical Africa). Climatic zone: subtrop. dry–moist, wmtemp. dry–moist. Riverine sites, perennial swamps and disturbed forest land (often associated with secondary forest), 0–500 m. In Uganda, for instance, it can replace semideciduous rainforest as a fire subclimax.

Region invaded Oceanic Is (Galápagos). Climatic zone: wmtemp. dry. In the *Scalesia* zone; its dense growth prevents regeneration of native species.

Notes Reproduction is mainly vegetative. It is tolerant of a wide range of environmental conditions, although it is easily killed by frost. It prefers rich well-drained soils. It is highly drought resistant.

References [72, 151, 178].

Pennisetum setaceum (Forsk.) Chiov. {*P. phalaroides* Schult.} (Gramineae)
Fountain grass

Description §Life form: herb, perennial §Seed disp.: wind §Seed prod.: low §Breeding syst.: hermaphrodite/monoecious §Invasive category **5.**

Region of origin Africa (North Africa) W. Asia. (Lebanon, Syria). Climatic zone: wmtemp. dry, trop. arid. Stony slopes and dry open places 300–1600 m.

Region invaded Pacific (Hawaii), Oceanic Is (St Helena – not invasive). Climatic zone: wmtemp. dry–moist, subtrop. moist–wet.

Notes Fountain grass is threatening species which are listed as endangered by the US wildlife service – *Gouania hillebradii, Haplostachys haplostacha, Kokia drynarioides, Lipochaeta venosa* and *Stenogyne angustifolia* var. *angustifolia*. It is a fire-adapted bunchgrass, and it both promotes fires and spreads as a result of them.

Control Physical control has prevented further spread of fountain grass in Hawaii. Biological control is hampered by opposition from the sugar cane industry.

References [235, 359, 361, 399, 422].

Peraserianthes falcataria (L.) Nielsen {*Albizia falcataria* (L.) Fosb.} (Leguminosae)

Batai wood, sau

Description tree

Region of origin Malesia

Region invaded Seychelles

Pereskia aculeata Miller (Cactaceae)

Barbados gooseberry

Description shrub

Region of origin S. America

Region invaded South Africa [207]

Persea americana Miller (Lauraceae)

Avocado pear, aguacate, alligator pear, palta

Description shrub

Region of origin C. America

Region invaded Galápagos

Pinus contorta Douglas (Pinaceae)

Lodge-pole pine

Description tree

Region of origin Western N. America

Region invaded New Zealand [188]

Figure 5.4 Throughout the world riparian vegetation is being replaced by alien species. In Namibia, some of the most important such invaders are Mesquite trees of the genus *Prosopis*. Ironically, these same species are often widely advocated for planting in land reclamation schemes in desertified areas. One possible solution to this potential conflict is the use of sterile forms. (Photo: H. Kolberg)

Pinus nigra Arnold (Pinaceae)

Black pine, Austrian pine

Description tree

Region of origin Europe

Region invaded New Zealand

Pinus patula Schiede & Deppe (Pinaceae)

Mexican weeping pine

Description §Life form: tree, evergreen §Seed disp.: wind §Seed prod.: high §Breeding syst.: monoecious §Invasive category **4**.

Region of origin C. America (Mexico). Climatic zone: wmtemp. arid. Highlands of the central and eastern states at 1500–3000 m elevation where *Pinus patula*, *P. pseudostrobus* and *Quercus reticulata* are characteristic species. In cool

183

subtropical climates with winter temperatures often below freezing.

Region invaded Pacific (Hawaii), Africa (Malawi – Mulanje Mt). Climatic zone: wmtemp. moist–wet, subtrop. dry–moist. It threatens the unique flora of Mulanje Mt by forming dense monotypic stands. It was originally planted as a nurse to promote regeneration of the native *Widdringtonia nodiflora*. It occurs between 1800 and 2400 m in tussock grassland, *Brachystegia* woodland and *Widdringtonia* forest.

Notes Capable of some self-fertilization, it is partly serotinous, with a high growth rate.

Control Cutting, followed by burning once the seeds have germinated, may be effective.

References [110, 111, 275, 276, 359].

Pinus pinaster Ait. (Pinaceae)

Cluster pine, maritime pine

Description §Life form: tree (to 40 m), evergreen §Seed disp.: wind §Seed prod.: high §Breeding syst.: monoecious §Invasive category **4.5**.

Region of origin Europe (Mediterranean). Climatic zone: wmtemp. arid–dry. In a wide range of habitats from Atlantic coastal dunes to mountain woodland, generally on sandy infertile soils.

Region invaded Africa (South Africa), Pacific (Hawaii – 1600–2200 m on Maui I.), Australasia (New Zealand – Abel Tasman National Park). Climatic zone: wmtemp. arid–moist, subtrop. wet. Replaces native vegetation in mountain and lowland fynbos on acid, leached soils, as it generally survives fynbos fires and has greater growth rate and longevity than native fynbos plants.

Notes It is susceptible to frost damage, although its thick bark gives it resistance to fires. It has been widely introduced around the world as a commercial forestry tree.

Control Cutting followed by burning to kill any regrowth has been effective. Any biological control must involve a seed predator so as not to interfere with commercial plantings.

References [188, 214, 215, 327, 359].

Pinus radiata D.Don (Pinaceae)

Monterey pine

Description tree

Region of origin California

Region invaded Australia, New Zealand, South Africa [329]. See main account in Chapter 4.

Pistia stratiotes L. (Araceae)

Water lettuce

Description herb

Region of origin Pantropical

Region invaded South Africa, Zambia, Malaysia, Philippines, Thailand, Australia

Pittosporum undulatum Vent. (Pittosporaceae)

Cheesewood

Description tree

Region of origin Australia

Region invaded Jamaica, South Africa, Australia, Lord Howe I., Hawaii [130, 328]. See main account in Chapter 4.

Poa annua L. (Gramineae)

Annual meadow-grass

Description herb

Region of origin North Temperate

Region invaded South Georgia

Prosopis glandulosa Torrey {*P.chilensis* (Molina) Stuntz var. *glandulosa* (Torrey) Standley} (Leguminosae)

Mesquite

Description §Life form: shrub or small tree (to 15 m), deciduous (spiny) §Seed disp.: mammals §Seed prod.: high §Breeding syst.: hermaphrodite §Invasive category **3.5**.

Region of origin N. America (SW United States), C. America (NE Mexico). Climatic zone: wmtemp. arid–dry, subtrop. moist. Occurs naturally in valleys and dry uplands, but invades pasture and disturbed land vigorously.

Region invaded Africa (South Africa, Namibia), Australasia (Australia – Queensland). Climatic zone: wmtemp. arid, subtrop. dry–moist. In South Africa it invades karoo and thornveld, and is beginning to spread into mountain fynbos.

Notes Seeds may remain dormant in soil for up to 10 years but germination is enhanced by passage through the digestive tract of herbivores. It is tolerant of extreme temperatures, severe drought and overgrazing. It has been widely cultivated for fodder. The most invasive type appears to be var. *torreyana* (Benson) Johnston rather than var. *glandulosa* [160].

Control Mesquite resprouts vigorously after cutting from dormant buds below the soil, so the roots must be grubbed out and the clearing operation must be followed up in successive years. If herbicides are to be used, these should be applied during the peak growing season. Biological control has recently been attempted using the seed-feeding beetle, *Algarobius prosopis*.

References [153, 357, 369].

Prosopis pallida (Willd.) Kunth (Leguminosae)

Description shrub

Region of origin South America

Region invaded Hawaii

Prosopis velutina Wooten (Leguminosae)

Velvet mesquite

Description shrub

Region of origin SW N. America

Region invaded South Africa [160].

Prunus serotina Ehrh. (Rosaceae)

Black cherry

Description §Life form: shrub or tree (to 30 m), deciduous §Seed disp.: bird §Seed prod.: high §Breeding syst.: hermaphrodite §Invasive category **4**.

Region of origin N. America (Ontario and Quebec southwards to Texas and Florida). Climatic zone: temp. wet. Woods and clearings, floodplains and thickets by roadsides.

Region invaded Europe (Britain, Netherlands). Climatic zone: temp. moist. Invades seminatural woodland on acid sandy soil.

Notes Regenerates by seed in gaps after disturbance. Resprouts vigorously after cutting.

Control Chondrostereum purpureum (fungus: Basidiomycetes) is under field evaluation as a biological control agent.

References [196, 407].

Psidium cattleianum Sabine (Myrtaceae)

Description shrub

Region of origin Tropical S. America

Region invaded Hawaii, tropical Polynesia, Raoul, Norfolk I., Mascarenes (Réunion). See main account in Chapter 4.

Psidium guajava L. (Myrtaceae)

Guava, guayaba (Galápagos)

Description §Life form: shrub/tree (to 10 m) §Seed disp.: mammal, bird §Seed prod.: high §Breeding syst.: hermaphrodite §Invasive category **4.5**.

Region of origin S. America (tropical and subtropical S. America). Climatic zone: trop. arid–dry.

Region invaded Oceanic Is (Galápagos), Pacific (Hawaii), Africa (Natal, E. Transvaal, Zimbabwe) Australasia (New Zealand). Climatic zone: wmtemp. moist, subtrop. dry–wet. In Galápagos invades forest edge communities, reducing regeneration of native evergreen trees, with an infestation covering some 40 000 ha. In Hawaii it invades *Acacia* forest at high elevations.

Notes It flowers and produces fruit nearly all the year round under favorable conditions, often being dispersed by introduced mammals and birds. It is intolerant of frost and deep shade but regenerates and grows quickly in gaps.

Control As *Psidium guajava* regenerates readily from underground parts by suckering, it is extremely difficult to kill. Penetration by herbicides is limited by the waxy cuticle.

References [65, 162, 183, 243, 251, 347].

Pueraria lobata (Willd.) Ohwi. {*P.thunbergiana* (Sieb. & Zucc.) Benth., *P.hirsuta* Schnied.} (Leguminosae)

Kudzu vine, Japanese arrowroot

Description §Life form: climber, perennial §Seed disp.: bird, mammal §Seed prod.: high §Breeding syst.: hermaphrodite §Invasive category **3.5**.

Region of origin N. Asia (China, Korea and Japan). Climatic zone: temp. wet. Climber of forest margins (up to 2000 m), common in tropical and subtropical regions.

Region invaded Africa (Eastern Transvaal), Australasia (Papua New Guinea), Pacific (Hawaii and Western Pacific Is), N. America (S. United States), C. America. Climatic zone: subtrop. dry, wmtemp. moist. In the United States it has been widely planted as an erosion control and green fodder. It has spread to forest margins where it is able to smother whole trees.

Notes It appears to be outbreeding and bee-pollinated. The aerial parts are damaged by frost but it is drought tolerant. It grows poorly where the temperature and humidity are high.

Control In Hawaii the herbicide Garlon 4 appears to be effective in controlling Kudzu vine.

References [61, 211, 408].

Ravenala madagascariensis Sonn. (Strelitziaceae)
Traveller's palm

Description §Life form: tree §Seed disp.: birds (fleshy aril) §Invasive category **4**.

Region of origin Malagassia (Madagascar). Climatic zone: trop. arid, subtrop. moist. characteristic of secondary forest.

Region invaded Malagassia (Mauritius) Climatic zone: subtrop. moist. Dominates large areas of marsh, mountain slopes and valleys, forming dense stands.

Notes Reproduces vegetatively by suckering. It was introduced to Mauritius in 1768.

References [238, 239].

Reynoutria japonica Houtt. {*Polygonum cuspidatum* Siebold & Zucc, *P. compactum* Hook f.} (Polygonaceae)
Japanese knotweed

Description §Life form: herb, perennial §Seed disp.: wind, water §Seed prod.: low §Breeding syst.: dioecious §Invasive category **3.5**.

Region of origin N. Asia (Japan, Korea and northern China). Climatic zone: temp. wet, wmtemp. moist–wet, subtrop. moist. Characteristically a colonist of bare volcanic soils, very common in open places on hills and high mountains, and in *Miscanthus sinensis* grassland.

Region invaded Europe (British Isles and N. Europe to E. Germany). Climatic zone: temp. moist, temp. wet. River and railway embankments, roadsides and waste ground. Introduced to Britain in 1825 and cultivated as an ornamental.

Notes Fruit is rarely seen in the British Isles, where spread is by vigorous rhizomatous growth. The flowers are functionally dioecious. It is intolerant of extreme frost, drought or high temperatures. A variable species.

Control Control can prove extremely difficult as *Reynoutria japonica* is fairly resistant to herbicides; when cleared manually, all the rhizomes must be removed to prevent resprouting. This usually proves impossible.

References [71, 76, 307, 383].

Reynoutria sachalinensis (Petrop.) Nakai (Polygonaceae)

Description herb

Region of origin Asia

Region invaded N. and C. Europe

Rhododendron ponticum L. (Ericaceae)

Description shrub

Region of origin Turkey, Spain, Portugal

Region invaded British Isles. See main account in Chapter 4.

Ricinus communis L. (Euphorbiaceae)

Castor bean, castor oil plant, jarak (Malaya)

Description §Life form: shrub, evergreen §Seed disp.: mammal (including humans), bird §Seed prod.: high §Breeding syst.: monoecious §Invasive category **3.5**.

Region of origin Africa (Tropical Africa). Climatic zone: trop. arid–dry.

Region invaded Africa (South Africa – Natal), N./W. Asia (Israel), Pacific (Hawaii), Australasia (Queensland, New South Wales), C. America (Antigua). Climatic zone: wmtemp. moist, subtrop. dry–moist. In South Africa it is widespread in the native vegetation of Natal where disturbance has occurred. In Hawaii it is present in dry disturbed habitats from sea level to 1200 m on all the major islands.

Notes A fast growing thicket-forming plant, widely cultivated for its oil and naturalized in most subtropical countries, easily destroyed by fire. It fruits all the year round under favorable conditions.

References [79, 251, 359].

Robina pseudacacia L. (Leguminosae)

False acacia, black locust

Description shrub

Region of origin N. America

Region invaded Netherlands, France, Germany, Switzerland, Hungary, Greece, Turkey, Cyprus, Israel, Australia, New Zealand

Reference [210].

Rosa rubiginosa L. (Rosaceae)
Sweet-briar

Description shrub

Region of origin Europe

Region invaded Argentina (Nahuel Huapi National Park)

References [94, 100].

Rubus argutus Link {R. *penetrans* Bailey} (Rosaceae)
Florida prickly blackberry

Description §Life form: shrub (thorny), perennial §Seed disp.: bird §Seed prod.: medium §Breeding syst.: possibly dioecious §Invasive category **3**.

Region of origin N. America (Prince Edward I. to Georgia and Alabama). Climatic zone: temp. moist, wmtemp. moist, subtrop. moist. Common in well drained sites.

Region invaded Pacific (Hawaii). Climatic zone: wmtemp. moist. Disturbed mesic to wet forests, 1000–2300 m, forming impenetrable thickets by tip rooting of the arching stems.

Notes Resprouts from basal and subterranean shoots after fire. Dispersed by alien birds in Hawaii.

Control Some biological control agents previously introduced to control blackberry in Hawaii have had adverse affects on two native *Rubus* species. Two rust diseases are under investigation for their biological control potential.

References [357, 359, 361].

Rubus cuneifolius Pursh. (Rosaceae)

Description shrub

Region of origin SE Asia

Region invaded South Africa

Rubus ellipticus Sm. {*R. flavus* Hamilton ex D.Don} (Rosaceae)

Yellow Himalayan raspberry

Description §Life form: shrub, semideciduous §Seed disp.: bird, mammal §Seed prod.: medium §Breeding syst.: hermaphrodite §Invasive category **4**.

Region of origin S. Asia (India, Sri Lanka, Burma, Tropical China, Philippines). Climatic zone: wmtemp. moist–wet. Evergreen oak–laurel forest, generally at altitudes of 1700–2300 m, in places associated with *Castanopsis* species; spreading into cultivated areas.

Region invaded Pacific (Hawaii), Africa (Mulanje Mt). Climatic zone: wmtemp. moist, subtrop. dry–moist. In Hawaii it invades pig-disturbed wet forest at 700–1700 m. In Malawi it often forms impenetrable thickets at forest margins or in clearings, having been introduced to the Zomba Botanic Garden between 1898 and 1900.

Notes It spreads rapidly by root suckers and regenerates from underground shoots after fire or cutting.

Control If physical clearance is undertaken, the roots must be grubbed out and burned. Alternatively, cut stumps may be treated with a herbicide: 2,4,5-T or a mixture of 2,4,5-T and 2,4-D are generally employed against *Rubus* species.

References [110, 111, 152, 359].

Rubus fruticosus L. sensu lato (Rosaceae)

Bramble, blackberry

Description shrub

Region of origin Europe

Region invaded Australia

Rubus moluccanus L. {Rubus alceifolius} (Rosaceae)

Description shrub

Region of origin Himalayas – Australasia

Region invaded Mauritius, Réunion

Rudbeckia lacinata L. (Compositae)

Coneflower, black-eyed Susan

Description herb

Region of origin N. America

Region invaded Japan

Saccharum spontaneum L. (Gramineae)

Description herb

Region of origin E. Africa

Region invaded Guam

Salix fragilis L. (Salicaceae)

Crack willow

Description shrub

Region of origin Eurasia

Region invaded New Zealand.

Salvinia molesta D.Mitch. (Salviniceae)

Water fern, kariba weed, African payal

Description herb

Region of origin Brazil

Region invaded Africa (South Africa, Namibia), India, Sri Lanka, Singapore, Malaysia, Philippines, Indonesia, New Guinea, Australia, New Zealand, Fiji, French Polynesia [148]. See main account in Chapter 4.

Sapium sebiferum (L.) Roxb. (Euphorbiaceae)
Chinese tallow

Description tree

Region of origin China, Japan

Region invaded SE USA (Louisiana to South Carolina) [75]

Scaevola plumieri (L.) Vahl. (Goodeniaceae)

Description shrub

Region of origin Australia

Region invaded Paraguay, Venezuela

Schinus terebinthifolia Raddi (Anacardiaceae)
Schinus, Florida holly, Brazilian pepper, Christmas berry

Description §Life form: shrub/tree, perennial, evergreen §Seed disp.: bird §Seed prod.: high §Breeding syst.: dioecious §Invasive category **3.5**.

Region of origin S. America (Brazil, Paraguay, Argentina). Climatic zone: wmtemp. dry, subtrop. dry, trop. arid–dry. Occurs in riverine forest and on damp soils, it thrives on disturbance but, in its native region, does not become a serious pest.

Region invaded N. America (Florida), Pacific (Hawaii), Australasia (Norfolk I.), Malagassia (Mauritius), Oceanic Is (St Helena). Climatic zone: wmtemp. dry, subtrop. moist–wet. Where invasive it has a broad ecological tolerance, for instance in Florida from mangrove swamp to pinelands. It thrives in disturbed habitats resulting from drainage or farming.

Notes A stand of *Schinus terebinthifolia* casts deep shade and lacks an herbaceous understorey. After burning, it resprouts from the base. It is said to release allelopathic chemicals. In Florida it is pollinated by flies (*Palpada vinetorum*). It produces its berries in winter, unlike many of the native Florida trees.

Control Several native pests from Brazil have been released in Hawaii to control *Schinus* but have not as yet been successful. Two have become established (*Bruchus atronotatus*–Coleoptera, *Episimus utilus*–Lepidoptera), but their effects are negligible.

References [21, 115, 196].

Sesbania punicea (Cav.) Benth. (Leguminosae)

Description shrub

Region of origin Argentina, Uruguay, Brazil

Region invaded South Africa [114, 173, 315]. See main account in Chapter 4.

Setaria palmifolia (Koenig) Stapf (Gramineae)

Description herb

Region of origin India

Region invaded Hawaii

Solanum mauritianum Scop. {syn. *Solanum auriculatum* Ait.} (Solanaceae)

Description shrub

Region of origin Tropical Asia

Region invaded South Africa, Uganda (Mabira Forest Reserve), Réunion, Norfolk I., New Zealand, Polynesia

Solidago canadensis L. (Compositae)

Canadian goldenrod

Description herb

Region of origin N. America

Region invaded C. Europe

Sorghum halepense (L.) Pers. (Gramineae)

Johnson grass

Description herb

Region of origin Mediterranean

Region invaded USA (Mediterranean climate areas)

Spartina anglica C.E.Hubbard (Gramineae)

Common cord-grass

Description herb

Region of origin uk. c. 1890 from *s.* x *townsendii* Groves and J. Groves

Region invaded British Isles [393], Netherlands, France, C. Europe, China, New Zealand.

Swietenia macrophylla King {*S. candollei* Pittier, *S. belizensis* Lundell} (Meliaceae)

Honduras mahogany, broadleaved mahogany

Description §Life form: tree, deciduous §Seed disp.: wind §Seed prod.: high §Breeding syst.: hermaphrodite §Invasive category **4**.

Region of origin C. America (S. Mexico to British Honduras), S. America (Panama, Colombia, Venezuela, Peru, Bolivia, Brazil). Climatic zone: subtrop. moist–wet, trop. dry.

Region invaded S. Asia (Sri Lanka). Climatic zone: trop. dry. It has been planted along skid-trails in previously logged tropical forest, in order to reforest areas of the Sinharaja Biosphere Reserve. Most of this reserve is relatively undisturbed, but *Swietenia macrophylla* threatens to dominate in places if not cleared, as it has a higher growth rate than native trees and is long-lived.

Notes Swietenia macrophylla becomes reproductively mature after 15 years and has a high seed production with good germination success.

Control Swietenia macrophylla has been logged near to roads in order to eradicate it. Further logging is planned. It has some ability to sprout after cutting.

References [79, 101].

Tamarindus indica L. (Leguminosae)

Tamarind

Description tree

Region of origin Tropical Africa

Region invaded Antigua

Tamarix aphylla (L.) Karsten (Tamaricaceae)
Athel pine (Australia)

Description tree

Region of origin Mediterranean

Region invaded Australia, Hawaii

Tamarix ramosissima Ledeb. (Tamaricaceae)

Description shrub

Region of origin Eurasia

Region invaded California, Australia (Finke River south of Alice Springs)

Thunbergia grandiflora (Rottler) Roxb. (Acanthaceae)

Blue thunbergia (Australia)

Description climber

Region of origin India

Region invaded Singapore, Australia (Queensland rainforest)

Tradescantia fluminensis Vell. Conc. (Commelinaceae)

Wandering jew

Description §Life form: herb §Seed prod.: low §Breeding syst.: hermaphrodite §Invasive category 3.

Region of origin S. America (Brazil). Climatic zone: trop. arid–dry.

Region invaded Australasia (New Zealand). Climatic zone: wmtemp. dry=moist. Invades disturbed lowland forest, producing a dense mat up to 60 cm deep, inhibiting the regeneration of native trees.

Notes Spreads rapidly by vegetative growth. Grows most rapidly under high light and is inhibited by deep shade.

Control Ecosystem management to keep the forest canopy unbroken and casting deep shade is probably the best preventive measure against this plant. Paraquat at 2 kg active ingredient per hectare has been used in New Zealand and reduces the standing crop of *Tradescantia fluminensis* by over 50% within 10 weeks [198], but it also damages native plants.

References [197, 198, 445].

Ugni molinae Turcz. (Myrtaceae)

Chilean guava

Description shrub

Region of origin Chile

Region invaded Juan Fernandez Is, Chatham Is

Ulex europaeus L. (Leguminosae)

Gorse, whin, furze

Description §Life form: shrub (spiny), evergreen §Seed disp.: Explosive dehiscence §Seed prod.: high §Breeding syst.: hermaphrodite §Invasive category **3.5**.

Region of origin Europe (western Europe). Climatic zone: temp. dry–wet. Heaths, upland pasture on mildly acidic soil.

Region invaded Pacific (Hawaii), S. Asia (India, Japan), N. America (USA), Australasia (New Zealand, Australia), Africa (South Africa). Climatic zone: temp. moist, wmtemp. moist, subtrop. moist. In New Zealand it vigorously invades pasture, wasteland, roadsides, fernland, scrubland and montane snow tussock. In Hawaii it is regarded as a threatening invader of open habitats.

Notes Although it is not tolerant of shade, in open areas it forms dense monospecific thickets. It regenerates rapidly after fire, both by resprouting and by growing from a persistent soil seed-bank.

Control Biological control using the weevil, *Exapion ulicis* Forster (Coleoptera: Apionidae), has been attempted in New Zealand, Australia and Chile, but has not been successful despite widespread establishment [196]. In New Zealand gorse has been suggested to help re-establishment of native trees as a nurse, in the anticipation that the gorse will disappear in the course of succession over a period of 50–60 years. Seed set of gorse is facilitated in Hawaii by the introduced honey-bee. The weevil, *Apion sculletone*, and the butterfly, *Lampides boeticus*, have been tested as gorse biological control agents in Hawaii.

References [196, 228, 235, 359].

Undaria pinnatifida (Harvey) Suringer (Phaeophyta: Laminariales)

Japanese kelp (Australia), japweed (British)

Description seaweed (kelp)

Region of origin Japan

Region invaded Tasmania, France, Britain

Vinca major L. (Apocynaceae)

Greater periwinkle

Description perennial herb, apparently only spreading vegetatively

Region of origin Europe

Region invaded New Zealand

Appendices

Appendix A – Use of herbicides: some environmental cautions

WWF policy is that herbicide use should be kept to a minimum. WWF cannot endorse the use of any particular herbicide, and it advises all potential users to check on current knowledge about the environmental consequences of particular formulations. Summary notes about the adverse effects of some common herbicides are given below to illustrate some of the issues. No claim is made that all toxic effects are listed.

2,4-D and 2,4,5-T
These substances are very poisonous to humans and there is evidence that they are carcinogenic. Dioxin, a contaminant of 2,4,5-T, is known to be a potent carcinogen. Fears have also been raised of a link between 2,4,5-T and birth defects. 2,4,5-T is persistent in the soil. It has been banned in over 10 countries; its use should be completely avoided (Martlew and Silver, 1991). 2,4-D is toxic to insects (Hurst *et al.*, 1991).

Bromacil (Hyvar X; Krovar II; Uragan)
A soil-acting herbicide. Care must be taken to keep it out of the rooting zone of native plants. Bromacil can be irritating to the skin, eyes and respiratory system. It is considered unlikely to cause an acute hazard to humans under normal use. However, it is dangerous to wildlife (Watterson, 1988).

Dalapon (as sodium/magnesium salts) (Dowpon; Radapon; Basfapon)
Effective for the control of grasses and usually applied as a foliar spray. It leaches readily but breaks down fairly rapidly in soil, persisting some two to four weeks. There is some evidence that Dalapon can cause skin and eye irritation, and it has moderate oral toxicity. It is said to be unlikely to be an acute health hazard to humans under normal use, although possible adverse effects on animal kidneys have been reported (Watterson, 1988).

Dicamba (Banvel; Fallowmaster)
Supplied as an amine or sodium salt, it is effective against woody plants and may be applied as foliar spray or on cut surfaces. It is mobile in soils and fairly persistent, with a half-life in soil of two weeks or more. It is a mammalian skin and eye irritant of moderate toxicity.

Diquat (as dibromide salt) (*Reglone; Reglox*)
It is highly toxic, although less toxic, and less effective as a herbicide, than paraquat. It is inactivated quickly in soils by adsorption onto clay minerals. Diquat can be fatal if swallowed, inhaled or absorbed through the skin, and can induce serious adverse effects on mammals (Watterson, 1988). There is no known antidote. It may also cause severe eye and skin irritation (Martlew and Silver, 1991).

Endothall (*Des-I-Cate; Aquathol*)
Highly toxic to people and wildlife, but less so to fish and occasionally used as an aquatic herbicide.

Glyphosate (*Round-up* (Monsanto); *Tumbleweed*)
A glycine derivative supplied as an amine salt. Herbaceous species are susceptible to foliar treatment, and some woody species are susceptible to cut-surface treatment or bark injection. It is inactivated immediately in soil by adsorption and decomposes relatively quickly with a half-life of some 60 days. It is said that it is unlikely to represent an acute health hazard to humans under normal use. Can cause nausea and eye irritation. Harmful to fish (Watterson, 1988).

Paraquat (as dichloride salt) (*Gramoxone, Dextrone X; Pillarxone*)
Foliar contact herbicide, highly toxic to plants and people. In soil it is immediately adsorbed to clay minerals in a persistent but biologically inactive form. This is an acutely poisonous chemical responsible for a significant number of deaths. It may also encourage Parkinson's disease, and can be fatal to small mammals (Watterson, 1988; Hurst *et al.*, 1991). There is no known antidote. The use of Paraquat is not advised (Martlew and Silver, 1991).

Picloram (as potassium salt) (*Tordon 22K* (Dow); *Amdan; Grazon*)
It is highly effective against woody plants and may be applied as foliar spray on cut surfaces or as soil-active granules. However, as it is persistent and mobile in soils, it should be used with caution. It is said that it is unlikely to present an acute health hazard to humans under normal use. Of moderate oral toxicity, it is an irritant to the skin, eyes and respiratory tract. High doses produce chronic renal disease in animals. Carcinogenic in animals and harmful to fish (Watterson, 1988).

Triclopyr (as amine salt or ester) (*Garlon* (Dow); *Crossbow*)
Forbs and woody plants are susceptible but grasses are tolerant. However, it is mobile in soil and persistent over a half-life of *c.* 46 days. Mild eye and skin irritant, slightly hazardous to humans and dangerous to fish (Watterson, 1988).

Sources:

Hurst, P., Hay, A. and Dudley, N. (1991) *The Pesticide Handbook*, Journeyman, London (UK) and Concord (USA).

Martlew, G. and Silver, S. (1991) *The Green Home Handbook*, Fontana, London.

Watterson, A. (1988) *Pesticide User's Safety Handbook*, Gower Technical, Aldershot (UK).

Appendix B – A selection of relevant addresses

Australia

CSIRO Division of Entomology,
Long Pocket Laboratories,
Private Bag 3, PO,
Indooroopilly,
Queensland 4068

CSIRO Division of Plant Industry,
GPO Box 1600,
Canberra,
ACT 2601

CSIRO Tropical Ecosystems Research Centre,
PMB 44,
Winellie,
Darwin NT08,
Northern Territory

Netherlands

Research Institute for Nature,
Kemperbergerweg 67,
6816 RM Arnhem

Dept of Vegetation Science, Plant Ecology and Weed Science,
Agricultural University,
Bornsesteeg 69,
6708 PD Wageningen

New Zealand

Land Care Research NZ Ltd,
PO Box 40,
Lincoln,
Canterbury

Department of Conservation,
PO Box 12–240,
Wellington

South Africa

National Botanical Institute,
Private Bag X101,
Pretoria 0001

Percy Fitzpatrick Institute of Ornithology,
University of Cape Town,
Private Bag,
Rondebosch 7700

Switzerland

Plants Officer,
IUCN – The World Conservation Union
Rue Mauverney 28
CH-1196 Gland

WWF International
Avenue du Mont Blanc
CH-1196 Gland

United Kingdom

Royal Botanic Gardens,
Kew,
Richmond,
Surrey TW9 3AA

Botanic Garden Conservation International,
Descanso House,
199 Kew Road,
Richmond,
Surrey TW9 3BW

International Institute for Biological Control,
Silwood Park,
Buckhurst Road,
Ascot,
Berkshire
SL5 7TA

WWF Plants Conservation Officer,
Panda House,
Weyside Park,
Godalming,
Surrey GU7 1XR

World Conservation Monitoring Centre,
219 Huntingdon Road,
Cambridge CB3 0DL

USA

WWF – United States
1250 24th Street NW
Washington DC 20037-1175
USA

USA – Hawaii

Hawaii Volcanoes National Park,
US National Park Service,
PO Box 52,
Honolulu,
Hawaii 96712

Cooperative Parks Study Unit,
University of Hawaii at Manoa,
90 Maile Way,
Honolulu,
Hawaii 96822

National Tropical Botanic Garden,
PO Box 340,
Lawai,
Kauai,
Hawaii 96765

Glossary

adventive	an immigrant, a plant recently arrived and not yet established
agamospermy	seed production without sexual fusion, a type of apomixis (q.v.)
alien	not native, exotic (adj.), a non-native organism (noun)
allelopathy	release by plants of chemicals into soil which inhibit the growth of others
apomixis	asexual reproduction
autecology	study of the relationship between an organism and its environment
biological control	control of pests by use of their predators or pathogens
biotype	a biologically distinct population or variant
cross-fertilization	fusion of the sex-cells of different organisms
cross-pollination	transfer of pollen from a different plant onto the stigma
database	source of organized information, generally held on a computer
dicotyledon	one of the two divisions of flowering plants, characteristically with reticulate venation and two seed-leaves, e.g. Leguminosae, Compositae
dioecy	separation of sexes, with male and female flowers on different plants
dispersal	process by which seeds, or other reproductive units, are carried away from the parent plant
endemic	(of an organism) restricted to a particular place, usually used of organisms which are not widespread
eutrophication	increase of availability or mineral nutrients (usually nitrates and phospates) in an ecosystem
evapotranspiration	loss of water from soil by evaporation from surface and from plant leaves (transpiration)
exotic	not native, alien (adj.), a non-native organism (noun)
facilitation	change in environmental conditions aiding or enabling invasion
facultative apomixis	reproduction sexual or asexual depending on conditions

feral	naturalized, not native
focus	point of introduction from which spread occurs (plural: foci)
fynbos	open scrubland of Cape Province, South Africa
genotype	the particular genetic constitution of an organism
germplasm	living material capable of propagation
grassveld	grassland of large perennial grasses characteristic of parts of South Africa
herbivore	plant-eating animal
hermaphrodite	having both male and female organs in each flower
inbreeding	reproduction by self-fertilization (q.v.)
indigenous	native
invasion	spread of an organism, without human assistence, into natural or seminatural habitats to produce a significant ecosystem change (of composition, structure or processes)
karoo	dry open vegetation of low shrubs and succulents, characteristic of parts of Cape Province, South Africa
mattock	implement with blade and handle for removal of small tree roots
monocotyledon	one of the two divisions of flowering plants, characteristically with parallel leaf-veins and one seed-leaf, e.g. grasses, orchids, palms (cf. dicotyledons)
monoecy	separation of male and female organs in different flowers but on the same plant
mycoherbicide	fungal spores in suspension used as a weedkiller
native	naturally occurring in an area
naturalization	process of establishment of an introduction in native vegetation from some point of introduction (e.g. a garden or plantation). An early stage in the process of invasion
obligate apomixis	inability to reproduce sexually
outbreeding	reproduction by cross-fertilization (q.v.)
polyploid	having more than two sets of chromosomes
propagule	a plant part or plant organ, such as a seed or stem fragment, capable of reproducing the plant after dispersal
remote sensing	collection of ground information by satellite or aircraft
ruderal	plant associated with human activity or disturbed sites
savanna	tropical grassland of tall grasses and sparse low trees, characteristic of much of Africa but also occurring in N. Australia and in Venezuela, Colombia ('ilanos') and Brazil ('campos')
sclerophyll	plant with tough, rigid leaves, such as *Eucalyptus*. Such plants are typical of Mediterranean-type ecosystems
seed-bank	reservoir of viable seed in soil capable of remaining dormant for some time, and usually germinating after disturbance

self-compatible	able to self-fertilize
self-fertilization	fusion of the sex-cells of the same organism
self-incompatible	not able to self-fertilize because of genetic and physiological factors
self-pollination	transfer of pollen from the same plant onto the stigma
serotiny	delayed seed release (often stimulated by fire)
subdioecy	incomplete separation of sexes between different plants
succession	change of plant communities over time following an environmental perturbation
syndrome	a collection of characteristics of a plant that are associated with particular ecological behavior
tannin	type of plant chemical that deters grazing and is used to tan leather
translocation	(of organism) movement of germplasm (q.v.) from one place to another; (of herbicide) movement of chemical from one part of the plant to another
triploid	having three sets of chromosomes, one set being unpaired

References

1. Abbott, R.J. (1992) Plant invasion, interspecific hybridization and the evolution of new plant taxa, *Trends in Ecology and Evolution*, **7**, 401–5.
2. Adams, C.D. (1972) *Flowering Plants of Jamaica*, University of the West Indies, Jamaica.
3. Allen, R.B. (1991) A preliminary assessment of the establishment and persistence of *Berberis darwinii* Hook., a naturalised shrub in secondary vegetation near Dunedin, New Zealand, *New Zealand Journal of Botany*, **29**, 353–60.
4. Amor, R.L. and P.L. Stevens (1975) Spread of weeds from a roadside into sclerophyll forests at Dartmouth, Australia, *Weed Research*, **16**, 111–18.
5. Andres, L.A. (1981) Conflicting interests and the biological control of weeds, in *Proceedings of the Fifth International Symposium on Biological Control of Weeds*, (ed. E.S. Del Fosse), CSIRO, Brisbane, Australia, pp. 11–20.
6. Anon. (1985) Ecology of Biological Invasions, *SCOPE Newsletter*, **23**, 1–5.
7. Arthington, A.H. and D.S. Mitchell (1986) Aquatic invading species, in *Ecology of Biological Invasions*, (ed. R.H. Groves and J.J. Burdon), Cambridge University Press, Cambridge, pp. 34–56.
8. Ashton, D.H. (1981) Tall open forests, in *Australian Vegetation*, (ed. R.H. Groves), Cambridge University Press, Cambridge, pp. 121–51.
9. Ashton, P.J. and D.S. Mitchell (1989) Aquatic plants: patterns and modes of invasion, attributes of invading species and assessment of control programs, in *Biological Invasions: a global perspective*, (ed. J.A. Drake *et al.*), John Wiley & Sons, Chichester, pp. 111–54.
10. Auld, B.A. (1969) The distribution of *Eupatorium adenophorum* Spreng. on the far north coast of New South Wales, *Journal and Proceedings of the Royal Society of New South Wales*, **102**, 159–61.
11. Auld, B.A. (1970) *Eupatorium* weed species in Australia, *PANS*, **16**, 82–6.
12. Auld, B.A. (1975) The autecology of *Eupatorium adenophorum* Spreng. in Australia, *Weed Research*, **15**, 27–31.
13. Auld, B.A. (1977) The introduction of *Eupatorium* species to Australia, *Journal of the Australian Institute of Agricultural Science*, September/December, 146–7.
14. Auld, B.A. (1981) Invasive capacity of *Eupatorium adenophorum*, in *Proceedings of the 8th Asian-Pacific Weeds Science Conference*, pp. 145–7.
15. Auld, B.A. and B.G. Coote (1980) A model of a spreading plant population, *Oikos*, **34**, 287–92.
16. Auld, B.A. and B.G. Coote (1981) Prediction of pasture invasion by *Nassella trichotoma* (Gramineae) in south east Australia, *Protection Ecology*, **3**, 271–7.
17. Auld, B.A., J. Hosking and R.E. McFadyen (1982/1983) Analysis of the spread of tiger pear and parthenium weed in Australia, *Australian Weeds*, **2**, 56–60.
18. Auld, B.A., K.M. Menz and N.M. Monaghan (1978/1979) Dynamics of weed spread: implications for policies of public control, *Protection Ecology*, **1**, 141–8.
19. Auld, B.A., K.M. Menz and C.A. Tisdell (1987) *Weed Control Economics*, Academic Press, London.

20. Auld, B.A., D.T. Vere and B.G. Coote (1982) Evaluation of control policies for the grassland weed, *Nassella trichotoma*, in south-east Australia, *Protection Ecology*, **4**, 331–8.

21. Austin, D.F. (1978) Exotic plants and their effects in south east Florida, *Environmental Conservation*, **5**, 25–34.

22. Baker, H.G. (1965) Characteristics and modes of origin of weeds, in *The Genetics of Colonizing Species*, (ed. H.G. Baker and C.L. Stebbins), Academic Press, New York, pp. 147–69.

23. Baker, H.G. (1986) Genetic characteristics of invasive plants, in *Ecology of Biological Invasion of North America and Hawaii*, (ed. H.A. Mooney and J.A. Drake), Springer-Verlag, New York, pp. 147–68.

24. Balloch, G.M., S. Ullah and A.A. Shah (1980) Some promising insects for the biological control of *Hydrilla verticillata* in Pakistan, *Tropical Pest Management*, **26**, 194–200.

25. Bannister, M.H. (1965) Variation in the breeding system of *Pinus radiata*, in *The Genetics of Colonizing Species*, (ed. H.G. Baker and G.L. Stebbins), Academic Press, New York, pp. 353–73.

26. Barneby, R.C. (1989) Reflections on typification and application of the names *Mimosa pigra* L. and M. *asperata* L. (Mimosaceae), in *The Davis and Hedge Festschrift*, (ed. K. Tan), Edinburgh University Press, Edinburgh, pp. 137–47.

27. Barrett, S.C.H. (1988) Evolution of breeding systems in *Eichhornea* (Pontederiaceae): a review, *Annals of the Missouri Botanical Garden*, **75**, 741–60.

28. Barrows, E.M. (1976) Nectar robbing and pollination of *Lantana camara* (Verbenaceae), *Biotropica*, **8**, 132–5.

29. Bass, D.A. (1990) Dispersal of an introduced shrub (*Crataegus monogyna*) by the brush-tailed possum (*Trichosurus vulpecula*), *Australian Journal of Ecology*, **15**, 227–9.

30. Bawa, K.S. (1980) Evolution of dioecy in flowering plants, *Annual Review of Ecology and Systematics*, **11**, 15–39.

31. Bean, W.J. (1976) *Trees and Shrubs Hardy in the British Isles*, 8th edn, John Murray, London, pp. 741–4.

32. Beard, J. (1976) Alien shrub runs riot in Hawaii, *New Scientist*, **1708**, 30.

33. Becker, D. (1988) *Management techniques: the control and removal of Rhododendron ponticum on RSPB reserves in England and Wales*, The Royal Society for the Protection of Birds. Sandy, UK.

34. Bennett, F.D. and V.P. Rao (1968) Distribution of an introduced weed *Eupatorium odoratum* Linn. (Compositae) in Asia and Africa and possibilities of its biological control, *Pest Articles and News Summaries, Section C*, **14**, 277–81.

35. Bennett, K.D. (1986) The rate of spread and population increase of forest trees during the postglacial, *Philosophical Transactions of the Royal Society, London, Series B*, **314**, 523–9.

36. Benson, L. (1982) *The Cacti of the United States and Canada*, Stanford University Press, Stanford.

37. Bentham, G. (1875) Revision of the suborder Mimoseae, *Transactions of the Linnean Society*, **30**, 335–664.

38. Binggeli, P. (1989) The ecology of *Maesopsis* invasion and dynamics of the evergreen forest of the East Usambaras and their implications for forest conservation and forestry practices, in *Forest Conservation in the East Usambara Mountains Tanzania*, (ed. A.C. Hamilton and R. Bensted-Smith), IUCN, Gland, pp. 269–300.

39. Binggeli, P. and A.C. Hamilton (1990) Tree species invasions and sustainable forestry in the East Usambaras, in *Research for Conservation of Tanzanian Catchment Forests*, (ed. I. Hedberg and E. Persson), Uppsala Universitet Reprocentralen HSC, Uppsala, pp. 39–47.

40. Birks, H.J.B. and J. Deacon (1973) A numerical analysis of the past and present flora of the British Isles, *New Phytologist*, **72**, 877–902.

41. Booth, T. *et al.* (1987) Grid matching: a new method of homoclime analysis, *Agricultural and Forest Meteorology*, **39**, 241–55.
42. Booth, T.H. (1985) A new method for assisting species selection, *Commonwealth Forestry Review*, **64**, 241–50.
43. Booth, T.H. and T. Jovanovic (1988) Assaying natural climatic variability in some Australian species with fuelwood and agroforestry potential, *Commonwealth Forestry Review*, **67**, 27–34.
44. Bottrell, D.G. and R.F. Smith (1982) Integrated Pest Management, *Environmental Science and Technology*, **16**, 282A–8A.
45. Boucher, C. (1980) Black wattle, in *Plant Invaders: beautiful but dangerous*, 2nd edn, (ed. C.H. Stirton), The Department of Nature and Environmental Conservation of the Cape Provincial Administration, Cape Town, pp. 48–51.
46. Boucher, C. and C.H. Stirton (1980) Long-leaved wattle, in *Plant Invaders: beautiful but dangerous*, 2nd edn, (ed. C.H. Stirton), Department of Nature and Environmental Conservation of the Cape Provincial Administration., Cape Town, pp. 44–7.
47. Boucher, C. and C.H. Stirton (1980) Port Jackson, in *Plants Invaders: beautiful but dangerous*, 2nd edn, (ed. C.H. Stirton), Department of Nature and Environmental Conservation of the Cape Provincial Administration, Cape Town, pp. 60–3.
48. Boucher, C. and C.H. Stirton (1980) Rooikrans, in *Plant Invaders: beautiful but dangerous*, 2nd edn, (ed. C.H. Stirton), Department of Nature and Environmental Conservation of the Cape Provincial Administration, Cape Town, pp. 40–3.
49. Boyette, C.D., G.E. Templeton and L.R. Oliver (1984) Texas gourd (*Cucurbita texana*) control with *Fusarium solani* f. sp. *cucurbitae*, *Weed Science*, **32**, 649–55.
50. Bradley, J. (1988) *Bringing Back the Bush*, Landsdowne Press, Sydney.
51. Braithwaite, R.W., W.M. Lonsdale and J.A. Esthbergs (1989) Alien vegetation and native biota in tropical Australia: the impact of *Mimosa pigra*, *Biological Conservation*, **48**, 189–210.
52. Bramwell, D. and Z.I. Bramwell (1974) *Wild Flowers of the Canary Islands*, Stanley Thornes, London.
53. Brenan, J.P.M. (1959) *Leguminosae Subfamily Mimosoideae, Flora of Tropical East Africa*, (ed. C.E. Hubbard and E. Milne-Redhead), Crown Agents, London.
54. Bridgewater, P.B. and D.J. Backshall (1981) Dynamics of some Western Australian ligneous formations with special reference to the invasion of exotic species, *Vegetatio*, **46**, 141–8.
55. Brockie, R.E. *et al.* (1988) Biological invasions of island nature reserves, *Biological Conservation*, **44**, 9–36.
56. Brockway, L.H. (1979) *Science and Colonial Expansion*, Academic Press, New York.
57. Brown, C.J., I.A.W. Macdonald and S.E. Brown, ed. (1985) *Invasive Alien Organisms in South West Africa/Namibia*, CSIR, Pretoria.
58. Buckley, R. (1981) Alien plants in central Australia, *Botanical Journal of the Linnean Society*, **82**, 369–80.
59. Burdon, J.J. and G.A. Chilvers (1977) Preliminary studies on a native eucalypt forest invaded by exotic pines, *Oecologia*, **31**, 1–12.
60. Burrell, J.P. (1981) Invasion of coastal heaths of Victoria by *Leptospermum laevigatum* (J.Gaertn.) F. Muell., *Australian Journal of Botany*, **29**, 747–64.
61. Burrows, J.E. (1989) Kudzu vine – a new plant invader of South Africa, *Veld & Flora*, **75**, 116–17.
62. Buxton, J.M. (1985) The potential for biological control of *Clematis vitalba* L., M.Sc. Thesis, Imperial College, Ascot.
63. Campbell, M.H. (1982) The biology of Australian weeds 9. *Nassella trichotoma* (Nees) Arech., *The Journal of the Australian Institute of Agricultural Science*, **48**, 76–84.
64. Cassani, J.R., T.W. Miller and M.L. Beach (1981) Biological control of aquatic weeds in southwest Florida, *Journal of Aquatic Plant Management*, **19**, 49–50.

65. Cattley, W. (1821) *Psidium cattleianum*, *Transactions of the Royal Horticultural Society*, **4**, 314–17.
66. Chapman, V.J. (1967) Conservation of maritime vegetation and the introduction of submerged freshwater aquatics, *Micronesica*, **3**, 31–5.
67. Charudattan, R. (1986) Integrated control of water hyacinth (*Eichhornea crassipes*) with a pathogen, insects and herbicides, *Weed Science*, **34**, (Suppl.1), 26–30.
68. Chilvers, G.A. and J.J. Burdon (1983) Further studies on a native Australian eucalypt forest invaded by exotic pines, *Oecologia*, **59**, 239–45.
69. Cilliers, C.J. (1991) Biological control of water fern, *Salvinia molesta* (Salviniaceae), in South Africa, *Agriculture, Ecosystems and Environment*, **37**, 219–24.
70. Cilliers, C.J. and S. Neser (1991) Biological control of *Lantana camara* (Verbenaceae) in South Africa, *Agriculture, Ecosystems and Environment*, **37**, 57–75.
71. Clapham, A.R., T.G. Tutin and E.F. Warburg (1962) *Flora of the British Isles*, 2nd edn, Cambridge University Press, Cambridge.
72. Clayton, W.D. and S.A. Renvoize (1982) Gramineae, part 3, in *Flora of Tropical East Africa*, (ed. R.M. Polhill), A.A. Balkema, Rotterdam.
73. Coblentz, B.E. (1990) Exotic organisms: A dilemma for conservation biology, *Conservation Biology*, **4**, 261.
74. Coffey, B.T. and J.S. Clayton (1988) Changes in the submerged macrophyte vegetation of Lake Rotoiti, central North Island, New Zealand, *New Zealand Journal of Marine and Freshwater Research*, **22**, 215–23.
75. Conner, W.H. and G.R. Askew (1993) Impact of saltwater flooding on red maple, redbay, and Chinese tallow seedlings, *Castanea*, **58**, 214–19.
76. Conolly, A.P. (1977) The distribution and history in the British of some alien species of *Polygonum* and *Reynoutria*, *Watsonia*, **11**, 291–311.
77. Cooper, R.C. (1956) The Australian and New Zealand species of *Pittosporum*, *Annals of the Missouri Botanical Garden*, **43**, 87–188.
78. Corlett, R.T. (1988) The naturalized flora of Singapore, *Journal of Biogeography*, **15**, 657–63.
79. Corner, E.J.H. (1988) *Wayside Trees of Malaya*, Vol. 1, 3rd edn, The Malayan Nature Society, Kuala Lumpur.
80. Correll, D.S. and M.C. Johnston (1970) *Manual of the Vascular Plants of Texas*, Series 1, (ed. C.L. Lundell), Texas Research Foundation., Texas.
81. Cowie, I.D. and P.A. Werner (1993) Alien plant species in the Kakadu National Park, tropical northern Australia, *Biological Conservation*, **62**, 127–35.
82. Cox, P. and P. Hutchinson (1963) Rhododendrons in north east Turkey, *Rhododendron Yearbook*, **17**, 64–7.
83. Crawley, M.J. (1987) What makes a community invasible?, in *Colonization, Succession and Stability*, (ed. M.J. Crawley, P.J. Edwards and A.J. Gray), Blackwell, Oxford, pp. 629–54.
84. Cronk, Q.C.B. (1986) The decline of the St Helena Ebony *Trochetiopsis melanoxylon*, *Biological Conservation*, **35**, 159–72.
85. Cronk, Q.C.B. (1987) Combating the invasive plant threat: the feasibility, structure and role of an action-oriented database, IUCN: unpublished report.
86. Cronk, Q.C.B. (1989) The past and present vegetation of St Helena, *Journal of Biogeography*, **16**, 47–64.
87. Crosby, A.W. (1986) *Ecological Imperialism: The biological expansion of Europe, 900–1900*, Cambridge University Press, Cambridge.
88. Cross, J.R. (1973) The ecology and control of *Rhododendron ponticum* L., Ph.D. Thesis, University of Dublin.
89. Cross, J.R. (1975) Biological Flora of the British Isles, no. 137. *Rhododendron ponticum* L., *Journal of Ecology*, **63**, 345–64.
90. Cross, J.R. (1982) The invasion and impact of *Rhododendron* in native Irish vegetation, in *Studies on Irish Vegetation*, (ed. J. White), Royal Dublin Society, Dublin, pp. 209–20.

91. Cruz, F.J. Cruz and J. Laweson (1986) *Lantana camara* L., a threat to native plants and animals, *Noticias de Galapagos*, **43**, 10–11.

92. D'Antonio, C.M. (1993) Mechanisms controlling invasion of coastal plant communities by the alien succulent *Carpobrotus edulis*, *Ecology*, **74**, 83–95.

93. D'Antonio, C.M. and B.E. Mahall (1991) Root profiles and competition between the invasive exotic perennial, *Carpobrotus edulis*, and the two native shrub species in California coastal scrub, *American Journal of Botany*, **78**, 885–94.

94. Damascos, M.A. and G.G. Gallopin (1992) Ecología de un arbusto introducido (*Rosa rubiginosa* L. = *Rosa eglanteria* L.): riesgo de invasión y efectos en las comunidades vegetales de la región andino-patagónica de Argentina, *Revista Chilena de Historia Natural*, **65**, 395–407.

95. Das, R. (1969) A study of reproduction in *Eichhornia crassipes*, *Tropical Ecology*, **10**, 195–8.

96. Dassanayake, M.D. and F.R. Fosberg, ed. (1985) *A Revised Handbook to the Flora of Ceylon*, Vol. 5, Amerind Publishing Company, New Delhi.

97. Davis, C.J., E. Yoshioka and D. Kageler (1992) Biological control of lantana, prickly pear, and hamakua pamakani in Hawaii: a review and update, in *Alien Plant Invasions in Native Ecosystems of Hawaii: management and research*, (ed. C.P. Stone, C.W. Smith and J.T. Tunison), University of Hawaii Cooperative National Park Resources Studies Unit, Honolulu, pp. 411–31.

98. Davis, D.R. *et al.* (1991) Systematics, morphology, biology, and host specificity of *Neurostrota gunniella* (Busck) (Lepidoptera: Gracillariidae), an agent for the biological control of *Mimosa pigra* L., *Proceedings of the Entomological Society of Washington*, **93**, 16–44.

99. Dawson, F.H. and E.A. Warman (1987) *Crassula helmsii* (T.Kirk) Cockayne: Is it an aggressive alien aquatic plant in Britain?, *Biological Conservation*, **42**, 247–72.

100. De Pietri, D.E. (1992) Alien shrubs in a national park: can they help in the recovery of natural degraded forest?, *Biological Conservation*, **62**, 127–30.

101. de Zoysa, N.D., C.V.S. Gunatilleke and I.A.U.N. Gunatilleke (1986) Vegetation studies of a skid-trail planted with mahogany in Sinharaja, *The Sri Lanka Forester*, **17**, 142–7.

102. Dennill, G.B. (1990) The contribution of a successful biocontrol project to the theory of agent selection in weed biocontrol – the gall wasp *Trichilogaster acaciaelongifoliae* and the *Acacia longifolia*, *Agriculture, Ecosystems and Environment*, **31**, 147–54.

103. Denton, G.R.W., and R. Muniappan (1991) Status and natural enemies of the weed, *Lantana camara*, in Micronesia, *Tropical Pest Management*, **37**, 338–44.

104. Devine, W.T. (1977) A programme to exterminate introduced plants on Raoul Island, *Biological Conservation*, **11**, 193–207.

105. Dickson, J.H., J.C. Rodriguez and A. Machado (1987) Invading plants at high altitudes on Tenerife especially in the Teide National Park, *Botanical Journal of the Linnean Society*, **95**, 155–79.

106. Downward, P. (1986) Herbicides for control of old man's beard, in *Proceedings of the 39th New Zealand Weed and Pest Control Conference*, pp. 108–9.

107. Du Cane, G. (1870) *Natural History of the Azores*, John van Voorst, London.

108. Duggen, K.J. and L. Henderson (1981) Progress with a survey of exotic woody plant invaders of the Transvaal, in *Proceedings of the Fourth National Weeds Conference of South Africa*, (ed. H.A. van de Venter and M. Mason), A.A. Balkema, Capetown, pp. 7–20.

109. Dyer, C. and D.M. Richardson (1992) Population genetics of the invasive Australian shrub *Hakea sericea* (Proteaceae) in South Africa, *South African Journal of Botany*, **58**, 117–24.

110. Edwards, I.D. (1982) Plant invaders on Mulanje Mountain, *Nyala*, **8**, 89–94.

111. Edwards, I.D. (1985) Conservation of plants on Mulanje Mountain, Malawi, *Oryx*, **19**, 86–90.

112. Ellenberg, H. (1974) Indicator values of vascular plants in central Europe, *Scripta Geobotanica*, **9**, 1–86.
113. Elton, C.S. (1958) *The Ecology of Invasions by Animals and Plants*, Chapman & Hall, London.
114. Erb, H.E. (1979) The natural enemies and distribution of *Sesbania punicea* (Cav.) Benth. in Argentina, in *Proceedings of the Third National Weeds Conference of South Africa, Pretoria*, pp. 205–10.
115. Ewel, J.J. (1986) Invasibility: lessons from South Florida, in *Ecology of Biological Invasion of North America and Hawaii*, (ed. H.A. Mooney and J.A. Drake), Springer-Verlag, New York, pp. 214–30.
116. Fairley, A. and P. Moore (1989) *Native Plants of the Sydney District: An Identification Guide*, Kangaroo Press, New South Wales.
117. Farell, T.P. (1979) Control of *Salvinia molesta* and *Hydrilla verticillata* in Lake Moondarra, Queensland, in *Australian Water Resources Council Seminar on Management of Aquatic Weeds*, pp. 57–71.
118. Flanagan, G.J., C.G. Wilson and J.D. Gillett (1990) The abundance of native insects on the introduced weed *Mimosa pigra* in Northern Australia, *Journal of Tropical Ecology*, **6**, 219–30.
119. Floyd, A.G. (1990) *Australian Rainforests in New South Wales*, Vol. 1, Surrey Beatty & Sons Pty Limited, New South Wales.
120. Forno, I.W. and K.L.S. Harkey (1979) The occurrence of *Salvinia molesta* in Brazil, *Aquatic Botany*, **6**, 185–7.
121. Forsyth, A.A. (1954) British poisonous plants, *Bulletin of the Ministry of Agriculture, Fisheries and Food, London*, No. 161.
122. Fuller, J.L. (1991) The threat of invasive plants to natural ecosystems, M.Phil. Thesis, University of Cambridge.
123. Fuller, R.M. and L.A. Boorman (1977) The spread and development of *Rhododendron ponticum* L. on dunes at Winterton, Norfolk, in comparison with invasion by *Hippophae rhamnoides* L. at Saltfleeby, Lincolnshire, *Biological Conservation*, **12**, 83–94.
124. Gardner, D.E. and V.A.D. Kageler (1982) *Herbicidal control of firetree in Hawaii Volcanoes National Park: a new approach, Ecological Services Bulletin*, No. 7, US Department of the Interior, Washington.
125. Gardner, D.E. and C.W. Smith (1985) Plant biocontrol quarantine facility at Hawaii Volcanoes, *Park Science*, **6**, 3–4.
126. Geldenhuys, C.J. (1982) The management of the southern Cape forests, *South African Forestry Journal*, **121**, 1–7.
127. Geldenhuys, C.J., P.J. Le Roux and K.H. Cooper (1986) Alien invasion in the indigenous evergreen forest, in *Ecology and Management of Biological Invasions in South Africa*, (ed. I.A.W. Macdonald, F.J. Kruger and A.A. Ferrar), Oxford University Press, Cape Town, pp. 275–82.
128. Gill, L.T. (1989) Perspectives on environmental education in Hawaii, in *Conservation Biology in Hawaii*, (ed. C.P. Stone and D.B. Stone), University of Hawaii Cooperative National Park Resources Studies Unit, Honolulu, pp. 177–8.
129. Gleadow, R.M. (1982) Invasion by *Pittosporum undulatum* of the forests of Central Victoria. 2, Dispersal, germination and establishment, *Australian Journal of Botany*, **30**, 185–98.
130. Gleadow, R.M. and D.H. Ashton (1981) Invasion by *Pittosporum undulatum* of the forests of Central Victoria. 1, Invasion patterns and morphology, *Australian Journal of Botany*, **29**, 705–20.
131. Gleadow, R.M., and K.S. Rowan (1982) Invasion by *Pittosporum undulatum* of the forests of Central Victoria. 3, Effects of temperature and light on growth and drought resistance, *Australian Journal of Botany*, **30**, 347–57.
132. Gleason, H.A. (1939) The genus *Clidemia* in Mexico and Central America, *Brittonia*, **3**, 97–140.

133. Godwin, Sir H. (1975) *The History of the British Flora*, 2 edn, Cambridge University Press, Cambridge.

134. Golley, F.B. (1965) Structure and function of an old-field broomsedge community, *Ecological Monographs*, **35**, 113–37.

135. Gooding, E.G.B., A.R. Loveless and G.R. Proctor (1965) *Flora of Barbados*, Her Majesty's Stationary Office, London.

136. Goodland, T. (1990) A report on the spread of an invasive tree species *Pittosporum undulatum* into the forests of the Blue Mountains, Jamaica, B.Sc. report, University College of North Wales.

137. Gopal, B. (1987) *Water Hyacinth: The most troublesome weed in the world*, Hindasia Publishers, Delhi.

138. Gowda, M. (1951) The genus *Pittosporum* in the Sino-Indian Region, *Journal of the Arnold Arboretum*, **32**, 263–343.

139. Graaf, J.L. and J. van Staden (1984) The germination characteristics of two *Sesbania* species, *South African Journal of Botany*, **3**, 59–62.

140. Griffen, J.L., V.H. Watson and W.F. Strachan (1988) Selective broomsedge (*Andropogon virginicus* L.) control in permanent pastures, *Crop Protection*, **7**, 80–3.

141. Griffin, G.F. *et al.* (1989) Status and implications of the invasion of Tamarisk (*Tamarix aphylla*) on the Finke River, Northern Territory, Australia, *Journal of Environmental Management*, **29**, 297–315.

142. Grime, J.P. (1979) *Plant Strategies and Vegetation Processes*, John Wiley & Sons, Chichester.

143. Groves, R.H. (1985) Invasion of weeds in mediterranean ecosystems, in *Resilience in Mediterranean Climatic Ecosystems*, (ed. B. Dell), Dr.W.Junk, The Hague.

144. Groves, R.H. (1986) Plant invasion of Australia: an overview, in *Ecology of Biological Invasions*, (ed. R.H. Groves and J.J. Burdon), Cambridge University Press, Cambridge, pp. 137–49.

145. Groves, R.H. and J.J. Burdon, ed. (1986) *Ecology of Biological Invasions*, Cambridge University Press, Cambridge.

146. Groves, R.H. and J.D. Williams (1975) Growth of Skeleton Weed (*Chondrilla juncea* L.) as affected by growth of subterranean clover (*Trifolium subterraneum* L.) and infected by *Puccinia chondrillina* Bubak & Syd., *Australian Journal of Agricultural Research*, **26**, 975–83.

147. Grubb, P.J. and E.V.J. Tanner (1976) The montane forest and soils of Jamaica: a re-assessment, *Journal of the Arnold Arboretum*, **57**, 313–68.

148. Guillarmod, A.J. (1980) Kariba weed, in *Plant Invaders: beautiful but dangerous*, 2nd ed, (ed. C.H. Stirton), Department of Nature and Environmental Conservation of the Cape Provincial Administration, Cape Town, pp. 132–5.

149. Guillarmod, A.J. (1980) Parrot's feather, in *Plant Invaders: beautiful but dangerous*, 2nd ed, (ed. C.H. Stirton), The Department of Nature and Environmental Conservation of the Cape Provincial Administration, Cape Town, pp. 96–9.

150. Haggar, J.P.C. (1988) The structure, composition and status of the cloud forests of the Pico Island in the Azores, *Biological Conservation*, **46**, 7–22.

151. Hamann, O. (1984) Changes and threats to the vegetation, in *Key Environments: Galapagos*, (ed. R. Perry), Pergamon Press, Oxford, pp. 115–31.

152. Hara, H. (1966) *Flora of the Eastern Himalaya*, University of Tokyo Press, Tokyo.

153. Harding, G. (1980) Mesquite, in *Plant Invaders: beautiful but dangerous*, (ed. C.H. Stirton), Department of Nature and Environmental Conservation for the Cape Provincial Administration, Cape Town, pp. 128–31.

154. Harper, J.L. (1977) *Population Biology of Plants*, Academic Press, New York.

155. Harris, P. (1979) Cost of biological control of weeds by insects, *Weed Science*, **27**, 242–50.

156. Harris, P. (1984) Current approaches to biological control of weeds, in *Biological*

Control Programmes Against Insects and Weeds, (ed. J.S. Kelleher and M.A. Hulme), CAB, Slough, pp. 95–104.

157. Harris, P. (1984) *Euphorbia esula-virgata* complex, Leafy Spurge and *E. cyparissias* L., Cypress spurge (Euphorbiaceae), in *Biological Control against Insects and Weeds in Canada 1969–1980*, (ed. J.S. Kelleher and M.A. Hulme), CAB, Slough, pp. 159–69.

158. Haslewood, E.L. and G.G. Motter, ed. (1984) *Handbook of Hawaiian Weeds*, University of Hawaii Press, Honolulu.

159. Hathaway, D.E. (1989) *Molecular Mechanisms of Herbicide Selectivity*, Oxford University Press, Oxford.

160. Henderson, L. (1991) Invasive alien woody plants of the northern Cape, *Bothalia*, **21**, 177–89.

161. Henderson, L. (1992) Invasive alien woody plants of the eastern Cape, *Bothalia*, **22**, 119–43.

162. Henderson, L. and K.J. Musil (1984) Exotic woody plant invaders of the Transvaal, *Bothalia*, **15**, 297–313.

163. Hengeveld, R. (1987) Theories on biological invasions, *Proceedings of the Koninklijke Nederlandse Akademie van Wetenschappen*, **C90**, 45–9.

164. Hengeveld, R. (1988) Mechanisms of biological invasions, *Journal of Biogeography*, **15**, 819–28.

165. Heywood, V.H. (1987) The changing role of the botanic garden, in *Botanic Gardens and the World Conservation Strategy*, (ed. D. Bramwell *et al.*), Academic Press, London, pp. 3–18.

166. Heywood, V.H. (1989) Pattern, extent and modes of invasions by terrestrial plants, in *Biological Invasion: a global perspective*, (ed. J.A. Drake *et al.*), John Wiley & Sons, Chichester, pp. 31–60.

167. Higashino, P.K., W. Guyer and C.P. Stone (1983) The Kilauea Wilderness marathon and Crater Rim runs: sole searching experiences, *Newsletter Hawaiian Botanical Society*, **22**, 25–8.

168. Hislop, E.C. (1987) Requirements for effective and efficient pesticide application, in *Rational Pesticide Use*, (ed. K.J. Brent and R.K. Atkin), Long Ashton Symposium Series (No. 9), Cambridge University Press, Cambridge.

169. Hobbs, R.J. (1989) The nature and effects of disturbance relative to invasions, in *Biological Invasions: a global perspective*, (ed. J.A. Drake *et al.*), John Wiley & Sons, Chichester, pp. 389–406.

170. Hodges, C.S. and D.S. Gardner (1985) Myrica faya: *potential biological contral agents*, Cooperative National Park Resources Studies Unit, University of Hawaii at Manoa, Department of Botany, Honolulu.

171. Hoffman, M.T. and D.T. Mitchell (1986) The root morphology of some legume spp. in the south-western Cape and the relationship of vesicular-arbuscular mycorrhizas with dry mass and phosphorus content of *Acacia saligna* seedlings, *South African Journal of Botany*, **52**, 316–20.

172. Hoffmann, J.H. (1990) Interactions between three weevil species in the biocontrol of *Sesbania punicea* (Fabaceae): the role of simulation models in evaluation, *Agriculture, Ecosystems, and Environment*, **32**, 77–87.

173. Hoffmann, J.H. and V.C. Moran (1988) The invasive weed *Sesbania punicea* in South Africa and prospects for its control, *South African Journal of Science*, **4**, 740–3.

174. Hoffmann, J.H. and V.C. Moran (1989) Novel graphs for depicting herbivore damage on plants: the biological control of *Sesbania punicea* (Fabaceae) by an introduced weevil, *Journal of Applied Ecology*, **26**, 353–60.

175. Hoffmann, J.H. and V.C. Moran (1991) Biocontrol of a perennial legume, *Sesbania punicea*, using a florivorous weevil, *Trichapion lativentre*: weed population dynamics with a scarcity of seeds, *Oecologia*, **88**, 574–6.

176. Holdridge, L.R. (1967) *Life Zone Ecology*, Tropical Science Center, Costa Rica.

177. Holm, L. *et al.* (1979) *A Geographical Atlas of World Weeds*, John Wiley & Sons, New York.
178. Holm, L.G. *et al.* (1977) *The World's Worst Weeds*, University Press of Hawaii, Honolulu.
179. Holmes, P.M., H. Dallas and T. Phillips (1987) Control of *Acacia saligna* in the south western Cape – are clearing treatments effective?, *Veld & Flora*, **73**, 98–100.
180. Holmes, P.M., I.A.W. Macdonald and J. Juritz (1987) Effects of clearing treatment on seed banks of the alien invasive shrubs *Acacia saligna* and *Acacia cyclops* in the Southern and South Western Cape, South Africa, *Journal of Applied Ecology*, **24**, 1045–51.
181. Horn, C.N. (1987) Pontederiaceae, in *Flora del Paraguay*, (ed. R. Spichiger), Missouri Botanical Garden, pp. 1–28.
182. Howard-Williams, C. and J. Davies (1988) The invasion of Lake Taupo by the submerged water weed *Lagarosiphon major* and its impact on the native flora, *New Zealand Journal of Ecology*, **11**, 13–19.
183. Huenneke, L.F. and P.M. Vitousek (1990) Seedling and clonal recruitment of the invasive tree *Psidium cattleianum*: implications for management of native Hawaiian forests, *Biological Conservation*, **53**, 199–211.
184. Hughes, C.E. and B.T. Styles (1989) The benefits and risks of woody legume introductions, *Monographs in Systematic Botany*, Missouri Botanic Garden, 505–31.
185. Hulbert, L.C. (1955) Ecological studies of *Bromus tectorum* and other annual brome-grasses, *Ecological Monographs*, **25**, 181–213.
186. Humbert, H., ed. (1951) *Flore de Madagascar et des Comores: Mélastomatacées*, Muséum National d'Histoire Naturelle, Paris.
187. Humphries, S.E., R.H. Groves and D.S. Mitchell (1991) *Plant Invasions of Australian Ecosystems: a status review and management directions*, Endangered Species Program, Project No. 58, Australian National Parks and Wildlife Service, Canberra.
188. Hunter, G.G. and M.H. Douglas (1984) Spread of exotic conifers on South Island Rangelands, *New Zealand Journal of Forestry*, **29**, 78–96.
189. IUCN (1987) *The IUCN Position Statement on Translocation of Living Organisms*, IUCN, Gland.
190. Janzen, D.H., ed. (1983) *Costa Rican Natural History*, The University of Chicago Press, Chicago.
191. Jarvis, P.J. (1977) The ecology of plant and animal introductions, *Progress in Physical Geography*, **3**, 187–214.
192. Joenje, W., K. Bakker and L. Vlijm, ed. (1986) *The Ecology of Biological Invasions*, Proceedings of the Koninklijke Nederlandse Akademie van Wetenschappen, **C90**.
193. Johnson, C. (1980) Australian Myrtle, in *Plant Invaders: beautiful but dangerous*, 2nd ed, (ed. C.H. Stirton), The Department of Nature and Environmental Conservation of the Cape Provincial Administration, Cape Town.
194. Johnstone, I.M. (1986) Plant invasion windows: a time-based classification of invasion potential, *Biological Reviews*, **61**, 369–94.
195. Jones, W.I. (1974) *A Rhododendron eradication trial*, Nature Conservancy Council, Information Paper No. 1.
196. Julien, M.H., ed. (1987) *Biological Control of Weeds: a world catalogue of agents and their target weeds*, CAB International Institute of Biological Control, Ascot.
197. Kelly, D. and J.P. Skipworth (1984) *Tradescantia fluminensis* in a Manawatu (New Zealand) forest I, *New Zealand Journal of Botany*, **22**, 393–7.
198. Kelly, D. and J.P. Skipworth (1984) *Tradescantia fluminensis* in a Manawatu (New Zealand) forest II, *New Zealand Journal of Botany*, **22**, 399–402.
199. Kelly, D.L. (1981) The native forest vegetation of Killarney, south-west Ireland: an ecological account, *Journal of Ecology*, **69**, 437–72.
200. Kennedy, P.C. (1984) The general morphology and ecology of *Clematis vitalba*, in *The Clematis vitalba Threat*, DSIR, Wellington, pp. 26–35.

201. Kenney, D.S. (1986) De Vine – The way it was developed – An industrialist's view, *Weed Science*, **34**, (Suppl. 1), 15–16.
202. Khoshoo, T.F. and C. Mahal (1967) Versatile reproduction of *Lantana camara*, *Current Science*, **36**, 201–3.
203. Killip, E.P. (1938) The American species of Passifloraceae, *Publications of the Field Museum Natural History, Botanical Series*, **19**, 613.
204. Kloot, P.M. (1983) The role of the common iceplant (*Mesembryanthemum crystallinum*) in the deterioration of mesic pastures, *Australian Journal of Ecology*, **8**, 301–6.
205. Kloot, P.M. (1987) The invasion of Kangaroo Island by alien plants, *Australian Journal of Ecology*, **12**, 263–6.
206. Kluge, R.L. (1983) Progress in the fight against hakea, *Veld & Flora*, **69**, 136–8.
207. Kluge, R.L. and P.M. Caldwell (1991) Alarming new records of *Pereskia*, *Veld & Flora*, **77**, 39.
208. Kluge, R.L. and S. Neser (1991) Biological control of *Hakea sericea* (Proteaceae) in South Africa, *Agriculture, Ecosystems and Environment*, **37**, 91–113.
209. Knight, R.S. and I.A.W. Macdonald (1991) Acacias and korhans: an artificially assembled seed dispersal system, *South African Journal of Botany*, **57**, 220–5.
210. Kohler, A., and H. Sukopp (1964) Uber die soziologische Struktur einiger Robinienbestände im Stadtgebiet von Berlin, *Sonderdruck aus Sitzungsberichte der Gesellschaft Naturforschender Freunde zu Berlin*, **4**(2), 74–88.
211. Koopowitz, H. and H. Kaye (1990) *Plant Extinction: A Global Crisis*, Christopher Helm, London.
212. Kornas, J. (1982) Man's impact upon the flora: processes and effects, *Memorabilia Zoologica*, **37**, 11–30.
213. Kornberg, H. and M.H. Williamson, ed. (1986) Quantitative aspects of the ecology of biological invasions, *Philosophical Transactions of the Royal Society, London, Series B*, **341**.
214. Kruger, F.J. (1977) Invasive woody plants in the Cape fynbos with special reference to the biology and control of *Pinus pinaster*, Proceedings of the Second National Weeds conference in South Africa, Balkema, Cape Town, pp. 57–74.
215. Kruger, F.J. (1980) Cluster Pine, in *Plant Invaders: beautiful but dangerous*, 2nd ed, (ed. C.H. Stirton), The Department of Nature and Environmental Conservation of the Cape Provincial Administration, Cape Town, pp. 124–7.
216. Kruger, F.J. *et al.* (1989) The characteristics of invaded mediterranean-climate regions, in *Biological Invasions: a global perspective*, (ed. J.A. Drake *et al.*), John Wiley & Sons, Chichester, pp. 181–214.
217. Kruger, F.J., D.M. Richardson and B.W. van Wilgen (1986) Processes of invasion by plants, in *The Ecology and Management of Biological Invasions in South Africa*, (ed. I.A.W. Macdonald, F.J. Kruger and A.A. Ferrar), Oxford University Press, Cape Town, pp. 145–55.
218. Kugler, H. (1980) Zur Bestäubung von *Lantana camara* L. (On the pollination of *Lantana camara* L.), *Flora*, **169**, 524–9.
219. Kushwaha, S.P.S., P.S. Ramakrishnan and R.S. Tripathi (1981) Population dynamics of *Eupatorium odoratum* in successional environments following slash and burn agriculture, *Journal of Applied Ecology*, **18**, 529–35.
220. Laroche, F.B. and A.P. Ferriter (1992) The rate of expansion of *Melaleuca* in South Florida, *Journal of Aquatic Plant Management*, **30**, 62–5.
221. LaRosa, A. (1987) Note on the identity of the introduced passion flower vine 'banana poka' in Hawaii, *Pacific Science*, **39**, 369–71.
222. LaRosa, A.M. (1984) The biology and ecology of *Passiflora mollissima* in Hawaii, Technical Report No.50, Cooperative National Park Resources Studies Unit, Hawaii.
223. LaRosa, A.M. (1992) The status of Banana Poka in Hawaii, in *Alien Plant Invasions in Native Ecosystems of Hawaii: management and research*, (ed. C.P. Stone, C.W. Smith

and J.T. Tunison), University of Hawaii Cooperative National Park Resources Studies Unit, Honolulu, pp. 000–000.

224. LaRosa, A.M., C.W. Smith and D.E. Gardner (1985) Role of alien and native birds in the dissemination of fire tree (*Myrica faya* Ait. – Myricaceae) and associated plants in Hawaii, *Pacific Science*, **39**, 372–7.

225. Lawton, J.H. (1988) Biological control of bracken in Britain: constraints and opportunities, *Philosophical Transactions of the Royal Society, London, Series B*, **318**, 335–54.

226. Leader-Williams, N., R.I.L. Smith and P. Rothery (1987) Influence of introduced reindeer on the vegetation of South Georgia: results from a long-term exclusion experiment, *Journal of Applied Ecology*, **24**, 801–22.

227. Lee, M.A.B. (1974) Distribution of native and invader plant species on the island of Guam, *Biotropica*, **6**, 158–64.

228. Lee, W.G., R.B. Allen and P.W. Johnson (1986) Succession and dynamics of gorse (*Ulex europaeus* L.) communities in the Dunedin ecological district, South Island, New Zealand, *New Zealand Journal of Botany*, **24**, 279–92.

229. Lewis, G.P. (1987) *Legumes of Bahia*, Royal Botanic Gardens, Kew.

230. Lock, M. (1989) *Legumes of Africa: a check-list*, Royal Botanic Gardens, Kew.

231. Lonsdale, W.M. (1988) Litterfall of an Australian population of Mimosa pigra an invasive tropical shrub, *Journal of Tropical Ecology*, **4**, 381–92.

232. Lonsdale, W.M. and D.G. Abrecht (1989) Seedling mortality in Mimosa pigra, an invasive tropical shrub, *Journal of Ecology*, **77**, 372–85.

233. Lonsdale, W.M., K.L.S. Harley and J.D. Gillet (1988) Seed bank dynamics in Mimosa pigra, an invasive tropical shrub, *Journal of Applied Ecology*, **25**, 963–76.

234. Lonsdale, W.M., I.L. Miller and I.W. Forno (1989) The biology of Australian weeds 20. Mimosa pigra L., *Plant Protection Quarterly*, **4**, 119–31.

235. Loope, L.L., O. Hamann and C.P. Stone (1988) Comparative conservation biology of oceanic archipelagos: Hawaii and the Galapagos, *Bioscience*, **38**, 272–82.

236. Loope, L.L. and D. Mueller-Dombois (1989) Characteristics of invaded islands with special reference to Hawaii, in *Biological Invasions: a global perspective*, (ed. J.A. Drake et al.), John Wiley & Sons, Chichester, pp. 257–80.

237. Loope, L.L. et al. (1988) Biological invasions of arid land nature reserves, *Biological Conservation*, **44**, 95–118.

238. Lorence, D.H. (1978) The pteridophytes of Mauritius (Indian Ocean): ecology and distribution, *Botanical Journal of the Linnean Society*, **76**, 207–47.

239. Lorence, D.H. and R.W. Sussman (1986) Exotic species invasion into Mauritius wet forest remnants, *Journal of Tropical Ecology*, **2**, 147–62.

240. Lorence, D.H. and R.W. Sussman (1988) Diversity, density, and invasion in a Mauritian wet forest, *Monographs of Systematics of the Missouri Botanical Garden*, **25**, 187–204.

241. Lucas, G. and H. Synge (1978) *The IUCN Red Data Book*, IUCN, Switzerland.

242. Mabberley, D.J. (1987) *The Plant-Book*, Cambridge University Press, Cambridge.

243. Macdonald, I.A.W. (1983) Alien trees, shrubs and creepers invading indigenous vegetation in the Hluhluwe-Umfolozi Game Reserve Complex in Natal, *Bothalia*, **14**, 949–59.

244. Macdonald, I.A.W. (1986) Range expansion in the pied barbet and the spread of alien tree species in southern Africa, *Ostrich*, **57**, 75–94.

245. Macdonald, I.A.W. (1987) Banana poka in the Knysna forest, *Veld & Flora*, **73**, 133–4.

246. Macdonald, I.A.W. (1987) State of the art in the science of wildlife management: a North–South comparison, *South African Journal of Science*, **83**, 397–9.

247. Macdonald, I.A.W. (1988) The history, impacts and control of introduced species in the Kruger National Park, South Africa, *Transactions of the Royal Society South Africa*, **46**, 252–76.

248. Macdonald, I.A.W. and G.W. Frame (1988) The invasion of introduced species into nature reserves in tropical savannas and dry woodlands, *Biological Conservation*, **44**, 67–93.

249. Macdonald, I.A.W. and W.P.D. Gertenbach (1988) A list of alien plants in the Kruger National Park, *Koedoe*, **31**, 137–50.

250. Macdonald, I.A.W. *et al.* (1988) Introduced species in nature reserves in Mediterranean-type climatic regions of the world, *Biological Conservation*, **44**, 37–66.

251. Macdonald, I.A.W. and M.L. Jarman (1985) *Invasive Alien Plants in the Terrestrial Ecosystems of Natal, South Africa*, South African National Scientific Programmes Report No. 118, CSIR, Pretoria.

252. Macdonald, I.A.W., F.J. Kruger and A.A. Ferrar, ed. (1986) *The Ecology and Management of Biological Invasions in Southern Africa*, Oxford University Press, Cape Town.

253. Macdonald, I.A.W. *et al.* (1989) Wildlife conservation and the invasion of nature reserves by introduced species: a global perspective, in *Biological Invasions: a global perspective*, (ed. J.A. Drake *et al.*), John Wiley & Sons, Chichester, pp. 215–56.

254. Macdonald, I.A.W. and T.B. Nott (1987) Invasive alien organisms in central south west Africa/Namibia: results of a reconnaisance survey conducted in November 1984, *Madoqua*, **15**, 21–34.

255. Macdonald, I.A.W. *et al.* (1988) The invasion of highlands in Galápagos by the red quinine tree *Cinchona succirubra*, *Environmental Conservation*, **15**, 215–20.

256. Macdonald, I.A.W. and C. Wissel (1992) Determining optimal clearing treatments for the alien invasive shrub *Acacia saligna* in the southwestern Cape, South Africa, *Agriculture, Ecosystems and Environment*, **39**, 169–86.

257. Mack, R.N. (1981) Invasion of *Bromus tectorum* L. into western North America: an ecological chronicle, *Agro-Ecosystems*, **7**, 145–65.

258. Mack, R.N. (1984) Invaders at home on the range, *Natural History*, **2**, 40–7.

259. Mack, R.N. (1985) Invading plants: their potential contribution to population biology, in *Studies on Plant Demography: a Festschrift for John L. Harper*, (ed. J. White), Academic Press, London, pp. 127–41.

260. Madsen, J.D. *et al.* (1991) The decline of native vegetation under dense Eurasian watermilfoil canopies, *Journal of Aquatic Plant Management*, **29**, 94–9.

261. Maheshwari, J.K. (1965) Alligator weed in Indian lakes, *Nature*, **206**, 1270.

262. Makepeace, W. (1981) Polymorphism and the chromosomal number of *Hieracium pilosella* L. in New Zealand, *New Zealand Journal of Botany*, **19**, 255–7.

263. Markin, G.P. (1989) Alien plant management by biological control, in *Conservation Biology in Hawaii*, (ed. C.P. Stone and D.B. Stone), Cooperative National Park Resources Studies Unit, Hawaii, pp. 70–3.

264. Martin, F.W. and H.Y. Nakasone (1970) The edible species of *Passiflora*, *Economic Botany*, **24**, 333–43.

265. Maslin, B.R. (1974) Studies in the genus *Acacia*, 3. The taxonomy of *Acacia saligna* (Labill.) H.Wendl., *Nuytsia*, **2**, 332–40.

266. Mason, R. (1960) Three waterweeds of the family Hydrocharitaceae in New Zealand, *New Zealand Journal of Science*, **3**, 383–95.

267. Mather, L.J. and P.A. Williams (1990) Phenology, seed ecology, and age structure of Spanish heath (*Erica lusitanica*) in Canterbury, New Zealand, *New Zealand Journal of Botany*, **28**, 207–15.

268. Mathur, G. and H.H. Mohan Ram (1978) Significance of petal colour in thrips-pollinated *Lantana camara* L., *Annals of Botany*, **42**, 1473–6.

269. McClune, B. (1988) Ecological diversity in North American pines, *American Journal of Botany*, **75**, 353–68.

270. McLoughlin, L. and J. Rawling (1990) *Making Your Garden Bush Friendly: how to recognise and control garden plants which invade Sydney's bushland*, McLoughlin-Rawling, Killara, NSW.

271. Meurk, C.D. (1977) Alien plants in Campbell Island's changing vegetation, *Mauri Ora*, **5**, 93–118.

272. Miller, I.L. (1983) The distribution and threat of *Mimosa pigra* in Australia, International Plant Protection Center, Corvallis, Oregon, USA.

273. Miller, I.L. and W.M. Lonsdale (1987) Early records of *Mimosa pigra* in the Northern Territory, *Plant Protection Quarterly*, **2**, 140–2.

274. Milton, S.J. (1980) Australian Acacias in the south western Cape: preadaption, predation and success.

275. Miranda, F. and A.J. Sharp (1950) Characteristics of the vegetation in certain temperate regions of eastern Mexico, *Ecology*, **31**, 313–33.

276. Mirov, N.T. (1967) *The Genus* Pinus, The Ronald Press Company, New York.

277. Mitchell, D.S. (1972) The kariba weed: *Salvinia molesta*, *British Fern Gazette*, **10**, 251–2.

278. Mitchell, D.S. (1980) The water fern *Salvinia molesta* in the Sepick River, Papua New Guinea, *Environmental Conservation*, **7**, 115–22.

279. Mitchell, D.S. and B. Gopal (1991) Invasion of tropical freshwaters by alien aquatic weeds, in *Ecology of Biological Invasion in the Tropics*, (ed. P.S. Ramakrishnan), International Scientific Publications, New Delhi, pp. 139–56.

280. Mitchell, D.S. and N.N. Tur (1975) The rate and growth of *Salvinia molesta* (*S. auriculata*) in laboratory and in natural conditions, *Journal of Applied Ecology*, **12**, 212–25.

281. Mohan Ram, H.Y. and G. Mathur (1984) Flower colour changes in *Lantana camara*, *Journal of Experimental Botany*, **35**, 1656–62.

282. Moll, E.J. (1980) Blackwood, in *Plant Invaders: beautiful but dangerous*, 2nd ed, (ed. C.H. Stirton), Department of Nature and Environmental Conservation of the Cape Provincial Administration, Cape Town, pp. 52–5.

283. Mollison, D. (1986) Modelling biological invasion: chance, explanation, prediction, *Philosophical Transactions of the Royal Society, London, Series B*, **314**, 675–92.

284. Moody, M.E. and R.N. Mack (1988) Controlling the spread of plant invasions: the importance of nascent foci, *Journal of Applied Ecology*, **25**, 1009–21.

285. Mooney, H.A. and J.A. Drake, ed. (1986) *Ecology of Biological Invasions of North America and Hawaii*, Springer-Verlag, New York.

286. Moran, V.C. and J.H. Hoffmann (1989) The effects of herbivory by a weevil species acting alone and unrestrained by natural enemies on growth and phenology of the weed *Sesbania punicea*, *Journal of Applied Ecology*, **26**, 967–77.

287. Moriarty, A. (1982) *Outeniqua Tsitsikamma & Eastern Little Karoo*, Botanical Society of South Africa, Kirstenbosch.

288. Morris, J.M. (1982) Biological control of hakea by a fungus, *Veld & Flora*, **68**, 51–2.

289. Morris, M.J. (1982) Gummosis and die-back of *Hakea sericea* in South Africa, in *Proceedings of the Fourth National Weeds Conference of South Africa*, (ed. H.A. van der Venter and M. Mason), Balkema, Cape Town, pp. 51–4.

290. Motooka, P., G. Nagai and L. Ching (1982) Weed and brush control in pasture and ranges of Hawaii 1–5, *HITAHR Brief*, 16–20.

291. Mueller-Dombois, D. (1973) A non-adapted vegetation interferes with water removal in a tropical rainforest area in Hawaii, *Tropical Ecology*, **14**, 1–16.

292. Mueller-Dombois, D. *et al.* (1980) Ohi'a rainforest study: investigation of the Ohi'a dieback problem in Hawaii, *Hawaii Agricultural Experimental Station Miscellaneous Publications*, **183**, 64.

293. Mueller-Dombois, D. and L.D. Whiteaker (1990) Plants associated with *Myrica faya* and two other pioneer trees on a recent volcanic surface in Hawaii Volcanoes National Park, *Phytocoenologia*, **19**, 29–41.

294. Munton, P. (1988) *The Role, Impact and Management of the Translocation of Living Organisms in the Context of Wildlife Resources (Draft Guidelines)*, Report of the Introduction Specialist Group of the Species Survival Commission,

223

295. Munz, P.A. (1959) *A California Flora*, University of California Press, Berkeley.
296. Myers, R.L. (1983) Site susceptibility to invasion by the exotic tree *Melaleuca quinquenervia* in southern Florida, *Journal of Applied Ecology*, **20**, 645–58.
297. Nakahara, L.M., R.M. Burkhart and J.Y. Funasaki (1992) Review and status of biological control of *Clidemia* in Hawaii, in *Alien Plant Invasions in Native Ecosystems of Hawaii: management and research*, (ed. C.P. Stone, C.W. Smith and J.T. Tunison), University of Hawaii Cooperative National Park Resources Studies Unit, Honolulu.
298. Natarajan, A.T. and M.R. Ahuja (1947) Cytotaxonomical studies of the genus *Lantana*, *Journal of the Indian Botanical Society*, **36**, 35–45.
299. National Trust, NSW (1991) *Bush Regenerators Handbook*, National Trust of Australia (NSW), New South Wales.
300. Naveh, Z. (1967) Mediterranean ecosystems and vegetation types in California and Israel, *Ecology*, **48**, 443–59.
301. Neser, S. (1980) Rock hakea, in *Plant Invaders: beautiful but dangerous*, 2nd ed, (ed. C.H. Stirton), The Department of Nature and Environmental Conservation of the Cape Provincial Administration, Cape Town, pp. 72–5.
302. Neser, S. (1980) Sweet hakea, in *Plant Invaders: beautiful but dangerous*, 2nd ed, (ed. C.H. Stirton), The Department of Nature and Environmental Conservation of the Cape Provincial Administration, Cape Town, pp. 80–3.
303. Neser, S. and S.R. Fugler (1980) Silky hakea, in *Plant Invaders: beautiful but dangerous*, 2nd ed, (ed. C.H. Stirton), The Department of Nature and Environmental Conservation of the Cape Provincial Administration, Cape Town, pp. 76–7.
304. Newbold, C. (1975) Herbicides in aquatic systems, *Biological Conservation*, **7**, 97–118.
305. Newbold, C. (1977) Aquatic herbicides: possible future developments, in *Ecological Effects of Pesticides*, (ed. F.H. Perring and K. Mellanby), Academic Press, London, Linnean Society Symposium Series, Number 5, pp. 000–000.
306. Noble, I.R. (1989) Attributes of invaders and the invading process: terrestrial and vascular plants, in *Biological Invasions: a global perspective*, (ed. J.A. Drake *et al.*), John Wiley & Sons, Chichester, pp. 301–14.
307. Ohwi, J. (1965) *Flora of Japan*, Smithsonian Institution, Washington DC.
308. Osmond, C.B. and J. Monro (1981) Prickly pear, in *Plants and Man in Australia*, (ed. D.J. Carr and S.G.M. Carr), Academic Press, Sydney, pp. 194–222.
309. Parham, J.W. (1964) *Plants of the Fiji Islands*, Government Press, Suva.
310. Parsons, W.T. (1973) *Noxious Weeds of Victoria*, Inkata Press, Melbourne.
311. Pathak, P.S., R. Debroy and P. Rai (1974) Autecology of *Leucaena leucocephala* (Lam.) de Wit 1. Seed production and germination, *Tropical Ecology*, **15**, 1–11.
312. Pedley, L. (1986) Derivation and dispersal of *Acacia* (Legumisosae), with particular reference to Australia, and the recognition of *Senegalia* and *Racosperma*, *Botanical Journal of the Linnean Society*, **92**, 219–54.
313. Penfound, W.T. (1948) The biology of the water hyacinth, *Ecological Monographs*, **18**, 447–72.
314. Pickard, J. (1984) Exotic plants on Lord Howe Island: distribution in space and time, 1853–1981, *Journal of Biogeography*, **11**, 181–208.
315. Pienaar, K. (1980) Sesbania, in *Plant Invaders: beautiful but dangerous*, 2nd ed, (ed. C.H. Stirton), The Department of Nature and Environmental Conservation of the Cape Provincial Administration, Cape Town, pp. 136–9.
316. Pienaar, K.J. (1977) *Sesbania punicea* (Cav.) Benth. The handsome plant terrorist, *Veld & Flora*, **63**, 17–18.
317. Popay, A.I. (1984) Prospects for herbicidal control of *Clematis vitalba*, in *The Clematis vitalba Threat*, DSIR, Wellington, pp. 43–6.
318. Popay, A.I. (1986) *Old Man's Beard* Clematis vitalba: *control measures*, Farm Production and Practice Report, Ministry of Agriculture and Fisheries, New Zealand.

319. Proctor, V.W. (1968) Long-distance dispersal of seeds by retention in the digestive tract of birds, *Science*, **160**, 321–2.

320. Ramakrishnan, P.S. (1991) Biological invasion in the tropics: an overview, in *Ecology of Biological Invasions in the Tropics*, (ed. P.S. Ramakrishnan), International Scientific Publications, New Delhi, pp. 1–20.

321. Ramakrishnan, P.S. and P.M. Vitousek (1989) Ecosystem-level processes and the consequences of biological invasions, in *Biological Invasions: a global perspective*, (ed. J.A. Drake *et al.*), John Wiley & Sons, Chichester, pp. 281–300.

322. Reiche, C. (1896) *Flora de Chile*, Vol. 1, Bandera, Santiago de Chile.

323. Reimer, N.J. (1988) Predation on *Liothrips urichi* Karny (Thysanoptera: Phlaeothripidae): a case of biotic interference, *Environmental Entomology*, **17**, 132–4.

324. Rejmanek, M. (1989) Invasibility of plant communities, in *Biological Invasions: a global perspective*, (ed. J.A. Drake *et al.*), John Wiley & Sons, Chichester, pp. 369–88.

325. Richards, A.J. (1986) *Plant Breeding Systems*, George Allen & Unwin, London.

326. Richards, P. W. and A. J. E. Smith (1975) A progress report on *Campylopus introflexus* (Hedw.) Brid. and C. *polytrichoides* De Not. in Britain and Ireland. *Journal of Bryology*, **8**, 293–8.

327. Richardson, D.M. and W.J. Bond (1991) Determinants of plant distribution: evidence from pine invasions, *The American Naturalist*, **137**, 639–68.

328. Richardson, D.M. and M.P. Brink (1985) Notes of *Pittosporum undulatum* in the south western Cape, *Veld & Flora* **71**, 75–7.

329. Richardson, D.M. and P.J. Brown (1986) Invasion of mesic mountain fynbos by *Pinus radiata*, *South African Journal of Botany*, **52**, 529–36.

330. Richardson, D.M. and R.M. Cowling (1992) Why is mountain fynbos invasible and which species invade?, in *Fire in South African Mountain Fynbos*, (ed. B.W. van Wilgen *et al.*), Springer-Verlag, Berlin, pp. 161–81.

331. Richardson, D.M., R.M. Cowling and D.C. Le Maitre (1990) Assessing the risk of invasive success in *Pinus* and *Banksia* in South African mountain fynbos, *Journal of Vegetation Science*, **1**, 629–42.

332. Richardson, D.M. and P.T. Manders (1985) Predicting pathogen-induced mortality in *Hakea sericea* (Proteacea) an aggressive alien plant invader in South Africa, *Annals of Applied Biology*, **106**, 243–54.

333. Richardson, D.M. and B.W. van Wilgen (1984) Factors affecting the regeneration success of *Hakea sericea*, *South African Forestry Journal*, **131**, 63–8.

334. Ridings, W.H. (1986) Biological control of strangler vine in citrus – A researcher's view, *Weed Science*, **34**(Suppl. 1), 31–2.

335. Ridley, H.N. (1923) *Flora of the Malay Peninsula*, Vol. 1, L. Reeves, London.

336. Room, P.M. (1983) 'Falling apart' as a lifestyle: the rhizome architecture and population growth of *Salvinia molesta Journal of Ecology*, **71**, 349–65.

337. Room, P.M. *et al.* (1981) Successful biological control of the floating weed *Salvinia*, *Nature*, **294**, 78–80.

338. Room, P.M. and P.A. Thomas (1986) Population growth of the floating weed *Salvinia molesta*: field observation and a global model based on temperature and nitrogen, *Journal of Applied Ecology*, **23**, 1013–28.

339. Rotherham, I.D. (1986) The introduction, spread and current distribution of *Rhododendron ponticum* in the Peak district and Sheffield area, *The Naturalist*, **111**, 61–7.

340. Rouw, A. De (1991) The invasion of *Chromolaena odorata* (L.) King & Robinson (ex *Eupatorium odoratum*), and competition with the native flora, in a rain forest zone, south-west Cote d'Ivoire, *Journal of Biogeography*, **18**, 13–23.

341. Sagar, G.R. (1974) On the ecology of weed control, in *Biology in Pest and Disease Control*, (ed. D. Price Jones and M.E. Solomon), Symposium of the British Ecological Society (13), Blackwell Scientific Publications, Oxford.

342. Sale, P.J.M. *et al.* (1985) Photosynthesis and growth rates on *Salvinia molesta* and *Eichhornia*, *Journal of Applied Ecology*, **22**, 125–37.

343. Sankaran, T. and E. Narayanan (1971) Occurrence of the alligator weed in South India, *Current Science*, **40**, 641.

344. Santos, G.L. *et al.* (1986) *Herbicidal control of selected alien plant species in Hawaii Volcanoes National Park*, Cooperative National Park Resources Studies Unit, University of Hawaii at Manoa, Technical Report 60, Honolulu.

345. Sauer, J.D. (1988) *Plant Migration: the dynamics of geographic patterning in seed plant species*, University of California Press, Berkeley.

346. Schofield, E.K. (1973) Galápagos flora: the threat of introduced plants, *Biological Conservation*, **5**, 48–51.

347. Schofield, E.K. (1989) Effects of introduced plants and animals on island vegetation: examples from the Galapagos archipelago, *Conservation Biology*, **3**, 227–38.

348. Schreiber, M.M. (1982) Modelling the biology of weeds for integrated weed management, *Weed Science*, **30** (Suppl.), 13–16.

349. Scott, D., J.S. Robertson and W.J. Archie (1990) Plant dynamics of New Zealand tussock grassland infested with *Hieracium pilosella*. 1. Effects of seasonal grazing, fertiliser, and overdrilling, *Journal of Applied Ecology*, **27**, 224–34.

350. Scowcroft, P.G. and K.T. Adee (1991) *Site Preparation Affects Survival, Growth of Koa on Degraded Montane Forest Land*, Pacific Southwest Research Station, Forest Service, USDA, Research Paper PSW-205, Berkeley, California.

351. Scowcroft, P.G. and P.G. Nelson (1976) *Disturbance After Logging Stimulates Regeneration of Koa*, Pacific South West Forest and Range Experimental Station, USDA Forest Research Notes 306.

352. Seratna, J.E. (1943) *Salvinia auriculata* Aublet – a recently introduced free-floating water weed, *Tropical Agriculture Magazine*, **99**, 146–9.

353. Shaughnessy, G. (1980) Historical Ecology of Alien Woody Plants in the Vicinity of Cape Town, South Africa, Ph.D. Thesis, Research Report No. 23, School of Environmental Studies, University of Cape Town, Cape Town.

354. Shaw, M.W. (1984) *Rhododendron ponticum* – ecological reasons for the success of an alien species in Britain and features that may assist in its control, *Aspects of Applied Biology*, **5**, 231–42.

355. Shimizu, Y. and H. Tabata (1985) Invasion of *Pinus luchuensis* and its influence on the native forest on a Pacific island, *Journal of Biogeography*, **12**, 195–207.

356. Sinha, S. and A. Sharma (1984) *Lantana camara* L. – a review, *Feddes Repertorium*, **95**, 621–33.

357. Small, J.K. (1913) *Flora of the South-Eastern States*, Published by the Author, New York.

358. Smathers, G.A. and D.E. Gardner (1979) Stand analysis of an invading firetree (*Myrica faya* Aiton) population, Hawaii, *Pacific Science*, **33**, 239–55.

359. Smith, C.W. (1985) Impact of alien plants on Hawaii's native biota, in *Hawaii's Terrestrial Ecosystems: preservation and management*, (ed. C.P. Stone and J.M. Scott), Cooperative National Park Resources Studies Unit, University of Hawaii, Honolulu.

360. Smith, C.W. (1989) Controlling the flow of non-native species, in *Conservation Biology in Hawaii*, (ed. C.P. Stone and Stone D.B.), University of Hawaii Cooperative National Park Resources Studies Unit, Honolulu, pp. 139–45.

361. Smith, C.W. (1989) Non-native plants, in *Conservation Biology in Hawaii*, (ed. C.P. Stone and D.B. Stone), University of Hawaii Cooperative National Park Resources Studies Unit, Honolulu, pp. 60–9.

362. Smith, C.W. (1992) Distribution, status, phenology, rate of spread, and management of *Clidemia* in Hawaii, in *Alien Plant Invasions in Native Ecosystems of Hawaii: management and research*, (ed. C.P. Stone, C.W. Smith and J.T. Tunison), University of Hawaii Cooperative National Park Resources Studies Unit, Honolulu.

363. Smith, D. (1984) *Clematis vitalba* – a paper on past control, in *The Clematis vitalba Threat*, DSIR, Wellington, pp. 47–9.

364. Smith, L.S. and D.A. Smith (1982) *The Naturalised* Lantana camara *Complex in*

Eastern Australia, Queensland Botany Bulletin, Report No. 1, Department of Primary Industries, Brisbane.

365. Soulé, M.E. (1990) The onslaught of alien species, and other challenges in the coming decades, *Conservation Biology*, **4**, 233–9.

366. Specht, R.L. (1970) Vegetation, in *The Australian Environment*, 4th ed, (ed. G.W. Leeper), CSIRO, Melbourne, pp. 44–67.

367. Spongberg, S.A. (1990) *A Reunion of Trees*, Harvard University Press, Cambridge, Mass.

368. Sprankle, P., W.F. Meggitt and E. Penner (1975) Rapid inactivation of glyphosate in the soil, *Weed Science*, **23**, 224–8.

369. Stanley, T.D. and T.M. Ross (1983) *Flora of South-eastern Queensland*, Vol. 1, Queensland Department of Primary Industries, Brisbane.

370. State Forester, Hawaii (*c.* 1980) *The Banana Poka Caper*, Department of Land and Natural Resources, Hawaii, Honolulu.

371. Stebbins, G.L. (1971) Adaptive radiation of reproductive characteristics in angiosperms, II: Seeds and seedlings, *Annual Review of Ecology and Systematics*, **2**, 237–60.

372. Steenis, C.G.G.J., ed. (1954) *Flora Malesiana*, Vol. 4, Noordhoff-Kolff N.V., Djakarta.

373. Steenis, C.G.G.J., ed. (1972) *Flora Malesiana*, Vol. 6, Wolters-Noordhoff Publishing Company, Groningen.

374. Steyermark, J.A. (1963) *Flora of Missouri*, The Iowa State University Press, Iowa.

375. Stirton, C.H. (1977) Some thoughts on the polyploid complex *Lantana camara* L. (Verbenaceae), in *Proceedings of the Second National Weeds Conference of South Africa*, Balkema, Cape Town, pp. 321–40.

376. Stirton, C.H. (1980) *Lantana*, in *Plant Invaders: beautiful but dangerous*, 2nd ed, (ed. C.H. Stirton), The Department of Nature and Environmental Conservation of the Cape Provincial Administration, Cape Town, pp. 88–91.

377. Stirton, C.H., ed. (1980) *Plant Invaders: beautiful but dangerous*, Department of Nature and Environmental Conservation of the Cape Provincial Administration, Cape Town.

378. Stockard, J., B. Nicholson and G. Williams (1985) An assessment of a rainforest regeneration program at Wingham Brush, *Victorian Naturalist*, **103**, 85–93.

379. Stone, C.P., L.W. Cuddihy and J.T. Tunison (1992) Responses of Hawaiian ecosystems to removal of feral pigs and goats, in *Alien Plant Invasions in Native Ecosystems of Hawaii: management and research*, (ed. C.P. Stone, C.W. Smith and J.T. Tunison), University of Hawaii Cooperative National Park Resources Studies Unit, Honolulu, pp. 666–704.

380. Strahm, W. (1988) The Mondrain Reserve and its conservation management, *Proceedings of the Royal Society of Arts and Science, Mauritius*, **5**, 139–77.

381. Strahm, W. (1990) Conservation of Endemic Plants of Mauritius and Rodrigues, Unpublished Progress Report of Project 3149, WWF/IUCN.

382. Sukopp, H. (1962) Neophyten in naturlichen Pflanzengesellschaften Mitteleuropas, *Sonderabdruck aus den Berichten der Deutschen Botanischen Gesellschaft*, **75**(6), 193–205.

383. Sukopp, H. and U. Sukopp (1988) *Reynoutria japonica* Houtt. in Japan und in Europa, *Veröffentlichungen des Geobotanisches Institut, Eidgenössiche Technische Hochschule, Stiftung Rübel, Zürich*, **98**, 354–72.

384. Sutherst, R.W. and G.F. Maywald (1985) A computerised system for matching climates in ecology, *Agriculture, Ecosystems and Environment*, **13**, 281–99.

385. Swarbrick, J.T., C.M. Finlayson and A.J. Cauldwell (1981) The biology of Australian weeds 7. *Hydrilla verticillata* (L.f.) Royle, *The Journal of the Australian Institute of Agricultural Science*, **47**, 183–90.

386. Tabbush, P.M. and D.R. Williamson (1987) *Rhododendron ponticum* as a forest weed, *Forestry Commission Bulletin*, **73**, 1–7.

387. Thaman, R.R. (1974) *Lantana camara*: its introduction, dispersal and impact on islands of the tropical Pacific Ocean, *Micronesica*, **10**, 17–39.
388. Thomas, K.J. (1975) Biological control of *Salvinia molesta* by the snail *Pila globosa* Swainson, *Biological Journal of the Linnean Society*, **18**, 263–78.
389. Thomas, K.J. (1979) The extent of *Salvinia* infestation in Kerala (South India): its impact and suggested methods of control, *Environmental Conservation*, **6**, 63–9.
390. Thomas, K.J. (1981) The role of aquatic weeds in changing the pattern of ecosystems in Kerala, *Environmental Conservation*, **8**, 63–6.
391. Thomas, P.A. and P.A. Room (1986) Taxonomy and control of *Salvinia molesta*, *Nature*, **320**, 581–4.
392. Thompson, D.Q., R.L. Stuckey and E.B. Thompson (1987) Spread, impact, and control of purple loosestrife (*Lythrum salicaria*) in North American wetlands, *Fish and Wildlife Research*, **2**, 1–55.
393. Thompson, J.D. (1991) The biology of an invasive plant: what makes *Spartina anglica* so successful?, *Bioscience*, **41**, 393–401.
394. Timmins, S. and P.A. Williams (1989) Reserve design and management for weed control, in *Alternatives to the Chemical Control of Weeds*, (ed. C. Bassett, L.J. White-house and J. Zabkiewicz), FRI Bulletin 155, Ministry of Forestry, Rotorua, New Zealand, pp. 133–8.
395. Timmins, S.M. and P.A. Williams (1987) Characteristics of problem weeds in New Zealand's protected natural areas, in *Nature Conservation: the role of remnants of native vegetation*, (ed. D.A. Saunders et al.), Surrey Beatty/CSIRO/CALM, pp. 241–7.
396. Tisdell, C.A., B.A. Auld, and K.M. Menz (1984) On assessing the value of biological control of weeds, *Protection Ecology*, **6**, 169–79.
397. Trewick, S. and P.M. Wade (1986) The distribution and dispersal of two alien species of *Impatiens*, waterway weeds in the British Isles, in *7th Symposium on Aquatic Weeds*, pp. 351–6.
398. Trujillo, E.E. (1986) *Colletotrichum gloeosporioides*, a possible control agent for *Clidemia hirta* in Hawaiian forests, *Plant Disease*, **70**, 974–6.
399. Tunison, J.T. (1992) Fountain grass control in Hawaii Volcanoes National Park: management considerations and strategies, in *Alien Plant Invasions in Native Ecosystems of Hawaii: management and research*, (ed. C.P. Stone, C.W. Smith and J.T. Tunison), University of Hawaii Cooperative National Park Resources Studies Unit, Honolulu.
400. Tunison, J.T. and C.P. Stone (1992) Special ecological areas: an approach to alien plant control in Hawaii Volcanoes National Park, in *Alien Plant Invasions in Native Ecosystems of Hawaii: management and research*, (ed. C.P. Stone, C.W. Smith and J.T. Tunison), University of Hawaii Cooperative National Park Resources Studies Unit, Honolulu, pp. 781–98.
401. Turner, D.R. and P.M. Vitousek (1987) Nodule biomass of the nitrogen-fixing alien *Myrica faya* Ait. in Hawaii Volcanoes National Park, *Pacific Science*, **41**, 186–90.
402. Turrill, W.B. (1929) *Plant Life of the Balkan Peninsula*, Clarendon Press, Oxford.
403. Tutin, T.G. et al., ed. (1972) *Flora Europaea*. Vol. 3, Cambridge University Press, Cambridge.
404. Usher, M.B. (1986) Invasibility and wildlife conservation: invasive species on nature reserves, *Philosophical Transactions of the Royal Society, London, Series B*, **314**, 695–709.
405. Usher, M.B. (1988) Biological invasions of nature reserves: a search for generalisations, *Biological Conservation*, **44**, 119–35.
406. Usher, M.B. (1991) Biological invasion into tropical nature reserves, in *Ecology of Biological Invasion in the Tropics*, (ed. P.S. Ramakrishnan), International Scientific Publications, New Delhi, pp. 21–34.
407. van den Tweel, P.A. and H. Eijsackers (1987) Black cherry, a pioneer species or 'forest pest', *Proceedings of the Koninklijke Nederlandse Akademie van Wetenschappen*, **C90**, 59–66.

408. van der Maesen, L.J.G. (1985) Revision of the genus *Pueraria* DC. with some notes on *Teylreia* Backer, *Agricultural University Wageningen Papers*, **85**, 3–132.
409. Van Wilgen, B.W. and D.M. Richardson (1985) The effects of alien shrub invasions on vegetation structure and fire behaviour in South African fynbos shrublands: a simulation study, *Journal of Applied Ecology*, **22**, 955–66.
410. van Wilgen, B.W. and W.R. Siegfried (1986) Seed dispersal properties of three pine species as a determinant of invasive potential, *South African Journal of Botany*, **52**, 546–8.
411. Veblen, T.T. (1975) Alien weeds in the tropical highlands of Western Guatemala, *Journal of Biogeography*, **2**, 19–26.
412. Vere, D.T. and B.A. Auld (1982) The cost of weeds, *Protection Ecology*, **4**, 29–42.
413. Vitousek, P.M. (1988) Diversity and biological invasions of oceanic islands, in *Biodiversity*, (ed. E.O. Wilson), National Academy Press, Washington, DC, pp. 181–92.
414. Vitousek, P.M. (1989) Biological invasion by *Myrica faya* in Hawaii: plant demography, nitrogen fixation, ecosystem effects, *Ecological Monographs*, **59**, 247–65.
415. Vitousek, P.M. (1990) Biological invasions and ecosystems processes: towards an integration of population biology and ecosystem studies, *Oikos*, **57**, 7–13.
416. Vitousek, P.M., L.L. Loope and C.P. Stone (1987) Introduced species in Hawaii: biological effects and opportunities for ecological research, *Trends in Ecology and Evolution*, **2**, 224–7.
417. Vitousek, P.M. *et al.* (1987) Biological invasion by *Myrica faya* alters ecosystem development in Hawaii, *Science*, **238**, 802–4.
418. Vivrette, N.J. and C.H. Muller (1977) Mechanisms of invasion and dominance of coastal grasslands by *Mesembryanthemum crystallinum*, *Ecological Monographs*, **47**, 301–18.
419. Waage, J.K. (1990) Ecological theory and the selection of biological control agents, in *Critical Issues in Biological Control*, (ed. M. Mackauer, L.E. Ehler and J. Roland), Intercept, Andover, pp. 135–57.
420. Wace, N. (1986) The arrival, establishment and control of alien plants on Gough Island, *South African Journal of Antarctic Research*, **16**, 95–101.
421. Wager, V.A. (1927) The structure and life history of South African *Lagarosiphons*, *Transactions of the Royal Society of South Africa*, **16**, 191–212.
422. Wagner, W.L., D.R. Herbst and R.S.N. Yee (1984) Status of the native flowering plants of the Hawaiian Islands, in *Hawaii's Terrestrial Ecosystems: preservation and management*, (ed. C.P. Stone and J.M. Scott), Cooperative National Park Resources Studies Unit, University of Hawaii, Honolulu, pp. 23–74.
423. Walter, H. (1968) *Die vegetation der Erde*, Gustav Fischer Verlag, Stuttgart.
424. Walter, H. and H. Lieth (1960) *Klimmadiagramm-Weltatlas*, Gustav Fischer Verlag GmbH, Jena.
425. Warshauer, F.R. *et al.* (1983) The Distribution, Impact and Potential Management of the Introduced Vine, *Passiflora mollissima* (Passifloraceae) in Hawaii, Cooperative National Park Resources Studies Unit, Hawaii, Report No. 48.
426. Waterhouse, B.M. (1988) Broom (*Cytisus scoparius*) at Barrington tops New South Wales, *Australian Geographical Studies*, **26**, 239–48.
427. Watson, A.K. and M. Clement (1986) Evaluation of rust fungi as biological control agents of weedy centaurea in North America, *Weed Science*, **34**(Suppl. 1), 7–10.
428. Webb, C.J., W.R. Sykes and P.J. Garnock-Jones (1988) *Flora of New Zealand: naturalised pteridophytes, gymnosperms, dicotyledons*, Vol. 4, DSIR, Christchurch.
429. Weiss, P.W. and I.R. Noble (1984) Interactions between seedlings of *Chrysanthemoides monilifera* and *Acacia longifolia*, *Australian Journal of Ecology*, **9**, 107–15.
430. Weiss, P.W. and I.R. Noble (1984) Status of coastal dune communities invaded by *Chrysanthemoides monilifera*, *Australian Journal of Ecology*, **9**, 93–8.

431. Wells, M.J. (1980) Nassella Tussock, in *Plant Invaders: beautiful but dangerous*, 2nd ed, (ed. C.H. Stirton), The Department of Nature and Environmental Conservation of the Cape Provincial Administration, Cape Town, pp. 140–3.

432. Wells, R.D.S. and J.S. Clayton (1991) Submerged vegetation and spread of *Egeria densa* Planchon in Lake Rotorua, central North Island, New Zealand, *New Zealand Journal of Marine and Freshwater Research*, 25, 63–70.

433. West, C.J. (1991) *Literature Review of the Biology of* Clematis vitalba *(Old Man's Beard)*, Vegetation Report, DSIR Land Resources, Report No. 725, New Zealand.

434. West, R.G. (1988) A commentary on Quaternary cold stage floras in Britain, *Journal of Biogeography*, 15, 523–8.

435. Wester, L.L. and H.B. Wood (1977) Koster's curse (*Clidemia hirta*), a weed pest in Hawaiian forests, *Environmental Conservation*, 4, 35–41.

436. Westman, W.E. (1990) Park management of exotic plant species: problems and issues, *Conservation Biology*, 4, 251–60.

437. Whiteaker, L.D. and D.E. Gardner (1985) *The Distribution of* Myrica faya Ait. *in the State of Hawaii*, Technical Report No. 55, Cooperative National Park Resources Studies Unit, University of Hawaii at Manoa, Honolulu.

438. Whiteaker, L.D. and D.E. Gardner (1992) Firetree (*Myrica faya*) distribution in Hawaii, in *Alien Plant Invasions in Native Ecosystems of Hawaii: management and research*, (ed. C.P. Stone, C.W. Smith and J.T. Tunison), University of Hawaii Cooperative National Park Resources Studies Unit, Honolulu.

439. Whiteman, J.B. and P.M. Room (1991) Temperatures lethal to *Salvinia molesta* Mitchell, *Aquatic Botany*, 40, 27–35.

440. Whitmore, T.C. (1991) Invasive woody plants in perhumid tropical climates, in *Ecology of Biological Invasion in the Tropics*, (ed. P.S. Ramakrishnan), International Scientific Publications, New Delhi, pp. 35–40.

441. Williams, P.A. (1981) Aspects of the ecology of broom (*Cytisus scoparius*) in Canterbury, New Zealand, *New Zealand Journal of Botany*, 19, 31–43.

442. Williams, P.A. (1983) Secondary vegetation succession on the Port Hills Banks Peninsula, Canterbury, New Zealand, *New Zealand Journal of Botany*, 21, 237–47.

443. Williams, P.A. (1984) Woody weeds and native vegetation – a conservation problem, in *Protection and Parks. Essays in the Preservation of Natural Values in Protected Areas*, (ed. P.R. Dingwall), Proceedings of Section A4e, 15th Pacific Science Congress, Dunedin, February 1983, Department of Lands and Survey, Wellington, NZ, pp. 61–6.

444. Williams, P.A. and R.P. Buxton (1985) Hawthorn (*Crataegus monogyna*) populations in mid-Canterbury, *New Zealand Journal of Ecology*, 9, 11–17.

445. Williams, P.A. and S.M. Timmins (1990) *Weeds in New Zealand Protected Natural Areas: a Review for the Department of Conservation*, Science and Research Series, Report No. 14, Department of Conservation, Wellington.

446. Williamson, M.H. and K.C. Brown (1986) The analysis and modeling of British invasions, *Philosophical Transactions of the Royal Society, London, Series B*, 314, 505–21.

447. Willson, B.W. (1985) The biological control of *Acacia nilotica* in Australia, in *Proceedings of the VI International Symposium on Biological Control of Weeds*, Agriculture Canada, Ottawa, pp. 849–53.

448. Wilson, C.G. and G.J. Flanagan (1991) Establishment of *Acanthoscelides quadridentatus* (Schaeffer) and *A. puniceus* Johnson (Coleoptera: Bruchidae) on *Mimosa pigra* in northern Australia, *Journal of the Australian Entomological Society*, 30, 279–80.

449. Wilson, J.B. and M.T. Sykes (1988) Some tests for niche limitation by examination of species diversity in the Dunedin area, New Zealand, *New Zealand Journal of Botany*, 26, 237–44.

450. Witkowski, E.T.F. (1991) Effects of invasive alien acacias on nutrient cycling in the coastal lowlands of the Cape Fynbos, *Journal of Applied Ecology*, 28, 1–15.

451. Witkowski, E.T.F. (1991) Growth and competition between seedlings of *Protea repens*

(L.) L. and the alien invasive *Acacia saligna* (Labill.) Wendl. in relation to nutrient availability, *Functional Ecology*, **5**, 101–10.

452. Wriggley, J.W. and M. Fagg (1979) *Australian Native Plants: a manual for their propagation, cultivation, and use in landscaping*, Collins, Sydney.

453. Zimmerman, H.G. (1980) Prickly pear, in *Plant Invaders: beautiful but dangerous*, 2nd ed, (ed. C.H. Stirton), The Department of Nature and Environmental Conservation of the Cape Provincial Administration, Cape Town, pp. 112–15.

INDEX

معذور عورتوں کے لئے صحت کی کتاب

ڈاکٹر شیر شاہ سیّد

رحمٰن ضیاء

پاکستان نیشنل فورم آن ویمنز ہیلتھ

معذور عورتوں کے لئے صحت کی کتاب

جملہ حقوق انگریزی اشاعت بحق ہیسپرین فاؤنڈیشن
اور
اُردو اشاعت بحق پاکستان نیشنل فورم آن ویمنز ہیلتھ محفوظ ہیں

Urdu Edition	اُردو ایڈیشن
Translation	ترجمہ
Mazoor Aurtoon Kayliay Sehat Ki Kitab	معذور عورتوں کے لئے صحت کی کتاب
●	●
Dr. Sher Shah Syed, Rahman Zia	ڈاکٹر شیر شاہ سیّد، رحمٰن ضیاء
●	●
Pakistan National Forum on Women's Health	پاکستان نیشنل فورم آن ویمنز ہیلتھ
Karachi, Pakistan	کراچی، پاکستان
English Edition	انگلش ایڈیشن
A Health Hand Book for Women with Disabilities	اے ہیلتھ بک فار وومین ودھ ڈس ایبلیٹیز
●	جین میکس ویل، جولیا واٹس بیلسر اینڈ ڈارلینا ڈیوڈ
Jane Maxwell, Julia Watts Belser	
and	●
Darlena David	ہیسپرین فاؤنڈیشن
●	برکلے کیلیفورنیا، یو ایس اے
Hesperian Foundation,	
Berkeley California, USA	

تکنیکی مشیر برائے اُردو اشاعت	:	ڈاکٹر حامد منظور، ڈاکٹر سجاد احمد صدیقی
لے آؤٹ/کمپوزنگ	:	ایس کے پریس میڈیا سروس
گرافک ڈیزائنر	:	خالد مرزا
پہلی اشاعت	:	جنوری 2009ء
طابع	:	فہیم آرٹ پرنٹرز
قیمت	:	600/- روپے

پاکستان نیشنل فورم آن ویمنز ہیلتھ

انتساب

عورتوں کے نام

جنہیں سماج ان کے پیدائشی اور صنفی حقوق دینے کے لئے تیار نہیں ہے۔

اس سماج کے نام

جو معذور عورتوں کے وجود کو تسلیم نہیں کرتا ہے۔

اور

ان مردوں کے نام

جو سماج میں اس تسلسل کے ذمہ دار ہیں۔

اظہارِتشکر

پاکستان نیشنل فورم آن ویمنز ہیلتھ (PNFWH)

شکر گزار ہے:

ہیسپیرین فاؤنڈیشن کا جس نے دیگر کتابوں کی طرح اس کتاب کی اشاعت کی اجازت دی اور حتی الامکان مدد کی تا کہ فورم اس کتاب کو منظرِ عام پر لا سکے۔

ان پاکستانی اداروں کا جو خاموشی کے ساتھ معذور لوگوں کی خدمتِ مسلسل میں لگے ہوئے ہیں۔

شکر گزار ہے جناب ڈاکٹر حامد منظور اور ڈاکٹر سجاد احمد صدیقی کا جنہوں نے وقت نکال کر کتاب کی پروف ریڈنگ میں بھی مدد کی۔

جناب خالد مرزا کا جنہوں نے ہمیشہ کی طرح کتاب کے چھپنے سے پہلے کے مراحل میں ساتھ دیا۔

جناب فاروق بکالی کا جو معذور لوگوں کے حقوق کے لئے انتھک کاوشوں میں لگے ہوئے ہیں اور جنہوں نے اس کتاب کے لئے اپنے قیمتی مشوروں سے نوازا اور کچھ مقامی معلومات مہیا کیں۔

جناب طارق نثار کا جن کی عنایت اس کتاب میں شامل رہی۔

ادارہ شکر گزار ہے پاکستان میڈیکل ایسوسی ایشن کے بہت سارے ممبران کا جنہوں نے ملک بھر سے معلومات مہیا کرنے میں مدد کی اور قابل قدر مشوروں سے نوازا۔ ادارہ ان لوگوں کا بھی شکر گزار ہے جن کا نام اس فہرست میں شامل نہیں ہے مگر جن کی درپردہ کاوشیں اس کتاب کی تیاری میں شامل رہی ہیں۔

Acknowledgement

The Pakistan National Forum on Women's Health (PNFWH) acknowledges the support and contribution of the Hesperian Foundation USA, for giving permission and encouraging PNFWH for the Urdu translation.

PNFWH also acknowledges the support of a number of individuals and organisations who supported, helped and contributed in numerious ways.

 Pakistan National Forum on Women's Health
National Secretariat, PMA House, Sir Aga Khan III Road,
Karachi, Pakistan

یہ کتاب پاکستان نیشنل فورم آن ویمنز ہیلتھ نے اردو میں پہلی بار ہیسپرین فاؤنڈیشن کی اجازت سے شائع کی ہے۔

پاکستان نیشنل فورم آن ویمنز ہیلتھ اور کتاب کے مترجم اور مرتبین اس کتاب میں موجود معلومات استعمال کرنے کے ضمن میں کوئی ذمہ داری نہیں رکھتے ہیں۔ اگر آپ خود کو درپیش مسئلے کے حوالے سے کچھ کرنے کے بارے میں پُریقین نہ ہوں تو اپنی کمیونٹی کے زیادہ تجربہ کار افراد یا مقامی طبی کارکنوں یا صحت حکام سے مشورہ اور مدد حاصل کیجئے۔

یہ کتاب معذور یوں کا شکار عورتوں کو صحت مند رہنے میں معاون بنیادی معلومات فراہم کرتی ہے اور ان لوگوں کی بھی مدد کرے گی جو اچھی دیکھ بھال کے لئے معذور عورتوں کی مدد کرتے ہیں۔ لہٰذا اگر آپ کسی معذوری کا شکار ہیں یا کسی معذور کی دیکھ بھال کرنے والے ہیں یا اس کتاب اور معذور عورتوں کی صحت کی بہتری کے لئے خیالات یا تجاویز رکھتے ہیں تو برائے کرم ہمیں ان سے آگاہ کیجئے۔ ہم آپ کے تجربات اور اس حوالے سے کئے گئے کام کے بارے میں جاننا چاہتے ہیں۔ آپ ہمیں اپنے خطوط پاکستان نیشنل فورم آن ویمنز ہیلتھ کے پتہ پر ارسال کر سکتے ہیں۔

پاکستان نیشنل فورم آن ویمنز ہیلتھ مقامی ضروریات کے مطابق اس کتاب کے کسی حصے یا تمام حصوں بشمول اشکال وغیرہ کی نقل، از سر نو تشکیل یا اخذ کی حوصلہ افزائی کرتا ہے بشرطیکہ طریقہ کار کاروباری فائدے کے لئے نہیں بلکہ مفت یا محض لاگت پر تقسیم کئے جائیں۔ کوئی بھی ادارہ یا فرد جو اس کتاب کے کسی حصے یا تمام حصوں کو کاروباری مقاصد سے استعمال یا شائع کرنا چاہتا ہے تو اسے پاکستان نیشنل فورم آن ویمنز ہیلتھ سے اجازت لینی ہوگی۔

اس کتاب کے کسی اور زبان میں ترجمے یا تلخیص سے پہلے تجاویز، کتاب میں موجود معلومات میں کسی اضافے اور کوششوں کی تکرار سے بچنے کے لئے پاکستان نیشنل فورم آن ویمنز ہیلتھ سے رابطہ کیجئے۔ اس کتاب کے مضامین کسی بھی صورت میں شائع کرنے والے ایک شمارہ پاکستان نیشنل فورم آن ویمنز ہیلتھ اور ہیسپرین فاؤنڈیشن کو ضرور بھیجیں۔

Pakistan National Forum on Women's Health

National Secretariat, Sir Agha Khan III Road

Karachi - PAKISTAN

CREDITS

ہیسپرین فاؤنڈیشن کی اس منفرد اور مفید کتاب کی تیاری اور انتظامات میں مندرجہ ذیل افراد نے حصہ لیا۔

Art coordination:
- Jane Maxwell

Community review coordination:
- Jane Maxwell and Sarah Constantine

Project support:
- Soo Jung Choi,
- Michelle Funkhauser,
- Tawnia Queen,
- Heather Rickard,
- Karen Wu

Design and production:
- Jacob Goolkasian,
- Shu Ping Guan,
- Christine Sienkiewicz,
- Sarah Wallis

Cover design:
- Iñaki Fernández de Retana,
- Jacob Goolkasian,
- Sarah Wallis

Additional writing:
- Pam Fadem,
- Judith Rogers,
- Edith Friedman

Copy editing:
- Kathleen Vickery,
- Todd Jailer

Indexing:
- Victoria Baker

Proofreading:
- Sunah Cherwin

Medical review:
- Lynne Coen,
- Suzy Kim,
- Melissa Smith,
- Susan Sykes,
- Sandra Welner

Editorial management:
- Darlena David

Editorial oversight:
- Sarah Shannon

Production management:
- Todd Jailer

Artists:
- Namrata Bali
- Heidi Broner
- Barbara Carter
- Regina Faul-Doyle
- Shu Ping Guan
- Haris Ichwan
- Delphine
- Joyce Knezevitch
- Naoko Miyamoto
- Mabel Negrete
- Connie Panzarini
- Petra Röhr-Rouendaal
- Ryan Sweere
- Lihua Wang
- Mary Ann Zapalac
- Sara Boore
- May Florence Cadiente
- Gil Corral
- Sandy Frank
- Jesse Hamm
- Anna Kallis
- Kenze
- Sacha Maxwell
- Lori Nadaskay
- Gabriela Nuñez
- Kate Peatman
- Carolyn Shapiro
- Sarah Wallis
- David Werner

Cover photo locations and photographers (left to right, counter-clockwise):

Location	Photographer
Uganda	Jan Sing
World Bank/Cambodia	Masaru Goto
Mexico	Suzanne C. Levine
India	Amy Sherts
Bulgaria	Sean Sprague/SpraguePhoto.com
World Bank/Uzbekistan	Anatoliy Rakhimbayev

Back cover

Uganda, UMCOR-ACT International, Paul Jeffrey
Bangladesh, Jean Sack/ICDDRB,
Courtesy of Photoshare

Permissions:

We thank the following organizations for permission to use their illustrations:
Breast Health Access for Women with Disabilities at the Alta Bates Summit Medical Center (for a drawing on page 130); Pearl S. Buck International, Vietnam (for sign language drawings on pages 369-370); Sahaya International, USA (for sign language illustrations drawn from photographs in The Kenyan's Deaf Peer Education Manual, on pages 369-370); and Jun Hui Yang (for Chinese Sign Language illustrations on pages 369-370).

THANKS

معذور عورتوں کے لئے لکھی جانے والی اس کتاب کے لئے چالیس سے زائد ممالک کی معذور اور غیر معذور عورتوں اور انسان دوستوں نے درد مندی سے کام کیا۔ ہیسپیرین فاؤنڈیشن نے ان سب کا شکر یہ ادا کیا ہے۔ پاکستان نیشنل فورم آن ویمنز ہیلتھ بھی ان سب کا ممنون ہے۔

It is impossible to adequately thank all the people who helped make A Health Handbook for Women with Disabilities a reality. It started 10 years ago as a good idea shared by 2 women, and grew into a remarkable international collaboration between women with disabilities and their friends in more than 40 countries.

Listing a person's name does not begin to say how much her efforts and ideas helped create this book. Every staff member, intern, and volunteer here at Hesperian also helped bring this book into the world, including those who raise funds, manage finances, publicize our materials, and pack and ship them around the world.

Along with our tireless medical editors, we called on a few reviewers over and over again, and they deserve a special mention and our sincerest thanks: Naomy Ruth Esiaba, Kathy Martinez, Gail McSweeney, Janet Price, Judith Rogers, Andrea Shettle, Ekaete Judith Umoh, and Veda Zachariah.

Many thanks to the following groups of people with disabilities who contributed so much of their hearts, time, and personal experience to help us make sure the material in this book would be useful to women with disabilities all over the world:

Afghanistan: the National Association of Women with Disabilities of Afghanistan (NAWDA)

Cambodia: the Women with Disabilities Committee of the Disability Action Council

China: MSI Professional Services Columbia: the Columbian Association for Disabled Peoples (ASCOPAR)

El Salvador: La Asociación Cooperativa de Grupo Independiente Pro Rehabilitación Integral (ACOGIPRI)

Fiji: the Support Group for Women with Disabilities

Finland: the Abilis Foundation, and The National Council on Disability

Republic of Georgia: the Gori Disabled Club

India: the Amar Jyoti Charitable Trust, Blind People Association, Catholic Relief Services (CRS), Disabled People's International, Humane Trust, and Sanjeevini Trust

Jamaica: Combined Disabilities Association

Kenya: The Bob Segero Memorial Project, and Hope

Laos: the Lao Disabled People's Association, and the Lao Disabled Women Development Center

Lebanon: the Arab Organization of Disabled People, and the National Association for the rights of Disabled People Lebanon (NARD)

Lesotho: the Lesotho National Federation of Organizations of Disabled

Mauritius: the Association of Women with Disability

Nepal: the Nepal Disabled Women Society, and Rural Health Education Services Trust (RHEST)

Nigeria: the Family-Centered Initiative for Challenged Persons (FACICP)

Palau: the Organization of People with Disabilities (Omekasang)

Philippines: Differently Abled Women's Network (DAWN), Disabled People's Internationa (DPI), and KAMPI

Russia: Perspektiva (the Regional Society of Disabled People)

South Korea: Korean Differently Abled Women United

Tanzania: The National Council for People with Disabilities

Thailand: Disabled People's International-Asia Pacific Trinidad/Tobago: the Tobago School for the Deaf, Speech and Language Impaired

Uganda: the Disabled Women's Network and Resource Organisation (DWNRO), Mobility Appliances by Disabled Women Entrepreneurs (MADE), and the National Union of Disabled Persons of Uganda

USA: Mobility International USA (MIUSA), Through the Looking Glass, Women Pushing Forward, and the World Institute on Disability (WID)

Vietnam: the Vietnam Veterans of America Foundation

Yemen: the Arab Human Rights Foundation

Zimbabwe: Disabled Women Africa (DIWA), the National Council on Disabled Persons of Zimbabwe, the Southern Africa Federation of the Disabled (SAFOD), and Women with Disabilities Development (ZWIDE)

Our heartfelt gratitude to everyone who gave so generously of their time and knowledge. Your commitment to health care for women with disabilities brought this book into the world.

Caroline Agwanda
Fatuma Akan
Firoz Ali
Janet Connatser
  Allem
Eric Anderson
Soc Balingit
Florence Baingana
Monica Bartley
Denise Bergez
Kim Best
Bimala Sharma
  Bhandari
Cheri Blauwet
Joan Bobb-Alleyne
Claire Borkert
Tina Bregvadze
Ron Brouillette
Arlene Calinao
Cynthia
  Carmichael
Susan Canas
Silvia Casey
Phonesavanh
  Chandavong
Sivila Chanpheng
Sujith J. Chandy
Gladys Charowa
Farai Cherera
Rosemary Ciotti
Alicia Contreras
Ann Cupola
  Freeman
John Day
Kathryn Day
Roshni Devi
Tara Dikeman

Lori Dobeus
Pamela Dudzik
Shalini Eddens
Sana Ali El-Saadi
Nancy Ferreyra
Anne Finger
Lee Gallery
Monica Gandhi
Katherine Gergen
Anita Ghai
Eileen Girón
  Batres
Nora Groce
Heba Hagrass
Maria Harkins
Sari Heifetz
Karen Heinicke-
  Motsch
Taija Heinonen
Susan Heller
Kevin Henderson
Judith Heumann
Rachael Holloway
Rob Horvath
Ralf Hotchkiss
Honora Hunter
Venus Ilagan
Namita Jacob
Lisa Jensen
Usha Jesudasan
Kathy Al Ju'beh
Rachel Kachaje
James G. Kahn
Wendy Kahn
Deborah Kaplan
Manali Kasbekar
Susan Kaur

Christie Keith
Jennifer Kern
Jahda Abou Khalil
Jackie Ndona Kingolo
Pat Kirkpatrick
Kristi L. Kirschner
Justine Kiwanuka
Mari Koistinen
Kathleen Lankasky
BA Laris
Ye Ja Lee
Anne Leitch
Cindy Lewis
Gertrude Likopo
  Lesoetsa
Rebecca C. Lim
Hoang Cam Linh
Sari Loijas
Lizzie Longshaw
Josephine Lyengi
Annie Malinga
Peggy Martinez
Rajaa Masabi
Melissa May
Katherine
  McLaughlin
Lemnis Geraldo Mendez
Ruth Miller
Linda D. Misek-
  Falkoff
Sruti Mohaptra
Linda Mona
Winifred Mujesia
Frank Mulcahy
Irene Busolo
  Mwenesi

Dorothy
  Musakanya
James Mwanda
Safia Nalule
Sucheta Narang
Kanika Sophak
  Nguon
Papa Djibril Niang
Cathy Noble
Corbett O'Toole
Deborah
  Ottenheimer
Judy Panko Reis
Lauri Paolinetti
Rafael Peck
Elizabeth Pearl
  Penumaka
KP Perkins
Minh Hang Pham
Allison Phillips
Judith Pollack
Jureeratana
  Pongpaew
Zohra Rajah
Barbara Ridley
Pia Rockhold
Denise Roza
Laura Ruttner
Mariana Ruybalid
Robert Sampana
Beatriz Elena
  Satizabal
Marsha Saxton
Estelle Schneider
Rosemary Segero
Lonny Shavelson
Maya Shaw
Julia Shelby

A.Shivasanthakumar
Caroline Signore
Meenu Sikand
Julia Simonova
Kathy Simpson
Jan Sing
Judith Smith
Florence Nayiga
  Ssekabira
Yvette Swan
Susan Sygall
Michael Tan
Supattraporn
  "Mai"
  Tanatikom
Carolyn
  Thompson
Uma Tuli
Meldah B.
  Tumukunde
Doralee Uchel
James Ullman
Nance Upham
Aruna Uprety
Elizabeth
  Valitchka
Koen Van Rompay
Jyoti Chandulal
  Vidhani
Zainab K. Wabede
Jessica Mak Wei-E
Ann Whitfield
Amy Wilson
Dayna Wolfe
Lin Yan

We also want to thank and remember the following women who contributed so much, not only to this book, but to the community of women with disabilities around the world. Sadly, they died before the book was published: Hellen Winifred Akot, Tanis Doe, Ana Malena Alvarado, Connie Panzarini, Nanette Tver, Barbara Waxman-Fiduccia, and Sandra Welner.

We also thank the following foundations and individuals for their generosity in financially supporting this project: Alexandra Fund; Chaim Tovim Tzedakah Fund of the Shefa Fund; Christopher Reeve Paralysis Foundation; Displaced Children and Orphans Fund/ Leahy War Victims Fund, U.S. Agency for International Development (under terms of JHPIEGO contract no. 06-TSC-022); Flora Family Foundation; Ford Foundation; Global Fund for Women; James R. Dougherty Jr. Foundation; Jennifer Kern; Kadoorie Charitable Foundation; Margaret Schink; Marguerite Craig; Marji Greenhut; May and Stanley Smith Charitable Trust; Norwegian-Dutch Trust Fund for Gender Mainstreaming/World Bank; Swedish International Development Agency; and the West Foundation.

فہرست مضامین

ایک معذور عورت کا حق

عورت کی صحت، سماج کی صحت ہے۔ پاکستان نیشنل فورم آن ویمنز ہیلتھ، عورت کی صحت ممکنہ حد تک یقینی بنانے کے لئے سرگرمِ عمل ہے۔ گزرے برسوں میں پاکستان نیشنل فورم آن ویمنز ہیلتھ نے اس مقصد کے لئے دستیاب ہر موقع پر عورت کے اس بنیادی حق کے لئے نہ صرف آواز اٹھائی بلکہ عملی اقدام بھی کئے۔ کتاب ''جہاں عورتوں کے لئے ڈاکٹر نہ ہو'' کی اشاعت (2001ء) اس کی ایک واضح مثال ہے۔ ہسپرین فاؤنڈیشن نے یہ کتاب انگریزی میں (1997ء) میں شائع کی تھی۔ اسی ادارے نے 2007ء میں معذور عورتوں کی صحت کے لئے اے ہیلتھ بک آف ویمنز ودھ ڈس ایبلٹی شائع کی ہے۔ ہم فاؤنڈیشن کی اجازت سے یہ کتاب پہلی بار اردو میں شائع کر رہے ہیں۔

معذور عورتیں دنیا بھر میں موجود ہیں۔ جن معاشروں میں ایک عام اور صحت مند عورت اپنی روزمرہ زندگی میں بے شمار مسائل سے دوچار ہے، معذور عورتوں کی حالتِ زار کا اندازہ لگایا جاسکتا ہے۔ صحت مند اور مطمئن زندگی ایک معذور عورت کا بھی حق ہے۔ وہ بھی ایک عام یا غیر معذور عورت کی طرح زندہ رہنے، خواب دیکھنے اور اپنے خیالات کے مطابق ممکنہ حد تک نارمل زندگی گزارنے کا حق رکھتی ہے۔

ہیسپرین فاؤنڈیشن کی یہ کتاب دس برسوں کی تحقیق اور تجزیوں کا نتیجہ ہے۔ قابل ذکر بات یہ ہے کہ اس کتاب کی ابتدا دو عورتوں نے کی اور یہ سلسلہ چالیس سے زیادہ ممالک تک وسیع ہو کر ایک بین الاقوامی منصوبہ بن گیا۔ اس کتاب سے نہ صرف معذور عورتوں بلکہ عام لوگوں کو بھی معذوری کے مسائل، ان کے حل اور بہتری کے لئے ضروری اقدام کے تعین میں مدد ملے گی۔ یہ منفرد کتاب لوگوں کو یہ احساس بھی دلاتی ہے کہ معذور عورتیں بھی اسی دنیا کا حصہ ہیں، انہیں نظر انداز کرنا یا غیر حقیقت پسندانہ نظریات کی بھینٹ چڑھانا ہر اعتبار سے قابل مذمت ہے۔ ہم معذور عورتوں کے مسائل جان اور سمجھ کر انہیں ایک باوقار اور مفید زندگی گزارنے کے لئے بھرپور مدد فراہم کر سکتے ہیں۔ ہم نے کوشش کی ہے کہ معذور عورتوں کے بنیادی مسائل جوں کے توں شامل کئے جائیں تاہم انتہائی ناگزیر صورتوں میں مقامی ماحول اور طرزِ زندگی کے مطابق رد و بدل کیا ہے۔ ہمیں امید ہے کہ یہ کتاب اپنی نوعیت اور افادیت کے لحاظ سے دنیا بھر کی معذور عورتوں کی طرح ہمارے معاشرے کی معذور عورتوں کے لئے بھی بے حد مفید اور معاون ثابت ہوگی۔ اس کے ساتھ ہم ہمیشہ کی طرح خواتین، معذور خواتین، عورتوں کی صحت کے مسائل سے دلچسپی رکھنے والے اداروں، ڈاکٹروں، صحت کارکنوں اور تنظیموں سے یہ توقع رکھتے ہیں کہ وہ ہمیں اس کتاب کے بارے میں اپنی آراء، تبصروں اور تجاویز سے ضرور آگاہ کریں گے تا کہ ہم اس کتاب کی آئندہ اشاعت کو مزید بہتر بنا سکیں۔

ڈاکٹر شیر شاہ سیّد

تعارف :

معذور عورتو! آؤ ہم اپنے حق کے لئے اٹھ کھڑے ہوں اور اس جدوجہد میں شامل ہوجائیں۔ **عمل کا یہی وقت ہے۔**

معذور عورتوں کی صحت کے بارے میں کتاب کیوں ضروری ہے؟

معذوریاں رکھنے والی عورتوں کو اچھی صحت کی ضرورت ہے۔ اچھی صحت، کوئی بیماری نہ ہونے سے زیادہ اہم ہے۔ ایک معذور عورت کی اچھی صحت کا مطلب یہ ہے کہ وہ اپنے جسم، دماغ اور روح کی بہتری محسوس کرتی ہو۔

معذور عورتیں خود اپنی صحت کے لئے اس وقت ہی اختیار سنبھال سکتی ہیں جب انہیں ایسی معلومات حاصل ہوں جو ان کے جسموں اور صحت کی ضروریات کے خود ان کے تجربات کی تصدیق کرتی ہوں۔ وہ اپنی معلومات، معذوری کے بارے میں لوگوں کے سوچنے کے انداز کو بدلنے کے لئے بھی استعمال کرسکتی ہیں۔ جیسے ہی معذوریاں رکھنے والی تمام عورتیں اپنی زندگیوں پر اختیار حاصل کریں گی وہ اپنے اپنے سماجوں میں احترام اور تائید بھی حاصل کریں گی۔

اگر چہ معذوری بذاتِ خود صحت کا مسئلہ نہیں ہوسکتی ہے، اکثر اوقات معذور عورتوں کو درپیش صحت کے مسائل کا علاج نہیں کیا جاتا ہے ۔ کسی معذور عورت کی صحت کے عام سے مسئلے کا علاج نہ ہو تو اس کا نتیجہ یہ بھی نکل سکتا ہے کہ وہ اس کی زندگی کو خطرے میں ڈالنے والا مسئلہ بن جائے۔

ہمیں وہ تمام رکاوٹیں دور کرنی چاہئیں جو معذور عورتوں کو اچھی صحت حاصل کرنے سے محروم رکھتی ہیں۔

میرے ملک میں معذور افراد خصوصاً معذور عورتوں کو بچہ سمجھا جاتا ہے کیونکہ بچوں کو کوئی ذمہ داری نہیں سونپی جاتی ہے اس لئے معذور عورتوں کو زندگی کی ہر سہولت مثلاً تعلیم، صحت، زمین کی ملکیت کے حصول وغیرہ وغیرہ سے محروم کردیا جاتا ہے۔

ہمیں معذور عورتوں کو معلومات فراہم کرنے کی ضرورت ہے تاکہ وہ بذات خود اپنی صحت کی بہتر دیکھ بھال کرنا سیکھ سکیں اور یہ بھی کہ وہ ایک کمیونٹی کے طور پر ڈاکٹروں، نرسوں اور اسپتال کے منتظمین کے رویوں کو تبدیل کرنے کے لئے اور صحت کی سہولتوں تک زیادہ رسائی اور دستیابی کے لئے کیا کرسکتی ہیں۔

لیزی لانگ شا،
نیشنل کونسل آف ڈس ایبلڈ پرسنز
آف زمبابوے

صحت کی اچھی دیکھ بھال میں رکاوٹیں

بیشتر عورتوں کی طرح معذور عورتیں بھی عموماً اپنی ضرورت کے مطابق یا جب انہیں ضرورت ہو، صحت کی دیکھ بھال کے حصول میں مشکلات سے دوچار ہوتی ہیں۔ حد تو یہ ہے کہ ایک عورت کسی مرکزِصحت کے قریب رہتی ہو اور علاج یا سہولیات کی ادائیگی کے لئے مناسب رقم بھی رکھتی ہو تو بیشتر کلینک، مراکزصحت اور اسپتال اس طرح نہیں بنے ہیں کہ ان میں جانا ہر ایک کے لئے آسان ہو۔ معذور عورتیں اس وقت رکاوٹوں سے دوچار ہوتی ہیں جب مراکزِصحت میں وہیل چیئر استعمال کرنے والوں کے لئے ریمپ نہ ہوں، نابینا یا بصارت میں نقائص رکھنے والوں کے لئے بریل یا آڈیوکیسٹ میں معلومات نہ ہوں۔ نہ سُن سکنے والی عورتوں کے لئے علامتی یا اشاروں کی زبان سمجھنے یا ان کی سمجھنے اور سیکھنے کی مشکلات حل کرنے والے نہ ہوں۔

صحت کی ضروریات

ایک اور مسئلہ یہ ہے کہ ڈاکٹر اور دیگر صحت کارکن عام طور پر ان ضرورتوں کو سمجھنے کی تربیت نہیں رکھتے ہیں جو معذور عورتوں کو لاحق ہوسکتی ہیں، اسی لئے ممکن ہے کہ صحت کارکن معذوری کے حوالے سے ایسے نظریات رکھتے ہوں جو معذور عورتوں کے لئے صحت کی اچھی دیکھ بھال کا حصول پریشان کن اور مشکل بنا دیتے ہوں۔ جب معذور عورتیں وسائل، تعلیم اور دیگر سہولتوں تک رسائی نہ رکھتی ہوں اور غربت، استحصال اور زیادتیوں کا زیادہ آسان ہدف ہوتی ہیں۔ خوداعتمادی اور اپنے حقوق سے آگہی کے بغیر عموماً وہ سماجی طور پر الگ تھلگ رہنے پر مجبور ہوتی ہیں۔ یہ صورت صحت کی دیکھ بھال تک رسائی میں ان کے لئے زیادہ رکاوٹوں کا سبب بنتی ہے۔

یہ کتاب کس کے لئے ہے؟

یہ کتاب دنیا بھر میں موجود ان کروڑوں معذور عورتوں کے لئے لکھی گئی ہے جو اس لئے مشکلات جھیل رہی ہیں اور غیر ضروری طور پر مر جاتی ہیں کہ وہ آبرو مندانہ اور مناسب صحت کی دیکھ بھال سے محروم ہیں۔ یہ کتاب معذوریاں رکھنے والی بیشتر عورتوں کے لئے اپنی بہتر دیکھ بھال، عمومی صحت کی بہتری، اپنی صلاحیتوں اور خود انحصاری میں اضافے اور اپنی کمیونٹی میں زیادہ موثر طور پر شریک عمل ہونے کی صلاحیت کو بڑھانے میں معاون ثابت ہو سکتی ہے یہ کتاب بحالی کا مینوئل نہیں ہے اور نہ ہی یہ مختلف اقسام کی بیماریوں، امراض یا معذوریوں کی تشخیص اور علاج کے لئے ضروری معلومات رکھتی ہے اس کتاب کے مقاصد دوسرے ہیں۔

یہ کتاب آگاہ کرتی ہے کہ کس طرح معذوری ایک معذور عورت کی صحت کی ضرورتوں کو کوئی معذوری نہ رکھنے والی کسی عورت سے مختلف بناتی ہے۔ اس کتاب کی معلومات دوسروں سے بہتر دیکھ بھال حاصل کرنے میں معذور عورتوں کی مدد کرے گی۔

یہ کتاب صحت کارکنوں، افراد خانہ اور دیکھ بھال کرنے والوں کو سکھائے گی کہ معذوری کا مطلب بیماری نہیں بلکہ معذوری رکھنے والی ایک عورت، جیسے نابینا عورت، وہیل چیئر استعمال کرنے والی عورت بھی، معذوریاں نہ رکھنے والی عورتوں کی مانند ایچ آئی وی/ایڈز، ملیریا جیسی بیماری میں مبتلا ہو سکتی ہے۔

یہ کتاب، گھرانوں، دوستوں، سماجی صحت کارکنوں اور معذور عورتوں کی معاونت کرنے والوں کی مدد کرے گی کہ وہ دیکھ بھال میں موثر طور پر شریک رہیں۔ اس کتاب میں معذوری کی سماجی وجوہ بھی ہیں اور اس میں ان خیالات اور عقائد کو تبدیل کرنے کے طریقے بھی تجویز کئے گئے ہیں جو معذوریاں رکھنے والی عورتوں کی صحت، ان کے گھرانوں اور ان کی کمیونٹیز کے لئے نقصان دہ ہیں۔ اس کتاب کو ممکنہ حد تک مفید بنانے کے لئے دنیا بھر کی معذور عورتوں نے ہمیں اپنی صحت کی ضرورتوں سے آگاہ کیا اور ان کی ضرورتیں، تجربات اور کہانیوں نے اس کتاب کو مرتب کرنے اور لکھنے میں ہماری مدد کی اور اس کتاب کے ہر صفحے پر ان کی عکاسی ہوتی ہے۔

باب - 1

معذوری اور کمیونٹی

معذوریاں رکھنے والی عورتیں اچھی صحت کا حق رکھتی ہیں۔ اچھی صحت کا انحصار کھانے کے لئے مناسب قوت بخش غذا، باقاعدہ جسمانی مصروفیات اور صحت کے مسائل خصوصاً تولیدی صحت کے مسائل سے تحفظ اور ان کے علاج کے لئے معلومات اور سہولتیں حاصل کرنے کی رسائی پر ہے۔ اس کے ساتھ معذوریاں رکھنے والی لڑکیوں اور عورتوں کو اپنی صلاحیتوں کی بھر پور نشوونما کے لئے اچھی تعلیم، روزگار اور اپنی اپنی کمیونٹیز میں عملی شراکت کے مواقعوں کی ضرورت ہوتی ہے۔

دس عورتوں میں ایک عورت کوئی نہ کوئی معذوری رکھتی ہے جو اس کی روزمرہ زندگی متاثر کرتی ہے۔

جب ہمیں یکساں مواقع ملیں گے تو ہم دوسری عورتوں کی طرح اپنے گھرانوں اور کمیونٹیز کے لئے مددگار ثابت ہوسکتے ہیں۔

معذوری کیا ہے؟

معذوری کا شکار بہت سی عورتیں اپنی انفرادی معذوریوں کے لئے بندش یا رکاوٹ کی اصطلاح استعمال کرتی ہیں۔ ان بندشوں یا رکاوٹوں میں اندھاپن، بہرہ پن یا کوئی ایسی صورت شامل ہو سکتی ہے جو ان کے لئے چلنا یا بولنا مشکل یا ناممکن بنائے یا ان کے لئے سمجھنا یا سیکھنا مشکل تر بنائے اور جو کسی اچانک دورے یا رکاوٹ کا سبب بن سکتی ہے۔

ایک معذور عورت، کوئی معذوری نہ رکھنے والی عورت کے مقابلے میں مختلف انداز سے حرکت کر سکتی ہے، دیکھ، سن یا سیکھ اور سمجھ سکتی ہے۔ جب وہ بات کرتی ہے، کھانا کھاتی ہے، غسل کرتی ہے، لیٹی ہوئی حالت سے اٹھتی ہے اور لباس پہنتی ہے اور اپنے بچے کو اٹھاتی یا دودھ پلاتی ہے تو وہ اپنی روزمرہ زندگی کے کاموں میں احتیاط اور خصوصی توجہ سے کام لیتی ہے۔ اپنی حدود یا معذوریوں سے مطابقت اس کی زندگی کا ایک عام سا حصہ ہے۔

ہمارا کم تر معیار زندگی ہماری معذوری کے باعث نہیں بلکہ سماجی حقائق کے باعث ہے، حل ہمارے جسموں میں پوشیدہ نہیں ہے۔

اپنی معذوری کے باعث درپیش مسائل کے حل ڈھونڈنے کی صلاحیت کے باوجود ہر عورت سماجی، جسمانی، ثقافتی اور اقتصادی رکاوٹوں کا سامنا بھی کرتی ہے جو اسے صحت کی دیکھ بھال، تعلیم، ووکیشنل تربیت اور روزگار حاصل کرنے سے روک سکتی ہیں۔

رویّے رکاوٹیں پیدا کرتے ہیں

ایک معذور عورت کیا کر سکتی ہے اور کیا نہیں کر سکتی ہے اس حوالے سے رویّے اور غلط نظریات ایک معذور عورت کو بھرپور اور صحت مند زندگی گزارنے اور اپنی کمیونٹی کے کاموں میں حصہ لینے سے روک سکتے ہیں۔ یہ رویّے رکاوٹیں پیدا کر کے اس کی معذوری میں اضافہ کرتے ہیں کیونکہ یہ اسے تعلیم یا روزگار اور سماجی زندگی میں حصہ لینے سے روک سکتے ہیں۔

مثال کے طور پر ایک استانی یہ سمجھ سکتی ہے کہ ایک لڑکی اس لئے کچھ سیکھ نہیں سکتی ہے کہ وہ اندھی یا بہری ہے لیکن ایک لڑکی کا نہ دیکھ سکنا یا بہرہ پن مسئلہ نہیں ہے۔ ایک اندھی لڑکی سن کر یا اپنے دوسرے حواس جیسے

معذوریاں رکھنے والی بہت سی عورتیں چھپی ہوئی ہیں ہم کمیونٹی کی سر گرمیوں میں شامل نہیں ہیں کیونکہ دوسرے لوگ سمجھتے ہیں کہ ہم معذوری نہ رکھنے والی عورتوں کی بہ نسبت کم مفید اور کم اہمیت رکھتے ہیں۔

سونگھنے یا چھونے سے سیکھ سکتی ہے، وہ زیادہ بہتر طور پر سیکھ سکتی ہے اگر بریل میں کتابیں یا آڈیوکیسٹ میں معلومات موجود ہوں۔ اس طرح ایک ایسی لڑکی جو بہری ہے، اشاروں کی زبان یا تدریس کے بصری طریقوں کے استعمال سے سیکھ اور سمجھ سکتی ہے۔

ایک عورت جو چل نہیں سکتی ہے، ممکن ہے کہ وہ کوئی گاڑی چلا سکتی ہو اور اپنے گھر کے لئے آمدنی حاصل کر سکتی ہو لیکن اس کا گھرانہ یا کمیونٹی اس کے چلنے کے انداز پر شرمندگی محسوس کرتی ہو اور اسے چھپا کر رکھنا چاہتی ہو تو ان کا یہ رویہ (شرمندگی کے جذبات) ہے جو اسے معذور بنائے گا۔

تمام کمیونٹیز میں معذور افراد شامل ہیں۔ یہ ایک نارمل بات ہے۔ یہ کسی شخص کے لئے یہ نارمل بات نہیں ہے کہ اس کے ساتھ امتیاز برتا جائے اور اسے عام زندگی سے اس لئے خارج کر دیا جائے کہ وہ معذور ہے۔ یہ بذاتِ خود معذور کرنے والی بات ہے۔

معذوری کا طبی فہم

بیشتر ڈاکٹر اور صحت کا رکن صرف اس معذوری کو دیکھتے ہیں جو کوئی بھی ظاہری طور پر رکھتا ہے۔ وہ معذور فرد کو ایک مکمل فرد یا عورت کی حیثیت سے نہیں دیکھتے ہیں۔ وہ سمجھتے ہیں کہ ''معذوری'' رکھنے والے لوگ کچھ نہ کچھ ''نقص'' رکھتے ہیں اور ان کا علاج، بحالی یا تحفظ ہونا چاہئے۔ جب تضحیک یا بُرا رویہ اسپتالوں یا دیگر پبلک مقامات یا عوامی سہولتوں کو ہر ایک کے لئے نا قابل استعمال بنا دیتا ہے تو پھر یہ طبی نظام ہے جس میں کوئی ''خرابی'' ہے اور اس کا علاج یا تدارک ہونا چاہئے۔ ان تمام معاملات میں یہ عورت کی معذوری نہیں بلکہ معذوری کا ناقص طبی فہم ہے جو ایک معذور عورت کے لئے صحت مند اور بھر پور زندگی بسر کرنا ناممکن بنا دیتا ہے۔

> ہم اپنی معذوریوں سے نمٹ لیں گے لیکن صرف آپ ہی اس سماجی امتیاز کو روک سکتے ہیں جس کا ہمیں سامنا ہے۔

> یقیناً کسی معذوری سے ہمیں کوئی نقصان ہوسکتا ہے لیکن ہم رویوں اور معاشرے میں اپنی شرکت کی راہ میں سماجی، ثقافتی، اقتصادی اور ماحولیاتی رکاوٹوں کے باعث مسلط بندشوں سے زیادہ مجروح ہیں۔

> ہم اپنی زندگیوں کے بارے میں خود فیصلے کرسکتے ہیں۔ ہم اپنی عملی کی حالت میں دیکھ بھال یا خیرات قبول کرنا نہیں چاہتے ہیں۔

معذوری زندگی کا ایک فطری حصہ ہے

یقیناً ہمیشہ ایسے لوگ بھی رہیں گے جو معذوری کے ساتھ پیدا ہوئے ہوں۔ حادثات اور بیماریوں کے باعث معذور ہونے والے بھی موجود ہیں گے لیکن حکومتیں اور کمیونٹیز معذوری کے سماجی اسباب اور معذور افراد کے لئے تبدیل نہ ہونے والے سماجی، ثقافتی، اقتصادی اور طبی رویوں کے باعث ان کی سماج میں شراکت کی راہ میں حائل رکاوٹوں کو دور کرنے کے لئے کام کرسکتی ہیں۔

معذوریاں رکھنے والی عورتوں کی جسمانی اور ذہنی صحت اس وقت بہتر ہوگی جب کمیونٹیز ان کے لئے رسائی بہتر بنائیں گی، تعصب کو مسترد کریں گی اور روزگار کے مواقع پیدا کریں گی۔

معذور عورتیں بنگلور (بھارت) میں راہ دکھاتی ہیں

جنوبی بھارت کے شہر بنگلور میں جسمانی معذوری رکھنے والی چار معذور عورتیں شاہانہ، نوری، دیوا کی اور چندرما دوسری معذور عورتوں کے لئے بحالی کے آلات اور اشیاء بناتی اور پہناتی ہیں۔ یہ چاروں ری ہیبلیٹیشن ایڈز ورک شاپ بائی ومین ودھ ڈس ایبلیٹیز (RAWWD) میں کام کرتی ہیں جن کی ابتداء 1997ء میں ایسی آٹھ معذور عورتوں نے کی جن کو حرکاتی اعانتیں بنانے والی موبلیٹی انڈیا نامی این جی او نے تربیت دی تھی۔

ہر چند کہ وہاں دوسری سہولتیں تھیں لیکن RAWWD کے آغاز میں امدادی اشیاء کے لئے ناپ لینے اور بنانے والے مرد ٹیکنیشنز ہی دستیاب تھے اور معذور عورتیں ان کے پاس جانے سے کتراتی تھیں۔ وہ مردوں کے ناپ لینے اور آلات فٹ کرنے سے گھبراتی تھیں۔ اسی لئے بہت سی عورتیں ایسے آلات استعمال نہیں کرتی تھیں جو انہیں متحرک رکھ سکتے تھے۔

RAWWD اب کہنیوں، پیروں اور گھٹنوں کے لئے بحالی کے آلات کی وسیع رینج بناتی ہے ان میں بیساکھیاں، واکر، جوتے، بیلٹ، سہارے اور مصنوعی اعضا (مصنوعی ٹانگیں، پیر) شامل ہیں۔

جوں جوں اس ادارے میں عورتیں اپنا اعتماد اور مہارتیں بڑھاتی گئیں انہوں نے معذور افراد کے لئے کام کرنے والی تنظیموں کو خدمات فراہم کرنا شروع کر دیں۔ اب یہ عورتیں بنگلور کے بہت سے اسپتالوں اور پرائیوٹ ڈاکٹروں کے لئے کام کر رہی ہیں۔ یہ عورتیں امدادی اشیاء بنانے کے لئے میٹریل حاصل کرتی ہیں، کلائنٹس کا ریکارڈ رکھتی ہیں اور بعد ازاں باقاعدہ معائنوں کا اہتمام کرنے کے ساتھ اپنے بزنس کو سنبھالتی ہیں۔ RAWWD معذوری رکھنے والی دوسری عورتوں کو ٹیکنیشن بنانے پر مائل کرتی ہے اور انہیں امدادی اشیاء و آلات بنانے اور ان کی مرمت کرنے کی تربیت دیتی ہے۔ اس سے معذور عورتوں کے لئے خصوصاً ان عورتوں کے لئے مواقع بڑھے ہیں جنہیں ان کے گھر والوں نے الگ تھلگ چھوڑ دیا تھا اور یہ کہ ان کے لئے ذریعہ معاش بھی فراہم ہوتا ہے۔

وسائل اور مواقع

بہت سی کمیونٹیز میں عورتیں مردوں کی بہ نسبت کم تر وسائل اور مواقع رکھتی ہیں۔ مردوں اور عورتوں کے درمیان یہ عدم مساوات ان لوگوں کے لئے بھی ہے جو معذوریاں رکھتے ہیں۔ وہیل چیئرز، مصنوعی اعضاء، علامتی زبان کی کلاسیں، بریل (جو اندھی عورتوں کو پڑھنے کے قابل بناتی ہے) اور دیگر وسائل یا ذرائع عموماً مہنگے، معذور مردوں کی بہ نسبت معذور عورتوں کے لئے کم دستیاب ہیں۔ ان معاون اشیاء کے بغیر معذور لڑکیاں اور عورتیں تعلیم حاصل کرنے اور اپنے لئے کچھ کرنے میں مشکلات جھیلتی ہیں۔ نتیجے میں وہ روزگار حاصل کرنے، اپنی زندگیوں پر اختیار حاصل کرنے اور اپنی اپنی کمیونٹیز کی سرگرمیوں میں فعال کردار ادا کرنے کے قابل نہیں ہوتی ہیں۔

> زیادہ وسائل کے ساتھ ہم دیکھ اور سن سکتے ہیں اور اپنی صحت پر کنٹرول حاصل کرسکتے ہیں۔

طبیعی مادّی رکاوٹیں

بہت سی معذور عورتیں سماجی سہولیات مثلاً بینک یا اسپتالوں سے فائدہ نہیں اٹھاسکتی ہیں کہ بیش تر عمارتوں میں ڈھلان سطحیں، ہینڈ ریل، ایلیویٹر یا لفٹ وغیرہ نہیں ہیں۔ طبیعی رکاوٹیں معذوریاں رکھنے والی عورتوں کے لئے اپنے طور پر آزادانہ نقل و حرکت مشکل بنا دیتی ہیں۔ جب عورتیں ان رکاوٹوں کے باعث بندشوں کا شکار ہوتی ہیں تو عموماً یہ ضرورت کے مطابق اچھی غذا، مناسب ورزش یا صحت کی دیکھ بھال کی سہولت حاصل نہیں کر پاتی ہیں۔

بہت سے لوگ بشمول صحت کارکن یہ باور کر سکتے ہیں کہ جب وہیل چیئر استعمال کرنے والی کوئی عورت اس لئے کسی عمارت میں داخل نہیں ہوسکتی ہیں کہ وہاں صرف سیڑھیاں ہیں تو وہ پیروں کے لئے سہارے یا بیساکھیوں کا استعمال سیکھے یا اس کے ساتھ کوئی ہو جو اسے عمارت میں اٹھا کر لے جائے۔ یہ اس عورت کی معذوری نہیں بلکہ وہ مادّی رکاوٹیں ہیں جو اس کے لئے کسی عمارت میں داخل ہونا ناممکن بناتی ہیں۔ اگر وہاں ڈھلان سطح ہوتی تو وہ اپنی وہیل چیئر خود دھکیل کر عمارت میں لے جاتی اور کوئی مسئلہ نہ ہوتا۔

> میں جسمانی معذوری رکھنے والی ایک ماں ہوں۔ میرا ایک بیٹا بھی جسمانی طور پر معذور ہے۔ ہر بار جب ہم کسی ریسٹورنٹ یا سپر مارکیٹ یا کہیں اور جاتے ہیں۔ ہمیں اٹھا کر سیڑھیوں سے اوپر پہنچایا اور اتارا جاتا ہے۔ یہ حقیقتاً تحقیر آمیز ہے اور ہمیں کم تر انسان ہونے کا احساس دلاتا ہے۔

ترقیاتی کام کرنے والے گاؤں آئے تو وہ اپنے ساتھ مختلف پروجیکٹ لائے۔ انہوں نے وہاں موجود تمام عورتوں کے لئے کام لیا مگر وہاں ایک معذور عورت بھی تھی۔ ان لوگوں کے منصوبوں میں پانی کی فراہمی بھی شامل ہے لیکن یہ منصوبے معذور عورت کے لئے کوئی سہولت نہیں رکھتے ہیں۔ معذور عورت کو بھی پانی کی ضرورت ہوگی، وہ پانی کس طرح نکالے گی اس بارے میں انہوں نے بالکل بھی نہیں سوچا۔

معذور عورتیں سماجی اور مادّی رکاوٹوں کے خاتمے کا حق رکھتی ہیں۔

معذوری کے اسباب

کچھ عورتیں پیدائشی معذور ہوتی ہیں۔ کچھ عورتیں خاصا وقت صحت مند زندگی گزارنے کے بعد معذور ہوجاتی ہیں۔ کچھ عورتیں اچانک حادثے یا کسی بیماری کے باعث معذور ہوجاتی ہیں۔

تمام معذوریوں سے تحفظ فراہم کرنا ممکن نہیں ہے۔ کچھ بچے رحم کے اندر ہی مختلف طور پر تشکیل پاتے ہیں اور کوئی نہیں جانتا ہے کہ کیوں؟

لیکن بچوں کی بہت سی معذوریاں عورتوں کی زندگی کے نقصان دہ حالات کے باعث ہوتی ہیں اگر عورتوں کو مناسب غذائیت والی اشیاء کھانے کو ملیں، وہ خود کو اور ہر لمحے ماؤوں سے بچا سکیں اور صحت کی اچھی دیکھ بھال جس میں بچے کی پیدائش کے وقت دیکھ بھال بھی شامل ہے تو پھر بہت سی معذوریوں سے بچا جا سکتا ہے۔

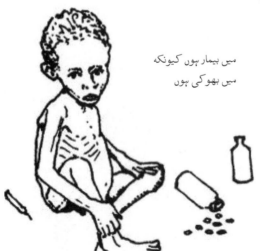

میں بیمار ہوں کیونکہ میں بھوکی ہوں

غربت اور ناقص غذائیت

غربت معذوری کا سب سے بڑا سبب ہے۔ غریب لوگ معذوری کا انتہائی نشانہ ہوتے ہیں کیونکہ وہ ناقص سینی ٹیشن، پرہجوم، تعلیم، صاف پانی یا مناسب غذا کے بغیر غیر محفوظ ماحول میں رہنے اور کام کرنے پر مجبور ہوتے ہیں۔

یہ سب ٹی بی اور پولیو جیسی بیماریوں اور ان کے باعث سنگین معذوریوں کو انتہائی عام بنا دیتا ہے کیونکہ ایسی بیماریاں ایک فرد سے دوسرے فرد میں نہایت آسانی سے منتقل ہو جاتی ہیں۔

جب کٹے ہوئے ہونٹ اور تالو والی اس لڑکی کی ماں حاملہ تھی تو اس کو کھانے کے لئے کافی مقدار میں فولک ایسڈ اور کیلشیم پر مشتمل غذا (جیسے سبز پتوں والی سبزیاں، پھلیاں اور انڈے وغیرہ) نہیں ملی تھی۔

غریب گھرانوں میں جنم لینے والے بہت سے بچے معذوریوں کے ساتھ پیدا ہو سکتے ہیں یا شیر خواری کے زمانے میں مر سکتے ہیں۔ ایسا اس لئے ہو سکتا ہے کہ ماں حاملہ تھی تو اسے مناسب مقدار میں کھانے کو نہ ملا ہو یا اس لئے کہ جب وہ لڑکی تھی تو اسے ضرورت کے مطابق کھانے کو نہ ملا ہو۔ بچپن میں لڑکی کو عام طور پر لڑکے کی بہ نسبت کم کھانے کو دیا جاتا ہے۔ نتیجے میں وہ بہت سست روی سے پروان چڑھتی ہے اور ممکن ہے اس کی ہڈیاں درست طور پر نشو و نما نہ پائی ہوں جو بعد میں بچے کی ولادت کے دوران مشکل پیدا کر سکتی ہیں خصوصاً اس صورت میں جب اسے صحت کی اچھی دیکھ بھال نہ ملے۔

اگر کسی چھوٹے یا نو عمر بچے کو کھانے کے لئے مناسب اور اچھی غذا نہ ملے تو وہ اندھا ہو سکتا ہے یا اسے سیکھنے اور سمجھنے میں مشکل پیش آ سکتی ہے۔

جنگ

اس زمانے کی جنگوں میں فوجیوں کے مقابلے میں شہری افراد زیادہ تعداد میں مارے جاتے ہیں یا معذور ہو جاتے ہیں اور ان میں بیش تر عورتیں اور بچے ہوتے ہیں۔ بم دھماکے لوگوں کو بہرہ، اندھا اور اعضاء سے محروم کرنے کے علاوہ دوسرے زخموں کا سبب بنتے ہیں۔ تشدد سے ان کی ذہنی صحت بھی بری طرح متاثر ہوتی ہے۔ ان تصادموں اور جنگوں کے نتیجے میں گھروں، اسکولوں، مراکز صحت اور زندگی گزارنے کے ذریعوں کی تباہی مزید معذوریوں، افلاس اور بیماریوں کا سبب بنتی ہے۔

جنگوں میں استعمال ہونے والی بارودی سرنگیں، کلسٹر بم، گولیاں اور کیمیائی مادے آج کے دنیا کی کسی اور وجہ کی بہ نسبت زیادہ معذوریوں کا سبب بنتے ہیں۔ یہ اکثر روزمرہ زندگی کے کام جیسے فارمنگ، پانی اور لکڑی وغیرہ جمع کرنے والی خواتین کو معذور کرتے ہیں۔

بم دھماکے اور بارودی سرنگیں بڑی تعداد میں ٹانگوں اور بازوؤں کو مجروح کرنے کا سبب بنتی ہیں اور اکثر کسی بچے یا عورت کی ٹانگ کاٹنی پڑتی ہے لیکن چار میں سے صرف ایک ہی معذور کو اس کی ضائع ہونے والی ٹانگ کے بدلے میں مصنوعی ٹانگ ملتی ہے، اس لئے کہ یہ عام طور پر مہنگی یا حاصل کرنا مشکل ہوتی ہے، ہکتی بازو اور جے پور پیر اچھی کوالٹی، کم قیمت والے دو مصنوعی اعضاء ہیں جو بھارت میں بنائے جاتے ہیں۔ ان کے بارے میں مزید معلومات کے لئے دیکھئے صفحہ 377

بارودی سرنگوں کو ممنوع قرار دینے والا بین الاقوامی معاہدہ بہت سی زندگیوں کو بچا اور معذوریوں سے تحفظ فراہم کر سکتا ہے لیکن کچھ حکومتیں اب بھی اس پر دستخط کرنے سے انکار کر رہی ہیں۔ اگر آپ کے ملک نے اب تک اس پر دستخط نہیں کئے تو حکومت پر اس کے لئے دباؤ ڈالیں۔

جوہری حادثے

بہت سے لوگ تابکاری کی بھاری مقدار کی زد میں آ کر نقصان اٹھاتے ہیں۔ ایسا 1979ء میں تھری مائل لینڈ امریکہ، 1986ء میں چرنوبل اور یوکرائن میں جوہری حادثوں کے بعد ہوا۔ جب امریکہ نے 1945ء میں جاپان پر جوہری بم گرائے تو ایسا ہی ہوا۔

یہ واقعات تابکاری اثرات کے باعث بڑے پیمانے پر تباہی اور موت کا سبب بنے۔ ان حملوں میں جولوگ بچ گئے زیادہ تر مختلف اقسام کے کینسر کا نشانہ بنے، خواہ وہ جسم کے مختلف حصوں خصوصا تھائیرائڈ گلینڈ میں ٹیومر کی شکل میں ہوا ہو یا لیوکیمیا (خون کے کینسر)، سب ہی وقت سے پہلے موت کا نشانہ بنے۔ ان قوموں میں جہاں جوہری سانحات پیش آئے وہاں ایسے بچوں کی تعداد میں اضافہ ہوا جو پیدائشی طور پر آموزشی مشکلات مثلاً ڈاؤن سنڈروم کا شکار تھے۔

صحت کی دیکھ بھال کے لئے ناقص رسائی

صحت کی اچھی دیکھ بھال بہت سی معذوریوں سے بچا سکتی ہے۔ دشوار زچگی اور ولادت سیریبرل پالسی (Cerebral Palsy) جیسی معذوری رکھنے والے بچے کی پیدائش کا سبب بن سکتی ہے۔ تربیت یافتہ برتھ اٹینڈنٹ جو خطرات کی شناخت کرکے ایمرجنسیوں سے نمٹ سکتے ہوں، پیدا ہونے والے بہت سے بچوں کو معذوری سے بچا سکتے ہیں۔ امیونائزیشن بھی بہت سی معذوریوں سے بچاتی ہے لیکن بیشتر ویکسین دستیاب نہیں ہوتی ہے یا لوگ جو بہت غریب ہوں یا شہری علاقوں سے دور رہتے ہوں یا اخراجات برداشت نہیں کر پاتے ہیں یا ہر ایک کے لئے ویکسین موجود نہیں ہوتی ہے۔

بیماری

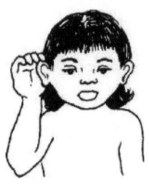

اگر حمل کے ابتدائی تین ماہ میں کوئی عورت جرمن خسرہ میں مبتلا ہوجاتی ہے تو اس کا نومولود بچہ بہرہ ہوسکتا ہے۔

کچھ بیماریاں جن میں حاملہ عورت بتلا ہوسکتی ہے، پیدا ہونے والے بچے کے لئے جسمانی یا آموزشی مسائل کا سبب بن سکتی ہے۔ پیدائشی نقائص کا سبب بننے والی بیماریوں میں جرمن خسرہ (rubella) شامل ہیں جو مولود بچوں میں بہرے پن کا ایک عام سبب ہے لیکن جرمن خسرہ کی ویکسین لینے والی عورت کو امیونائزیشن کے ایک ماہ بعد تک حاملہ نہیں ہونا چاہئے۔

سفلس (Syphilis) (دیکھئے صفحہ 163، ہرپیز (Herpes) (دیکھئے صفحہ 165) اور ایچ آئی وی (دیکھئے صفحہ 169) بھی ماں سے اس کے بچے میں منتقل ہوسکتے ہیں اور پیدائشی نقائص کا سبب بن سکتے ہیں اس لئے جب رحم میں بچہ پروان چڑھ رہا ہوتو جنسی طور پر منتقل ہونے والے امراض کے اثرات سے بچاؤ کے لئے عورتوں کے معائنے اور علاج کی ضرورت ہے۔

کچھ بیماریاں جن میں نومولود یا چھوٹا بچہ بتلا ہوسکتا ہے، جیسے Menigitis، پولیو اور خسرہ بچے کے لئے معذوری کا سبب بن سکتی ہیں۔ تحفظ کے لئے ضروری ہے کہ نومولود بچوں کو حفاظتی ٹیکے لگائے جائیں۔ (دیکھئے صفحہ 276) وہ مقامات جہاں جذام عام ہے، بچوں کو جس قدر جلد ممکن ہو ٹیسٹ کیا جانا چاہئے۔

ادویات اور انجیکشنز

جب درست استعمال ہو تو انجکشن کے ذریعے دی جانے والی ادویات، جیسے کچھ ویکسینز صحت کے تحفظ اور معذوری سے بچاؤ کے لئے اہم ہوتی ہیں تاہم پوری دنیا میں غیر ضروری انجکشنوں کا استعمال عام ہے۔

گندی سوئی یا سرنج سے لگائے جانے والے انجکشن انفیکشن کا ایک عام سبب ہیں اور ایسے وائرس دوسروں تک منتقل کر سکتے ہیں جو ایچ آئی وی/ایڈز یا ہپاٹائٹس جیسی سنگین بیماریوں کا سبب بنتے ہیں۔ ہر سال یہ غیر ضروری انجکشن لاکھوں افراد خصوصاً بچوں کو بیمار ڈالتے ہیں، ہلاک یا معذور کرتے

غیر ضروری انجکشنوں سے بچیں

ہیں۔ غیر صاف شدہ انجکشن بھی ایسے انفیکشن کا سبب بنتے ہیں جو فالج یا ریڑھ کی ہڈی کی خرابی یا موت کا سبب بن سکتے ہیں۔ انجکشن کے ذریعے دی جانے والی کچھ ادویات ماں کے رحم میں موجود بچے کے لئے خطرناک الرجک ردِعمل، زہریلے پن اور بہرے پن کا سبب بن سکتی ہیں۔

سوئی یا سرنج ہر بار جراثیم سے پاک کئے بغیر ایک سے زیادہ فرد کو انجکشن لگانے کے لئے استعمال نہ کی جائے۔

بعض اوقات حمل کے دوران لی جانے والی کچھ ادویات یا نشہ آور اشیاء رحم میں پروان چڑھنے والے بچے میں معذوری کا سبب بن سکتی ہیں۔ ولادت کے عمل یا ماں کی زچگی کو قوت دینے والی آکسی ٹوسن (Oxytocin) ادویات کی زائد مقدار ولادت کے دوران بچے کو آکسیجن سے محروم کر سکتی ہیں۔ یہ دماغ کو پہنچنے والے نقصان کا ایک اہم سبب ہے۔ حمل کے دوران الکحل اور تمباکو کا استعمال بھی نشوونما پاتے ہوئے بچے کو نقصان پہنچا سکتا ہے۔

ہر ایک کسی بھی دوا کے استعمال کے ممکنہ خطرات اور فوائد پر غور کرے۔ ڈاکٹر، نرسیں، دیگر صحت کارکن، ادویات فروش سب ہی ادویات خصوصاً انجکشنوں کے غلط استعمال اور بے جا استعمال کو روکیں۔ غیر ضروری انجکشنوں کے خطرے کے بارے میں تدریس کے نظریات کے لئے **ہیلپنگ ہیلتھ ورکرز لرن** کا باب 19،18 اور 27 دیکھئے

خطرناک حالات میں کام

ایسی عورتیں جو مناسب آرام کے بغیر مسلسل کام کرتی ہیں حادثات کا امکان زیادہ رکھتی ہیں۔ فیکٹریوں، کانوں اور زرعی کاشت کے لئے کام کرنے والی خواتین مشینوں، آلات یا کیمیائی مادّوں کی زد میں آ سکتی ہیں۔ حادثات، کام کی زیادتی اور کیمیائی مادّوں کی زد میں رہنا سب ہی معذوری کا سبب بن سکتے ہیں۔

عورتوں کی بڑھتی ہوئی تعداد کام کرتے ہوئے تشدد کے باعث مستقلاً مجروح ہو رہی ہے بعض اوقات عورتوں سے تندہی اور تیز رفتاری سے کام لینے کے لئے نگران پُرتشدد اور دھمکی آمیز طریقے اپناتے ہیں۔ کچھ صورتوں میں انتظامیہ عورتوں کو غیر محفوظ حالات میں کام کرنے پر ہڑتال یا احتجاج سے روکنے کے لئے فوج یا پولیس کو استعمال کرتی ہے۔

حادثات

بہت سی عورتیں اور بچے گھروں میں چولہوں کی آگ سے یا گر کر، سڑکوں پر حادثات، زہریلے کیمیائی مادّوں میں سانس لینے یا انہیں پینے سے معذور کرنے والی خرابیوں میں مبتلا ہو جاتے ہیں۔ کام کی جگہ ہونے والے حادثات خصوصاً تعمیرات، زراعت، کان کنی اور چھوٹے کاروباروں جیسے کم تر باضابطہ شعبوں میں معذوری کا ایک عام ذریعہ ہیں۔

زہر اور پیسٹی سائیڈز

رنگوں میں سیسہ جیسے زہر، چوہے مارنے کے زہر جیسے پیسٹی سائیڈز اور دوسرے کیمیائی مادّے لوگوں میں معذوریوں اور رحمِ مادر میں پروان چڑھنے والے بچوں میں پیدائشی نقائص کا سبب بن سکتے ہیں۔ حمل کے دوران تمباکو کھانا اور سگریٹ نوشی، دھواں نگلنا، الکحل پینا بھی بچے کو پیدا ہونے سے پہلے نقصان پہنچا سکتا ہے۔

یہ عورت ایک فارم ورکر تھی اور دوران حمل خطرناک کیمیائی مادّوں میں کام کرتی رہی؛ جس نے اس کے بچے کو اس وقت متاثر کیا جب وہ رحم میں تھا۔ یہ بچہ معذوری کے ساتھ پیدا ہوا۔

محنت کش عموماً اپنی ملازمت کی نوعیت کے اعتبار سے بغیر یہ سیکھے کہ انہیں محفوظ طریقے سے کس طرح استعمال کیا جائے یا یہ جانے بغیر یہ خطرناک ہیں، کیمیائی مادّے استعمال کرتے ہیں۔ فیکٹریوں میں حادثات زہریلے مادّوں کے ہوا، پانی یا زمین پر اخراج کا سبب بن سکتے ہیں اور اس صورت میں صحت کے خوفناک مسائل کا جن میں مستقل معذوری بھی شامل ہے سبب بن سکتے ہیں۔

توارثی معذوریاں

بعض معذوریاں جیسے اسپائنل مسکیولر اٹروپی اور مسکیولر ڈس تھروپی (مسلز اور نرووز کی بیماریاں) توارثی کہلاتی ہیں۔ ایسی عورتیں جن کے ایک یا دو بچے توارثی معذوری کا شکار ہوں اسی معذوری کا شکار ایک اور بچے کے زیادہ امکانات رکھتی ہیں۔ دوسری معذوریاں خون کا قریبی رشتہ رکھنے والوں (جیسے بھائیوں، بہنوں، فرسٹ کزنز، والدین اور بچے) کے بچوں میں ہو سکتی ہیں، چالیس برس یا اس سے زائد عمر کی ماؤں کے بچے ڈاؤن سنڈروم کا شکار ہونے کے امکانات رکھتے ہیں **تاہم بیشتر معذوریاں توارثی نہیں ہوتی ہیں** بیش تر صورتوں میں معذور پیدا ہونے والے بچوں کے والدین معذوری کا سبب بننے کے ذمے دار نہیں ہوتے ہیں لہذا ان کو الزام نہیں دینا چاہیئے۔

ڈاؤن سنڈروم کے جسمانی نقائص

آنکھیں جو اوپر کی طرف گھومی ہوئی ہیں، بعض اوقات ترچھی آنکھیں یا ناقص بصارت

ڈھکے ہوئے کان

چھوٹا منہ، کھلا رہتا ہے منہ کا اوپری حصہ اونچا اور تنگ ہے۔ زبان باہر لٹکتی رہتی ہے۔

معذوری کے بارے میں غلط نظریات اور مفروضے

معذوری کے بارے میں مقامی رواج اور عقائد میں غلط اور نقصان دہ خیالات شامل ہو سکتے ہیں، کچھ لوگ سمجھتے ہیں کہ کوئی عورت اس وقت معذوری کا شکار ہوتی ہے جب اس نے یا اس کے والدین یا اس کے آباؤ اجداد یا والدین میں سے کسی ایک کو ناراض کیا ہو یا غیر جائز جنسی تعلق قائم کیا ہو۔ عام طور پر لوگ ماں کو ذمے دار قرار دیتے ہیں لیکن **بچے کی معذوری کی ذمہ دار مائیں نہیں ہوتی ہیں۔** معذوروں کا کسی کو ذمہ دار قرار دینا کوئی مدد نہیں کرتا ہے۔

معذوری کے حوالے سے ایک اور تکلیف دہ خیال یہ عقیدہ ہے کہ کوئی بھی جو "مختلف" ہے اسے نکال دیا جائے، اس کی تضحیک اور توہین کی جانی چاہیے۔ کچھ لوگ سمجھتے ہیں کہ معذوری رکھنے والا فرد بدی کی علامت ہے یا بدقسمتی کا سبب بنے گا۔ معذور یوں کا شکار عورتیں عموماً برے سلوک کا نشانہ بنتی ہیں یا گزر بسر کے لئے بھکاری بننے یا جسم فروشی کے لئے مجبور ہوتی ہیں۔ بعض اوقات معذور عورتیں اس لئے جنسی زیادتی کا نشانہ بنتی ہیں کہ لوگ یہ سمجھتے ہیں کہ یہ ایچ آئی وی/ایڈز سے محفوظ ہیں یا کسی معذور عورت کے ساتھ جنسی عمل سے ایچ آئی وی/ایڈز کا علاج ہو سکتا ہے۔

لیکن سچ یہ ہے کہ کسی بھی معذور عورت کے ساتھ زیادتی نہیں ہونی چاہیے۔ معذوری کوئی سزا نہیں ہے۔ معذوری کسی جادوگری یا بددعا کے ذریعے نہیں ہوتی ہے۔ معذوری متعدی یا دوسروں کو منتقل ہونے والی بیماری نہیں ہے لہذا یہ دوسروں تک نہیں پھیلتی ہے۔

ممکن ہے کہ لوگ نہ سمجھتے ہوں کہ ایک معذور عورت کیا کر سکتی ہے اور کیا نہیں کر سکتی ہے۔ وہ تسلیم نہیں کرتے ہوں کہ

> کچھ حاملہ عورتیں میری دکان سے دور رہتی ہیں کیونکہ وہ سمجھتی ہیں کہ میرے قریب آنے سے ان کا بچہ بھی میری طرح بہرہ پیدا ہوگا۔

- آپ ایک بالغ فرد ہیں اور خود فیصلے کر سکتی ہیں۔
- آپ کو تعلیم کی ضرورت ہے۔
- آپ کو صحت کی دیکھ بھال کی ضرورت ہے۔
- آپ کینسر، ایچ آئی وی/ایڈز جیسے امراض میں بھی مبتلا ہو سکتی ہیں۔
- آپ کو ترس اور ہمدردی نہیں مواقعوں اور احترام کی ضرورت ہے۔
- آپ کام کر سکتی ہیں، آپ بھی پیشہ ور ماہر بن سکتی ہیں اور کوئی روزگار حاصل کر سکتی ہیں۔
- آپ سرمایہ حاصل کر سکتی ہیں، جائیداد رکھ سکتی ہیں، اپنے گھر کو تشکیل دے سکتی ہیں اور اس کی معاونت کر سکتی ہیں۔
- آپ سوچتی، محسوس کرتی ہیں اور جذبات رکھتی ہیں۔
- آپ ڈانس اور ورزش کر سکتی ہیں۔
- آپ ذمہ داریاں اٹھا سکتی ہیں، فیصلے کر سکتی ہیں اور قائدانہ کردار سنبھال سکتی ہیں اور اپنی کمیونٹی کے کاموں میں شامل ہو سکتی ہیں۔
- آپ کسی سے ازدواجی تعلق رکھ سکتی ہیں، آپ کسی سے محبت کر سکتی ہیں یا کوئی غیر معذور یا معذور فرد آپ سے محبت کر سکتا ہے۔

> میں بچہ نہیں ہوں اور مجھے ضرورت نہیں ہے کہ آپ میرے لئے سوچیں یا کچھ کریں۔

- آپ جنسی خواہشات رکھتی ہیں اور جنسی طور پر فعال ہوسکتی ہیں ۔
- آپ شادی کرسکتی ہیں اور اولاد پیدا کرسکتی ہیں ۔
- آپ جنسی تعلق کی اہل ہیں لیکن ممکن ہے ایسا نہ چاہتی ہوں ۔
- اگر آپ سیکھنے یا سمجھنے میں دشواری محسوس کرتی ہیں تو آپ دوسری عورتوں سے نہ تو زیادہ نہ ہی کم جنسی ضروریات رکھتی ہیں ۔
- آپ دوسری عورتوں کی طرح بیش تر ایسے بچے پیدا کرسکتی ہیں جو معذوری نہ رکھتے ہوں ۔
- ایک اچھی ماں بن سکتی ہیں ۔
- اگر آپ کوئی جسمانی خرابی آموزشی معذوری رکھتی ہیں تو آپ دماغی طور پر بیمار یا غیر مستحکم نہیں ہیں ۔
- آپ لوگوں یا بچوں کو بددعا نہیں دیتی ہیں اور کوئی ایسا بُرا شگون یا علامت نہیں ہیں جس سے بچا جائے ۔

یہ بہت ہی بُری بات ہے کرانتی نے دیوتاؤں کو ناراض کیا ہوگا

میں نے دیوتاؤں کو ناراض نہیں کیا ہے میں جب چھوٹی تھی تو مجھے آلودہ سوئی سے انجیکشن لگایا گیا تھا یہی وجہ ہے کہ میری ٹانگ مفلوج ہے -

تبدیلی کے لئے کام

ایک عورت کی معذوری صرف اسے ہی متاثر نہیں کرتی ہے ۔ یہ بہت سے لوگوں، اس کے گھر انے، اس کے دوستوں اور سب سے زیادہ اس کی کمیونٹی پر اثر انداز ہوتی ہے ۔ ایک معذور عورت زیادہ صحت مند بن سکتی ہے جب اس کے اِرد گرد رہنے والے اسے اہمیت دیں اور اس کی مدد کریں، جس طرح معذوریاں نہ رکھنے والی عورتوں کے ساتھ پیش آیا جاتا ہے، یہ تبدیلی لانا مشکل کام ہے لیکن یہ ناممکن نہیں ہے ۔

میں ایک ایسے دن کا خواب دیکھتی ہوں جب دنیا بھر کے لوگ یہ سمجھ لیں گے کہ معذور ہونا، بیمار ہونے کی طرح نہیں ہے اور یہ کہ ہم عام طور پر انتہائی صحت مند ہوتے ہیں اور دوسری تمام عورتوں کی طرح صحت مند رہنا ہماری بھی ضرورت ہے -

معذور عورتیں کیا کر سکتی ہیں

آپ اپنے حقوق کے لئے پیروکاری کرکے اپنی آواز یں قابل سماعت اور یہ یقینی بنائیے کہ معذوری کے امور ایک ترجیح بن جائیں۔

- ایک جگہ محدود ہونا مسترد کر دیں، ہم جو بہیں اور مختلف تجربات کا خیر مقدم کریں۔
- کاروباری مہارتیں سیکھیں، خود کو معاشی طور پر بااختیار بنائیں۔

مارکیٹ میں سیکیورٹی حاصل کرنا

اوپا نڈ لو کا تعلق زمبابوے سے ہے وہ وہیل چیئر استعمال کرتی ہے اور اپنی کمیونٹی کی ایک قابل احترام شخصیت ہے۔ اس نے سبزیاں اور ٹماٹر فروخت کرکے ایک کامیاب منصوبے کی ابتداء کی۔ اب اس کے اردگرد کے سب لوگ اس سے سبزی خریدتے ہیں۔ اپنی مستحکم آمدنی کے باعث اوپا نے ایک گھر بھی خرید لیا ہے۔

اپنی کہانی سنائیے

- کمیونٹی کی ہر سطح پر شراکت پر زور دیجئے۔
- لڑکیوں اور عورتوں کے رول ماڈل بنیں۔
- اپنی معذوریوں کے بارے میں اظہار کریں۔
- دوسری معذور لڑکیوں اور عورتوں کو جو کہیں جانا چاہیں، ساتھ جانے کی پیشکش کیجئے۔
- کھیلوں میں حصہ لیجئے۔

اولمپکس معیار کے ایتھلٹس

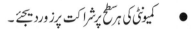

پیرا اولمپک گیمز میں حصہ لینے والی عورتوں کی تعداد مسلسل بڑھ رہی ہے۔ یہ کھیلوں کے بین الاقوامی مقابلے ایسے ایتھلٹس کے لئے ہیں جو نقل و حرکت، کٹے اعضاء، بصری معذوریوں اور سیریبرل پالسی سمیت تمام دیگر معذوریاں رکھتے ہیں۔ پیرا اولمپک گیمز، اولمپک گیمز کے بعد ہر چار سال میں ہوتے ہیں۔ بہت سے لوگوں کے لئے چشم کشا ہے کہ وہ معذور عورتوں کو اعتماد اور مہارت کے ساتھ مقابلہ کرتا ہوا دیکھیں۔

بالرجس نے مفروضوں کو غلط ثابت کیا

یوگنڈا میں نابینا بالرکونسٹینس سپنڈا کو 'سپورٹس پرسن آف دا ایئر' نامزد کیا گیا، اس نے ان مفروضوں کی تردید کی ہے کہ اس جیسی معذور عورت دنیا کے لئے غیر ضروری اور غیر مفید ہے۔ کونسٹینس ایک عالمی مقابلے میں دو گولڈ میڈل حاصل کر چکی ہے اس نے بلائنڈ بالنگ میں بہت سے میڈل حاصل کئے ہیں جنوبی افریقہ، اسکاٹ لینڈ اور برطانیہ میں ہونے والے مختلف مقابلوں میں شرکت کرنے والی کونسٹینس دنیا بھر کی عورتوں اور لڑکیوں کو اپنی چھپی ہوئی صلاحیتیں ڈھونڈ نے پر مائل کرتی ہے۔

آپ مل کر طے کر سکتی ہیں کہ آپ کی کمیونٹی میں کون سی چیزیں سب کے لئے زندگی بہتر بنانے کے لئے تبدیل کی جاسکتی ہیں ۔ مثلاً آپ مندرجہ ذیل کام کر سکتی ہیں۔

- ایسی عورتوں کے لئے جو پڑھ یا لکھ نہ سکتی ہوں، خواندگی کا آغاز کرنا۔

- مل کر دستکاری، آلات وغیرہ بنانے اور فروخت کرنے کا کاروبار شروع کرنا۔

- سماجی خدمات سے متعلق معلومات میں ایک دوسرے کو شریک کرنا اور انہیں زیادہ قابل رسائی بنانا۔

- خواہ کم شرح سے قرض ہو یا عطیہ حاصل کرنے کی کوشش کرے، آمدنی والا یا کوئی منصوبہ شروع کرنا یا کمیونٹی کو زیادہ قابل رسائی بنانا۔

- معذوریوں کے بارے میں آگہی بڑھانا اور خود مختاری کے بارے میں سوچنے کے نئے انداز کے لئے نمائندگی کرنا۔

- مقامی رہنماؤں یا حکومت کے ساتھ مل کر معذور عورتوں کے بہتر علاج کے لئے کام کرنا۔

آپ سماجی گروپوں کی مدد بھی کر سکتی ہیں

- ان مسائل صحت کے مختلف اسباب کا جائزہ لیں جو آپ اور دیگر معذور عورتوں کو درپیش ہیں پھر فیصلہ کریں کہ ان میں سے کن کو کمیونٹی تبدیل کر سکتی ہے۔

- پیروکاری کے ذریعے ایسی سرگرمیوں اور خدمات کے لئے حرکت میں آ ئے، جو معذور افراد سمیت سب کے لئے صحت کی بہتر دیکھ بھال، تعلیم اور آمدورفت کی سہولتوں کی فراہمی سے زندگی بہتر بنائے، قابل رسائی سہولتوں پر زور دیں۔

- ایک چھوٹا گروپ تشکیل دیں، کیونکہ ادارے کی آواز ایک فردکی آواز سے زیادہ طاقتور ہوتی ہے۔ یہ طے کریں کہ گروپ کے مقاصد کیا ہیں اور ان مقاصد کے لئے گروپ کیا اقدام کرے گا۔

- ایسی پالیسیوں اور قوانین کے خلاف آواز اٹھا ئیں جو آپ کے خلاف اور امتیازی ہیں۔

ایک وقت تھا کہ یوگنڈا میں بہرے لوگوں کو گاڑی چلانے کی اجازت نہیں تھی لیکن ہم نے اس پر احتجاج کیا۔ اب ہم کار خرید سکتے ہیں اور ہمیں گاڑی چلانے کی اجازت حاصل ہے۔

سکھانا کہ لوگ اپنا حق کس طرح حاصل کریں

بھارت کے ریاستی دارالحکومت بنگلور کی وہیل چیئر استعمال کرنے والی ڈورتھی نے اس عمارت میں جہاں وزیراعلیٰ کے دفاتر ہیں کوئی ریمپ نہیں پایا۔ عمارت کا داخلی راستہ بھی اتنا چھوٹا تھا کہ وہ وہیل چیئر سمیت اندر داخل نہیں ہو سکتی تھی۔ اس نے گارڈز سے اس بارے میں بات کی اور ان پر زور دیا کہ وہ وزیراعلیٰ سے ملاقات کا وقت حاصل کرنے کے لئے اس کی مدد کریں۔

بعد میں ڈورتھی نے لوگوں کو یہ بتانے کے لئے کہ اس کے ساتھ کیا پیش آیا ہے سینکڑوں ای میل بھیجے۔ اس سے حکومت پر دباؤ بڑھا کہ وہ تبدیلیاں لائے۔

ایک اور موقع پر ڈورتھی ایک کرکٹ میچ دیکھنے گئی جہاں پولیس نے اس سے پوچھا ''تم یہاں کیوں آنا اور براہ راست میچ دیکھنا چاہتی ہو، تم گھر میں رہ کر آرام سے ٹی وی پر میچ دیکھ سکتی ہو۔''

اس نے جواب دیا کہ ''دوسروں کی طرح وہ بھی میچ براہ راست دیکھنا چاہتی ہے۔''

گھرانے کیا کر سکتے ہیں؟

وہ انداز جس سے معذوریاں رکھنے والی خواتین کے گھر والے، دوست اور دوسرے لوگ ان کی معاونت کرتے ہیں، بہت بڑا فرق ڈالتا ہے۔ اکثر اوقات معذور لڑکی کو احمق، محتاج، اپنی اور دوسروں کی مدد کرنے کے لئے نا قابل سمجھا جاتا ہے لہٰذا وہ کسی سہولت کی حقدار نہیں ہوتی۔ بعض اوقات کچھ گھرانے اسے چھپانے کے قابل شرمناک بوجھ کے بطور دیکھتے ہیں اور وہ اس کے سننے یا جانے یا خود فیصلہ کرنے کے حق سے انکار کرتے ہیں۔ اگر کسی گھرانے یا کمیونٹی میں ایسا ہوتا ہے تو خرابی معذور لڑکی یا عورت میں نہیں بلکہ اس کے اِرد گرد موجود لوگوں میں ہے۔

اعتماد پروان چڑھانا

جب کرسٹائن تیرہ برس کی تھی، اس کی ٹانگ ایک بیماری کی وجہ سے کاٹ دی گئی۔ ابتداء میں اس نے سوچا کہ یہ اس کے خوابوں کی موت ہے لیکن اس کے والدین اس نے اس کے ساتھ بہت اچھا برتاؤ کیا اور کرسٹائن کا اعتماد بحال ہوگیا۔ پہلے پہل کرسٹائن کے والدین نے اس کو حد سے زیادہ تحفظ دیا لیکن اس نے اپنے والدین پر زور دیا کہ وہ اس کے ساتھ بھی اس کے دوسرے بہن بھائیوں کی طرح پیش آئیں۔ کرسٹائن نے کالج کی تعلیم حاصل کی اور اپنی برتری کے لئے ایوارڈ پائے۔ کرسٹائن میں تبدیلی اس کے گھر والوں اور کمیونٹی کے یہ تسلیم کرنے سے آئی کہ کرسٹائن کی کٹی ہوئی ٹانگ اسے اپنے خوابوں کی تکمیل سے دور نہیں رکھ سکے گی۔

وہ واحد چیز جوان رویوں کو تبدیل کر سکتی ہے سماجی آگہی ہے۔ معذور عورتوں اور لڑکیوں کو اچھی غذا، تعلیم، صحت کی دیکھ بھال اور جسمانی و سماجی سرگرمیوں میں شامل ہونے کے مواقعوں کی ضرورت ہوتی ہے۔ ''دماغی صحت'' اور دیکھ بھال کرنے والوں کی تائید و حمایت کے لئے دیکھئے باب نمبر 3 اور 15۔

آپ کئی مہارتیں پروان چڑھا سکتی ہیں

لاؤس کی ہونگ ہا دو برس کی عمر میں پولیو کا شکار ہوگئی تھی۔ اپنے خاندان کی مدد کے ساتھ اس نے یونیورسٹی سے فرانسیسی میں گریجویشن کی ڈگری حاصل کی۔ جب ہونگ ہا کو اپنے لئے ملازمت نہ ملی تو اس نے کپڑے سینا سیکھا اور اپنے گھر میں دکان کھول لی۔ اس کے ساتھ اس نے ایک بار پھر تعلیم کا سلسلہ شروع کیا اور انگریزی سیکھنے لگی۔ اس نے اپنی ایک سہیلی کے ساتھ مل کر اپنے گھر میں چھوٹا سا انگلش ٹریننگ سینٹر کھول لیا۔ وہ معذوری کے ایک پروگرام کی کوآرڈینیٹر بھی ہے۔

ابتدائی معاونت

اپنی زندگی کے اولین برسوں میں تمام بچے بقیہ عمر کے کسی بھی حصے سے زیادہ بہتر اور آسانی سے جسمانی، ذہنی، ابلاغی اور سماجی مہارتیں سیکھتے ہیں۔ کیونکہ ایک بچہ پیدا ہونے کے بعد ہی سیکھنا شروع کر دیتا ہے لہٰذا یہ نہایت ہی اہم ہے کہ گھر والے جس قدر جلد ممکن ہو، معذوری رکھنے والے بچوں کی مدد پر اضافی توجہ دینا شروع کر دیں۔

یہ اس لئے بھی اہم ہے کہ ہر نئی مہارت جو بچہ سیکھتا ہے اس کی بنیاد وہ مہارتیں ہوتی ہیں جو وہ پہلے حاصل کر چکا ہو۔ ہر نئی مہارت اس کے لئے یہ بھی ممکن بناتی ہے کہ وہ زیادہ مشکل مہارتیں سیکھ سکے۔ اسی لئے جب کوئی بچہ ابتدائی مہارت سیکھ نہیں پاتا ہے تو وہ ایسی دوسری مہارتیں بھی نہیں سیکھ سکے گا جن کا انحصار اس پر ہو۔

معذور بچوں کے والدین کے لئے تائیدی گروپ کی ابتداء کرنا

معذور بچوں کی ماؤں کو اکثر ان کے شوہر چھوڑ دیتے ہیں اور وہ خود ان کی پرورش کرنے پر مجبور ہوتی ہیں۔ والدین کے تائیدی گروپ ان کی مدد کر سکتے ہیں۔ معذور بالغ عورتیں ان مسائل کی اقسام پر مشاورت فراہم کر سکتی ہیں جو معذور لڑکیوں کو آئندہ بڑے ہونے پر پیش آ سکتے ہیں۔ اس سے معذور لڑکیوں کی ماؤں کو مدد مل سکتی ہے کہ وہ انہیں بہتر طور پر مدد دے سکیں۔

آپ نو عمر معذور لڑکیوں کے لئے بھی تائیدی گروپ شروع کر سکتی ہیں تا کہ وہ خود بھی ایک دوسرے کی مدد اور معاونت کر سکیں۔

کمیونٹیز کیا کر سکتی ہیں؟

> جب صحت کارکن، اساتذہ، سماجی رہنما اور ہمارے گھر والے اور پڑوسی معذوری کے حوالے سے اپنے رویئے بدلیں گے تو ہم اپنا کام کر سکیں گے، ٹھوس تعلقات اور ربط رکھتے ہوئے ہم اپنی کمیونٹیز کو خوشحال اور مضبوط بنا سکتے ہیں۔

کمیونٹی گروپ، حکومت، صحت کارکنوں، اساتذہ، کمیونٹی بنیاد بحالی کارکنوں اور سماجی رہنماؤں کو معذوری کے مسائل سے آگاہ کر سکتے ہیں۔ یہ عوام کو بھی عوامی تھیٹر، تبادلہ خیال اور دوسرے طریقوں سے آگاہ کر سکتے ہیں کہ معذور عورتیں بھی دوسرے غیر معذور افراد کی طرح تعلیم، صحت کی دیکھ بھال اور ٹرانسپورٹ کی سہولتوں کا حق رکھتی ہیں۔ کمیونٹیز معذور عورتوں کے لئے روزگار کے مواقعے پیدا کر سکتی ہیں اور معذور عورتوں کے لئے خدمات کے حوالے سے معلومات فراہم کر سکتی ہیں۔

جب والدین اور گھرانے انہیں پسند اور قبول کر لیتے ہیں اور وہ تعلیم، روزگار اور صحت کی دیکھ بھال کی سہولتیں حاصل کر سکتی ہوں تو معذور لڑکیاں اور عورتیں پُر اعتماد اور خود پر بھروسہ کرنے والی بن جاتی ہیں اور اپنی بھر پور صلاحیتوں کے ساتھ پروان چڑھتی ہیں۔ تمام سماجی وسائل اور ذرائع جیسے اسکول، بینک، مذہبی مقامات، اسپتال اور کلینک بھی ہر فرد کے لئے قابل رسائی ہونا چاہئیں۔

تعلیم

معذور لڑکیاں، معذور لڑکوں کے مقابلے میں اسکول جانے کے کمتر مواقع رکھتی ہیں۔

معذوریاں رکھنے والی لڑکیوں کے لئے تعلیم انتہائی اہم ہے۔ اس میں بہری یا اندھی لڑکیوں کے لئے اشاروں کی زبان اور بریل یا آڈیوکیسٹ کے ذرائع بھی شامل ہیں۔

بہت سے غریب ممالک میں معذور لڑکیاں اسکول جانے اور تعلیم حاصل کرنے سے قاصر ہوتی ہیں، ممکن ہے کہ وہ بالغ ہونے کے بعد گزر بسر کے لئے بھیک مانگنے پر مجبور ہوجائیں اگر پورا معاشرہ معذور افراد کے تعلیمی حقوق کے لئے کام کرتا ہے تو اس سے نمایاں فرق پڑ سکتا ہے۔ سماج بنیادی گروپ مسائل پر تبادلہ خیال کر سکتا ہے اور دوسرے بچوں سمیت ہر ایک کو مائل کر سکتا ہے کہ وہ معذوریاں رکھنے والی لڑکیوں کا خیر مقدم اور احترام کریں۔ یہ گروپ معذوروں کے لئے ابتدائی تعلیم کے مواقعوں کا اہتمام کر سکتا ہے یا سرکاری امداد یا دیگر ذرائع سے مدد فراہم کر سکتا ہے۔ معذور لڑکیاں تعلیم حاصل کرکے اپنی اپنی کمیونٹیز کی مدد کر سکتی ہیں اور انہیں خوشحال بنا سکتی ہیں۔

پڑھنا اور لکھنا سیکھنے سے میرے لئے یہ ممکن ہوا کہ میں وقار کے ساتھ اپنی گزر بسر کر سکوں۔

کمیونٹیز کو ہر ایک کے لئے قابل رسائی بنانا

دنیا بھر میں معذور عورتیں کلینک، اسکولوں، مارکیٹوں، شہری سڑکوں، بسوں اور سماج کو معذور افراد کے لئے زیادہ قابل رسائی بنانے کے لئے کام کر رہی ہیں۔

میں خود کو بے بس اور لاچار محسوس کرتی ہوں۔ میں ہمیشہ گھر سے نکلنے کے لئے دوسرے لوگوں کی مدد پر انحصار کرنے پر مجبور ہوں۔ اگر میرے گھر میں اور دوسری عوامی عمارتوں میں ریمپ ہوتے تو میں خود ہی اپنے طور پر ارد گرد کے مقامات آ جا سکتی۔ میں جب چاہتی باہر جا سکتی تھی اور مجھے مدد کے لئے دوسرے لوگوں کا انتظار نہ کرنا پڑتا۔

کمیونٹی یہ بات یقینی بنا سکتی ہے اس کے بجائے اس بعد میں تبدیلیاں لائی جائیں جب عمارتیں اور سڑکیں تعمیر ہو رہی ہوں تو انہیں قابل رسائل بنایا جائے۔ اس طرح تمام عوامی سہولتیں، عمر، صلاحیت یا صورتحال سے قطع نظر ممکنہ حد تک زیادہ سے زیادہ لوگوں کے لئے قابل استعمال ہوں گی۔ یہ مثالی یا دشوار حالات میں سب لوگوں کے کام آئیں گی خواہ وہ جوان ہوں یا بوڑھے، شاندار صلاحیتوں کے مالک ہوں یا محدود صلاحیتیں رکھتے ہوں۔ (دیکھئے صفحہ 38 تا 40)

جب مقامات قابل رسائی ہوں تو ہر ایک کے لئے آسان ہوتا ہے کہ وہ سماجی سرگرمیوں میں زیادہ سے زیادہ شامل ہو سکے لیکن رسائی، ریمپ جیسی ماڈی سے زیادہ چیز ہے۔ رسائی حاصل ہونے کا مطلب یہ بھی ہے کہ ہر فرد یہ اظہار کر سکے اور سمجھ سکے کہ کیا ہو رہا ہے۔ اس صورت میں ایک معذور عورت اپنے لئے زیادہ کام کر سکتی ہے اور اس صورت میں زیادہ لوگ دیکھیں گے کہ معذوری زندگی کا ایک فطری حصہ ہے۔ جب معذور عورتیں سماج کا قابل قدر حصہ بنیں گی، سماج بھی معذوری کے حوالے سے مختلف انداز سے سوچنا شروع کرے گا۔

حکومت کو چاہیئے کہ وہ آمدورفت کے ذرائع، عمارتوں، عوامی پروگراموں اور سہولتوں کو بشمول، معذور عورتوں کے، ہر ایک کے لئے قابل استعمال بنانے کے لئے ضروری وسائل فراہم کرے۔ ان لوگوں کو بھی کچھ سزا دی جائے جو تعاون کرنے سے انکار کریں۔

یہ چند مثالیں ہیں کہ کس طرح معذوریاں رکھنے والی عورتوں نے اپنی زندگی میں تبدیلیاں ممکن بنائیں۔

حکومت کو تبدیلیوں کے لئے آمادہ کرنا

رکاوٹ کے بغیر رسائی پر ورکشاپ میں شریک ہونے کے بعد، لاؤ ڈس ایبلڈ ومنز ڈیولپمنٹ سینٹر نے معذور افراد کے لئے رکاوٹوں کے بغیر رسائی کے موضوع پر ایک ویڈیو بنائی۔ انہوں نے حکومت کے مختلف شعبوں سے معذوریاں رکھنے والے افراد کی معاشرے میں شراکت کی سہولت فراہم کرنے کے لئے مذاکرات شروع کئے۔ ان کے خیالات وزیراعظم کے علاوہ ابلاغ، مواصلات، ڈاک اور تعمیرات، محنت، سماجی بہبود اور امور خارجہ کی وزارتوں نے قبول کر لئے۔ انہوں نے دارالحکومت وینٹائین میں 47 مقامات پر ریمپ تعمیر کرنے کے لئے فنڈ حاصل کر لئے۔

رسائی ممکن بنانے کے لئے عورتوں کی کوشش

جب بچپن میں میکسیکو کی ایلیسیا کونٹریاس پولیو کے باعث معذور ہوئی تو اس کے والدین نے یہ تصور بھی نہیں کیا تھا کہ وہ اسکول جا سکے گی۔ ایلیسیا کے بارے میں فزیکل تھراپسٹ نے اندازہ لگایا کہ وہ کتنی ذہین ہے۔ اس نے ایلیسیا کے والدین پر زور دیا کہ وہ اسے اسکول میں داخل کرائیں۔ اس کی پسندیدہ کلاس تیسری منزل پر ہوتی تھی۔ وہ وہیل چیئر استعمال کرتی تھی لہٰذا ہر روز اسے اپنی کلاس میں جانے کے لئے ہاتھوں اور گھٹنوں کے بل تین منزلہ سیڑھیاں طے کرنی پڑتی تھیں۔ یونیورسٹی کے آخری سال میں، ایلیسیا کی ملاقات معذور عورتوں کے ایک گروپ ''Free Access'' سے ہوئی جو معاشرے میں تبدیلیوں کے لئے کام کر رہا تھا۔ گروپ کا یقین تھا کہ وہ بھی دوسروں کی طرح یکساں حقوق رکھتا ہے لہٰذا وہ اپنے معاشروں کو زیادہ قابل رسائی بنانے کے لئے کام کر رہا تھا۔ اس گروپ سے وابستہ خواتین، اداروں کے سربراہوں کے پاس گئیں اور انہیں معذور افراد کی سہولت کے لئے زندگی کی بہتر بنانے کے لئے تبدیلیوں کا قائل کرنے کی کوشش کی ۔ مثلاً انہوں نے اپنے شہر کے ڈائریکٹر ٹرانسپورٹ سے بات کی ۔ انہوں نے واضح کیا کہ معذور افراد کے لئے کہیں جانا کس قدر دشوار ہے ۔ ڈائریکٹر ان سے متاثر ہوا اور اس نے بہت سی بسوں کو معذور افراد کے لئے قابل رسائی بنایا۔

ایلیسیا نے بھی تبدیلی کے لئے کام کرنے کا فیصلہ کیا۔ وہ یونیورسٹی ڈائریکٹر کے پاس گئی اور اس نے کہا کہ اس کی کلاس تیسری منزل سے پہلی منزل میں منتقل کی جائے۔ ڈائریکٹر نے اس پر فوراً ہی آمادگی ظاہر کر دی۔ ایلیسیا کہتی ہے ''ڈائریکٹر نے کبھی یہ نہیں سوچا کہ معذور طلبہ بلندی پر واقع اپنی کلاسوں میں نہیں جا سکتے ہیں اور نہ ہی میں نے کبھی انہیں یہ بتانے کا سوچا۔'' اس کے بعد اسے سیڑھیاں چڑھنے کی زحمت برداشت نہیں کرنی پڑی۔

پالیسیوں میں تبدیلیاں لانا آسان نہیں ہے ۔ یہ جدوجہد طویل اور پیچیدہ ہو سکتی ہے۔ اس میں کئی برس لگ سکتے ہیں اور بہت سے لوگوں کو مسلسل کوشش کرنی پڑ سکتی ہے۔ آپ اس کے اخراجات، متاثر افراد اور ان پالیسیوں کی نوعیت سمجھیں۔ تبدیلی کے لئے کام کرتے ہوئے مایوسی کا احساس عام سی بات ہے ۔ اگر آپ اس دوران خود کو مغلوب یا مایوس محسوس کریں تو اپنے ملک میں حتیٰ کہ ملک سے باہر موجود کسی اور معذور عورت سے مشورہ لینے کی کوشش کریں اور یاد رکھیں کہ آپ بھی عوامی سہولیات کو استعمال کرنے کا حق رکھتی ہیں ۔ آپ معاشرے کو قابل رسائی بنا سکتی ہیں ۔

یہ ان معذور لوگوں کی کہانی ہے جن کے تشکیل شدہ گروپ نے ان کے شہر میں شاندار تبدیلیاں یقینی بنائیں۔

ایک شہر کو قابل رسائی بنانا

اکٹیر نبرگ (Ekaterin burg) روس میں ''دی فریڈم آف موومنٹ سوسائٹی'' شہری حکومت کے تعاون سے شہر کو زیادہ سے زیادہ قابل رسائی بنانے کے لئے کام کر رہی ہے۔ روسی قانون کے مطابق معذور افراد کو عوامی عمارتوں اور بسوں کے استعمال کرنے کی سہولت ملنی چاہئے لیکن بہت سے مقامات اب بھی نا قابل رسائی ہیں۔ اکٹیر نبرگ کی شہری حکومت نے عمارتوں کو قابل رسائی بنانے کے لئے ایک ڈس ایبلٹی پروگرام تشکیل دیا۔

لیکن معذور افراد کے ایک گروپ نے جو وہیل چیئرز یا بیساکھیاں استعمال کرتا ہے یہ اندازہ لگایا کہ ہر چند حکومت مدد کرنے کی کوشش کرتی رہی ہے بہت سے مقامات تبدیلیوں کے بعد بھی معذور افراد کے استعمال کے لئے دشوار ہیں۔ انہوں نے اندازہ لگایا کہ حکومت معذور افراد کی مدد کے بغیر یہ کام نہیں کر سکتی ہے اس لئے انہوں نے ''فریڈم فار موومنٹ سوسائٹی'' تشکیل دی۔ انہوں

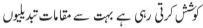

نے اپنے کام کی ابتداء شہر کے انتہائی اہم مقامات کو قابل رسائی بنانے کے لئے فہرست مرتب کرنے سے کی۔ وہ شہری حکام سے ملے اور انہیں اپنی فہرست دکھائی۔ حکام نے تسلیم کیا کہ انہیں معذور افراد سے مشاورت کی ضرورت ہے۔ سوسائٹی نے یہ بھی یقینی بنایا کہ اس کے اراکین اس شہری کمیٹی میں شامل ہوں جو رسائی بہتر بنانے کی ذمہ دار ہے۔

شہری کمیٹی کے کسی بھی طے شدہ پروجیکٹ پر عمل کے لئے معذور افراد کی منظوری ضروری تھی۔

فریڈم فار موومنٹ سوسائٹی نے وہ رہنما خطوط تشکیل دیئے جو ماہر تعمیرات عمارتوں کو قابل رسائی بنانے کے لئے استعمال کر سکتے ہیں۔ انہوں نے ان عمارتوں کی تصاویر کھینچیں جن میں وہ درستگی چاہتے تھے پھر ایسی واضح اشکال بنائیں کہ تبدیلیاں کس طرح عمل میں لائی جائیں۔ اب پرانی عمارتیں آہستہ آہستہ تبدیل ہو رہی ہیں جبکہ نئے رہنما خطوط پر شہر بھر میں عمل ہو رہا ہے۔

فریڈم فار موومنٹ سوسائٹی کی جدوجہد کے باعث تمام سرکاری عمارتیں اور بہت سی دوسری عمارتیں معذور افراد کے لئے قابل رسائی ہیں۔ نیا شہری مال معذور افراد کے لئے باآسانی قابل استعمال ہے۔ سوسائٹی نے شہر کے بہت سے اسکولوں اور فلمی تھیٹر کو بھی معذور افراد کے لئے قابل رسائی بنا دیا ہے۔

معذور عورتوں کے لئے کچھ اور آئیڈیے ہیں جو کمیونٹیز میں اقدامات کے لئے ان کی مدد کرسکتے ہیں۔

کرنے والے کام

- ایسی سماجی سرگرمیوں کا اہتمام کریں جو معذور لڑکیوں کو گھروں سے باہر آنے اور معذوریاں رکھنے والے دوسرے افراد اور اپنی ہم عمر لڑکیوں سے ملنے کے مواقع فراہم کریں۔
- کام کی تلاش یا ذرائع پیدا کرنے کے حوالے سے معلومات میں شراکت۔
- ایسی خواتین کی مدد جن کے ساتھ گھر یا عوام میں یا کام کی جگہوں پر بُرا سلوک کیا جاتا ہو۔
- قیادت اور سماجی مہارتوں میں تربیت کی پیشکش۔

یوگنڈا میں تبدیلی کے لئے وسائل کی فراہمی

یوگنڈا ڈس ایبلڈ ومنز ایسوسی ایشن نے معذور عورتوں کے لئے قرض کا ایک گردشی پروگرام شروع کیا تا کہ وہ اپنا کاروبار شروع کر سکیں۔ ایسوسی ایشن معذور عورتوں کو تعلیم، حرکاتی اعانتیں فراہم کرتی ہے اور معذور عورتوں کے بارے میں عوامی آگہی بڑھانے کے لئے ڈرامہ گروپ بھی چلاتی ہے۔ ایسوسی ایشن معذور عورتوں کے حقوق اور بہتری میں اضافہ کرنے، معذور بچوں کے لئے بہتر تعلیم، خود مختار زندگی کی مہارتیں سکھانے، تولیدی صحت کی معلومات فراہم کرنے نیز غربت، تغافل، سماجی تفاوت اور بیماریوں سے جنگ کے لئے جدوجہد کرتی ہے۔

عورتوں نے ایل سلواڈور میں تبدیلی یقینی بنائی

ایل سلواڈور میں معذوری کے حقوق کے لئے سرگرم عمل گروپ (ACOGIPRI) 1987ء سے پروگراموں کا اہتمام کر رہا ہے۔ یہ گروپ مختلف النوع پس منظر رکھنے والی معذور عورتوں کو

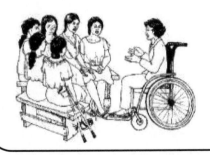

- جنسی مسائل اور دیگر امور پر تبادلہ خیال کے لئے جمع کرتا ہے۔
- خواندگی اور قیادت کی تربیت فراہم کر رہا ہے۔
- کسی بھی قسم کی معذوری رکھنے والی لڑکیوں اور عورتوں کے لئے امداد اور خدمات کی دستیابی ممکن بناتا ہے۔
- امتیازی سلوک اور تشدد روکتا ہے۔

جب ہر ایک شامل ہو تو کمیونٹیز مضبوط تر ہوتی ہیں

تعصب، تغافل اور امتیاز کی رکاوٹوں کے باوجود دنیا بھر کی معذور عورتیں اپنی خود انحصاری کے لئے اپنی مہارتیں بہتر بنا رہی ہیں

اور صحت اور تعلیم میں تبدیلی کے لئے جدوجہد کر رہے ہیں

ہم سماج میں برتر شمولیت کے لئے دباؤ ڈال رہے ہیں

میں نے ایک ایسے دن کا خواب دیکھا ہے جب معذوریاں رکھنے والی تمام عورتیں خود مختار، اپنا خاندان رکھنے والی اور عوامی سہولیات کو کسی بھی دوسرے فرد کی طرح استعمال کر سکنے والی ہوں گی

ہر سطح پر اور ہر معاملے میں فیصلے کرتے ہوئے معذوریاں رکھنے والی عورتوں کی آواز سنی جائے، صرف معذوری کے امور پر نہیں، کوئی بھی تمام سوالوں کے جواب نہیں رکھتا ہے۔ ہر ایک، عورتیں اور مرد، معذور اور غیر معذور، وہ لوگ جو انسانی حقوق، محنت کشوں کے حقوق، عورتوں کے وقار کے لئے کام کرنے والے، دنیا کے تمام حصوں میں معذوریاں رکھنے والی عورتوں کے لئے صحت مند، خود مختارانہ اور تعمیری زندگی یقینی بنانے کے لئے اکٹھے ہوں، یہ یقینی بنا کر کہ ہم ایک دوسرے کے لئے زندگی بہتر بنا کر اور ایک دوسرے سے سیکھ کر ہی اس دنیا کو ہر ایک کے لئے زیادہ منصفانہ بنا سکتے ہیں۔

آئیے بات کریں کہ ہمارے سماج میں کونسی تبدیلیاں ہر ایک کے لئے زیادہ شمولیت اور تمام لوگوں کے لئے بہتر صحت یقینی بنانے میں مدد کریں گی۔

ایک عام عورت مجبور ہے لیکن معذور عورت کا حال تو انتہائی بدتر ہے۔ اس میں ذمے دار معاشرہ بھی ہے تو قصور خود عورت کا بھی ہے۔ عورتوں میں شعور بیدار کرنے کی اشد ضرورت ہے۔ دنیا کی ہر عورت خواہ وہ معذور ہی کیوں نہ ہو، ممکنہ حد تک بہتر اور اپنی خواہشوں کے مطابق زندگی گزارنے کا حق رکھتی ہے۔ رہنے کی مناسب جگہ، صحت افزاء غذا، صحت کی سہولتیں اور اعتماد کے ساتھ جینے کا حق یہی تو ہر انسان کی ضرورت ہے۔ پھر ایک معذور عورت کو یہ حق کیوں نہ ملے۔ معذور عورتوں کے لئے صحت کی کتاب اسی سلسلے میں ایک کوشش ہے جس کے ذریعے ہم اس عالمی مہم میں شامل ہو رہے ہیں جو معذور عورتوں میں صحت مند زندگی کا جذبہ ابھارنے کے لئے سرگرم عمل ہے۔

ڈاکٹر عزیز خان ٹانک

صدر

پاکستان میڈیکل ایسوسی ایشن کراچی

29

باب -2

معذوری دوست، ہیلتھ کیئر کے لئے منظّم ہونا

معذوری کا شکار ہونے والی عورتیں صحت مند رہنے اوراچھی دیکھ بھال حاصل کرنے کا حق رکھتی ہیں لیکن چند ہی مراکزصحت، کلینک اوراسپتال اس طرح تعمیر کئے گئے ہیں کہ معذورعورتیں ان میں رسائی حاصل کرسکیں۔ یہ بہت مہنگے یا بہت دوربھی ہوسکتے ہیں اورممکن ہے آپ وہاں جانے کی،ادویات یا علاج کے لئے ادائیگی کی استطاعت نہ رکھتی ہوں یاصحت کارکنوں سے اپنے مسئلے کے اظہارکے قابل نہ ہوں۔

ہم اپنے حقوق اور یہ یقینی بنانے کے لئے کہ معذوری کے مسائل ایک ترجیح بن جائیں، اپنی آوازوں کو تائید اورنمائندگی کے ذریعے قابل سماعت بناسکتے ہیں۔

اس باب میں ہم ایک عورت ڈلفائن کی کہانی سنائیں گے کہ اس نے اپنی صحت کے ایک مسئلے کوحل کرنے کے لئے کس طرح دوسری عورتوں کیساتھ کام کیا۔ ڈلفائن اوراس کی سہیلیوں نے جان لیا کہ اس کے مسئلے کا پائیدارحل ڈلفائن کی موجودہ صورتحال سے ہٹ کرسوچنے میں ہے۔ایک معذورعورت کے مسائل صحت، دوسری تمام عورتوں کے مسائل صحت کی طرح صرف ایک عورت کے مسائل نہیں ہوتے ہیں۔اس کے مسائل صحت،سماجی مسئلہ ہیں۔

ڈلفائن اوراس کی سہیلیوں کی طرح آپ اور دیگر معذورعورتیں جنہیں آپ جانتی ہوں مل جل کرصحت کی اچھی دیکھ بھال، اپنی کمیونٹی میں مسائل کے بنیادی اسباب کی نشاندہی اورتبدیلیوں کے حصول کے لئے کام کرسکتی ہیں۔

ساؤتھ امریکہ کی ایک کہانی

معذوری کا شکار عورت، اپنی معذوری سے قطع نظر جذبات و احساسات اور فطری ضروریات کے لحاظ سے ایک عام عورت کی طرح ہوتی ہے۔ وہ بھی ایک عام عورت کی طرح مختلف مسائل سے دو چار ہو سکتی ہے۔ جنوبی امریکہ میں رہنے والی ڈلفائن کے ساتھ بھی ایسا ہی کچھ ہوا مگر اسے علاج کی سہولت کے لئے خاصی جدوجہد کرنی پڑی صرف اس لئے کہ مقامی صحت کارکن یہ ماننے پر تیار ہی نہیں تھے کہ ایک معذور عورت کسی عام مسئلے کا شکار ہو سکتی ہے یا یہ کہ اس کے طبی مسئلے کا سبب اس کی معذوری کے علاوہ کچھ اور ہو سکتا ہے۔ ہوا یہ کہ ایک دن ڈلفائن نے محسوس کیا کہ اس کی وجائنا سے رطوبت کا غیر معمولی اخراج

یہ عورت محفوظ جنسی عمل کے بارے میں معلومات کیوں حاصل کر رہی ہے

ہو رہا ہے۔ اس نے ابتداء میں اپنے طور پر علاج کیا مگر کوئی فائدہ نہیں ہوا۔ آخر وہ ایک کلینک گئی اور اپنی پریشانی بتائی۔ وہاں موجود کوئی بھی فرد اس کی بات پر یقین کرنے کے لئے آمادہ نہ تھا۔ ڈلفائن اپنے لائف پارٹنر کے ذریعے جنسی انفیکشن میں مبتلا ہو گئی تھی۔ وہ ایک معذور عورت تھی

لہٰذا اس کا لائف پارٹنر لوگوں کے سامنے آنے پر آمادہ نہ تھا۔ کلینک میں موجود صحت کارکنوں نے اس کے مسئلے کا سبب اس کی معذوری کو قرار دیا۔ کلینک میں اس کے بازو اور ٹانگیں کھینچی گئیں جس سے اس کے مسلز کی ٹینشن بڑھ گئی۔ اسے یہ ٹینشن دور کرنے کے لئے دوا دی گئی مگر اس کی تکلیف بڑھتی گئی۔ پیٹ کا درد شدید تر ہو گیا، اب تیز بخار کے علاوہ اسے پیشاب کرتے ہوئے بھی تکلیف ہو رہی تھی۔ ڈلفائن کو یاد آیا کہ اس کی ایک سہیلی نے اسے معذور عورتوں کے ایک گروپ کا پتہ بتایا تھا۔ ڈلفائن نے اس گروپ سے رابطہ کیا۔ انہی دنوں گروپ میں شامل عورتوں نے ہیسپرین فاؤنڈیشن کی کتاب ''ویئر وومین ہو نو ڈاکٹر'' پڑھی تھی۔ جس میں بتایا گیا ہے کہ کس طرح انفیکشن جنسی عمل کے دوران ایک فرد سے دوسرے فرد میں منتقل ہوتا ہے۔ گروپ کی دو معذور عورتوں نے ڈلفائن کے ساتھ کلینک جانے کا فیصلہ کیا۔ انہوں نے مل کر ڈاکٹر کو قائل کیا کہ ڈلفائن بھی ایک عورت ہے اور اپنا ایک لائف پارٹنر رکھتی ہے۔ ڈاکٹر نے اب ضروری ٹیسٹ کرائے اور مان لیا کہ ڈلفائن جنسی طور پر رحم میں منتقل ہونے والے سنگین انفیکشن کا شکار ہے جو گنوریا اور کلے مائیڈیا کے باعث ہوتا ہے (دیکھئے باب 8)۔ ڈاکٹر نے اسے تشخیص کے مطابق دوا دی اور اس کی حالت بہتر ہونے لگی۔ ڈاکٹر نے تاکید کی کہ اس کا پارٹنر بھی یہی دوائیں اور کنڈوم استعمال کرے ورنہ وہ ایک بار پھر اسی انفیکشن کا شکار ہو جائے گی۔

مسائل کے بنیادی اسباب کی تلاش

ڈلفائن نے اس گروپ میں شمولیت اختیار کر لی اور ان سے بہت کچھ سیکھا۔ یقیناً ڈلفائن کی کہانی بہت سے معاشروں سے مختلف ہے لیکن یہ جاننا دلچسپ ہوگا کہ جب ڈلفائن نے اس گروپ کی معذور عورتوں کے سامنے اپنا مسئلہ رکھا تو انہوں نے اس مسئلے کے باعث پیدا ہونے والی صورتحال کی نشاندہی اور ڈلفائن کی مدد کے لئے ایک مکالماتی کھیل کھیلنے کا فیصلہ کیا جس کا نام انہوں نے "لیکن کیونکہ" رکھا تھا۔

ڈلفائن کیوں گنوریا اور کلے مائیڈیا کا شکار ہوئی

کیونکہ وہ اپنے لائف پارٹنر کے ذریعے انفیکشن کی زد میں آئی

لیکن کلینک میں صحت کارکنوں نے میرے رطوبت کے اخراج کا علاج کرنے کے بجائے میرے بازو اور ٹانگیں کیوں کھینچیں۔

کیونکہ وہ سمجھتے تھے کہ تمہاری معذوری ہی صحت کا مسئلہ تھی۔ انھوں نے یہ یقین نہیں کیا کہ یہ ممکن ہے کہ تم جنسی عمل کر سکتی ہو۔

لیکن انھوں نے یہ یقین کیوں نہیں کیا کہ میرے لئے جنسی تعلق ممکن ہے

کیونکہ صحت کارکن کسی معذور فرد کو نارمل جذبات رکھنے والے ایک نارمل فرد کی طرح نہیں سمجھتے ہیں۔ وہ نہیں سمجھتے ہیں کہ معذوری، جنسی عمل کے لئے رکاوٹ نہیں ہے۔

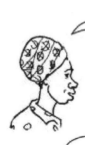

لیکن ڈلفائن نے صحت کارکنوں کو اپنے لائف پارٹنر کے بارے میں کیوں نہیں بتایا

کیونکہ وہ خوفزدہ تھی کہ پھر وہ اس سے ملنے نہیں آئے گا۔

لیکن وہ اسے مزید ملنے کے لئے کیوں نہیں آتا

کیونکہ وہ شرمندہ ہوتا۔ گاؤں کے لوگ کے ایک معذور عورت سے تعلق رکھنے پر اس کی ہنسی اُڑاتے۔

لیکن انہوں نے اس کی ہنسی کیوں اُڑاتے؟

کیونکہ وہ ہمیں حقیقی عورت نہیں سمجھتے ہیں۔ کیونکہ اس نے ایک معذور عورت سے تعلق قائم کیا لہٰذا لوگ سمجھتے کہ ضرور اس کے ساتھ کوئی گڑبڑ ہے۔

جب عورتوں نے وجوہات کی ایک طویل فہرست بنالی، تو انہوں نے وجوہ کی درجہ بندی کرکے ان کے گروپ بنالئے۔ اس طرح ان عوامل کا جائزہ لینا آسان تر بن سکتے ہیں جو مسائل صحت کا سبب بن سکتے ہیں اور ان پہلوؤں کے حل ڈھونڈے جاسکتے ہیں۔

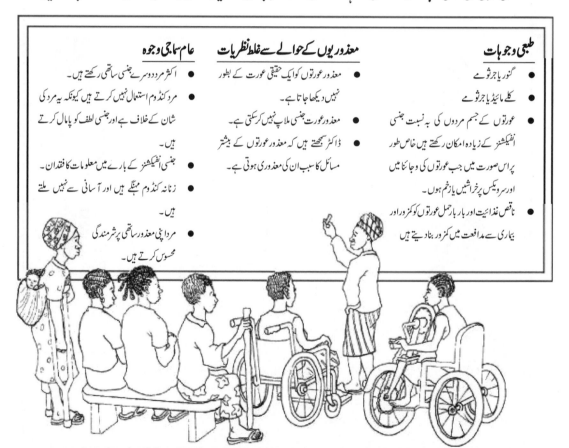

طبی وجوہات

- گنور یا جرثومے
- کلے مائیڈیا جرثومے
- عورتوں کے جسم مردوں کی یہ نسبت جنسی انفیکشنز کے زیادہ امکان رکھتے ہیں خاص طور پر اس صورت میں جب عورتوں کی وجائنا میں اور سرویکس پر خراشیں یا زخم ہوں۔
- ناقص غذائیت اور بار بار حمل عورتوں کو کمزور اور بیماری سے مدافعت میں کمزور بنادیتے ہیں

معذوریوں کے حوالے سے غلط نظریات

- معذور عورتوں کو ایک حقیقی عورت کے بطور نہیں دیکھا جاتا ہے۔
- معذور عورت جنسی ملاپ نہیں کرسکتی ہے۔
- ڈاکٹر سمجھتے ہیں کہ معذور عورتوں کے بیشتر مسائل کا سبب ان کی معذوری ہوتی ہے۔

عام سماجی وجوہ

- اکثر مرد دوسرے جنسی ساتھی رکھتے ہیں۔
- مرد کنڈوم استعمال نہیں کرتے ہیں کیونکہ یہ مردی شان کے خلاف ہے اور جنسی لطف کو پامال کرتے ہیں۔
- جنسی انفیکشنز کے بارے میں معلومات کا فقدان۔
- زنانہ کنڈوم مہنگے ہیں اور آسانی سے نہیں ملتے ہیں۔
- مردانی معذور ساتھی پر شرمندگی محسوس کرتے ہیں۔

صحت کی دیکھ بھال، سب کا انسانی حق ہے

صحت کی اچھی دیکھ بھال اُن مسائل سے بچاتی ہے جو معذوریوں کو بدتر بناتے ہیں۔ صحت کی اچھی دیکھ بھال ان مسائل صحت سے بھی تحفظ فراہم کرتی ہے جو معذوریوں کے باعث جنم لیتے ہیں۔ کسی عام سے مسئلے/خرابی کا بروقت حل مثلاً ایک ہی انداز سے مسلسل طویل مدت بیٹھنے یا لیٹنے کے باعث، دباؤ سے ہونے والے زخموں کا علاج، انہیں زندگی کے لئے خطرہ بننے والے بحران، میں بدلنے سے روکتا ہے۔

ہمیں اچھی غذائیت، جسمانی فعالیت، تولیدی صحت کی دیکھ بھال اور مسائل صحت سے تحفظ اور علاج کے ساتھ اچھی صحت کو فروغ دینا چاہئیے۔ ہمیں اپنی زندگی کے حالات (انداز) بھی بدلنا چاہئیں تاکہ ہمیں خود اپنی صحت پر اختیار حاصل ہو

صحت کی دیکھ بھال تمام معذور اور عورتوں کو ان کی سماجی حیثیت سے قطع نظر ملنی چاہئے۔ صحت کی اچھی دیکھ بھال میں مفت یا کم اخراجات پر صحت کی سہولیات، صحت کی دیکھ بھال کے لئے ادائیگی کے لئے نقد رقم یا انشورنس اور پبلک ٹرانسپورٹیشن شامل ہے، جن کا استعمال آسان ہو۔ یہ ان عورتوں کے لئے خصوصی طور پر اہم ہے جو الگ تھلگ رہتی ہیں یا غریب ہیں۔

میں جانتی ہوں کہ یہ گلٹی خطرناک ہوسکتی ہے لیکن میں کیا کروں؟ دوائیں بہت مہنگی ہیں اور کلینک بہت دور ہے۔ اس کے علاوہ سب ہی میری ہنسی اڑائیں گے۔

ہم تھائی لینڈ میں اپنی صحت کی دیکھ بھال کے لئے ریاستی اسپتالوں میں ایک امریکی ڈالر سے بھی کم ادا کرتے ہیں۔

غربت اور صحت

دنیا بھر میں اقتصادی اور کاروباری پالیسیوں نے افلاس بڑھایا ہے، صحت کی دیکھ بھال کے وسائل کم تر کئے ہیں اور لوگوں کے درمیان سماجی فرق بڑھایا ہے، ان عدم مساواتوں نے دنیا بھر میں عورتوں کے لئے مشکل کر دیا ہے کہ وہ خود اپنے لئے اور اپنے گھرانوں کے لئے صحت کی سہولتیں حاصل کرسکیں۔ نئی پالیسیاں جیسے صحت کی دیکھ بھال کے لئے فیس، صحت کی سہولتوں کے حصول کی راہ میں ایک اور رکاوٹ ہے، دیگر مالیاتی رکاوٹیں مثلاً ادویات کی قیمتیں، آنے جانے کے اخراجات، ذرائع صحت کی دیکھ بھال کے حصول کو ناقابل برداشت بناسکتے ہیں۔

معذور اور عورتوں کے لئے صحت کی سہولتیں حاصل کرنا بہت دشوار ہے۔ افریقہ کے بہت سے ممالک میں سو معذور افراد میں صرف ایک معذور افراد میں صرف ایک ہی اپنے لئے ضروری صحت کی خدمات حاصل کرسکتا ہے۔ خدمات کے فقدان کے ساتھ اخراجات، فاصلے، ماڈی رکاوٹیں اور نقصان دہ رویئے بھی آڑے آتے ہیں۔ اگر کوئی عورت کچھ سرمایہ رکھتی ہو تو اس صورت میں بھی دستیاب خدمات شاذ و نادر ہی معذور اور عورتوں کی صحت کی ضرورت (خصوصاً تولیدی صحت) کے مطابق ہوتی ہیں۔

طبی معائنہ مفت بھی ہے تو طبی معائنہ کے لئے جانے میں کیا فائدہ ہے، ادویات تو مفت نہیں ملیں گی۔ میرے گھر والے میرے لئے مزید دوا نہیں خریدسکیں گے۔

نائیجیریا میں رکاوٹوں کا خاتمہ

ای جوڈتھ امونائیجیریا کے قدرتی تیل کے مالا مال ڈیلٹا نائیجیر کی ہے۔ وہ پولیوزدہ ہے آج لوگ اسے صحت کی دیکھ بھال کے تمام پروگراموں کی منصوبہ بندی اور خدمات کے ہر مرحلے میں معذور لڑکیوں اور عورتوں کی شمولیت پر اصرار کے باعث ''ماماین اسٹریم'' کہتے ہیں۔

وہ جو جوتے پہنتی ہے جانتی ہے کہ یہ پیر کے کسی حصے میں زیادہ تکلیف پہنچاتے ہیں ۔وہ کہتی ہے :''ہم عورت ہیں اور ان تمام خدمات اور سہولتوں کے حق دار ہیں جو معاشرے میں دوسری عورتوں کے لئے فراہم کی جا رہی ہیں۔

2000ء میں جوڈتھ نے ایک این جی او فیملی سینٹر ڈانیشیٹیو فار چیلنجڈ پرسنز (FACICP) کی بنیاد رکھی جو معذور افراد خصوصاً عورتوں اور لڑکیوں کے حقوق اور ضروریات یقینی بنانے کے لئے کام کرتی ہے اور صحت کی دیکھ بھال اور بہتری کے تمام پروگراموں میں اپنا ایک مقام رکھتی ہے۔ وہ اپنی تنظیم (FACICP) کے ہیلتھ کیئر پروگرام کو رکاوٹوں سے پاک پروجیکٹ قرار دیتی ہے۔ اس کا کہنا ہے ''پروجیکٹ کا مقصد تولیدی صحت کی سہولیات بشمول ایچ آئی وی /ایڈز کے بارے میں معلومات کو معذور عورتوں کے لئے قابل حصول بنانا ہے۔ ہم نابینا خواتین کے لئے تولیدی صحت کے بارے میں عام معلومات کو بریل میں منتقل کر رہے ہیں اور جنسی تعلیم خصوصاً حمل، بچوں کی پرورش اور معذوری سے تعلق رکھنے والے امور پر تبادلہ خیال کی ماہانہ ملاقاتوں کا اہتمام کرتے ہیں۔

ہم عورتیں ہیں اور ان تمام خدمات اور سہولتوں کے حق ہیں جو معاشرے میں دوسری عورتوں کے لئے فراہم کی جا رہی ہیں۔

FACICP عورتوں کی صحت کے امور پر وسیع تر معلومات فراہم کرنے والی آرگنائزیشن ''سوسائٹی فار فیملی ہیلتھ (SFH)'' کی شراکت میں بھی کام کرتی ہے۔

''ایس ایف ایچ ہمیں اپنے ہر ایسے تربیتی پروگرام یا ورکشاپ میں مدعو کرنے پر آمادہ ہے جو وہ معذوریاں رکھنے والی عورتوں کی صحت کی ضروریات کے بارے میں آگہی بڑھانے کے لئے منعقد کرے گی۔'' ای جوڈتھ نے بتایا۔ ایف اے سی آئی سی پی مختلف مقامات پر وہیل چیئرز کی فراہمی کے ساتھ ایسی ورکشاپوں کے انعقاد میں ایس ایف ایچ کے ساتھ کام کر رہی ہے جہاں علامتی زبانوں سے ترجمانی کی جاتی ہے لہذا بہری عورتیں بھی ان میں بھرپور شرکت کر سکتی ہیں۔ ایس ایف ایچ کی ٹریننگ کے بعد معذور عورتیں اپنی اپنی کمیونٹیز میں فیملی ہیلتھ ایجوکیٹر بن سکتی ہیں۔

ای جوڈتھ اور اس کے ساتھی، حکومتوں، ہم پہلو آرگنائزیشنوں اور سول سوسائٹی پر تمام ترقیاتی کاموں میں Disability lens استعمال کرنے کے لئے زور دے رہی ہے۔

مثال کے طور پر انہوں نے تجویز دی ہے کہ ورلڈ بینک کے فراہم کردہ فنڈز سے چلنے والے منصوبوں میں معذور افراد کو تربیت، تیکنیکی معاونت، مشاورت، پروجیکٹ فنڈنگ اور ضروریات ہمیشہ توجہ کا مرکز رہیں اور بھلائی نہ جائیں۔ ای جوڈتھ ہمیں یاد دلاتی ہے معذور افراد ہر جگہ ہیں، وہ انہی حقوق اور مراعات کے حقدار ہیں جن سے کسی بھی کمیونٹی کے دوسرے افراد لطف اندوز ہوتے ہیں ۔

صحت کی سہولتوں کو استعمال کرنا/آسان بنانا

معذور عورتیں اور صحت کارکن مل کر صحت کی سہولتوں کو معذور عورتوں کے لئے بہتر بنا سکتی ہیں۔ وہ معذور خواتین کے لئے مراکز صحت میں جانے، آلات استعمال کرنے اور معذوریوں کے بارے میں معلومات بڑھانے کو آسان تر بنانے کے لئے اور معذور عورتوں کے لئے صحت کارکنوں کے رویوں کو بہتر بنانے کی راہیں تلاش کر سکتے ہیں۔ ان میں سے بیشتر تبدیلیاں لانا مشکل یا مہنگا نہیں ہے۔

یہ تبدیلیاں اور بہت سے لوگوں جیسے ان بوڑھے لوگوں کی بھی مدد کر سکتی ہیں جو اپنی جوانی کی طرح چل پھر نہیں سکتے ہیں یا ایسا کوئی فرد جو کسی حادثے کا شکار ہو کر عارضی طور پر اپنی ٹوٹی ہوئی ٹانگ یا بازو کے باعث معذور ہو گیا ہو۔

صحت کی سہولتوں کو زیادہ معذور دوست بنانے کے لئے تجاویز

- ایسے لوگ جو مراکز صحت سے دور رہتے ہیں ہفتہ وار یا ماہانہ ملاقاتوں کی پیشکش کریں۔
- معذور عورتوں کے لئے صحت کی مفت سہولیات پیش کی جائیں۔
- آلات کے استعمال کو آسان بنائیں۔
- مراکز صحت تک پبلک یا پرائیویٹ ذرائع آمدورفت فراہم کریں۔
- ذرائع آمدورفت ان لوگوں کے لئے سہولت رکھتے ہوں جو وہیل چیئر یا بیساکھیاں استعمال کرتے ہوں یا چلنے میں دشواری محسوس کرتے ہوں۔

رسائی کے لئے مزید معلومات حاصل کرنے کے لئے صفحہ 376 دیکھئے۔

صحت کی دیکھ بھال میں رکاوٹیں

- وہیل چیئر یا بیساکھیاں استعمال کرنے والی عورت کے لئے بیش تر مراکز صحت اور اسپتالوں میں جانا مشکل ہوتا ہے۔ یہ عموماً دور ہوتے ہیں اور ایک معذور عورت وہاں تک جانے کے لئے کسی بھی سواری میں کوئی سہولت نہیں رکھتی ہے۔
- کم بلند بیڈ یا اچھے معیار کے کینتھیٹر عام طور پر دستیاب نہیں ہوتے ہیں۔
- مراکز صحت کے کھلنے کے اوقات موزوں نہیں ہو سکتے ہیں، انہیں تبدیل کرنا چاہیئے۔
- بہت سی عورتیں مرد ڈاکٹروں کے پاس جاتے ہوئے ہچکچاتی ہیں اور وہاں چند خواتین ڈاکٹر ہی ہو سکتی ہیں۔
- صحت کارکن نہیں جانتے ہیں کہ کسی ایسے فرد سے کس طرح ابلاغ کیا جائے جو بہرا ہو اور نابینا عورتوں کے لئے صحت کی معلومات فراہم کرنے والا سامان نہیں ہوتا ہے۔
- صحت کارکن جن میں نرسیں اور ڈاکٹر بھی شامل ہیں، ممکن ہے پوری طرح تربیت یافتہ نہ ہوں یا معذوری کے بارے میں زیادہ نہ جانتے ہوں۔ ممکن ہے وہ معذوری کے بارے میں غلط نظریات رکھتے ہوں اور وہ معذور عورت کی بات نہ سنیں۔
- مراکز صحت مہنگے ہو سکتے ہیں اور ممکن ہے کہ آپ کو کسی صحت کارکن سے ملنے کے لئے کسی اور کو رشوت دینی پڑے۔ (بدعنوانی)

بیش تر صحت کارکن ہماری بات نہیں سنتے ہیں کہ وہ ہمیں ناکارہ سمجھتے ہیں اگر ہمیں معائنہ کرانے کا موقع مل جائے تو وہ خاموشی سے ٹیسٹ وغیرہ کرتے رہتے ہیں اور اگر ہم ان سے سوالات کریں تو وہ ہم پر چیخنے چلانے لگتے ہیں۔

کلینک اوراسپتالوں کا استعمال آسان بنانے کے لئے تجاویز

کلینک یا اسپتالوں کو لازمی طور پر

- قریب ہونا اوران تک پہنچنے کے لئے ٹرانسپورٹ دستیاب ہونا چاہئے ۔
- انہیں ایسے لوگوں کے لئے باسہولت ہونا چاہئے جو وہیل چیئر یا بیساکھیاں استعمال کرتے ہوں یا چلنے میں دشواری محسوس کرتے ہوں ۔
- سیڑھیوں کے علاوہ ریمپ یالفٹ ہونی چاہئے ۔
- ایسے ٹوائلٹ ہوں جنہیں معذور عورتیں استعمال کرسکیں ۔

کلینکوں اوراسپتالوں میں ایسا تربیت یافتہ عملہ بھی ہونا چاہئے جو بہرے یا نابینا اور سیر یبرل پالسی کے شکار افراد کے ساتھ موثر طور پر ابلاغ کر سکے اور جو یقینی بنا سکے کہ وہ عورتیں جو مشکلات کو سمجھنے میں دشواری محسوس کر رہی ہوں وہ سمجھیں کہ کلینک میں کیا ہو رہا ہے ۔

کلینک اوراسپتال یہ کر سکتے ہیں کہ

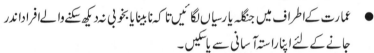

آپ کے مسائل صحت کی تحریر شدہ فہرست آپ کی مدد کرسکتی ہے لہٰذا آپ یہ فہرست بنانا نہ بھولیں ۔

- ہر ایک کو معذوری کے حوالے سے تربیت دیں ۔
- کلینک اوراسپتالوں میں معذوری رکھنے والی عورتوں کو بطور صحت کارکن اور اسٹاف کارکن شامل کریں ۔
- عمارت کے اطراف میں جنگلہ یا رسیاں لگائیں تا کہ نابینا یا بخوبی نہ دیکھ سکنے والے افراد اندر جانے کے لئے اپنا راستہ آسانی سے پاسکیں ۔
- معذوریاں رکھنے والے عورتوں اوران کی صحت کے حوالے سے سرگرمیاں منعقد کریں ۔
- معذوریاں رکھنے والی عورتوں کے لئے ماہانہ یا با قاعدہ مشاورتی سیشن کا اہتمام کریں ۔
- معذوریاں رکھنے والی عورتوں کے لئے آسان بنائیں کہ وہ جس دن بھی کلینک آئیں اسی روز انہیں اپنی ضرورت کے مطابق مختلف شعبوں میں طبی عملے سے ملاقاتوں کی سہولت ملے ۔ بعض مراکزصحت معذور عورتوں کے لئے یہ طے کرنے کے لئے مقامی ہیلتھ ورکرز کو ایسا کرنے کی سہولت دیتے ہیں ۔
- معلومات فراہم کریں کہ صحت کی سہولیات کس طرح آسانی سے حاصل کی اور بھی جاسکتی ہیں ۔
- مختلف زبانوں میں صحت کی معلومات فراہم کریں ۔
- نابینا عورتوں کو بریل یا آڈیوکیسٹ کے ذریعے صحت کی معلومات فراہم کریں ۔
- وہ عورتیں جو سیکھنے یا سمجھنے میں دشواری محسوس کرتی ہوں انہیں سمجھانے کے لئے صحت کارکنوں پر زور دیں کہ وہ سادہ، واضح زبان اور تصاویر استعمال کریں ۔
- صحت کارکنوں کو ان عورتوں کے ساتھ ابلاغ کی تربیت دیں جو صاف بولنے میں دشواری رکھتی ہوں ۔
- عملے کے ارکین کو علامتی زبان کی تربیت دیں تا کہ وہ بہری خواتین کو معلومات فراہم کرسکیں ۔

ایک کلینک بہری عورتوں کے لئے باسہولت ہوگا اگر وہاں موجود کوئی ایک صحت کارکن اس کمیونٹی میں رہنے والے بہرے افراد کی علامتی زبان جانتی ہو۔ کلینک میں کام کرنے والوں میں سے کوئی ایک فرد مقامی ڈیف ایسوسی ایشن یا قریب رہنے والے بہرے فرد سے یہ زبان سیکھ سکتا ہے۔ اگر دستیاب ہو تو مقامی علامتی زبان کی ڈکشنری کو استعمال کیا جا سکتا ہے۔ حتٰی کہ باضابطہ علامتی زبان استعمال کئے بغیر بھی صحت کارکن ابلاغ کے لئے اشارے استعمال کر سکتے ہیں۔ بہری عورتیں بذات خود صحت کارکنوں کو ابلاغ کی وہ قسم بتانے کے لئے بہترین ذریعہ ہوں گی جو ان کے لئے بہترین ہوگی۔

صحت سے تعلق رکھنے والی علامتی زبان کی کچھ تجاویز صفحہ 369 تا 371 دیکھئے۔

کمیونٹی صحت کارکن دیکھ بھال فراہم کر سکتے ہیں

بہت سے ممالک میں معذور عورتوں کی دیکھ بھال کے لئے ضروری مہارتوں کو اہم سمجھا جاتا ہے اور یہ خدمات صرف ڈاکٹروں کے ذریعے فراہم کی جاتی ہیں تاہم ان میں سے بہت سی خدمات تربیت یافتہ کمیونٹی صحت کارکنوں، اساتذہ اور بحالی صحت کارکنوں کے ذریعے کم تر اخراجات پر فراہم کی جا سکتی ہیں۔

معذور بچوں کے لئے طبی خدمات کا حصول

معذور بچوں کے اسپتال اور بحالی مرکز (کیوری، نیپال) کے فیلڈ ورکر پورے نیپال میں معذور بچوں کی مدد کرتے ہیں۔ یہ تربیت یافتہ فیلڈ ورکر معذور بچوں میں دباؤ سے بننے والے زخموں کا علاج کرتے ہیں۔ انہیں فزیوتھراپی کی سہولت دیتے ہیں اور متاثرہ مسلز کو تقویت پہنچانے اور انہیں سکڑنے سے بچانے کے لئے ورزش کراتے ہیں۔ یہ فیلڈ ورکر انہیں امدادی اشیاء بھی فراہم کرتے ہیں تا کہ یہ بچے اپنی کمیونٹیز میں زیادہ سہولت کے ساتھ آ جا سکیں۔

معذور افراد کے لئے کمیونٹی بنیاد پر طبی امداد کے بارے میں مزید معلومات کے لئے کتاب "Disabled village children" دیکھئے۔

مراکزِصحت اور اسپتالوں کی عمارتوں کو مریض دوست بنانے کے لئے تجاویز

عمارتیں سب لوگوں کو خوش آمدید کہنے یا کچھ لوگوں کو باہر رکھنے کے لئے ڈیزائن کی جاسکتی ہیں۔ یہ بات حیران کن ہے کہ چند داخلی راستے، ریمپ، جنگلے، کم بلندی رکھنے والی سیڑھیاں، لفٹ (ایلیویٹرز) کشادہ ٹوائلٹ اور ایسے فرش جو پھسلنے والے نہ ہوں ان لوگوں کے لئے بھی کسی عمارت میں داخل ہونے اور اسے استعمال کرنے کو آسان تر بنا دیتے ہیں۔

جنگلے (یا رسیاں)

عمارت تک جانے والے راستے اور اندرونی دیواروں کے ساتھ ساتھ جنگلے یا رسیاں ان لوگوں کی مدد کریں گی جو نابینا ہوں یا توازن رکھنے میں دشواری یا چلنے میں مشکل محسوس کرتے ہوں۔ ان کی مدد سے یہ لوگ عمارت میں آسانی سے داخل ہوکر اپنے لئے راستہ تلاش کر سکتے ہیں۔

رسیوں سے آراستہ راستہ اور کھردری بناوٹ والی سڑک مراکزِ صحت کے اطراف کے علاقے کو استعمال کرنے کے لئے آسان بنا دیتی ہے۔ کھردری ساخت ایسی عورتوں کو سہولت فراہم کرسکتی ہے جو نابینا ہوں یا بخوبی دیکھ نہ سکتی ہوں

دروازے

دروازے میں نصب گول دستے کے بجائے ہینڈل کو استعمال کرنا آسان ہے۔ وہ لوگ جو اپنے ہاتھوں کو آسانی سے حرکت نہیں دے سکتے ہیں، ہینڈل کو آسانی سے دبا سکتے ہیں۔ اور وہ شخص بھی جو اپنے ہاتھوں میں کچھ اُٹھائے ہو دروازے کے ہینڈل کو آسانی سے استعمال کر سکے گا۔

آپ گول دستے کے ساتھ دھاتی پلیٹ ویلڈ کر کے اسے آسانی سے گھومنے والے ہینڈل میں تبدیل کر سکتے ہیں۔

ہینڈل کو اس حد تک نچلی سطح پر رکھیں کہ اُسے چھوٹے قد یا وہیل چیئر استعمال کرنے والے بھی استعمال کر سکیں۔

اگر دروازہ کھولنا دشوار ہو تو قبضوں پر تیل، گریس یا موم لگا ئیں تا کہ دروازہ آسانی سے حرکت کر سکے۔

بعض اوقات دروازے میں اتنی گنجائش نہیں ہوتی ہے کہ وہ وہیل چیئر کمرے میں جا سکے اگر جگہ چھوٹی ہے جیسے کسی ٹوائلٹ میں تو پھر کوشش کریں کہ دروازہ کھلی جگہ میں یا کمرے میں کھلے۔ جب دروازہ چھوٹے کمرے میں کھلتا ہے تو یہ کسی کے لئے کمرے میں اندر جانا اور باہر آنا مشکل بنا سکتا ہے۔ اگر جگہ محدود ہو تو دروازے سلائیڈنگ بنائے جا سکتے ہیں۔

دروازہ اس قدر کشادہ ہونا چاہئے کہ اس سے وہیل چیئر گزر سکے۔ دروازے میں معذور عورت کی وہیل چیئر اور پہیوں پر رکھے جانے والے ہاتھوں کے لئے مناسب گنجائش ہونی چاہئے۔

اکثر صورتوں میں آپ دروازہ کھلنے کا رُخ بدل سکتے ہیں تاکہ وہ دوسرے انداز سے کھلے اور اس میں وہیل چیئر لے جانا ممکن ہو۔

ریمپ بنانا

ریمپ بہت سے لوگوں کے لئے عمارتوں اور پبلک مقامات جیسے مراکز صحت، اسکول اور لائبریریوں میں آنا اور جانا اور آسان بناتے ہیں ۔ ریمپ صرف وہیل چیئر استعمال کرنے والوں کو ہی سہولت نہیں فراہم کرتے ہیں بلکہ ایسے لوگوں کے لئے بھی معاون ہوتے ہیں جو چلنے میں دشواری محسوس کرتے ہیں یا عارضی زخموں کے باعث مشکل سے چل پاتے ہیں ۔

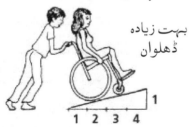

بہت زیادہ ڈھلوان

اچھا

بہت زیادہ ڈھلوان اپنی اونچائی سے چار گنا ہے ۔ یہ کم فاصلہ وہیل چیئر استعمال کرنے والوں کے علاوہ دیگر لوگوں کے لئے حد سے زیادہ ڈھلوان ہے ۔

ریمپ اپنی اونچائی کے مقابلے میں آٹھ سے بارہ گنا طویل ہوسکتے ہیں ۔ یہ ریمپ اپنی اونچائی سے بارہ گنا طویل ہے ۔ اس کی ڈھلان وہیل چیئر استعمال کرنے والوں کے لئے آسان تر ہے ۔

ٹوائلٹ

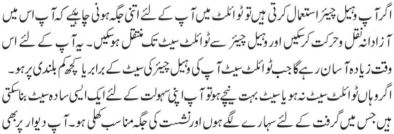

اگر آپ وہیل چیئر استعمال کرتی ہیں تو ٹوائلٹ میں آپ کے لئے اتنی جگہ ہونی چاہئے کہ آپ اس میں آزادانہ نقل و حرکت کر سکیں اور وہیل چیئر سے ٹوائلٹ سیٹ تک منتقل ہوسکیں ۔ یہ آپ کے لئے اس وقت زیادہ آسان رہے گا جب ٹوائلٹ سیٹ آپ کی وہیل چیئر کے سیٹ کے برابر یا کچھ بلندی پر ہو ۔ اگر وہاں ٹوائلٹ سیٹ نہ ہو یا سیٹ بہت نیچے ہو تو آپ اپنی سہولت کے لئے ایک ایسی سادہ سیٹ بنا سکتی ہیں جس میں گرفت کے لئے سہارے لگے ہوں اور نشست کی جگہ مناسب کھلی ہو ۔ آپ دیوار پر بھی سہارے یا گرفت کے لئے کوئی سلاخ یا ہینڈل ہولڈر لگا سکتی ہیں تا کہ آپ کے گرنے کا کوئی امکان نہ رہے ۔(دیکھئے صفحہ 123)

اسپتال کے بیڈ

صرف معذور افراد ہی نہیں بلکہ اکثر لوگ شکایت کرتے ہیں کہ اسپتال کے بیڈ استعمال کرنا مشکل ہوتا ہے ۔ یہ بیڈ عام طور پر گھروں میں استعمال ہونے والے بیڈ کے مقابلے میں فرش سے خاصے بلند ہوتے ہیں ۔ اگر بیمار افراد کے لئے اسپتال کے بیڈ کو نیچا قابل رسائی بنانا ممکن نہ ہو تو صحت کارکنوں کو ان کی مدد کرنی چاہئے ۔

لیکن جب کوئی بیمار یا معذور ہو تو اس کے لئے بیڈ پر چڑھنا انتہائی مشکل ہو سکتا ہے کیونکہ عام طور پر اسپتال کے بیڈز میں پہیے بھی نصب ہوتے ہیں، یہ ان کے لئے خطرناک ثابت ہو سکتے ہیں کیونکہ یہ بیڈ بیمار یا معذور شخص کے بیڈ پر منتقل ہونے کے دوران لڑھکنا شروع کر سکتے ہیں ۔

اگر مراکز صحت میں کچھ بیڈ پہیوں کے بغیر اور نچلی سطح کے ہوں تو ہر فرد اپنے لئے موزوں ترین بیڈ کا انتخاب کر سکے گا ۔

صحت کارکنوں کے لئے :

معذوری کے بارے میں جاننا

ڈاکٹر اور دیگر صحت کارکن عام طور پر صرف معذوری نہ رکھنے والے افراد کے علاج کی تربیت رکھتے ہیں یہ اپنی تعلیم کے دوران عموماً معذوری کے بارے میں بہت ہی کم معلومات حاصل کر پاتے ہیں، معذور افراد کے ساتھ ان کا پہلا اور واحد رابطہ ان کو لاحق معذوری کے علاج کے لئے ہوسکتا ہے۔

صحت کارکنوں کو معذوریوں کے بارے میں زیادہ سے زیادہ جاننے کی ضرورت ہے۔ انہیں سیکھنا/ جاننا چاہئے کہ کوئی مخصوص معذوری معذور عورت کی زندگی کے مختلف پہلوؤں، مثلاً اس کے حاملہ ہونے یا عمر رسیدہ ہونے پر کیسے اثر انداز ہوسکتی ہے۔

معذوری کے بارے میں زیادہ سے زیادہ جاننے کا ایک اچھا طریقہ یہ ہے کہ تربیتی پروگراموں میں معذور عورتوں کو شامل کیا جائے۔ صحت کارکن معذور عورتوں کے تجربات سے اعتماد حاصل کریں گے اور وہ سیکھیں گے کہ صحت کارکنوں کو کس طرح بہترین طور پر سکھایا جائے کہ وہ اپنی دیکھ بھال کو معذوری دوست بنائیں۔

میں خوش ہوں کہ تم نے زور دیا کہ میں تمہارے سینے کا معائنہ کروں۔ تمہاری سانس گھٹنے کا سبب یہ ہے کہ تمہیں دمہ ہے۔ اس کا تمہاری معذوری سے کوئی تعلق نہیں ہے۔

توجہ سے سنیں کہ ایک معذور عورت اپنی صحت کے حوالے سے کس تشویش کا اظہار کر رہی ہے۔ بعد میں آپ اس سے پوچھ سکتی ہیں کہ کیا وہ یہ سمجھتی ہے یا نہیں کہ اس کی معذوری اس کے مسائل صحت پر اثر انداز ہوتی ہے۔

صحت کارکن معذور عورتوں سے سیکھتے ہیں

یوگنڈا میں وزارت صحت نے یہ جاننے کے لئے کہ انہیں اپنا کام بہتر طور پر کرنے کے لئے کس قسم کی معلومات کی ضرورت ہے ملک بھر کی مڈوائفز اور روایتی دائیوں سے تبادلہ خیال اور سروے کیا۔

ان میں سے بیشتر نے کہا کہ انہیں معذور عورتوں کی مدد کے حوالے سے زیادہ معلومات کی ضرورت ہے۔ اب یوگنڈا کی وزارت صحت، معذور عورتوں کی صحت کے حوالے سے زیادہ معلومات حاصل کرنے کے لئے تربیتی سیشنوں کا انعقاد کر رہی ہے۔ معذور عورتیں ان تربیتی سیشنوں میں معاونت کر رہی ہیں۔ وہ صحت کارکنوں کو اپنے تجربات سے آگاہ کرکے ان سوالات کے جواب دے سکتی ہیں کہ معذور عورتوں کا علاج کرنے کے اچھے طریقے کیا ہوسکتے ہیں۔ اس طرح صحت کارکن معذور عورتیں ایک دوسرے سے سیکھ رہے ہیں۔

اگلے ہفتے نابینا عورتوں کی ایسوسی ایشن ہم سے ملاقات کرے گی۔

بہت خوب میں جاننا چاہتی ہوں کہ فیملی پلاننگ کے بارے میں معلومات حاصل کرنے کی خواہاں ایک نابینا عورت کی مدد کس طرح کی جائے۔

جب کوئی معذور عورت اپنی صحت کے کسی مسئلے کے لئے آپ کے پاس آتی ہے تو یاد رکھیں کہ وہ بھی دوسری تمام عورتوں کی طرح ایک عورت ہے۔ سب سے پہلے اس سے پوچھیں کہ وہ آپ کے پاس کیوں آئی ہے اور اس کی مدد کس طرح کی جا سکتی ہے۔ یہ فرض نہ کریں کہ اس کے مسئلے کا سبب اس کی معذوری ہے۔

اس کی ہمت بڑھائیں کہ وہ سوالات کرے اس طرح وہ اپنے مسائل اچھی طرح بیان کر سکتی ہے۔ اس کی رائے کا احترام کیجئے کیونکہ وہ کسی اور سے بہتر طور پر اپنی صحت کے مسائل سمجھتی ہے اور اپنے علاج کے لئے بہتر فیصلے کر سکتی ہے۔ اسے پرسکون ہونے میں مدد دیجئے اور اپنے سوالات کے جوابات کے لئے مہلت دیجئے۔

اس سے اسے مدد ملے گی اور وہ خوفزدہ نہیں ہوگی۔ بعض اوقات معذوری رکھنے والی عورت اس الجھن کے بارے میں پوچھنے کا

میرے تصور کے مطابق مثالی کلینک میں صحت کارکن کہے ۔ "آپ کی معذوری کے بارے میں کوئی ایسی بات ہے جو آپ کے خیال میں مجھے جاننا چاہئے؟ مجھے بتائیے کہ آپ کی معذوری آپ کی صحت کی دیکھ بھال پر کس طرح اثر انداز ہو سکتی ہے ۔" کیا آپ یہ سمجھتی ہیں کہ آپ کے لئے کیا موثر رہے گا۔"

اعتماد نہیں رکھتی ہے جو حقیقت میں اسے پریشان کر رہی ہو۔ اس کا اعتماد بحال کرنے میں مدد دیں اور مطلوبہ معلومات حاصل کریں اور اس کے لئے جو ضروری ہو سہولت فراہم کریں۔

معذور افراد سے یہ پوچھنا اہم ہے کہ وہ اپنے لئے کیا جاننا پسند کریں گے۔ جب وہ سوالات پوچھیں تو ضروری نہیں کہ آپ کے پاس ان سب کے جوابات ہوں۔ اچھی بات یہ ہے کہ جب آپ کوئی بات نہ جانتے ہوں تو اس کا اعتراف کریں۔

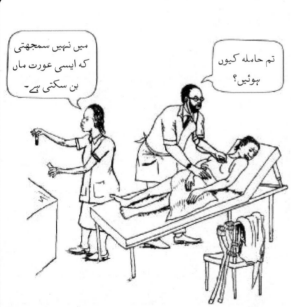

جب ڈاکٹر اور دوسرے صحت کارکن معذوری کے بارے میں کم جانتے ہوں تو ان کا رویہ ایک معذور عورت کے لئے توہین آمیز ہو سکتا ہے۔

معذور عورتوں کا احترام

ہر اس شخص کو جو معذور عورتوں کے صحت کے حوالے سے کوئی تعلق رکھتا ہے جاننا چاہئے کہ اُن کی بہتر طور پر دیکھ بھال کس طرح کی جائے۔ اُن کی دیکھ بھال کی سہولت فراہم کرنے والے ہر فرد کو ان کے ساتھ احترام اور وقار کے ساتھ پیش آنا چاہئے۔ بدقسمتی سے بعض اوقات لوگوں کو یہ یاد دلانے کی ضرورت پڑ سکتی ہے۔ عورت کو یہ بتانے پر آمادہ کیا جائے کہ ۔۔۔۔۔ وہ اپنے مسئلے کی کیا وجہ سمجھتی ہے اور اسے کس طرح حل کیا جانا پسند کرے گی۔ اس طرح ایک صحت کارکن مختلف معذوریوں کو سمجھ سکے گا۔ مل جل کر کام کرنا باہمی تنازعوں اور جھگڑوں کو کم کرتا ہے اور بہترین نتائج سامنے لاتا ہے۔

مخصوص معذوریاں رکھنے والی عورتوں کی مدد کرنا

ایک عورت کو جو اندھی ہے یا دیکھنے میں دشواری محسوس کرتی ہے

مجھے بتائیے کہ میں کہاں ہوں، مجھے کرسی یا معائنہ ٹیبل تک لے جائیے۔ مجھے کمرے کے بیچ میں مت چھوڑیئے۔

- اس وقت تک یہ بتائے بغیر اسے نہ چھوئیں کہ آپ کون ہیں، ماسوائے کہ کوئی فوری ضرورت ہو۔
- یہ نہ سوچیں کہ وہ آپ کو بالکل بھی نہیں دیکھ سکتی ہے۔
- اس سے عام آواز میں بات کریں۔
- اگر اس کے پاس چھڑی ہے تو وہ چھڑی کسی وقت اس سے دور نہ کریں۔
- اس سے دور جاتے ہوئے یا اس سے الگ ہوتے ہوئے خدا حافظ کہیں۔

اگر کوئی عورت بہری ہو یا سننے میں دشواری محسوس کرتی ہو تو

میری علامتی زبان کی توجیہہ کرنے والے یا میرے گھریلو اشارے سمجھنے والے میرے گھر والوں کی طرف نہیں بلکہ میری طرف دیکھئے۔

- گفتگو شروع کرنے سے قبل اس کی توجہ حاصل کرنا یقینی بنائیے۔ اگر وہ آپ کی طرف متوجہ نہ ہو تو اس کے کندھے کو نرمی سے چھوئیں۔
- زور سے نہ بولیں یا اپنی گفتگو کو طول نہ دیں۔
- براہ راست اس کی طرف دیکھیں اور اپنے منہ کو کسی چیز سے بھی نہ چھپائیں۔
- اس سے معلوم کریں کہ اس کے ساتھ ابلاغ کا بہترین طریقہ کیا ہے۔

جسمانی معذوریاں رکھنے والی عورت ہو تو

میرے گھر والوں یا میری دیکھ بھال کرنے والے سے نہیں بلکہ مجھ سے براہ راست بات کیجئے۔

- یہ فرض نہ کریں کہ وہ دماغی طور پر بھی کم تر ہے۔
- اگر ممکن ہو تو اس طرح بیٹھیں کہ آپ اس کی آنکھوں کی سطح پر ہوں۔
- عورت کی اجازت کے بغیر اور ان کی واپسی کی یقین دہانی کے بغیر بیساکھیاں، چھڑی، واکر یا وہیل چیئر اس سے الگ نہ کریں۔

ایک عورت جو واضح طور پر بول نہ سکتی ہو

اگر سمجھ نہ پائیں تو یہ مت ظاہر کیجئے کہ آپ میری بات سمجھ چکے ہیں

- ہر چند کہ اس کی گفتگو سست ہو یا سمجھنا مشکل ہو، اس کا یہ مطلب ہرگز نہیں کہ وہ سیکھنے یا سمجھنے میں کوئی دشواری رکھتی ہوگی۔
- اگر آپ کوئی بات نہ سمجھ پائیں تو اسے وہ بات دوہرانے کے لئے کہیں۔
- ایسے سوالات پوچھئے جن کے جواب وہ "ہاں" یا "نہیں" میں دے سکے۔
- صبر اور تحمل سے کام لیں اور اپنے مسئلے کو بیان کرنے میں اسے جس قدر وقت کی ضرورت ہو، اسے وہ وقت دیں۔

اگر ایک عورت سیکھنے یا سمجھنے میں مشکل رکھتی ہو

ایک وقت میں مجھے ایک بات بتائیے اور ضروری ہو تو اسے دوہرائیے۔

- سادہ الفاظ اور مختصر جملے استعمال کریں۔
- شائستہ اور صابر رہیں اور اس کے ساتھ بچوں جیسا برتاؤ نہ کریں۔

تبدیلی کے لئے کام

یہ چند تجاویز ہیں جو آپ صحت کی دیکھ بھال کی خدمات بہتر بنانے کے لئے ،صحت کارکنوں کے ساتھ مل کرکام کرنے کے لئے استعمال کرسکتے ہیں ۔

یہ سرگرمیاں مندرجہ مقاصد کے لئے استعمال کی جاسکتی ہیں :

● رسائی ،دستیابی اوران رویوں کے متعلق جومعذور عورتوں کی دیکھ بھال سے متعلق ہیں آگہی بڑھانا۔

● ان اقدامات کی نشاندہی جومعذور عورتوں کے لئے صحت کی دیکھ بھال بہتر بناسکتے ہیں ۔

صحت کی دیکھ بھال کی خدمات بہتر بنانے میں آپ کی مدد کے لئے چند رہنما خطوط ذیل میں بیان کئے گئے ہیں ۔ گروپ کی صورت میں صحت کی دیکھ بھال کی رکاوٹوں کے ذاتی تجربات میں ایک دوسرے کو شریک کرکے عورت میں اعتماد پیدا کیا جاسکتا ہے ۔

پہلا مرحلہ : ہر فرد پیش کرنے کے لئے کچھ نہ کچھ رکھتا ہے ۔

ہر فرد کومطمئن رکھنے اور یہ ظاہر کرنے کے لئے کہ ہر فرد گروپ سازی میں ایک حصہ رکھتا ہے آپ ہر عورت سے درخواست کرسکتے ہیں کہ وہ آپ کو بتائے کہ وہ کیا کام بہتر طور پر کرسکتی ہے یا اسے اپنی کس خوبی پر ناز ہے ۔ آپ انہیں ایک دوسرے کے بارے میں رائے دینے پر بھی آمادہ کرسکتے ہیں ۔

بطور مثال :

ماریا اپنی بہنوں کے مابین امن بحال رکھتی ہے ۔

کرانتی ایک اچھی باورچن ہے

ایڈی ٹون ایک اچھی داستان گو ہے ،ان کے گھر اور پڑوسیوں کے بچے اس سے کہانیاں سننا پسند کرتے ہیں ۔

رائنا ایک ماہر مڈوائف ہے اور سینکڑوں بچے ڈلیور کراچکی ہے

پاکستان میں بھی معذور افراد کے بارے میں آگاہی پھیلانے کی ضرورت ہے ۔ پیرامیڈیکس ،نرسنگ ، مڈوائفری اورطب کے طلباء و طالبات کومعذور افراد اور خاص طور پر معذور خواتین کی ضرورتوں کے بارے میں آگاہی نہیں دی جاتی ہے ۔ معذور افراد ہمارے سماج کے لئے قابل قدر خدمات انجام دے سکتے ہیں ۔ ہم سب مل جل کر معذور افراد کے لئے ہر طرح کی سہولتوں کی فراہمی کے لئے کام کرنا چاہیئے ۔

دوسرا مرحلہ: صحت کی دیکھ بھال کی رسائی کے حوالے سے تجربات سے آگاہ ہونا

ہر فرد سے درخواست کیجیے کہ وہ اپنے مشاہدے یا تجربے کے مطابق ایسی بات کے بارے میں بتائے جو کسی معذور عورت کے لئے صحت کی اچھی دیکھ بھال حاصل کرنے میں رکاوٹ بنی ہو۔ عورتوں کے بیان کردہ مشکلات کی فہرست بنائیے۔

تیسرا مرحلہ: صحت کی اچھی دیکھ بھال میں حائل رکاوٹوں کو جاننے کے لئے تمثیلی خاکے

انہوں نے مشکلات کی جو فہرست بنائی ہے ان کے بارے میں ہر ایک کی ہم آہنگی بڑھانے کے لئے تمثیلی خاکے استعمال کیجیے۔ آپ اس مقصد کے لئے گروپ کو مختلف ٹیموں میں تقسیم کریں جن میں صحت کارکن اور معذور عورتیں دونوں شامل ہوں۔ ہر ٹیم سے کہیں کہ وہ کسی معذور عورت کے حوالے سے ایسا تمثیلی خاکہ تیار کریں جو صحت کی اچھی دیکھ بھال حاصل کرنے میں دشواری یا رکاوٹ رکھتی ہے۔ حصہ لینے کے لئے ہر ایک کی حوصلہ افزائی کیجیے۔

تمثیلی خاکے (Role plays)

لوگوں کو حقیقی زندگی کے مسائل یا صورتحال کو سمجھانے کا ایک بہترین طریقہ یہ ہے کہ مسائل کو عملی طور پر پیش کیا جائے۔ جب باقاعدہ اور اجتماعی تبادلہ خیال کے بعد معاملات کو تمثیلی خاکے کے ذریعے پیش کیا جائے تو لوگوں کے لئے رویوں، رواج اور برتاؤ کو سمجھنا آسان ہو جاتا ہے۔ اس طرح سے یہ معلومات عورتوں کے صحت کے مسائل کو حل کرنے میں معاون ثابت ہو سکتی ہیں۔

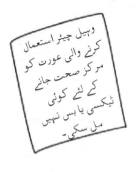

بہروپ بدلنا سماجی مسائل کے حوالے سے آگہی پروان چڑھانے اور متبادل حل تلاش کرنے کے لئے مفید ہو سکتا ہے۔

بہروپ سازی پر مزاح ہونی چاہئے لیکن اسے سنجیدگی سے کیا جانا چاہئے۔ حرکات اور کردار بعض اوقات حد درجہ مبالغہ آمیز ہو سکتے ہیں لیکن یہ بنیادی طور پر اسی طرح ہوں جس طرح معاملات اور لوگ حقیقت میں ہوتے ہیں۔ بہروپ سازی معمولی یا کسی مشق کے بغیر اور مکالمہ وغیرہ یاد کئے بغیر بھی کی جاسکتی ہے۔

(تمثیلی خاکوں اور دیگر تعلیمی تھیئٹر کے بارے میں مزید معلومات کے لئے دیکھئے کتاب ہیلپنگ ہیلتھ ورکر زلرن۔ باب 14 اور 27)

تمثیلی خاکے۔امکانات

اگر آپ کے گروپ کو اپنے خاکے سوچنے میں مشکلات درپیش ہوتو رہنمائی کے لئے چند تمثیلی خاکے یہ ہیں:

ایک دن سیریبرل پالسی کی شکار ایک عورت بیدار ہوئی تو وہ بخار، کپکپی اور ڈائریا میں مبتلا تھی۔ وہ اپنے گھر میں اس طرح بیمار ہونے والی چوتھی فرد تھی۔

وہ مقامی مرکز صحت گئی جہاں موجود صحت کار کر نے اس سے اس کی معذوری سے متعلق بہت سے سوالات کئے مگر ام کی بیماری کے بارے میں کچھ بھی نہیں پوچھا۔

ایک بہری عورت کو کلینک میں ایسا کوئی فرد نہیں ملا جو سمجھ سکتا کہ وہ کیا چاہتی ہے۔

وہیل چیئر استعمال کرنے والی عورت کو مرکز صحت جانے کے لئے کوئی ٹیکسی یا بس نہیں مل سکی۔

ہر تمثیلی خاکے کے بعد ادا کاروں سے کہیں کہ وہ گروپ میں واپس آ جائیں۔ گروپ سے کہیں کہ وہ ایک دوسرے سے تمثیلی خاکے کے بارے میں سوالات کریں اس سے معذور عورتوں کے ان مسائل کو اچھی طرح سمجھنے میں مدد ملے گی جو انہیں مراکز صحت میں پیش آتے ہیں۔

چوتھا مرحلہ: مراکز صحت کا دورہ

ایک گروپ کسی فرد کی بہ نسبت معاملات زیادہ سنجیدگی سے لے گا۔وقت سے پہلے ہی یہ طے کریں کہ گروپ کی نمائندگی کون کرے گا اور وہ کیا کہے گا۔ممکن ہے آپ کو دورہ کرنے سے پہلے اس کی اجازت لینے کی ضرورت ہو۔

صحت کی اچھی دیکھ بھال تک رسائی میں حائل چند عمومی رکاوٹوں کی نشاندہی کے بعد گروپ ان چیزوں کے مشاہدے کے لئے مقامی مراکز صحت کا دورہ کرسکتا ہے جو کسی معذور عورت کے لئے مسائل پیدا کرسکتی ہیں۔اگر گروپ میں لوگوں کی تعداد مناسب ہو تو گروپوں کو دو حصوں میں اس طرح بانٹا جاسکتا ہے کہ ہر گروپ میں ایک صحت کارکن ضرور ہو اور ہر گروپ ایک یا زیادہ مراکز صحت جاسکتا ہے (اگر ممکن ہو تو کسی مرکز صحت کے دورے کے لئے ایسے گروپ کو نہ بھیجا جائے جس میں اس مرکز صحت میں کام کرنے والا صحت کارکن شامل ہو) گروپ میں شامل ایک یا دو عورتوں سے کہا جائے کہ وہ سامنے آنے والے مسائل اور رکاوٹوں کو لکھتی رہیں۔ان سے یہ بھی کہیں کہ وہ ایسی باتیں بھی نوٹ کریں جو معذور یاں رکھنے والی عورتوں کے لئے معاون ہوں۔

مرکز صحت کا دورہ ایک مثال کے بطور بھی استعمال کیا جاسکتا ہے کہ کس طرح عورتیں ایک دوسرے کی مدد اور مسائل پر قابو پانے کے لئے ایک دوسرے کی قوت بن سکتی ہیں۔مثلاً وہیل چیئر استعمال کرنے والی عورتیں نابینا عورتوں کی رہنمائی کرسکتی ہیں۔نابینا عورتیں ان عورتوں کی مدد کرسکتی ہیں جنہیں چلنے میں مدد کی ضرورت ہو۔

پانچواں مرحلہ: آپ نے مرکز صحت میں کیا دیکھا

دورے سے واپسی پر ہر گروپ ان مسائل کو جو اس کے مشاہدے میں آئے اور وہ باتیں جو معذور یاں رکھنے والی عورتوں کے لئے مددگار تھیں، بیان کرے۔ہر گروپ سے یہ بھی پوچھیں کہ مرکز صحت کے ڈائریکٹر اور عملے نے ان کے ساتھ کیسا برتاؤ کیا۔آپ پائے جانے والے مسائل کی فہرست بنا سکتے ہیں یا ایک خاکہ کھینچ سکتے ہیں۔

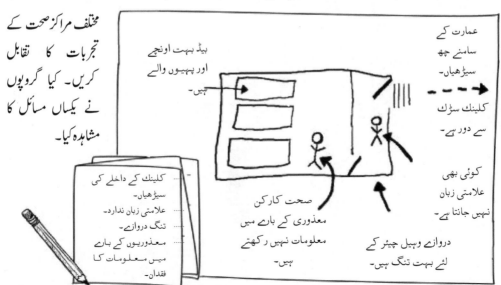

چھٹا مرحلہ: کون سے مسائل سب سے زیادہ اہم ہیں؟

مراکزِ صحت کے دورہ کرتے ہوئے سامنے آنے والے مسائل پر تبادلہ خیال کریں اور عورتوں سے پوچھیں کہ ان میں سے کن مسائل کو وہ سب سے زیادہ اہم سمجھتی ہیں یا وہ کس مسئلے کا حل سب سے پہلے چاہتی ہیں۔ مرکزِ صحت کی خدمات میں بہتری کے عمل کے لئے منصوبہ بندی اور وقت درکار ہوسکتا ہے۔ ممکن ہے کہ آپ کا گروپ فہرست کا جائزہ لینا چاہتا ہوتا کہ یہ فیصلہ کر سکے کہ کن چیزوں کو آپ فوری طور پر بدل سکتے ہیں اور کن باتوں کو تبدیل کرنے میں زیادہ مدت لگ سکتی ہے۔ آپ ان سے پوچھیں کہ وہ اپنے منتخب کئے گئے مسائل کو کیوں اہم سمجھتی ہیں کہ ان کے حل کے لئے کیا کام کیا جائے اور تبدیلی کے لئے ان کی امیدیں اور خواہشات کیا ہیں جوصورتحال کے بہتر ہونے پر پوری ہوسکتی ہیں۔

ساتواں مرحلہ: بہتری کے لئے عملی منصوبہ

جیسے ہی گروپ ایک دو ایسے مسائل چن لے جو معذور عورتوں کے لئے صحت کی اچھی دیکھ بھال کا حصول مشکل بنا رہے ہوں، گروپ ان مسائل کو حل کرنے کے منصوبے پر کام شروع کرسکتا ہے۔ گروپ سے کہیں کہ وہ ہر مسئلے کو حل کرنے کے مختلف طریقوں پر تبادلہ خیال کرے اور گروپ سے یہ بھی کہیں کہ وہ ان بہتریوں کے لئے معاون دوسرے لوگوں کے بارے میں بھی سوچے۔ وہ مراحل طے کریں جو ان بہتریوں کے لئے ضروری ہوں اور یہ بھی فیصلہ کریں کہ کس مرحلے کے لئے کون کس عمل کا ذمہ دار ہوگا، اس کے بعد کام شروع کردیں۔

باب- 3

ذہنی صحت

جب آپ اچھی دماغی صحت اور خودداری رکھتی ہیں تو آپ خود اپنا احترام کرتی ہیں اس صورت میں آپ اپنے وقار کا مضبوط احساس اور ٹھوس اعتماد رکھتی ہیں۔

جب آپ اپنے آپ پر اعتماد کرتی ہیں تو آپ اپنی اور دوسروں کی مدد کے لئے قوت حاصل کرسکتی ہیں۔

صحت مند جسم کے ساتھ ساتھ صحت مند ذہن اور روح کا وجود بھی ضروری ہے۔ جب آپ کا ذہن اور روح صحت مند ہوگی تو آپ کے اندر خود اپنی اور اپنے گھرانے کی دیکھ بھال، مسائل سمجھنے اور انہیں حل کرنے کے لئے بہترین کوشش کرنے، اپنے مستقبل کے لئے منصوبہ بندی کرنے اور دوسرے لوگوں کے ساتھ اطمینان بخش تعلقات قائم کرنے کی جذباتی قوت ہوگی۔ جب آپ دماغی طور پر صحت مند ہوں گی تو دوسرے لوگوں کی مدد بھی کرسکتی ہیں اور خود اپنی اہمیت بھی برقرار رکھ سکتی ہیں۔

بہت سی معذور عورتیں دماغی صحت کے مسائل کا شکار ہو جاتی ہیں جوان کے لئے زندگی میں درپیش چیلنجوں کا سامنا کرنا اور اپنی زندگی سے مطمئن رہنا یا سماج میں اپنا کردار ادا کرنا مشکل بنا دیتے ہیں۔ بعض اوقات دماغی صحت کے یہ مسائل عورت کی معذوری کی وجہ سے ہوتے ہیں لیکن عام طور پر ان کا سبب سماج کا وہ انداز ہوتا ہے جو وہ معذور عورتوں کے لئے رکھتا ہے۔

یہ باب ان چیلنجوں کے بارے میں بتاتا ہے جو معذوریوں کا شکار بہت سی عورتوں کو درپیش ہوتے ہیں۔ یہ باب عام دماغی صحت کے مسائل اور ایسی تجاویز بھی پیش کرتا ہے جو بتاتی ہیں کہ بہتری یا بہتری کے احساس کے لئے کس طرح کام کریں۔

یاد رکھئے کہ دماغی صحت کے مسائل کا فوری حل موجود نہیں ہیں۔ ایسے ہر فرد سے مختلف طریقے جواس قسم کا دعویٰ کرتا ہے۔

دماغی صحت کے مسائل

دباؤ، امتیازی رویہ، تنہائی اور اذیت ناک واقعات وہ مشکلات ہیں جن کا معذور عورتوں کو سامنا کرنا پڑتا ہے۔ یقیناً ان مسائل کا سامنا کرنے والی ہر عورت میں دماغی صحت کے مسائل پروان نہیں چڑھیں گے۔ مثلاً دباؤ صحت کا دماغی مسئلہ نہیں ہے لیکن جب آپ طویل مدت تک درپیش مسائل سے نمٹ نہ سکیں تو حد سے زیادہ دباؤ بذات خود مسئلہ بن جاتا ہے۔ آپ کی زندگی میں اذیت ناک واقعات ہمیشہ دماغی صحت کے مسائل کا سبب نہیں بنتے ہیں لیکن جب آپ انہیں سمجھنے کی کوشش اور حل کرنے میں مدد نہ رکھتی ہوں تو یہ عموماً دماغی صحت کا مسئلہ بن جاتے ہیں۔

جب ہم دماغی صحت کے مسائل کے بارے میں سوچ رہے ہوں تو یاد رکھیں

- زندگی کے واقعات پر عام ردِعمل اور دماغی مسائل کے مابین کوئی واضح حد نہیں ہے۔

- بیش تر لوگ اس باب میں بیان کی ہوئی چند علامات اپنی زندگی کے مختلف زمانے میں، وقت کی مناسبت سے درپیش مسائل کے مطابق محسوس کر سکتے ہیں۔

- دماغی صحت کے مسائل کی علامات ایک معاشرے سے دوسرے میں مختلف ہو سکتی ہیں۔ وہ رویے جو کسی اجنبی کو عجیب و غریب نظر آتے ہوں ممکن ہے کہ کسی کمیونٹی کی روایات یا اقدار کا ایک عام سا جزو ہوں۔

اگر آپ سمجھتی ہیں کہ کوئی دماغی صحت کے مسئلے سے دوچار ہے

اگر آپ کو کسی کے بارے میں شبہ ہے کہ وہ دماغی صحت کے کسی مسئلے سے دوچار ہے تو اس کے بارے میں جاننے کی بہتر کوشش کریں۔ سنیں کہ دوسرے لوگ اس کے رویوں کے بارے میں کیا کہہ رہے ہیں اور اس میں کس طرح تبدیلی آئی۔ کیونکہ دماغی صحت کے مسائل عموماً خاندان یا کمیونٹی سے پیوست ہوتے ہیں، غور کیجیے یہ کس طرح درپیش مسئلے کا سبب بن سکتے ہیں۔ لیکن ضروری نہیں کہ ہر دماغی صحت کے مسئلے کے اسباب یا وجوہات کی نشاندہی ہو سکے۔ بعض اوقات ہم یہ نہیں جان پاتے ہیں کہ کسی فرد میں دماغی صحت کا کوئی مسئلہ کیوں پروان چڑھا ہے۔

دباؤ

جب آپ کو روزانہ اور طویل عرصہ بہت سے دباؤ کا سامنا کرنا پڑتا ہے، تو آپ خود کو مغلوب ہوتا اور خود کو اس سے نمٹنے کے ناقابل محسوس کرتی ہیں۔ مسئلہ مزید سنگین ہو سکتا ہے جب آپ نے خود کو اہمیت دینا نہ سیکھا ہو اور آپ اپنی ضروریات کو نظر انداز کر رہی ہوں۔

کیا واقعی میں اعصابی خرابی سے دوچار ہوں

ممکن ہے آپ محسوس کریں کہ آپ کمزور یا بیمار ہیں لیکن بعض اوقات حقیقی مسائل ممکن ہے منصفانہ یا درست نہ ہوں۔

دباؤ کے باعث جسمانی تبدیلیاں اور بیماری

جب آپ دباؤ کا شکار ہوتی ہیں تو آپ کا جسم فوری طور پر ردِعمل کے لیے تیار ہو جاتا ہے اور دباؤ سے نجات کے لیے جنگ لڑتا ہے۔ اس صورت میں یہ تبدیلیاں ہو سکتی ہیں:

- دل کی دھڑکن تیز ہونا شروع ہو جاتی ہے۔

- خون کا دباؤ بڑھ جاتا ہے۔

- ایسا فرد سانس تیزی سے لیتا ہے۔

- ہاضمہ کا عمل سست پڑ جاتا ہے۔

اگر دباؤ اچانک اور شدید ہے تو آپ اپنے جسم میں یہ تبدیلیاں محسوس کرسکتی ہیں، جیسے دباؤ ختم ہوتا ہے آپ کا جسم نارمل حالت میں لوٹ جاتا ہے۔ لیکن اگر دباؤ کم شدت کا ہے اور آہستہ روی سے بڑھتا ہے تو ممکن ہے کہ علامات کے موجود ہونے کے باوجود آپ یہ اندازہ نہ لگا سکیں کہ دباؤ کس طرح آپ کے جسم کو متاثر کر رہا ہے۔

طویل عرصہ برقرار رہنے والا دباؤ ان جسمانی علامات کا سبب بن سکتا ہے جو انزائٹی اور ڈپریشن میں عام ہیں جیسے سردرد، آنتوں کے مسائل، توانائی کا فقدان وغیرہ۔ وقت گزرنے پر دباؤ مختلف بیماریوں مثلاً ہائی بلڈ پریشر کا سبب بھی بن سکتا ہے۔

سماجی رکاوٹیں دباؤ پیدا کرتی ہیں۔

وہ رکاوٹیں جو معذور عورتوں کو صحت کی سہولتیں حاصل کرنے سے روکتی ہیں بیشتر صورتوں میں ان کی روزمرہ کی زندگی میں دباؤ پیدا کرتی ہیں۔ کیونکہ وہ دباؤ کے بہت سے ذریعوں کا سامنا کرتی ہیں اس لئے یہ معذور عورتوں کے لئے خصوصی طور پر اہم ہے کہ وہ ایسی مدد تلاش کریں جس کی انہیں اپنی صلاحیتوں پر اعتماد بحال رکھنے اور خودداری برقرار رکھنے کے لئے ضرورت ہے۔

صنفی تضاد

صنف وہ طریقہ ہے جس سے کمیونٹی متعین کرتی ہے کہ مرد یا عورت ہونے کا کیا مطلب ہے۔ ان کمیونٹیز میں جو لڑکیوں کو لڑکوں کی طرح اہمیت نہیں دیتیں لڑکیاں زیادہ دباؤ میں رہتی ہیں۔ آپ کے بھائیوں کو زیادہ تعلیم یا خوراک دی جاسکتی ہے۔ آپ پر زیادہ تنقید ہوسکتی ہے اور آپ کے سخت محنت طلب کام نظر انداز ہوسکتے ہیں۔ معذور لڑکی کے ساتھ، ایک عام لڑکی یا معذور لڑکے کے مقابلے میں یہ سلوک

(speech bubble: یہ انصاف نہیں ہے یہ لوگ تو اسکول جائیں اور میں گھر میں رہوں اور کام کرتی رہوں)

بدر جہا زیادہ بدتر ہوسکتا ہے۔ بڑے ہونے پر ممکن ہے کہ آپ باور نہ کریں کہ آپ اپنے شریک حیات یا گھر والوں کے اچھے سلوک کی، بیمار ہونے پر صحت کی دیکھ بھال یا اپنی مہارتیں پروان چڑھانے کی حقدار ہیں۔ جب آپ اس انداز سے محسوس کریں گی تو ممکن ہے کہ آپ یہ سوچیں کہ گھر اور کمیونٹی میں آپ کو اہمیت نہ دیا جانا فطری اور درست ہے حالانکہ حقیقت میں یہ غیر منصفانہ اور غلط ہے۔

غربت

جب کوئی گھرانہ غریب ہو تو ایک معذور لڑکی یا عورت کے لئے وہ مہارتیں حاصل کرنا انتہائی مشکل ہوتا ہے جو کام کرنے کے لئے اس کی ضرورت ہیں۔ ممکن ہے وہ اسکول جانے کے لئے سماجی آلہ یا بیساکھیاں حاصل نہ کر پائے۔ اگر کوئی معذور لڑکی یا عورت اپنے گھرانے کی مدد کرنے کا موقع نہیں رکھتی ہوتو اس کو بوجھ سمجھا جاسکتا ہے۔ اگر گھر والوں کے پاس کم غذا ہے تو وہ فیصلہ کر سکتے ہیں کہ زیادہ غذا انہیں ملنی چاہیئے جو کام کرنے جاتے ہیں اور گھر والوں کی مدد کرتے ہیں۔

معذوری کے بارے میں رویئے

کمیونٹیز بھی ممکن ہے معذور لڑکیوں اور عورتوں سے زندگی میں کامیابی کی کمتر توقعات رکھتی ہوں، یہ جاننے کے بعد کہ ان سے کم امید کی جاتی ہے، معذور عورتیں خود کو کم اہمیت دینے لگتی ہیں۔ یہ عموماً کمیونٹی میں تبدیلی کے لئے نمائندگی کرنے کی خوداعتمادی سے محروم ہوتی ہیں۔

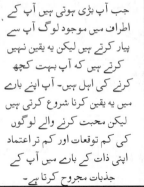

جب آپ بڑی ہوتی ہیں آپ کے اطراف میں موجود لوگ کیوں آپ سے پیار کرتے ہیں لیکن یہ یقین نہیں کرتے ہیں کہ آپ بہت کچھ کرنے کی اہل ہیں۔ آپ اپنے بارے میں یہ یقین کرنا شروع کرتی ہیں لیکن محبت کرنے والے لوگوں کی کم توقعات اور کم تر اعتماد اپنی ذات کے بارے میں آپ کے جذبات مجروح کرتا ہے۔

امتیازی رویہ، دباؤ اور عزت نفس

ہماری ایسوسی ایشن معذور عورتوں کی بہبود کو پروان چڑھانے میں مدد کے لئے معذور عورتوں کے ذریعے 1989ء میں وجود میں آئی۔ مختلف معذوریاں رکھنے والی ایکس عورتیں (بصارت، سماعت، بولنے اور حرکت کرنے میں دشواری) ایسوسی ایشن کی رکن ہیں۔ ہم مہینے میں ایک بار اپنے مسائل پر گفتگو اور ان کے حل ڈھونڈنے کے لئے میٹنگ کرتے ہیں۔

ہم سب متفق ہیں کہ معذور عورتوں کے ساتھ عموماً امتیازی سلوک کیا جاتا ہے کیونکہ :

● ہم عورت ہیں۔

● ہم معذوری رکھتے ہیں۔

● ہم میں سے بیش تر غریب ہیں۔

ہمیں شادی کے لئے موزوں نہیں سمجھا جاتا ہے یا کام کی جگہوں میں ''غلط'' تاثر دیا جاتا ہے۔ معذور لڑکیاں اور عورتیں عموماً اور اس صورت میں بھی تعلیم حاصل نہیں کر پاتی ہیں جب تعلیم دستیاب ہو مثلاً معذور بچوں کے خصوصی اسکولوں میں بھی عام طور پر لڑکوں کو ترجیح دی جاتی ہے۔

ہمیں کسی بھی قسم کے کام کی تربیت نہیں ملتی ہے۔ ہم جسمانی، جذباتی اور جنسی زیادتیوں سے دوچار ہوتے ہیں، ہمیں گھر میں یا کمیونٹی میں شاذ ہی فیصلے کرنے کی اجازت ملتی ہے۔

لیکن ایسوسی ایشن میں شامل ہم میں سے ہر ایک کے لئے سب سے بڑا مسئلہ خودداری کا فقدان ہے۔ ہمیں معاشرے نے سکھایا کہ ہم خود کو اہمیت نہ دیں۔ ہمیں شادی کرنے اور بچے پیدا کرنے اور کوئی مفید کام کرنے کے لئے نااہل سمجھا جاتا ہے۔ اسی لئے ہمیں بے وقعت سمجھا جاتا ہے۔ حد تو یہ ہے کہ ہمارے توسیع شدہ گھرانے ہم سے چاہتے ہیں کہ ہم خود کو ان کے لئے مفید ثابت کریں۔

تصورات

کمیونٹی ایک معذور کو کسی اور عورت کے مقابلے میں کمتر اہمیت کا حامل سمجھ سکتی ہے کیونکہ وہ کمیونٹی کی خوبصورت عورت کے تصور پر پوری نہیں اترتی ہے۔ لیکن معذور عورتیں اپنے اطراف میں اجسام اور رویوں کی مختلف قسمیں دیکھتی ہیں اور ان کے فرق کو سمجھ سکتی ہیں۔ وہ اپنے زخموں، بدہئیتی، اعضاء سے محرومی، غیر معمولی تاثرات اور حرکات، وہیل چیئر، بیساکھیوں، چھڑی، لاٹھیوں اور لوگوں کے درمیان اچانک دورے یا پیشاب پاخانہ خطا ہونے کے باوجود خود کو خوبصورت، خوش لباس، اہل اور مضبوط دیکھ اور سمجھ سکتی ہیں۔

جب میں ہم رنگ ساڑھی اور بلاؤز کے ساتھ اسی مناسبت سے چوڑیاں پہنتی ہوں اور اپنے ماتھے پر بندیا لگاتی ہوں تو خود کو بہت اچھا اور زیادہ پُراعتماد محسوس کرتی ہوں۔

میں نے اپنا تاثر کس طرح بدلا

میرا نام روز ہے اور میں کینیا سے آئی ہوں۔ میں نابینا ہوں اور اپنے گھر والوں اور دوستوں میں ایسے کئی افراد رکھتی ہوں جو روزمرہ دیکھ بھال میں میری مدد کرتے ہیں۔ میں ان کی مدد کو بہت زیادہ سراہتی ہوں۔ البتہ میں اس لئے دکھی اور مایوس تھی کہ مجھے اس پر زیادہ اختیار نہ تھا کہ میں کس طرح لباس پہنوں یا کس طرح کچھ کیا جاتا ہے۔ میں محسوس کرتی تھی کہ ہمیشہ میرے ساتھ بچوں جیسا سلوک کیا جاتا ہے کیونکہ کوئی بھی مجھ سے احترام کا برتاؤ کرتا محسوس نہیں ہوتا تھا۔ میں خود کو زیادہ آزاد اور خودمختار محسوس کرنا چاہتی تھی لہذا میں نے سوالات پوچھنا شروع کر دئے جب کوئی لباس بدلنے میں میری مدد کرتا تو میں پوچھتی کہ کپڑے کیسے لگ رہے ہیں اور میرے بال کس طرح سنوارے گئے ہیں۔ میں یہ بھی پوچھتی کہ میری عمر کی دوسری عورتیں کیسا لباس پہنتی ہیں اور اپنے بال کیسے سنوارتی ہیں۔

میں نے جلد ہی جان لیا کہ جب میرے مددگار مجھے لباس پہناتے اور میرے بال سنوارتے ہیں تو میں خود کو کسی بچے کی مانند محسوس کرتی ہوں، کوئی حیرت نہیں کہ لوگ مجھ سے احترام کا برتاؤ نہیں کرتے لیکن میں 25 برس کی ایک عورت تھی اور نہیں چاہتی تھی کہ میرے ساتھ بچوں جیسا برتاؤ کیا جائے اس لئے میں نے اپنے مددگاروں سے کہا کہ کیا وہ معاشرے کی دوسری عورتوں کی طرح مجھے بھی بال سنوارنا سکھا سکتی ہیں۔ وہ بھی اس سے خوش ہوئیں۔ اس سے قبل انہوں نے ایسا نہیں سوچا تھا لیکن وہ خود اپنی نوعمر بیٹیوں کے بال سنوارنے کی عادی تھیں لہذا میری مدد بھی اسی انداز سے کرتی تھیں۔ اب میری سہیلیاں معاشرے کی دوسری عورتوں کی طرح لباس پہننے میں میری مدد کرتی ہیں اور معاشرے کے دوسرے لوگ میرے ساتھ احترام سے پیش آتے ہیں۔

تنہائی

معذور لڑکیاں ممکن ہے کہ دوسرے بچوں سے الگ تھلگ پروان چڑھیں اور دوستیاں استوار کرنے کا موقع نہ رکھتی ہوں۔ ممکن ہے کہ وہ بالغ ہونے پر ٹھوس تعلقات بنانے کے لئے ضروری سماجی مہارتیں نہ سیکھ سکیں۔ تنہا، الگ تھلگ رہنا ذہنی دباؤ پیدا کرتا ہے۔ عزتِ نفس کے لئے دوست ہونا اور معاشرے یا سماج کا ایک حصہ ہونا اہم ہے۔ ایک نوعمر لڑکی جو کوئی معذوری رکھتی ہوا سے اپنے جسم کے حوالے سے اعتماد پروان چڑھانے کے لئے مدد کی ضرورت ہوتی ہے تاکہ وہ قریبی ذاتی اور جنسی تعلقات قائم کرسکے (دیکھئے صفحہ 142)۔

روزگار کی مہارتیں

معذور عورتیں کام کی تربیت کم ہی پاتی ہیں۔ اگر وہ روزگار کی مہارتیں حاصل کرنے کا موقع نہیں رکھتی ہوں تو ان کے لئے اپنے گھرانوں اور خود کو سہارا دینا مشکل تر ہوتا ہے۔

عام ذہنی مسائلِ صحت

ہر چند کہ ذہنی صحت کے مسائل کی کئی اقسام ہیں لیکن انتہائی عام انزائٹی، ڈپریشن، پُر اذیت واقعات پر ردعمل اور الکحل یا منشیات کا ناجائز استعمال ہے۔

ڈپریشن (انتہائی افسردگی یا کچھ بھی احساس نہ ہونا)

ڈپریشن ارد گرد کے معذوری نہ رکھنے والے دس میں سے دو افراد کے مقابلے میں معذوریاں رکھنے والی دس میں سے پانچ عورتوں کو متاثر کرتا ہے۔ یہ بات حیران کن نہیں ہے کیونکہ معذور لڑکیاں تعلیم حاصل کرنے یا اعتماد پروان چڑھنے کا کوئی موقع نہیں پاتی ہیں۔ یا یہ سیکھنے کے لئے اپنے لئے کچھ کس طرح کیا جائے، جوں جوں آپ کی عمر بڑھتی ہے سماجی رکاوٹیں اور آپ کی صحت میں آنے والی تبدیلیاں وہ کچھ کرنا زیادہ مشکل بنا دیتی ہیں جو آپ عام طور پر کر لیتی تھیں، یہ صورت زیادہ ناخوش اور یاسیت زدہ بناتی ہے۔

علامات

- زیادہ افسردہ محسوس کرنا۔
- سونے میں مشکل یا حد سے زیادہ سونا۔
- واضح طور پر سوچنے میں دشواری
- خوشی بخشنے والی سرگرمیوں، کھانے پینے یا جنسی تعلق میں دلچسپی ختم ہونا۔
- جسمانی مسائل جیسے سردرد، آنتوں کی تکالیف جو کسی بیماری کے باعث نہیں ہوتی ہیں۔
- روزمرہ کے کاموں کے لئے توانائی کی کمی۔
- موت یا خودکشی کے بارے میں سوچنا۔

ہر چند کہ جب آپ اس کی زد میں ہوں تو یہ یقین کرنا مشکل ہے لیکن ڈپریشن ہمیشہ نہیں رہتا ہے۔ ڈپریشن پر قابو پانے کے طریقوں کے لئے دیکھئے صفحہ 60 تا 69۔

جوں جوں آپ بوڑھی ہوتی ہیں

عمر کے ساتھ ساتھ آپ کا جسم بھی تبدیل ہوتا رہے گا۔ آپ کے روزمرہ کے کام زیادہ وقت لیں گی۔ بعض معذوریاں بدتر ہوجائیں گی اور آپ اپنے جسم کے مختلف حصوں کے زائد استعمال کے باعث ثانوی معذوریوں میں مبتلا ہوسکتی ہیں۔ آپ کو جسمانی طور پر زیادہ مشکلات پیش آسکتی ہیں اور آپ کو تواتر کے ساتھ اپنے کام کرنے کے لئے نئے طریقے اختیار کرنے کی ضرورت ہوگی۔ یہ مستقل تبدیلیاں آپ کو یہ سوچنے پر مجبور کرسکتی ہیں کہ آپ کبھی بھی حقیقی طور پر خودمختار نہیں ہوں گی اور آپ کو ہمیشہ دوسروں کی مدد پر انحصار کرنا ہوگا۔ یہ احساس کہ آپ کا دوسروں پر انحصار بڑھ رہا ہے، آپ کی عزت نفس پر اثر انداز ہوسکتا ہے۔ دیکھئے باب 13 ''معذوروں کے ساتھ عمر رسیدگی۔''

اگر آپ زیادہ تر افسردگی محسوس کرتی ہیں یا آپ سو نہیں پاتی ہیں یا اگر اپنے موڈ میں تبدیلیاں دیکھتی ہیں تو اپنے گھر میں جس پر آپ کو اعتماد ہو اس سے یا کسی صحت کارکن سے بات کیجئے۔

خودکشی

شدید ڈپریشن خودکشی (خود کو ہلاک کر لینے) کا سبب بن سکتا ہے۔ بہت سے لوگ زندگی میں کم از کم ایک بار خودکشی کے بارے میں سوچتے ہیں لیکن جب یہ خیالات بار بار اور اکثر آئیں یا انتہائی شدید ہوجائیں تو آپ کو فوری طور پر کسی تربیت یافتہ مشیر یا ذہنی صحت کارکن سے مدد حاصل کرنے کی ضرورت ہے۔

- کیا آپ خود کو تنہا اور اپنے خاندان یا دوستوں سے الگ تھلگ محسوس کر رہی ہیں؟
- کیا آپ زندہ رہنے کی امنگ کھو چکی ہیں؟
- کیا آپ الکحل یا منشیات با قاعدگی کے ساتھ استعمال کر رہی ہیں؟
- کیا آپ صحت کے کسی شدید مسئلے سے دوچار ہیں؟
- کیا آپ خود کو ہلاک کرنے کے خیالات رکھتی ہیں؟
- کیا آپ نے کبھی خود کو ہلاک کرنے کی کوشش کی؟

اگر ان سوالات میں سے کسی سوال کا جواب ''ہاں'' ہے تو آپ کسی قابل بھروسہ فرد کے ساتھ اپنے مسائل پر بات کرکے ہی بہتر محسوس کرسکتی ہیں۔ کچھ مشیر یا ڈاکٹر ڈپریشن کے علاج کے لئے ادویات بھی استعمال کرتے ہیں۔

اگر کوئی بھی جسے بھی آپ جانتی ہیں آپ سے خود کو ہلاک کرنے کی باتیں کرتی ہو تو کسی کو تاکید کریں کہ وہ اس کی سختی سے نگرانی کرے اور ہر وقت اس کے ساتھ رہے۔ گھر والوں سے کہیں کہ وہ اس کے اطراف سے خطرناک چیزیں ہٹا دیں۔ اگر آپ کی کمیونٹی میں دماغی صحت کی خدمات میسر ہوں تو کسی ایسے فرد کو تلاش کیجئے جو اس کے ساتھ باقاعدگی سے گفت و شنید کرسکے۔

انزائٹی (نروس ہونا یا فکر مندی محسوس کرنا)

انزائٹی کے دیگر عام نام، نروز، نروس، نروس اٹیک اور اذیت محسوس کرنا بھی ہیں۔

علامات

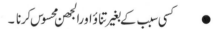

- کسی سبب کے بغیر تناؤ اور الجھن محسوس کرنا۔
- پسینے کی زیادتی، اختلاج قلب۔
- مسلسل جسمانی شکایات جو جسمانی بیماری کے باعث نہ ہوں اور اس وقت شدت اختیار کر لیتی ہوں جب آپ اپ سیٹ ہوں۔

خوف و ہراس کے حملے انزائٹی کی شدید قسم ہیں۔ یہ اچانک ہوتے ہیں اور کئی منٹ سے کئی گھنٹے جاری رہ سکتے ہیں۔ مندرجہ بالا علامات کے ساتھ آپ دہشت یا خوف محسوس کر سکتی ہیں اور اندیشہ رکھتی ہیں کہ آپ اپنے حواس کھو سکتی ہیں (بے ہوش ہو سکتی ہیں) یا مر سکتی ہیں، آپ سینے میں درد، سانس لینے میں دشواری بھی محسوس کر سکتی ہیں یا یہ محسوس کر سکتی ہیں کہ کوئی خوفناک بات ہونے والی ہے۔

جذباتی صدمہ

جب عورت کے ساتھ کوئی ہولناک بات پیش آتی ہے تو وہ جذباتی صدمے کا شکار ہو جاتی ہے، جذباتی صدمے یا دھچکے کی انتہائی عام وجوہ میں گھر میں تشدد، عصمت دری، جنگ، تشدد اور قدرتی آفات شامل ہیں۔ جذباتی صدمات عورت کی جسمانی یا ذہنی حالت (یا دونوں) کو نقصان پہنچاتے ہیں۔ نتیجے میں وہ خود کو غیر محفوظ، تحفظ سے محروم، بے یار و مددگار محسوس کرتی ہے اور دنیا یا ارد گرد کے لوگوں پر بھروسہ نہیں کر سکتی ہے۔ جذباتی صدمے سے بحال ہونے کے لئے ایک عورت کو طویل مدت لگ سکتی ہے خصوصاً اس صورت میں جب یہ کسی دوسرے فرد کے ذریعے پہنچے۔

جذباتی صدمے کے باعث معذوری

جب کوئی عورت جنگ، کسی حادثے یا بیماری کے باعث زندگی کے کسی مرحلے میں معذور ہوتی ہے تو یہ اچانک تبدیلی اس کے لئے انتہائی مشکل ہو سکتی ہے۔ کچھ عورتیں جو حال ہی میں معذور ہوئی ہوں خود اپنے لئے، اپنے گھر انوں اور کمیونٹیز کے لئے ساری وقعت کھو سکتی ہیں۔ وہ پہنچنے والے جذباتی صدمے کے باعث خوفزدہ یا ذہنی طور پر منتشر بھی ہو سکتی ہیں۔

عموماً اپنی زندگی کے کسی مرحلے میں معذور ہونے والی عورت اعتماد، اچھی تعلیم اور بہت سی مہارتوں کے ساتھ پروان چڑھتی ہے۔ وہ ہمیشہ دوسروں کے ساتھ ٹھوس تعلقات رکھنے والی اور دوسروں سے توقع کر سکتی ہے کہ وہ اس کے ساتھ احترام سے پیش آئیں۔ جب وہ معذور ہو جاتی ہے تو اسے اپنے جسم میں آنے والی تبدیلیوں سے مطابقت حاصل کرنے میں وقت لگ سکتا ہے۔ لوگ اسے کس طرح دیکھتے ہیں یا وہ خود کو کس طرح دیکھتی ہے، اس حوالے سے آنے والی تبدیلیوں کو قبول کرنا اس کے لئے دشوار تر ہو سکتا ہے۔

ایک عمر گزر نے کے بعد معذور ہونے والی بہت سی عورتیں کہتی ہیں کہ انہیں دستبردارانہ ہونے کے لئے فیصلہ کرنا پڑا ہر چند کہ انہوں نے دکھ اور صدمہ محسوس کیا تاہم انہوں نے تسلیم کیا کہ وہ اپنی زندگیاں گزارنے کے حوالے سے مواقع رکھتی ہیں۔ صفحہ 63 پر ڈاکٹر عینی کی کہانی پڑھیے جو نہایت مطمئن اور کامیاب زندگی گزار رہی تھیں کہ اچانک قوتِ سماعت سے محروم ہو گئیں۔

زیادتی ایک قسم کا جذباتی صدمہ ہے

معذور لڑکیاں خصوصی طور پر اپنے گھرانوں میں کسی نہ کسی کے ہاتھوں زیادتی یا تشدد کی زد میں ہوتی ہیں۔ اگر کوئی لڑکی کو جنسی طور پر چھوئے یا کوئی باپ، بھائی، کزن یا نگراں لڑکی پر جنسی تعلق کے دباؤ ڈالے تو زیادتی جنم لیتی ہے۔ زیادتی میں لڑکی کو مارنا یا مجروح کرنا، اس کی توہین کرنا، اس کی بے رحمی سے دیکھ بھال یا دیکھ بھال سے انکار بھی شامل ہو سکتا ہے۔ زیادتی ایک قسم کا جذباتی صدمہ ہوتی ہے جو لڑکی کی دماغی صحت کو سخت نقصان پہنچاتی ہے۔ اگر کسی عورت کے ساتھ بچپن میں زیادتی ہوئی ہو یا اسے مجروح کیا گیا ہو تو یہ برسوں اُسے متاثر رکھتی ہے۔

معذوریاں رکھنے والی بہت سی عورتیں جو مسلسل زیادتی کا شکار ہیں، شکایت نہیں کرتی ہیں کیونکہ وہ سمجھتی ہیں کہ وہ اچھے برتاؤ کی حقدار نہیں ہوتی ہیں۔ (زیادتی سے متعلق مزید معلومات کے لئے دیکھئے باب 14)

جذباتی صدمے کے ردعمل

اگر آپ جذباتی یا ذہنی صدمے کا شکار رہی ہوں تو آپ متعدد مختلف ردعمل سے دوچار ہو سکتی ہیں جیسے:

- **آپ کے ذہن میں تکلیف دہ واقعہ کی تکرار**۔ جب آپ بیدار ہوں تو ہونے والی ہولناک باتوں کی ذہن میں بار بار تکرار ہو سکتی ہے۔ رات سوتے ہوئے آپ کو انہی کے خواب نظر آئیں یا آپ سو ہی نہ سکیں کیونکہ آپ انہی کے بارے میں سوچ رہی ہوں۔

- **بے حسی یا پہلے کے مقابلے میں کم جذباتی کیفیات**۔ ممکن ہے آپ ان لوگوں یا مقامات سے گریز کریں جو آپ کو اس واقعے کی یاد دلاتے ہوں۔

- **حد سے زیادہ چوکنا ہو جانا**۔ اگر آپ مستقل طور پر خطرے کے اندیشے میں گھری رہیں تو آپ کے لئے پُرسکون رہنا اور سونا مشکل ہو سکتا ہے۔ آپ چونکانے والی باتوں پر بے جا ردعمل کا اظہار کر سکتی ہیں۔

- **جو کچھ ہوا اس پر انتہائی غصہ ہونا یا شرمندگی محسوس کرنا**۔ اگر آپ کسی ایسے اذیت ناک واقعے سے دوچار ہوئی ہوں جس میں دوسرے کئی افراد ہلاک یا شدید زخمی ہوئے ہوں تو آپ اس پر ندامت محسوس کر سکتی ہیں کہ دوسروں کو آپ سے زیادہ اذیت جھیلنی پڑی۔

- **دوسرے لوگوں سے الگ تھلگ اور دور ہونے کا احساس**

- **غیر معمولی یا پُر تشدد درویے پر شدید بڑھی بھی کا اظہار**۔ ہر چند کہ آپ ابہام میں ہوں کہ آپ کہاں موجود ہیں۔

ان میں سے کئی مشکل حالات میں نارمل ردعمل کی علامات ہیں۔ مثلاً یہ بات نارمل ہے کہ کسی اذیت ناک واقعے کے ہونے پر غصہ کیا جائے یا صورتحال پُر خطر ہو تو چوکنا رہا جائے۔ لیکن اگر علامات اتنی شدید ہوں کہ آپ روزمرہ کے کام نہ کر سکیں یا واقعے کے مہینوں بعد بھی علامات موجود ہوں تو آپ کو مدد کی ضرورت ہے۔

اذیت ناک واقعے سے دوچار افراد انتہائی فکرمند یا ڈپریشن زدہ بھی ہو سکتے ہیں یا وہ الکحل/منشیات کا حد سے زیادہ استعمال کرنے لگیں۔

جذباتی صدمے کے ردِعمل پر قابو میں مدد

اگر آپ کسی جذباتی صدمے سے دوچار ہوئی ہیں تو آپ کو اعتماد اور اپنے اوپر بھروسہ حاصل کرنے کے لئے دوسروں کی ضرورت پڑ سکتی ہے لیکن خیال رکھیں کہ

- دوسروں پر ازسرِ نو اعتماد کرنا ایک طویل عمل ہے۔

- واقعے سے قبل کی اور اپنے تجربات کے حوالے سے گفتگو بعض دفعہ مشکل ہو جاتی ہے۔ ہر چند زندگی بے حد تبدیل ہو چکی ہے لیکن بہت سی صورتوں میں آپ وہی ہیں جو پہلے تھیں۔

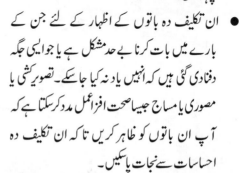

- ان تکلیف دہ باتوں کے اظہار کے لئے جن کے بارے میں بات کرنا بے حد مشکل ہے یا جو ایسی جگہ دفنا دی گئی ہیں کہ انہیں یاد نہ کیا جا سکے۔ تصویر کشی یا مصوری یا مساج جیسا صحت افزا عمل مدد کر سکتا ہے کہ آپ ان باتوں کو ظاہر کریں تا کہ ان تکلیف دہ احساسات سے نجات پا سکیں۔

- ان یاد دہانیوں کے لئے منصوبہ بنائیں جن سے آپ گریز نہیں کر سکتی ہیں۔ اگر آپ پیش آنے والے واقعے کی یاد دلانے والی باتوں پر خوفزدہ ہو جاتی ہیں تو ان باتوں سے نمٹنے کے لئے منصوبہ بنانا آپ کی مدد کرے گا جن سے آپ گریز نہیں کر سکتی ہیں۔ مثلاً آپ خود سے کہیں۔"اس کا چہرہ اس مرد کے چہرے جیسا ہے جس نے مجھ پر حملہ کیا تھا لیکن یہ ایک مختلف شخص ہے اور مجھے نقصان پہنچانا نہیں چاہتا ہے۔

- یاد رکھیں کہ کسی بھی طرح جنسی حملے یا مجروح ہونے کی صورت میں آپ نے جو کچھ کہا یا کیا، آپ اس کی ذمہ دار نہیں ہیں۔ سراسر قصور ان لوگوں کا ہے جنہوں نے آپ کو مجروح کیا۔ آپ کو مجروح کرنے والے لوگ آپ کے ساتھ ایسا سلوک کر سکتے ہیں کہ آپ خود کو پہلے جیسا نہ محسوس کر سکیں لیکن یہ سچ نہیں ہے۔

- اپنی نئی زندگی میں سوتے ہوئے آپ اپنے قریب کوئی حفاظتی چیز رکھیں۔ اس طرح اذیت ناک واقعہ کا خواب نظر آنے پر بیدار ہونے کے بعد اس حفاظتی چیز کی موجودگی آپ کو یہ یاد دلائے گی کہ اب آپ محفوظ ہیں۔

اگر آپ کی جاننے والی کوئی عورت جذباتی صدمے سے دوچار ہو

تو سب سے پہلے اس عورت کے دوستوں، گھر والوں اور دیکھ بھال کرنے والوں کے لئے ضروری ہے کہ وہ اس کے ساتھ مل جل کر روزمرہ کے کام کریں یا اس کے لئے وہ کام کریں جو کرنا چاہتی ہو۔ اسے یہ بتائیں کہ آپ اس کی باتیں سننا چاہتے ہیں اور اس کے گفتگو پر آمادہ ہونے تک انتظار کریں۔ بعد میں جب بہتر حالت میں محسوس ہو تو اسے ایسی سرگرمیوں میں حصہ لینے کے لئے آمادہ کریں جن سے وہ پہلے لطف اندوز ہوتی رہی تھی یا جو اس کی روزمرہ زندگی کا ایک حصہ تھیں۔

سنگین ذہنی امراض (سائیکوسس)

معذور عورتیں ذہنی بیماری کا خطرہ رکھتی ہیں اگر وہ

- ماضی میں ذہنی صحت کے کسی مسئلے سے دو چار رہی ہوں۔
- اپنے گھرانے کے افراد کھو چکی ہوں یا اپنے گھرانوں سے الگ تھلگ ہوں۔
- تشدد کا مشاہدہ کیا ہو یا پُر تشدد والدین رکھتی ہوں۔
- کم تر سماجی حیثیت رکھتی ہوں۔

ایک معذور عورت ذہنی طور پر بیمار ہو سکتی ہے اگر وہ ان میں سے کوئی علامات رکھتی ہو۔

- وہ ایسی آوازیں سنتی ہو یا ایسی غیر معمولی چیزیں دیکھتی ہو جو دوسرے لوگ نہیں سنتے ہوں یا دیکھ سکتے ہوں۔ (فریب خیال hullucinations)
- وہ ایسے غیر معمولی عقائد رکھتی ہو جو اس کی روزمرہ زندگی پر متاثر کریں۔ (واہمے ۔ delusions) مثلاً وہ سوچتی ہو کہ اس کے پڑوسی اسے ہلاک کرنے کی کوشش کر رہے ہیں۔
- وہ طویل مدت سے اپنی دیکھ بھال خود نہ کر رہی ہو۔ مثلاً وہ خود لباس نہ بدلتی ہو یا صاف ستھری نہ رہتی ہو یا کھاپی نہ سکتی ہو۔
- وہ عجیب و غریب انداز میں پیش آتی ہو جیسے کوئی مفہوم نہ رکھنے والی باتیں کرنا وغیرہ۔

ایسی ہی علامات چند دوسری بیماریوں زہر خورانی، ادویات، منشیات کی زیادتی یا دماغ کو نقصان پہنچنے سے بھی ہو سکتی ہیں۔ وہ لوگ جو ذہنی طور پر بیمار نہ ہوں بعض اوقات اس انداز سے کوئی کام کرتے ہیں جو ان کی ذہنی صحت کے بارے میں شبہات ابھارتے ہیں خصوصاً اس صورت میں جب ان رویوں کا تعلق عقائد یا روایات سے ہو جن میں بقیہ کمیونٹی شریک نہ ہو۔ مثلاً اگر کوئی عورت یہ کہتی ہے کہ اسے خواب کے ذریعے رہنمائی ملی ہے تو ممکن ہے کہ اس نے معلومات اور رہنمائی کے روایتی ذرائع استعمال کئے ہوں، وہ فریب خیال یا ذہنی عارضہ کا شکار نہ ہو۔ لیکن جب یہ باتیں اکثر و بیشتر ہونے لگیں اور اس حد تک طاقتور ہوں کہ کسی فرد کو اپنے روزمرہ کے کام نمٹانے میں دشواری ہو رہی ہو تو یہ علامات ذہنی عارضہ ہونے کا زیادہ امکان رکھتی ہیں۔

ذہنی بیماری کے لئے دیکھ بھال کی سہولت حاصل کرنا

ہر چند کہ بیش تر مقامات پر ایسے افراد کی دیکھ بھال جو ذہنی طور پر بیمار ہوں گھر والے کرتے ہیں، لیکن ایسے مریض کا علاج کسی تربیت یافتہ صحت کارکن کے ذریعے کرایا جانا بہترین عمل ہے۔ کچھ صورتوں میں دوائیں ضروری ہوتی ہیں لیکن یہ واحد علاج نہیں ہونا چاہئے۔

روایتی معالج اکثر ذہنی بیماری کے علاج میں اہم کردار ادا کرتے ہیں۔ ایک ایسا معالج جو اس کمیونٹی سے تعلق رکھتا ہو جس سے مریضہ کا تعلق ہو، ممکن ہے مریضہ اور اس کے گھرانے کو جانتا ہو، ممکن ہے وہ اس بات کو اچھی طرح سمجھتا ہو کہ مریضہ کو کس قسم کے دباؤ کا سامنا ہے۔ بعض معالج علاج کے مقامی طریقے یا روایتی ذرائع استعمال کرتے ہیں جو مسئلے پر قابو یا حل میں عورت کی مدد کر سکتے ہیں۔

اس سے غرض نہیں کہ کس طرح علاج کیا گیا، ذہنی بیمار فرد کا علاج ہمیشہ رحم دلی، احترام اور وقار کے ساتھ کیا جانا چاہئے۔

ذہنی عارضے کے لئے علاج کا فیصلہ کرنے سے پہلے یہ سوالات کیجئے:

- علاج میں ہر مرحلے کا مقصد کیا ہے؟

- کیا بات ہونے کی توقع ہے؟

- اگر بیمار بذاتِ خود یا دوسروں کے لئے خطرہ نہیں ہے تو کیا اس کا علاج اور دیکھ بھال گھر میں یا کمیونٹی میں دوسروں کے ساتھ رہتے ہوئے ہوسکتی ہے؟

- کیا افراد خانہ بھی علاج میں شامل ہوں گے؟

- کیا علاج کرنے والا فرد معاشرے میں احترام رکھتا ہے؟

- کیا کوئی علاج جسمانی نقصان یا شرمندگی کا سبب بن سکتا ہے؟

جینیتا تمہیں کسی بات پر بھی فکرمند ہونے کی ضرورت نہیں ہے میں بچوں کی دیکھ بھال کرلوں گا۔

کسی بھی علاج کا سب سے اہم حصہ گھر والوں اور دوستوں کی تائید اور دیکھ بھال ہے۔

اگر کسی کا علاج اسپتال میں ہونا چاہئے، تو مریضہ کو وہاں چھوڑنے سے پہلے ادارے کو دیکھنے کے لئے درخواست کریں۔ اس بات کا یقین کریں کہ اسپتال صاف ستھرا ہے، مریض محفوظ ہیں اور ان سے ملاقات کے لئے لوگ آ سکتے ہیں اور یہ کہ انہیں تربیت یافتہ ذہنی صحت کارکنوں کے ذریعے باقاعدہ علاج کی سہولت ملے گی۔ مریض جب تک اپنے لئے یا دوسروں کے لئے خطرہ نہ بنیں انہیں آزاد رکھا جائے گا۔ اس کے ساتھ یہ بھی معلوم کیجئے کہ بعد میں مریضہ کو اسپتال سے لے جانے کے لئے کیا کرنا ہوگا؟

ذہنی صحت کے ادارے میں بھی عموماً دوسری عمارتوں اور سہولتوں کی طرح وہی رکاوٹیں ہوسکتی ہیں جو معذور افراد کے لئے نقل و حرکت اور ابلاغ مشکل بناتی ہیں۔ تمام مراکزِ صحت تک رسائی بہتر بنانے کے حوالے سے معلومات کے لئے دیکھئے صفحہ 36 سے 40۔

ذہنی صحت کے لئے کام

بہتر زندگی کے لئے معذور عورتوں کو صحت، تعلیم اور آزادانہ طور پر نقل و حرکت کی اہلیت اور گزر بسر کے لئے آمدنی کی ضرورت ہوتی ہے۔ لیکن ان مقاصد کو حاصل کرنے میں مشکلات آپ کے لئے ذہنی صحت کے مسائل پیدا کر سکتی ہیں۔ عام طور پر آپ کو ڈپریشن، فکرمندی یا پریشانی یا عزتِ نفس مجروح ہونے کے احساسات پر قابو پانے کے لئے کسی تربیت یافتہ ذہنی صحت کارکن سے علاج کی ضرورت نہیں ہوتی ہے۔ ان طریقوں سے آپ اپنی مدد خود کر سکتی ہیں اور وہ طریقے بھی ہیں جن کے مطابق آپ کسی اور فرد یا گروپ کی مدد سے بہتر محسوس کرنا شروع کر سکتی ہیں۔

وہ کام جو آپ محدود وسائل سے کر سکتی ہیں

- دوستوں کے ساتھ، باغبانی، کھانا پکانے یا دیگر روزمرہ کے کاموں میں شریک ہو کر وقت گزاریں۔

- اپنے خیالات کا اظہار کریں۔ نظمیں، گیت اور کہانیاں لکھنا اس وقت معاون ثابت ہو سکتا ہے جب آپ دوسروں سے کچھ کہنے میں مشکل محسوس کر رہی ہوں۔ آپ کچھ کہے بغیر بھی اپنے احساسات رقص، ڈرائنگ، مصوری یا موسیقی کے ذریعے ظاہر کر سکتی ہیں، ضروری نہیں آپ ان کاموں میں ماہر ہوں۔

- اطراف کو خوشگوار بنائیے۔ اپنے رہنے کی جگہ کو اپنی پسند کے مطابق ترتیب دینے کی کوشش کیجئے، ممکنہ حد تک وہ جگہ روشن اور ہوادار ہو۔

- اپنے اطراف میں کچھ دلکشی پیدا کرنے کی کوشش کیجئے۔ اس کا مطلب یہ بھی ہو سکتا ہے کہ آپ اپنے کمرے میں کچھ پھول رکھیں، موسیقی سنیں یا کسی ایسی جگہ جائیں جو دلکش محسوس ہو۔

- ایسے روایتی طریقے اپنائیے جو اندرونی تقویت بڑھائیں، جسم اور دماغ کو پُرسکون رکھنے میں معاون ثابت ہوں۔

پُرسکون ہونا سیکھئے

- اپنی آنکھیں بند کریں اور ایسے محفوظ پُرسکون ماحول کا تصور کیجئے جہاں آپ ہونا پسند کریں۔ یہ جگہ کہیں بھی ہو سکتی ہے۔ کسی پہاڑی مقام پر، کسی جھیل یا سمندر کے کنارے یا کسی میدان میں۔

- ناک سے گہری سانسیں لیتے اور منہ سے خارج کرتے ہوئے اس جگہ کے بارے میں سوچنا جاری رکھیئے۔

- اگر اس سے مدد ملتی ہے تو مثبت خیال ذہن میں لائے۔ جیسے ''میں پُرسکون ہوں'' یا ''میں محفوظ ہوں''۔

- محفوظ جگہ یا مثبت خیال پر توجہ مرکوز کرتے ہوئے سانس اسی طرح لیتی رہیں۔ یہ عمل تقریباً پانچ منٹ جاری رکھیں (وہی وقت جو چاول اُبالنے کے لئے ضروری ہے)۔

- اگر آپ پُرسکون رہنے کی اس مشق کے دوران کسی بھی وقت بے اطمینانی یا خوف محسوس کریں، آنکھیں کھول دیں اور گہری سانس لیں۔

آپ یہ عمل گروپ کے ساتھ یا آپ سونے میں دشواری محسوس کرنے، تناؤ یا خوف محسوس کرنے پر، گھر میں کر سکتی ہیں، گہرے سانس لینا منتشر خیالات دور کر کے پُرسکون ہونے میں مدد فراہم کرتا ہے۔

مددگار تعلقات

مددگار تعلقات میں دو یا زائد افراد ایک دوسرے کو جاننے، سمجھنے اور ایک دوسرے کی مدد کرنے کا عہد/ وعدہ کرتے ہیں۔

مددگار تعلقات، مدد حاصل کرنے، احساسات کو جاننے اور ہیجانی ردِعمل پر قابو پانے میں معاون ثابت ہو سکتے ہیں۔ مددگار تعلقات دوستوں، افرادِ خانہ، معذور عورتوں، ایک جگہ کام کرنے والی عورتوں یا کسی اور مقصد کے لئے ملنے والے گروپوں میں قائم کئے جا سکتے ہیں۔

مددگار تعلقات کا فیصلہ کرتے ہوئے محتاط رہیں، تعلق ایسے ہی لوگوں سے استوار کریں جو آپ کے احساسات اور پرائیوسی کا احترام کریں۔ مددگار گروپوں کی تشکیل کے بارے میں معلومات کے لئے دیکھئے صفحہ 65 اور 66۔

اپنے آپ کو اہمیت دینا سیکھیں

جب ایک عورت اپنے خاندان، اسکول اور کمیونٹی کی تائید اور مدد کے ساتھ پرورش پاتی ہے تو اس میں خودداری کے احساسات نہایت ہی اعلیٰ ہوں گے خواہ وہ معذور ہو یا نہ ہو۔ لیکن اگر ایک عورت صرف اس لئے کہ وہ معذور ہے، دوسروں سے کم تر اور بے وقعت ہونے کے احساسات کے ساتھ پروان چڑھتی ہے تو اسے اپنے آپ کو اہمیت دینا سیکھنا پڑے گا۔

جب آپ اپنے بارے میں اچھا سوچتی ہیں، آپ اپنا سر بلند رکھ سکتی ہیں اور آپ جو کچھ ہیں اس پر فخر کر سکتی ہیں۔ آپ میں نئی باتیں کرنے کا حوصلہ اور اپنے آپ پر یقین کرنے کی قوت ہو گی۔ آپ خود اپنا احترام کریں، اس صورت میں بھی جب آپ غلطیاں کریں اور جب آپ اپنی عزت کرتی ہیں تو دوسرے لوگ بھی آپ کا احترام کرتے ہیں۔

جب آپ جانتی ہیں کہ آپ اہم ہیں، آپ اپنی زندگی کے بارے میں اچھے فیصلے کرتی ہیں، اپنے تحفظ، اپنے احساسات، اپنی صحت، اپنی پوری شخصیت کو اہمیت دیتی ہیں۔ عزتِ نفس یہ جاننے میں آپ کی مدد کرتی ہے کہ آپ کی ذات کے ہر حصے کے لئے دیکھ بھال اور تحفظ اہمیت رکھتا ہے۔

ذہنی صحت کے انتہائی اہم حصوں میں سے ایک عزتِ نفس ہے، عزتِ نفس یا خودداری کا اچھا احساس اسی وقت ہوتا ہے جب آپ جانتی ہوں کہ آپ احترام کے ساتھ برتاؤ کئے جانے کی حیثیت رکھتی ہیں۔ آپ جانتی ہیں کہ لوگ آپ کی باتیں سنتے ہیں اور آپ کی رائے کو اہمیت دیتے ہیں۔ آپ خود کو مشکلات اور چیلنج کا مقابلہ کرنے کے قابل سمجھتی ہیں۔

وہ لڑکیاں اور عورتیں جن کے خاندان، اسکول اور کمیونٹیز انہیں احترام دیتے ہیں، ان میں اچھی عزتِ نفس پروان چڑھتی ہے۔ خاندانوں اور کمیونٹیز کی تائید جس قدر زیادہ ہو گی اتنی ہی آپ بہترین زندگی گزار سکتی ہیں اور اسی قدر زیادہ اپنی اہمیت محسوس کریں گی۔ اچھی عزتِ نفس میں معاون دوسری چیزوں میں بامقصد کام، معاشی تحفظ، پیار بھرے تعلقات، جسمانی اور جنسی زیادتی سے تحفظ شامل ہیں۔

عزت نفس پروان چڑھانا

اپنے آپ کو اہمیت دینا سیکھنے اور عزت نفس پروان چڑھانے کا عمل اس وقت شروع ہوتا ہے جب آپ بلوغت کا دور طے کرتی ہیں اور یہ عمل بقیہ ساری عمر جاری رہتا ہے۔ لیکن اگر آپ کو بچپن میں اہمیت نہ دی جائے یا حد سے زیادہ تحفظ فراہم کیا جائے یا اعتماد پروان چڑھنے یا اپنے طور پر کچھ کرنے کا موقع نہ ملے تو آپ بالغ فرد کی حیثیت سے اس طرح نہیں رہ سکیں گی۔ آپ اپنے طور پر خود کو اہمیت اور احترام دے سکتی ہیں اور ویسی ہی نظر آ سکتی ہیں کہ جیسی کہ آپ ہیں۔ آخر کار آپ کے تجربے نے آپ کو اپنی معذوری سے مطابقت حاصل کرنا اور کام کرنا سکھا دیا ہے۔

دنیا بھر کی معذور عورتیں ازسرنو متعین کر رہی ہیں کہ ہم کون ہیں اور ایک دوسرے کی مدد کر رہی ہیں۔
ہم دل کش ہیں اور ہم جیسے بھی ہیں اس پر فخر کرتے ہیں، ہم خوبصورت ہیں۔

اچانک ہی معذور ہونے والی عورت اپنے دوستوں اور گھر والوں کی مدد اور تعاون سے اپنی معذوری سے نمٹنا سیکھ سکتی ہے۔ وہ کاموں کو اس طرح کرنا سیکھ سکتی ہے جو اس کی معذوری کے مطابق مؤثر ہو۔ لیکن اسے صرف اس لئے اپنی اقدار اور احترام کے طریقے میں تبدیلی کی ضرورت نہیں کہ اس کا جسم یا ذہن تبدیل ہو چکا ہے۔

یونگ یوک کیوان کی شخصیت حیران کن ہے وہ بارودی سرنگ کی زد میں آکر اپنی ٹانگ سے محروم ہونے سے قبل ایک بہترین اُستاد تھیں اور اب بھی ایک شاندار اُستاد ہیں۔

عینی کی کہانی

ڈاکٹر عینی اپنے شعبے کی ماہر، ایک بیوی اور ایک ماں۔ وہ ایک بیماری کے باعث بہری ہو گئیں اور قوت سماعت سے محروم ہونے کی وجہ سے اچانک معذور افراد کے زمرے میں آ گئیں۔ معذور افراد کی دنیا میں شامل ہونے کے بعد انہیں تنہائی کا تجربہ ہوا جو اکثر معذور عورتیں محسوس کرتی ہیں۔ ڈاکٹر عینی جانتی تھیں یا تو وہ زندگی کا پرانا طرز چھوڑ دیں یا پھر ایسے مواقع ڈھونڈیں جو انہیں جس حد تک ممکن ہو نارمل زندگی کے قابل بنا سکیں۔ انہوں نے ہونٹوں کی حرکات سمجھنا اور جب دوسرے لوگ ان کی بات نہ سمجھ سکیں تو تحریر کے ذریعے ابلاغ سیکھا اور خود کو الگ تھلگ نہ ہونے دیا۔ انتہائی ذاتی نقصان اور مصائب کے سامنے ڈاکٹر عینی کا وقار اور حوصلہ بہت سے لوگوں کے لئے ایک مثبت مثال ہے۔

نیلما کا انتخاب

جب نیلما نوجوان تھی تو اس نے تیزاب پی کر خودکشی کی کوشش کی۔ تیزاب نے اس کی غذائی نالی اور معدے کو مکمل طور پر جلا دیا۔ اس کی زندگی بچانے والے بھارتی ڈاکٹر نے اس سے کہا: اس کے پاس انتخاب کا ایک ہی موقع ہے: آپریشن کے بعد یا تو وہ بول سکے گی یا پھر چھوٹے چھوٹے نوالے نگل سکے گی۔ اب وہ دونوں میں سے صرف ایک کر سکتی ہے۔ نیلما نے دوسری بات کو ترجیح دی۔ آلۂ صوت نکال دیئے

جانے کے بعد نیلما ذہنی طور پر مضبوط رہی۔ اب وہ بول نہیں سکتی تھی۔ اپنی معذوری کے باوجود نیلما نے اپنا اسکول کا امتحان پاس کیا اور کھانا فراہم کرنے کا کام شروع کر دیا۔

وہ اچھا کھانا پکاتی ہے۔ اس نے گھر میں کھانا تیار کر کے فروخت کرنے کو اپنا ذریعہ روزگار بنا کر خوب نام کمایا ہے۔

اپنے طور پر خود کو اہمیت دینا ہمیشہ آسان نہیں ہوتا ہے لیکن یہ چھوٹے اقدامات کر کے کیا جا سکتا ہے۔

پہلا مرحلہ دوسرے لوگوں سے ملنا ہے۔ اگر آپ باہر جانے کی عادی نہیں ہیں تو آپ کو کوشش کرنی چاہئے کہ اپنے گھر کے دروازے پر بیٹھیں اور اپنے پڑوسیوں کا حال معلوم کریں۔ پھر اگر آپ مارکیٹ جانے کے قابل ہیں تو وہاں جائیں اور لوگوں سے گفتگو کریں۔ جوں ہی وہ آپ سے واقف ہوں گے وہ جان لیں گے کہ معذوریوں کے ساتھ یہ عورت دوسروں سے مختلف نہیں ہے۔ ہر بار جب آپ باہر جائیں گی آپ کے لئے لوگوں سے ملنا اور بات کرنا آسان تر ہو جائے گا۔

بعض اوقات ایک عورت کی معذوری، اس کے لئے دوسروں سے بات کرنا مشکل بنا دیتی ہے۔ وہ عورتیں جو بہری ہیں یا

کیا میں دو آلولے سکتی ہوں

اس ہفتے آلو بہت ہی اچھے آئے ہیں تم تین آلولے سکتی ہو۔

واضح طور پر بول نہیں سکتی ہیں، دوسروں تک اپنی بات پہنچانے کے لئے اشارے یا تصاویر استعمال کر سکتی ہیں۔ ایک بہری عورت اپنے پڑوسیوں کو کچھ اشاروں کی زبان بھی سکھا سکتی ہے۔ ابتداءً ان دو یا تین لوگوں سے کیجئے جن سے آپ گفتگو کرنا چاہتی ہیں۔ ایسے لوگ ڈھونڈنے کی کوشش کیجئے جو صابر ہوں اور آپ کے ساتھ تعاون کرنا چاہتے ہیں۔ مل جل کر آپ زیادہ سے زیادہ چیزوں کے بارے میں ابلاغ کے طریقے تلاش کر سکتی ہیں۔ پھر وقت کے ساتھ ساتھ آپ اور زیادہ لوگوں تک پہنچنے کے لئے اقدام کر سکتی ہیں۔

دوسرا مرحلہ: معذور عورتوں کے لئے ایک گروپ شروع کرنا یا ایسے گروپ میں شامل ہونا ہے۔ گروپ عورتوں کے لئے آزادانہ طور پر گفتگو کے لئے ایک محفوظ مقام فراہم کرسکتا ہے۔ دوسری عورتوں سے بات چیت آپ کی مندرجہ ذیل ابتداء کے لئے مدد کرسکتی ہے۔

- اپنی خوبیوں اور خامیوں کے بارے میں جاننا۔

- معذوری کے نتیجے میں درپیش خصوصی چیلنجوں کے حوالے سے خیالات اور تجربات میں شمولیت۔

- اپنے جسموں کو بخوبی قبول کرنے اور دیکھ بھال کے بارے میں گفتگو۔

- خوشی اور مشکلات کے زمانے میں ایک دوسرے کی مدد۔

- یہ سیکھنا کہ کس طرح خودمختار بنا جائے۔

- اپنے بارے میں اچھا محسوس کرنا اور اس احساس کو تبدیل کرنے والے معذوری کے کسی منفی احساس کو ذہن میں جگہ نہ دینا۔

خود کو قائل کرنا انتہائی دشوار تھا

ایک جرم کا ہدف بننے کے بعد معذور ہونے والی جارجیا کی ٹینا اپنے تجربے میں شریک کرتی ہے۔

جب میں نے جان لیا کہ میں معذور اور وہیل چیئر استعمال کرنے پر مجبور ہوں تو مجھے شدید صدمہ ہوا۔ پہلے میں نے سوچا کہ میں خود اس کی ذمہ دار ہوں لیکن دنوں گزرنے کے ساتھ میں نے خود سے کہا ''تمہارے بیٹے تم سے پیار کرتے ہیں اور تمہارے شوہر کو تمہاری ضرورت ہے۔ تم ایک ماہر آرائش ہو اور بہت سی عورتیں اپنے چہروں کو خوبصورت بنانے کے لئے تمہاری منتظر ہیں۔ لہٰذا تمہیں زندہ رہنا ہے۔''

میں نے زندہ رہنے اور ان کے لئے کام کرنے اور ان کے ساتھ رہنے کا فیصلہ کیا۔ اب میں دیکھ سکتی ہوں کہ میری زندگی بہتری کی جانب گامزن ہے۔

مددگار گروپ بنانا

دوسری معذور عورتوں سے باہمی ملاقاتیں ایک عورت کو زیادہ تقویت اور امید دے سکتی ہیں جو اس کی روزمرہ زندگی کے چیلنجوں سے نمٹنے میں اس کی مدد کرتی ہے۔

کسی مسئلے پر دوسروں سے گفتگو کرنے کے قابل ہونا بھی مددگار ثابت ہوگا۔ ایک عورت کے اپنی روداد سنانے کے بعد گروپ لیڈر دوسروں سے کسی ایسے تجربے کی بابت پوچھ سکتی ہے۔

> بعض اوقات ہم میٹنگ میں خراب احساسات کے ساتھ آتے ہیں۔ ہم بات کرنا نہیں چاہتے ہیں کہ ہم کوئی امنگ نہیں رکھتے ہیں۔ لیکن پھر گرمجوشی کا کوئی اظہار یا ایک قہقہہ سب کچھ بدل دیتا ہے اور ہم سب خود کو زیادہ مضبوط محسوس کرتے ہیں۔ صرف ایک ساتھ ہونا اور تنہا نہ ہونا ہمیں تقویت دیتا ہے۔

جب سب لوگ یہ تجربات سن چکیں تو پھر گروپ باہمی طور پر طے کرسکتا ہے کہ اُن کہانیوں میں کیا مشترک ہے۔ کیا مسائل جزوی طور پر سماجی حالات کا نتیجہ ہیں؟ اور اگر ایسا ہے تو ہمیں یہ حالات تبدیل کرنے کے لئے کیا کرنا چاہیئے؟

پھر عورتیں طے کرسکتی ہیں کہ کیا اس مسئلے کے لئے الگ الگ کام کیا جائے یا مل جل کر کام کرکے۔ مل کر کام کرنے والی عورتیں تنہا کام کرنے والی ایک عورت کے مقابلے میں زیادہ طاقتور ہوتی ہیں۔

مددگار گروپ کس طرح شروع کریں

1- دو یا زائد ایسی عورتیں تلاش کیجیے جو گروپ شروع کرنا چاہتی ہوں ۔

2- طے کیجیے کہ کہاں ملنا ہے۔اس کے لئے کوئی پُرسکون مقام جیسے کوئی اسکول ، مرکز صحت ، سماجی مرکز یا عبادت کی کوئی جگہ تلاش کرنے میں مدد ملتی ہے ۔

3- تبادلہ خیال کیجیے کہ آپ کیا کرنا چاہتی ہیں ، ایسے انتہائی اہم موضوع چنیے جن پر آپ مل کر گفتگو کرنا چاہتی ہیں ۔ عام طور پر مددگار گروپ اس وقت بہترین کام کرتے ہیں جب یہ معذور عورتوں کے ذریعے معذور عورتوں کے لئے چلائے جاتے ہوں ۔

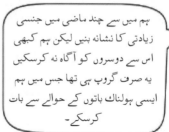

4- مشورہ دینے کی بجائے تائید کیجیے۔ ہر عورت اپنے لئے پسند کرے کہ وہ کس طرح خود کو درپیش چیلنج کا سامنا کرے گی ، کوئی اس سے یہ نہ کہے کہ اسے کیا کرنا چاہیے ۔

5- سب سے درخواست کیجیے کہ وہ اجتماعی تبادلہ خیال کو خود تک محدود رکھیں ۔

6- ہر ایک کو بات کرنے کا موقع دیجیے اور گفتگو کو مرکزی نکتے پر محدود رکھنا یقینی بنائیے ۔ ابتدائی چند ملاقاتوں کے بعد گروپ کے ارکان گروپ کی قیادت تبدیل کرنا چاہیں ایک سے زیادہ سربراہ رکھنا ، شرمیلی عورتوں کے لئے قیادت کا جذبہ اُبھارنے میں معاون ہو سکتا ہے ۔

احساسات کا اندازہ لگائیں

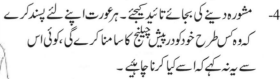
ہم میں سے چند ماضی میں جنسی زیادتی کا نشانہ بنیں لیکن کبھی اس سے دوسروں کو آگاہ نہ کر سکیں یہ صرف گروپ ہی تھا جس میں ہم ایسی ہولناک باتوں کے حوالے سے بات کر سکے ۔

بعض اوقات عورتیں اپنے احساسات چھپاتی ہیں (یا یہ بھی تسلیم نہیں کرتی ہیں کہ وہ کچھ احساسات رکھتی ہیں) کیونکہ وہ سمجھتی ہیں کہ یہ خراب ، خطرناک اور شرمناک ہیں ۔

کوئی کہانی ، ڈرامہ تخلیق کریں یا پینٹنگ بنائیں

آپ گروپ اراکین کے تجربات سے ملتی جلتی صورتحال کے حوالے سے کوئی کہانی سوچ سکتی ہیں ۔ احساسات کے بارے میں دوسروں کی گفتگو سننا ، خود اپنے احساسات سے نمٹنے میں ایک عورت کی مدد کر سکتا ہے ۔ گروپ لیڈر کہانی کی ابتداء کرتا ہے اور پھر دوسرا رکن اس کا دوسرا حصہ سنانا شروع کرتا ہے اور یہ سلسلہ اس وقت تک جاری رہتا ہے جب تک ہر ایک اس کہانی میں کچھ اضافہ نہ کرے اور اس طرح کہانی مکمل ہو جاتی ہے ۔ گروپ کہانی کے مطابق اداکاری بھی کر سکتا ہے یا کہانی کے مطابق تصویر پینٹ کر سکتا ہے ۔ یہ سوالات اپنے احساسات کے حوالے سے تبادلہ خیال میں گروپ کی مدد کر سکتے ہیں ۔

- اس کہانی میں کون سے احساسات یا تجربات سب سے زیادہ اہم ہیں؟
- یہ احساسات کس طرح وجود میں آئے۔ان احساسات سے عورت کس طرح نمٹ رہی ہے؟
- اس کی زندگی میں نیا توازن پروان چڑھانے کے لئے کیا چیز اس کی مدد کر سکتی ہے؟
- گروپ اس کی مدد کے لئے کیا کر سکتا ہے؟

مسئلے کے اسباب سمجھنا

ایک دوسرے سے گفتگو کرکے مختلف قسم کی معذور عورتیں یہ تسلیم کرنا شروع کردیتی ہیں کہ ان میں سے کئی یکساں نوعیت کے مسائل سے دوچار ہیں۔ اس سے مسائل کے بنیادی اسباب کی نشاندہی میں مدد مل سکتی ہے۔

میں اپنے بارے میں خراب تصور رکھتی تھی کہ جیسے میں ہی اپنے خاندان کی غربت کی ذمہ دار ہوں لیکن یہ میرا قصور نہیں تھا کہ میں معذور ہوں۔ دوسروں سے گفتگو نے مجھے یہ سمجھنے میں مدد دی کہ معذور لوگ کیوں اس طرح تکلیف جھیلتے ہیں جس طرح ہم جھیل رہے ہیں۔

اپنی کمیونٹی کی تصویر بنایئے

گروپ کے یکجا ہونے کے کچھ عرصہ بعد یہ عمل بہترین اثر رکھتا ہے۔ آپ کا گروپ، آپ کی کمیونٹی کی ایک تصویر بنا سکتا ہے۔ (گروپ لیڈر کے لئے یہ مددگار ہوگا کہ وہ ابتدا کرنے کے لئے سادہ تصویر بنائے) پھر دوسرے اس میں اضافہ کریں اور کمیونٹی کے ان حصوں کی تصویر بنائیں جو معذور عورتوں کی ذہنی صحت کے لئے اپنا کردار ادا کرتے ہیں اور وہ جو ذہنی مسائلِ صحت کا سبب بنتے ہیں۔ یہ سوالات عملی منصوبہ بنانے میں گروپ کی مدد کر سکتے ہیں۔

- ہم کس طرح کمیونٹی کے ان حصوں کو مضبوط بنا سکتے ہیں جو اس وقت معذور عورتوں کی اچھی ذہنی صحت کے لئے کام کر رہے ہیں؟
- کیا نئی چیزیں شروع کرنے کی ضرورت ہے؟
- ہم ان تبدیلیوں کے لئے کس طرح مدد کر سکتے ہیں؟

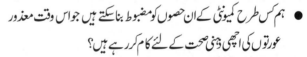

ہم نے کمیونٹی واٹر پروجیکٹ اور ٹوائلٹس تک رسائی کے حوالے سے آگہی بڑھانے کے لئے مل کر ویلیج کونسل جانے کا پروگرام بنایا۔ اگر ہم میں سے کوئی تنہا یہ کرنے کی کوشش کرے تو اس کے لئے یہ انتہائی مشکل ہوگا۔

یقینی بنائیں کہ معذور لڑکیاں اور عورتیں خاندان اور کمیونٹی کی تمام سرگرمیوں میں شریک ہوں۔

خاندان اور کمیونٹیز ذہنی صحت فروغ دے سکتی ہیں

وہ گھرانے یا خاندان جو معذور لڑکیوں اور عورتوں کی حوصلہ افزائی کرتے اور ان کی خوبیوں کو بڑھاتے ہیں، ان کی ذہنی صحت کو فروغ دیتے ہیں۔ جب گھر اور کمیونٹی آپ سے اچھی کارکردگی کی توقع کرتے ہیں اور چاہتے ہیں کہ آپ اپنی بہترین صلاحیتوں سے کام لیں تو آپ عزتِ نفس کے بھرپور احساس کے ساتھ پروان چڑھیں گی اور اپنے باطن میں مضبوط ہوں گی۔

خاندان اور کمیونٹی کے لئے ضروری ہے کہ وہ

- معذور عورتوں اور لڑکیوں کو مکمل حصہ دار ار کان کے بطور ان کے تسلیم کریں۔

- مثالوں کے ذریعے ظاہر کریں کہ وہ لڑکیوں کو بھی لڑکوں کی طرح اہمیت دیتے اور قبول کرتے ہیں اور معذور لڑکیوں اور عورتوں کو بھی دوسروں کی طرح سمجھتے ہیں۔

- معذور لڑکیوں کی دور بلوغت میں مدد کریں جب وہ لڑکی سے عورت میں بدلتی ہیں۔ یہ وہ وقت ہے کہ جب معذور اور غیر معذور لڑکیوں دونوں ہی میں ذاتی تاثر کے مسائل پیدا ہوتے ہیں۔ یہی وہ زمانہ ہے جب بیشتر نو جوان جس حد تک ممکن ہو، پُرکشش نظر آنا چاہتے ہیں۔

- معذوریاں رکھنے والی ان لڑکیوں اور عورتوں کی مدد کریں جو ذہنی صحت کے مسائل رکھتی ہیں۔ گھرانے/خاندان، اساتذہ، صحت کارکن اور دوسرے تمام لوگ بجائے اس کے کہ آپ کیا نہیں کرسکتی ہیں، آپ کیا کرسکتی ہیں پر توجہ دے کر مدد کر سکتے ہیں۔ جیسے۔

- یہ اعتماد کرنا کہ آپ ایک خوشگوار اور بھرپور زندگی گزار سکتی ہیں اور سماج میں اپنا کردار ادا کرسکتی ہیں۔ لوگ بے جا تحفظ فراہم کرنے اور آپ کے لئے کام کرنے کے بجائے، نئے کام کرنے میں آپ کی حوصلہ افزائی کریں۔

- یقینی بنائیں کہ آپ گھر کے کاموں میں ان کا ہاتھ بٹائیں اور خاندانی معاملات میں شریک ہوں۔

اس وقت سے جب میں ایک چھوٹی سی لڑکی تھی، میں ڈاکٹر بننا چاہتی تھی۔ میرے والدین کو مجھ پر اعتماد تھا۔ انہوں نے کہا کہ میں ہر وہ کام کرسکتی ہوں جو میں چاہتی ہوں اور انہوں نے میرے خواب کو پورا کرنے کے لئے میری بھرپور مدد کی۔

معذور لڑکیوں کو تعلیم دلانا

معذور لڑکیوں کو ضرورت ہوتی ہے کہ وہ اسکول جائیں اور دوسرے بچوں کے ساتھ سیکھیں۔ اگر ان کے گھر والے ان کے لئے اسکول جانے اور اسکول کو ان کے لئے ایک اچھی جگہ بنانے کا راستہ ڈھونڈ لیتے ہیں تو ایک معذور لڑکی کی مضبوط عزتِ نفس کی مالک بن سکتی ہے۔

اسکول تمام اقسام کے معذور بچوں کو قبول کر لیں، اس مقصد کے لئے دوسرے گھرانوں کے ساتھ مل کر کام کریں۔ اپنی بیٹی کی صلاحیتوں کو سمجھنے میں مدد کے لئے اور معذوریوں کے بارے میں ان کی آگہی بڑھانے کے لئے اساتذہ سے بات کریں۔

معذور لڑکیوں کو تعلیم اور ایسی مہارتیں سیکھنے کی ضرورت ہوتی ہے جو انہیں روزگار حاصل کرنے کے قابل بنائیں۔ اس صورت میں وہ خود اپنی کفالت کرنے اور اپنے گھرانوں اور کمیونٹیز کے لئے مفید کردار ادا کرنے کے قابل ہوں گی۔

این لن پہلے ہی بہت کچھ سیکھ چکی ہے۔ اگر وہ اسکول جاتی ہے تو وہ اپنے گھروالوں کی مدد کرنے کے زیادہ مواقعے حاصل کر لے گی اور خود اپنے مستقبل کے لئے بھی تیار ہو گی۔

معذور مرد ہوں یا عورتیں ان کی دیکھ بھال اور بہتر صحت کا اہتمام معاشرے کی ذمہ داری ہے۔ معاشرہ یہ ذمہ داری اس صورت پوری کر سکتا ہے جب وہ معذور افراد خصوصاً عورتوں کو اپنا ایک حصہ اور بحیثیت ایک انسان ان کے حقوق نہ صرف تسلیم کرے بلکہ انہیں یقینی بنانے کے لئے اپنا کردار ادا کرے۔ پوری دنیا کی طرح ہمارے ملک میں بھی معذور افراد کو نظر انداز کیا جاتا ہے۔ معذور عورتوں کی حالت انتہائی بدتر ہے۔ یہ ایسا ظلم ہے جس پر ہر باشعور انسان کو آواز بلند کرنی چاہیئے اور بہتر زندگی کی سہولتوں کے لئے معذور افراد کی مدد اور حوصلہ افزائی کرنی چاہیئے۔

ڈاکٹر حبیب الرحمٰن سومرو

سیکریٹری جنرل

پاکستان میڈیکل ایسوسی ایشن کراچی

باب -4

اپنے جسم کو سمجھنا

ان چار عورتوں کے جسم ایک دوسرے سے مختلف نظر آتے ہیں لیکن یہ چاروں یکساں تبدیلیوں سے گزرتی ہیں۔

یہ سمجھنا اہم ہے کہ آپ کا جسم کس طرح کام کرتا ہے۔ آپ اپنے جسم کے بارے میں جس قدر زیادہ جانیں گی بذاتِ خود اپنی دیکھ بھال کرنے کے اسی قدر بہتر قابل ہوں گی۔ جب آپ اپنے جسم اور اپنی عام تبدیلیوں کو سمجھیں گی تو آپ یہ جاننے کے قابل ہوں گی کہ آیا کوئی بات آپ کی معذوری کے باعث ہوئی ہے یا یہ ایک عام تبدیلی ہے جو تمام عورتوں میں ہوتی ہے۔ اس سے آپ کو اپنے لئے یہ فیصلہ کرنے میں بھی مدد ملے گی کہ دوسروں کا دیا ہوا مشورہ آپ کے لئے فائدہ مند ہے یا نقصان دہ۔ اس صورت میں آپ اعتماد کے ساتھ فیصلے کرسکتی ہیں۔

جب ایک لڑکی کا جسم تبدیل ہونا شروع ہوتا ہے (دورِ بلوغت)

ظاہری طور پر یہ کتنی ہی مختلف محسوس ہوں، بیش تر عورتوں کے جسم زندگی کے مختلف مراحل میں یکساں تبدیلیوں سے دو چار ہوتے ہیں۔

بعض اوقات نو سے پندرہ برس کی عمر کے درمیان ایک لڑکی کا جسم بڑھنا اور ایک عورت کے جسم میں بدلنا شروع ہوتا ہے۔ یہ بلوغت کہلاتی ہے۔ آپ کی معذوری آپ کے ساتھ ایسا ہونے میں رکاوٹ نہیں بنے گی۔ یہ تمام تبدیلیاں نارمل ہیں اور کسی بھی لڑکی میں خواہ وہ معذور ہو یا نہ ہو رونما ہوتی ہیں۔

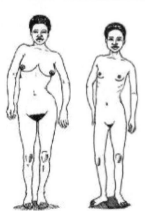

یہ وہ بنیادی تبدیلیاں ہیں جو دورِ بلوغت میں آپ کے علم میں آئیں گی۔

- آپ کی لمبائی اور گولائی میں افزائش ہوتی ہے۔
- آپ کی بغل کے نیچے اور ٹانگوں کے درمیان جنسی عضو پر بال اُگنے لگتے ہیں۔
- آپ کی چھاتیاں بڑھتی ہیں تا کہ وہ حمل کے بعد بچے کی ولادت پر اس کیلئے دودھ تیار کرسکیں۔
- آپ کے جسم کے اندر رحم (بچہ دانی)، بیضہ نالیاں، بیضہ دانی اور روجائنا نشو و نما پاتی ہیں اور اپنی پوزیشن بدلتی ہیں۔
- آپ کی وجائنا سے رطوبت کا اخراج شروع ہوتا ہے۔
- آپ کا ماہانہ اخراج خون (حیض، مینسز) شروع ہو جاتا ہے۔ ہر ماہ حیض شروع ہونے سے پہلے آپ کو پیٹ میں مروڑ، سر میں درد، کمر کے نچلے حصے یا چھاتیوں میں درد یا مزاج میں تبدیلی محسوس ہوسکتی ہے۔ مثلاً آپ حد سے زیادہ حساس ہوسکتی ہیں یا آسانی سے اپنا ضبط کھوسکتی ہیں۔
- آپ کے ذہن میں زیادہ جنسی خیالات آنا یا ہیجان شروع ہوسکتا ہے۔
- آپ کا چہرہ چکنا ہوسکتا ہے اور اس پر دانے اور دھبے پیدا ہو سکتے ہیں۔
- ممکن ہے آپ کو پسینہ زیادہ آئے اور اب آپ کے پسینے کی بو بلوغت سے قبل کی بو سے مختلف ہو۔

اگرچہ یہ تبدیلیاں فطری اور نارمل ہیں لیکن دورِ بلوغت کے یہ سال مشکل ثابت ہوسکتے ہیں۔ ممکن ہے آپ ایک لڑکی یا ایک عورت کی طرح محسوس نہ کریں۔ آپ کا جسم ان دو مرحلوں کے درمیان کہیں ہو۔ آپ معذور ہوں یا نہ ہوں، ان برسوں کے دوران یہ اہم ہے کہ آپ اپنی دیکھ بھال اچھی طرح کریں۔ صحت بخش غذا کھائیں (دیکھئے صفحہ 86) اور ماہانہ اخراج خون کے دوران صاف ستھری رہیں۔ (دیکھئے صفحہ 109) یہ آپ کے لئے نہایت ہی اہم ہے کہ آپ خود کو جنسی زیادتی سے بچائیں (دیکھئے باب - 14)۔

آپ کے جسم اور احساسات میں تبدیلی آپ کو آگاہ کرتی ہے کہ آپ ایک عورت میں تبدیل ہو رہی ہیں جو جنسی تعلق کے لئے موزوں ہوتی ہے اور حاملہ ہوسکتی ہے۔ بعض اوقات اس انداز کے باعث جس سے لوگ ایک معذور لڑکی کے ساتھ پیش آتے ہیں، معذور لڑکی خود ترسی کا شکار ہوسکتی ہے اور اپنے جسم کی تبدیلیوں پر پشیمانی محسوس کرسکتی ہے۔ وہ سرجھکانے اور دوسرے لوگوں سے ملنے سے گریزاں اور اپنے افرادِ خانہ پر زیادہ انحصار کرنے والی ہوسکتی ہے (عزتِ نفس اور ذہنی صحت کے لئے دیکھئے صفحہ 62 اور 63)۔

ہارمونز

ایک لڑکی کی بہت سی تبدیلیوں سے اس وقت گزرتی ہے جب اس کے جسم میں آنے والی تبدیلیاں ہارمونز کے ذریعے واقع ہوتی ہیں۔ ہارمونز آپ کے جسم میں بننے والے ایسے کیمیائی مادّے ہیں جو کنٹرول کرتے ہیں کہ کس طرح اور کب آپ کا جسم بڑھے۔ آپ کا پہلا ماہانہ اخراج خون (حیض) شروع ہونے سے کچھ پہلے آپ کا جسم ایسٹروجن اور پروجسٹرون ہارمون زیادہ بنانا شروع کردیتا ہے۔ یہ بنیادی ہارمون ہیں جو ایک عورت کے جسم میں تبدیلیوں کا سبب بنتے ہیں۔

ہارمونز یہ تعین کرتے ہیں کہ عورت کی بیضہ دانی کب ایک بیضہ (ہر ماہ ایک بیضہ) جاری کرے۔ ہارمونز یہ بھی کنٹرول کرتے ہیں کہ ایک عورت کب حاملہ ہوسکتی ہے۔ یہ بچے کی ولادت کے بعد بچے کو پلانے کے لئے اس کی چھاتیوں میں دودھ تیار کرتے ہیں۔ خاندانی منصوبہ بندی کے کئی طریقے عورت کے جسم میں ہارمونز کنٹرول کرکے حمل کو روکتے ہیں (دیکھئے صفحہ 196) ہارمونز حیض کے سلسلے کو با قاعدہ بھی بناتے ہیں (دیکھئے صفحہ 75)

چھاتیاں

ایک نوعمر لڑکی کی چھاتیاں نو سے پندرہ برس کی عمر کے درمیان بڑھنا شروع ہوتی ہیں۔ آپ کو اپنی چھاتیوں کے بارے میں شرمندہ یا حساس ہونے کی ضرورت نہیں۔ یہ اس بات کی علامت ہیں کہ آپ کا جسم ایک عورت کے جسم میں تبدیل ہو رہا ہے۔ آپ کی چھاتیوں میں ممکن ہے ایک دوسری سے قبل بڑھنا شروع ہو لیکن دونوں چھاتیاں ہمیشہ تقریباً یکساں ہو جاتی ہیں۔ اگر آپ کی چھاتیاں بالکل یکساں نظر نہ آتی ہوں تو خوفزدہ نہ ہوں۔ بہت سی عورتوں کی چھاتیاں سائز یا بناوٹ میں ایک دوسری سے قدرے مختلف ہوتی ہیں۔ اور اگر آپ کی چھاتیاں کسی اور لڑکی کی چھاتیوں سے مختلف نظر آتی ہیں تو یہ بھی کوئی غیر معمولی بات نہیں۔ چھاتیاں مختلف سائزوں اور بناوٹوں کی ہوتی ہیں۔

جیسے آپ کی چھاتیاں بڑی ہوتی ہیں یہ حمل کے بعد بچوں کی غذا کے لئے دودھ بنانے کے قابل ہو جاتی ہیں اور جب جنسی عمل کے دوران آپ کی چھاتیاں چھوئی جاتی ہیں تو آپ کا جسم آپ کی بھٹنیاں سخت اور ہیجان انگیز کر کے آپ کی وجائنا گیلی اور مباشرت کے لئے تیار کر کے ردِعمل کرتا ہے۔

ماہانہ حیض سے کچھ پہلے آپ کی چھاتیاں سوج سکتی ہیں اور پُر درد ہو سکتی ہیں یا آپ کی بھٹنیاں بعض اوقات مجروح ہو سکتی ہیں۔

جیسے ہی آپ کی چھاتیاں بڑھ جائیں آپ ایک مہینے میں ایک بار یہ یقینی بنانے کے لئے کہ یہ صحت مند ہیں اور ان میں کوئی غیر معمولی گومڑ وغیرہ تو نشو و نما نہیں پا رہا ہے، ان کا معائنہ ضرور کریں۔ عام طور پر ایک عورت یہ جاننے کے بعد کہ چھاتیوں کا جائزہ کس طرح لیا جائے، بذاتِ خود چھاتی کے غیر معمولی گومڑ وغیرہ کو ڈھونڈ سکتی ہے۔ بعض اوقات ختم نہ ہونے والا چھاتی کا گومڑ بریسٹ کینسر کی علامت بھی ہو سکتا ہے۔ باقاعدہ معائنہ صحت مسائل ابتداء ہی میں معلوم ہونے میں آپ کی مدد کرے گا۔ چھاتیوں کا معائنہ کس طرح کیا جائے۔ یہ جاننے کے لئے دیکھئے صفحہ 128۔

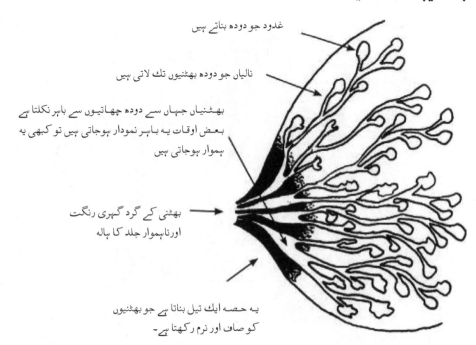

غدود جو دودھ بناتے ہیں

نالیاں جو دودھ بھٹنیوں تک لاتی ہیں

بھٹنیاں جہاں سے دودھ چھاتیوں سے باہر نکلتا ہے بعض اوقات یہ باہر نمودار ہو جاتی ہیں تو کبھی یہ ہموار ہو جاتی ہیں

بھٹنی کے گرد گہری رنگت اور ناہموار جلد کا ہالہ

یہ حصہ ایک تیل بناتا ہے جو بھٹنیوں کو صاف اور نرم رکھتا ہے۔

ماہانہ اخراج خون (پیریڈ، حیض)

معذوریاں رکھنے والی تقریباً تمام لڑکیاں عورت کی حیثیت سے غیر معذور عورتوں کی طرح حیض کا ماہانہ سلسلہ رکھتی ہیں۔ خون کا ماہانہ اخراج اس بات کی علامت ہے کہ آپ حاملہ ہوسکتی ہیں۔ کوئی بھی لڑکی یہ نہیں جان سکتی ہے کہ اس کا اولین حیض کب شروع ہوگا۔ یہ عام طور پر چھاتیاں نشو و نما پانے اور جسم پر بال اُگنا شروع ہونے کے بعد ہوتا ہے۔ اولین حیض سے کئی ماہ پہلے ممکن ہے و جائنا سے کچھ رطوبت خارج ہو۔ یہ رطوبت آپ کے کپڑوں پر داغ ڈال سکتی ہے۔ یہ عام سی بات ہے۔

اگر آپ نابینا ہیں یا دیکھنے میں دشواری محسوس کرتی ہیں یا آپ اپنے بازو اور ٹانگیں حرکت دینے میں مشکل محسوس کرتی ہیں تو اپنے گھر والوں یا سہیلیوں سے جن پر آپ اعتماد کرتی ہوں درخواست کرسکتی ہیں کہ وہ حیض کے دوران آپ کی مدد کریں۔ ایک ایسی لڑکی یا عورت کے لئے جو ماہانہ حیض کے بارے میں سمجھنے یا سیکھنے میں دشواری محسوس کرتی ہو دیکھئے صفحہ 110۔ یہ جاننے کے لئے کہ آپ کس طرح ماہانہ حیض کے دوران اپنی دیکھ بھال کریں دیکھئے صفحہ 109۔ جنسی عمل اور ماہانہ حیض کے بارے میں معلومات کے لئے دیکھئے صفحہ 182۔

جیسے ہی عورت بوڑھی ہوتی ہے اس کے حیض کا خاتمہ ہو جاتا ہے۔ بیش تر عورتوں میں یہ تبدیلی 45 سے 55 برس کی عمر کے درمیان آتی ہے۔ معلومات کے لئے دیکھئے صفحہ 282۔

دنیا بھر میں عورتیں اپنے ماہانہ اخراج خون (حیض)
کے لئے مختلف نام استعمال کرتی ہیں

ماہواری کا سلسلہ (حیض کا سلسلہ)

ماہواری کا سلسلہ ہر عورت کے لئے مختلف ہوتا ہے۔ بیشتر عورتوں میں مکمل ماہواری کا چکر چاند کے چکر کی طرح تقریباً 28 دن کا ہوتا ہے لیکن کچھ عورتوں کو ماہواری ہر بیس دن بعد آتی ہے یا کچھ عورتوں کی درمیانی مدت پینتالیس دن کی ہوتی ہے۔ ماہواری شروع ہونے کے بعد پہلے برس میں امکانی طور پر یہ ہر ماہ مختلف وقت میں ہوسکتی ہے یہ ایک معمول کی بات ہے۔ ماہواری کا باقاعدہ سلسلہ شروع ہونے سے پہلے کئی ماہ بھی لگ سکتے ہیں۔

ماہانہ چکر

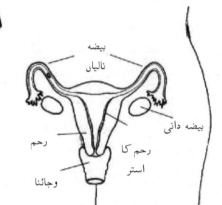

آپ کے ماہانہ چکر کے ختم ہونے سے تقریباً 14 دن پہلے بیضہ دانیوں میں سے ایک، ایک بیضہ جاری کرتی ہیں۔ یہ بیض ریزی (ovulation) کہلاتا ہے۔ اس وقت ہارمون پروجسٹرون رحم کا استر تیار کرتا ہے جس سے رحم کا استر دبیز اور حمل کے امکان کے لئے تیار ہوجاتا ہے۔

بیضہ — نالیاں

بیضہ دانی

رحم

رحم کا استر

وجائنا

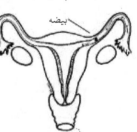

بیضہ

جب بیضہ دانی ایک بیضہ خارج کرتی ہے تو یہ بیض نالی کے ذریعے سفر کرتا ہوا رحم میں چلا جاتا ہے۔ اس دوران عورت بارور ہوسکتی ہے اور مرد سے جنسی ملاپ کی صورت میں مرد کا نطفہ عورت کے بیضے سے مل کر حمل کا آغاز کر سکتا ہے۔ یہ عمل باروری کہلاتا ہے۔

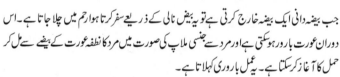

بیضہ

زیادہ تر بیضے بارور نہیں ہوتے ہیں لہٰذا رحم کے دبیز استر کی ضرورت نہیں رہتی ہے۔ یہ استر خونی سیال میں تبدیل ہوجاتا ہے جو حیض کے دوران عورت کی وجائنا سے خارج ہوتا ہے۔ اس کے بعد ماہانہ چکر دوبارہ شروع ہوجاتا ہے۔

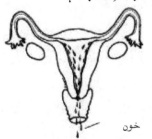

خون

جب آپ کا جسم تبدیل ہوتا ہے

یہ اہم ہے کہ آپ سوالات پوچھنے اور اپنے احساسات خصوصاً اپنے بدلتے ہوئے جسم اور جذبات کے حوالے سے اپنے ابہام اور اندیشے بیان کرنے کے قابل ہوں۔ یہ بات آپ کی پوری زندگی کے لئے درست ہے۔ جنسی صلاحیت، باروری، حمل اور سن یاس آپ

> ایک صحت مند عورت بننے کے لئے مجھے اپنے جسم اور یہ کیوں تبدیل ہوتا ہے کے بارے میں جس حد تک ممکن ہو زیادہ سے زیادہ جاننے کی ضرورت ہے۔

کے جسم اور آپ کی صحت کے لئے بڑی تبدیلیوں کی علامت بن سکتا ہے۔ جب آپ اپنی جسمانی نشوونما، جذبات اور جنسی احساس کو قبول کر لیتی ہیں تو آپ ایک عورت کی حیثیت سے اپنی دیکھ بھال اور خود اپنا احترام کر سکتی ہیں۔ آپ خود اپنے احساسات کو پرکھنے کے لئے کچھ وقت صرف کیجئے اور ان کے اظہار میں دوسروں کو شریک کیجئے۔

- اپنے جسم سے مطمئن رہیں اور اپنی معذوری کو اپنے جسم کا ایک جز تسلیم کیجئے۔
- اپنی صنفی خصوصیات اور اس حوالے سے ذمہ داریوں کے بارے میں جانیں، گھر کے بزرگ، صحت کارکن، کونسلر اور معذوری رکھنے والے دوسرے بالغ افراد معلومات کا بہترین ذریعہ ہو سکتے ہیں۔
- گھر والوں، احباب اور ہمدردی رکھنے والوں سے خیر سگالی اور محبت کے تعلقات پروان چڑھائیں۔ صحت مندی کے لئے مثبت تعلقات لازمی ہیں۔
- معذوریاں رکھنے والی لڑکیوں اور عورتوں خصوصاً ان عورتوں سے ملیں جو روزگار کرتی ہیں اور خاندان پروان چڑھا رہی ہیں۔
- ان لوگوں کے ساتھ وقت گزارنے سے گریز کریں جو آپ کے لئے ناخوشگوار تاثر بنتے ہیں۔
- اپنے گھر سے باہر مختلف سرگرمیوں میں حصہ لیں۔ ان میں اپنی شرکت کوئی جہتیں تلاش کرنے اور دوستیاں پروان چڑھانے کے مواقع سمجھیں اور ان کاموں میں حصہ لیں جو آپ بخوبی کر سکتی ہوں۔
- خود کو جنسی زیادتی سے محفوظ رکھیں (دیکھئے باب 14)

لڑکی کو عورت بننے میں مدد فراہم کرنا

ایک لڑکی کو اس کے جسم میں ہونے والی ان تبدیلیوں کے لئے ذہنی طور پر تیار کرنا نہایت ہی اہم ہے جن سے وہ عورت بنتے ہوئے دو چار ہوتی ہے۔

> نوبالغ ہونا بے حد دشوار ہے اگر آپ کے گھر والے اس پر زور دیں کہ آپ کا جسم اب بھی ایک چھوٹی سی لڑکی کی طرح ہے

یقینی بنائیے کہ وہ اپنی پہلی ماہواری سے قبل خون کے ماہانہ اخراج کے بارے میں جان لے اور جب اس کی ماہواری کا سلسلہ شروع ہو تو اسے ضرورت کے مطابق اقدام کے لئے تیار کریں۔

اسے یہ سمجھائیں کہ اس کی جسمانی اور جذباتی تبدیلیاں ایک نارمل بات ہیں۔

خاندان کی بزرگ عورتیں اور دیکھ بھال کرنے والے ایک لڑکی کو گفتگو کرنے اور اس سے نرم دلی کے ساتھ اس کے جسم کی تبدیلیوں کے بارے میں پوچھ کر اسے سوالات پوچھنے پر آمادہ کریں۔ اس طرح ایک معذور لڑکی کی اپنی بلوغت شروع ہونے سے پہلے ہی یہ جان لیتی ہے کہ اس کے قریبی افراد اس کے سوالات کا جواب دینے کے لئے موجود ہیں۔

گھرانے اور دیکھ بھال کرنے والے کیا کر سکتے ہیں؟

والدین اور گھر کے دوسرے افراد مندرجہ ذیل طریقوں سے مددگار ثابت ہو سکتے ہیں۔

- وہ یہ سمجھیں کہ یہ بھی دوسری لڑکیوں کی طرح تبدیلیوں سے گزر کر عورت بن رہی ہے۔
- اسے معذوریاں رکھنے والی دوسری لڑکیوں اور عورتوں سے ملنے کی سہولت فراہم کریں۔
- دوستی کرنے اور گھر سے باہر سرگرمیوں میں حصہ لینے کے لئے اس کی حوصلہ افزائی کریں۔ اس سے اس میں خوداعتمادی پیدا ہوگی اور اپنے ہونے کا احساس ہوگا۔
- اسے اچھی غذا اور بروقت صحت کی دیکھ بھال فراہم کریں۔
- اس کے علاوہ اس کی صنفی خصوصیات کے بارے میں گفتگو کریں۔ صنفی حوالوں سے سوالات پوچھنے اور اپنے احساسات کا اظہار کرنے کے لئے اس کی ہمت بندھائیں۔
- اس کو جنسی زیادتیوں سے تحفظ فراہم کریں۔

عمر کی مناسبت سے تقریبات

کچھ کمیونٹیز میں لڑکی کے بالغ ہونے پر تقریب کے اہتمام کا مقصد لوگوں کو یہ آگاہ کرنا ہوتا ہے کہ لڑکی جوان ہو چکی ہے اور شادی کے قابل ہے۔ اگر آپ کسی ایسی کمیونٹی میں رہتی ہیں جہاں لڑکی کے بالغ ہونے پر تقریب منعقد کی جاتی ہے تو آپ اپنی بیٹی کے لئے اس کا اہتمام ضرور کریں۔

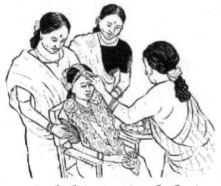

بھارت کے کچھ علاقوں میں لڑکی کے بالغ ہونے پر اسے غسل کرایا اور دلہن کی طرح سجایا جاتا ہے۔ اس کے بعد جشن منایا جاتا ہے اور تقریب میں شریک افراد لڑکی کو تحفے دیتے ہیں۔

صحت کارکن کیا کر سکتے ہیں؟

یقینی بنائیں کہ آپ لڑکیوں کے لئے ان کے جسم کے بارے میں معلومات اور صحت کی تعلیم کا جو بھی منصوبہ بنائیں اس میں معذور لڑکیوں کو ضرور شامل کریں۔ معذوریاں رکھنے والی لڑکیوں اور عورتوں کے گھر والوں اور اساتذہ کو سمجھائیں اور سکھائیں کہ معذور لڑکی یا عورت کا جسم بھی تقریباً ایسی لڑکی یا عورت کی طرح ہوتا ہے جو معذور نہ ہو۔

ایک عورت کا تولیدی نظام

کئی اعتبار سے ایک عورت کا جسم بھی خواہ وہ کوئی معذوری رکھتی ہو یا نہ ہو مرد کے جسم سے مختلف نہیں ہے۔ مثلاً عورتیں اور مرد دونوں کے دل، گردے، پھیپھڑے اور جسم کے دوسرے اعضاء ایک جیسے ہوتے ہیں لیکن ان کے جنسی یا تولیدی حصے، جو ایک مرد اور عورت کو بچہ جنم دینے کی سہولت دیتے ہیں یکسر مختلف ہوتے ہیں۔ بہت سی عورتوں کے مسائل صحت عورت کے جسم کے ان حصوں پر اثر انداز ہوتے ہیں۔

معذوریاں رکھنے والی عورتوں اور معذوری نہ رکھنے والی عورتوں کے یہ جنسی اور تولیدی حصے عام طور پر یکساں نظر آتے ہیں اور یکساں انداز سے کام کرتے ہیں۔ جسم کے یہ جنسی حصے تناسلی اعضاء (genitals) جبکہ اندرونی تولیدی اعضاء کہلاتے ہیں۔

ایک عورت کے تولیدی اعضاء

بعض اوقات اپنے جسم کے جنسی حصوں کے بارے میں بات کرنا مشکل ہوسکتا ہے خصوصاً اس صورت میں جب آپ شرمیلی ہوں یا جسم کے مختلف حصوں کے نام نہ جانتی ہوں۔ بہت سی جگہوں پر جسم کے تولیدی حصوں کو نجی معاملہ سمجھا جاتا ہے۔

یہ جاننا کہ آپ کے جنسی اور تولیدی اعضاء کس طرح کام کرتے ہیں آپ کو اس بات سے آگاہ کرے گا کہ کس طرح حاملہ ہوا جاسکتا ہے یا حمل سے محفوظ رہا جاسکتا ہے۔

بیرونی جنسی حصے

عورت کے جسم کے بیرونی اور ٹانگوں کے درمیان موجود جنسی حصوں کو مشترکہ صورت میں ولوا(vulva) کہتے ہیں۔ خاکہ بتاتا ہے کہ ولوا کیسا نظر آتا ہے اور اس کے ہر حصے کو کیا کہا جاتا ہے۔ کیونکہ ہر عورت کا جسم مختلف ہے لہٰذا ان حصوں کے سائز، بناوٹ اور رنگ خصوصاً بیرونی اور اندرونی تہوں میں فرق ہوتا ہے۔

بعض اوقات اس پورے حصے کے لئے فرج (وجائنا) کا لفظ استعمال کرتے ہیں۔ فرج ولوا کا ایک حصہ ہے جو ایک سوراخ کی شکل میں شروع ہوتا ہے اور جسم کے اندر بچہ دانی تک جاتا ہے۔ فرج کو بعض اوقات پیدائش کا راستہ بھی کہا جاتا ہے۔

جلد کی بیرونی اور اندرونی تہیں فرج کی حفاظت کرتی ہیں جنہیں لپ (Lip) بھی کہلاتی ہیں۔ جلد کی اندرونی تہیں نرم اور بالوں کے بغیر ہوتی ہیں اور حد درجہ حساس ہوتی ہیں۔ جنسی عمل کے دوران اندرونی تہیں پھول جاتی ہیں اور گہری رنگت کی ہو جاتی ہیں۔

پردہ بکارت فرج کے سوراخ کے فوراً بعد جلد کا پتلا سا ٹکڑا ہوتا ہے۔ سخت محنت، کھیل کود یا کسی اور عمل کے باعث یہ کھنچ یا پھٹ سکتا ہے اور اس حصے سے خون بہہ سکتا ہے۔ یہ اس وقت بھی ہوسکتا ہے جب عورت پہلی بار جنسی عمل سے گزرے۔ تمام پردہ بکارت ایک دوسرے سے مختلف ہوتے ہیں۔ کچھ عورتوں میں پردہ بکارت ہوتا ہی نہیں ہے اور ضروری نہیں ہے کہ تمام عورتوں میں پہلی بار جنسی عمل کے دوران خون بہے۔

بظر (clitoris) ایک چھوٹا سا گلی نما حصہ ہے۔ یہ ولوا کا چھوا جاسکنے والا سب سے حساس حصہ ہے اسے اور اس کے ارد گرد کے علاقے کو سہلانے سے عورت جنسی طور پر پُر جوش ہوسکتی ہے اور یہ عمل جنسی ملاپ کی انتہا کا سبب بن سکتا ہے۔

پیشاب کا اخراج فرج کے دہانے اور بظر کے درمیان موجود چھوٹے سے سوراخ سے ہوتا ہے جو پیشاب کی تھیلی تک ایک چھوٹی سی ٹیوب کی شکل میں جاتا ہے۔ یہ ٹیوب پیشاب کو مثانے سے جسم کے باہر لاتی ہے۔ مقعد بڑی آنت کے سر پر موجود سوراخ ہے جس سے فضلہ جسم سے خارج ہوتا ہے۔

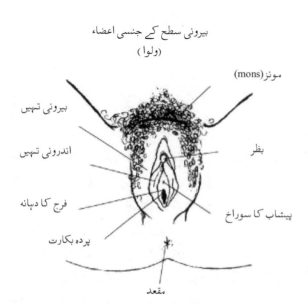

بیرونی سطح کے جنسی اعضاء (ولوا)

مونز (mons)

بیرونی تہیں

اندرونی تہیں

بظر

فرج کا دہانہ

پیشاب کا سوراخ

پردہ بکارت

مقعد

کولہے کی ہڈیاں

عورت کے جسم کے اندرونی جنسی حصے

عورت کے جسم کے اندرونی تولیدی حصے اس کے پیٹرو، کولہوں کے درمیانی حصے میں ہوتے ہیں۔ آپ اپنے کولہوں کی ہڈیوں کو اپنی کمر کے نیچے محسوس کرسکتی ہیں۔ اگر آپ کے کولہے کی ہڈیاں ہموار شکل میں نہیں ہیں تو آپ کے تولیدی حصوں پر کوئی اثر نہیں پڑے گا۔

رحم

بیضہ دانیاں

بیض نالی

سرویکس

فرج یا پیدائش کا راستہ بیضہ

ایک عورت کے رحم کے اطراف میں دو بیضہ دانیاں ہوتی ہیں، ہر بیضہ دانی، بادام یا انگور کے دانے کے برابر ہوتی ہے بیضہ دانیاں ہر ماہ عورت کی بیض نالی میں ایک بیضہ خارج کرتی ہیں۔

بیض نالیاں بیضے کو بیضہ دانی سے عورت کے رحم میں لے جاتی ہیں جو ایک کھوکھلا عضو ہے جو عورت کے حاملہ ہونے پر پھیل اور بڑھ جاتا ہے۔

ایک مرد کے جنسی حصے

ایک مرد کے جنسی اعضاء، عورت کے جنسی اعضاء کی بہ نسبت زیادہ تر اس کے بیرونی حصے میں ہوتے ہیں۔ خصیے اہم مردانہ ہارمون ٹیسٹوسٹیرون بناتے ہیں۔ جب ایک لڑکے کا جسم تبدیل ہونے لگتا ہے تو خصیے زیادہ مقدار میں ٹیسٹوسٹیرون بنانے لگتے ہیں۔ اس سے وہ تبدیلیاں ہوتی ہیں جو لڑکے کو مرد کی طرح بناتی ہیں۔ یہ تبدیلیاں ویسی ہی ہوتی ہیں جو ایک لڑکی کے جسم میں زیادہ مقدار میں زنانہ ہارمونز زیادہ مقدار میں بننے پر ہوتی ہیں۔ خصیے مرد کا نطفہ بھی بناتے ہیں۔ نطفہ خصیوں سے ایک نالی کے ذریعے عضو تناسل میں پہنچتا ہے جہاں سے خصوصی غدود سے بننے والی رطوبت میں مل جاتا ہے اس رطوبت اور نطفے کا آمیزہ ''منی'' کہلاتا ہے۔

جنسی عمل کے دوران کیا ہوتا ہے؟

جنسی عمل کے دوران مرد کی منی مرد کے انزال پر اس کے عضو تناسل سے باہر نکل جاتی ہے۔ منی کے ہر قطرے میں لاکھوں نطفے ہوتے ہیں جو اس قدر چھوٹے ہوتے ہیں کہ انہیں دیکھا نہیں جا سکتا ہے۔ جب مرد، عورت کے فرج یا تولیدی اعضاء کے نزدیک انزال ہوتا ہے تو نطفہ رحم کے منہ کے ذریعے رحم کے اندر داخل ہو سکتے ہیں۔

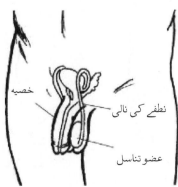

خصیہ

نطفے کی نالی

عضو تناسل

فرج ایک خاص قسم کی جلد سے بنی ہے جو جنسی عمل (اور بچے کی ولادت کے وقت) آسانی سے پھیل سکتی ہے۔ فرج ایک رطوبت یا نمی بناتی ہے جو اسے صاف اور انفیکشن سے محفوظ رہنے میں معاون ہوتی ہے۔ فرج جنسی عمل کے دوران معمول سے زیادہ رطوبت بناتی ہے جو عضو تناسل کو فرج میں آسانی سے اندر جانے دیتی ہے اور فرج کو رگڑ سے محفوظ رکھتی ہے اور نطفے کو رحم تک پہنچنے میں مدد دیتی ہے۔

نو ماہ بعد

حاملہ ہونا

ہر ماہ حیض شروع ہونے کے تقریباً چودہ دن کے بعد جب رحم کا استر تیار ہوتا ہے، کسی ایک بیضہ دانی سے ایک بیضہ نکلتا ہے۔ یہ باروری کہلاتی ہے۔ بیضہ ایک نالی سے گزرتا ہوا رحم کے اندر پہنچتا ہے اس وقت عورت زرخیز ہوتی ہے اور حاملہ ہوسکتی ہے۔ اگر اس نے اسی دوران کسی مرد کے ساتھ جنسی عمل کیا ہے تو مرد کا نطفہ رحم کے منہ (cervix) کے ذریعے رحم میں داخل ہو کر بیضے میں مل سکتا ہے۔ یہ صورت تولیدی عمل کہلاتا ہے اور یہ حمل کی ابتدا ہے۔ اگر بیضہ میں مرد کا نطفہ شامل نہیں ہوتا ہے تو حمل نہیں ٹھہرے گا اور رحم کا استر ماہانہ اخراج خون (حیض) سے جھڑ جاتا ہے۔

ہر عورت کو جنسی عمل کے بارے میں کیا جاننا چاہیے؟

- عورتیں مرد کے ساتھ پہلی بار مباشرت پر حاملہ ہوسکتی ہیں۔
- عورتیں خاندانی منصوبہ کے طریقے پر عمل کے بغیر جنسی عمل سے کسی بھی وقت حاملہ ہوسکتی ہیں۔ (خواہ جنسی عمل ایک بار ہی کیا گیا ہو)۔
- عورتیں حاملہ ہوسکتی ہیں خواہ مرد یہ سوچتا ہو کہ وہ اپنا نطفہ باہر نہیں نکلنے دے گا۔
- عورتیں جنسی طور پر منتقل ہونے والے انفیکشن (STI) یا ایچ آئی وی کا شکار ہوسکتی ہیں۔ اگر انہوں نے متاثرہ شریک حیات سے مباشرت کے دوران کنڈوم استعمال نہ کیا ہو (کسی شخص کو دیکھ کر یہ نہیں کہا جاسکتا ہے کہ وہ انفیکشن زدہ ہے یا نہیں، دیکھئے صفحہ 172)۔
- جنسی عمل کے دوران ایک عورت کا مرد کے ذریعے ایس ٹی آئی یا ایچ آئی وی میں مبتلا ہونا، مرد میں اس کے ذریعے یہ بیماریاں منتقل ہونے سے کئی گنا زیادہ آسان ہے کہ مرد کی منی اس کی فرج میں طویل عرصہ موجود رہتی ہے۔
- کسی لڑکی یا عورت کے بارے میں یہ جاننا مشکل ہے کہ وہ کسی ایس ٹی آئی میں مبتلا ہے، اس لئے کہ انفیکشن کی علامات عموماً اس کے جسم کے اندر ہوتی ہیں۔

ایس ٹی آئی اور ایچ آئی وی/ ایڈز سے بچاؤ کے لئے مردانہ یا زنانہ کنڈوم استعمال کیا جائے ہر چند کہ منی اور انفیکشن پیدا کرنے والے جرثومے بہت ہی چھوٹے ہوتے ہیں لیکن یہ کنڈوم درست طور پر استعمال کرنے پر پلاسٹک یا لیٹکس کے پار نہیں جاسکتے ہیں۔ دیکھئے صفحہ 190 اور 191۔

خود کو انفیکشن سے بچانے کے لئے دیکھئے باب نمبر 8۔ غیر مطلوب حمل سے تحفظ کے لئے دیکھئے باب نمبر 9۔

عورتوں کے لئے کنڈوم
(زنانہ کنڈوم)

مردوں کے لئے کنڈوم

جب عورتیں بچہ پیدا کرنے کے قابل نہ ہوں (بانجھ پن)

معذوری بانجھ پن کا سبب نہیں ہوتی ہے۔ کچھ معذور عورتیں بانجھ ہوں گی لیکن ان کی تعداد ان بانجھ عورتوں سے زیادہ نہیں جو معذور نہیں ہوتی ہیں۔ ایک معذوری رکھنے والی کوئی عورت بانجھ ہے تو عام طور پر یہ اس کی معذوری کے سبب نہیں ہوتا ہے۔

بانجھ پن کیا ہے؟

ہم ایک جوڑے، ایک مرد اور ایک عورت کو اس وقت بانجھ قرار دیتے ہیں جب وہ سال بھر میں ہر مہینے چند بار، خاندانی منصوبہ بندی کا کوئی بھی طریقہ استعمال کئے بغیر مباشرت (جنسی عمل) کرتے ہیں اور عورت حاملہ نہیں ہوتی ہے۔ ایک جوڑا اس وقت بھی اس مسئلے سے دو چار ہو سکتا ہے جب وہ تین یا زیادہ بار اسقاط حمل سے دوچار ہو چکا ہو۔

ایسا جوڑا جو اولاد رکھتا ہو، وہ بھی بانجھ ہو سکتا ہے۔ آخری بچے کی ولادت کے بعد گزری مدت میں ان میں یہ خرابی پیدا ہو سکتی ہے۔ بعض اوقات مسئلہ صرف مرد یا عورت میں نہیں بلکہ دونوں میں ہی ہوتا ہے اور بعض اوقات دونوں فریق صحت مند محسوس ہوتے ہیں اور کوئی بھی ڈاکٹر یا ٹیسٹ یہ تعین نہیں کر سکتا ہے کہ بانجھ پن کی وجہ کیا ہے؟

بانجھ پن کے اسباب کیا ہیں؟

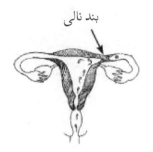

بند نالی

عورت میں بانجھ پن

عورت میں بانجھ پن کی بنیادی وجوہات مندرجہ ذیل ہیں۔

1- بیض نالی میں زخم یا رحم ۔ نالی میں زخم یا موٹی کھردری جلد بیضے کو نالی سے گزرنے یا نطفے کو بیضے تک پہنچنے سے روک سکتی ہے۔ رحم کے اندر زخم بار بار آور انڈے کو رحم کی دیوار سے جڑنے سے روک سکتے ہیں۔ بعض اوقات عورت خراشوں یا زخموں کا شکار ہو جاتی ہے لیکن وہ خود کو بیمار محسوس نہ کرنے کے باعث ان سے لاعلم رہتی ہے لیکن برسوں بعد اسے معلوم ہوتا ہے کہ وہ بانجھ ہے۔

زخم یا خراشوں کے اسباب مندرجہ ذیل ہو سکتے ہیں۔

- جنسی عمل کے دوران غیر علاج شدہ ایس ٹی آئی سے منتقل ہونے والا انفیکشن جو رحم یا نالیوں تک پہنچ گیا ہو (پیٹ کی سوجن یا پی آئی ڈی) یا پیٹ کی ٹی بی جس کے باعث بیضہ نالیوں یا رحم میں زخم پیدا ہو گئے ہوں۔

- غیر محفوظ اسقاط حمل یا بچے کی ولادت کی پیچیدگیاں جو رحم میں انفیکشن کا سبب بنتی ہیں۔

- حمل روکنے کا آلہ (آئی یو ڈی) رکھتے ہوئے صفائی کا خیال نہ رکھنا۔ آئی یو ڈی ایک چھوٹا سا آلہ ہے جو حمل روکنے کے لئے رحم کے اندر نصب کیا جاتا ہے۔ اس صورت میں یہ انفیکشن کا سبب بن جاتا ہے۔

- فرج، رحم، نالیوں یا بیضہ دانیوں کے آپریشن کے باعث مسائل۔

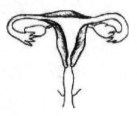

بیضہ دانی بیضہ تیار نہیں کرتی ہے

2- بیضے بننے میں رکاوٹیں

اگر ایک بانجھ عورت کا ماہانہ اخراج خون 21 دن سے کم یا 35 دن سے زیادہ مدت میں ہو تو ممکن ہے وہ بیضے پیدا نہ کر سکتی ہو۔ یہ اس وقت ہوتا ہے جب اس کا جسم مناسب مقدار میں ہارمونز نہ بنا رہا ہو یا درست وقت پر ہارمونز نہ بن رہے ہوں۔ بعض اوقات یہ عورت کے بوڑھا ہونے اور سن یاس کے قریب ہونے پر بھی ہوتا ہے۔ کچھ عورتیں تیزی سے موٹی یا دبلی ہو جائیں یا بہت موٹی یا دبلی عورتوں یا بیمار عورتوں میں بھی بیضے نہیں بنتے ہیں۔

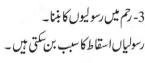

رسولی

3- رحم میں رسولیوں کا بننا۔

رسولیاں اسقاط کا سبب بن سکتی ہیں۔

مرد میں بانجھ پن

مرد میں بانجھ پن کے بنیادی اسباب یہ ہیں۔

1- مطلوبہ مقدار میں نطفے نہ پیدا ہونا۔

2- اس کے خصیے صحت مند نطفے نہ بناتے ہوں۔ یہ اس صورت میں ہوتا ہے جب وہ ایسے تنگ کپڑے پہنتا ہو جو اس کے خصیے کو اس کے جسم سے متصل رکھتے ہوں، یا وہ بوائلر، بھٹیوں یا انجن کے قریب رہ کر کام کرتا ہو، خاص طور پر جب وہ وقفے کے بغیر گھنٹوں گاڑی چلاتا ہو۔ یہ سارا دن بیٹھے رہنے یا جنسی عمل سے قبل دیر تک ہاتھ باتھ لینے پر بھی ہوتا ہے۔

3- اس کا عضو تناسل سخت نہ ہوتا ہو۔ کیونکہ اس کی ٹیوب میں ایس ٹی آئی کے باعث زخم ہوں یا اس کی ریڑھ کی ہڈی مجروح ہو۔

مردوں اور عورتوں میں بانجھ پن

مردوں اور عورتوں دونوں میں بانجھ پن کے مندرجہ اسباب ہو سکتے ہیں۔

1- بیماریاں۔ گلسوئے کی بیماری، ذیابیطس، ٹی بی اور ملیریا۔

2- شراب نوشی، سگریٹ نوشی یا تمباکو کو کھانا یا منشیات کا استعمال۔

3- ناقص غذائیت، حد سے زیادہ دباؤ، کام کی زیادتی یا مضر کیمیائی مادوں کی زد میں رہنا۔

بچہ گود لے کر گھرانے کی تشکیل

معذوریاں رکھنے والی کچھ عورتیں بچے گود لے کر اپنا خاندان تشکیل دینے کا فیصلہ کرتی ہیں۔ ایک عورت خود اپنے یا اپنے شوہر کے بانجھ ہونے یا ان مسائل صحت کے باعث ایسا کرتی ہے جو بچے کی پیدائش میں رکاوٹ بنتے ہیں یا وہ محض اس لئے یہ فیصلہ کرتی ہے کہ اس کے خیال میں ماں بننے اور گھر بنانے کا یہ ایک اچھا طریقہ ہے۔

امریکہ کی ایک معذور لڑکی کی کہانی

میں جب جوان ہوئی تو امریکہ کی بیشتر لڑکیوں کی طرح میں نے بھی ایک شریک حیات اور گھر کا خواب دیکھا۔ لیکن دوسری نوجوان لڑکیوں کے برعکس مجھے اس خواب کے پورا ہونے کا یقین نہ تھا کیونکہ میں وہیل چیئر استعمال کرتی تھی اور میرے سامنے ایسی عورتوں کی کوئی مثال نہیں تھی جو میری طرح وہیل چیئر استعمال کرتی ہوں اور ماں بنی ہوں۔ میری کبھی حوصلہ افزائی نہیں کی گئی کہ کبھی میں اپنا گھر یا خاندان رکھ سکوں گی۔ جب میں پہلی بار اپنے شوہر سے ملی تو مجھے یقین آ گیا کہ یہی شخص میرے لئے موزوں ہے۔ اس نے بچے گود لے کر گھر بنانے کے میرے پوشیدہ خواب کی تائید کی۔ مجھے معلوم تھا کہ اپنے گھرانوں سے محروم بہت سے بچے گھروں کے منتظر ہیں۔ میں یہ یقین رکھتی تھی کہ ہم ایک موزوں بچے کے لئے اچھا خاندان بن سکتے ہیں۔

پہلے پہل میرے والدین نے محسوس کیا کہ میرے شوہر پر بچے کی دیکھ بھال کا سارا بوجھ ڈالنا غیر منصفانہ بات ہوگی۔ ان کا خیال تھا کہ میں بچے کی دیکھ بھال نہیں کر سکوں گی۔ ہر چند کہ میں خود بھی نروس تھی لیکن میں نے اندازہ لگا لیا تھا کہ وہ بہت سے کام کس طرح کروں گی جو دوسروں کے مطابق میں نہیں کر سکتی تھی۔ میں اپنے گھر کے مختلف کام کر لیتی تھی، اپنے لئے روزگار رکھتی تھی اور اپنے دوستوں کے بچوں کی دیکھ بھال کر چکی تھی۔ میں جانتی تھی کہ میں اور میرا شوہر یہ سب کر لیں گے۔

ہم نے کئی اداروں سے بچہ لینے کی کوشش کی حتیٰ کہ ہمیں وہ ادارہ مل گیا جس نے ہمارے اس منصوبے کی تائید کی، لیکن یہ فوراً ہی نہیں ہوا۔ ہمیں اندازہ تھا کہ ہم لوگوں کے اندیشوں کو تبدیل نہیں کر سکتے ہیں لہذا اگر کوئی ادارہ یہ سمجھتا ہے کہ ہمارا منصوبہ کامیاب نہیں رہے گا تو اس سے ہمیں مایوس ہونے کی ضرورت نہیں۔ ہم نے ایک کے بعد دوسرے ادارے سے رابطہ کیا اور آخر کار ہمیں ایک ایسا ادارہ مل گیا جو ہمارے خیال کا حامی تھا۔ ہم نے انہیں دکھایا کہ میں جو کچھ نہیں کر سکتی ہوں اس سے ہٹ کر کس قدر اچھی ماں بن سکتی ہوں۔

آخر ہمیں وہ بچی مل گئی جو ہماری امیدوں اور خوابوں کو پورا کر سکتی تھی۔ وہ بھی میری طرح وہیل چیئر استعمال کرتی تھی۔ ہم فکرمند تھے کہ ہمیں بچہ گود لینے کی قانونی منظوری دینے والا جج میری معذوری کے باعث انکار بھی کر سکتا تھا لیکن جب اس نے ہمیں دیکھا تو ہمیں وہ بچی گود لینے کی اجازت دے دی۔

میں اپنی بیٹی کی اور اس کے اس قدر دل کش اور اہل بننے پر جیسی کہ اب وہ ہے، گزرے برسوں میں رہنمائی اور مدد پر فخر محسوس کرتی ہوں۔

کیرین بریٹ میئر

دنیا کی کل آبادی کا دس فیصد حصہ معذور افراد پر مشتمل ہے۔ 80 فیصد معذور افراد کا تعلق ترقی پذیر اور پسماندہ ممالک سے ہے۔ پاکستان میں معذور افراد کل آبادی کا 7 فیصد ہیں جن میں 40 فیصد جسمانی اور 20 فیصد ذہنی معذور ہیں بقیہ 20 فیصد بصارت جبکہ 10 فیصد سماعت سے محروم ہیں۔ 10 فیصد افراد ایک سے زائد معذوریوں کا شکار ہیں۔

سندھ میں 3.05 فیصد پنجاب میں 2.48 فیصد، بلوچستان میں 2.23 فیصد اور سرحد میں 2.02 فیصد افراد معذور ہیں۔ مرد 2.8 فیصد جبکہ خواتین میں 02.02 فیصد معذور ہیں۔

معذوروں کے عالمی دن پر روز نامہ جنگ کی رپورٹ، 3/ دسمبر 2008ء

باب -5

اپنے جسم کی دیکھ بھال

کچھ لوگ سمجھتے ہیں کہ معذوری کا مطلب یہ ہے کہ آپ بیمار ہیں۔ یہ درست نہیں ہے۔ لیکن معذوری کے باعث یہ ممکن ہے کہ آپ کو اپنے روزمرہ معمولات کے مطابق صحت مند رہنے کے لئے زیادہ احتیاط کی ضرورت ہو۔ ایک معذور عورت کی حیثیت سے آپ اپنے جسم کو کسی اور کی بہ نسبت بہتر طور پر جانتی اور سمجھتی ہیں۔ لیکن آپ ہمیشہ ہی دوسروں کو اپنی تکالیف سے آگاہ کرنے پر انحصار نہیں کر سکتی ہیں۔ آپ کو روزانہ احتیاط اور با قاعدگی کے ساتھ اپنے جسم، خاص طور پر جسم کے ان حصوں کا جائزہ لینے کی ضرورت ہے جو آپ محسوس نہیں کر سکتی یا دیکھ نہیں سکتی ہیں۔ یا اگر آپ کو کسی غیر معمولی بات یا جسمانی رد عمل یا کسی جگہ درد یا زخموں یا بیماری کا احساس ہو، آپ جس قدر جلد ممکن ہو یہ معلوم کرنے کی کوشش کریں کہ اس کا سبب کیا ہو سکتا ہے؟ جب ضروری ہو آپ اپنے گھر کے کسی فرد، دوست یا جس پر آپ اعتماد کر سکتی ہوں اس سے اپنی مدد کے لئے کہیں۔

اس باب میں دی گئی معلومات کا مقصد یہ ہے کہ آپ صحت مند اور بہت سے مسائل صحت سے محفوظ رہیں۔ اگر آپ کو اپنی روزمرہ دیکھ بھال کے لئے مدد کی ضرورت ہے تو اس باب میں آپ کے گھر والوں یا دیکھ بھال کرنے والوں یا معاونت کے مطلوب کے لئے ضروری معلومات موجود ہیں۔

اچھی صحت کے لئے اچھی غذا کھایئے

تمام عورتوں کو اپنے روزمرہ کے کاموں، بیماری سے بچاؤ، محفوظ اور صحت مند ولادتوں کے لئے اچھی غذا کی ضرورت ہے لیکن غریب ممالک کی عورت میں ناقص غذائیت اور معذور کرنے والے صحت کے مسائل انتہائی عام ہیں۔ جب گھر یا کمیونٹی میں غذا کی مساوی تقسیم نہ ہو تو یہ عام طور پر عورتیں خصوصاً معذور عورتیں ہی ہوتی ہیں جنہیں ضروری مقدار میں غذا نہیں ملتی ہے۔

بچپن ہی سے عموماً لڑکی کو لڑکے کے مقابلے میں کم مقدار میں غذا دی جاتی ہے۔ نتیجے میں اس کی افزائش سست ہوسکتی ہے، ممکن ہے اس کی ہڈیاں مناسب طور پر نشوونما نہ پائیں اور یہ آئندہ زندگی میں اس کے لئے معذوری کا سبب بن جائے۔ ایک ایسی لڑکی کی جو کسی معذوری کے ساتھ پیدا ہوئی ہو، یہ صورت اس کی معذوری کو خراب تر بنا سکتی ہے اور جب ایک عورت کو مناسب مقدار میں کھانے کو نہ ملے تو وہ بیمار ہوجاتی ہے۔ ایسی عورت سنگین پیچیدگیوں میں مبتلا ہونے کے زیادہ امکانات رکھتی ہے۔

بہتر صحت کے لئے گھر کے ہر فرد کو بشمول معذور لڑکیوں اور عورتوں کے مناسب اور اچھی غذا کی ضرورت ہے

صحت افزا غذا

صحت مند رہنے کے لئے ضروری نہیں کہ آپ صفحہ 87 پر دی گئی تمام غذائیں کھائیں۔ آپ وہ بنیادی غذائیں کھا سکتی ہیں جن کی آپ عادی ہیں، پروٹین آمیز غذائیں جلد اور عضلات کو مضبوط رکھنے کے لئے بہترین ہیں۔ جن غذاؤں میں کیلشیم موجود ہے (دودھ یا دودھ کی مصنوعات، سبز پتوں والی ترکاریاں، سویابین اور شیل فش) ہڈیوں کو مضبوط رکھنے کے لئے خاص طور پر اچھی ہیں۔

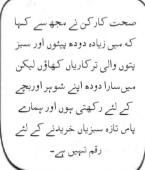

صحت کارکن نے مجھ سے کہا کہ میں زیادہ دودھ پیوں اور سبز پتوں والی ترکاریاں کھاؤں لیکن میں سارا دودھ اپنے شوہر اور بچے کے لئے رکھتی ہوں اور ہمارے پاس تازہ سبزیاں خریدنے کے لئے رقم نہیں ہے۔

اس سلسلے میں کچھ مشورے درج ذیل ہیں:

- کم قیمت کی بنیادی غذا جیسے چاول، مکئی، باجرہ، گندم، کساوا☆، آلو وغیرہ
- کچھ پروٹین آمیز غذائیں جانوروں سے جیسے دودھ، دہی، چیز، انڈے، مچھلی یا گوشت (جو جسم کی نشوونما کرتی ہیں)۔
- پروٹین کے دیگر ذرائع جیسے پھلیاں، دالیں، گری دار میوہ، سمندری کائی، سویا۔
- وٹامن اور معدنیات سے بھر پور پھل اور سبزیاں (جو جسم کی حفاظت اور مرمت کرتی ہیں)
- چکنائی اور شکر کی کم مقدار (جو توانائی فراہم کرتی ہیں)

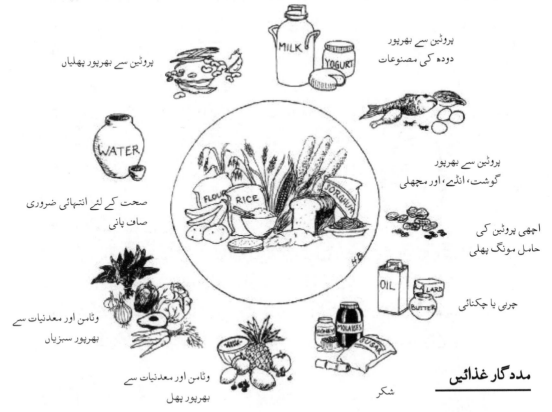

پروٹین سے بھرپور
دودھ کی مصنوعات

پروٹین سے بھرپور پھلیاں

پروٹین سے بھرپور
گوشت، انڈے، اور مچھلی

صحت کے لئے انتہائی ضروری
صاف پانی

اچھی پروٹین کی
حامل مونگ پھلی

چربی یا چکنائی

وٹامن اور معدنیات سے
بھرپور سبزیاں

وٹامن اور معدنیات سے
بھرپور پھل

مددگار غذائیں

شکر

انیمیا سے بچاؤ (کمزور خون)

مناسب مقدار میں اچھی غذا کے بغیر کوئی بھی لڑکی یا عورت عمومی خراب صحت کا شکار ہوسکتی ہے اور وہ انیمیا میں مبتلا ہوسکتی ہے۔ یہ اس وقت ہوتا ہے جب آپ آئرن آمیز غذائیں مناسب مقدار میں نہ کھائیں۔ انیمیا عورتوں میں خاص طور پر حاملہ اور دودھ پلانے والی عورتوں میں عام ہے۔ یہ انتہائی تھکن، انفیکشن اور بیماری کے خلاف عورت کی مزاحمت میں کمی کا سبب بنتا ہے۔ بچے کی ولادت کے دوران بھاری مقدار میں خون بہنے سے بھی خون ملیریا اور پیٹ کے کیڑوں (ہک وارم) کی طرح انیمیا ہوسکتا ہے۔ (کسی صحت کارکن سے معلوم کیجئے کہ ملیریا سے بچاؤ اور اس کا علاج کس طرح کیا جائے)

انیمیا کی مندرجہ ذیل علامات ہیں:

- آنکھ کے پپوٹوں کا اندرونی طور پر، ناخنوں اور ہونٹوں کے اندرونی حصوں کا زرد ہونا۔

- کمزوری اور انتہائی تھکن محسوس کرنا۔

- سر چکرانا خاص طور پر بیٹھ کر یا لیٹ کر اُٹھتے وقت۔

- بے ہوشی (حواس گم ہونا)

- دم گھٹنا۔

- دل کی تیز دھڑکن۔

انیمیا سے تحفظ اور نجات کے لئے روزانہ ایسی غذائیں کھائیں جن میں آئرن موجود ہو جیسے سبز پتوں والی تر کاریاں (پالک ہیسکس کے پتے، پالک، ڈرم اسٹک کے پتے، ٹارو کے پتے، کساوا کے پتے) ۔ انڈے، دودھ، کشمش، دال اور گوشت بھی کھائیے۔

زیادہ مقدار میں آئرن حاصل کرنے کے طریقے :

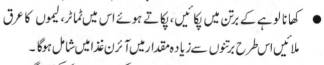

- آئرن سے بھر پور غذائیں ٹماٹر یا آم، پپیتا، سنگتروں، لیموں کے ساتھ کھائیں۔ ان سب میں وٹامن 'سی' ہوتا ہے جو غذا میں موجود آئرن استعمال کرنے میں آپ کے جسم کی مدد کرتا ہے۔

- کھانا لوہے کے برتن میں پکائیں، پکاتے ہوئے اس میں ٹماٹر، لیموں کا عرق ملائیں اس طرح برتنوں سے زیادہ مقدار میں آئرن غذا میں شامل ہوگا۔

- کھانا پکانے کے برتن میں لوہے کا صاف ستھرا ٹکڑا مثلاً لوہے کی کیل یا گھوڑے کی نعل وغیرہ ڈالیں۔ یہ یقین کریں کہ ڈالی گئی چیز خالص لوہے کی بنی ہے اور اس میں دوسری دھاتیں شامل نہیں ہیں۔ کچھ دھاتیں مثلاً سیسہ نقصان دہ ہوتی ہیں اور پیدائشی نقائص کا سبب بنتی ہیں۔

- چند گھنٹے لیموں کے عرق میں خالص لوہے کا ٹکڑا جیسے لوہے کی کیل وغیرہ رکھیں پھر اس سے لیمونیڈ بنا کر پئیں بہت سے مقامات پر مراکز صحت حاملہ عورتوں کو انیمیا سے بچانے کے لئے آئرن پلز (فیرس سلفیٹ) دیتے ہیں۔

اپنے جسم کو متحرک رکھیئے

تمام عورتوں کو اپنے جسم مضبوط، لچکدار اور صحت مند رکھنے کے لئے ورزش کی ضرورت ہوتی ہے۔ ورزش آپ کے عضلوں، آپ کے دل اور آپ کے پھیپھڑوں کو مضبوط رکھنے میں مدد دیتی ہے یہ آپ کا بلڈ پریشر بڑھنے، ہڈیاں کمزور ہونے اور قبض ہونے سے بچاتی ہے۔ ورزش آپ کو حد سے زیادہ موٹا ہونے سے بچانے میں مدد کرتی ہے۔ حد سے زیادہ موٹا ہونا صحت مند ہونا نہیں ہے یہ آپ کے لئے روزمرہ کے کام زیادہ مشکل بنا دے گا۔

بعض اوقات عورت کی معذوری اسے اپنا جسم یا اس کا کوئی حصہ استعمال کرنے یا حرکت میں لانے سے روکتی ہے، بہتری کے لئے اسے ورزش کی ضرورت ہوتی ہے۔ عضلات اگر باقاعدگی سے استعمال نہ ہوں تو کمزور ہو جاتے ہیں یا اینٹھ جاتے ہیں۔ جوڑ اگر پوری طرح حرکت نہ کریں تو وہ جکڑ جاتے ہیں اور مکمل طور پر سید ھے نہیں ہوتے ہیں یا مڑتے نہیں ہیں۔ اگر آپ کوئی معذوری رکھتی ہوں جو آپ کے جسم کو متاثر کرتی ہو تو آپ اپنے جسم کے تمام حصوں کو بھر پور انداز سے حرکت دے کر انہیں فعال رکھنے کی یقینی کوشش کریں۔

ورزش ان عورتوں کی بھی مدد کرتی ہے جو خود کو ڈپریشن زدہ محسوس کرتی ہیں۔ کچھ ورزشیں درد میں کمی کے احساس میں معاون ہوتی ہیں۔ بہت سے لوگ باقاعدگی سے ورزش کریں تو انہیں اچھی نیند آتی ہے جب آپ کا جسم مضبوط اور صحت مند ہو تو آپ میں زیادہ توانائی ہوگی، آپ خود کو بہتر محسوس کریں گی اور کم زخمی ہوں گی۔

زیادہ تر عورتیں روزمرہ کے کاموں جیسے کھانا پکانے، صفائی، کھیتوں میں کام کرنے، لکڑی اور پانی لانے اور بچے سنبھالنے سے ضروری ورزش کر لیتی ہیں۔ جس حد تک بھی ممکن ہو معذور عورتیں بھی اسی طرح ورزش یقینی بنا سکتی ہیں۔

اگر آپ کے لئے اپنے جسم کو حرکت دینا بہت ہی مشکل ہے تو آپ دن میں کئی بار اپنی پوزیشن تبدیل کرنے کی کوشش کریں۔ اگر آپ پورا دن بیٹھی رہتی ہیں تو کچھ دیر لیٹ کر اپنی پوزیشن تبدیل کرتی رہیں۔

اپنے سینے کے عضلات پھیلانے کی کوشش کریں

اگر آپ زیادہ آگے کی طرف جھکی رہتی ہیں۔۔۔۔۔۔

آپ کو بہتری کے لئے پُرمشقت ورزش نہیں کرنا چاہیے۔ بہتر ہے کہ آہستہ روی سے ابتداء کریں، خصوصاً اگر آپ فی الحال زیادہ حرکت نہ کر سکتی ہوں یا اگر آپ اپنے جسم کے کسی حصے کو حرکت نہ دے پاتی ہوں، یا کوئی عضو کمزور ہو یا تکلیف دیتا ہو یا آپ زیادہ تر وقت ایک ہی پوزیشن میں گزارتی ہوں۔ مناسب حرکت نہ کرنا، جوڑوں اور عضلات کو سخت اور تکلیف دہ کر دیتا ہے۔ جوں جوں آپ کا جسم حرکت کا عادی ہو جائے گا آپ اتنا ہی زیادہ حرکت کرنے کے قابل ہوں گی۔

ورزش پُرلطف بھی ہو سکتی ہے

ایسی ورزش ڈھونڈئے جو پُرلطف ہو۔ بعض ممالک میں عورتیں گدھے یا خچر پر سواری کرنا پسند کرتی ہیں۔ جانور کو کنٹرول کرنا، اس کی حرکات کے مطابق جسم کا رِدّعمل اور اپنا توازن برقرار رکھنا ہر قسم کی ورزش پر مشتمل ہے۔

کسی اور فرد کے ساتھ مل کر ورزش کرنے کی کوشش کریں، آپ زیادہ تر ورزش جاری رکھنا اس وقت پسند کریں گی جب آپ اپنا وقت کسی سہیلی یا دوست کے ساتھ گزارتی ہوں۔ ورزش کے دوران کسی ایسے فرد کا ساتھ ہونا بھی اچھا ہے جو ضرورت پڑنے پر آپ کی مدد کر سکے۔

یا کوئی کھیل پسند کرتی ہیں   کچھ عورتیں رقص سے لطف اندوز ہوتی ہیں

معذوریاں رکھنے والی بہت سی عورتوں کے لئے تیرا کی یا پانی میں متحرک رہنا ورزش کرنے کا ایک بہترین طریقہ ہے۔ پانی میں آپ کا جسمانی وزن کم ہو جاتا ہے لہذا وہ عورتیں جن کے لئے اپنے جسم کو حرکت دینا یا چلنا دشوار ہوتا ہے پانی میں زیادہ سہولت سے حرکت کر سکتی ہیں یا پانی میں وہ کم تکلیف محسوس کرتی ہیں۔ تیرا کی آرتھرائٹس میں مبتلا افراد کے لئے بہترین ورزش ہے۔ اگر آپ وہیل چیئر استعمال کرتی ہیں تو اسے اپنے حلقے میں آنے جانے کے لئے خود کو دھکیلنے کی کوشش کریں۔

اطمینان کرلیں کہ پانی زیادہ ٹھنڈا نہ ہو۔ ٹھنڈ سے عضلات زیادہ آسانی سے مجروح ہوجاتے ہیں۔

بھاری چیزیں بار بار اٹھانے سے آپ کے عضلات اور ہڈیاں مضبوط ہوسکتی ہیں۔

اگر یہ ممکن نہ ہو تو مختلف اشیاء (جیسے پتھر، کھانوں کے ڈبے یا پانی سے بھری بوتلیں وغیرہ) بار بار اٹھانے کی کوشش کریں۔اس سے آپ کو کندھوں اور بازوؤں کے عضلات مضبوط رکھنے میں مدد ملے گی۔

وزن کس طرح اٹھائیں: وزن اٹھانے سے پہلے آپ کے لئے جس حد تک ممکن ہو، سیدھی ہوکر بیٹھیں، گہری سانس لیں اور خارج کریں جونہی آپ سانس خارج کریں وزن اٹھاتے ہوئے اپنے شانوں کی ہڈیاں اپنی ریڑھ کی ہڈی کی طرف لے جائیں۔جیسے ہی وزن اٹھالیں ایک اور گہری سانس لیں اور پھر وزن کو آہستگی سے واپس رکھتے ہوئے سانس خارج کریں۔

اپنے عضلات کھینچیں

عضلات کو کھینچنا انہیں زیادہ لچکدار بناتا ہے لہذا آپ انہیں زیادہ سہولت سے موڑ اور گھما سکتی ہیں۔

معذوریاں رکھنے والی بہت سی عورتوں کے لئے باقاعدگی سے عضلات کو پھیلانے کا مطلب کم تر درد محسوس کرنا ہے۔ اسٹریچنگ زخموں سے بچنے میں بھی مدد دیتی ہے۔

محنت کا کوئی کام یا ورزش کرنے سے پہلے ہمیشہ اسٹریچ ضرور کریں۔ اسٹریچنگ رفتہ رفتہ شروع کرنا آپ کو اور آپ کے عضلات کو مجروح ہونے سے محفوظ رہنے میں مدد دے گا۔ ورزش یا محنت کا کام کرنے کے بعد بھی اسٹریچ کرنا ایک اچھا خیال ہے۔ اسٹریچنگ آپ کے جسم کو لچکدار رکھتی ہے اور جوں آپ بوڑھی ہوتی ہیں درد اور کمزوری سے تحفظ فراہم کرتی ہے۔

کسی عضلے کو کھینچنا (Stretching)

1- ایسی جگہ تلاش کیجیے جہاں آپ خود کو محفوظ محسوس کریں اور گریں نہیں ۔ کھچاؤ کا عمل نرمی سے ہونا چاہیے ۔ یہ تکلیف نہ پہنچائے مثلاً اپنی کمر کے نچلے حصے میں کھچاؤ کے لیے کسی چٹائی وغیرہ پر اس طرح لیٹیے کہ آپ کا چہرہ اوپر ہو،اپنے گھٹنے موڑئیے اور دونوں پیروں کو اپنے سینے کی طرف اس حد تک کھینچیے کہ یہ آپ کے لیے تکلیف کا سبب نہ بنے ۔

2- اپنے جسم کو اسی انداز میں رکھیے اور تیس تک گنیے (یا تین بار دس تک گنیے)۔اپنے جسم کو آگے اور پیچھے حرکت مت دیں ۔

3- کھچاؤ کے دوران سانس لینا یاد رکھیں ۔اگر کھچاؤ تکلیف دینے لگے تو اس حصے کو حرکت دینے کی کوشش کیجیے جسے آپ کھینچ رہی ہیں تا کہ یہ عمل زیادہ بہتر طور پر ہو سکے ۔اگر اس سے درد ندر کے تو مختلف انداز آزمائیں ۔

محدود حرکت رکھنے والی عورتیں مختلف عضلات کو کھینچنے کے تجربات کریں ۔بعض اوقات آپ کو دوسرے فرد کی مدد کی ضرورت پڑسکتی ہے۔اگر کوئی اور آپ کی مدد کر رہا ہو تو یہ بات یقینی بنائیں کہ آپ کے عضلات آہستگی سے حرکت کریں اور جب آپ کھچاؤ تکلیف دہ محسوس کریں تو آپ مددگار کو رکنے کے لیے کہہ سکتی ہیں ۔

ثنا کا پائوں پولیو کی وجہ سے مفلوج ہے۔ وہ کھانے پکانے کی تیاری کے دوران اپنے پیر کو ایک پوزیشن میں رہ کر جام ہونے سے بچانے کے لیے اسے اسٹریچ کرتی رہتی ہے۔

کچھ لوگ اسٹریچنگ سے پہلے اپنے عضلات پر برف، گرم کپڑا یا تولیہ یا ہیٹ پیک (اگر دستیاب ہو) رکھنا پسند کرتے ہیں،آپ بھی یہ دیکھنے کے لیے کہ کیا آپ کا جسم اس سے بہتر محسوس کرتا ہے،اسے آزما سکتی ہیں ۔سخت عضلات رکھنے والی بہت سی عورتیں ہر صبح کام کا آغاز کرنے سے پہلے جسم کو اسٹریچ کرتی ہیں تا کہ وہ دن میں مجروح نہ ہوں ۔رات کے وقت وہ ایک طویل دن کے بعد بہتر نیند اور کم درد کے لیے دوبارہ اسٹریچ کرتی ہیں ۔

بہت سی عورتیں جان لیتی ہیں کہ وہ کوئی اور کام کرتے ہوئے بھی اپنے عضلے کو اسٹریچ کرسکتی ہیں ۔ آپ بھی ایسا کرسکتی ہیں ۔روزمرہ کے کام نمٹاتے ہوئے ایسا کرنے کے طریقے تلاش کیجیے۔

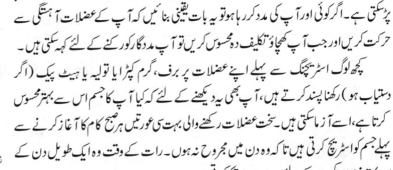

ماریا سیریبرل پالسی کا شکار ہے وہ اپنے عضلات کو روزمرہ کے کام کرنے کے دوران اسٹریچ کرتی ہے۔ سنگی ٹکڑا اس کی ٹانگوں کو ایک دوسرے سے دور رکھتا ہے جس سے اسے اپنی ٹانگوں کے اندرونی عضلات کو اسٹریچ کرنے میں سہولت ملتی ہے۔ اس سے اسے عضلاتی جکڑاؤ سے تحفظ ملتا ہے ، وہ اپنے بازو، پیر اور گردن اسٹریچ کرتے ہوئے اپنی کمر جس حد تک ممکن ہو سیدھی رکھتی ہے۔

اگر آپ کے عضلات جکڑے ہوں، سیریبرل پالسی یا ریڑھ کی ہڈی مجروح ہونے سے مفلوج ہو یا جوڑوں میں درد ہو

وہ عورتیں جن کے جوڑوں میں تکلیف یا عضلات جکڑے ہوئے ہوں ہیں انہیں دوڑنے یا بھاری وزن اٹھانے کی ورزش کرتے ہوئے محتاط رہنا چاہیے ۔ اس قسم کی ورزشیں عضلات یا جوڑوں پر حد سے زیادہ دباؤ ڈال سکتی ہے ۔ یہ آپ کے عضلات کو مضبوط بنانے کے بجائے انہیں مجروح کر سکتی ہے ۔

جکڑے (اینٹھے ہوئے) عضلات کو نرم کرنا

سیریبرل پالسی، مخلوط اعصابی اکڑن یا ریڑھ کی ہڈی کے زخموں میں اکثر عضلات نہایت ہی سخت ہوتے ہیں اور اکڑ جاتے ہیں ۔ بعض اوقات کوئی عضلہ انتہائی سخت ہو جاتا ہے لرزنے لگتا ہے اور عورت اسے کنٹرول نہیں کر پاتی ہے ۔ عضلات سخت اینٹھے ہوئے ہوں تو

- اینٹھے ہوئے عضلے کو براہ راست مخالف سمت میں کھینچیے مت یا دبائیے یہ اکڑن یا اینٹھن کو بڑھاتا ہے ۔
- اینٹھے عضلات پر مساج مت کریں ۔ سہلانا یا مساج کرنا انہیں مزید سخت کر دیتا ہے ۔
- اینٹھے ہوئے عضلات کو بہتر بنانے کے لئے ایسی پوزیشن معلوم کیجئے جس میں جسم کو آرام ملے اسے آہستہ آہستہ اطراف میں حرکت دینا معاون ثابت ہو سکتا ہے ۔ بعض اوقات جسم کے کسی اور حصے کو حرکت دینے سے بھی اینٹھے ہوئے عضلات میں نرمی آتی ہے ۔ آپ گرم کپڑا (گیلا یا خشک) بھی اینٹھے عضلات کو نرم کرنے کے لئے استعمال کر سکتی ہیں ۔

اگر آپ پہیہ گاڑی، بیساکھیاں یا وہیل چیئر استعمال کرتی ہوں

اگر آپ پہیہ گاڑی، بیساکھیاں یا وہیل چیئر استعمال کرتی ہیں تو اپنے بازو حد سے زیادہ استعمال کرنے کے باعث کندھوں اور کلائیوں کی تکلیف میں مبتلا ہو سکتی ہیں ۔ آپ کے بازو اور کندھے آسانی سے مجروح یا ناقابل استعمال ہو سکتے ہیں ۔ اس سے بچاؤ کے لئے آپ اپنے بازوؤں اور کندھوں کو بار بار اسٹریچ کریں مثلاً وہیل چیئر استعمال کرنے والی عورتیں عموماً مضبوط بازوؤں کی مالک ہوتی ہیں ۔ لیکن صرف ان عضلات کو ہی نہیں جو آپ وہیل چیئر دھکیلنے کے لئے استعمال کرتی ہیں بلکہ بازوؤں کے تمام عضلات اور کندھوں کو مضبوط رکھنا ضروری ہے اپنے بازوؤں اور کندھوں کے حد سے زیادہ استعمال سے بچنے کے لئے ایک ہی کام طویل مدت تک کرنے کی کوشش نہ کریں مثلاً چیزوں کو اٹھانے کا انداز بدلیں ۔ پہلے اپنا الٹا ہاتھ استعمال کریں پھر سیدھا ۔ اپنے کندھوں کے دوسرے عضلات مضبوط بنانے کا ایک اچھا طریقہ یہ ہے کہ اپنی وہیل چیئر کو پیچھے کی طرف دھکیلیں ۔

غیر معمولی استعمال سے تکالیف

جوڑ جسم کے وہ حصے ہیں جہاں ہڈیاں آپس میں ملتی ہیں۔ان جوڑوں پر نسیجیں جیسے عضلات کو ہڈیوں سے منسلک کرتی ہیں۔اگر آپ کوئی حرکت جیسے وہیل چیئر یا پہیہ گاڑی کو دھکیلنا، بار بار دوہرائیں یا بیساکھیوں کے سہارے کے سہارے چلیں تو آپ کی کلائیوں کے پٹھوں کو نقصان پہنچ سکتا ہے۔ اس صورت میں آپ اپنے ہاتھوں سے اس جگہ کو (کلائی کو) نرمی سے تھپکیں گی تو درد محسوس کریں گی۔

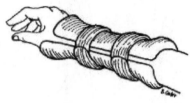

علاج:

- **آرام:** اپنی کلائیوں اور ہاتھوں کو جس قدر ممکن ہو آرام دہ پوزیشن میں رکھیں۔اگر آپ کو مسلسل حرکت میں رہنا ہے یا وہیل چیئر دھکیلنا ہے تو اپنے ہاتھوں اور کلائیوں کو جس حد تک ممکن ہو ساکت رکھنے کے لئے اسپلنٹ پہنیے۔

- **اسپلنٹ:** نرم اسپلنٹ بنانے کے لئے اپنی کلائی اور بازو کے نچلے حصے میں کپڑا لپیٹیے تاکہ یہ حرکت نہ کر سکیں کپڑا لپیٹے ہوئے پہلے پتلی لکڑی کا ٹکڑا رکھنے سے جوڑ کو سیدھا رکھنے میں مدد مل سکتی ہے کپڑے کو اس حد تک سختی سے لپیٹیے کہ آپ کی کلائی حرکت نہ کر سکے لیکن یہ اس قدر سختی سے نہ لپیٹیں کہ دوران خون رک جائے یا وہ حصہ سن ہو جائے۔اگر ممکن ہو تو ارد گرد گھومتے ہوئے اور آرام کرتے ہوئے یا سوتے ہوئے بھی اسپلنٹ پہنیں۔

- **پانی:** ایک برتن میں گرم پانی اورایک برتن میں ٹھنڈا پانی بھریں۔اپنے ہاتھوں اور کلائیوں کو پہلے ایک منٹ ٹھنڈے پانی میں اور پھر چار منٹ گرم پانی میں رکھیں۔ یہ عمل پانچ مرتبہ کریں۔اختتام گرم پانی سے کریں۔ یہ عمل دن میں کم از کم دو (اور ہو سکے تو زیادہ) بار دوہرائیں۔ گرم پانی کا برتن ہمیشہ آخری ہونا چاہیے جس میں آخری بار آپ اپنے ہاتھ رکھیں۔

- **ورزش:** آبی علاج کے بعد ہر بار ہاتھوں اور کلائیوں کی ورزش کریں۔اس سے آپ کے پٹھوں کو زیادہ نقصان پہنچنے سے بچاؤ میں مدد ملے گی۔اپنے ہاتھوں کو نیچے دیے گئے انداز سے رکھتے ہوئے ہر انداز میں پانچ تک گنیں اگر آپ ان میں سے کسی انداز میں درد محسوس کریں تو اس انداز کو اپنے لئے زیادہ آرام دہ بنانے کے لئے انداز میں قدرے تبدیلی کریں۔ یہ حرکت دس مرتبہ دوہرائیں۔

- **ادویات:** اگر آپ کے ہاتھوں یا کلائیوں میں درد ہو یا یہ سوجے ہوئے ہوں تو اسپرین یا درد دور کرنے والی کوئی ایسی دوا کھائیں جو سوجن اور سوزش دور کرتی ہے(دیکھئے صفحہ 335)

- **آپریشن:** اگر درد چھ ماہ بعد مستقل ہو،اگر آپ اپنے ہاتھوں کو کمزور محسوس کریں یا اگر آپ چھونے کی حس گنوا دیں یا اپنے ہاتھوں میں جھنجھناہٹ محسوس کریں تو طبی مدد حاصل کریں۔ممکن ہے آپ کی کلائیوں کو انجکشن کے ذریعے دوا دینے یا آپریشن کی ضرورت ہو۔

احتیاط

- اگر آپ کے لئے ممکن ہو تو وہیل چیئر اس انداز سے دھکیلنے یا حرکت کرنے کی کوشش کریں کہ آپ کے ہاتھ اور کلائیاں کم سے کم مڑیں اور ان پر کم دباؤ پڑے۔

- اگر ممکن ہو تو اپنے ہاتھوں اور بازوؤں کو آرام دینے کے لئے کسی اور سے اپنی وہیل چیئر یا پہیہ گاڑی دھکیلنے کی درخواست کریں۔

- ہر گھنٹے بعد اپنے ہاتھوں اور کلائیوں کو تمام تر ممکنہ انداز سے گھما کر ورزش کریں۔ یہ پٹھوں اور عضلات کو اسٹریچ اور مضبوط کرے گی۔ اگر ورزش درد کا سبب بنے، آہستگی اور نرمی سے حرکت دیں۔ اگر آپ کے ہاتھ اور کلائیاں سرخ یا گرم ہیں تو یہ انفیکشن زدہ ہو سکتے ہیں۔ فوری طور پر کسی صحت کار کن سے ملیں۔

بیساکھیوں کا استعمال

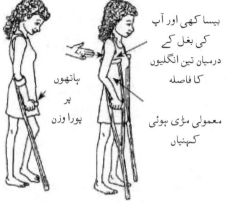

بیساکھی اور آپ کی بغل کے درمیان تین انگلیوں کا فاصلہ

ہاتھوں پر پورا وزن

معمولی مڑی ہوئی کہنیاں

بیساکھی ...... کہنی کی بیساکھی

اگر آپ بیساکھیاں استعمال کرنا چاہتی ہوں تو خیال رکھئے کہ یہ آپ کے لئے مناسب حد تک موزوں ہوں۔ جب آپ بیساکھیاں استعمال کرتی ہیں تو آپ کے جسم کا بیشتر بوجھ آپ کے ہاتھوں پر پڑے گا لہٰذا اپنے ہاتھوں کو پہنچنے والے نقصان سے بچنے کے لئے صفحہ 93 پر دیئے گئے مشورے پر عمل کریں اگر ممکن ہو تو اپنی بغلوں کے اعصاب کو پہنچنے والے نقصان سے بچنے کے لئے کہنی والی بیساکھی استعمال کیجے لیکن اگر آپ کی ترجیح لمبی بیساکھی ہے تو یہ یقینی بنائیں کہ بیساکھیاں آپ کی بغلوں کو نقصان نہ پہنچائیں بیساکھی استعمال کرتے ہوئے آپ کی کہنیاں ہلکی سی مڑی ہوئی ہوں، آپ کی بغل اور بیساکھی کے درمیان تین انگلیوں کے برابر فاصلہ ہونا چاہئے۔ اگر لمبی بیساکھیاں آپ کی بغلوں کو دباتی ہیں تو ان پر پڑنے والا دباؤ وقت گزرنے پر آپ کے ہاتھوں کو مفلوج کرنے کا سبب بن سکتا ہے۔

سکڑاؤ

سکڑاؤ

بازو یا پاؤں طویل عرصے تک مڑا رہے تو وہ ایک پوزیشن میں جم جاتا ہے (سکڑ جاتا ہے) کچھ مسلز چھوٹے ہو جاتے ہیں لہٰذا بازو یا پیر پوری طرح سیدھا نہیں ہو پاتا ہے۔ یا چھوٹے ہونے والے مسلز جوڑ کو سیدھا رکھتے ہیں اور اسے موڑا نہیں جا سکتا ہے بعض اوقات سکڑاؤ تکلیف کا سبب بنتا ہے۔

اگر آپ کا سکڑاؤ برسوں پرانا ہے تو نرم روی سے ہلانے جلانے اور اسٹریچنگ سے جوڑ کو بد تر ہونے سے بچایا جا سکتا ہے۔ جوڑوں اور مسلز کو پوری طرح سیدھا کرنا مشکل ہو گا لیکن مناسب ورزشیں سختی کو کسی حد تک کم اور آپ کے مسلز کو مضبوط بنا سکتی ہیں۔ سکڑاؤ سے بچنے اور مسلز کو مضبوط رکھنے کے لئے روزانہ بازوؤں اور پیروں کی ورزشیں کیجے اگر ضروری ہو تو کسی ایسے فرد کا تعاون حاصل کیجے جو آپ کے جسم کے مختلف حصوں کو حرکت دینے میں آپ کی مدد کر سکے۔

وہ چند ورزشیں جو بعض اقسام کے سکڑاؤ سے آپ کو بچاتی ہیں
اور مسلز کو مضبوط رکھنے میں مدد کرتی ہیں

پیر کے اوپر کے سامنے والے
حصے کی ورزش کے لئے

2- سیدھا کیجئے　　　　1- موڑئیے

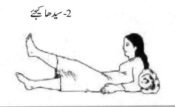

پیر کے اوپر کے حصے کی
پشت کی ورزش کے لئے

2- سیدھا کیجئے　　　　1- موڑئیے

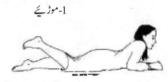

نچلے پیر کی ورزش کے لئے

2- پھر آرام کیجئے　　　1- انگلیوں کا رخ اوپر کی جانب کیجئے

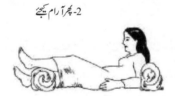
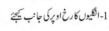

بازوؤں کی ورزش کے لئے

موڑئیے

اوپر کی طرف سیدھا کیجئے　　　　سیدھا کیجئے

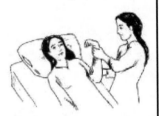

اہم بات:　　　اگر ایک جوڑ طویل عرصے سے مڑا ہوا ہے تو نرمی سے کام لیجئے،
اسے زبردستی (قوت لگا کر) سیدھا کرنے کی کوشش مت کیجئے۔

اپنے پیروں کو زخم اور انفیکشن سے بچانے کے لئے خصوصی ا

عام مسائل صحت سے بچاؤ

کیونکہ آپ کسی بھی دوسرے فرد کی بہ نسبت اپنے جسم کو بہتر طور پر جانتی اور سمجھتی ہیں لہٰذا آپ اپنے گھر والوں دوستوں اور دیکھ بھال کرنے والوں کو سکھا سکتی ہیں کہ وہ کس طرح آپ کی بہترین مدد کر سکتے ہیں اگر آپ کسی مسئلے سے دوچار ہیں تو ان سے مدد کی درخواست کرنے میں خوفزدہ یا شرمندہ نہ ہوں اگرچہ ہمیشہ یہ ممکن نہیں ہوتا ہے کہ بیماری سے بچا جاسکے، بیشتر صحت کے مسائل کا ابتداء ہی میں علاج کر لیا جائے تو وہ سنگین نہیں ہوتے ہیں۔ (دیکھئے باب نمبر 6)

روزمرہ دیکھ بھال

آپ اپنے جسم کو روزانہ دھو کر اور اپنی جلد کا با قاعدگی سے دیکھ بھال کر کے صحت مندرہ سکتی ہیں اور انفیکشنز سے بچ سکتی ہیں۔ اگر آپ زیادہ تر بیٹھی رہتی ہیں اور دن بھر میں کوئی خاص حرکت نہیں کرتی ہیں تو آپ کو اپنی جلد صحت مندرکھنے کے لئے اضافی اور محتاط دیکھ بھال کی ضرورت ہوگی (دیکھے صفحہ 114 تا 117)۔ سوجن یا انفیکشن کی دوسری علامات پر نظر رکھئے۔ اگر آپ خراشیں، زخم یا سوزش دیکھیں تو انہیں دھو کر ڈھک لیں یا پٹی باندھ لیں تا کہ وہ خراب نہ ہوں۔ آپ اپنے جسم کے ان حصوں کا جائزہ لینے کے لئے جنہیں دیکھنا مشکل ہے آئینہ استعمال کر سکتی ہیں۔ بہت سی نابینا عورتیں سونگھ کر یا چھو کر زخموں یا دیگر تنبیہی علامات کا اندازہ لگا لیتی ہیں۔

اپنی جلد کا روزانہ جائزہ لیں

آپ بال با قاعدگی سے دھوئیں اور ان میں جوؤں کے لئے بھی جائزہ لیتی رہیں اپنے سر کی جلد پر زخموں یا گھر نڈ کا جائزہ بھی لیں روزانہ صاف ستھرے کپڑے خاص طور پر زیر جامے اور موزے پہننے کی کوشش کریں۔

کچھ معذور عورتوں کو جب انہیں صحت کا کوئی مسئلہ درپیش ہو تو چھوٹی چھوٹی علامتوں پر توجہ دینے اور بیان کرنے کی ضرورت ہوتی ہے مثلاً اپنے رحم میں انفیکشن رکھنے والی عورت ممکن ہے اس کی وجہ سے تکلیف محسوس نہ کرے لیکن وہ اپنی فرج سے غیر معمولی اخراج یا بو کو محسوس کر سکتی ہے۔ ایک نابینا عورت ایسا زخم نہیں دیکھ سکتی ہے جو سنگین انفیکشن بن رہا ہو لیکن وہ اس کے باعث کچھ درد یا سوجن محسوس کر سکتی ہے۔

آپ کی فرج سے اخراج کی بو میں تبدیلی کا مطلب یہ ہوسکتا ہے آپ انفیکشن کا شکار ہیں

پیروں اور ہاتھوں کی دیکھ بھال

اگر آپ اپنے ہاتھوں اور پیروں میں محسوس ہونے کی زیادہ صلاحیت نہیں رکھتی ہیں تو ان کے تحفظ کے لئے محتاط رہیں۔ روزانہ ان کا جائزہ لیں۔ اگر آپ محسوس نہیں کر سکتی ہیں تو یہ آسانی سے جل سکتے ہیں یا آپ کو محسوس بھی نہ ہو اور ان میں کوئی گھاؤ یا زخم موجود ہو۔ اگر آپ کو کوئی زخم یا خراش لگ جائے تو زخم بھرنے تک اسے صاف اور ڈھک کر رکھیں۔

اپنے جسم کے ان حصوں کی حفاظت کیجئے جو حرارت یا ٹھنڈک محسوس نہ کرتے ہوں۔ کوئی بھی گرم چیز اٹھاتے ہوئے دبیز دستانے پہنیں یا ہاتھوں پر کپڑ الپیٹ لیں اور اگر کسی سرد جگہ میں ہیں تو ہاتھوں اور پیروں کی حفاظت کے لئے انہیں ڈھانپ کر رکھیں۔

اپنے تلووں کو دیکھنے کے لئے آئینہ استعمال کریں یا کسی سے مدد کی درخواست کریں کہ وہ دیکھے۔

- سرخی، سوجن، گرم جلد یا انفیکشن کی دیگر علامات
- چٹخن، خراشیں یا پھٹی ہوئی جلد
- پس، خون کا اخراج یا بو
- بے ربط بڑھتے ہوئے پیر کی انگلیوں کے ناخن (ناخن انگلیوں کی جلد میں گھس رہے ہوں)۔

اگر درد، جھنجھناہٹ، سوزش محسوس کرتی ہوں یا آپ کے پیروں میں سُن پن ہو تو کسی صحت کارکن سے رجوع کریں۔ ممکن ہے آپ کو کوئی انفیکشن ہو اور اس کے علاج کے لئے ادویات ضروری ہوں۔ انفیکشن سے بچنے کے لئے روزانہ اپنے پیر صابن اور گرم پانی سے دھوئیں۔ دھونے سے پہلے پانی کی حدت اپنی کہنی ڈبو کر چیک کریں یا پھر کسی صحت مند فرد سے کہیں کہ وہ چیک کر کے بتائے کہ پانی زیادہ گرم تو نہیں ہے اپنے پیروں کو خشک کرتے ہوئے انگلیوں کی درمیانی جگہ اچھی طرح صاف اور خشک کریں۔

اگر آپ کے پیروں کی جلد خشک ہو جاتی ہے یا چٹخنے لگی ہے تو اپنے پیر روزانہ بیس منٹ پانی میں رکھیں پھر ان پر نباتاتی تیل یا پیٹرولیم جیلی یا کوئی لوشن ملیں۔

پیروں کو محفوظ رکھنے کے دوسرے طریقے

- ننگے پاؤں نہ پھریں۔
- اپنے ناخن گولائی میں نہیں بلکہ کناروں سے سیدھے کاٹیں اس طرح ناخن کے سرے اطراف کی کھال میں نہیں دھنسیں گے ناخنوں کو زیادہ بڑھنے نہ دیں کہ وہ تتر خنے لگیں۔ اگر ضروری ہو تو کسی سے اپنی مدد کے لئے کہیں۔
- یقینی بنائیں کہ جوتے آپ کے پیروں کی مناسبت سے ہوں، جلد سے رگڑ نہ کھائیں، پیروں میں خراشوں یا سوزش کا ذریعہ نہ بنیں۔
- جوتے پہننے سے پہلے ہر بار جائزہ لیں کہ ان میں کوئی ایسی چیز (جیسے کنکر، مٹی یا کوئی کیڑا وغیرہ) نہ ہو جو آپ کے پیروں کے لئے تکلیف دہ بن سکتی ہے۔
- پیروں کو کراس (x) کر کے مت بیٹھیں اس سے آپ کے پیروں کی طرف خون کا بہاؤ مشکل ہو جاتا ہے۔
- اپنے پیروں سے گومڑ، مردہ کھال مت کاٹیں۔ یہ انفیکشن کا سبب بن سکتا ہے۔
- ایسے موزے پہنیں جو ہموار ہوں اور آپ کے پیروں پر رگڑ نہ ڈالیں۔ اگر آپ کو اپنے موزوں کے سوراخ وغیرہ بند کرنے کی ضرورت ہو تو کوشش کیجئے کہ ٹانکے نہایت ہی ہموار ہوں۔
- اگر آب و ہوا گرم ہو تو دن میں جس قدر ممکن ہو پیروں کو ڈھکے بغیر رکھیں۔ اس سے دوران خون بہتر رکھنے میں مدد ملتی ہے اور پیر کی انگلیوں کے بیچ میں انفیکشن سے تحفظ ملتا ہے۔

جذام میں مبتلا عورتیں (Hansen's Disease) اپنے پیروں کو زخم اور انفیکشن سے بچانے کے لئے خصوصی احتیاط سے کام لیں۔ جذام، ٹانگوں اور پیروں کی حس ختم کر دیتا ہے لہٰذا جذام میں مبتلا عورتیں درپیش مسئلے کو ابتدا ءہی میں جب اس کا علاج آسان ہوتا ہے، بہت کم محسوس کر پاتی ہیں۔

جذام میں مبتلا عورتوں کو چیزیں گرفت کرنے میں مشکل پیش آتی ہے۔ چیزوں پر آسان گرفت اور زخموں سے بچنے کے لئے چوڑے ہموار ہینڈل والے آلات یا برتن استعمال کیجئے یا ہینڈل کے گرد موٹا کپڑا لپیٹ لیں۔

سہولت کے مطابق ہینڈل بنانا

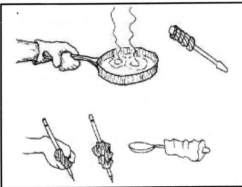

آپ متعلقہ فرد کی گرفت کے مطابق ہینڈل مولڈ کرسکتی ہیں۔ اس مقصد کے لئے ایپوکسی یا گلو آمیز پلاسٹر آف پیرس استعمال کرسکتی ہیں۔ اسے ہینڈل پر جماکر متعلقہ فرد سے کہیں کہ وہ اس پر گرفت کرے۔ پھر اسے سخت ہونے دیں۔

آپ مٹی یا موٹے پتوں (جیسے کیلے کے پتے) کی کئی تہہ لپیٹ کر یا بُنٹے کے بال لپیٹ کر بھی ہینڈل بناسکتی ہیں۔

منہ اور دانتوں کی دیکھ بھال

ایسی عورتیں جنہیں اپنے منہ اور زبان کے مسلز کنٹرول کرنے یا حرکت دینے میں مشکل پیش آتی ہے یا جو اپنے ہاتھوں اور بازوؤں کو استعمال کرنے میں دشواری محسوس کرتی ہیں ان کے لئے اپنے دانتوں اور مسوڑھوں کو صاف کرنا مشکل ہوسکتا ہے۔ اگر دانتوں کو باقاعدگی سے صاف نہ کیا جائے اور مسوڑھوں پر غذائی ذرے چپکے رہیں تو اس سے مسوڑھے خراب ہوسکتے ہیں۔ اگر ضروری ہو تو کسی قابل بھروسہ فرد سے اپنی مدد کے لئے کہیں۔

"بعض اوقات ڈینٹسٹ سیریبرل پالسی کے مریضوں کا علاج کرنے سے انکار کردیتے ہیں لیکن یہ تمام لوگوں کے لئے ضروری ہے کہ انہیں اپنے دانتوں کی صحت کے لئے سہولت ملے"

مرگی میں مبتلا عورتیں (دورے پڑنا یا جھٹکے کا مرض)

اگر آپ دوروں سے بچنے کے لئے Phenytoin(diphenylrhdanatoin) استعمال کرتی ہیں تو یہ آپ کے مسوڑھوں میں سوجن یا ان کے پھولنے کا سبب بن سکتی ہے۔ دانتوں اور مسوڑھوں کی اچھی دیکھ بھال اس سوجن سے بچاسکتی ہے۔

ہر کھانے کے بعد اپنے دانت اچھی طرح صاف کرنے کی کوشش کریں اور صاف پانی سے اپنے منہ کو صاف کریں۔ دانتوں کی درمیانی جگہ صاف کرنے کا خصوصی طور پر خیال رکھیں انگلی مسوڑھوں پر رگڑ نا بھی مددگار ثابت ہوتا ہے۔

"مرگی کی ادویات مسوڑھوں کی سوجن اور تکلیف کا سبب بن سکتی ہے جو دانتوں پر بھی اثر انداز ہوتی ہے دانتوں کو صاف رکھ کر آپ اس سے بچ سکتی ہیں"

دانتوں کی صفائی کے لئے ٹوتھ پیسٹ ضروری نہیں ہے۔ کچھ لوگ اس کے بجائے کونلہ یا نمک استعمال کرتے ہیں اگر آپ کے پاس ٹوتھ برش ہے تو یہ اس کے بال ہیں جو دانتوں کی صفائی کرتے ہیں۔ اس صفائی کے لئے پانی ہی کافی ہے۔ نرم بالوں والا برش استعمال کریں۔ سخت اور کھر درا برش مسوڑھوں کو زخمی کرے گا، انہیں فائدہ نہیں پہنچائے گا۔ اگر آپ مسواک استعمال کرتی ہیں تو احتیاط سے کام لیں کچھ لکڑیاں سخت ہوتی ہیں اور مسوڑھوں کو زخمی کر سکتی ہیں یا نقصان پہنچا سکتی ہیں۔ نیم کی مسواک (جو بہت سے گرم ممالک میں ہوتا ہے) بہتر کام کر سکتی ہے۔ دانتوں کو باری باری صاف کرنے کے لئے آپ باریک لکڑی یا ٹوتھ پک کے سرے پر صاف کپڑا لپیٹ سکتی ہیں۔

آنکھوں کی دیکھ بھال

اپنا چہرہ روزانہ مناسب صابن اور صاف پانی سے دھوئیں۔ اس سے آپ کو آنکھوں کی سوزش (Conjunctivitis) جیسے انفیکشنز سے بچنے میں مدد ملے گی۔ اس انفیکشن سے ایک یا دونوں آنکھوں میں سرخی، پس اور قدرے سوزش ہوتی ہے بیدار ہونے پر آنکھوں کے پپوٹے چپکے ہوئے ہو سکتے ہیں۔ بیشتر آشوب چشم تیزی سے ایک فرد سے دوسرے کو لگنے والے ہوتے ہیں۔ اگر کسی کی آنکھوں میں انفیکشن ہو تو اس کا تولیہ یا کپڑے وغیرہ استعمال نہ کریں۔ ہر بار اپنی آنکھوں کو چھونے سے پہلے اور بعد میں اپنے ہاتھ دھوئیں۔ مکھیوں سے بچیں خاص طور پر انہیں آنکھوں پر نہ بیٹھنے دیں۔ مکھیاں، انفیکشن ایک فرد سے دوسرے فرد تک منتقل کر سکتی ہیں۔

علاج

سب سے پہلے ابال کر ٹھنڈا کئے پانی سے نم اور صاف کپڑے سے آنکھوں کا پس صاف کریں پھر ہر آنکھ میں آئی آئنٹمنٹ لگائیں (دیکھئے صفحہ 343) نچلے پپوٹے کو قدرے کھینچ کر اندرونی حصے میں آئنٹمنٹ لگائیں۔ آنکھ کے بیرونی حصے پر آئنٹمنٹ لگانا کوئی فائدہ نہیں دیتا ہے۔

ہدایت: ٹیوب کو آنکھ کے ساتھ نہ لگنے دیں۔

اگر آپ کو جذام ہو

بعض معذوریاں جیسے جذام میں مبتلا فرد کے لئے بینائی کے مسائل زیادہ امکان رکھتے ہیں یا وہ آنکھوں کے انفیکشن میں آسانی سے مبتلا ہو سکتا ہے۔

اگر آپ کو جذام ہے تو آنکھوں کے گرد کے مسلز کمزور ہو سکتے ہیں یا ممکن ہے ان میں حس باقی نہ رہے۔ اس کا مطلب یہ ہے کہ آپ کی آنکھیں اپنے طور پر جھپک نہیں سکتی ہیں۔ پلکوں کا نہ جھپکنا آنکھوں کی خشکی اور انفیکشن کا سبب بن سکتا ہے۔ اگر آپ کی پلکیں نہیں جھپکتی ہیں یا آنکھیں سرخ ہوں تو آپ مندرجہ ذیل باتوں پر عمل کر سکتی ہیں۔

- دھوپ کا چشمہ پہنیں خصوصاً ایسا چشمہ جو آپ کی آنکھوں کے گرد کے حصے کو ڈھانپ لے۔
- ایسا ہیٹ پہنیں جو آپ کی آنکھوں پر سایہ کرے۔
- دن میں جتنی بار ممکن ہو آنکھوں کو سختی سے بند کریں۔
- آنکھیں سختی سے بند کر کے آنکھ کے ڈھیلوں کو اتنا تر سے اوپر کی طرف گھمائیں۔
- آنکھوں کے اطراف کی جلد میں کئی بار دھوئیں۔

اگر پس (مواد) بن رہا ہے تو ''پنک آئی'' کے لئے دیا گیا علاج کریں (دیکھئے 99)
آنکھ کو جس حد تک ممکن ہو بند رکھیں اگر ضروری ہو تو آنکھ کو ڈھانک کر رکھیں ۔

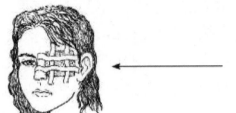

آپ آنکھ کو ڈھانپنے کے لئے کسی بھی صاف کپڑے یا روئی سے گدی یا پٹی سے آنکھ پر اس طرح باندھیں کہ وہ اس پر مستقل رہے۔

یا

آنکھ پر پٹی یا گدی شکل کے مطابق ٹیپ سے لگا لیں۔

آنکھ کو مسلیں یا ملیں نہیں

اگر آپ آنکھوں کو بند نہیں کر پاتی ہیں (Lidlag) تو آنکھ پر صاف اور نرم کپڑا رکھ کر
اسے ڈھانپ کر رکھیں تا کہ وہ خشکی اور انفیکشن سے محفوظ رہ سکے۔ اپنی آنکھوں کو نم رکھنے اور انفیکشن سے بچنے کے لئے ہر آنکھ میں
نمکین پانی کے چند قطرے ڈالیئے (ایک کپ یا ایک گلاس صاف پانی میں ایک چٹکی نمک ملائیے)۔

پیشاب اور پاخانے کا اخراج

کچھ معذور عورتیں اپنے پیشاب یا پاخانے کے مکمل طور پر اخراج پر قابو نہیں رکھتی ہیں یہ خاص طور پر ان عورتوں کا مسئلہ ہے
، جن کے نچلے دھڑ کے مسلز معذوری (جیسے پولیو یا ریڑھ کی ہڈی مجروح ہونے) کے باعث مفلوج ہو جاتے ہیں۔ اگر آپ تناسلی
اعضا کو دھو نہیں پاتی ہیں تو اپنے گھر کے کسی فرد یا کسی مدد گار سے ان کی صفائی اور انہیں خشک کرنے کے لئے مدد حاصل کریں۔ اگر
آپ نیپی یا کپڑا استعمال کرتی ہیں تو ریشنز (Rashes) انفیکشنز اور زخموں سے بچنے کے لئے انہیں مناسب وقفے سے بدلتی
رہیں (دیکھئے صفحہ 114)۔

جب آپ باہر جائیں اور ممکن ہو تو تبدیل کرنے کے لئے اضافی کپڑے ساتھ لے کر جائیں۔ اس صورت میں آپ
بے قابو اخراج کے باعث خراب ہونے والے کپڑے تبدیل کر سکتی ہیں اور شرمندگی کے ساتھ ساتھ انفیکشن سے بھی بچ سکتی
ہیں ۔

مثانے پر کنٹرول

اگر آپ بار بار پیشاب کرتی ہیں یا آپ کا پیشاب نکل جاتا ہے تو آپ اپنے کمزور مسلز کو مضبوط بنانے کے لئے بھینچنے والی ورزش کریں۔ یہ ورزش آپ کے مسلز کو مضبوط رکھنے کے علاوہ بڑھاپے میں بھی آپ کے کام آئے گی کہ آپ بڑھاپے میں پیشاب خود بخود نکلنے کے عارضے میں بھی مبتلا ہونے سے بچ سکتی ہیں۔

بھینچنے والی ورزش

پہلے پیشاب کرتے ہوئے اس کی مشق کریں، جیسے ہی پیشاب خارج ہونے لگے، اپنی فرج کے مسلز بھینچ کر یا سکیڑ کر اسے روک لیں، دس تک گنیں اور پھر پیشاب خارج ہونے دیں یہ مشق پیشاب کرتے ہوئے کئی بار دوہرائیں جب آپ یہ ورزش کرنا سمجھ لیں تو دن میں عام صورت میں اسے دوہرائیں، کسی کو بھی پتہ نہیں چلے گا کم از کم دن میں چار بار یہ ورزش کریں، ہر بار پانچ سے دس بار اپنے مسلز کو بھینچے کچھ عورتوں میں پیشاب کا غیر اختیاری اخراج روکنے کے لئے سرجری کی ضرورت ہوتی ہے اگر آپ کا پیشاب زیادہ مقدار میں خارج ہوا اور اس ورزش سے کوئی فائدہ نہ ہو، کسی صحت کارکن سے مشورہ کریں جو عورتوں کی صحت کی تربیت رکھتی ہو۔ بھینچنے والی ورزش تمام عورتوں کے لئے مفید ہے یہ ورزش روزانہ کریں اس سے مسلز طاقتور رہتے ہیں اور آئندہ کے مسائل سے تحفظ ملتا ہے۔

میں اپنی بھینچنے والی ورزش کر رہی ہوں اور امانا یہ کرنا نہیں جانتی

مثانہ خالی کرنا

اگر آپ کی معذوری کسی معاونت کے بغیر پیشاب کرنے میں حائل ہوتی ہے تو آپ کو اپنا مثانہ خالی کرنے کے لئے کوئی اور طریقہ اپنانے کی ضرورت ہے۔ کچھ عورتیں پیشاب کا اخراج ممکن بنا سکتی ہیں اور اپنا مثانہ خالی کر کے سکتی ہیں اگر وہ

- اپنی ناف کے عین نیچے مثانے کے اوپر اور پیٹرو کی ہڈی کے اوپر دبائیں۔
- اپنے ہاتھوں سے مثانے سے اوپر پیٹ کے نچلے حصے کو دبائیں۔
- نچلے پیٹ پر گھونسا بنا کر رکھیں اور جسم کے اوپری حصے کو موڑ کر نرمی سے اسے دبائیں۔
- پیٹ کے مسلز کو سخت کر کے پیشاب کے اخراج کے لئے دباؤ ڈالیں۔

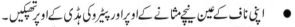

آپ یہ تدابیر اس صورت میں استعمال کریں جب پیشاب مناسب زور کے ساتھ آسانی سے خارج ہو جائے۔ اگر آپ کے مسلز نرم ہیں اور پیشاب خارج نہیں ہونے دیتے ہیں تو مثانے کو دبانے سے پیشاب گردوں میں واپس جا سکتا ہے اور انہیں نقصان پہنچا سکتا ہے۔

اگر ان میں سے کوئی بھی طریقہ کام نہ آئے تو آپ کو ربر یا پلاسٹک کی ایک ٹیوب کی ضرورت ہے جسے کیتھیٹر کہتے ہیں۔ کیتھیٹر اس وقت تک استعمال نہ کریں جب تک اس کے علاوہ پیشاب کے اخراج کی کوئی اور صورت نہ ہو۔ کیتھیٹر کا غیر محتاط استعمال بھی مثانے اور گردے میں انفیکشن کا سبب بن سکتا ہے۔

معیاری کیتھیٹر کا استعمال

کیتھیٹر لچکدار ربڑ سے بنی ٹیوب ہے جو مثانے سے پیشاب باہر نکالنے کے لئے استعمال ہوتی ہے۔ بہت سی عورتیں جن کے لئے کیتھیٹر کا استعمال ضروری ہوتا ہے اُن کے لئے مثانے کو خالی کرنے کے لئے ہر چار سے چھ گھنٹے بعد صاف یا جراثیم سے محفوظ کیتھیٹر مثانے میں لگاتی ہیں۔ ایک عورت دن میں جس قدر پانی پیئے گی اسے اتنی ہی بار کیتھیٹر استعمال کرنے کی ضرورت ہوتی ہے۔

کچھ عورتیں زیادہ پانی نہیں پیتی ہیں کیونکہ وہ بار بار کیتھیٹر استعمال کرنا نہیں چاہتی ہیں لیکن یہ دوسرے مسائل کا سبب بن سکتا ہے۔ اگر آپ مناسب مقدار میں مائع نہیں پئیں گی تو آپ کے مثانے یا گردوں میں انفیکشن ہوسکتا ہے یا قبض کی شکایت ہوسکتی ہے۔ یہ بات اہم ہے کہ آپ کا مثانہ بھرا ہوا ہو تو اس سے Dysreflexia ہوسکتا ہے (دیکھئے 117) اور اس صورت میں پیشاب گردوں میں واپس جا کر انہیں نقصان پہنچا سکتا ہے۔

بہت سی عورتیں ٹوائلٹ یا پاٹ پر بیٹھے ہوئے کیتھیٹر کا استعمال سیکھ لیتی ہیں۔ عورتیں وہیل چیئر پر بیٹھے ہوئے بھی کیتھیٹر کا استعمال کر کے ٹوائلٹ یا کسی بوتل میں پیشاب خارج کر سکتی ہیں۔ وہی طریقہ اپنائیے جو آپ کے لئے درست اور موزوں ہو۔ جب آپ بیٹھی ہوں تو کیتھیٹر کا استعمال سیکھنے کے لئے مشق کی ضرورت ہے لیکن بہت سی عورتیں اپنی روزمرہ مصروفیات کے دوران کیتھیٹر کے استعمال کا آسان طریقہ سیکھ لیتی ہیں۔ بیشتر عورتوں کے لئے 16-a سائز کا کیتھیٹر بہترین ہوتا ہے۔ بہت چھوٹی جسامت کی عورت کے لئے 14 سائز کا کیتھیٹر بہتر ہوسکتا ہے۔

کیتھیٹر استعمال کرنے والا فرد کسی اور فرد کے مقابلے میں یورینری انفیکشن کا زیادہ امکان رکھتا ہے یہ عام طور پر کیتھیٹر کے صاف نہ ہونے کے باعث جراثیموں کے مثانے میں جانے سے ہوتا ہے۔ یورینری انفیکشن سے بچنے کا بہترین طریقہ کیتھیٹر کو احتیاط کے ساتھ صاف کرنا ہے۔ کیتھیٹر کو چھونے سے پہلے ہر بار اپنے ہاتھ مناسب صابن سے یا گرم پانی سے دھوئیں اور کیتھیٹر استعمال کرنے سے پہلے اور بعد میں بھی ہاتھ دھوئیں۔ کیتھیٹر کو استعمال نہ کرنے کی صورت میں صاف جگہ میں رکھیں۔

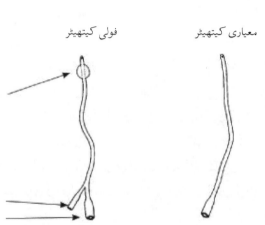

فولی کیتھیٹر میں سرے سے نزدیک چھوٹا سا (غبارہ) ہوتا ہے یہ کیتھیٹر اس انداز سے اس لئے بنایا جاتا ہے کہ یہ مثانے میں زیادہ عرصہ تک رہ سکے۔ جب یہ مثانے میں ہو تو غبارے کو پانی سے بھرا جاتا ہے تاکہ یہ باہر نہ نکلے۔ فولی کیتھیٹر کو مثانے کے اندر رکھنے کے لئے 5cc پانی عام طور پر کافی ہوتا ہے لیکن یہ باہر نکل جاتا ہے تو پانی کی مقدار 12cc سے 15cc تک بڑھائی جائے۔

فولی کیتھیٹر   معیاری کیتھیٹر

غبارے میں پانی اس جگہ سے داخل کیا جاتا ہے۔

پیشاب اس جگہ سے خارج ہوتا ہے۔

کیتھیٹر کس طرح رکھا جائے

1۔ کیتھیٹر (اور سرنج یا استعمال میں آنے والے کسی بھی آلے کو) بیس منٹ پانی میں اُبالیں یا انہیں کم از کم ابال کر ٹھنڈا کئے ہوئے پانی سے اچھی طرح دھوئیں اور صاف ستھرا رکھیں۔

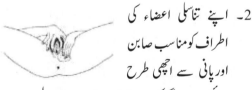

2۔ اپنے تناسلی اعضاء کی اطراف کو مناسب صابن اور پانی سے اچھی طرح دھوئیں۔ اس جگہ کو دھونے میں احتیاط سے کام لیں جہاں سے پیشاب خارج ہوتا ہے۔ اگر آپ کے پاس صابن نہ ہو تو صاف پانی استعمال کریں۔ تیز صابن آپ کی جلد کو نقصان پہنچا سکتا ہے۔

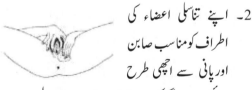

3۔ اپنے ہاتھ دھوئیں۔ ہاتھ دھونے کے بعد صرف انہی اشیاء کو چھوئیں جو صاف ستھری ہوں۔

4۔ اگر ممکن ہو آپ ایسی جگہ بیٹھیں جہاں آپ کے تناسلی اعضاء کسی چیز سے نہ چھوئیں جیسے کرسی کا اگلا حصہ یا صاف ٹوائلٹ کی نشست وغیرہ۔ اگر آپ زمین پر یا کسی سخت سطح پر بیٹھیں تو اس پر صاف کپڑا بچھا لیں۔

5۔ اگر آپ کو کوئی اور چیز چھونا پڑے تو اپنے ہاتھوں کو دوبارہ مناسب صابن اور پانی سے دھولیں۔

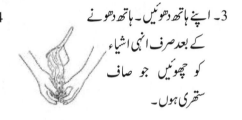

6۔ کیتھیٹر پر جراثیم سے پاک پانی آمیز چکنائی (تیل یا پیٹرولیم جیلی نہیں) لگائیں۔ اس سے آپ کے تناسلی اعضاء کی نرم جلد اور پیشاب کی نالی (یوریتھرا) کو تحفظ ملتا ہے۔ اگر آپ کے پاس ایسی چکنائی موجود نہیں ہے تو اس بات کا یقین کرلیں کہ یہ اس پانی سے نم ہے جس میں آپ نے اسے اُبالا یا دھویا تھا جب آپ اسے مثانے میں رکھیں تو اسے خاطر خواہ نرم ہونا چاہیئے۔

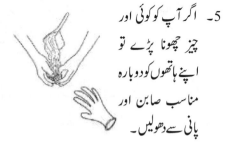

اگلے صفحہ پر ملاحظہ فرمائیے

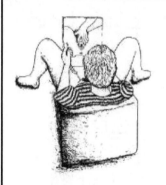

7- اگر آپ خود کیتھیٹر رکھ رہی ہیں تو پیشاب کا سوراخ دیکھنے کے لئے آئینہ استعمال کریں۔ فرج کی اطرافی جلد کو کھلا رکھنے کے لئے اپنی دوسری اور تیسری انگلی استعمال کریں۔ پیشاب کا سوراخ تقریباً فرج کے دہانے کے اوپر اور بظر کے نیچے ہوتا ہے (دیکھئے صفحہ 78) یہ عمل چند بار کرنے کے بعد آپ جان لیں گے کہ مطلوبہ سوراخ کہاں ہے اور آپ کو آئینہ استعمال کرنے کی ضرورت نہیں ہوگی۔

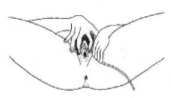

یقینی بنائیے کہ کیتھیٹر کا دوسرا سرا نچلے رخ پر ہو تاکہ پیشاب باہر خارج ہوسکے

8- پھر بیچ کی انگلی سے اپنے بظر کے نیچے چھوئیں آپ کو چھوٹا سا ڈینٹ یا گڑھا سا محسوس ہوگا اس کے ساتھ ہی پیشاب کا سوراخ ہے۔ اپنی بیچ کی انگلی اس جگہ سے بالکل اوپر رکھیں اور اپنے دوسرے ہاتھ سے صاف شدہ کیتھیٹر کو اگلے سرے سے چار پانچ انچ فاصلے سے پکڑیں اندر لے جائیں حتیٰ کہ پیشاب باہر نکلنے لگے محتاط رہئے کہ کیتھیٹر کے سرے کو اس وقت تک انگلیوں یا ہاتھ سے نہ چھوئے۔ اگر کیتھیٹر پیشاب کی نالی کے بجائے فرج میں جائے گا تو آپ کو باآسانی معلوم ہو جائے گا کیونکہ یہ نہ صرف آسانی سے اندر چلا جائے بلکہ پیشاب کا اخراج بھی نہیں ہوگا۔ جب آپ کیتھیٹر باہر نکالیں گی تو اس میں فرج سے نکلنے والی رطوبت (Mucus) ہوگی لہٰذا کیتھیٹر کو صاف پانی سے دھو کر دوبارہ کوشش کریں۔ اگر آپ کو مثانے یا گردے کا انفیکشن ہو جائے تو کسی صحت کارکن سے رجوع کریں۔ ممکن ہے آپ کی فرج میں انفیکشن ہو۔

اہم بات: کیتھیٹر کے استعمال کے انفیکشن سے تحفظ کے لئے ضروری ہے کہ آپ صاف ستھری رہیں اور جراثیم سے پاک کیتھیٹر استعمال کریں۔ اگر ہر بار ایسا ممکن نہ ہو تو کیتھیٹر کو انتہائی صاف ہونا چاہیئے۔

پیشاب کے نظام کے انفیکشن کا علاج اور بچاؤ

مثانے کا انفیکشن

بودار پیشاب
انفیکشن کی علامت ہے

بیشتر عورتیں اس وقت کہہ سکتی ہیں کہ وہ مثانے کے انفیکشن میں مبتلا ہیں جب وہ پیشاب کرنے کے بعد درد یا جلن محسوس کریں یا انہیں پیشاب کرنے کے فوراً بعد پیٹ کے نچلے حصے میں درد ہو۔ اگر آپ اپنے پیٹ پر محسوسات نہیں رکھتی ہوں تو آپ کو دوسری علامات پر توجہ دینی ہوگی۔

- بار بار پیشاب کی حاجت
- دھندلا نظر آنے والا پیشاب
- بدبودار پیشاب
- پیشاب میں خون یا پیپ
- پسینہ آنا یا حدت محسوس کرنا (رد عمل میں کمی۔ دیکھئے صفحہ 117)

اکثر مثانے کے انفیکشن کا علاج چائے یا دیگر نباتاتی اشیاء سے ہو جاتا ہے۔ اپنی کمیونٹی کی عمر رسیدہ خواتین سے پوچھئے کون سی نباتات مفید ہوں گی۔

مثانے کے انفیکشن کا علاج

جونہی علامات علم میں آئیں انفیکشن کا علاج شروع کر دیں۔

- خوب پانی پئیں ہر تین منٹ کے بعد کم از کم ایک کپ صاف پانی پئیں۔ اس سے آپ کو پیشاب بار بار آئے گا اور انفیکشن کے سنگین ہونے سے قبل جراثیموں کے خارج ہونے میں مدد مل سکتی ہے۔
- چند دن کے لئے یا علامات ختم ہونے تک جنسی عمل روک دیں۔

اگر آپ ایک یا دو روز میں بہتر محسوس نہ کریں تو پانی کا زیادہ استعمال نباتاتی علاج بند کر دیں اور دوا لینا شروع کر دیں۔ اگر اگلے دو روز میں آپ بہتری محسوس نہ کریں تو صحت کارکن سے رجوع کیجیے ممکن ہے آپ جنسی طور پر منتقل ہونے والے انفیکشن میں مبتلا ہوں (دیکھئے صفحہ 158)

کب اور کس طرح لیں	کتنی مقدار میں لیں	دوا
مثانے کے انفیکشن کی ادویات		
منہ کے ذریعے، دن میں دو مرتبہ، تین دن تک	480mg کی دو ٹیبلٹ	کوٹرائی موکسازول (Cotrimoxazole)
	160mg ٹرائی میتھو پرام (Trimethoprim)	
	800mg سلفامیتھوا ایکسازول (Sulfamethoxazole)	
		یا
منہ کے ذریعے، دن میں دو مرتبہ، تین دن تک	100mg	نائٹروفیورنٹوان (Nitrofurantoin)

گردے کا انفیکشن

بعض اوقات مثانے کا انفیکشن پیشاب کی نالی کے ذریعے گردوں تک پہنچ جاتا ہے گردے کا انفیکشن مثانے کے انفیکشن سے زیادہ خطرناک ہے۔

گردے کے انفیکشن کی علامات

- کمر کے درمیان میں یا نچلے حصے میں درد، اکثر شدید جو سامنے سے اطراف تک اور پھر پشت تک جا سکتا ہے۔
- متلی اور قے
- بخار اور سردی کا احساس
- مثانے کے انفیکشن کی کوئی بھی علامت

مثانے اور گردے کے انفیکشن کی علامات ہوں تو آپ کو امکانی طور پر گردے کا انفیکشن ہو سکتا ہے۔ جب کسی عورت کے دونوں گردوں میں انفیکشن ہو تو وہ عام طور پر شدید درد محسوس کرتی ہے اور سخت بیمار ہو سکتی ہے اور اسے فوری طور پر طبی مدد کی ضرورت ہوتی ہے۔ گھریلو علاج کافی نہیں ہوتا ہے۔ یہ دوائیں فوری طور پر لینا شروع کر دیں لیکن جب دو دن کے بعد بھی بہتری محسوس نہ ہو تو کسی صحت کارکن سے رجوع کریں۔

دوا	کتنی مقدار میں لیں	کب اور کس طرح لیں
گردے کے انفیکشن کی ادویات		
سیپروفلاکسن (Ciprofloxain)	500mg	منہ کے ذریعے، دن میں دو مرتبہ، دس دن تک
		(اگر آپ بچے کو دودھ پلا رہی ہوں تو یہ دوا استعمال نہ کریں)
یا سیفکزائم (Cefexime)	500mg	منہ کے ذریعے، دن میں دو مرتبہ، دس دن تک
یا کوٹرائی موکسازول (Cotrimoxazole)	480mg کی دو گولیاں	منہ کے ذریعے، دن میں دو مرتبہ، دس دن تک
ٹرائی میتھو پرام (Trimethomptinm)	160mg	
سلفا میتھو ایکسازول (Selfamethoxazole)	800mg	
اگر آپ متلی اور قے کے باعث دوا نگل نہ پائیں تو کسی صحت کارکن سے ملیں۔ آپ کو انجکشن کے ذریعے دوا دینا ہوگی۔		

پیشاب کے انفیکشن سے کس طرح محفوظ رہا جا سکتا ہے

اپنے تناسلی اعضاء کو صاف رکھیں۔ جراثیم تناسلی اعضاء اور خاص طور پر مقعد سے آپ کے پیشاب کے اخراجی راستے میں پہنچ کر انفیکشن پیدا کر سکتے ہیں۔ اپنے تناسلی اعضاء روزانہ دھوئیں اور پاخانہ کرنے کے بعد ہمیشہ سامنے سے پیچھے کی طرف صفائی کریں۔ صفائی کرتے ہوئے سامنے کی طرف آنے میں مقعد کے جراثیم پیشاب کے اخراجی راستے میں جا سکتے ہیں اسی طرح جنسی عمل سے پہلے اور بعد میں تناسلی اعضاء دھونے کی کوشش کریں۔ ماہواری کے لئے پیڈ زانتہائی صاف اور خشک استعمال کریں۔

- یقینی بنائیے کہ آپ کا کیتھیٹر مڑا ہوا یا بل کھایا ہوانہ ہوتا کہ پیشاب با آسانی خارج ہو سکے۔
- جنسی عمل کے بعد پیشاب کریں، اس سے پیشاب کی نالی صاف ہونے میں مدد ملتی ہے۔
- پانی خوب پئیں اور اپنے مثانے کو با قاعدگی سے خالی کریں۔
- دن بھر لیٹی نہ رہیں، جس قدر ممکن ہو متحرک رہیں۔

بیشتر عورتیں دوا اسی وقت لیتی ہیں جب تک انفیکشن کی علامات ہوں لیکن کچھ عورتیں زیادہ تر اپنی ماہواری شروع ہونے پر بار بار انفیکشن میں مبتلا ہوتی ہیں زیادہ تر اپنی ماہواری شروع ہونے پر، لہٰذا انہیں اس کے ساتھ ہی دوا کھانا شروع کر دینا چاہیئے۔

باؤل کنٹرول

کوشش کریں کہ رفع حاجت روزانہ یا ہر دوسرے دن ایک مقررہ وقت پر ہو۔ جب آپ مقررہ وقت کی پابندی کریں گی تو آخرکار آپ کا جسم اس سے مطابقت حاصل کر لے گا اور اس وقت آپ کے جسم سے فضلہ آسانی سے مقررہ وقت پر خارج ہونے لگے گا۔

رفع حاجت کے لئے bisacodyl یا گلیسرین جیسی اشیاء استعمال کی جاسکتی ہیں bisacodyl کی کارتوس نما گولیاں مقعد میں رکھی جاتی ہیں جو آنتوں کو متحرک کرتی ہیں جس سے فضلہ خارج ہوجاتا ہے اگر آپ فضلہ خارج کرنے کے لئے اپنے جسم کے نچلے حصے کے مسلز استعمال نہیں کرسکتی ہیں تو آپ فضلہ نکالنے کے لئے انگلی استعمال کرسکتی ہیں آپ یہ طریقہ قبض ہونے کی صورت میں بھی استعمال کرسکتی ہیں جب آپ بیٹھی ہوں تو فضلہ عام طور پر آسانی سے خارج ہوجاتا ہے لہذا فضلے کو نکالنے کی کوشش اس وقت کریں جب آپ ٹوائلٹ یا Pot پر بیٹھی ہوں۔ اگر آپ بیٹھ نہیں سکتی ہیں تو الٹی طرف لیٹی حالت میں کوشش کریں۔ اگر ضروری ہو تو کسی سے مدد کے لئے کہیں۔ احتیاط رکھیں کہ فضلہ آپ کی فرج یا پیشاب کے سوراخ میں نہ لگے فضلے میں موجود نقصان دہ جراثیم انفیکشن پیدا کرسکتے ہیں۔

فضلہ کس طرح نکالا جائے

اپنی انگلی کو صاف رکھنے کے لئے شکل کے مطابق اس پر ربڑ یا کوئی نرم تہہ والی چیز یا "finger cot" چڑھائیں

1- اپنے ہاتھ کو صاف پلاسٹک یا ربڑ کے دستانے یا پلاسٹک بیگ سے ڈھکیں۔ اپنی پہلی انگلی یا جو انگلی آپ بہتر طور پر استعمال کرسکتی ہوں اس پر نباتاتی یا معدنی تیل لگائیں۔

2- اپنی انگلی مقعد میں لگ بھگ دو سینٹی میٹر (ایک انچ) چڑھائیں۔

3- انگلی کو دائرے میں تقریباً ایک منٹ گھمائیں تاکہ مسلز نرم پڑیں اور فضلہ خارج ہوجائے۔

4- اگر فضلہ خود خارج نہ ہو تو اسے جس حد تک ممکن ہو اپنی انگلی سے باہر نکالیں، نرمی سے کام لیں تاکہ مقعد کے اندرونی حصے کی جلد پر خراشیں نہ پڑیں۔

5- مقعد اور اس کی اطراف کی جلد کو اچھی طرح صاف کریں اور اپنے ہاتھ دھولیں۔

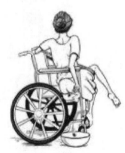

کچھ عورتیں وہیل چیئر پر بیٹھ کر فضلہ خارج کرسکتی ہیں۔ ایسا کرنے کے لئے زمین میں ایک گڑھا کیجئے یا کوئی برتن رکھ لیجئے۔ پھر سیٹ پر آگے کی طرف سرکیں اور خود کو ایک طرف جس حد تک بھی ممکن ہو اٹھا لیں۔ اپنی ٹانگ اور پیر کو مڑا ہوا رکھنے کے لئے کوئی اسٹریپ یا بیلٹ استعمال کریں تاکہ آپ اپنا ہاتھ مقعد تک با آسانی پہنچا سکیں۔ آپ اپنی ٹانگ کو مناسب جگہ رکھنے کے لئے اسٹریپ یا بیلٹ کو اپنی کرسی کے دوسرے سرے پر باندھ سکتی ہیں۔

قبض (فضلہ خارج کرنے میں دشواری)

سیربرل پالسی اور ریڑھ کی ہڈی کی خرابی میں مبتلا عورتوں کو اگر قبض ہو جاتا ہے یا وہ کئی کئی دن پاخانہ نہیں کر پاتی ہیں۔ اس سے سنگین مسائل پیدا ہو سکتے ہیں جیسے فضلہ بڑی آنت کے نچلے حصے میں سخت ٹکڑوں کی صورت اختیار کر لیتا ہے۔ سخت ٹکڑوں کی صورت میں فضلہ مقعد کو نقصان پہنچا سکتا ہے (دیکھئے صفحہ 117 سے 119)

قبض سے بچنے کے لئے

- روزانہ کم از کم آٹھ گلاس مائع لیجیے۔ پانی بہترین ہے اگر دستیاب ہو۔
- پھل، سبزیاں اور ریشہ دار اشیاء خوب کھانے کی کوشش کیجیے جیسے اجناس، کساوا، پھلی دار سبزیاں یا ایسی غذائیں جو ریشے سے بھر پور ہوں۔
- اپنے جسم کو متحرک رکھئے اور جس قدر ممکن ہو ورزش کیجیے۔
- با قاعدہ باؤل پروگرام برقرار رکھئے۔
- اپنے پیٹ کا مساج کیجیے۔
- پکا ہوا پپیتا یا آم یا سبز کیلے کھائیے۔
- دن میں دو بار ایک گلاس پانی میں ایک چمچہ اسپغول ملا کر پئیں۔
- اگر چار دن یا اس سے زیادہ دن پاخانہ نہ آئے تو آپ کوئی مسہل دوا جیسے ملک آف میگنیشیاء لے سکتی ہیں لیکن اگر آپ کے پیٹ میں درد ہو تو مسہل نہ لیں۔ گلیسرین بھی قبض دور کرنے کے لئے استعمال کی جاسکتی ہے۔

مقعد کے گرد درد انگیز سوجن (رگوں کا پھولنا، بواسیر)

مقعد کے گرد وریدوں کی سوجن بواسیر کہلاتی ہے۔ یہ عموماً تکلیف اور سوزش کا سبب بنتی ہیں یا ان سے خون خارج ہونے لگتا ہے۔ قبض اس صورت میں مزید خرابی کا سبب بنتا ہے۔ وہیل چیئر استعمال کرنے والی عورتیں، وہ جو زیادہ بیٹھی رہتی ہیں یا سیربرل پالسی میں مبتلا عورتیں عمر بڑھنے کے ساتھ بواسیر جیسے مسائل میں مبتلا ہونے کا زیادہ امکان رکھتی ہیں۔ اگر آپ فضلہ ہاتھ سے صاف کرتی ہیں تو جائزہ لیں کہ اس میں خون تو نہیں لگا ہوا ہے۔ یہ بواسیر کی عام علامت ہے۔

اگر آپ کو بواسیر ہو تو کیا کریں

ٹھنڈے پانی میں بیٹھ کر بواسیر کو کم تکلیف دہ بنایا جاسکتا ہے۔

- درد میں آرام کے لئے ٹھنڈے پانی کے برتن میں بیٹھیں۔
- اس صفحہ پر درج قبض سے بچاؤ کے لئے مشوروں پر عمل کریں۔
- کوئی صاف کپڑا ہیزل (ایک نباتاتی سیال) میں بھگو کر درد والے حصے میں رکھیں۔
- گھٹنوں کے بل جھکیں اس سے بھی درد میں آرام مل سکتا ہے۔

ماہواری

ماہانہ اخراج خون کے دوران زیادہ تر عورتیں اور لڑکیاں فرج سے نکلنے والے خون کو جذب کرنے کے لئے کئے کپڑے کی گدی یا روئی کی تہہ استعمال کرتی ہیں۔ یہ بیلٹ، پن یا زیر جامہ کے ذریعے درست جگہ رکھی جاتی ہیں۔ یہ گدی یا تہہ دن میں کئی بار تبدیل ہونا چاہیئے اور اگر اسے دوبارہ استعمال کیا جائے تو انہیں صابن اور پانی سے خوب اچھی طرح دھولینا چاہیئے۔

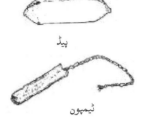

پیڈ

ٹیمپون

اگر ممکن ہو تو انہیں دھونے کے بعد سورج کی روشنی میں سکھایا جائے یا خوب گرم استری سے استری کیا جائے۔ حرارت انہیں نہ صرف خشک کرے گی بلکہ جراثیموں کو بھی ہلاک کرے گی اور ان کے دوبارہ استعمال کی صورت میں انفیکشن سے بھی تحفظ ملے گا۔ ماہواری کے بعد انہیں صاف، خشک جگہ، مٹی، گرد اور حشرات الارض سے دور رکھا جائے۔

کچھ عورتیں اپنی فرج میں کاٹن، کپڑے یا اسفنج سے بنائی ہوئی کوئی چیز رکھتی ہیں جسے ٹیمپونز (Tampons) کہتے ہیں اگر آپ ٹیمپونز استعمال کرتی ہیں تو یہ روزانہ دن میں کم از کم تین بار ضرور تبدیل کریں۔ ایک ٹیمپون کو ایک دن سے زائد رکھنے سے سنگین انفیکشن ہو سکتا ہے۔

اپنے تناسلی اعضاء سے خون وغیرہ صاف کرنے کے لئے انہیں پانی سے اچھی طرح دھویں۔ اگر ممکن ہو مناسب صابن استعمال کریں اگر آپ پیشاب کے اخراج کے لئے کیتھیٹر استعمال کرتی ہیں تو ماہواری کے دوران پیشاب کے اخراجی حصے کو خصوصی توجہ سے صاف کریں۔ اگر کیتھیٹر کی ٹیوب میں خون چلا جائے تو اسے فوری طور پر دھولیں، خون ٹیوب کا راستہ بند کرکے پیشاب خارج ہونے سے روک سکتا ہے۔

کچھ معذور عورتوں کو ماہواری کے زمانے میں اضافی مدد کی ضرورت ہوتی ہے۔

اگر ماہواری کے دوران آپ کے کپڑوں یا بستر پر خون لگ جائے تو شرمندگی محسوس مت کیجیے۔

یہ کبھی نہ کبھی تمام عورتوں کے ساتھ ہو جاتا ہے۔ اگر آپ کو پیشاب اور پاخانے کے اخراج میں مدد کی ضرورت ہوتی ہے تو پیڈز اس وقت تبدیل کئے جاسکتے ہیں اگر رات کے دوران پیڈ تبدیل کرنا آپ کے لئے مشکل ہو تو اپنے نیچے تولیہ یا کوئی کپڑا بچھالیا کریں تا کہ اس پر خون لگے تو آپ اسے آسانی سے دھوسکیں۔

اگر آپ نابینا ہیں

کیونکہ آپ خون نہیں دیکھ سکتی ہیں اس لئے جب پہلی بار آپ کا ماہانہ اخراج خون ہوتو آپ کے لئے بتانا مشکل ہوسکتا ہے کہ آپ کی ماہواری شروع ہوچکی ہے۔ لیکن چند ماہ بعد جب یہ آپ کی زندگی کا معمول بن جائے گا تو آپ اپنے جسم میں ایسی علامات محسوس کرسکتی ہیں جو آپ کو یہ بتائیں گی۔ ماہواری کے زمانے میں جس قدر ممکن ہو پیڈ یا ٹیمپونز کی تبدیلی یقینی بنائیں۔ پیڈ تبدیل کرنے یا یہ جائزہ لینے کے بعد کہ کیا خون خارج ہورہا ہے، اپنے ہاتھ ضرور دھویں۔ اپنے گھر کے کسی فرد یا جس پر آپ بھروسہ کرتی ہوں یہ جائزہ لینے کے لئے کہیں کہ آپ کے کپڑوں پر خون تو نہیں لگا ہے۔ اگر ایسا ہو تو اس سے اپنی مدد کے لئے کہیں تا کہ آپ اپنے کپڑوں کو دھوکر خون کے دھبے صاف کرسکیں۔

اگر آپ ماہواری میں کسی عورت کی مدد کرتی ہیں تو اپنی جلد پر خون لگنے سے بچاؤ کے لئے پلاسٹک کے دستانے پہننا یا پلاسٹک بیگ چڑھا لینا بہترین تدبیر ہے۔اگرچہ ماہواری کے دوران ایک عورت سے دوسری عورت میں بیماری منتقل ہونے کا امکان بہت ہی کم ہے تاہم بہتر یہ ہے کہ ہیپا ٹائٹس اور ایچ آئی وی ایڈز سے انفیکشن منتقل ہونے کے امکان سے بچا جائے۔

ان لڑکیوں کی مدد کرنا جو سیکھنے اور سمجھنے میں دشواری محسوس کرتی ہیں

اگر کوئی لڑکی مشکل سے سمجھتی ہے تو اسے اپنی روزمرہ دیکھ بھال کے لئے مدد کی ضرورت ہے بڑی بہن، خالہ یا ماں اسے بتا سکتی ہے کہ وہ کس طرح ماہواری میں اپنا خیال رکھے۔

- آپ ماہواری کے لئے وہی پیڈ یا کپڑا استعمال کریں جو لڑکی کی استعمال کرتی ہو۔
- اسے بتائیں کہ پیڈ یا کپڑا کہاں رکھا ہے۔
- اسے بتائیں کہ پیڈ یا کپڑا کہاں پھینکتے ہیں یا انہیں دوبارہ استعمال کرنا ہے تو انہیں کس طرح دھویا جاتا ہے۔
- اس کے زیرجامے میں پیڈ یا کپڑا رکھیں تا کہ وہ اس کی مشق کر سکے اور اس کے استعمال کی عادی ہو سکے۔
- اسے بتائیں کہ ماہواری کے دوران وہ گہرے رنگ کے کپڑے پہن سکتی ہے تا کہ خون کے دھبے دوسروں کو نظر آنے کے گھٹ جائیں۔

- اس کے زیرجامے میں پیڈ یا کپڑا رکھیں تا کہ وہ اس کی مشق کر سکے اور اس کے استعمال کی عادی ہو سکے۔
- اسے بتائیں کہ ماہواری کے دوران وہ گہرے رنگ کے کپڑے پہن سکتی ہے تا کہ خون کے دھبے دوسروں کو نظر آنے کے امکانات گھٹ جائیں۔

ماہواری کے دوران بے اطمینانی

ماہواری کے دوران استر کو باہر دھکیلنے کے لئے رحم سکڑتا ہے۔ یہ سکڑاؤ پیٹ کے نچلے حصے یا کمر کے نچلے حصے میں درد پیدا کر سکتا ہے جسے بعض اوقات مروڑ بھی کہا جاتا ہے، یہ درد یا تکلیف ماہواری شروع ہونے سے کچھ پہلے شروع ہو سکتی ہے۔ پیٹ کی سنکائی مروڑ میں کمی لا سکتی ہے۔ کسی بوتل یا برتن میں گرم پانی بھریں اور اسے پیٹ یا کمر کے نچلے حصے پر رکھیں یا گرم پانی میں ڈوبا ہوا موٹا کپڑا استعمال کریں۔ اگر حرارت سے فائدہ نہ ہو تو Ibuprofen جیسی کوئی درد دور کرنے والی دوائیں (دیکھئے صفحہ 345) استعمال کریں۔ ماہواری آپ کے مسلز بھی سخت کر سکتی ہے یا آپ معمول سے زیادہ تھکن محسوس کر سکتی ہیں۔ آپ کی معذوری کی عمومی علامت بھی ماہواری کے دوران سنگین ہو سکتی ہے۔ ماہواری کے دوران کچھ عورتوں کی چھاتیاں سوج جاتی ہیں اور تکلیف کا باعث بنتی ہیں۔ اور کچھ عورتیں ایسے جذبات محسوس کرتی ہیں جو نہایت ہی شدید ہوتے ہیں۔
ماہواری کے بارے میں مزید معلومات کے لئے (دیکھئے صفحہ 74)

ماہواری یا اخراج خون کی زیادتی

کچھ عورتوں کی ماہواری ہر ماہ شدید ہوتی ہے۔ یہ بات کچھ عورتوں کے لئے نارمل ہوسکتی ہے لیکن دوسری عورتوں کے لئے خون کی کمی (انیمیا) کا سبب بن سکتی ہے (دیکھئے صفحہ 87) ماہواری اس وقت شدید ہوتی ہے جب آپ کا پیڈ یا کپڑے کی تہہ تین گھنٹے سے کم وقت میں بھر جائے۔ اگر ایسا ہو تو Ibuprofen لیں (دیکھئے صفحہ 345)۔ اس سے خون کے اخراج کی رفتار کم ہوسکتی ہے اور آپ انیمیا سے بچ سکتی ہیں اگر اس سے فائدہ نہ ہو یا آپ کو تین سے چار ہفتوں میں ایک سے زیادہ بار ماہواری آتی ہو تو کسی صحت کارکن سے رجوع کیجئے۔

یسٹ (Yeast) کے باعث انفیکشنز

یسٹ، فنگس کے ذریعے ہونے والا ایک عام انفیکشن ہے۔ یہ عام طور پر جنسی اعضاء میں یا ان جگہوں پر ہوتا ہے جو طویل عرصہ تک ڈھکی رہتی ہیں (پیشاب کے اخراج یا پسینے کے باعث) یسٹ عام طور پر جنسی عمل سے منتقل ہونے والا انفیکشن نہیں ہے۔

رطوبت کی بو یا رنگ میں تبدیلی کا مطلب یہ ہے کہ آپ انفیکشن میں مبتلا ہوسکتی ہیں۔

فرج سے رطوبت کا اخراج (یسٹ، سفید پانی، سفید پریڈ، کینڈیڈا، تھرش) فرج میں نمی یا رطوبت کا معمولی مقدار میں اخراج نارمل بات ہے۔ یہ فرج کو صاف ستھرا رکھنے اور تحفظ فراہم کرنے کا قدرتی طریقہ ہے لیکن آپ کی فرج سے اخراج کی مقدار، رنگ یا بو میں تبدیلی کا مطلب یہ ہوسکتا ہے کہ آپ انفیکشن زدہ ہیں لیکن خارج ہونی والی رطوبت سے یہ بتانا مشکل ہوسکتا ہے کہ آپ کس قسم کے انفیکشن میں مبتلا ہیں۔

کوئی بھی عورت فرج کے یسٹ انفیکشن میں مبتلا ہوسکتی ہے خاص طور پر اگر وہ زیادہ تر بیٹھی رہتی ہو یا وہیل چیئر استعمال کرتی ہو۔ یسٹ انفیکشن زیادہ ان عورتوں کو بھی ہوسکتا ہے جنہیں ذیابیطس ہو یا جو اینٹی بایوٹک ادویات استعمال کر رہی ہوں۔ ایک حاملہ عورت کے لئے بہتر ہے کہ بچے کی پیدائش سے قبل اس کا علاج کیا جائے ورنہ بچہ بھی تھرش کہلانے والے یسٹ انفیکشن میں مبتلا ہوسکتا ہے۔

جلد کا انفیکشن

یسٹ انفیکشن ہمیشہ فرج میں ہی نہیں ہوتے ہیں۔ عورتیں جلد کے یسٹ انفیکشن میں بھی مبتلا ہوسکتی ہیں خاص طور پر چھڑوں کے گرد جلد کی تہوں، رانوں کے اندرونی حصوں میں جہاں جلد باہم ملی رہتی ہے یا چھاتیوں کے نیچے۔

جلد کا یسٹ انفیکشن زخم بن سکتا ہے۔ یہ فضلے یا پیشاب سے گندہ ہوسکتا ہے اور اس طرح دوسرے سنگین انفیکشن شروع ہو کر جسم کے دوسرے حصوں تک پھیل سکتا ہے محدود و نقل و حرکت رکھنے والی عورتوں کے لئے جو گھنٹوں بیٹھی رہتی ہیں، یہ بہت خطرناک ہے کیونکہ یہ ریڑھ کی ہڈی کے انتہائی نچلے سرے پر ہڈیوں میں منتقل ہوسکتا ہے۔

یسٹ انفیکشن کی علامات

- آپ اپنی فرج کے اندر یا بیرونی حصے میں انتہائی خارش محسوس کرتی ہیں۔
- فرج کے اندر یا باہر، جلد کی دوسری تہوں یا رانوں کے اندرونی حصوں میں سرخی جس سے بعض اوقات خون بہتا ہے۔
- پیشاب کرتے ہوئے جلن۔
- دودھ، دہی یا چھاج جیسا سفید اور گاڑھا اخراج۔
- ایک خاص قسم کی بدبو۔

یسٹ کا علاج عموماً قدرتی جڑی بوٹیوں سے ہوسکتا ہے۔ایساہی ایک علاج یہ ہے کہ ایک لیٹر اُبال کرٹھنڈا کئے ہوئے پانی میں تین کھانے کے چمچے (ٹیبل اسپون) سرکہ ملا کراس آمیزے میں صاف روئی بھگو کررات کے وقت فرج میں رکھیں۔یہ عمل تین رات کریں صبح ہوتے ہی روئی فرج سے باہر نکال لیں۔

یسٹ انفیکشن کے لئے ادویات

روئی جینیئان وائلٹ ایک فیصد(Gentian Violet 1%)میں بھگوئیں اور ہررات اپنی فرج میں رکھیں یہ عمل تین رات کریں۔صبح روئی نکال دیں۔یا پھر نیچے دی گئی ادویات میں سے کوئی دوا استعمال کریں۔ادویات سے بنی یہ کریم فرج کے بیرونی حصے کولہوں یا ٹانگوں کی سرخی میں بھی استعمال کی جاسکتی ہے۔کریم نرمی سے متاثرہ حصہ پر ملیں۔

کب اور کس طرح	کتنی مقدار میں	دوا
فرج میں اوپر تک، ہررات، تین راتوں تک	200ملی گرام	میکونیزول (Miconezole)
فرج میں اوپر تک، ہررات، تین راتوں تک	100,000یونٹ	نیسٹاٹن (Nystatin)
فرج میں اوپر تک، ہررات، تین راتوں تک	100ملی گرام	کلوٹرائی میزول (Clotrimazole)

اگر دستیاب ہوتو آپ اینٹی فنگل پاؤڈر جیسے نیسٹاٹن یا بغیر خوشبووالا بے بی پاؤڈر بھی استعمال کرسکتی ہیں۔پاؤڈر متاثرہ حصے پر کم مقدار میں لگائیے۔

یسٹ انفیکشن سے بچاؤ

یسٹ گرم اور مرطوب حصوں میں پیدا ہوتی ہے۔یسٹ انفیکشن سے بچنے کا بہترین طریقہ اپنی فرج اس کے ارد گرد کے حصے اور چھاتیوں کی اطرافی جلد کو صاف اور خشک رکھنا ہے۔اس سلسلے میں چند تجویزیہ ہیں۔

- اگر آپ کا پیشاب خارج ہوتا رہتا ہے تو اپنے زیر جامے جتنی بار ممکن ہو تبدیل کریں۔آپ صاف کپڑا یا پیڈ (ماہواری کے پیڈ کی طرح) بھی استعمال کرسکتی ہیں اوران میں دن میں کئی بار بدلتی رہیں۔

- اگر آپ زیادہ تر بیٹھی رہتی ہیں تو اپنی پوزیشن میں ہر گھنٹے میں بدلنے کی کوشش کریں۔دن میں یہ جتنی زیادہ بار ہوبہتر ہے۔ اس کے ساتھ دن میں کم از کم دوبار، ہر بار پندرہ منٹ اپنی جگہ سے اٹھ کر اس طرح لیٹیں کہ آپ کی ٹانگیں کھلی رہیں۔اس سے دباؤ کے زخموں سے بھی تحفظ ملے گا (دیکھئے صفحہ 114)

- اگر آپ اپنے بدن کے نچلے حصے میں حس نہیں رکھتی ہیں، یہ جائزہ لینے کے لئے کہ آپ کی فرج میں یا اس کے ارد گرد غیر معمولی سرخی تو نہیں ہے آئینہ استعمال کریں اگر آپ ایسا نہیں کرسکتی ہیں تو کسی ایسے فرد سے جس پر آپ کو بھروسہ ہو اپنی مدد کے لئے کہیں خصوصاًاس صورت میں جب آپ اپنے جنسی اعضاء کی غیر معمولی بو محسوس نہ کرسکتی ہوں۔

- صاف ستھرے اور خشک سوتی زیر جامے پہنیں (کیونکہ یہ رطوبت کو جذب کر لیتے ہیں) یہ آپ کے جسم پر قدرے ڈھیلے ہوں تا کہ آپ کے جنسی اعضاء تک ہوا پہنچ کر انہیں خشک رکھے۔

- آپ سونے کے لئے لیٹیں تو زیر جامہ مت پہنیں اس سے آپ کے جنسی اعضاء کو خشک رہنے میں مدد ملے گی۔

- ماہواری کے دوران دن میں کئی بار پیڈیا کپڑا تبدیل کریں۔ اگر یہ دوبارہ استعمال کرنا ہوں تو انہیں صابن اور پانی سے اچھی طرح دھولیں اور سورج کی روشنی میں مکمل طور پر سکھائیں۔

- اپنی فرج میں (روئی ، کپڑے یا اسفنج کے بنے) ٹیمپون استعمال کریں تو انہیں دن میں کم از کم تین بار تبدیل کریں۔ ٹیمپون کو ایک دن سے زیادہ فرج میں رکھنے سے سنگین انفیکشن ہوسکتا ہے (ماہواری کے بارے میں مزید معلومات کے لئے دیکھئے صفحہ 109)

بیکٹیریل وجائی نوسس (Bacterial Vaginosis)

بیکٹیریل وجائی نوسس ایک انفیکشن ہے جو فرج سے رطوبت کے اخراج کا سبب بنتا ہے۔ یہ جنسی طور پر منتقل نہیں ہوتا ہے اگر آپ حاملہ ہوں تو یہ بچے کی جلد ولادت کا سبب بن سکتا ہے۔

علامات

- معمول سے زیادہ رطوبت کا اخراج
- مچھلی کی جیسی بو، خاص طور پر جنسی عمل کے بعد
- ہلکی ہلکی سوزش

علاج

دوا	کتنی مقدار میں لیں	کب اور کس طرح لیں
بیکٹیریل وجائی نوسس کے علاج کے لئے ادویات		
میٹرونیڈازول (Metronidazole)	400 سے 500 ملی گرام	منہ کے ذریعے، دن میں دوبار، سات دن تک
یا میٹرونیڈازول (Metronidazole) حمل کے ابتدائی تین ماہ میں میٹرونیڈازول مت کھائیں	200 / 2 گرام ملی گرام	منہ کے ذریعے، ایک خوراک
یا کلینڈامائی سن (Clindamycin)	300 ملی گرام	منہ کے ذریعے، دن میں دوبار، سات دن تک
یا کلینڈامائی سن (Clindamycin)	5 گرام 2 فیصد کی کریم	فرج کے اندر لگائیں، سوتے وقت سات دن تک
آپ کے شریک حیات کو بھی صرف ایک بار میٹرونیڈازول کی دو گرام کی خوراک لینی چاہیئے		
اہم : جب آپ دوا لے رہی ہوں اس دوران الکحل نہ پئیں۔		

اہم فرج کے دوسرے انفیکشن یا رطوبتوں کا اخراج جنسی طور پر منتقل ہونے والے امراض (ایس ٹی آئی) کے باعث ہوسکتا ہے (مزید معلومات کے لئے دیکھئے صفحہ 158)

دباؤ کے زخم

دباؤ کے زخم خاص طور پر ان عورتوں میں عام ہیں جو وہیل چیئر استعمال کرتی ہیں یا بستر پر لیٹی رہتی ہیں اور اپنے جسم کو متحرک نہیں رکھتی ہیں۔ دباؤ کے زخم اس وقت شروع ہوتے ہیں جب جسم کے استخوانی حصوں کی کھال کرسی یا بستر پر دباؤ میں رہتی ہے ان صورتوں میں خون کی رگیں رکاوٹوں کا شکار ہوجاتی ہیں اور کھال کو مطلوبہ خون نہیں ملتا ہے نتیجے میں کھال کے اس حصے پر گہرا یا سرخ دھبہ نمودار ہوجاتا ہے اگر دباؤ جاری رہے تو کھلا زخم ہوسکتا ہے جو جسم کے اندر بڑھنے لگتا ہے یا زخم جسم کے قریب ہڈی کے اندر شروع ہوکر بتدریج کھلے حصے کی طرف بڑھنے لگتا ہے اگر دباؤ کے زخم کا علاج نہ کیا جائے تو انفیکشن پورے جسم میں پھیل سکتا ہے اور فرد کو ہلاک کر سکتا ہے۔

کیونکہ ایک دبلی پتلی عورت کی ہڈیوں پر گوشت کم ہوتا ہے لہٰذا وہ دباؤ کے زخموں کے زیادہ امکانات رکھتی ہے آپ بھی دباؤ کے زخم کا شکار ہوسکتی ہیں اگر۔

- آپ وہیل چیئر استعمال کرتی ہیں یا زیادہ وقت بیٹھی یا بستر پر لیٹی رہتی ہیں۔
- آپ کا پیشاب قابو میں نہیں رہتا ہو۔
- آپ مسلز کے کھنچاؤ کا شکار رہتی ہوں جو جسم کی بستر یا کپڑوں سے رگڑ کا سبب بنتا ہے۔

علامات

- گرم، سرخ یا سیاہ جلد ہو، دبانے پر ہلکے رنگ کی نہ ہو۔
- سوجن یا کھال پر کھلا زخم۔

جب آپ کے علم میں دباؤ کے زخم کی اولین علامات آئیں

- گھنٹہ بھر میں کم از کم ایک بار اپنی پوزیشن بدلیں۔
- متعلقہ حصے کو دباؤ سے بچانے کے لئے اضافی پیڈنگ استعمال کریں۔
- اس جگہ کا جائزہ لیتی رہیں کہ وہ بہتر ہو رہی ہے یا خراب تر۔

اپنی جلد کا روزانہ جائزہ لیں

اگر آپ دباؤ کے زخم رکھتی ہیں تو

- زخمی حصے کو ہر قسم کے دباؤ سے محفوظ رکھیں۔
- زخم اور اردگرد کی جلد دن میں دوبار صاف، اُبال کر ٹھنڈا کئے ہوئے پانی سے دھوئیں۔ سب سے پہلے زخم کے سروں کو دھوئیں، پھر نئے صاف کپڑے یا روئی سے زخم کے بیرونی حصے سے اندر کی طرف صاف کریں۔
- زخم صاف کرنے کے بعد صاف کپڑے یا روئی پر کوئی مرہم لگا کر زخم پر رکھیں۔ آپ کوئی اینٹی بائیوٹک کریم یا پیٹرولیم جیلی جیسی چیز استعمال کر سکتی ہیں۔ اس سے آپ کی جلد خشک نہیں ہوگی اور زخم دھول، مٹی، مکھیوں اور دیگر حشرات الارض سے محفوظ رہے گا۔
- خشک جلد آسانی سے تڑخ اور پھٹ سکتی ہے لہذا دن میں ایک بار اس پر کوئی لوشن لگائیں۔
- دباؤ کے زخم کے اطراف کی جلد کو رگڑیں یا اس پر مساج مت کریں اس سے جلد کمزور ہو کر پھٹ سکتی ہے جس سے زخم سنگین ہو جاتا ہے۔

اگر زخم گہرا ہو اور اس میں مردہ کھال اور گوشت موجود ہو

کھلا زخم

متاثرہ حصہ سے بدبو آتی ہے اور مردہ کھال سرمئی، کالی، ہری یا پیلی ہو سکتی ہے

- اس صورت میں زخم کو دن میں تین بار صاف کیا جانا چاہیئے۔
- عموماً ایسا زخم گہرا ہوتا ہے اور جلد کی تہوں کے نیچے تک جا سکتا ہے۔ جب زخم صاف کیا جائے تو احتیاط کے ساتھ مردہ ٹشوز اور کھال بھی صاف کی جائے۔ مردہ کھال رفتہ رفتہ نکالی جائے حتی کہ صحت مند سرخ گوشت نظر آنے لگے یا ہڈی دکھائی دے۔
- جب بھی مردہ اجزاء صاف کئے جائیں زخم کو صابن اور پانی سے دھوئیں۔ اگر دستیاب ہو تو مائع سرجیکل صابن استعمال کریں اس کے بعد زخم کو صاف یا اُبال کر ٹھنڈا کئے پانی سے دھوئیں۔

دباؤ کے زخموں کے گھریلو علاج

پپیتا……اس میں ایسے کیمیائی اجزاء ہوتے ہیں جو دباؤ کے زخموں میں موجود مردہ حصے کو نرم بناتے ہیں، اس طرح انہیں صاف کرنا آسان ہو جاتا ہے۔ پپیتے کے درخت کے تنے یا کچے پپیتے سے نکالے ہوئے ''دودھ'' میں جراثیم سے پاک کپڑے یا روئی کو بھگو کر زخم پر رکھیں یہ عمل دن میں تین بار دہرائیں۔

شہد اور شکر……یہ جراثیم ہلاک کریں گے انفیکشن سے تحفظ دے کر زخموں کے تیزی سے بھرنے میں معاون ہوں گے۔ شہد اور شکر کو ملا کر گاڑھا پیسٹ بنائیں اس آمیزے کو زخم میں بھریں اور اسے صاف سے کپڑے یا بینڈ یج سے ڈھانپ لیں (راب یا خام شکر بھی استعمال کی جا سکتی ہے) دن میں زخم کو دوبار صاف کرکے اس آمیزے سے بھریں اگر شہد اور نمک کے آمیزے کے زخم میں سے خارج ہونے والا سیال شامل ہوتا رہے تو اس سے جراثیم ہلاک ہونے کے بجائے پروان چڑھیں گے۔

اگر دباؤ کا زخم عفونت زدہ ہو گیا ہو

اگر دباؤ کے زخم سے بدبو آرہی ہو یا وہ سوجا ہوا، سرخ یا گرم ہو یا آپ کو بخار ہو رہا ہو یا سردی لگتی ہو تو اس کا مطلب ہے کہ زخم عفونت زدہ (انفیکشن زدہ) ہوگیا ہے۔ اس صورت میں سب سے بہتر کسی صحت کار کن سے رجوع کیا جائے جو پتہ چلا سکے کہ کس قسم کے جراثیم عفونت پیدا کر رہے ہیں اور کون سی دوا موثر ترین رہے گی۔ اگر یہ ممکن نہیں ہے تو آپ ڈوکسی سائیکلین (doxycycline) ، اریتھرومائسین (erythromycin)، ڈائی کلوک سیلین (dicloxcellin) استعمال کرسکتی ہیں (دوائیں استعمال کرنے کی معلومات کے لئے سبز صفحات دیکھئے)

- دباؤ کے زخم اندر سے باہر کی طرف ٹھیک ہوتے ہیں لہذا آپ محسوس کریں گی زخم بتدریج بھرنا شروع ہو رہا ہے۔ یہ زخم تیزی سے نہیں بھریں گے لہذا تحمل اور برداشت سے کام لیں۔

- اگر ضروری ہو تو درد کے لئے پیراسیٹامول لیں (دیکھئے صفحہ 350)۔

اگر اپنے جسم کے کسی حصے میں محسوس کرنے کی حس کھو چکی ہیں تو آپ کے لئے اور آپ کے گھر والوں یا دیکھ بھال کرنے والوں کے لئے ضروری ہے کہ وہ دباؤ کے زخموں کے بارے میں جس قدر جلد رجلد ہو جان لیں کہ ان کا علاج کس طرح کیا جائے اور آپ کو ان سے کس طرح بچایا جائے۔ ریڑھ کی ہڈی کے مریضوں میں دباؤ کے زخم عام ہوتے ہیں۔ اکثر یہ زخم اسپتال میں ہی علاج کے دوران ہی شروع ہو جاتے ہیں کیونکہ ایسا فرد دباؤ سے نجات کے لئے ایک پوزیشن سے دوسری پوزیشن میں حرکت نہیں کر پاتا ہے۔ اگر مناسب توجہ دی جائے تو کوئی بھی دباؤ کے زخموں میں مبتلا نہ ہو۔

دباؤ کے زخموں سے تحفظ

اگر آپ اپنے جسم کو نمایاں طور پر حرکت دینے سے قاصر ہوں تو پھر بھی کوشش کیجیے کہ کم از کم ہر دو گھنٹے اپنے جسم کو حرکت دیں یا اپنا بوجھ ایک جگہ سے دوسری جگہ منتقل کریں۔ اگر آپ سارا وقت لیٹی رہتی ہیں یا جسم کو آسانی سے حرکت نہیں دے سکتی ہیں تو اپنی پوزیشن بدلنے کے لئے کسی کی مدد حاصل کریں۔

اگر آپ سارا دن بیٹھی رہتی ہیں تو اپنے نچلے حصے تکیہ یا نرم کمبل وغیرہ رکھیں۔

یا اپنے جسم کو باری باری دائیں بائیں جھکائیں

اور بازوؤں کے درمیان کے سہارے اوپر کی طرف اٹھائیں

رگڑ کھانے والی جگہوں مثلاً اپنے گھٹنوں کے درمیان یا اپنے سر اور بازوؤں کے درمیان تکیہ یا نرم کمبل وغیرہ رکھیں۔ آپ کسی نرم چیز پر بیٹھ یا لیٹ سکتی ہیں جس سے آپ کے جسم کے ہڈی والے حصوں پر دباؤ کم ہو سکے۔ کشن یا ایسا سلیپنگ پیڈ جس میں ہڈی والے حصوں کے اطراف میں کھو کھلے خانے سے ہوں آپ کی مدد کرے گا۔ آپ پلاسٹک کے بیگ میں کچے دانے یا چاول بھر کر بھی کشن یا سلیپنگ بیگ بنا سکتی ہیں۔ اسے مہینے میں ایک بار نئے دانوں یا چاول سے بھرا جائے۔ اگر آپ وہیل چیئر استعمال کرتی ہیں تو کوشش کریں کہ آپ نرم اور آرام دہ کشن پر بیٹھیں۔

روزانہ اپنے پورے جسم کا احتیاط سے جائزہ لیں۔ آپ اپنی پشت کا جائزہ لینے کے لئے آئینہ استعمال کرسکتی ہیں۔ اگر جسم کے کسی حصے میں سیاہ یا سرخ نشان دیکھیں تو اس حصے پر مزید دباؤ ڈالنے سے گریز کریں حتٰی کہ آپ کی جلد دوبارہ نارمل ہوجائے۔

اپنے جسم کو روزانہ صابن اور صاف پانی سے دھونے کی کوشش کریں۔ اپنی جلد کو خشک رکھیں مگر اسے رگڑیں مت۔ اپنی جلد کو خشکی سے بچانے کے لئے جو آپ کی جلد کو آسانی سے چٹخا سکتی ہے دن میں ایک بار معمولی مقدار میں کوئی لوشن ملیں۔ **اپنی جلد پر الکحل مت لگائیں۔** الکحل جلد کو خشک کرکے اسے کمزور بنا سکتی ہے۔

پھل سبزیاں، پروٹین اور آئرن والے پھل اور سبزیاں جیسے (پھلیاں، مٹر، گوشت (خصوصاً جگر، دل اور گردے) مچھلیاں یا مرغی خوب کھائیں۔ یہ چیزیں آپ کی جلد اور مسلز کو صحت مند اور مضبوط بنائیں گی جس سے آپ کو دباؤ کے زخموں سے تحفظ ملے گا۔

اچانک شدید ضرب لگاتے سردرد کے ساتھ خون کے دباؤ میں اضافہ (ڈس ریفلیکسیا)

ریڑھ کی ہڈی کے T6 مُہرے کے اوپر کے حصے کے زخم اچانک ہی شدید سردرد کے ساتھ ہائی بلڈ پریشر (ڈس ریفلیکسیا) کا سبب بن سکتے ہیں۔ یہ جسم کا کسی ایسی چیز پر ردِعمل ہوتا ہے جو عام صورت میں درد بے اطمینانی کا سبب بنتی ہے لیکن اسے ایسا فرد زخم کے باعث محسوس نہیں کرپاتا ہے۔ ڈس ریفلیکسیا اس صورت میں ہوسکتا ہے جب کوئی چیز جسم کے کسی اندرونی حصے جیسے bowel، تولیدی اعضاء، مثانے یا آنتوں یا نچلے جسم کی جلد یا چھاتیوں کو چھوری ہوی ان میں ہیجان پیدا کررہی ہو۔

ڈس ریفلیکسیا کی عام وجوہات

ڈاکٹر ریڑھ کی ہڈی کے مُہروں کی نشاندہی کے لئے حروف تہجی اور نمبر استعمال کرتے ہیں۔ T6 کہلانے والا مُہرہ اس جگہ ہوتا ہے۔

- انتہائی بھرا ہوا مثانہ۔ اس کا سبب ایسا کیتھیٹر بھی ہوسکتا ہے جو مُڑ گیا ہو یا بلاک کھا گیا ہو۔

- مثانے کا انفیکشن یا مثانے یا گردے میں پتھری۔ (دیکھئے صفحہ 105)

- جسم میں حد سے زیادہ فضلہ (قبض) (دیکھئے صفحہ نمبر 108)

- دباؤ کے زخم، جلا ہوا جسم یا جلد کی سوزش جسے آپ محسوس نہ کرسکتی ہوں

- دباؤ کے زخموں کے لئے (دیکھئے صفحہ نمبر 114)

- جلد کو محسوس نہ ہونے والی ٹھنڈک یا حرارت (جیسے معائنے کی ٹھنڈی میز)

- ماہواری یا بچے کی ولادت کے دوران رحم کا سکڑاؤ یا کھچاؤ

- جنسی عمل

ڈس ریفلیکسیا کی علامات

1- چہرے، بازوؤں یا سینے پر زیادہ پسینہ آنا

2- ریڑھ کی ہڈی کے زخم کے اوپر سرخ یا سیاہ یا سوجی ہوئی جلد

3- بازوؤں یا سینے پر goose bumps یا مہاسے/دانے

4- دھندلاہٹ یا دھبے نظر آنا

5- بند ناک

6- شدید ضرب لگا تا سر درد

7- بیمار محسوس کرنا/متلاہٹ

8- اچانک بلڈ پریشر بڑھنا (150 / 240 تک)

ان میں سے کوئی بھی ایک یا زیادہ علامات ڈس ریفلیکسیا کی نشاندہی کرسکتی ہے۔ اگر آپ سمجھتی ہیں کہ آپ ڈس ریفلیکسیا کا شکار ہیں تو آپ کو فوری طور پر مدد کی ضرورت ہے۔ اگر بلڈ پریشر اچانک بڑھ جائے تو گھر کے کسی فرد یا دیکھ بھال کرنے والے کی مدد حاصل کریں جو جانتا ہو کہ ایسی صورت میں کیا کرنا چاہیے۔ آپ یہ معلومات اپنے مددگار یا صحت کارکن کو بتانے کے لئے استعمال کرسکتی ہیں کہ اس صورت میں آپ کی مدد کس طرح کی جائے۔

اہم بات : ڈس ریفلیکسیا میڈیکل ایمرجنسی ہے۔ ہائی بلڈ پریشر، دورے یا دماغ میں مہلک جریان خون کا سبب بن سکتا ہے۔ دیکھ بھال کرنے والے ڈس ریفلیکسیا کے مریض کو تنہا نہ چھوڑیں۔

ڈس ریفلیکسیا کی علامات پر ہمیشہ توجہ دیں۔ کچھ علامات ایمرجنسی کی نشاندہی نہیں کرتی ہیں لیکن طریقے ایسے موجود ہیں کہ ریڑھ کی ہڈی کی مجروح عورت جان سکتی ہے کہ اس کے جسم کے ساتھ کوئی غیر معمولی بات ہو رہی ہے مثلاً اگر آپ خود کو قدرے گرم اور پسینے میں ڈوبا محسوس کرنا شروع کریں یا اگر آپ اپنی جلد میں جھنجھناہٹ محسوس کریں یہ اس لئے بھی ہو سکتا ہے کہ آپ کا جوتا بل کھایا ہوا ہو یا آپ کے پیروں کے ناخن جلد میں گھس رہے ہوں۔ اگر آپ اپنے مسئلے سے نمٹ سکتی ہوں تو ڈس ریفلیکسیا کی علامات رفع ہو جائیں گی۔

ڈس ریفلیکسیا کا علاج

- اگر لیٹی ہوئی ہیں تو بیٹھ جائیں اور علامات دور ہونے تک بیٹھی رہیں۔

- تنگ کپڑوں کو (بشمول تنگ موزے یا زیر جامے) وغیرہ ڈھیلے کریں۔

- اگر یہ باؤ درجہ حرارت کے باعث ہے تو باؤ ختم کرنے یا گرم یا ٹھنڈی سطح سے الگ ہونے کے لئے اپنی پوزیشن تبدیل کریں۔

- جلد سے رگڑ کھانے والی کوئی بھی چیز الگ کر دیں۔

- یہ دیکھیے کہ کیا آپ کا مثانہ بھرا ہوا تو نہیں، اپنے پیٹ کے نچلے حصے کو محسوس کریں۔

اگر آپ کا پیشاب خارج نہیں ہو رہا ہو:

- کیتھیٹر لگائیں اور مثانہ خالی کریں۔
(دیکھئے صفحہ 103 اور 104)

اگر آپ پہلے ہی کیتھیٹر استعمال کر رہی ہوں:

- دیکھیے کہ کیتھیٹر میں خم یا بل تو نہیں ہیں، انہیں درست کیجیے تاکہ پیشاب باہر نکل سکے۔
- کیا کیتھیٹر میں رکاوٹ ہے؟ کیتھیٹر تبدیل کریں یا اس میں ابال کر ٹھنڈا کیا ہوا 30cc پانی یا جراثیم سے پاک سیلائن محلول داخل کریں تاکہ کیتھیٹر کی رکاوٹ دور ہو سکے۔

آپ کو پیشاب کے انفیکشن کی علامات ہوں

- دیکھیے صفحہ نمبر 105 تا 106۔ اگر وجہ ایسی ہی ہوں تو مثانے میں کیتھیٹر کے ذریعے anesthetic محلول داخل کریں۔ ابالے ہوئے 20cc پانی میں ایک فیصد لیگنوکین (Lignocaine) کے 10cc ملا کر استعمال کریں۔ کیتھیٹر کو بیس منٹ بند رکھنے کے بعد کھلا چھوڑ دیں۔ انفیکشن کی صورت میں اینٹی بایوٹک ادویات سے علاج کی ضرورت بھی ہوتی ہے۔

اگر آپ کو قبض ہو:

- اگر آپ کو پاخانہ نہ ہوئے طویل مدت گزر چکی ہے تو اپنی انگلی کو ڈھک کر اس پر Lignocaine جیلی رکھیں اور اپنی مقعد میں داخل کریں۔ کیا bowel بھرا ہوا ہے؟ اگر وہ سخت فضلے سے بھرا مقعد میں مزید Lignocaine جیلی داخل کریں اور پھر پندرہ منٹ سردرد کم ہونے تک انتظار کریں اور پھر انگلی سے فضلہ نکالیں (دیکھئے صفحہ 107)۔ اگر علامات دس منٹ میں ختم نہ ہوں تو دوا استعمال کریں۔ Nifedipine بلڈ پریشر کو پانچ سے دس منٹ میں کم کر دے گی۔

دوا	کتنی مقدار میں اور کس طرح لی جائے؟
ڈس ریفلیکسیا کے لئے ادویات	
Nifedipine	10 ملی گرام کا کیپسول دانتوں سے کاٹ کر نگل لیں
یا nifedipine	10 ملی گرام کی گولی پیس کر تھوڑے سے صاف پانی میں پیسٹ بنائیں اور یہ پیسٹ زبان کے نیچے رکھ لیں

جسم کے اندر سے زیادہ پیشاب یا پاخانے کی موجودگی ڈس ریفلیکسیا کا سبب بن سکتی ہے۔ روزانہ مقررہ وقت پر پاخانہ کرنے کے اپنے پروگرام پر عمل کا خیال رکھیں۔ خوب پانی پئیں اور ایسی غذائیں کھائیں جو فضلے کے اخراج میں معاون ہوں۔ اس کے ساتھ دن میں کئی بار پیشاب کریں۔ اگر آپ کیتھیٹر استعمال کرتی ہیں تو خیال رکھیں کہ یہ مڑے یا بل نہ کھائے۔

درد میں آرام کے لئے

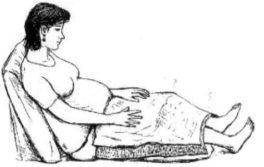

بعض معذوریاں جیسے آرتھرائٹس مسلز اور جوڑوں میں درد کا سبب بنتا ہے۔ بعض اوقات عورتوں کو اپنے جسم کے مخصوص حصوں میں درد ہوتا ہے۔ یا وہ پورے بدن میں درد کا شکار ہو سکتی ہیں۔ درد میں آرام کے لئے بہت سی ترکیبیں موجود ہیں جو آپ آزما سکتی ہیں۔

زخموں، جکڑے ہوئے جوڑوں اور مسلز کے لئے حدت یا حرارت عام طور پر موثر ہوتی ہے۔ گرم پانی میں کپڑا بھگوئیں اور درد والے حصوں پر رکھیں۔ پانی زیادہ گرم نہیں ہو۔ آپ اس میں اطمینان سے اپنا ہاتھ رکھ سکیں، دوسری صورت میں آپ اپنی جلد جلا سکتی ہیں۔

پُرسوزش جوڑوں یا زخموں کے لئے ٹھنڈک عام طور پر بہترین ہوتی ہے۔ جب آپ کے جسم کا حصہ کوئی پُرسوزش ہو تو وہ آپ کو گرم محسوس ہوگا، یہ سرخ اور سوجا ہوا بھی ہو سکتا ہے۔ برف کسی کپڑے یا تولیے میں لپیٹ کر تکلیف دہ حصے پر رکھیں۔ برف براہِ راست جلد پر نہ رکھیں۔ دس سے پندرہ منٹ کے بعد برف لپٹے کپڑے کو ہٹائیں اور اپنی جلد کو گرم ہونے دیں۔ جب آپ کی جلد گرم ہو جائے تو اس پر آپ دوبارہ برف لپٹا کپڑا رکھ سکتی ہیں۔

تکلیف زدہ حصے کو آرام دینے کی کوشش کریں۔ مسلز اور جوڑوں پر بوجھ نہ ڈالیں اور بھاری کاموں یا ان حصوں کو زیادہ استعمال کرنے سے بچیں تا کہ ان میں کھچاؤ پیدا نہ ہو۔

اکثر مدھم حرکات معاون ثابت ہوتی ہیں۔ آپ کے جوڑوں اور مسلز کو متحرک رکھنے کی ایسی صورتیں ہیں جو درد میں کمی لا سکتی ہیں۔

- درد والے حصے کو نرمی سے سہلائیں۔

  نرمی کے ساتھ اپنے مسلز اسٹریچ کریں۔

- کسی سے اپنے مسلز پر مساج کرائیں۔

- صاف، گرم پانی میں تیریں یا پانی میں رہ کر کچھ وقت گزاریں۔

درد میں، درد دور کرنے والی کوئی دوا جیسے پیراسیٹامول (acetaminophen) معاون ہو سکتی ہے لیکن اس سے سوجن کم نہیں ہوگی۔ اسپرین اور ibuprofen جوڑوں کا درد اور سوجن کنٹرول کرنے میں معاون ہوں گی، درد میں آرام کے لئے مزید معلومات حاصل کرنا ہوں تو ادویات کے خصوصی صفحات دیکھیے۔

اہم بات : اگر آپ کے کان بجنا شروع ہوں یا آپ آسانی سے مجروح ہو جاتی ہوں تو اسپرین کم لیں۔

اگر آپ جوڑوں کی سوجن کی وجہ سے اسپرین یا ibuprofen لے رہی ہوں تو درد ختم ہونے کے بعد اس وقت تک کھاتی رہیں جب تک جوڑوں کی سوجن کم نہ ہو۔ اسپرین اور ibuprofen یہ دونوں دوائیں چار گھنٹے کے اندر ایک کے بعد دوسری دوا مت لیں۔

تبدیلی کے لئے کام

اگرچہ بہت سے لوگ معذور عورتوں کی دیکھ بھال کو اہم سمجھتے ہیں لیکن حقیقت میں بہت سی معذور عورتیں مناسب دیکھ بھال یا معلومات نہیں حاصل کر پاتی ہیں، انہیں صحت مند اور فعال زندگی کے لئے رہنمائی کی ضرورت ہے۔

گھرانے اور دیکھ بھال کرنے والے کیا کر سکتے ہیں؟

گھر کے افراد اور مدد کرنے والے ہماری زندگی کو مختلف طریقوں سے خوشگوار اور

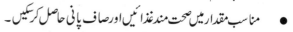

ہم اپنی مدد کے لئے آپ پر اعتماد کرتے ہیں

بہتر بنا سکتے ہیں۔ وہ ہماری اس طرح بھی مدد کر سکتے ہیں کہ ہمیں زیادہ سے زیادہ خودمختار بنانے کے لئے ہماری حوصلہ افزائی کریں تا کہ اپنے جسموں کی دیکھ بھال جس حد تک ممکن ہو خود کر سکیں۔ یوں بھی معذوریوں کے ہوتے ہوئے ہمیں اضافی توجہ اور مدد درکار ہوتی ہے کہ ہم

- مناسب مقدار میں صحت مند غذائیں اور صاف پانی حاصل کر سکیں۔
- اپنے جسموں کو مضبوط اور لچکدار بنانے کے لئے ورزش کر سکیں۔
- غسل کر سکیں اور اپنے دانت صاف کر سکیں۔
- پاخانے یا پیشاب کے اخراج، کپڑے بدلنے یا ماہواری میں خون جذب کرنے کے لے پیڈ زوغیرہ استعمال کر سکیں۔
- دباؤ کے زخموں کے جائزے۔
- صفائی اور علاج کے لئے۔ دباؤ کے زخموں سے بچاؤ کے لئے ہر دو گھنٹے کے بعد معذور عورت کی پوزیشن بدل کر یا اس کا وزن دوسرے انداز میں منتقل کر کے اس کی مدد کی جائے۔ اسے نرم سطح پر بٹھایا یا لٹایا جائے جس سے ان کے ہڈیوں والے حصوں پر دباؤ کم ہوتا ہے۔ اگر دباؤ کے زخموں کا علاج نہ کیا جائے تو وہ موت کا سبب بن سکتے ہیں۔
- ادویات اور ضروریات کی دیگر اشیاء کی گھر میں دستیابی خصوصاً اس صورت میں جب طبی سہولت کا مرکز دور ہو۔ ادویات میں درد دور کرنے والی دوائیں، پیشاب یا جلد کے انفیکشن کے لئے کوئی اینٹی بایوٹک، صاف پٹیاں اور ایسی کوئی دوا جو عورت کی معذوری کی مناسبت سے با قاعدہ استعمال کی جاتی ہو۔

ہم میں سے بہت سی عورتیں جو نابینا یا بہری ہیں، اپنی جسمانی ضرورت کی تکمیل اور دیکھ بھال خود کر سکتی ہیں۔ لیکن ممکن ہے کہ ہمیں خود کو صحت مند رکھنے کے لئے معلومات کی اب بھی ضرورت ہو۔ مثلاً ایک نابینا فرد کو ضرورت ہو سکتی ہے کہ کوئی اسے صحت کے بارے میں معلومات پڑھ کر سنائے، خواہ کچھ معلومات پیچیدہ یا سمجھنے میں مشکل ہی کیوں نہ ہوں۔ اور ایک بہری فرد کی ضرورت ہے کہ اسے کوئی ریڈیو یا صحت کار کن سے سنی صحت افزاء باتیں اشارے کی زبان میں سمجھائے۔

کمیونٹیز کیا کرسکتی ہیں؟

کمیونٹیز حالات کو بہتر بنانے کے لئے بہت کچھ کرسکتی ہیں جس سے ہمیں اپنی دیکھ بھال کی سہولت ملے گی اور ہم صحت مندرہ سکیں گے۔ بہت سی معذور عورتیں غریب ہیں اور ان میں سے کچھ تنہا زندگی گزار رہی ہیں۔ دوسروں کی طرح ہمیں بھی دیکھ بھال، پُر غذائیت غذا، صاف پانی، سینی ٹیشن اور رہنے کے لئے محفوظ جگہ کی ضرورت ہے۔ ہمیں اپنے ہمسایوں کے ساتھ بہتر تعلقات اور احترام کی ضرورت بھی ہے۔ ہم سے اور ہمارے گھر والوں سے معلوم کریں کہ کمیونٹی ہماری اور ہماری صحت کی دیکھ بھال کے لئے کیا کرسکتی ہے۔

- ہم میں سے کچھ کو اپنے گھر والوں اور مددگاروں سے روزانہ دیکھ بھال کی ضرورت ہوسکتی ہے۔ کمیونٹی رہنما اور ہمسایہ گروپ اس کا بندوبست کرسکتے ہیں کہ ہماری روزمرہ کی ضروریات پوری ہوتی رہیں اور ہمارے گھر والوں یا مستقل مدد کرنے والوں کو وقفہ مل سکے۔

- معذوری رکھنے والی بہت سی بوڑھی عورتیں انتہائی غریب ہیں، تنہا رہتی ہیں یا انہیں کسی مدد حاصل کرنے میں سخت دشواری پیش آتی ہے۔ ان کے لئے مددگار یا کچھ وقت ساتھ گزارنے والوں کا اہتمام کرکے یا انہیں دوسرے طریقوں سے عملی مدد فراہم کرکے، کمیونٹی معذور عورتوں کی زندگی میں غیر معمولی بہتری لاسکتی ہے۔

- کمیونٹی ہماری مدد کرسکتی ہے کہ وہ اپنے طور پر مناسب مقدار میں اچھی غذا حاصل کرسکیں۔

- معذور عورتوں کو صاف پانی تک رسائی حاصل ہو۔

- کمیونٹی کو آمادہ کریں کہ وہ بیت الخلاء اور ٹوائلٹ کو تعمیر یا تبدیل کرے تا کہ معذوری رکھنے والے افراد آسانی سے اور محفوظ طور پر استعمال کرسکیں (دیکھئے صفحہ 123)

یہ جاننے کے لئے کمیونٹیز اور گھرانے کس طرح ہمیں صحت مند اور محفوظ رکھ سکتے ہیں، کتاب ''جہاں عورتوں کے لئے ڈاکٹر نہ ہو'' کا باب نمبر 10 اور کتاب ''سینی ٹیشن، صفائی اور زندگی کے لئے پانی'' ملاحظہ کیجئے۔

ہمیں باقاعدگی سے ایسے صحت کارکنوں کے ذریعے چیک اپ کی ضرورت ہے جو ہماری بہتری کے لئے دلی جذبہ رکھتے ہوں۔

ٹوائلٹس اور بیت الخلاء کو استعمال کے لئے آسان بنانا

معذور بچوں اور بالغ افراد کے استعمال کے لئے ٹوائلٹس کو موزوں بنانے کے بہت سے طریقے ہیں۔ کمیونٹی کو اپنی ضروریات کے مطابق تبدیلیوں سے آگاہ کرنے کے لئے مستعد اور تخلیقی رہیئے، گھٹنے موڑنے میں دشواری رکھنے والے فرد کے لئے دستی سہارا یا بیٹھنے کے لئے اونچی جگہ بنائی جائے۔ اگر ٹوائلٹ زمین میں ہے تو کسی اسٹول یا کرسی میں سوراخ کرکے اسے ٹوائلٹ کے اوپر رکھیئے۔

بٹھائی جاسکنے والی
سلاخ اگر ضرورت ہو تو
شامل کی جاسکتی ہے

اگر کسی فرد کو اپنا جسم کنٹرول کرنے میں دشواری ہوتی ہے تو اس کی پشت، اطراف اور پیروں کے لئے سہارے اور سیٹ بیلٹ بنایئے۔

گھر سے ٹوائلٹ تک جانے کے لئے نابینا افراد کے لئے رسی یا جنگلے کے ذریعے راستے کی رہنمائی کا اہتمام کیجئے۔ اگر کسی فرد کو اپنے کپڑے اُتارنے میں دشواری ہوتی ہے تو اس کے کپڑے ممکنہ حد تک ڈھیلے بنائیں اور لاسٹک استعمال کریں۔ اس کے لیٹنے اور کپڑے بدلنے کے لئے صاف ستھری اور خشک جگہ کا بندوبست کیجئے۔ ایسے فرد کے لئے جسے بیٹھنے میں مشکل ہوتی ہے آپ دستی سہارے یا حرکت کرنے والی سیڑھیاں لگایئے۔

وہیل چیئر استعمال کرنے والوں کے لئے ٹوائلٹس

گھنٹی یا کوئی ایسی آواز پیدا کرنے والی چیز جسے ضرورت پڑنے پر مدد کے لئے بجایا جاسکے

مدد کے لئے سہارے

وہیل چیئر سے ٹوائلٹ تک آسانی سے جانے کے لئے دستی سہارے

ٹوائلٹ سیٹ اور وہیل چیئر کی یکساں سطح

آسانی سے داخل ہونے کے لئے باہر کھلنے والا چوڑا دروازہ

اتنی جگہ کہ وہیل چیئر آسانی سے اس میں سماسکے

ٹوائلٹ جانے یا نکلنے کے لئے ریمپ جو نقل و حرکت کو آسان بناتا ہے

کمیونٹی کے لوگوں کو احساس دلائیں کہ معذوری رکھنے والے فرد کو اسی طرح پرائیویسی اور احترام کی ضرورت ہوتی ہے جس طرح دوسرے کسی بھی فرد کو۔ انہیں بھی یقینی طور پر پرائیویسی ملنا چاہیئے۔

”جہاں عورتوں کے لئے ڈاکٹر نہ ہو“

عورتوں کی صحت کے لئے ہیسپرین فاؤنڈیشن کی کتاب

”ویئر ویمن ہیونو ڈاکٹر“ کا اُردو ترجمہ

کتاب سندھی زبان میں بھی دستیاب ہے

دنیا بھر کی عورتوں کے لئے بیش قیمت کتاب

وہ باتیں جن کا جاننا ہر عورت کے لئے ضروری ہے

باب -6

معائنہ صحت

معائنہ صحت معذور عورتوں کی ضرورت ہے

اکثر لوگ سوچتے ہیں کہ صحت کے حوالے سے ایک معذور عورت کی وجہ پریشانی اس کی معذوری ہوتی ہے اور اسے صحت کے دیگر معائنوں کی ضرورت نہیں ہے۔ لیکن یہ سچ نہیں ہے۔ ہر دو سے تین برس میں کسی صحت کارکن سے چیک اپ، خواہ وہ خود کو ٹھیک ہی محسوس کر رہی ہو، ایک عورت کی صحت کے مسائل ابتداء ہی میں جاننے کا موثر طریقہ ہے، اس صورت میں ایسی عورتوں کا علاج بھی بہترین طور پر ہوسکتا ہے۔

معذور عورتیں عموماً صحت کی جانچ کے لئے دشواریوں سے دوچار ہوتی ہیں۔ ممکن ہے آپ صرف اس لئے اپنے معائنے سے کتراتی ہوں کہ آپ اپنے جسم کے بارے میں شرمندگی کے جذبات کے ساتھ پروان چڑھی ہوں یا آپ چاہتی ہوں کہ کوئی آپ کے جسم کو نہ چھوئے۔ یا پھر آپ اتنی بار طبی معائنے اور آپریشن کرا چکی ہوں کہ اب کسی صحت کارکن کو دیکھنا بھی نہیں چاہتی ہوں۔

لیکن باقاعدگی سے طبی معائنہ دوسری تمام عورتوں کی طرح معذور عورتوں کے لئے بھی اہم ہے۔ اس بارے میں اس کتاب اور دوسری صحت کی کتابوں سے جس قدر زیادہ ممکن ہو سیکھیں اور پھر آپ مقامی صحت کارکنوں سے پوچھ سکتی ہیں اور اسپتال کے ڈائریکٹروں اور وزرائے صحت سے مطالبہ کر سکتی ہیں کہ وہ ان خدمات کو آپ کے لئے اور دوسری معذور عورتوں کے لئے فراہم کریں۔

اس باب میں چھاتیوں کے معائنے (دیکھئے صفحہ 128) اور پیڑو کے معائنے (دیکھئے صفحہ 130) کے بارے میں معلومات فراہم کی گئی ہیں۔ یہ دو معائنے صحت مند رہنے کی خواہاں ہر عورت کے لئے اہم ہیں۔ صحت کے دیگر ٹیسٹوں کے بارے میں معلومات کے لئے صفحہ 135 دیکھئے۔

باقاعدگی سے معائنہ آپ کو کیا بتا سکتا ہے

صحت کے ایسے بہت سے مسائل ہیں جو باقاعدہ چیک اپ سے معلوم ہو سکتے ہیں۔ بعض اوقات ایک فرد بیمار ہو سکتا ہے مگر ممکن ہے کہ اسے مسئلے کے انتہائی سنگین ہونے تک اس کا اندازہ ہی نہیں ہو اور اس کا علاج مشکل ہو جائے۔ کچھ صحت کے مسائل ایسے ہیں جو ابتداءً ہی علم میں آ جائیں تو علاج میں سہولت رہتی ہے مثلاً اینیمیا (کمزور خون) ٹی بی، ایچ آئی وی / ایڈز اور دیگر جنسی طور پر منتقل ہونے والے انفیکشنز، ملیریا، کچھ اقسام کے کینسر، ہائی بلڈ پریشر، کیڑے اور آنتوں کے دیگر پیراسائٹس اور ذیابطیس۔ ہر عورت خواہ وہ معذور ہو یا نہ ہو ان مسائل کا شکار ہو سکتی ہے۔

معائنہ صحت کے مسئلے کو کیمونٹی کے سامنے لانا

لیزی لانگ شا جانتی تھی کہ زمبابوے میں اس کی کیمونٹی کی بیش تر معذور عورتیں پیڑو یا چھاتیوں کے معائنے سے محروم رہتی ہیں۔ کلینک جہاں یہ معائنے ممکن ہیں معذور عورتوں کے لئے بہت دور اور بہت مہنگے تھے۔ لیکن وہ جانتی تھی کہ یہ معائنے معذور عورتوں کے لئے بھی کس قدر ضروری ہیں۔ کیونکہ یہ معائنے کرانا آسان نہ تھا لہٰذا بہت سی عورتیں اپنے مسائل صحت کے بارے میں اس وقت تک نہیں جان پاتی تھیں جب تک وہ سنگین نہ ہو جائیں، بہت سی عورتیں اسی لئے کینسر کے باعث مر جاتی تھیں۔

لیزی نے جو خود بھی معذور تھی، معذور عورتوں کا ایک گروپ تشکیل دیا۔ ان عورتوں نے اکٹھا ہو کر جس قدر بھی ممکن تھا کینسر اور صحت کے دوسرے مسائل اور یہ کہ کس طرح معائنے مسائل کو جلدی جاننے میں معاون ہو سکتے ہیں، کے بارے میں معلومات حاصل کیں۔ پھر اس گروپ نے وزارت صحت کے ایک نمائندے سے مل کر اسے آگاہ کیا کہ معذور عورتوں کو کیا مسائل درپیش ہیں۔ اس گروپ نے واضح کیا کہ معذور عورتوں کو کلینکوں تک جانے اور صحت کی سہولتوں کے حصول کے لئے ادائیگیوں میں کن مشکلات کا سامنا ہے۔ نمائندہ اس بات سے انتہائی متاثر ہوا کہ یہ عورتیں کس قدر زیادہ جانتی ہیں۔ اس نے سرکاری طور پر اس کیمونٹی کی معذور عورتوں کے لئے مہینے میں ایک بار کینسر اسکریننگ اور خاندانی منصوبہ بندی کی خدمات کے لئے مفت موبائل کلینک کا اہتمام یقینی بنایا۔

چھاتیوں کا معائنہ اور پیڑو کا معائنہ ایسے دو انتہائی اہم معائنے ہیں جن کی سہولت ہر عورت کو ملنی چاہیے۔ عورتوں میں پروان چڑھنے والے دو عام کینسر چھاتیوں اور سرویکس کے کینسر ہیں۔ یہ معائنے اور ٹیسٹ ان کی جلد نشاندہی کر کے علاج میں معاون ثابت ہو سکتے ہیں۔

چھاتیوں اور پیڑو کے معائنوں کے لئے کس طرح تیار رہوں

کیا ہونے والا ہے،اس بارے میں پیشگی جان کر آپ چھاتیوں یا پیڑو کے معائنے کے لئے تیار ہوسکتی ہیں۔صحت کارکن سے معائنے کے ہر مرحلے کے بارے میں بتانے کے لئے کہیں اور اگر آپ کی سمجھ میں کوئی بات نہ آئے تو اس کی وضاحت کے لئے کہیں۔ اس سے پوچھے جانے والے سوالات کے بارے میں سوچنے سے مدد دلکتی ہے۔

معذوری رکھنے والی ایک عورت کی حیثیت سے ممکن ہے کہ معائنے کے دوران آپ کی ضروریات مختلف ہوں،اپنے ساتھ کسی ایسی سہیلی کے فرد کو لے جائیں جو اس دوران ہمہ وقت آپ کے ساتھ رہے۔معائنے سے قبل صحت کارکن سے اپنی خصوصی ضروریات کے بارے میں بات کیجئے تا کہ وہ معائنہ اس طرح کرے جو آپ کے لئے محفوظ تر اور سہل ہو۔

اگر آپ بہری ہیں یا ٹھیک سے سن نہیں پاتی ہیں تو اپنے ہمراہ کسی ایسے فرد کو رکھیں جو آپ کے اور صحت کارکن کے درمیان ابلاغ کے لئے علامتی یا اشاروں کی زبان استعمال کر سکتا ہو۔

اگر آپ نابینا ہیں یا اچھی طرح دیکھ نہیں پاتی ہیں تو معائنے کی وضاحت اور تفصیل جاننے کے لئے کسی معاون کو ساتھ لے جائیں۔صحت کارکن سے کہیے کہ وہ اچھی طرح بتائے کہ وہ کیا کر رہی ہے اور آپ کیا نہیں دیکھ سکتی ہیں۔

اگر آپ حرکت سے تعلق رکھنے والی معذوری رکھتی ہیں یا چل نہیں پاتی ہیں تو کسی کو ساتھ لائیں یا وقت سے پہلے سوچ لیں کہ آپ کلینک یا مرکز صحت میں کس طرح داخل ہوں گی۔

اگر آپ سمجھنے یا سیکھنے میں دشواری رکھتی ہوں اور چھاتیوں یا پیڑو کا معائنہ آپ کو خوفزدہ ،نروس یا بے چین کر دیتا ہو تو جس پر آپ کو اعتماد ہو اس سے درخواست کریں کہ وہ معائنے کے دوران آپ کے ساتھ رہے۔ گھر کے افراد اور دیکھ بھال کرنے والے ایسی معذور عورتوں کی مدد کر سکتے ہیں جو سیکھنے اور سمجھنے کی صلاحیت پر اثر انداز ہوتی ہیں۔

- **معائنے کے بارے میں پیشگی بات کیجے۔**گھر کا کوئی فرد یا دوست، سمجھنے میں دشواری رکھنے والی عورت کو معائنوں کے بارے میں سمجھا سکتا ہے۔ آپ اسے سمجھائیں کہ یہ معائنے اس کی صحت کے لئے کس قدر ضروری ہیں۔یہ بتائیں کہ معائنے کے دوران کیا ہوگا؟اس کے سوالوں کے جواب دیں۔اگر ممکن ہو تو اسے بتائیں کہ معائنہ کون کرے گا۔

- **اگر ممکن ہو تو معائنے سے پہلے کلینک جائیں۔** معائنے سے ایک دن پہلے معذور عورت کے ساتھ اس جگہ جائیں جہاں معائنہ کیا جائے گا۔

- **اس کے ساتھ ایسے فرد کو رہنے دیں جس پر وہ بھروسہ کرتی ہو۔**اگر وہ چاہتی ہو تو معائنے کے دوران اس کے ساتھ اس کی کوئی سہیلی یا گھر کا کوئی فرد ٹھہر سکتا ہے۔ اگر معائنہ کرنے والا صحت کارکن مرد ہو تو یقینی بنائیں کہ اس کے ہمراہ معائنے کے دوران کوئی ایسی عورت رہے جس پر وہ اعتماد کرتی ہو۔

کیا آپ مجھے بتائیں گی کہ آپ میری چھاتیوں کا معائنہ کس طرح کریں گی

صحت کا رکن اس طرح مدد کر سکتے ہیں کہ وہ

- **معائنے سے قبل معائنے کی وضاحت کریں۔** یہ بتائیں کہ معائنہ شروع ہونے سے پہلے کیا ہوگا۔اس سے پوچھیں کہ کیا اس کے ذہن میں کوئی سوال ہے۔اگر وہ معائنہ شروع ہونے سے پہلے سوالات کر سکتی ہو تو وہ ممکنی طور پر کم تر خوفزدہ ہوگی۔

- **اسے وہ آلات دکھائیں جو آپ استعمال کریں گے جیسے** speculum۔ پیڑو کے معائنے سے پہلے یقینی بنائیں کہ وہ جان لے کہ speculum کیا ہے تا کہ وہ معائنے کے دوران پریشان نہ ہو۔ اگر وہ آلے کو چھونا چاہتی ہے تو اسے ایسا کرنے دیں۔

- **معائنے کے دوران اس سے باتیں کریں۔** ہر مرحلے پر بتائیں کہ کیا ہو رہا ہے۔ اسے آگاہ کریں کہ اگلے مرحلے میں آپ کیا کریں گے۔ اس سے پوچھیے کہ کیا وہ تیار ہے،اس کی آمادگی کے لئے انتظار کیجئے۔اس طرح وہ معائنے کے دوران خود پر قابو رکھ سکے گی۔

چھاتیوں کا معائنہ

چھاتیوں کا باقاعدگی سے معائنہ اس بات کا یقین کرنے کا ایک اچھا طریقہ ہے کہ آپ چھاتیوں کے کینسر کی کوئی علامت نہیں رکھتی ہیں۔ بیش تر عورتوں کی چھاتیوں میں چھوٹے چھوٹے گومڑ (lumps) ہوتے ہیں۔ یہ گومڑ ان کی ماہواری کے دوران عموماً اپنی جسامت اور ساخت بدل لیتے ہیں۔ یہ کبھی کبھار ماہواری شروع ہونے سے پہلے انتہائی نرم ہو سکتے ہیں۔ چھاتی کا ایسا گومڑ جو ختم نہیں ہوتا ہو، چھاتی کے کینسر کی علامت ہو سکتا ہے۔ بہت سی عورتیں چھاتی کے کینسر میں مبتلا ہوتی ہیں جس کا علاج نہ ہو تو انہیں ہلاک کر سکتا ہے۔ چھاتیوں کا باقاعدگی سے معائنہ یقینی بناتا ہے کہ کینسر ابتداء ہی میں علم میں آ جائے اور اس کا فوری علاج کیا جا سکے۔

جب بھی آپ کے پیڑو کا معائنہ ہو تو کوئی تربیت یافتہ صحت کا رکن آپ کی چھاتیوں کا معائنہ بھی کرے۔ صحت کا رکن اس باب میں بیان شدہ معائنے کے طریقے استعمال کرے گا۔

ہر چند کہ ایک صحت کا رکن ہر سال یا دو سال میں آپ کی چھاتیوں کا معائنہ کر سکتا ہے لیکن آپ خود بھی بار بار اپنی چھاتیوں کا جائزہ لے سکتی ہیں۔

اگر آپ خود یہ نہ کر سکیں تو کسی ایسے فرد کی خدمات حاصل کریں جس پر آپ کو اعتماد ہو۔ بہتر یہ ہے کہ ہر بار ایک ہی فرد سے معائنہ کرائیں اس طرح آپ کی مدد کرنے والا آسانی سے جان لے گا کہ کوئی تبدیلی تو نہیں آئی ہے۔

آپ اپنی ماہواری کے دوران ہر ماہ ایک مخصوص دن اپنے چھاتیوں کے معائنہ کی کوشش کریں (دیکھئے صفہ 75)۔ اگر ممکن ہو تو معائنہ ہر ماہ اپنی ماہواری شروع ہونے کے بعد ساتویں دن کریں۔ اگر باقاعدگی سے ایسا کریں گی تو آپ جان لیں گی کہ آپ کی چھاتیاں عام طور پر کیسی محسوس ہوتی ہیں اور اگر کوئی گڑ بڑ ہونے پر آپ آسانی سے جان لیں گی۔ اس کے ساتھ آپ اپنی چھاتیوں کا معائنہ اس وقت کریں جب آپ کے پاس آرام کے لئے اور اچھی طرح معائنے کے لئے کافی وقت ہو۔

اگر آپ خود چھاتیوں کا جائزہ نہیں لے سکتی ہیں تو آپ کی بہن یا کوئی سہیلی آپ کی چھاتیوں کا معائنہ کر سکتی ہے۔

اگر آپ کی چھاتیاں بڑی ہیں تو انہیں چار حصوں میں تقسیم کر کے باری باری ہر حصے کا جائزہ لیں آپ اس طرح شکل بنا کر کسی بھی حصے میں پائے جانے والے گومڑ کی نشاندہی بھی کر سکتی ہیں۔

اگر آپ کو گومڑ ملے تو کیا کریں

اگر گومڑ ہموار اور ملائم ہے اور دبانے پر آپ کی جلد کے نیچے حرکت سی کرتا ہے تو اس کے لئے فکر مند مت ہوں لیکن ہر ماہ اسے چیک کرتی رہیں۔ اور اگر یہ سخت اور غیر ہموار ساخت رکھتا ہوا اور یہ درد کے بغیر سائز میں بڑھ رہا ہو تو اس پر نگاہ رکھیں۔ خصوصاً اس وقت جب گومڑ ایک چھاتی میں ہو اور دبانے پر حرکت نہیں کرتا ہو۔ اگر گومڑ اگلی بار بھی اسی جگہ موجود ہو تو کسی صحت کار کن سے ملیں۔ یہ کینسر کی علامت بھی ہو سکتا ہے۔ آپ اس صورت میں طبی مدد حاصل کریں جب آپ کے نپل سے خون یا مواد جیسا سیال خارج ہو رہا ہو۔

آپ کسی بھی گومڑ کے دریافت ہونے پر کسی تجربہ کار صحت کار کن سے چیک کرنے کی درخواست کریں خواہ گومڑ ہموار ہو یا غیر ہموار۔ ماہواری بند ہونے (سن یاس) کے بعد بھی اپنی چھاتیوں کا معائنہ با قاعدگی سے جاری رکھیں۔

ایک طریقہ جس سے آپ اپنی چھاتیوں کا معائنہ کر سکتی ہیں

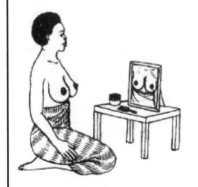

آپ اپنی چھاتیوں کا معائنہ کس طرح کریں

آئینے میں اپنی چھاتیوں کو دیکھیں۔ اپنے بازو سر سے اوپر تک اٹھائیں، اپنی چھاتیوں کی ساخت میں کسی تبدیلی یا سوجن، جلد یا نپل میں تبدیلیوں کا جائزہ لیں۔ اب اپنے بازو اپنے اطراف میں رکھ کر اپنی چھاتیوں کو دوبارہ غور سے دیکھیں۔

اپنی چھاتی کے ہر حصے کو اچھی طرح چیک کریں۔ اس طرح ہر ماہ ایک ہی انداز سے معائنے کی سہولت ملتی ہے۔

اگر ممکن ہو تو لیٹ جائیں، اپنا ایک بازو سر کے نیچے رکھیں، اپنی انگلیوں کو سیدھار کھتے ہوئے اپنی چھاتی کو دبائیں اور گومڑ وغیرہ کی موجودگی کو محسوس کریں۔ دوسری چھاتی کا جائزہ لینے کے لئے بازوؤں کی ترتیب بدل لیں۔

دوسرے طریقے جن سے آپ اپنی چھاتیوں کو چیک کر سکتی ہیں

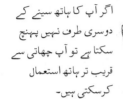

اگر آپ کا ہاتھ سینے کے دوسری طرف نہیں پہنچ سکتا ہے تو آپ چھاتی سے قریب تر ہاتھ استعمال کر سکتی ہیں۔

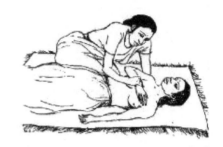

اگر آپ کے مسلز کمزور ہیں یا ہاتھ لرزتے ہیں، آپ اپنی انگلیوں کو استعمال کرنے کے لئے دوسرا ہاتھ استعمال کر سکتی ہیں یا کوئی اور اس میں آپ کی مدد کر سکتا ہے۔ مدد کرنے والا آپ کے ہاتھ کو تھام کر آپ کی چھاتی پر رکھ کر آپ کی انگلیوں کو درست جگہ رکھ سکتا ہے۔

یاد رکھیئے: اگر آپ تھکی ہوئی ہیں تو وقفہ کیجیے۔ ضروری نہیں کہ آپ ایک بار میں مکمل معائنہ کر لیں۔

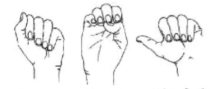

اگر آپ انگلیوں سے اچھی طرح محسوس نہیں کر سکتی ہیں تو اپنے ہاتھ کے دوسرے حصے استعمال کریں۔ آپ اپنا انگوٹھا، ہتھیلی یا اپنی انگلیوں کی پشت استعمال کر سکتی ہیں۔ اپنے چھاتی کے ہر حصے کو چھونا اور محسوس کرنا یقینی بنائیں۔

پیڑو کا معائنہ

پیڑو کے معائنے سے آپ کو پتہ چل سکتا ہے کہ کیا آپ

- اپنے تناسلی اعضاء کے گرد گومڑ، سوجن، زخم وغیرہ تو نہیں رکھتی ہیں۔ ان میں کچھ خطرناک ہو سکتے ہیں لہٰذا ان کے علاج کی ضرورت ہو سکتی ہے۔

- آپ کے رحم، بیض نالیوں، بیضہ دانی یا فرج میں کوئی انفیکشن ہے۔ انفیکشن کا علاج نہ کیا جائے تو وہ خطرناک ہو سکتے ہیں۔

- آپ رحم کے منہ، بیضہ دانی یا رحم کے کینسر میں مبتلا ہیں۔

- آپ رحم یا بیضہ دانیوں کے دیگر مسائل جیسے فائبرائڈ ٹیومر، اینڈومیٹروسز (endometriosis یا cysts) کا شکار ہیں جو کینسر کے باعث نہیں ہوتے ہیں (دیکھئے صفحہ 81اور 82)

اگر آپ چلتے ہوئے لنگڑاتی ہیں یا چھڑی، بیساکھی یا وہیل چیئر استعمال کرتی ہوں؟

اگر آپ کوئی جسمانی معذوری رکھتی ہیں تو آپ بہتر طور پر جانتی ہوں گی کہ ایک پوزیشن سے دوسری پوزیشن میں حرکت کس طرح کی جاسکتی ہے۔ اپنی سہیلی یا صحت کارکن سے مدد کے لئے کہیں۔ پیڑو کا معائنہ شروع ہونے سے پہلے یقینی بنائیے کہ آپ اچھی طرح متوازن، محفوظ اور آرام دہ پوزیشن میں ہیں (دیکھئے صفحات 133اور 134)۔

معائنے سے پہلے جس قدر ممکن ہو فضلہ اور پیشاب خارج کرنے کی کوشش کریں۔ پیڑو کا معائنہ مسلز کو ڈھیلا کر سکتا ہے جس سے پیشاب یا فضلہ خارج ہو سکتا ہے۔ اگر آپ سارا وقت کیتھیٹر استعمال کرتی ہیں تو اسے ہٹانے کی ضرورت نہیں ہے یہ معائنے پر اثر انداز نہیں ہوگا۔ اگر آپ نے اپنی ٹانگ سے پیشاب کی تھیلی باندھ رکھی ہے تو اسے الگ کر دیں اور اسے اپنے یا بیڈ کے ساتھ پیٹ کے پار اس طرح رکھیں کہ ٹیوب نہ مڑے اور پیشاب کا اخراج ہوتا رہے۔

پیڑو کے معائنے کے مراحل

1- صحت کارکن آپ کے بیرونی تناسلی اعضاء کا جائزہ لے گی کہ وہاں سوجن، گومڑ، زخم یا رنگ میں تبدیلی تو نہیں آئی۔

2- عام طور پر صحت کارکن آپ کی فرج میں Speculum رکھے گی۔ یہ دھات یا پلاسٹک کا چھوٹا سا آلہ ہے جو فرج کے اندرونی حصے کو کھلا رکھتا ہے اور صحت کارکن آسانی سے فرج کی اندرونی دیواروں اور سرویکس کا جائزہ لیتی ہے کہ کہیں سوجن، گومڑ، زخم یا رطوبت کا اخراج تو نہیں ہے۔ Speculum کے

اب میں تمہاری فرج کے اندرونی حصے کا جائزہ لینے کے لئے یہ آلہ استعمال کرنے والی ہوں

استعمال کے دوران آپ قدرے دباؤ یا بے اطمینانی محسوس کر سکتی ہیں لیکن اس سے کوئی نقصان نہیں پہنچے گا۔ اگر آپ کے مسلز ڈھیلے اور مثانہ خالی ہو تو معائنہ زیادہ آرام دہ ہوگا۔

3- اگر کلینک میں لیبارٹری کی سہولت ہے تو صحت کارکن کینسر کے لئے Pap test اور ضروری ہوا تو جنسی انفیکشنز کے لئے ٹیسٹ کرائے گی۔ Pap test کے لئے صحت کارکن سرویکس سے ٹشو کھرچنے کے لئے ایک چھوٹی سی گول ڈنڈی استعمال کرے گی۔ یہ تکلیف دہ عمل نہیں ہے۔ آپ صرف معمولی سا دباؤ محسوس کریں گی۔ ٹشو کا نمونہ لیبارٹری بھیجا جائے گا جہاں اس میں کینسر کی علامات چیک ہوں گی۔ اگر سرویکس میں کینسر پایا جائے اور اس کا فوری علاج ہو تو ہمیشہ اس کا علاج ہو سکتا ہے۔

4- Speculum ہٹانے کے بعد صحت کارکن پلاسٹک کا صاف دستانہ پہن کر اپنی دو انگلیاں آپ کی فرج میں ڈالے گی۔ وہ دوسرے ہاتھ سے آپ کے پیٹ کے نچلے حصے کو دبائے گی اس طرح وہ آپ کے رحم، ٹیوبز، بیضہ دانیوں کا سائز، بناوٹ اور جگہ محسوس کر سکتی ہے۔ معائنے کا یہ حصہ تکلیف دہ نہیں ہونا چاہیئے۔ اگر ایسا ہو تو اس کا مطلب یہ ہے کہ کچھ نہ کچھ گڑ بڑ ضرور ہے۔

5- کچھ مسائل کے لئے صحت کارکن کو ریکٹل معائنے کی ضرورت ہو سکتی ہے۔ وہ ایک انگلی آپ کی مقعد اور ایک فرج میں ڈالے گی۔ اس معائنے سے صحت کارکن کو فرج، رحم، ٹیوبز اور بیضہ دانیوں کے امکانی مسائل کے بارے میں مزید معلومات مل سکتی ہیں۔ ریکٹم معائنہ آسان ہوگا اگر آپ صحت کارکن کی انگلی مقعد پر محسوس کرتے ہی خود بھی زور لگائیں۔ اس سے آپ کی ریکٹم کے اطراف مسلز نرم پڑیں گے اور معائنہ کم تر تکلیف دہ ہوگا۔

صحت کارکنوں کے لئے

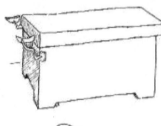

بہت سے اسپتالوں اور کلینکوں میں معائنہ ٹیبل اونچی ہوتی ہیں جو اپنی ٹانگوں کو حرکت دینے میں دشواری محسوس کرنے والی عورتوں کے لئے استعمال کرنا مشکل ہوتی ہیں۔ زمین سے قریب تر ٹیبل معذوریوں والی بیشتر عورتوں کے لئے بہترین ہوتی ہیں۔ لیکن آپ کو پیڑو کے معائنے کے لئے کسی خصوصی ٹیبل کی ضرورت نہیں۔ صحت کارکن یہ معائنہ کسی بھی صاف ستھری ہموار جگہ پر حد تو یہ ہے کہ صاف فرش پر صاف کپڑا اچھا بھی کر سکتی ہے۔

جب فرش پر کسی کا معائنہ کریں تو speculum کو اس طرح استعمال کیا جاسکتا ہے کہ اس کا اوپری حصہ نیچے کی طرف رہے۔

فرش پر معائنہ کرنے کے لئے Speculum کا ہینڈل اس طرح گھمائیے کہ اس کا رخ عورت کی فرج میں رکھتے ہوئے اوپر کی طرف رہے، دوسری صورت میں Speculum کو کھولنا مشکل ہوگا۔ یہ یقینی بنانے کے لئے کہ Speculum فرش کو نہ چھوئے، عورت کے کولہوں کے نیچے تہہ کیا ہوا کپڑا بچھا دیں تا کہ وہ قدرے اوپر رہیں۔

بہت سی عورتیں پہلی بار Speculum دیکھ کر خوفزدہ ہو جاتی ہیں، وہ سمجھتی ہیں کہ جب یہ ان کی فرج میں جائے گا تو انہیں زخمی کردے گا۔ جب آپ کسی ایسی عورت کا معائنہ کررہی ہوں جس نے اس سے قبل پیڑو کا معائنہ نہیں کرایا ہوتا ہو تو اسے چھوٹے سائز کا Speculum دکھائیں خواہ آپ نے بڑا سائز ہی استعمال کرنا ہو۔

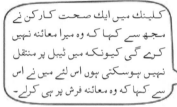

کلینک میں ایک صحت کارکن نے مجھ سے کہا کہ وہ میرا معائنہ نہیں کرے گی کیونکہ میں ٹیبل پر منتقل نہیں ہوسکتی ہوں اس لئے میں نے اس سے کہا کہ وہ معائنہ فرش پر ہی کرلے۔

اطمینان کریں کہ وہ معائنہ سے پہلے پُرسکون ہو، اسے نرمی سے چھوئیں اور بتائیں کہ آپ کیا کرنا چاہتی ہیں۔ جب معائنہ مکمل ہوجائے تو معائنے کے دوران تعاون پر اس کا شکریہ یہ ادا کیجئے۔

مختلف معذوریاں رکھنے والی عورتوں کے معائنوں کو سہل بنانے کے لئے مزید تجاویز دیکھئے صفحہ 133، 134 پر۔

ڈس ریفلیکسیا (بلڈ پریشر میں اچانک اضافہ اور تیز سر درد) سے تحفظ کے لئے احتیاطی تدابیر اختیار کریں۔

ریڑھ کی ہڈی کے زخموں والے افراد میں ڈس ریفلیکسیا عام ہے۔ یہ اس چیز پر جسم کا ردِعمل ہے جو عام طور پر درد یا بے اطمینانی کا سبب بنے گی لیکن ایسا فرد اپنے زخم کے باعث اسے محسوس نہیں کرتا ہے۔

عورت کے پیڑو کے معائنے کے دوران ڈس ریفلیکسیا مندرجہ ذیل وجوہ کے باعث ہوسکتا ہے۔

- عورت کا جسم سخت معائنہ ٹیبل یا سطح پر ہو (خواہ وہ اسے محسوس نہ کر سکتی ہو)
- معائنہ کرنے والے فرد کے ہاتھوں یا استعمال کئے جانے والے آلے کے باعث فرج یا ریکٹم پر دباؤ خصوصاً جب سردی ہو۔
- کلینک میں یا معائنے کی جگہ ٹھنڈک (کم درجہ حرارت)
- کیتھیٹر کی ٹیوب مڑی ہو یا خم کھائی ہو۔

اہم بات اگر آپ مجروح ریڑھ کی ہڈی والی عورت کا معائنہ کر رہی ہوں تو ڈس ریفلیکسیا کی علامت پر نظر رکھیں اور معائنہ روکنے کے لئے تیار رہیں۔ **ڈس ریفلیکسیا میڈیکل ایمرجنسی ہے۔** بلند خون کا دباؤ دوروں کا سبب بن سکتا ہے یا اس سے دماغ میں خون بہہ سکتا ہے۔ ڈس ریفلیکسیا کے مریض کو تنہا نہ چھوڑیں۔ ڈس ریفلیکسیا کی علامات اور علاج کے لئے دیکھئے صفحہ 117 تا 119۔

پیڑو کے معائنے کے لئے پوزیشن

اگر آپ اپنی ٹانگوں کو آسانی سے الگ نہیں رکھ سکتی ہیں تو اس کا مطلب یہ نہیں ہے کہ آپ کا معائنہ نہیں ہوسکتا ہے۔ صحت کار کن سے مختلف پوزیشنوں کے بارے میں بات کیجئے جو آپ کے جسم کے لئے بہتر ہوں۔ درج ذیل کچھ پوزیشنیں جسمانی معذوریاں رکھنے والی بہت سی عورتوں کے لئے استعمال ہوسکتی ہیں۔

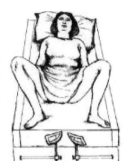

کولہوں کے مسائل رکھنے والی عورتوں کے لئے یہ پوزیشن آسان ہوسکتی ہے۔ یہ پوزیشن اس صورت میں اچھی ہے جب آپ کے پیروں کو پکڑنے کے لئے کوئی موجود نہ ہو۔ کیونکہ بہت سی عورتیں اپنے مسلز اکڑائے بغیر ٹانگوں کو اس پوزیشن میں رکھ سکتی ہیں۔

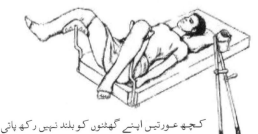

کچھ عورتیں اپنے گھٹنوں کو بلند نہیں رکھ پاتی ہیں تو وہ اپنے گھٹنوں کو سہارا دینے کے لئے فٹ ریسٹ استعمال کرتی ہیں۔

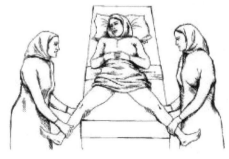

کچھ معائنہ ٹیبلوں میں عورت کے پیروں کے رکھنے کی جگہ ہوتی ہے۔ معذوریوں والی بہت سی عورتیں ان کو استعمال نہیں کرتی ہیں۔

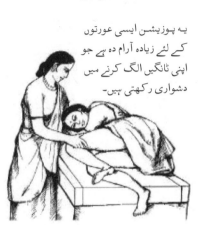

یہ پوزیشن ایسی عورتوں کے لئے زیادہ آرام دہ ہے جو اپنی ٹانگیں الگ کرنے میں دشواری رکھتی ہیں۔

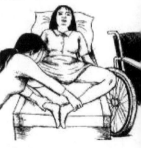

یہ پوزیشن ان عورتوں کے لئے ہے جو اپنے پیروں کو خود حرکت نہیں دے سکتی ہیں۔ یہ ان کے لئے بھی اچھی ہے جو اپنے گھٹنے موڑنے میں دشواری محسوس کرتی ہیں۔

اگر آپ کے مسلز اکڑے ہوئے یا سخت ہیں

معائنے کے دوران مسلز اچانک ہی اکڑ یا سخت ہوسکتے ہیں۔ یہ عام طور پر ریڑھ کی ہڈی کی چوٹ یا سیریبرل پالسی کی صورت میں ہوتا ہے۔ مسلز میں اچانک کھچاؤ اس وقت ہوسکتا ہے جب

- آپ معائنہ ٹیبل پر لیٹ رہی ہوں۔
- تکلیف دہ پوزیشن میں ہوں۔
- کوئی آلہ جیسے speculum فرج میں رکھا ہوا ہو۔
- بائی مینول یا ریکٹم معائنے کے لئے انگلیاں فرج یا مقعد میں رکھی گئی ہوں۔

معائنے کے دوران آپ کی سہیلی یا گھر کا کوئی فرد آپ کی ٹانگوں کو سہارا دے سکتا ہے۔ اس صورت میں آپ اپنے مسلز کو ڈھیلا رکھ سکتی ہیں جو معائنے کو آسان بناتا ہے۔

اگر آپ کے مسلز سخت ہیں تو صحت کارکن سے کہیں کہ وہ اپنا کام آہستگی سے کرے تا کہ آپ کو آرام کے لئے زیادہ وقت ملے۔ اگر معائنے کے دوران کھچاؤ محسوس ہو تو صحت کارکن سے رکنے اور اس وقت تک انتظار کرنے کے لئے کہیں جب تک آپ کے مسلز پہلے کی طرح نرم نہ ہو جائیں۔ اکڑے مسلز کے خلاف براہ راست کھچاؤ یا دباؤ نہ ڈالیں اس سے کھچاؤ کی کیفیت اور شدید اور شدید ہو جائے گی۔ کوئی فرد مسلز کے نرم پڑنے تک متاثرہ حصے کو تھام کر یا سہارا دے کر رکھ سکتا ہے۔

آپ کمبل کو گول تہہ کر کے یا تکیوں کو اپنی ٹانگوں کو سہارا دینے کے لئے استعمال کرسکتی ہیں

معائنہ اس صورت میں آسان ہوگا اگر آپ ایسی آرام دہ پوزیشن معلوم کریں جس میں آپ کو آرام دہ حالت میں رکھنے کے لئے اپنے مسلز سخت نہ کرنے پڑیں یا پھر اپنی کسی سہیلی یا گھر کے کسی فرد سے کہیں کہ وہ معائنے کے دوران آپ کے جسم کو سہارا دے۔ اگر یہ ممکن نہ ہو تو آپ کمبل کو گول تہہ کر کے اپنے گھٹنوں کے نیچے رکھ سکتی ہیں۔

اہم بات اکڑے یا بل کھائے مسلز پر نہ مساج کریں اور نہ ہی انہیں رگڑیں۔ مساج مسلز کو اور زیادہ سخت کرے گا۔

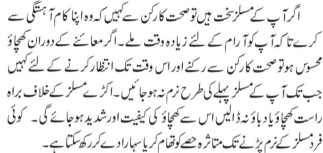

دھکیلنے کی ورزش

سیریبرل پالسی یا اکڑے ہوئے مسلز رکھنے والی خواتین، اپنے مسلز کو آرام دہ صورت میں لانے میں دشواری محسوس کرتی ہیں۔ پیڑو کے مسلز کو آرام دہ کرنے کے لئے آپ دھکیلنے والی ورزش کرسکتی ہیں۔ پہلے نیچے کی طرف دھکیلنے کی مشق اس طرح کریں جیسے آپ فضلہ خارج کرنے کی کوشش کر رہی ہوں۔ کچھ عورتیں تصور کرتی ہیں کہ جیسے وہ انڈا دینے کی کوشش کر رہی ہیں جو فرج سے باہر آئے گا۔

کچھ دیر بعد جب آپ دوبارہ دھکیلنے کی ورزش شروع کریں گہرا سانس لینے کی کوشش کریں۔ معائنے سے پہلے اس ورزش کی مشق کیجیے۔ آپ یہ ورزش معائنے کے دوران بھی کرسکتی ہیں۔ خصوصاً اس وقت جب صحت کارکن آپ کی فرج پر انگلی یا آلہ رکھے۔ جیسے ہی صحت کارکن معائنہ شروع کرے، آپ دھکیلنا بند کر کے اور اپنے پیٹ کے مسلز کو ڈھیلا چھوڑ کر اس کی مدد کریں گی۔ اگر آپ کے مسلز سخت ہوں تو صحت کارکن آپ کے اندرونی حصے کی صورتحال کو محسوس نہیں کرسکے گی۔

صحت مند رہنے کے لئے دیگر معائنے

تمام عورتیں اپنی صحت کا بہتر طور پر خیال رکھ سکتی ہیں اگر وہ یہ جانتی ہوں کہ بیماری کی علامات کا جائزہ کس طرح لیا جائے۔ مثال کے طور پر ایک ایسا فرد جس پر آپ کو اعتبار ہو وہ آپ کے چھاتیوں کے معائنے میں آپ کی مدد کرسکتی ہے۔ چھاتی کے کینسر کی علامات دیکھنے کے لئے صفحہ 128 تا 130 پر دی گئی معلومات استعمال کرسکتی ہیں۔

ایسے ٹیسٹ بھی ہیں جو گھر میں آپ خود یا گھر کا کوئی فرد یا آپ کی دیکھ بھال کرنے والا فرد کرسکتا ہے لیکن کچھ ٹیسٹ کسی کلینک یا اسپتال میں صحت کارکن کے ذریعے ہونا ہی بہترین ہے۔

آپ اپنا معائنہ خود کرنے، مختلف بیماریوں کی علامات سیکھنے یا عام طور پر کلینکوں میں کئے جانے والے ٹیسٹوں کے بارے میں زیادہ معلومات کے لئے ایک عام ہیلتھ گائیڈ استعمال کرسکتی ہیں۔ آپ نیچے دی گئی کتابوں سے بھی معلومات حاصل کرسکتی ہیں۔

جہاں عورتوں کے لئے ڈاکٹر نہ ہو (WWHND - Where Women have no doctor)

درسی کتاب برائے مڈوائف (M.W - A book for Midwives)

جہاں کوئی ڈاکٹر نہ ہو (WTND - Where there is no doctor)

انگریزی زبان میں یہ کتابیں ہسپرین فاؤنڈیشن نے شائع کیں ہیں۔ پہلی دو کتابوں کے اُردو اور سندھی تراجم پاکستان نیشنل فورم آن ویمینز ہیلتھ نے شائع کئے ہیں۔ "جہاں عورتوں کے لئے ڈاکٹر نہ ہو" کا اُردو ترجمہ پاکستان نیشنل فورم آن ویمینز ہیلتھ نے شائع کردیا ہے۔

- پیٹ کا معائنہ ۔ درد اور غیر معمولی گومڑ/ اُبھار چیک کرنے کے لئے

  [WWHND - page 334]

- حمل کے دوران مسائل جاننے کے لئے

  [MW - page 109]

- نبض چیک کرنا۔ یہ یقین کرنے کے لئے رفتار مستحکم ہے [WTND - page 32-33]

- بلڈ پریشر [WWHND - page 462] اور ٹمپریچر [WTND - page 30]

- انیمیا کی علامات [WTND - page 124]

- ہیپاٹائٹس کی علامات [WTND - page 172]

- نظر کا ٹیسٹ (WTND - page 33]

معائنے/ٹیسٹ جو کلینک یا اسپتال میں ہوتے ہیں

- کینسر کے لئے رحم کے دہانے (سرویکس) کا Pap Test
- گنوریا اور کلے مائیڈیا کے لئے ٹیسٹ
- انیمیا کے لئے بلڈ ٹیسٹ
- سفلس کے لئے بلڈ ٹیسٹ
- ایچ آئی وی کے لئے بلڈ ٹیسٹ
- ہیپاٹائٹس اے، بی اور سی کے لئے بلڈ ٹیسٹ

- ملیریا کے لئے بلڈ ٹیسٹ (خصوصاً حاملہ عورتوں کے لئے اہم)
- ذیابیطس کے لئے یورین ٹیسٹ
- کیڑوں اور پیراسائٹس کے لئے اسٹول ٹیسٹ
- ٹی بی کے لئے بلغم ٹیسٹ
- حمل کے لئے یورین یا بلڈ ٹیسٹ
- مثانے یا گردے کے انفیکشن کے لئے یورین ٹیسٹ

تبدیلی کے لئے کام

معذور عورتیں کیا کر سکتی ہیں

اس کتاب اور دیگر صحت کی کتابوں سے معائنہ صحت کے بارے میں جس قدر زیادہ ممکن ہو سیکھیں۔

جب ہم کسی صحت کارکن سے ملیں تو ہم اس سے چھاتی اور پیڑو کے معائنے کے لئے کہیں۔ ہم گروپ کی صورت اکٹھا ہو کر عورتوں کے لئے ضروری معائنوں کے بارے میں جس قدر بھی زیادہ ممکن ہو معلومات حاصل کرنے کے لئے اس کتاب اور صحت کی دوسری کتابوں کا مطالعہ کر سکتے ہیں۔ پھر ہم مقامی صحت کارکنوں اور اسپتال یا کلینک چلانے والوں سے کہہ سکتے ہیں کہ وہ ان خدمات کو ہمارے لئے دستیاب بنائیں۔ ایک گروپ کے بطور ہم وزارت صحت پر زور دے سکتے ہیں کہ معذوری رکھنے والی عورتوں کے لئے یہ معائنے کس قدر اہم ہیں۔

وہ لوگ جنہوں نے پہلی بار صحت کے کتابچوں کو سمجھنا آسان بنایا

1997ء میں سیکھنے کی مشکلات سے دوچار بہت سی عورتوں نے ایک ویمینز گروپ 'پیوپل فرسٹ لیورپول' بنایا تا کہ وہ مل جل کر عورتوں کی صحت کی دیکھ بھال کے بارے میں زیادہ سے زیادہ سیکھیں۔ انہوں نے عورتوں کے ایک ہیلتھ کلینک کے ساتھ مل کر ایسے کئی کتابچے تخلیق کئے جو صحت کے معائنوں کو سمجھنا آسان بناتے ہیں، ان کتابچوں کے بارے میں جاننے کے لئے دیکھئے صفہ 381۔

جنسی صلاحیت اور باروری کے بارے میں جاننا

کرانتی اور سبالا بھارت کی غریب ترین آبادیوں میں سے ایک کی صحت کارکن ہیں، جہاں معذور اور غیر معذور بیش تر عورتیں کسی بھی قسم کی صحت کی سہولت حاصل نہیں کر سکتی ہیں۔

انہوں نے عورتوں کو سکھایا کہ وہ کس طرح اپنے جسموں کا جائزہ لیں اور انہیں سمجھیں۔ انہوں نے فرج سے غیر معمولی اخراج جیسے مسئلوں کا جائزہ لیا اور باروری اور جنسی صلاحیت پر آگہی کے لئے خیال بادلہ کیا۔ انہوں نے مسائل کے ایسے حل تلاش کرنے پر توجہ دی جن پر عورتیں خود عمل کر سکیں اور جن کے لئے بے تحاشہ وسائل کی ضرورت نہیں پڑتی ہے۔

گھر اور دیکھ بھال کرنے والے کیا کر سکتے ہیں؟

معذور عورتوں کے احباب اور گھر والے معذور عورتوں سے تبادلہ خیال کر کے معائنے کے سلسلے میں ان کی مدد کر سکتے ہیں۔ وہ بیان کرنا سیکھیں کہ معائنے کے دوران کیا ہوگا اور نتائج جاننا کیوں اہم ہے۔ وہ معائنوں اور ان طریقوں کے بارے میں جو صحت کارکن ان کی ضروریات کے مطابق معائنوں کے لئے اختیار کر سکتی ہیں، اپنی معلومات سے آگاہ کریں۔

> معائنہ صحت کے لئے میرے ساتھ جانے اور اگر میں کہوں تو میرے ساتھ رکنے کی پیشکش

معذور عورتوں کو معائنوں سے دور یا محروم رکھنے والی کلینکوں یا اسپتالوں کی ان رکاوٹوں کے بارے میں بات کیجئے اور یہ بھی کہ وہ معائنوں کو آسان بنانے کے لئے کیا کر سکتے ہیں۔ جیسے ہی آپ کی معذور بیٹی ایک لڑکی سے عورت میں تبدیل ہوتی ہے تو اسے بتائیں کہ وہ معائنوں سے خوفزدہ نہ ہو۔ آپ اور آپ کی بیٹی مل کر کلینکوں کو قابل رسائی، صحت کارکنوں کو تربیت یافتہ اور ذرائع آمد ورفت کو دستیاب بنانے کے لئے کام کر سکتے ہیں۔

صحت کارکن کیا کر سکتے ہیں

> پہلی بار کوئی معائنہ خصوصاً پیڑو کا معائنہ کراتے ہوئے بہت سی عورتیں پریشان یا خوفزدہ ہوتی ہیں

صحت کارکن کسی بھی معائنے سے پہلے عورت سے گفتگو کر کے ابتداء کر سکتے ہیں، وہ بتائیں کہ کیا ہونے والا ہے، وہ اس کے سوالات کے جواب دیں اور اس سے کہیں کہ وہ معائنے کے دوران بھی کوئی سوال کر سکتی ہے۔

معذور عورتوں کو یہ سمجھائیں کہ پیڑو اور چھاتی کے معائنوں سمیت صحت کے دیگر معائنے ان کے لئے کیوں اہم ہیں۔ آپ ان پر واضح کر سکتے ہیں کہ یہ معائنے تمام عورتوں کے لئے ضروری ہیں۔ وضاحت کیجئے کہ ایک معذور عورت بھی ان معائنوں کی سہولت حاصل کر سکتی ہے خواہ اس کے لئے اپنے بازوؤں اور پیروں کو حرکت دینا مشکل ہی کیوں نہ ہو۔ بتائیے کہ معذور عورتوں اور ان کے صحت کارکنوں نے ایسی بہت سی پوزیشن دریافت کر لی ہیں جو عورتیں ان معائنوں کے لئے اپنا سکتی ہیں۔ یاد رکھیے کہ ایک معذور عورت اپنے جسم کو دوسروں سے بہتر سمجھتی ہے لہٰذا اس سے کہیں کہ وہ آپ کو آگاہ کرے کہ وہ کس حد تک حرکت کر سکتی ہے اور یہ کہ کیا اسے مدد کے لئے کسی فرد کی ضرورت ہے۔

عورتوں کو عموماً سکھایا جاتا ہے کہ وہ اپنے جسموں کو نہ چھوئیں اور نہ ہی شکایت کریں۔ اس لئے بہت سی عورتیں چھاتیوں کا معائنہ کے دوران بے اطمینانی محسوس کرتی ہیں یا کسی کو یہ بتاتے ہوئے جھجکتی ہیں کہ وہ اپنے پیٹ میں غیر معمولی درد محسوس کر رہی ہیں۔ اکثر عورتیں جنس یا جنسی اعضاء کے بارے میں بات کرتے ہوئے گھبراتی ہیں لہٰذا ان کے لئے فرج سے رطوبت کے اخراج کے بارے میں کسی سے بات کرنا مشکل ہو سکتا ہے۔ صحت کارکن اپنی کمیونٹی میں ایسی عورتوں کی حوصلہ افزائی کر کے انہیں درپیش کسی بھی مسئلے کے بارے میں گفتگو کے لئے مائل کر کے مدد کر سکتے ہیں۔

معذور عورت سے ہمیشہ براہ راست بات کریں اور اس سے اس کے مسائل پوچھیں خواہ کمرے میں اس کی مدد کے لئے کوئی اور بھی موجود ہو۔ آپ اس سے اسی طرح بات کریں جس طرح دوسرے لوگوں سے کرتی ہیں چاہے اسے آپ سے بات کرتے ہوئے دشواری ہی کیوں نہ ہو۔

جب آپ کسی ایسی عورت کا معائنہ کریں جو نابینا ہو یا اچھی طرح دیکھ نہ سکتی ہو

نابینا عورت کے لئے کلینک جیسی کسی غیر مانوس جگہ جانا پریشان کن ہوسکتا ہے۔ وہ نہیں جانتی ہے کہ کس جگہ کیا چیز ہے، یا کہاں جانا ہے۔ بعض اوقات لوگ نابینا عورت سے ناشائستہ انداز سے پیش آتے ہیں۔ یہ کسی بھی طرح مناسب نہیں ہے۔

جب آپ کسی نابینا عورت کی رہنمائی کریں تو اس کا بازو یا ہاتھ نہ تھامیں۔ بہت سی نابینا عورتیں ''چھوکر'' دیکھنے کے لئے اپنے ہاتھوں پر انحصار کرتی ہیں اس کے بجائے اسے اپنا بازو پیش کریں اور اسے اپنا بازو پکڑنے دیں یا اس کا ہاتھ اپنے بازو پر رکھیں اسے بتائیں کہ کس جگہ کیا چیز ہے اور آپ اسے کہاں لے جا رہی ہیں۔ اس طرح وہ سیکھ لے گی کہ وہ اس جگہ کس طرح اپنے طور پر بہتر انداز سے پھر سکتی ہے۔ اس طرح وہ اپنے معائنے کے دوران زیادہ اطمینان محسوس کرے گی۔

جب آپ کسی ایسی عورت کا معائنہ کریں جو بہری ہو یا ٹھیک سے سن نہ سکتی ہو

بہری عورتوں کے لئے کسی کلینک جانا انتہائی پریشان کن ہوسکتا ہے جب وہاں کوئی علامتی زبان استعمال کرنے والا نہ ہو۔ بعض اوقات بہری عورت اپنے ساتھ کسی ایسے فرد کو لائے گی جو سن سکتا ہو اور علامتی زبان بھی جانتا ہو اور اس کے لئے ترجمانی کرسکتا ہو۔ اگر وہ ایسا کرتی ہے تو اس سے بات کرتے ہوئے یا جب

وہ بات کرے تو آپ ترجمان کے بجائے بہری عورت کی طرف دیکھیں۔ آپ یہ اس وقت بھی کریں جب آپ اس کے ترجمان کی بات سن رہے ہوں۔ آپ صرف بہری عورت کی طرف متوجہ رہیں۔ ترجمان وہاں مدد کرنے کے لئے آیا ہے جبکہ بہری عورت وہ شخصیت ہے جو آپ کے پاس اپنی صحت کی بہتری کے لئے آئی ہے۔

جب آپ کسی ایسی عورت کا معائنہ کریں جو سیکھنے یا سمجھنے میں دشواری رکھتی ہو

ایسی عورتیں جو سیکھنے یا سمجھنے میں دشواری محسوس کرتی ہیں انہیں بھی اپنی صحت کے بارے میں معلومات حاصل کرنے اور اپنے جسموں کے بارے میں فیصلے کرنے کے لئے مدد کی ضرورت ہے۔ ممکن ہے آپ کو سمجھنے کی دشواری رکھنے والی عورت کو سمجھانے میں زیادہ وقت لگے۔ بجائے اس کے کہ آپ اس سے پوچھیں کہ کیا اس کی سمجھ میں آ گیا، اس سے کہیں کہ وہ اپنے الفاظ میں بتائے کہ اس نے کیا سیکھا۔

میں پیڑو یا چھاتی کے معائنوں کے بارے میں معذور عورتوں کے سوالات کا انتظار کرتی تھی لیکن بیش تر معذور عورتیں یہ جان کر حیران ہوتی ہیں کہ وہ بھی یہ معائنے کرا سکتی ہیں یا وہ ان کے بارے میں بات کرتے ہوئے گھبراتی تھیں۔ اب میں ان سے یہ پوچھنا یقینی بناتی ہوں کہ کیا وہ ان معائنوں کے بارے میں جانتی ہیں جو تمام عورتوں کو کرانے چاہئیں۔

باب - 7

جنسی جذبہ/جنسیت

جنسی جذبہ زندگی کا ایک فطری حصہ ہے۔ بہت سی عورتوں کے لئے جنسی تعلق خوشی محسوس کرنے، اپنے شریک حیات کے لئے محبت اور طلب ظاہر کرنے یا مطلوبہ بچوں کے لئے حاملہ ہونے کا ایک طریقہ ہے۔ لیکن جنسی جذبہ جنسی تعلق سے زیادہ اہم ہے کیونکہ وہ طریقہ جس سے ایک عورت اپنے شریک حیات سے تعلق قائم کرتی ہے وہ انداز جس سے وہ خود اپنے بدن سے ہم آہنگی پیدا کرتی ہے اور ایک عورت کی حیثیت سے اپنے بارے میں سوچتی ہے، اس کی جنسیت کے جزو ہیں۔

معذور عورتیں بھی قریبی، محبت آمیز جنسی تعلق کا حق رکھتی ہیں یا ایسا چاہتی ہیں یا رکھ سکتی ہیں۔ بعض اوقات کمیونٹی کے رویئے معذور عورتوں کے اس حق پر اثر انداز ہوتے ہیں یا اسے محدود کرتے ہیں کہ اس کے نزدیک یہ کس طرح جنسی جذبے کو محسوس کرسکتی ہیں۔ اگر کسی کمیونٹی کے لوگ یہ سمجھتے ہیں اور اس سے اتفاق کرتے ہیں کہ معذور عورتیں بھی دوسرے تمام لوگوں کی طرح محبت، جنس اور خاندان کی ضرورت رکھتی ہیں تو پھر ایک معذور عورت

ہمیں حق ہے کہ ہم جیسے ہیں، اسی طرح ہمیں پسند کیا جائے، اس کے لئے ناپسند نہ کیا جائے جو ہم نہیں کرسکتے ہیں یا ہم کیسے نظر آتے ہیں

- اپنے جنسی جذبے کو اس طرح ظاہر کر سکتی ہے کہ اسے خوشی ملے۔
- اپنا ازدواجی شریک چن سکتی ہے۔
- انتخاب کرسکتی ہے کہ وہ کب حاملہ ہو۔
- جنسی طور پر منتقل ہونے والے انفیکشنز سے بچاؤ کرسکتی ہے۔
- جنسی تشدد بشمول جبری تعلق سے آزاد رہ سکتی ہے۔

معذور عورتوں کی جنسیت کے بارے میں نقصان دہ عقائد

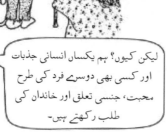

انہیں کوئی حق نہیں ہے

لیکن کیوں؟ ہم یکساں انسانی جذبات اور کسی بھی دوسرے فرد کی طرح محبت، جنسی تعلق اور خاندان کی طلب رکھتے ہیں۔

معذور عورتوں کی جنسیت یا جنسی جذبے کے بارے میں کچھ نقصان دہ عقائد نیچے بیان کئے گئے ہیں یہ منفی رویئے اور عقائد انہیں قریبی، محبت آمیز جنسی تعلقات سے لطف اندوز ہونے سے محروم کرتے ہیں۔ بعض اوقات غلط عقائد عورتوں کو ایسے وفادار شریک حیات تلاش کرنے سے روکتے ہیں جو ان کا احترام کریں اور انہیں اہمیت دیں۔

ضرر رساں عقیدہ: معذور عورتوں کے جسم شرمناک ہیں

بیش تر کمیونیٹیز میں عورتوں کو ان کی جسمانی کشش کے معیار کی مناسبت سے اہمیت دی جاتی ہے۔ اگر ایک عورت مروجہ معیار سے انتہائی مختلف دکھائی دیتی ہے تو لوگ اسے شادی کے قابل نہیں سمجھتے ہیں۔

گیتا ہمیں شادی کے لئے ایک مکمل ساتھی کے بارے میں بتاتی ہے

بھارت میں مسئلہ یہ ہے کہ ہم مکمل جسموں کے مالک مکمل لوگوں کو تلاش کرتے ہیں۔ جسمانی خدوخال کو اس حد تک اہمیت دی جاتی ہے کہ ایک ایسے فرد کے لئے کوئی گنجائش نہیں ہے جو ناقص / ناممکن ہو۔ ذرا ایک نظر ضرورت رشتہ کے اشتہاروں پر ڈالئے۔ تمام مرد ایسی عورت چاہتے ہیں جو دبلی پتلی، خوبصورت، صاف رنگ والی، تعلیم یافتہ اور اچھے خاندان سے تعلق رکھتی ہو۔ یہاں یہ عقیدہ بھی پایا جاتا ہے کہ اگر کوئی لڑکا کسی ایسے خاندان میں شادی کرتا ہے جہاں کوئی فرد کسی معذوری مثلاً پولیو کا شکار ہو تو اس کے بچے بھی اسی معذوری کے ساتھ جنم لیں گے۔ ایسا میرے ساتھ بھی ہوا لہٰذا میں یہ اچھی طرح جانتی ہوں، اس کا ایک انتہائی ٹھوس صنفی پہلو بھی ہے۔ یہ عورت ہے جس کے ہر اعتبار سے مکمل ہونے کی توقع کی جاتی ہے، دوسری طرف دولہا معذور، بدصورت یا مرگی کا مریض، نشے کا عادی ہی کیوں نہ ہو وہ پھر بھی مکمل دولہا ہے۔

ضرر رساں عقیدہ : معذور عورتیں جنسی جذبات نہیں رکھتی ہیں

دوسری تمام عورتوں کی طرح معذور عورتیں بھی جنسی جذبات رکھتی ہیں لیکن ان کے خاندان یا کمیونٹی کے افراد یہ سوچ سکتے ہیں کہ اسے جنسی جذبات نہیں رکھنا چاہئیں یا وہ جنسی جذبات نہیں رکھتی ہے۔

> میرا بہرہ پن مجھے جنسی احساسات سے محروم نہیں کرتا اسی طرح جیسے عینک پہننے سے آپ پیاس یا بھوک کے احساس سے بچ نہیں سکتے۔

بعض اوقات معذور عورت کے بارے میں یہ سمجھا جاتا ہے کہ وہ جنسی احساسات نہیں رکھتی ہے لہٰذا کوئی بھی اس پر اس لحاظ سے توجہ نہیں دیتا ہے اور اس سے اپنے بچوں کا طلب گار نہیں ہوتا ہے۔ اگر کوئی عورت معذوری کے ساتھ پیدا ہوئی ہو یا وہ نوعمری میں معذور ہوئی ہو اُسے یہ باور کرانے میں سخت محنت کا سامنا کرنا پڑتا ہے کہ وہ بھی عمر کی مناسبت سے جذبات رکھتی ہے۔ اگر کوئی نو جوان عورت اس وقت معذور ہوتی ہے جب وہ اپنے جنسی جذبے کو دریافت کرنا شروع کر رہی ہوتی ہے تو اس کا ذاتی تصور مجروح ہو سکتا ہے اور ممکن ہے کہ وہ یہ یقین ہی نہ کرے کہ وہ کبھی جنسی تعلق رکھ پائے گی۔

ایک بڑی عمر کی عورت بھی معذور ہونے کے بعد نشو و نما پانے اور اپنے جنسی جذبے کا تجربہ ہونے کے بعد بھی اس تبدیلی سے دو چار ہو سکتی ہے کہ وہ اپنے جسم کے بارے میں کس طرح محسوس کرتی ہے۔ وہ سوچ سکتی ہے کہ شاید اب وہ جنسی طور پر پرکشش نہیں رہی یا یہ محسوس کرے کہ اب اس کے لئے جنسی عمل مختلف ہوگا۔ ممکن ہے کہ وہ یہ تسلیم ہی نہ کرے کہ اب بھی وہ جنسی عمل سے لطف اندوز ہو سکتی ہے۔

ضرر رساں عقیدہ : معذور عورتیں ہر وقت جنسی عمل چاہتی ہیں

کچھ لوگ سمجھتے ہیں کہ معذور عورتیں، خصوصاً وہ عورتیں جو سیکھنے اور سمجھنے میں دشواری رکھتی ہیں ہمیشہ جنسی عمل کی خواہاں رہتی ہیں لیکن یہ خیال درست نہیں ہے۔

عورتوں کے کسی بھی گروپ کی طرح معذور عورتیں اس بارے میں مختلف جذبات رکھتی ہیں۔ کچھ عورتیں اکثر جنسی عمل چاہتی ہیں اور کچھ اسے بالکل بھی پسند نہیں کرتی ہیں۔ کسی بھی عورت کی طرح معذور عورت کو بھی جب وہ چاہے جنسی عمل کے انتخاب کا حق ہونا چاہئے اور ہر عورت کو جب وہ نہ چاہے جنسی عمل کے لئے انکار کا حق ہونا چاہئے۔

سیکھنے یا سمجھنے میں دشواری رکھنے والی لڑکیوں اور عورتوں کی مدد

بہت سی معذور لڑکیاں اور عورتیں خاطر خواہ توجہ پانے یا ضرورت کے مطابق از دواجی تعلق قائم کرنے کے مواقع نہیں رکھتی ہیں۔ اگر آپ گھر میں تنہا ہیں یا آپ کے گھر والے آپ کو نظر انداز کرتے ہیں تو آپ تنہائی کا شکار ہو سکتی ہیں اور ممکن ہے کسی ایسے دوست یا فرد کو تلاش کریں جو آپ پر توجہ دے۔ دوسرے لوگ اس طلب کو غلط سمجھ سکتے ہیں۔ یا ممکن ہے وہ اس لئے فائدہ اٹھانے کی کوشش کریں کہ آپ تنہا ہیں یا انہیں یقین ہو کہ آپ کو ضرر پہنچایا تو کوئی بھی پرواہ کرنے والا نہیں ہے۔

آپ حقیقتاً کیا چاہتی ہیں، یہ سمجھنا بھی مشکل ہو سکتا ہے خصوصاً اس صورت میں جب آپ نے جنس کے بارے میں کچھ نہ سیکھا ہو۔ آپ کے لئے اپنے جسم اور جنسی جذبے کے بارے میں سیکھنا اہم ہے۔ ایسا کوئی فرد جس پر آپ اعتماد کریں آپ کو تحفظ فراہم کرنے اور جنس کے حوالے سے اچھے فیصلے کرنے میں آپ کی مدد کر سکتا ہے۔

جنسی جذبے/جنسیت کے بارے میں سیکھنا

بہت سی معذور لڑکیاں پہلی بار جنس کے بارے میں اس وقت زیادہ جانتی ہیں جب کوئی ان کے ساتھ زیادتی کرتا ہے یا انہیں جنسی عمل کے لئے ورغلاتا ہے۔ صحت کارکن، گھر والے، دوست اور مددگار معذور لڑکیوں اور عورتوں کو جنسی جذبے یا جنسیت اور نسوانیت کے بارے میں گفتگو میں شامل کر سکتے ہیں۔ معذور لڑکیوں کو جنسی تعلیم کے پروگراموں میں شامل کریں اور انہیں جنسیت کے بارے میں سکھائی گئی باتیں لڑکیوں کو ان مردوں سے تحفظ فراہم کرنے میں معاون ثابت ہوں گی جو ان سے فائدہ اٹھانا چاہتے ہیں جنسی زیادتی پر مزید معلومات کے لئے دیکھیے باب 14۔

کچھ کمیونٹیز میں لڑکیوں کو عورت بننے کے بارے میں کمیونٹی کی کسی بزرگ عورت کے ذریعے سکھایا جاتا ہے، دوسری کمیونٹیز میں لڑکیاں جب عورت بنتی ہیں تو ان کے لئے خصوصی تقریبات کا اہتمام کیا جاتا ہے اور کچھ کمیونٹیز میں لڑکیاں ایک دوسرے کے تجربے سے آگاہ ہونے کے لئے جمع ہوتی ہیں۔ وہ جب ایک دوسرے سے بالوں اور کپڑوں کے نئے انداز، جنس مخالف کے دوستوں اور اپنے پسندیدہ پاپ گلوکاروں کے بارے میں باتیں کرتی ہیں تو بہت کچھ سیکھتی ہیں۔ ایسی محفلوں میں معذور لڑکیوں کو شامل کرنے کی ضرورت ہے۔

> مجھے جنسی تعلیم بالکل بھی نہیں ملی، میں اپنی ماں کو دوسری عورتوں سے باتیں کرتے ہوئے سنتی ہوں لیکن انہوں نے کبھی مجھ سے اس موضوع پر بات نہیں کی۔ انہیں قطعاً توقع نہیں ہے کہ میری شادی ہوگی لہٰذا وہ سمجھتی ہیں کہ مجھے کچھ جاننے کی ضرورت بھی نہیں ہے۔

> میری ماں نے مجھے کبھی تفصیل سے نہیں بتایا۔ اگر میں نے کبھی جاننا بھی چاہا تو انہوں نے مجھے مارنے کی دھمکی دی، میں نے زیادہ تر باتیں اپنی سہیلیوں سے سیکھی ہیں۔

> میں نے جنسی جذبے / جنسیت کے بارے میں دوسری عورتوں کی باتیں سن کر سیکھا۔ میری خواہش ہے کہ کوئی مجھ سے براہ راست اس موضوع پر بات کرے تاکہ میں اس سے کچھ پوچھ سکوں جو میں چاہتی ہوں۔

> میری آنٹی جانتی ہیں کہ میری شادی ہو سکتی ہے۔ انہوں نے دوسری لڑکیوں کی طرح مجھے عورت ہونے کے بارے میں سمجھایا۔

> معذور لڑکیوں اور عورتوں کو عورت بننے، جنسیت اور رشتوں کے بارے میں جاننے کی ضرورت ہے۔

چاہنے والا ایک ساتھی ڈھونڈنا

دوسری عورتوں کی طرح آپ بھی ایک ایسے ساتھی کی حقدار ہیں جو آپ کا احترام کرے اور آپ کی دیکھ بھال کرے۔ آپ ایک ایسے ساتھی کی حقدار ہیں جو آپ کی باتیں سنے اور آپ کے ساتھ اچھا برتاؤ کرے۔ آپ ایسا ساتھی پانے کا حق رکھتی ہیں جو آپ کی حدوں کے مطابق آپ کی مدد کرے۔ آپ ایک ایسے ساتھی کی حقدار ہیں جو آپ کی مدد، آپ کا اعتماد اور آپ کے پیار کا طلب گار ہو۔

بہت سی عورتیں جنہیں ایک چاہنے والا ساتھی ملا ہے، کہتی ہیں کہ انہوں نے معذوری کے بارے میں غلط خیالات پر یقین کرنا چھوڑ دیا ہے۔ انہوں نے مہارتیں حاصل کیں اور اپنے گھر انوں کے لئے معاونت کے طریقے تلاش کئے اور خود اپنا احترام کرنا سیکھا۔ جب آپ اپنی ذات اور اقدار کا احترام کرتی ہیں تو آپ اپنے لئے ایسا ساتھی بھی چاہتی ہیں جو آپ کا بھی احترام کرے۔ عزت نفس پر مزید معلومات کے لئے دیکھئے صفحہ 62 تا 65۔

حادثے سے دوچار ہونے کے بعد میری پشت پر کُب اُبھر گیا اور مجھے چلنے پھرنے میں مشکل پیش آنے لگی لیکن مجھے ایسا شخص مل گیا جو میری شخصیت اور حوصلے سے متاثر تھا۔ اس نے اس انداز کو پسند کیا کہ میں اپنی مشکلات کو ہنس کر جھیلتی ہوں۔ جب اس نے مجھ سے شادی کے لئے کہا تو میں رضامند ہو گئی۔ اب ہمارا ایک خوبصورت بچہ بھی ہے۔

اس سے پہلے کہ میرا ساتھی مجھ سے شادی کے لئے کہتا، ہم نے کھلے دل سے لاحق معذوری کے حوالے سے تبادلہ خیال اس نے مجھ سے کہا کہ میں جیسی بھی ہوں بشمول میری حدود کے، اسے قبول ہوں۔ انہوں نے وعدہ کیا کہ وہ کبھی مجھ پر شرمندہ نہیں ہوں۔ ہمارے درمیان شادی کی بنیاد اعتماد اور احترام ہے اور ہمارے دو بچے بھی ہیں۔

شادی سے پہلے میں اور میرے شوہر ایک ساتھ کام کرتے تھے، ہم نے ایک دوسرے کو پسند کیا، ہمارے والدین نے مل کر ہماری شادی کا اہتمام کیا۔ میری ساس کا سلوک میرے ساتھ بہت اچھا تھا لیکن میرے شوہر کے کچھ رشتہ دار مجھے قبول کرنا نہیں چاہتے تھے، میرے شوہر ایک سرکاری اہلکار ہیں اور وہ رشتہ دار سمجھتے تھے کہ وہ ایک معذور بیوی کو اپنے ہمراہ نہیں لے جاسکیں گے۔ ابتداء میں ان کے خیالات سے میرے جذبات مجروح ہوئے لیکن اپنے شوہر کی تائید کے ساتھ میں ان کے ہمراہ سفر کرنے لگی اور ہمارے درمیان ٹھوس رفاقت استوار ہو گئی۔

جب ایک معذور لڑکی روزگار کی مہارتیں اور تعلیم حاصل کرتی ہے تو وہ اپنے لئے ایک چاہنے والا ساتھی پسند کرتی ہے۔ روزگار اسے زیادہ مواقع اور عزت نفس دیتا ہے اور اس طرح دوسرے لوگوں کے لئے یہ باور کرنا آسان ہو جاتا ہے کہ ایک معذور عورت جو روزگار بھی رکھتی ہے اپنے گھر کی کفالت میں بھی ہاتھ بٹا سکتی ہے۔

بدکلامی کرنے والا شریک زندگی

بعض اوقات ایک عورت سوچتی ہے کہ وہ کسی سے بھی نبھا کرلے خواہ وہ اس سے زیادتیاں ہی کیوں نہ کرے یا اس کی دیکھ بھال نہ کرے یا اس کی اور گھر والوں کی کفالت بھی نہ کرے۔ بعض اوقات ایک عورت کسی ایسے فرد کو قبول کرلیتی ہے جواس کے گھر والوں کی کفالت کے لئے سرمایہ فراہم کرے۔ یا ایک عورت اپنی گزر بسر کے لئے جنس کی تجارت بھی کرسکتی ہے۔ کچھ صورتوں میں یہ عورت ہوتی ہے جو کام کرتی ہے اور مرد اس سے اس کا سرمایہ لے جاتا ہے۔ وہ اس سے کہتا ہے کہ اسے اس بات پر ممنون ہونا چاہیئے کہ وہ اس کا ساتھ رکھتی ہے۔

جب کسی عورت کو اس کے خاندان، کمیونٹی اور بذات خود اپنی نظر میں اہمیت حاصل ہوتی ہے تو وہ ایسے فرد کا ساتھ قبول کرتی ہے جو اس سے اچھا سلوک کرے۔ وہ ایسا ساتھی قبول نہیں کرے گی جو اسے زد وکوب کرے یا اس کے ساتھ بدکلامی کرے۔

طے شدہ شادیاں

بعض کمیونٹیز میں گھر والے اپنی بیٹیوں کی شادیاں کرتے ہیں۔ جب ایسا ہوتا ہے تو ایک معذور عورت کو دوسری عورتوں کے مقابلے میں کم اہمیت دی جاتی ہے۔ اس کے گھر والوں کے لئے ممکن ہے اس کی شادی کسی ایسے فرد سے کر دیں جو محض اسے قبول کرے خواہ وہ شخص اس کے ساتھ اچھا سلوک نہ کرے۔ ایک عورت ایسی شادی کے لئے خود بھی آمادہ ہوسکتی ہے کیونکہ وہ خود کو کوئی اہمیت نہ دیتی ہو یا سوچتی ہو کہ اس کے علاوہ کوئی بھی اس سے شادی نہیں کرے گا۔

بعض اوقات مرد لڑکی کے گھر والوں سے زیادہ جہیز یا دوسری ادائیگیاں چاہتے ہیں کیونکہ وہ کوئی معذوری رکھتی ہے۔ یا معذور عورت دوسری بیوی بنتی ہے لیکن اس سے غیر معذور بیوی جیسا بہتر سلوک نہیں کیا جاتا ہے۔ اگر کوئی مرد عورت کی معذوری کے عوض نقد سرمایہ یا تحائف طلب کرتا ہے تو یہ علامت ہے کہ وہ اس کی ایک عورت کی حیثیت سے عزت نہیں کرتا ہے۔ اکثر وہ دوسرے انداز سے اس سے ناروا سلوک رکھے گا یا بدکلامی کرے گا (دیکھئے باب 14)

نصف شب کے شوہر

کچھ ممالک میں مرد معذور عورت کے ساتھ جنسی تعلق کے لئے رشتہ جوڑتے ہیں اور صرف رات میں ہی ملتے ہیں اور پھر صبح ہونے سے پہلے اس وقت رخصت ہوجاتے ہیں جب اندھیرا پھیلا ہوا ہو۔ ایسے مردوں کو نصف شب کے شوہر کہا جاتا ہے۔ یہ عام طور پر اس وقت آنا چھوڑ دیتے ہیں جب عورت حاملہ ہوجائے، عام طور پر ایسے مرد بچے کے لئے بھی کسی قسم کی مالی مدد فراہم نہیں کرتے ہیں۔

جنس سے لطف اندوز ہونا

یہ فطری ہے کہ ایک عورت اپنے ساتھی کے ساتھ جنسی آسودگی میں شرکت چاہے لیکن بعض اوقات ممکن ہے کہ ایک عورت جنسی عمل سے خوشی محسوس نہ کرے۔ اس کی بہت سی وجوہات ہوسکتی ہیں۔ اگر اس کا ساتھی ایک ناسمجھ مرد ہے تو ممکن ہے کہ وہ یہ تسلیم نہ کرتا ہو کہ ایک عورت کا جسم جنسی لمس پر ایک مرد کے جسم سے مختلف ردعمل کا اظہار کرتا ہے۔ ممکن ہے کہ عورت کو یہ سکھایا گیا ہو کہ عورت کو مرد کے مقابلے میں جنسی عمل سے کم تر لطف اندوز ہونا چاہئے یا یہ کہ وہ اپنے ساتھی سے یہ نہ کہے کہ وہ کیا پسند کرتی ہے۔

ایک عورت کی معذوری اس کے لئے خوشی یا لطف محسوس کرنا مشکل بنا سکتی ہے۔ اسے جنسی عمل کے لئے ایسے مختلف طریقہ تلاش کرنے کی ضرورت ہوسکتی ہے کہ وہ اچھا محسوس کرے۔ خصوصاً اس وقت جب وہ حال ہی میں معذور ہوئی ہو۔ جس طرح وہ اپنی ذاتی صحت وصفائی کے لئے ذمہ داریاں سنبھالتی ہے اسی طرح وہ جنسی عمل کے لئے بھی وہ طریقے تلاش کرے جو اس کی معذوری کے ساتھ اس کے لئے خوشی کا سبب بنیں۔

جنسی لطف پر جسم کس طرح ردعمل کرتا ہے

عورتیں اور مرد دونوں جنسی جذبہ رکھتے ہیں لیکن ان کے جسم جنسی خیالات اور لمس پر مختلف انداز سے ردعمل کرتے ہیں۔ جب مرد اور عورتیں جنسی خیالات رکھتے ہوں یا ایک دوسرے کو جنسی انداز سے چھوئیں تو وہ سنسنی (ہیجان) محسوس کرتے ہیں۔ مزید تصورات اور لمس جسم کو زیادہ ہیجان انگیز بناتے ہیں۔

مرد میں جنسی ہیجان محسوس کیا جاسکتا ہے لیکن عورت میں یہ ہیجان دیکھنا زیادہ مشکل ہے۔ اس کی بظر (cliloris) سخت ہوجاتی ہے اور سوج سکتی ہے۔ اس کے Labia اور فرج کی دیواریں حد درجہ حساس ہوجاتی ہیں۔ فرج شفاف اور چکنے سیال سے گیلی ہوجاتی ہے۔ اس کی بھٹنیاں (Nipples) بھی حساس ہوسکتی ہیں اگر جنسی لمس اور خیالات جاری رہیں تو جنسی تناؤ اس حد تک بڑھ جاتا ہے کہ وہ لطف کے عروج پر پہنچ جاتی ہے اس کی انتہا (Climax) پر پہنچ جاتی ہے۔ جب ایسا ہوتا ہے تو اس کے جسم کی توانائی اور تناؤ انتہا پر ہوتا ہے جو خارج ہوجاتا ہے تو وہ خود کو پرسکون اور انتہائی پُرلطف محسوس کرتی ہے۔

جب ایک مرد اپنے لطف سے عروج پر پہنچتا ہے تو اس کا عضو تناسل مادہ حیات (Semen) خارج کرتا ہے جس میں منی (Sperm) شامل ہوتی ہے۔ انتہا پر پہنچنے کے بعد اس کا جسم پرسکون ہوجاتا ہے اور اس کا عضو تناسل دوبارہ نرم ہوجاتا ہے۔

بظر کو چھونا ایک عورت کے عروج پر پہنچنے کا انتہائی عام ذریعہ ہے ایسا اس کی فرج کو اندرونی طور پر رگڑنے، عضو تناسل یا انگلیوں سے چھونے پر بھی ہوسکتا ہے۔ اس کی مقعد بھی چھونے کے لئے حساس ہوسکتی ہے۔ عموماً ایک عورت مرد کی بہ نسبت عروج یا انتہا پر پہنچنے کے لئے زیادہ وقت لیتی ہے۔

بیرونی حصے میں ایک عورت کے تولیدی حصے

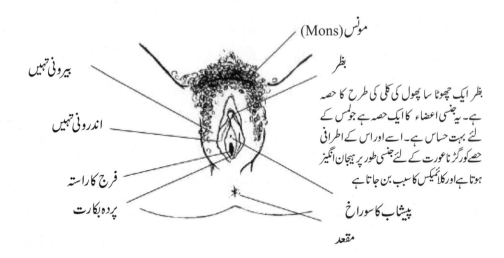

بیشتر عورتوں کے لئے جنسی ہیجان کی انتہا تک پہنچنا ممکن ہے لیکن ہر عورت کا جسم اور تجربہ مختلف ہوتا ہے۔ وہ عورتیں جو اپنے نچلے دھڑ میں کمتر محسوسات رکھتی ہیں، ممکن ہے انہیں عروج پر پہنچنے کے لئے بظر پر زیادہ دباؤ اور سختی سے چھوئے جانے اور بعض اوقات اس کے ساتھ فرج کے اندر دباؤ کی ضرورت ہو۔ ریڑھ کی ہڈی کی خرابی اور اپنے جسموں میں احساسات نہ رکھنے والی عورتیں بھی عروج پر پہنچ سکتی ہیں ممکن ہے وہ مختلف طور پر محسوس کرتی ہوں۔

مختلف عورتیں اپنے جسموں میں مختلف مقامات پر چھوا جانا پسند کرتی ہیں اور لطف اندوز ہوتی ہیں۔ زیادہ تر لوگ سمجھتے ہیں کہ تولیدی اعضاء یا چھاتیاں عورت کے جسم کے جنسی اعضاء ہوتے ہیں لیکن ایک عورت کے جسم کے ہاتھ، گردن، چہرہ اور پیٹ بھی اس کے جسم کے حساس حصے ہوتے ہیں جو چھوئے جانے پر وہ خوشی اور لطف محسوس کرتی ہے۔ وہ عورتیں جو مفلوج ہوں یا ریڑھ کی ہڈی کی چوٹ رکھتی ہوں یہ دوسرے حصے بھی حساس ہو سکتے ہیں کہ وہ انہیں چھونے پر عروج جیسی خوشی محسوس کر سکتی ہیں۔

یہ تمام حواس، لمس، بو، سماعت، ذائقہ اور بصارت اہم اور جنسی لطف محسوس کرنے کے اطمینان بخش ذرائع ہو سکتے ہیں۔ ہر فرد، مرد ہو یا عورت، معذوری کے بغیر ہو یا معذوری کے ساتھ، اپنے اپنے طور پر ان حواس سے لطف اندوز ہوتے ہیں۔ ایک نابینا عورت کے لئے لمس، بو یا سننا انتہائی اہم ہو سکتا ہے۔ ایک بہری عورت کے لئے چھوا جانا اور دیکھنا جنسی عمل کا سب سے زیادہ اطمینان بخش حصہ ہو سکتا ہے۔

ایک عورت اپنے جسم کے حساس حصوں کے بارے میں سیکھ سکتی ہے اور یہ کہ وہ کس طرح خود کو چھو کر عروج پر پہنچ سکتی ہے یا اپنے ساتھی کو آگاہ کر سکتی ہے کہ اسے کیا اچھا محسوس ہوتا ہے۔ یہ خاص طور پر ایسی عورت کے لئے مددگار ثابت ہوگا جو حال ہی میں معذور ہوئی ہو۔ اس سے اس کے شوہر کو یہ معلوم ہو سکتا ہے کہ وہ کس طرح اسے خوش کرے۔

جنسی عمل کے مختلف طریقے

مختلف طریقے ہیں جن کے مطابق لوگ جنسی عمل کرتے ہیں اور اس سے لطف اندوز ہوتے ہیں۔ عموماً لوگ جنسی عمل کو ایک مرد اور عورت کے درمیان تعلق کو سمجھتے ہیں جس میں مرد کے تناسلی عضو کا عورت کی فرج میں دخول ہوتا ہے۔ لیکن جنسی عمل کے اور بہت سے طریقے ہیں جو لوگ اپناتے ہیں اور دوسرے فرد سے اپنی محبت کا اظہار کرتے ہیں۔ بوسہ لینا، لپٹانا اور گفتگو بھی جنسی عمل کے جز ہیں کسی کسی فرد کے چہرے، ہاتھوں، پشت اور گردن کو چھونا بھی جنسی عمل کے اچھے طریقے ہیں اور اورل سیکس بہت سے لوگوں کے لئے پر لطف ہو سکتا ہے ایک دوسرے کے تولیدی اعضا کو چھونا اور مسلنا (باہمی مشت زنی) بھی جنسی لطف حاصل کرنے اور فراہم کرنے کا ایک اور طریقہ ہو سکتا ہے۔

لطف اندوز ہونے کے لئے کسی کا خود کو چھونا (خود لذتی)

آپ خود کو اس طرح چھو سکتی ہیں کہ اس سے آپ کو جنسی لذت ملے۔ یہ اپنے جسم کے بارے میں اور یہ جاننے کا ایک اچھا طریقہ ہے کہ آپ کس قسم کا جنسی لمس بہترین محسوس کرتی ہیں۔ یہ آپ کے لئے اپنے خود بے جنسی جذبے، حیثیت کے بارے میں زیادہ پراعتماد اور بہتر محسوس کرنے میں مددگار ہو سکتا ہے۔ بہت سی کمیونٹیز کے خیال میں آپ کو اپنے خود کو چھونا غلط یا نقصان دہ ہوتا ہے لہٰذا بعض اوقات لوگ ایسا کرتے ہوئے ندامت محسوس کرتے ہیں لیکن خود کو چھونا نقصان کا سبب نہیں بنتا ہے۔ یہ لطف اندوز ہونے یا اپنی خواہش کی تسکین کا ایک اچھا ذریعہ ہو سکتا ہے۔

از دواجی تعلق اور جنسی عمل

بیش تر عورتیں کسی ایسے فرد کے ساتھ قریبی اور پیار بھرا تعلق چاہتی ہیں جو ان کا خیال رکھے۔ یہ تعلق کسی اور معذور فرد یا غیر معذور فرد کے ساتھ ہو سکتا ہے۔ کچھ معذور عورتیں شادی شدہ ہوتی ہیں اور کچھ نہیں۔ بہت سی معذور عورتیں مرد کے ساتھ اور کچھ دوسری عورتوں کے ساتھ تعلق رکھتی ہیں کچھ مائیں ہیں اور کچھ نہیں ہیں۔ مغد وریاں رکھنے والی عورتوں اور ان کے پارٹنرز کے لئے عموماً جنسی عمل اس صورت میں زیادہ پُر لطف ہو سکتا ہے جب وہ تجربہ کرنے اور نئے طریقے تلاش کرنے کے خواہاں ہوں جو آپ کرنے نہیں سکتی ہیں۔ جو کرنا آپ کے لئے مشکل ہو، اس پر توجہ دینے کے بجائے ان باتوں پر توجہ دیں جو آپ اور آپ کے از دواجی ساتھی کے لئے خوشی اور لطف کا سبب بنے۔

اپنے ساتھی سے گفتگو

بہت سی عورتیں اپنی خواہشات یا احساسات کے بارے میں گفتگو کرتے ہوئے شرمندگی محسوس کرتی ہیں لیکن طلب محسوس کرنا اور ایسا تعلق قائم کرنا جو آپ کو خوش کرے ایک فطری عمل ہے اگر آپ از دواجی زندگی گزارنے کی سہولت رکھتی ہیں تو وقت سے پہلے محفوظ جنسی تعلق وغیرہ جیسے موضوعات (دیکھئے صفحہ 180 تا 182) اور دیگر ضروریات اور توقعات پر بات کرنا معاون ثابت ہوگا۔

اپنی نقل و حرکت کی حد اور ان طریقوں کے بارے میں جن سے آپ کا جسم جنسی آمادگی کے ردِعمل کا اظہار کر سکتا ہے، اپنے شریک حیات سے بات کیجئے۔ بعض اوقات دوسرا فریق یہ سوچ کر فکر مند ہوتا ہے کہ جنسی عمل عورت کی معذوری کے باعث اسے مجروح کر سکتا ہے یا خطرناک ہو سکتا ہے یہ سوچ جنسی خواہش کے خاتمہ کا سبب بن سکتی ہے۔ جب ہر فریق ایک دوسرے کے بارے میں بہتر طور پر جانتا ہوتو دونوں فریق زیادہ لطف اندوز ہو سکتے ہیں۔ ہر فرد کی طلب اور خواہشات مختلف ہیں لہٰذا یہ جاننے کا کہ دوسرے فرد کو کیا پسند ہے، بہترین طریقہ آپس میں گفتگو اور تجربات ہیں۔ گفتگو کے لئے موزوں باتیں مندرجہ ذیل ہیں۔

- جنسی عمل کہاں سہل ہوگا مثلاً بیڈ پر، وہیل چیئر میں یا کسی کرسی پر یا فرش پر۔
- کون سی پوزیشن تکلیف دہ ہے اور کونسی زیادہ آرام دہ ہو سکتی ہے۔
- آپ کی معذوری آپ کی جسمانی کارکردگی پر کس طرح اثر انداز ہوتی ہے۔
- آپ کس طرح ایک دوسرے کو خوشی دے سکتے ہیں۔
- اگر آپ جلد تھک جاتی ہیں تو دن یا ہفتے کا کونسا وقت ایسا ہے جب آپ سب سے زیادہ توانائی رکھتی ہیں۔

کیا اچھا محسوس ہوتا اور کیا اچھا محسوس نہیں ہوتا ہے اس بارے میں بات کیجئے۔ اگر آپ کا ساتھی آپ کی دیکھ بھال کرنے والا بھی ہے تو اس سے آپ دیکھ بھال کے دوران گزرے وقت اور از دواجی شریک کی حیثیت سے گزرے ہوئے دورانیے کے مابین محسوس کئے جانے والے فرق پر بات کیجئے۔

اہم بات: ہر عورت کے لئے ضروری ہے کہ جنسی طور پر منتقل ہونے والے انفیکشنز بشمول ایچ آئی وی سے بچاؤ کے لئے جنسی عمل کے محفوظ تر طریقے اپنائے۔ خصوصاً متاثرہ علاقوں میں جنسی عمل سے پہلے اپنے شوہر کے ساتھ جا کر ایچ آئی وی ٹیسٹ کرانا بھی اہم ہے (مزید معلومات کے لئے دیکھئے صفحہ 172)۔ اگر آپ جنسی تعلق کے ساتھ حاملہ نہیں ہونا چاہتی ہیں تو فیملی پلاننگ کے بارے میں معلومات کے لئے باب 9 اور ایمر جنسی انسدادِ حمل کے لئے صفحہ 105 دیکھئے۔

ان عورتوں کے لئے جنسی جذبہ/جنسیت جو معذور ہو جاتی ہیں

کسی حادثے یا بیماری کے باعث معذور ہونے والی عورت کے جنسی احساسات میں تبدیلی آسکتی ہے۔ کچھ عورتیں کمتر جنسی احساسات رکھتی ہیں یا کچھ عرصے جنسی عمل میں دلچسپی نہیں رکھتی ہیں۔ بعض اوقات عورتیں سوچتی ہیں کہ وہ اب جنسی زندگی سے لطف اندوز ہونے کے قابل نہیں رہی ہیں۔ ایسی ہر عورت کو ایسی معلومات کی ضرورت ہے جس سے اسے پتہ چلے کہ اس کی معذوری اس کی جنسیت کو کس طرح متاثر کرتی ہے۔ اگر اس کا کوئی ساتھی ہے تو دونوں کے لئے معذوری کے باعث جنسیت کس طرح متاثر ہو سکتی ہے۔ انہیں اس بارے میں معلوم ہونا چاہئے۔

مجھے امید ہے کہ ہیم کیری مجھ سے شادی برقرار رکھے گا صحت کا رکن کا کہنا ہے کہ میں اب بھی جنسی تعلق قائم کر سکتی ہوں اور بچے پیدا کر سکتی ہوں۔

مجھے بہت دکھ ہے کہ سوک چھم بارودی سرنگ کی زد میں آکر اس طرح شدید زخمی ہو گئی لیکن میرے گھر والے چاہتے ہیں میں دوسری شادی کرلوں۔ وہ کہتے ہیں کہ اب یہ عورت بدقسمتی کا سبب بنے گی اور بچے بھی پیدا نہیں کر سکے گی۔

اگر آپ معذوری سے قبل از دواجی تعلق رکھتی تھیں تو ممکن ہے کہ آپ دوبارہ جنسی تعلق قائم کرنے سے خوفزدہ ہوں۔ آپ فکر مند ہو سکتی ہیں کہ آپ کا شوہر ممکن ہے معذوری کے بعد آپ کو جنسی طور پر پرکشش نہ پائے یا آپ اور آپ کا شوہر یہ سوچیں کہ آپ دونوں اب ایک دوسرے کو مطمئن کرنے کے قابل نہیں رہے۔ آپ دونوں کے لئے بہتر ہوگا آپ اپنے احساسات اور ان میں تبدیلیوں کے حوالے سے گفت و شنید کریں جن کی آپ کو ضرورت پڑ سکتی ہے۔ آپ کے جنسی عمل کا انداز مختلف ہو سکتا ہے لیکن دوسرے بیش تر جوڑوں کی طرح آپ ایک دوسرے کو خوش اور مطمئن رکھنے کے طریقے ڈھونڈ سکتے ہیں اگر آپ معذوری سے پہلے قابل بھروسہ تعلق اور اچھا ربط رکھتی تھیں تو یہ بات انتہائی ممکن اور درست ہے۔

تخلیہ (Privacy)

جنسی عمل کے لئے مناسب وقت اور جگہ پانا معذور عورتوں کے لئے دشوار ہوسکتا ہے خصوصاً اس صورت میں جب انہیں اس کی تیاری کے لئے کسی اور کی مدد کی ضرورت ہو۔ یہ ان عورتوں کے لئے مشکل مرحلہ ہوسکتا ہے جو اپنے خاندان کے ساتھ رہتی ہوں۔

اس مسئلے کا کوئی بھی آسان حل نہیں ہے خصوصاً اس وقت جب عورت کے گھر والے یا دیکھ بھال کرنے والے یہ نہیں سوچتے ہوں کہ جنسی تعلق اس کی بھی ضرورت ہے بعض اوقات ان لوگوں سے بات کرنا معاون ہوتا ہے جو آپ کی مدد کر رہے ہوں یا آپ کسی ایسے قابلِ بھروسہ فرد سے بات کرسکتی ہیں جو آپ کے گھر والوں یا مدد کرنے والوں سے بات کرسکے۔ دوسری معذور عورتوں سے بات کرنا اور ان کے تجربات سے آگاہ ہونا بھی معاون ثابت ہوتا ہے۔ کچھ لوگ خود اپنے طریقے طے کر لیتے ہیں مثلاً ایک عورت اپنی وہیل چیئر پر رہ کر بھی جنسی عمل کرسکتی ہے اس صورت میں اسے وہیل چیئر سے اتر کے بیڈ تک جانے کے لئے کسی کی مدد کی ضرورت نہیں ہوتی ہے اور کچھ لوگ اپنے مدد گاروں کو اپنی ضروریات کے لئے حد درجہ حساس پاتے ہیں۔

آرام دہ پوزیشن تلاش کرنا

اگر آپ کی نقل و حرکت محدود ہے تو ممکن ہے آپ کو جنسی عمل کے لئے آرام دہ پوزیشن معلوم کرنے کے لئے تجربات کی ضرورت ہو۔ سیریبرل پالسی، کمزور عضلات، سخت اکڑے عضلات یا آرتھرائٹس یا ایسی عورتیں جن کی معذوری درد یا کمزوری کا سبب بنتی ہو انہیں آرام دہ پوزیشن کے لئے اپنے ساتھی کی مدد کی ضرورت پڑسکتی ہے۔ آپ کی ٹانگوں یا کولہوں کو سہارا دینے کے لئے تکیوں یا گولائی میں لپٹے کپڑے کی ضرورت پڑسکتی ہے اگر شوہر کا وزن درد یا تکلیف کا سبب بن رہا ہو تو کوئی اور پوزیشن آزمائیں جس میں دونوں فریق کروٹ کے بل لیٹے ہوں یا کرسی پر بیٹھے ہوں۔

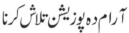

اگر آپ مفلوج ہیں تو یقیناً آپ دونوں جانتے ہوں گے کہ آپ کس قدر حرکت کرسکتی ہیں یا کونسی جسمانی پوزیشن آپ کے لئے ممکن ہیں۔ پیڑوں کے معائنے کی پوزیشنیں جنسی عمل کے لئے استعمال کی جاسکتی ہیں۔
دیکھئے صفحہ 137۔

جنسی عمل کے دوران ممکنہ مسائل

مباشرت کے دوران مدد

جنسی عمل کو نقصان دہ نہیں ہونا چاہئے بعض اوقات درد عورت کی معذوری کے باعث ہوتا ہے لیکن عموماً اس کے دوسرے اسباب بھی ہوتے ہیں اور عام طور پر ایسے طریقے ہوسکتے ہیں کہ جن پر عورت اور اس کا شوہر عمل کریں تو عورت کم درد محسوس کرسکتی ہے بعض اوقات ایک عورت اس وقت درد محسوس کرتی ہے جب مرد کا عضو تناسل، انگلی یا ہاتھ عورت کی فرج میں داخل ہوتا ہے ایسا اس وقت ہوتا ہے جب

- جب عورت کا ساتھی عجلت کرتا ہے اور عورت کے پرسکون ہونے اور اس کی فرج کے گیلے ہونے کا انتظار نہیں کرتا ہے۔
- عورت غلطی یا پشیمانی محسوس کرتی ہے یا جنسی عمل نہیں چاہتی ہے۔
- اس کی فرج یا نچلے پیٹ میں انفیکشن یا کوئی افزائش ہو۔
- عورت کی ختنہ کی گئی ہوں یا اس کے جنسی اعضا کوسی دیا گیا ہو۔ کچھ عورتوں کے عضلات سخت یا اکڑے ہوئے ہوتے ہیں اور کچھ عورتیں جن کی ریڑھ کی ہڈی مجروح ہوتو ان کے لئے اپنی فرج میں کسی چیز کا دخول تکلیف دہ ہوسکتا ہے اگر ایسا ہو تو جنسی عمل کے لئے ایسا کوئی اور طریقہ ڈھونڈا جائے جو دونوں فریقوں کے لئے خوشی اور طمانیت کا باعث بنے۔

خشک فرج

جنسی عمل تکلیف دہ ہوسکتا ہے اگر فرج بہت زیادہ خشک ہو۔ جب ایک عورت جنسی طور پر ہیجان میں ہوتو عموماً اس کی فرج گیلی ہوجاتی ہے۔ یہ فطری ہے لیکن بعض اوقات عورت کی معذوری اس کی فرج کا گیلا ہونا کم بناتی ہے ایسا بہت سی عورتوں کے ساتھ ہوسکتا ہے لیکن خاص طور پر Rheumatoid آرتھرائٹس اور رر یڑھ کی ہڈی مجروح ہوتو ایسا ہونا عین ممکن ہے۔

فرج کو گیلا کرنے کا ایک عام طریقہ یہ ہے کہ تیاری میں زیادہ وقت صرف کیا جائے تا کہ جسم اپنے طور پر خود مطلوبہ رطوبت بنائے آپ فرج کو چکنا بنانے کے لئے کوئی چکناہٹ استعمال کرسکتی ہیں تا کہ کھال پر خراشیں وغیرہ نہ پڑیں۔

اہم بات: اگر جنسی عمل میں لیٹکس کنڈوم استعمال کیا جائے تو فرج کو چکنا کرنے کے لئے تیل، پیٹرولیم جیلی، ویسلین، مکھن، معدنی تیل یا لوشن وغیرہ استعمال نہ کیا جائے اس سے کنڈوم پھٹ سکتا ہے کنڈوم کے ساتھ پانی میں بنی چکنائی جیسے کے وائی جیلی استعمال کریں۔

کچھ علاقوں میں لوگ جنسی عمل کو اس وقت ترجیح دیتے ہیں جب فرج بالکل خشک ہو لہٰذا ان علاقوں کی بہت سی عورتیں جنسی عمل سے پہلے اپنی فرج میں جڑی بوٹیاں یا پاؤڈر رکھ لیتی ہیں یا انہیں دھولیتی ہیں لیکن جب فرج خشک ہوتو یہ جنسی عمل کے دوران خراش زدہ ہوسکتی ہے جس سے عورت ایچ آئی وی یا دیگر STIs سے متاثر ہونے کا زیادہ امکان رکھتی ہے۔

تکلیف دہ عضلات اور جوڑ

بعض اوقات کوئی معذوری جیسے آرتھرائٹس، عورت کے لئے جب وہ معمول سے زیادہ چلے پھرے کا سبب بن سکتی ہے۔ اگر آپ کے ساتھ ایسا ہو تو حرارت درد میں کمی لاسکتی ہے۔ جنسی عمل سے پہلے گرم پانی میں بھیگا ہوا کپڑا درد والے یا سوجے ہوئے جوڑوں پر رکھیں یا گرم پانی سے غسل کریں اس سے آپ کے جسم کو آرام ملے گا اور آپ جنسی عمل سے زیادہ بہتر طور پر لطف اندوز ہوسکتی ہیں۔ اگر آپ درد کے لئے دوا استعمال کرتی ہیں تو اس روز دن میں وہ دوا استعمال کریں اس سے آپ کو جنسی عمل کے لئے تیار ہونے میں مدد ملے گی۔

عضلاتی اینٹھن

سیریبرل پالسی یا فالج کے باعث معذور عورت جنسی طور پر ہیجان کا شکار ہوتی ہے تو اس کے عضلات اچانک ہی سخت ہوسکتے ہیں اگر زیادہ عرصہ نہ رہے تو عضلاتی اینٹھن خطرناک یا نقصان دہ نہیں ہے اس باعث جنسی عمل روکنے کی ضرورت نہیں ہے بعض اوقات سخت ہونے یا اینٹھنے والے حصے کو نرمی سے دبانے سے آرام آسکتا ہے بعض اوقات عضلے کو نرمی سے اسٹریچ کرنے سے بھی آرام آسکتا ہے لیکن عضلے کو سختی سے کھینچیں نہیں یا اینٹھن کو روکنے کی کوشش نہ کریں اگر آپ عضلاتی اینٹھن کے لئے دوا استعمال کرتی ہیں تو مباشرت سے پہلے دوا لینے سے مدد مل سکتی ہے۔

مثانے اور معدے کا دھیان

مباشرت سے پہلے پیشاب اور پاخانہ کرلیں۔ اگر آپ پیشاب کے اخراج کے لئے کیتھڑ بیگ والی لیگ بیگ استعمال کرتی ہوں تو اسے خالی کرلیں۔ مباشرت کے دوران اتفاقی طور پر پیشاب یا پاخانے کے اخراج سے بچنے کے لئے مباشرت سے کچھ پہلے کھانے پینے سے گریز کریں اگر یورین کیتھڑ مستقل لگا رہتا ہے تو اسے مباشرت کے دوران احتیاط سے ٹیپ سے چپکا کر یا کسی اور ذریعے سے الگ رکھا جاسکتا ہے۔ خیال رکھیں کہ اس دوران اس کی ٹیوب مڑے یا خم نہ کھائے انفیکشن سے بچنے کے لئے انتہائی احتیاط سے کام لیا جائے (دیکھئے صفہ 102 تا 104)۔ یہ بہتر ہے کہ مباشرت دوران کیتھڑ کو لگا ہی رہنے دیا جائے۔ اگر کوئی عورت جنسی عمل کے دوران کیتھڑ نکال دیتی ہے تو اس بات کا امکان ہے کہ اس دوران اس کا پیشاب خارج ہوجائے کیونکہ کیتھڑ استعمال کرنے کے باعث اس کا مثانہ پیشاب کو روکنے کا عادی نہیں رہتا ہے لہٰذا اس کا پیشاب خارج ہوجائے گا اس کے ساتھ ہی اس دوران فضلہ خارج ہونے کا امکان بھی ہے اگر ایسا ہوجائے تو اسے صاف کرنے کے لئے صاف ستھرا کپڑا بھی قریب ہی رکھیئے۔ وقت سے پہلے اپنے شوہر سے ممکنہ مسائل پر بات کرلینا معاون ثابت ہوتا ہے یہ ایک مشکل کام ہے اور ہر عورت اس حوالے سے مختلف انداز سے بات کرے گی۔ کچھ عورتیں اسے زندگی کے دوسرے معمولات کی طرح لیتی ہیں کچھ عورتیں مزاح سے کام لیتی ہیں اور اس حوالے سے ہنسی کا موقع ڈھونڈ لیتی ہیں۔

اہم بات : کیتھڑ کے ساتھ رگڑ کھانے پر کنڈوم پھٹ یا ٹوٹ سکتا ہے۔ اس سے بچنے کے لئے کنڈوم کی بیرونی سطح پر پانی میں بنی چکنی جیلی استعمال کریں یا یہ فرج کے اندر لگائیں۔

اہم بات: جنسی عمل کے دوران ریڑھ کی ہڈی کی خرابی رکھنے والی کچھ عورتوں کا بلڈ پریشر تیز سر درد کے ساتھ اچانک بڑھ سکتا ہے ان کی جلد سرخ ہوسکتی ہے یا دل کی دھڑکن تیز ہوسکتی ہے۔ یہ صورت dysreflexia کہلاتی ہے اور ایک سنگین مسئلہ صحت ثابت ہوسکتی ہے اس سے بچنے کے لئے یقینی بنائیں کہ جنسی عمل سے پہلے آپ پیشاب اور پاخانہ کرلیں۔ اس ضمن میں مزید معلومات کے لئے دیکھئے صفحہ 117 تا 119۔

حد سے زیادہ تھکن محسوس کرنا

ممکن ہے کہ بعض اوقات آپ اپنی معذوری کی وجہ سے دن کا بیش تر حصہ خود کو بے حد تھکا تھکا محسوس کریں یا ممکن ہے کہ آپ ایسی دوائیں لے رہی ہوں جو آپ میں تھکن کا احساس پیدا کرتی ہوں۔ یہ تھکن آپ میں جنسی عمل کے لئے کمتر دلچسپی کا سبب بن سکتی ہے۔

جنسی عمل کے لئے دن کے ایسے وقت کا انتخاب کیجئے جب آپ خود کو سب سے کم تھکا ہوا محسوس کرتی ہوں اگر یہ عملاً ممکن نہ ہو تو اپنے صحت کار کن سے پوچھئے کہ کیا دوا دن میں کسی اور وقت لینا آپ کے لئے مناسب اور محفوظ ہے۔ اگر آپ جلدی تھک جاتی ہیں یا آپ کے عضلات بہت مضبوط نہیں ہیں تو آپ کے لئے بہتر یہ ہوسکتا ہے کہ جنسی عمل آہستگی کے ساتھ ہو کیونکہ اس طرح نسبتاً کم توانائی صرف ہوگی۔ اگر آپ بھرپور جنسی عمل کے لئے خود کو حد سے زیادہ تھکا محسوس کرتی ہیں تو آپ اپنے شریک حیات کو خوش کرنے کے طریقے ڈھونڈ سکتی ہیں اور آپ اس سے کہہ سکتی ہیں کہ اکثر محبت بھرا لمس عورت اور اس کے ساتھی کے لئے انتہائی پرلطف ثابت ہوتا ہے۔

خواہش کا فقدان

بہت سی باتیں ایک عورت کے لئے جنسی خواہش میں کمی یا جنسی عمل کے لطف کو کم کرسکتی ہیں۔ کچھ باتوں کا تعلق عورت کی معذوری سے ہوسکتا ہے اور کچھ کی وجہ دیگر ہوسکتی ہیں۔ آپ کی جنسی خواہش کم ہوسکتی ہے اگر۔

- آپ سخت محنت کے باعث تھکی ہوئی ہوں، آپ نے ڈھنگ سے کھانا نہ کھایا ہو، آپ کے ہاں حال ہی میں ولادت ہوئی ہو یا آپ کی معذوری اس کا سبب ہو۔
- آپ کا ساتھی آپ کو پسند نہ ہو یا وہ آپ سے بُرا برتاؤ کرتا ہو۔
- آپ کی معذوری درد کا سبب بنتی ہو یا ہلنا جلنا تکلیف پہنچاتا ہو۔
- آپ اپنے جسم کے لئے نامناسب سوچ رکھتی ہوں یا اپنی معذوری پر شرمندہ رہتی ہوں۔
- آپ ڈپریس ہوں یا زیادہ تر افسردہ رہتی ہوں۔
- ماضی میں جنسی عمل کے لئے کسی نے جبر کیا ہو یا آپ کو مجروح کیا ہو۔
- آپ حاملہ ہونے یا جنسی امراض میں مبتلا ہونے سے خوفزدہ ہوں۔
- جب ایک عورت میں جنسی جذبے کی کمی ہو تو اس کا جسم قدرتی نمی کم بناتا ہے اور اسے جنسی عمل کے تکلیف دہ نہ ہونے سے بچاؤ کے لئے کوئی چکنائی استعمال کرنے کی ضرورت پڑ سکتی ہے۔ (دیکھئے صفحہ 151)

جنسی زیادتی یا جبر کے بعد جنسی عمل

اگر کسی عورت کے ساتھ جنسی زیادتی ہوئی یا اس کی بالجبر عصمت دری کی گئی ہو تو عام طور پر اسے اپنے جسم کے بارے میں بہتر محسوس کرنے اور جنسی عمل کے لئے ذہنی طور پر آمادہ ہونے میں وقت لگتا ہے۔ تشدد کی جسمانی علامات ختم ہونے کے بعد بھی عورت جذباتی طور پر مجروح اور بری یادوں میں گھری رہتی ہے۔ جنسی زیادتی یا جبر کے بعد جنسی عمل کے بارے میں مزید معلومات کے لئے دیکھئے صفحہ 305۔

اگر آپ جنسی خواہش کی کمی کا شکار ہوں تو ایسی چیزوں کا تصور کرنے کی کوشش کریں جو آپ کو جنسی خوشی دیتی ہیں۔ جنسی خیالات اور تصورات آپ کو بذاتِ خود ایسی عورت کے روپ میں دیکھنے میں مدد دیں گے جو جنسی خواہشات رکھتی ہو۔ ان تصورات سے جنسی ہیجان پروان چڑھنے میں مدد ملتی ہے اور جنسی عمل سے مزید لطف اندوز ہونے میں مدد ملتی ہے، تصورات تنہارا کر بھی ابھارے جاسکتے ہیں اور ان میں اپنے پارٹنر کو بھی شریک کیا جاسکتا ہے۔

تبدیلی کے لئے کام

معذور عورتوں کے لئے جنسی عمل کے حق کے بارے میں کمیونٹیز کے عقائد اور رویوں میں تبدیلی میں طویل وقت لگ سکتا ہے لیکن یہ رویئے وقت کے ساتھ تبدیل ہو سکتے ہیں۔ معذور عورت کے لئے یہ اہم ہے کہ وہ اپنے بارے میں اچھا محسوس کرے۔ ایک ایسی عورت جو خود کو اہمیت دیتی ہے وہ اپنے پارٹنر سے بھی اپنے لئے احترام چاہے گی۔

معذور عورتیں کیا کرسکتی ہیں

کسی بھی ایسے فرد سے تعلق رکھنے میں انکار کردو جو آپ کے وقار کو احترام نہ دے

- خود کو جنسیت اور اپنے احساسات کے بارے میں سکھائیں

- جنسیت کے بارے میں مزید جاننے کے لئے دوسری معذور عورتوں یا دوسری ایسی عورتوں سے جن پر آپ بھرپور بھروسہ کرسکتی ہوں تبادلہ خیال کریں۔ اگر ہم نے اپنی بلوغت کے زمانے میں جس کے بارے میں نہیں سیکھا تو اب ہم اس بارے میں اور زیادہ بہتر سیکھ سکتے ہیں۔

- ایک دوسرے کو اپنے جنسی جذبے کو سمجھنے اور اس سے لطف اندوز ہونے کے طریقوں کے بارے میں مشورے دیں۔

- ایک دوسرے کی جنسی اور خاندانی زندگی کی تائید اور افادیت منوانے کے لئے دوسری معذور عورتوں کا گروپ بنائیں یا کسی گروپ میں شامل ہو جائیں۔

- معذور عورتوں کو جنسیت کے بارے میں تعلیم میں شامل کئے جانے کی تائید کیجئے۔

- تمام عورتوں کی جنسی ضروریات کے احترام کے لئے نمائندگی کیجئے۔

- اپنے جنسی جذبے اور عورت ہونے کی حیثیت سے اظہار کے طریقے تلاش کیجئے۔

ہماری کمیونٹی میں معذور عورتیں خود کو عورتوں کی حیثیت سے بااختیار بنانے کے لئے اکٹھی ہو رہی ہیں۔ ہم مل جل کر اپنی نسوانیت اور جنسیت ظاہر کرنے کے طریقے ڈھونڈ رہے ہیں۔ ہم میں سے کچھ اپنے بالوں میں پھول یا ماتھے پر بندی لگاتی ہیں بعض اوقات ہم اپنے ہاتھوں اور پیروں پر مہندی سے نقش و نگار بناتے ہیں ہاتھوں اور پیروں میں زیورات پہنتے ہیں یا روایتی غسل کرتے ہیں یہ سب باتیں ہماری جنسیت دریافت کرنے اور اپنے بارے میں اچھا محسوس کرنے میں معاون ہوتی ہیں ہم خود کو ایسی عورتوں کی طرح دیکھنا سیکھ رہے ہیں جیسا کہ ہم بننا چاہتے ہیں۔

تامل ناڈو بھارت کی معذور عورتوں کا ایک گروپ

گھر والے اور دیکھ بھال کرنے والے کیا کر سکتے ہیں

- معذور لڑکی سے وہی سلوک کیجئے جو اس عمر کے دوسرے بچوں سے کیا جاتا ہے اس سے اس میں اپنے بارے میں یا اپنے جسم اور اپنے جذبات کے بارے میں اچھے خیالات پروان چڑھانے میں مدد ملے گی

> یقینی بنائیں کہ معذور لڑکیوں اور عورتوں کو بھی بھرپور سماجی زندگی پروان چڑھانے کے یکساں مواقع ملیں۔

جب وہ بڑی ہوگی تو اس کے لئے اپنے شریک حیات کے ساتھ ایک محبت بھرا قابل احترام تعلق استوار کرنا آسان ہوگا۔

- دورِ بلوغت میں جب وہ ایک لڑکی سے ایک عورت میں تبدیل ہو رہی ہو لڑکی کی مدد کیجئے۔ اگر معذور لڑکی کو جنسیت کے بارے میں اچھی معلومات ملتی ہیں، اسے خود کو پرکشش بنانے کی سہولت اور اجازت ہے اور وہ جیسی ہے اسے اس حوالے سے کسی امتیازی رویے کا سامنا نہیں ہے تو وہ اپنے جسم، اپنی جنس کے بارے میں بہترین انداز سے سوچے گی اور اس کی خودداری بھی ٹھوس انداز سے پروان چڑھے گی۔
- یقینی بنائیں کہ معذور لڑکیوں اور عورتوں کو باہمی تبادلہ خیال اور تقاریب میں شریک کیا جائے۔
- ساتھی چننے اور فیصلہ کرنے میں معذور عورتوں کی مدد کیجئے۔
- اگر رشتے کا طالب مرد یا اس کے گھر والے اس معذور لڑکی کے گھر والوں سے عورت کی معذوری کے عوض نقد رقم یا کچھ طلب کرتے ہیں تو ایسی شادی مت کیجئے۔
- کمیونٹی کے لوگوں کو مائل کریں کہ وہ معذور عورتوں کو دوسری عورتوں کی طرح احساسات اور ضروریات رکھنے والا سمجھیں اور ان کے ساتھ اسی احترام سے پیش آئیں۔

پیار بھرے تعلق کی طلب تمام لوگوں کے لئے نارمل ہے معذوری رکھنے والی ایک عورت بھی ان سے مختلف نہیں ہوتی ہے۔

کمیونٹی کیا کر سکتی ہے؟

دوسری لڑکیوں اور عورتوں کو جنسی تعلیم فراہم کرنے کے لئے معذور عورتوں کو صحت کارکن کی حیثیت سے تربیت دیں۔

- یقینی بنائیں کہ معذوری والی عورتیں جنسیت اور نسوانیت کے حوالے سے کمیونٹی کی تقریبات میں شامل رہیں اور اہم کردار ادا کریں۔

- تمام عورتوں اور لڑکیوں کی جنسیت کے احترام کی پیروکاری کریں۔

- معذور لڑکیوں اور عورتوں کی شمولیت کے لئے جنسی تعلیمی پروگراموں کا اہتمام کریں۔ مثلاً ایک نابینا عورت کو کنڈوم محسوس کرنے دیں اور اسے سکھائیں کہ یہ کس طرح استعمال کیا جاتا ہے۔ بہری عورتوں اور ایسی عورتوں کو جو پڑھنے یا سمجھنے میں دشواری رکھتی ہوں، سمجھانے کے لئے تصاویر اور ماڈل استعمال کیجئے۔

- معذور لڑکیوں اور عورتوں کی نگرانی کیجئے خصوصاً ان کی جن کی معذوری سمجھنے اور سیکھنے پر اثر انداز ہوتی ہے اکثر ایسی لڑکیاں اور عورتیں جنسی زیادتی کا آسان ہدف ہوتی ہیں عورت کا تحفظ پوری کمیونٹی کی ذمہ داری ہے۔ زیادتی کا شکار ہونے والی عورتوں کی مدد کے لئے دیکھئے باب 14۔

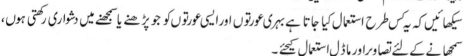

صحت کارکن کیا کر سکتی ہیں۔

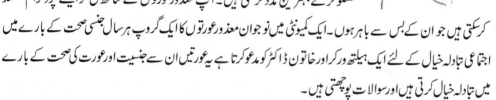

معذوریاں رکھنے والی بہت سی عورتیں جنسیت کے بارے میں سوالات رکھتی ہیں لیکن اکثر وہ اپنے سوالات پوچھتے ہوئے خوفزدہ یا شرمندہ ہوتی ہیں ایک صحت کارکن کے بطور آپ مختلف نئے طریقے سکھانے کے علاوہ معذور عورتوں سے ان کی فکر مندیوں اور توقعات کے بارے میں گفتگو کرکے بہترین مدد کر سکتی ہیں۔ آپ معذور عورتوں کے ساتھ مل کر ایسے پروگرام منعقد کر سکتی ہیں جو ان کے بس سے باہر ہوں۔ ایک کمیونٹی میں نوجوان معذور عورتوں کا ایک گروپ ہر سال جنسی صحت کے بارے میں اجتماعی تبادلہ خیال کے لئے ایک ہیلتھ ورکر اور کراور خاتون ڈاکٹر کو مدعو کرتا ہے یہ عورتیں ان سے جنسیت اور عورت کی صحت کے بارے میں تبادلہ خیال کرتی ہیں اور سوالات پوچھتی ہیں۔

آپ لڑکیوں اور نو عمر عورتوں کو ان کی جسمانی تبدیلیوں کے بارے میں معلومات فراہم کرکے بھی ان کی مدد کر سکتی ہیں۔ آپ بیان کر سکتی ہیں ایک عورت ہونے کا مقصد کیا ہے؟ آپ انہیں جنسیت اور رشتوں کے حوالے سے سوالات کے جواب فراہم کر سکتی ہیں ایک صحت کارکن کی حیثیت سے آپ ان عقائد اور رویوں کو بدل سکتی ہیں جو معذور لڑکیوں اور عورتوں کے لئے مشکل بناتے ہیں کہ وہ اپنے جسموں اور جنسیت کے بارے میں اچھا سوچ سکیں۔

باب- 8

جنسی صحت

اچ آئی وی/ ایڈز سمیت
جنسی طور پر منتقل ہونے والی
بیماریوں سے بچاؤ

جنسی صحت، صحت مند ہونے کی جسمانی اور جذباتی وہ حالت ہے جو ہمیں اپنے جنسی احساسات کے مطابق لطف اندوز ہونے اور عمل کرنے کی سہولت فراہم کرتی ہے۔ ہم اپنے جسموں کے بارے میں، کیا چیز ہمیں خوشی دیتی ہے، غیر مطلوبہ حمل اور جنسی عمل کے دوران انفیکشنز کے خطرے کو کم کرنے کے طریقے جان کر خود کو جنسی طور پر صحت مند رکھ سکتے ہیں۔

بہت سی کمیونیٹیز میں عورت کے حوالے سے نقصان دہ عقائد وہاں کی عورتوں کے لئے صحت مند جنسی عمل کو مشکل بنا سکتے ہیں۔ کیونکہ عورتیں جنسی عمل کے حوالے سے بہت معمولی اختیار رکھتی اور عموماً جنسی عمل کے لئے انکار نہیں کر سکتی ہیں اس لئے دنیا بھر میں لاکھوں عورتیں ہر سال اچ آئی وی اور جنسی طور پر منتقل ہونے والے دوسرے انفیکشنز (STIs) میں مبتلا ہو جاتی ہیں۔

اکثر لوگ سمجھتے ہیں کہ ایک ایسی عورت جو کوئی معذوری رکھتی ہو کسی انفیکشن میں مبتلا نہیں ہو سکتی۔ یہ بات درست نہیں ہے۔ معذور عورتیں بھی غیر معذور عورتوں کی طرح انفیکشنز میں مبتلا ہو سکتی ہیں۔ حقیقت یہ ہے کہ معذور لڑکیاں اور عورتیں غیر معذور عورتوں کی یہ نسبت جنسی طور پر منتقل ہونے والے انفیکشنز میں مبتلا ہونے کا زیادہ خطرہ رکھتی ہیں۔ ان کے لئے جنسی صحت کے بارے میں معلومات حاصل کرنا ہی مشکل نہیں ہوتا ہے بلکہ وہ اس بات کا بہت کم اختیار رکھتی ہیں کہ وہ کس طرح جنسی تعلق کو محفوظ رکھیں۔ یہ صورت انہیں اچ آئی وی سمیت دیگر جنسی طور پر منتقل ہونے والے انفیکشنز کا آسان ہدف بناتی ہے۔

معذور عورتوں کے ساتھ جنسی زیادتیوں کے بارے میں مزید معلومات کے لئے دیکھئے باب 14۔

یہ باب اچ آئی وی/ ایڈز اور دیگر جنسی طور پر منتقل ہونے والے انفیکشنز اور ان سے بچنے کے طریقوں کے بارے میں معلومات فراہم کرتا ہے۔ معیاری معلومات کے ساتھ عورتیں خود کو ان عوارض سے بچانے اور صحت مند جنسی عمل سے لطف اندوز ہونے کے لئے اقدام کر سکتی ہیں۔

جنسی طور پر منتقل ہونے والے انفیکشنز کیا ہیں؟

جنسی طور پر منتقل ہونے والے انفیکشنز (STIs) وہ انفیکشنز ہیں جو ایک فرد سے دوسرے فرد کو جنسی عمل کے دوران منتقل ہوتے ہیں۔ یہ انفیکشنز ایک فرد سے دوسرے فرد کو کسی بھی قسم کے جنسی تعلق سے منتقل ہوسکتے ہیں۔ یہ فطری مباشرت، غیر فطری مباشرت یا اورل سیکس کوئی بھی طریقہ ہوسکتا ہے۔ جنسی طور پر منتقل ہونے والے انفیکشنز صرف متاثرہ عضو تناسل یا فرج کو دوسرے فرد کے تناسلی اعضاء سے رگڑنے سے بھی منتقل ہوسکتے ہیں۔

علاج نہ ہونے کی صورت میں یہ انفیکشنز خطرناک ہوتے ہیں۔

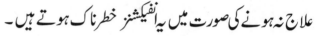

بیشتر STIs کا علاج ادویات سے ہوسکتا ہے۔ اگر ان کا ابتداء ہی میں علاج نہ کرایا جائے تو یہ مرد اور عورتوں دونوں میں بانجھ پن پیدا کرسکتے ہیں۔ انفیکشن ہونے کی صورت میں بچے وقت سے پہلے، بہت چھوٹے یا مردہ بھی پیدا ہوسکتے ہیں۔ حمل ٹیوبز میں ٹھہر سکتا ہے۔ پیٹ کے نچلے حصے میں مستقل درد، رحم (سرویکس) کے منہ کے کینسر اور شدید انفیکشن کے باعث موت ہوسکتی ہے۔

ایس ٹی آئی کی علامات

اگر مندرجہ ذیل میں سے ایک یا زائد علامات ہوں تو آپ ایس ٹی آئی میں مبتلا ہوسکتی ہیں۔

- فرج سے غیر معمولی رطوبت کا اخراج
- فرج سے غیر معمولی بو
- پیٹ کے نچلے حصے میں خاص طور پر فطری طریقے سے مباشرت کے دوران درد یا غیر معمولی کیفیات
- جنسی اعضاء میں سوزش، سرخ دھبے، ابھار یا درد

معذوری کے اعتبار سے یہ آپ کے لئے مشکل ہوسکتا ہے کہ آپ کسی کو یہ بتا سکیں کہ آپ ان میں سے کوئی علامات رکھتی ہیں۔ ممکن ہے آپ کو ضرورت ہو کہ کسی قابل بھروسہ فرد سے کہیں کہ وہ ایس ٹی زی آئی کی علامات کا جائزہ لینے میں آپ کی مدد کرے۔

رطوبت کے اخراج میں تبدیلیاں

فرج میں معمولی مقدار میں نمی یا اخراج نارمل بات ہے۔ یہ فرج کو صاف اور محفوظ رکھنے کا فطری طریقہ ہے۔ اخراج آپ کی ماہواری کے دوران تبدیل ہوتا ہے۔ یہ ماہواری شروع ہونے سے تقریباً چودہ دن پہلے گاڑھا، نہایت شفاف اور چکنا ہوجاتا ہے۔

دوسری صورت میں آپ کی فرج سے اخراج کی مقدار، رنگ یا بو میں تبدیلی کا مطلب یہ ہوسکتا ہے کہ آپ کسی انفیکشن میں مبتلا ہیں لیکن آپ کے اخراج سے یہ کہنا مشکل ہوسکتا ہے کہ انفیکشن کس قسم کا ہے۔ ایسے اخراج کے بارے میں جاننے کے لئے جو جنسی طور پر منتقل نہ ہوا ہو (یسٹ یا بیکٹیر یا کی رطوبتیں) دیکھیے صفحہ 111 تا 113۔

ایس ٹی آئی کی علامات کس طرح چیک کریں؟

جب بھی آپ اپنے تناسلی اعضاء دھوئیں تو کسی غیر معمولی اخراج، ابھار یا سوجن کو محسوس کرنے کے لئے اپنی انگلیاں استعمال کریں۔ ہفتے میں ایک بار یہ جائزہ ضرور لیں۔ آپ اگر روزانہ ایسا کریں گی تو کسی تبدیلی کا اندازہ لگانا آپ کے لئے مشکل ہوگا۔

اگر آپ ہاتھوں سے یا ہاتھوں کے استعمال سے معذور ہیں۔ اگر آپ اپنے تناسلی اعضاء میں کسی تبدیلی کو محسوس کرنے کے لئے انگلیاں استعمال نہیں کر سکتی ہیں تو جائزہ لینے کے لئے آئینہ استعمال کرنے کی کوشش کریں۔ اگر آپ آئینہ پکڑ نہیں سکتی ہیں تو اسے فرش پر رکھ کر اس طرح جھکیں کہ جائزہ لے سکیں۔

اگر آپ ریڑھ کی ہڈی کی خرابی رکھتی ہیں۔ اگر آپ اپنے تناسلی اعضاء کو محسوس کر سکتی ہیں اور ان کا جائزہ لے سکتی ہیں تو ہفتے میں ایک بار غسل کرتے ہوئے ایسا کریں۔ اگر آپ اس سے قاصر ہیں تو کسی قابل بھروسہ فرد سے مدد طلب کریں۔ ممکن ہے کہ آپ اپنے پیٹ کے نچلے حصے میں درد یا تناسلی اعضاء میں سوزش محسوس نہ کر سکتی ہوں لیکن اگر آپ انفیکشن میں مبتلا ہیں اور اس کا علاج نہ ہو تو آپ dysreflexia کا شکار ہو سکتی ہیں جو خطرناک ہوتا ہے۔ علاج کے لئے دیکھئے صفحہ 117 تا 119۔

اگر آپ اپنے پیروں کی حرکت محدود یا بالکل بھی نہیں رکھتی ہوں۔ اگر ممکن ہو تو آپ ایسی پوزیشن ڈھونڈیں کہ آپ دھوتے ہوئے اپنی انگلیوں سے تناسلی اعضاء کو محسوس کر سکیں یا انہیں دیکھنے کے لئے آئینہ استعمال کریں۔ اگر ضروری ہو تو اپنے پیروں کو گرفت کرنے کے لئے کسی سے مدد حاصل کریں۔

ٹرائیکوموناس (Trichomonas)

ٹرائیکوموناس ایک انتہائی پریشان کن انفیکشن ہے۔ مردوں میں عام طور پر اس کی کوئی علامات نہیں ہوتی ہے لیکن ان کے عضو تناسل میں یہ انفیکشن ہو سکتا ہے جو جنسی ملاپ کے دوران عورت میں منتقل ہو سکتا ہے۔

علامات

- تناسلی اعضاء اور فرج کا سرخ اور خارش زدہ ہونا
- سرمئی یا زرد رنگ کی بلبلہ دار رطوبت کا اخراج
- پیشاب کرتے ہوئے درد یا جلن ہونا
- بدبودار رطوبت کا اخراج

اگر آپ ٹرائیکوموناس کی تصدیق کے لئے ٹیسٹ کرا سکتی ہیں تو تصدیق ہونے پر صفحہ 160 پر دی گئی ادویات میں سے کوئی دوا لے سکتی ہیں۔ اگر ٹیسٹ ممکن نہیں تو پھر بھی صفحہ 162 پر درج ادویات لینا بہتر ہے کیونکہ انفیکشن دیگر وجہ سے بھی ہو سکتا ہے۔

ٹرائیکوموناس کے لئے ادویات

دوا	کتنی مقدار میں لی جائے	کب اور کس طرح لی جائے
میٹرونیڈازول (Metronidazole)	400 تا500 ملی گرام	منہ کے ذریعے، دن میں دوبار سات دن تک
یا میٹرونیڈازول (Metronidazole) (حمل کے ابتدائی 3 ماہ میں میٹرونیڈازول مت لیں)	2 گرام (2000 ملی گرام)	منہ کے ذریعے۔ ایک خوراک
یا کلینڈامائی سین (Clindamycin)	300 ملی گرام	منہ کے ذریعے۔ دن میں دوبار سات دن تک

اہم بات میٹرونیڈازول کھانے کے دوران الکحل مت پئیں، میٹرونی ڈے زول کے بارے میں مزید معلومات کے لئے دیکھئے صفحہ 347۔ آپ کے شوہر کو بھی یہی دوا کھانی ہوگی۔

آتشک (کلیپ، گونوریا، وی ڈی) اور کلے مائیڈیا

آتشک اور کلے مائیڈیا دونوں ہی سنگین انفیکشن ہیں۔ اگر ان کا علاج ابتداء ہی میں ہو تو یہ آسانی سے ٹھیک ہو سکتے ہیں۔ اگر نہیں تو یہ مردوں اور عورتوں دونوں ہی میں شدید انفیکشنز اور بانجھ پن کا سبب بن سکتے ہیں۔ مرد میں ان کے علامات متاثر عورت سے مباشرت کے دو سے پانچ دن بعد سامنے آتی ہیں۔ عورتوں میں علامات ہفتوں بلکہ مہینوں بعد نمودار ہوتی ہیں۔ یہ بھی ممکن ہے کہ مرد اور عورت دونوں انفیکشن زدہ ہوں اور کوئی علامت نہ ہو۔ علامات نہ ہونے پر بھی ایک فرد آتشک اور کلے مائیڈیا دوسرے فرد میں منتقل کر سکتا ہے۔

مرد کی علامات
- عضو تناسل سے رطوبت کا اخراج
- پیشاب کرتے ہوئے درد یا جلن
- خصیوں میں درد یا سوجن
- کوئی بھی علامت نہیں

عورتوں کی علامات
- فرج یا مقعد سے زرد یا سبز رطوبت کا اخراج
- پیشاب کرتے ہوئے درد یا جلن
- بخار
- پیٹ کے نچلے حصے میں درد
- مباشرت کے وقت درد یا اخراج خون
- کوئی بھی علامت نہیں

علاج

آتشک یا کلے مائیڈیا کی کوئی بھی علامت ہو اور ممکنہ طور پر متاثر فرد سے غیر محفوظ مباشرت کی گئی ہو تو انفیکشن کی تشخیص کے لئے ٹیسٹ کرایا جائے تا کہ اسی کے مطابق دوا دی جائے۔

آتشک کے لئے ادویات

دوا	کتنی مقدار میں لی جائے	کب اور کس طرح لی جائے
سیفکسائم (Cefixime)	400 ملی گرام	منہ کے ذریعے، ایک بار
یا سپروفلوکساسن (Ceprofloxacin)	500 ملی گرام	منہ کے ذریعے ایک بار

کلے مائیڈیا کے لئے ادویات

دوا	کتنی مقدار میں لی جائے	کب اور کس طرح لی جائے
ازتھرومائیسین (Azithromycin)	ایک گرام	منہ کے ذریعے، دن میں ایک بار
ڈوکسی سائیکلین	100 ملی گرام	منہ کے ذریعے، دن میں دو بار، سات دن تک
ٹیٹرا سائیکلین	500 ملی گرام	منہ کے ذریعے، دن میں چار بار، سات دن تک
اریتھرومائیسین (Erythromycin)	500 ملی گرام	منہ کے ذریعے، دن میں چار بار، سات دن تک

بدقسمتی سے ٹیسٹ ہر جگہ اور ہر بار ممکن نہیں ہوتے ہیں لہذا یہ بہتر رہتا ہے کہ ایک سے زیادہ انفیکشن کے لئے ادویات لی جائیں۔ ایک وقت میں کئی انفیکشن ہو سکتے ہیں جن کا سبب صرف آتشک یا کلے مائیڈیا ہی نہیں بلکہ ٹرائیکوموناس (دیکھئے صفحہ 159) بیکٹیریائی ویجائی نوسز (دیکھئے صفحہ 113) بھی ہو سکتے ہیں۔ صفحہ 162 پر دی گئی ادویات ان تمام انفیکشنز کا علاج کریں گی۔

پیٹرو کی سوجن کا عارضہ

پیٹرو کی سوجن (PID) عورت کے پیٹ کے نچلے حصے کے تولیدی حصوں میں سے کسی میں بھی انفیکشن کو کہا جاتا ہے۔ اسے پیٹرو کا انفیکشن بھی کہا جاتا ہے۔ پیٹرو کا انفیکشن جنسی طور پر منتقل ہونے والے کسی بھی انفیکشن خصوصاً آتشک یا کلے مائیڈیا کے ذریعے ہو سکتا ہے۔

علامات

- پیٹ کے نچلے حصے میں درد
- تیز بخار
- بیمار سا یا تھکا ہوا محسوس کرنا
- فرج سے سبز یا زرد رطوبت کا اخراج
- مباشرت کے وقت درد یا اخراج خون

علاج

یہ انفیکشن عموماً ملے جلے جرثوموں کی وجہ سے ہوتا ہے لہذا اس کے علاج کے لئے ایک سے زیادہ ادویات استعمال کی جانی چاہئیں۔ صفحہ 162 پر دی گئی ادویات استعمال کریں۔

آتشک، کلے مائیڈیا، ٹرائیکوموناس، بیکٹر یائی وجائی نوسز اور پیڑو کے انفیکشن کے لئے ادویات

اگر آپ میں ان انفیکشنز کی علامات ہوں اور انفیکشن کے تشخیص کے لئے ٹیسٹ ممکن نہ ہو تو آپ یہ ادویات لیں۔

دوا	کتنی مقدار میں لی جائے	کب اور کس طرح لی جائے
سیفیکسائم (Cefixcime) .... 400 ملی گرام	.... منہ کے ذریعے، ایک خوراک
یا سپر وفلوکساسن (Ceprofloxacin) .... 500 ملی گرام	.... منہ کے ذریعے، ایک خوراک

اگر آپ حاملہ ہوں یا بچے کو اپنا دودھ پلا رہی ہوں تو سپر وفلوکساسن مت لیں

| **اور** |

| از تھرومائسن (Azithromycin) ..... ایک گرام (ایک ہزار ملی گرام) ..... منہ ذریعے ایک خوراک |
| یا اریتھرومائسن (Erythromycin) ... 500 ملی گرام .... منہ کے ذریعے دن میں چار بار، سات دن تک |
| یا اموکسی سلین (Amoxycillin) .... 500 ملی گرام .... منہ کے ذریعے، دن میں تین بار، سات دن تک |
| یا ڈوکسی سائیکلین (Doxicycline) ... 100 ملی گرام .... منہ کے ذریعے، دن میں دو بار، سات دن تک |

(اگر آپ حاملہ ہوں یا بچے کو اپنا دودھ پلا رہی ہوں تو ڈوکسی سائیکلین مت لیں)

| یا ٹیٹرا سائیکلین .... 500 ملی گرام .... منہ کے ذریعے، دن میں چار بار، سات دن تک |

| **اور** |

| میٹرونیڈازول .... 400 تا 500 ملی گرام .... منہ کے ذریعے، دن میں دو بار، سات دن تک |

(Metronidazole)

(حمل کے ابتدائی تین ماہ میں میٹرونیڈازول مت لیں، اس کے بجائے کلینڈامائسن اور ٹینی ڈازول دونوں دوائیں لیں)

| **یا** |

| کلینڈامائسن (Clindamycin) .... 300 ملی گرام .... منہ کے ذریعے، دن میں دو بار، سات دن تک |
| **یا** 2 فیصد کی 5 گرام کریم .... سوتے وقت فرج میں لگائیں، سات دن تک |
| ٹینی ڈازول (Tinidazole) .... 2 گرام (2000 ملی گرام) .... منہ کے ذریعے، ایک خوراک |
| **یا** 500 ملی گرام .... منہ کے ذریعے، دن میں دو بار، 5 دن تک |

اہم بات میٹرونیڈازول یا ٹینی ڈازول لینے کے دوران الکحل استعمال نہ کریں۔ آپ کے شوہر کو بھی یہی ادویات لینی ہوں گی۔

تناسلی اعضاء کے زخم

تناسلی اعضاء کے بیشتر زخم یا السر جنسی طور پر منتقل ہوتے ہیں لیکن دباؤ کے زخم، بال ٹوٹڑیا چوٹیں وغیرہ بھی تناسلی اعضاء پر زخموں کا سبب بن سکتی ہیں۔ تناسلی عضو کے ہر زخم کو صابن اور پانی سے دھو کر صاف رکھا جائے، اسے احتیاط سے خشک کریں، ہر کپڑے کو استعمال کرنے سے پہلے دھو کر خشک کر لیں۔

تنبیہہ تناسلی اعضاء پر کوئی زخم ہو تو اس کے ذریعے دوسرے انفیکشنز، خاص طور پر ایچ آئی وی اور ہیپاٹائٹس بی آسانی سے منتقل ہوسکتے ہیں۔ انفیکشن سے بچنے کے لئے ایسے زخموں کے ٹھیک ہونے تک جنسی ملاپ سے گریز کریں۔

سوزاک (Saphilis)

سوزاک (سفلس) ایک سنگین انفیکشن ہے جو پورے جسم کو متاثر کرتا ہے۔ یہ برسوں رہ سکتا ہے اور بدسے بدتر صورت اختیار کر سکتا ہے۔ سفلس کا علاج ممکن ہے اگر ابتداء ہی میں شروع ہو جائے۔

علامات

1- اس کی پہلی علامت درد کے بغیر چھوٹا سا زخم ہے جو کسی مہا سے، دانے یا ہموار گیلے مہا سے یا کھلے زخم کی طرح ہوسکتا ہے۔ زخم چند دن یا چند ہفتے رہنے کے بعد خود ہی غائب ہوجاتا ہے لیکن بیماری پورے جسم میں پھیلتی رہتی ہے۔

2- کئی ہفتوں یا مہینوں کے بعد متاثرہ فرد، گلے کے درد، بخار، منہ کے زخموں، جوڑوں کی سوجن یا خاص طور پر ہتھیلیوں یا تلووں میں سرخ اُبھار کی زد میں آ جاتا ہے۔ اس دوران وہ دوسرے افراد کو بھی متاثر کر سکتا ہے۔

3- عام طور پر یہ ساری علامات خود بخود ختم ہو جاتی ہیں لیکن بیماری موجود رہتی ہے۔ علاج نہ ہونے پر سفلس، عارضہ قلب، فالج، دماغی بیماری اور موت کا سبب بن سکتا ہے۔

سوزاک

علاج

اگر علامات سال بھر سے کم مدت موجود رہیں تو بینزاتھائین پنسیلین (benzathine pencillin) کے 2.4 ملین یونٹ دو مساوی حصوں میں کر کے انجکشن کے ذریعے باری باری ہر کولہے میں لگائیے۔ پنسیلین سے الرجی رکھنے والے افراد 500 ملی گرام ٹیٹرا سائیکلین (Tetracycline) دن میں چار بار پندرہ دن تک لے سکتے ہیں۔

● ایک علامت ایک سال سے زائد مدت سے ہے تو تین ہفتے تک، ہفتے میں ایک بار 2.4 ملین یونٹ بینزاتھائین (Benzathine) نصف ہر کولہے میں انجکشن کے ذریعے داخل کریں۔ پنسیلین سے الرجی ہونے پر تیس دن تک، روزانہ چار بار 500 ملی گرام ٹیٹرا سائیکلین فی بار لیں۔

نوٹ پنسیلین سے الرجی رکھنے والی حاملہ یا بچے کو اپنا دودھ پلانے والی عورتیں اریتھرومائسن (erythromycin) لے سکتی ہے۔ (دیکھئے صفحہ 343)۔ یہی علاج آپ کے شوہر کا بھی ہونا چاہئے۔

حمل اور سوزاک

حاملہ عورت سے سوزاک، اس کے نشوونما پاتے بچے میں منتقل ہوسکتا ہے جس کی وجہ سے وہ وقت سے بہت پہلے ناقص صورت میں یا مردہ جنم لے سکتا ہے۔ آپ حمل کے دوران ٹیسٹ کے ذریعے تشخیص اور علاج کے ذریعے ایسی صورت سے محفوظ رہ سکتی ہیں۔ اگر آپ اور آپ کے شوہر کے بلڈٹیسٹ سے سوزاک ہونے کی تصدیق ہو تو آپ دونوں کو تین ہفتے، ہفتے میں ایک بار انجکشن کے ذریعے 2.4 ملین یونٹس بینزاتھائن پینسیلین لینی پڑے گی۔

شنکرائڈ (Chancroid)

شنکرائڈ بھی جنسی طور پر منتقل ہونے والا انفیکشن ہے جو تناسلی اعضاء پر زخموں کا سبب بنتا ہے۔ ابتداءً ہی میں اس کا ادویات سے علاج ہوسکتا ہے۔ یہ سوزاک سے ملتا جلتا ہے اور اسے سمجھنے میں مغالطہ ہوسکتا ہے۔

علامات

شنکرائڈ

- تناسلی اعضاء یا مقعد پر ایک یا زیادہ، نرم پر درد زخم جن سے با آسانی خون بہنے لگے۔
- جانگھ میں بڑے اور تکلیف دہ غدود ہوسکتے ہیں۔
- ہلکا بخار۔

شنکرائڈ کے لئے ادویات

دوا	کتنی مقدار	کب اور کتنی بار لی جائے
ازیتھرومائسن (Azithromycin)	ایک گرام	منہ کے ذریعے، ایک بار
یا اریتھرومائسن (Erythromycin)	500 ملی گرام	منہ کے ذریعے، دن میں چار بار، سات دن تک
یا سپروفلوکساسن (Ceprofloxacin) (اگر آپ حاملہ ہیں تو سپروفلاکساسن مت لیں)	500 ملی گرام	منہ کے ذریعے، دن میں دو بار، تین دن تک

نوٹ اگر یہ یقین سے کہنا ممکن نہ ہو کہ آپ کے زخم شنکرائڈ کے باعث ہیں یا اس کا ٹیسٹ نہیں ہوسکتا ہو تو امکانی طور پر بہترین یہی ہے کہ آپ سوزاک کی ادویات (دیکھئے صفحہ 163) لیں۔

تناسلی ہرپز (Genital Herpes)

تناسلی ہرپز وائرس کے ذریعہ ہونے والا انفیکشن ہے۔اس میں تناسلی اعضاء پر چھوٹے چھوٹے دانے نمودار ہوتے ہیں۔ یہ جنسی عمل کے دوران ایک فرد سے دوسرے فرد میں منتقل ہوتا ہے ۔اورل سیکس کی صورت میں یہ منہ پر بھی ہوسکتا ہے۔(یہ منہ پر ہونے والے عام ہرپز سے مختلف ہوتا ہے جو جنسی عمل سے نہیں پھیلتا ہے)۔

ہرپز وائرس کے زخم مہینوں یا برسوں ہوتے اور ٹھیک ہوتے رہتے ہیں ۔ ہرپز کا کوئی علاج نہیں البتہ یہ ممکن ہے کہ آپ خود کو بہتر محسوس کریں۔

ہرپز

علامات

● تناسلی اعضاء کی جلد یا رانوں پر کھجلی، خارش یا تکلیف کا احساس۔

● چھوٹے تکلیف دہ چھالے جو جلد پر پانی کے قطروں کی طرح نظر آئیں۔ یہ چھالے پھٹ کر تکلیف دہ کھلے زخم بن سکتے ہیں۔

پہلی بار ہرپز کے زخم تین ہفتے یا اس سے زائد مدت رہ سکتے ہیں۔ اس دوران آپ کو بخار، سردرد یا جسم میں درد، کپکپاہٹ ہو سکتی ہے یا جانگھ میں لمف کی گٹھیں سوج سکتی ہیں۔ تاہم زخم ختم ہوجاتے ہیں لیکن انفیکشن ختم نہیں ہوتا ہے۔ دوسرا حملہ نسبتاً ہلکا ہوگا۔

علاج: acyclovir استعمال کریں، دیکھئے صفحہ 333۔

حمل اور ہرپز

ہرپز انفیکشن میں مبتلا حاملہ عورت کو زخم ولادت کے وقت ہوں تو وائرس اس کے بچے میں منتقل ہوسکتے ہیں ۔اگر ماں کو ہرپز اس سے پہلے ہوتا ہو تو بچے کو ماں سے ہرپز منتقل ہونے کے امکانات کم ہوتے ہیں۔ اگر آپ کو ہرپز کے زخم ہوں تو کوشش کیجئے کہ بچے کی ولادت اسپتال میں ہو۔

اگر ماں پہلی بار ہرپز میں مبتلا ہو تو بچے کی ولادت کے لئے ڈاکٹر آپریشن کر سکتے ہیں (سی سیکشن، دیکھئے صفحہ 244) تا کہ بچے زخموں کے باعث انفیکشن زدہ نہ ہوں یا پھر بچے کو ولادت کے بعد دوائیں دی جا سکیں۔

تناسلی مہاسے (HPV)

تناسلی مہاسے یا ابھار ہیومن پاپیلوماوائرس (HPV) کے ذریعے ہوتے ہیں ۔ یہ جسم کے دوسرے حصے کے مہاسوں کی طرح نظر آتے ہیں ۔ یہ ممکن ہے کہ آپ ایچ پی وی زدہ ہوں تو آپ کو معلوم نہ ہو۔ خاص طور پر اس وقت مہاسے فرج کے اندر یا مردانہ عضو تناسل کے سرے پر اندر ہوں۔ کچھ لوگوں کے ایچ پی وی نہ نہ ہوں۔ مہاسے کے علاج کے بغیر بھی ختم ہوجاتے ہیں لیکن اس میں طویل وقت لگتا ہے۔ عام طور پر یہ بدتر ہوتے رہتے ہیں اور ان کا علاج کرانا چاہیے کیونکہ یہ جنسی عمل کے دوران دوسرے فرد کو آسانی سے منتقل ہوجاتے ہیں۔

اہم بات اگر تناسلی اعضاء پر موجود مہاسوں کا علاج نہ کیا جائے تو کچھ مہاسے رحم کے منہ کے کینسر بن سکتے ہیں۔ اگر آپ تناسلی مہاسوں میں مبتلا ہوں تو PAP ٹیسٹ کرانے کی کوشش کریں (دیکھئے صفحہ 131) تا کہ معلوم ہو سکے کہ کہیں آپ کے رحم کے منہ پر ایچ پی وی یا کینسر میں تو مبتلا نہیں ہیں۔

ایچ پی وی کی علامات

- خارش

- درد کے بغیر سفید یا بھورے کھر درے اُبھار

مردوں میں یہ اُبھار عام طور پر مرد کے عضو تناسل پر یا اس کے منہ کے اندر، خصیوں یا مقعد میں ہوتے ہیں

عورتوں میں یہ اُبھار عام طور پر فرج کے منہ کے گرد جلد کی تہوں یا مقعد میں ہوتے ہیں۔

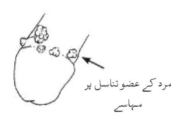

مرد کے عضو تناسل پر مہاسے

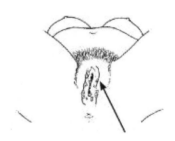

عورت کے مہاسے

علاج

مہاسوں کے علاج کی یہ اشیاء عام طور پر دوا فروشوں یا کیمسٹ کی دکان سے مل جاتی ہیں۔

1- صحت مند جلد کو محفوظ رکھنے کے لئے ہر اُبھار (مہاسے) کے گرد جلد پر پیٹرولیم جیلی (ویسلین) یا کوئی چکنی چیز لگا ئیے۔

2- کسی تیلی سی تیلی یا ٹوتھ پک کے ذریعے 80 سے 90 فیصد ٹرائی کلوروایسٹک ایسڈ (TCA) یا بائی کلوروایسٹک ایسڈ محلول (BAC) بہت ہی کم مقدار مہاسوں کے اندر لگا ئیے (دیکھئے صفحہ 354) جب تک یہ سفید نہ ہوں، یہ عمل کرتی رہیں۔

3- لگ بھگ دو گھنٹے کے بعد یا اس سے پہلے اگر سوزش یا جلن انتہائی تکلیف دہ ہو تو تیزابی محلول کو صاف کر لیں۔

یا

مہاسوں کے براؤن ہونے تک اسی طرح 20 فیصد پوڈوفائلین (Podophyllin) محلول لگا ئیں (دیکھئے صفحہ 351) پوڈوفائلین محلول کو چھ گھنٹے بعد صاف کر لیا جائے۔

تیزاب اب مہاسوں کو جلا دے گا اور اس کی جگہ تکلیف دہ زخم رہ جائے گا۔ زخموں کو صاف اور خشک رکھیے۔ زخم ایک سے دو ہفتے میں ٹھیک ہو جائیں گے۔ ان پر نظر رکھیے کہ وہ انفیکشن زدہ نہ ہوں۔ جب تک زخم نہ بھریں جنسی عمل مت کیجیے۔ اگر اس دوران جنسی ملاپ کیا جائے تو آپ کا شوہر کنڈوم استعمال کرے۔ تمام مہاسوں کے خاتمے کے لئے عام طور پر کئی اقسام کے علاج کی ضرورت ہوتی ہے (اہمیت اس کی نہیں ہے کہ آپ کونسا محلول استعمال کریں) آپ ہفتہ بھر بعد علاج دوہرا سکتی ہیں۔ کوشش کریں کہ اس جگہ ایسڈ نہ لگے جہاں کبھی مہاسہ موجود ہوا کرتا تھا۔ اگر بہت زیادہ سوزش ہو تو اگلی بار علاج کے لئے کچھ مزید انتظار کر لیں۔

حمل اور مہاسے

اگر آپ حاملہ ہوں تو پوڈوفائلین استعمال مت کریں۔ یہ آپ کی جلد میں جذب ہو جائے گا اور نشوونما پاتے بچے کو نقصان پہنچا سکتا ہے۔ حمل کے دوران مہاسے بڑھ سکتے ہیں اور ان سے خون بہہ سکتا ہے لیکن مہاسے بذات خود بچے کو کوئی نقصان نہیں پہنچائیں گے۔ بعض اوقات حمل کے بعد مہاسے بہت ہی چھوٹے ہوجاتے ہیں۔

ہیپاٹائٹس (یرقان، زرد آنکھیں)

ہیپاٹائٹس جگر کا ورم ہے جو عام طور پر ایک وائرس کے ذریعے ہوتا ہے۔ اس کا سبب بیکٹیریا، الکحل یا کیمیائی زہر بھی بنتا ہے۔ ہیپاٹائٹس کی تین اہم اقسام (اے، بی اور سی) ہیں۔ مرض کی علامات بھی نہ ہوں تو یہ ایک فرد سے دوسرے فرد کو منتقل ہوسکتا ہے۔

ہیپاٹائٹس اے

یہ عموماً بچوں میں معمولی نوعیت کا اور بڑی عمر کے افراد اور حاملہ عورتوں میں زیادہ شدید ہوتا ہے۔

ہیپاٹائٹس بی

ہر ایک کے لئے خطرناک ہے۔ یہ جگر کو مستقل نوعیت کا نقصان (Cirrhosis) پہنچا سکتا ہے، جگر کے کینسر اور موت کا سبب بن سکتا ہے۔ اگر آپ حاملہ ہیں اور ہیپاٹائٹس بی کی علامات رکھتی ہوں تو طبی مشاورت کیجئے۔

ہیپاٹائٹس سی

یہ بھی انتہائی خطرناک ہے اور جگر کے مستقل انفیکشن کا سبب بن سکتا ہے۔ ہیپاٹائٹس سی، ایچ آئی وی/ایڈز میں مبتلا افراد کی موت کا ایک بڑا سبب ہے۔

علامات

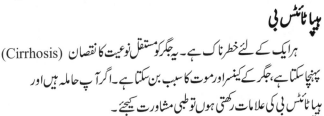

* پیٹ میں درد یا متلی
* گہرے رنگ کا پیشاب اور سفید محسوس ہونے والا پاخانہ
* بعض اوقات بخار
* کوئی بھی علامات نہیں

* تھکن اور کمزوری کا احساس
* عدم اشتہا (بھوک محسوس نہ ہونا)
* زرد آنکھیں اور/ یا جلد (خصوصاً ہتھیلیاں اور تلوے)

علاج

ایسی کوئی دوا نہیں جو مدد کرے۔ حقیقت یہ ہے کہ کچھ دوائیں آپ کے جگر کو مزید نقصان پہنچا سکتی ہیں۔ آرام کریں اور مائعات خوب پئیں۔ اگر آپ کچھ کھانا نہیں چاہتی ہیں تو پھلوں کا رس یا سبزیوں کا جوس پینے کی کوشش کریں۔ تے پر قابو پانے کے لئے کولا یا جنجر کے چند گھونٹ لیں۔ جڑی بوٹیوں کی چائے بھی معاون ہو سکتی ہے۔ اپنے اِردگرد موجود عمر رسیدہ عورتوں سے پوچھیں کہ کون سی جڑی بوٹی اچھی رہے گی۔

جب آپ کو کھانے کی خواہش ہو تو حیوانی پروٹین (گوشت، انڈے، مچھلی) وغیرہ زیادہ مقدار میں نہ کھائیں کیونکہ یہ نقصان دہ جگر کے کام کو بہت مشکل بناتے ہیں۔ حیوانی چکنائی یا وجیٹیبل آئل میں پکا کھانا بھی نہ کھائیں۔ ان کے بجائے پھل اور تازہ یا بھاپ میں پکی سبزیاں اور معمولی مقدار میں پروٹین آمیز غذائیں کھائیں۔ کم از کم چھ ماہ الکحل مت پئیں۔

بچاؤ

ہپاٹائٹس بی اور سی کے وائرس جنسی ملاپ کے دوران، غیر صاف شدہ سوئی سے لگائے گئے انجکشنوں، متاثرہ خون کی منتقلی اور متاثرہ ماں سے بچے کو منتقل ہو سکتے ہیں۔ ہپاٹائٹس سے دوسرے لوگوں کو بچانے کے لئے جنسی ملاپ کے دوران کنڈوم استعمال کریں (دیکھئے صفحہ 181 سے 182 اور 189 سے 192)۔ یقینی بنائیں کہ سوئیاں، سرنج، جلد کو کاٹنے یا گودنے والے آلات نقش بنانے، ختنہ کرنے، چیرا لگانے کے)استعمال سے قبل اُبال لئے جائیں۔

ہپاٹائٹس اے کا وائرس آلودہ پانی یا غذا کے ذریعے منتقل ہوتا ہے۔ دوسروں کو بیماری سے تحفظ دینے کے لئے ضروری ہے کہ بیمار فرد کا فضلہ غذا یا پانی کو آلودہ نہ کرے اور بیمار فرد صاف ستھرا رہے۔ بیمار افراد، اس کے گھر والے، دیکھ بھال کرنے والے سب صاف ستھرے رہیں اور اپنے ہاتھ وقفہ وار دھوتے رہیں۔

ہپاٹائٹس اے اور بی کی ویکسین موجود ہے لیکن یہ کچھ مقامات پر مہنگی ہو سکتی ہیں یا ممکن ہے دستیاب ہی نہ ہوں۔ اگر آپ حمل کے دوران ویکسین حاصل کر سکتی ہیں تو اس سے بچے میں آپ کے ذریعے وائرس منتقل نہیں ہوگا۔

کوئی جنسی مرض ہو تو آپ کیا کریں؟

اگر آپ یا آپ کا شوہر کسی جنسی مرض میں مبتلا ہو تو

● اس مرض کا فوری طور پر علاج شروع کر دیں۔ ابتداء ہی میں علاج کو تاخیر سے ہونے والے سنگین مسائل سے تحفظ دے گا اور انفیکشن مزید نہیں پھیلے گا۔

● اگر ممکن ہو تو ٹیسٹ کرائیں۔ اپنی بیماری کی تشخیص کے لئے کسی کلینک یا اسپتال جائیں جہاں ٹیسٹ ہو سکے اس صورت میں آپ کو وہ دوائیں لینے کی ضرورت نہیں ہوگی جو غیر ضروری ہوں۔ اگر ٹیسٹ ممکن نہ ہو تو علاج کے لئے کسی تجربہ کار صحت کارکن سے مشورہ لیں۔

● اپنے علاج کے ساتھ اپنے شوہر کا علاج بھی کرائیں، ورنہ وہ جنسی ملاپ کے دوران آپ کو پھر اسی انفیکشن میں مبتلا کر دے گا۔ اسے مناسب دوا لینے یا کسی صحت کارکن سے ملنے پر مجبور کریں۔

● یقینی بنائیں کہ آپ تمام دوائیں لیں خواہ علامات ختم ہونا شروع کر دیں۔ دوا ادھوری مقدار میں مت لیں۔ جب آپ اور آپ کا شوہر پوری دوا نہیں لے گا تو آپ دونوں صحت مند نہیں ہوں گے (دیکھئے صفحہ 327)

● جنسی ملاپ کا محفوظ طریقہ اپنائیں۔ اگر آپ احتیاط نہیں کریں گی تو آپ کسی بھی جنسی مرض میں مبتلا ہو سکتی ہیں (دیکھئے صفحہ 180 تا 182)۔

ایچ آئی وی / ایڈز کیا ہے؟

ایچ آئی وی (ہیومن امیونو ڈیفیشینسی وائرس جسم کے مدافعتی نظام (جسم کے انفیکشنز اور بیماریوں سے لڑنے والے حصے کو کمزور کرنے والا) ایک چھوٹا سا جرثومہ ہے جسے آپ دیکھ نہیں سکتیں ہیں۔ ایچ آئی وی زیادہ تر جنسی ملاپ کے دوران ایک فرد سے دوسرے فرد میں منتقل ہوتا ہے۔ اگر ایک مرد حاملہ عورت میں ایچ آئی وی منتقل کر دے یا عورت حاملہ ہونے سے پہلے ہی ایچ آئی وی رکھتی ہو تو حمل کے دوران ولادت کے وقت یا بریسٹ فیڈنگ کے دوران وائرس بچے میں منتقل ہو سکتا ہے۔ ایچ آئی وی سے تحفظ کے طریقے جاننے کے لئے دیکھئے صفحہ 170، 171۔

ایڈز (اکوئرڈ امیون ڈیفیشینسی سنڈروم) ایک ایسی بیماری ہے جو کسی فرد کے ایچ آئی وی سے متاثر ہونے کے کچھ عرصے بعد پروان چڑھتی ہے۔ ایک ایسے فرد کو ایڈز زدہ کہا جاتا ہے جب وہ معمول سے زیادہ صحت کے عام عارضوں میں مبتلا ہونے لگتا ہے۔ وزن کم ہونا، صحیح نہ ہونے والے زخم، بدترین کھانسی، رات کے وقت پسینہ آنا، ڈائریا، جلد کے دھبے، بخار، فرج سے رطوبت کا اخراج یا سارا وقت انتہائی تھکن محسوس کرنا ایڈز کی چند علامات ہیں لیکن ان کی دوسری وجوہ بھی ہو سکتی ہیں۔ خصوصی بلڈ ٹیسٹ کے بغیر کسی فرد کے بارے میں یقین سے نہیں کہا جا سکتا ہے کہ وہ ایچ آئی وی / ایڈز میں مبتلا ہے (دیکھئے صفحہ 172)۔

ایچ آئی وی سے متاثرہ مدافعتی نظام ہر بیماری کے بعد کمزور سے کمزور تر ہونے کے باعث ایسے فرد کا جسم بیماری سے لڑنے اور صحت یاب ہونے کی صلاحیت کھونے لگتا ہے حتیٰ کہ اس کا جسم اس حد تک کمزور ہو جاتا ہے کہ اس کی بقا ممکن نہیں رہتی اور وہ مر جاتا ہے۔ فرد معذور ہو یا غیر معذور ایچ آئی وی/ ایڈز میں یکساں طور پر مبتلا ہو سکتا ہے۔

کچھ لوگ ایچ آئی وی ہونے کے بعد بہت بہت جلد ایڈز سے مر جاتے ہیں لیکن بہت سے لوگ ایڈز ہونے سے پہلے کئی برس گزار لیتے ہیں۔ اس کا مطلب یہ ہے کہ ایک شخص ایچ آئی وی سے متاثر ہو سکتا ہے لیکن اسے یہ معلوم نہیں ہوتا کیونکہ وہ خود کو صحت مند محسوس کرتا ہے۔ قطع نظر اس کے کہ وہ جو بھی محسوس کرتا ہو وہ دوسرے لوگوں میں ایچ آئی وی منتقلی کر سکتا ہے۔ یہ جاننے کے لئے کہ آپ کو ایچ آئی وی تو نہیں، واحد طریقہ بلڈ ٹیسٹ ہے۔ یہ ٹیسٹ بہت سے کلینکوں اور اسپتالوں میں کیا جاتا ہے۔

اے آر وی (antiretrovirals) کہلانے والی ادویات ایچ آئی وی/ ایڈز میں مبتلا افراد کو زیادہ عرصہ صحت مند رکھنے میں معاون ہو سکتی ہیں۔ یہ ادویات ایچ آئی وی کو ہلاک نہیں کرتی یا ایڈز کا علاج نہیں کرتی ہیں لیکن فرد کے لئے زندگی آسان بناتی ہیں۔ حاملہ عورت کا اے آر وی علاج بچے کو ماں سے ایچ آئی وی کی منتقلی سے تحفظ فراہم کر سکتا ہے۔ بدقسمتی سے اے آر وی ادویات کچھ ممالک میں مہنگی اور ان کا حاصل کرنا مشکل ہو سکتا ہے۔ اے آر وی ادویات کے بارے میں مزید جاننے کے لئے دیکھئے صفحہ 176۔

ایچ آئی وی/ ایڈز کس طرح پھیلتا ہے

ایچ آئی وی/ ایڈز کا سبب بنے والا وائرس ایچ آئی وی میں مبتلا افراد کی جسمانی رطوبتوں مثلاً خون، مرد کی منی، عورت کی فرج کی رطوبتوں میں رہتا ہے۔ یہ وائرس ان رطوبتوں کے کسی اور فرد کے جسم میں داخل ہونے پر اسے بھی اپنا ہدف بنا لیتا ہے۔ ایچ آئی وی مندرجہ ذیل طریقوں سے پھیل سکتا ہے۔

| اگر کنڈوم استعمال نہ کیا جائے تو ایچ آئی وی میں مبتلا فرد کے ساتھ جنسی ملاپ سے۔ | جلد میں سوراخ کرنے، کاٹنے والے غیر صاف شدہ آلات اور سوئیوں کے ذریعے۔ | کٹی ہوئی جلد، کھلے زخم میں متاثرہ خون شامل ہونے سے۔ | حمل، ولادت یا بریسٹ فیڈنگ کے ذریعے متاثرہ ماں سے بچے میں۔ |

جہاں خون کا ایچ آئی وی ٹیسٹ نہیں ہوتا ہے لوگ متاثرہ خون کی منتقلی سے بھی ایچ آئی وی میں مبتلا ہو سکتے ہیں۔

ایچ آئی وی/ ایڈز سے تحفظ کی مزید معلومات کے لئے دیکھئے صفحہ 179 تا 182۔

ایچ آئی وی/ایڈز کس طرح نہیں پھیلتا ہے

ایچ آئی وی انسانی جسم کے باہر چند منٹ سے زیادہ زندہ نہیں رہتا ہے۔ یہ ہوا یا پانی میں زندہ نہیں رہتا ہے۔ اس کا مطلب یہ ہے کہ کسی کو بھی ایچ آئی وی مندرجہ ذیل طریقوں سے نہیں ہوسکتا ہے۔

- چھونے، لپٹانے یا بوسہ لینے سے
- ساتھ کھانے سے
- ساتھ سونے سے
- کپڑوں، بستر یا ٹوائلٹ کے مشترکہ استعمال سے
- حشرات کے کاٹنے سے

چھونے سے ایچ آئی وی/ایڈز منتقل نہیں ہوتا ہے

ایچ آئی وی/ایڈز عورتوں کو کس طرح متاثر کرتا ہے؟

مردوں کی بہ نسبت ایچ آئی وی عورتیں زیادہ جلدی ایڈز میں مبتلا ہوتی ہیں۔ ناقص غذائیت اور بچوں کی ولادت عورتوں کو بیماری سے لڑنے کے قابل نہیں رہنے دیتی ہے۔ اس لئے عورتیں مردوں کی بہ نسبت آسانی سے ایچ آئی وی سے متاثر ہوجاتی ہیں۔ جب جنسی ملاپ کے دوران مرد کی منی عورت کے جسم میں داخل ہوتی ہے تو ایچ آئی وی اس کی فرج یا رحم کے منہ کے ذریعے اس کے خون میں آسانی سے منتقل ہوسکتے ہیں۔ خصوصاً اس وقت جب کوئی کٹاؤ یا زخم ہو۔ عورت معذور ہو یا نہ ہو ایسا ہوسکتا ہے۔

معذور عورتوں اور ایچ آئی وی/ایڈز کے بارے میں خطرناک تصورات

ایچ آئی وی/ایڈز کے حوالے سے ایک انتہائی نقصان دہ اور غلط تصور یہ ہے کہ اگر ایچ آئی وی/ایڈز میں مبتلا شخص کسی ایسی عورت کے ساتھ مباشرت کرے جس نے کبھی مباشرت نہ کی ہو تو وہ صحت یاب ہوجائے گا۔ اس غلط تصور کے باعث ایچ آئی وی/ایڈز زدہ شخص کسی معذور عورت کو اپنا نشانہ بنا سکتا ہے اگر وہ یہ سمجھتا ہے کہ معذور ہونے کے باعث وہ کنواری ہوگی اور اس کا علاج ثابت ہوسکتی ہے۔ یہ خیال درست نہیں ہے۔

کنواری لڑکی یا دوشیزہ سے جنسی ملاپ سے ایچ آئی وی/ایڈز ہی پھیل سکتا ہے۔ یہ عمل کسی ایچ آئی وی ایڈز زدہ کو دوبارہ صحت مند نہیں بنائے گا۔

کچھ عرصہ پہلے میں ایک ایڈز زدہ شخص سے ملی جس نے کہا کہ اگر وہ شادی کے بغیر کسی عورت کے ساتھ رہنا پسند کرے گا تو وہ میں ہوں۔ جب میں نے اس سے سبب پوچھا تو اس نے صاف گوئی سے کہا کہ اسے یقین ہے کہ میں نہ تو حاملہ ہوسکتی ہوں اور یہ کہ میں ایچ آئی وی سے محفوظ ہوں۔ احمق شخص!

یہ جاننا کہ کیا آپ ایچ آئی وی ہیں

ایچ آئی وی ٹیسٹ

ایچ آئی وی جسم میں داخل ہوتا ہے تو جسم کا مدافعتی نظام وائرس سے لڑنے کے لئے فوری طور پر اینٹی باڈیز بنانا شروع کر دیتا ہے۔ دو سے چار ہفتوں میں ایچ آئی وی ٹیسٹ خون میں موجود ان اینٹی باڈیز کا پتہ لگا سکتا ہے۔

ایچ آئی وی پازیٹیو ٹیسٹ کا مطلب یہ ہے کہ آپ وائرس زدہ ہیں اور آپ کے جسم نے ایچ آئی وی کے لئے اینٹی باڈیز بنا لی تھیں۔ اگر آپ خود مکمل طور پر بہتر اور صحت مند محسوس کریں تو جب بھی آپ دوسروں کو ایچ آئی وی منتقل کر سکتی ہیں۔

ایچ آئی وی نگیٹیو ٹیسٹ کے دو مطلب ہو سکتے ہیں۔

- آپ ایچ آئی وی سے متاثر نہیں ہیں۔
- آپ حال ہی میں ایچ آئی وی میں زدہ ہوئی ہیں اور آپ کے جسم نے اب تک اتنی اینٹی باڈیز نہیں بنائی ہیں جو ایچ آئی وی ٹیسٹ میں ظاہر ہو سکیں۔

اگر آپ کا ایچ آئی وی ٹیسٹ نگیٹیو ہے مگر آپ سمجھتی ہیں کہ آپ ایچ آئی وی سے متاثر ہو سکتی ہیں تو آپ کو تقریباً چھ ہفتے کے بعد دوبارہ یہ ٹیسٹ کرانا چاہیے۔ بعض اوقات پازیٹیو ٹیسٹ بھی دوبارہ کرانے کی ضرورت ہوتی ہے، ایک تجربہ کار صحت کارکن یہ فیصلہ کرنے میں آپ کی مدد کر سکتا ہے۔

نوٹ ایچ آئی وی کے لئے جانچ اور مشاورت ایک ساتھ کی جاتی ہے اور اب یہ زیادہ دستیاب ہے۔ کسی صحت کارکن سے معلوم کیجیے کہ آپ کے قرب و جوار میں یہ ٹیسٹ کہاں ہو سکتا ہے۔ اب بہت سے اسپتالوں اور کلینکوں میں یہ ٹیسٹ بہت کم قیمت پر یا مفت ہوتا ہے۔ آپ کو ٹیسٹ کا نتیجہ اسی روز مل سکتا ہے۔ کچھ ٹیسٹ مراکز بریل میں یا کچھ علامتی زبان میں معلومات فراہم کرتے ہیں۔

اہم بات متاثر ہونے کی صورت میں خواہ آپ خود کو صحت مند محسوس کرتی ہوں یا نظر آتی ہوں، جب بھی آپ ایچ آئی وی دوسرے افراد میں منتقل کر سکتی ہیں۔ آپ کسی فرد کو صرف دیکھ کر نہیں کہہ سکتی ہیں کہ وہ ایچ آئی وی سے متاثر ہے۔ کسی کے ایچ آئی وی ہونے کے بارے میں جاننے کا واحد ذریعہ ایچ آئی وی ٹیسٹ ہے۔

مشاورت

ایچ آئی وی ٹیسٹ ہونا چاہئے۔

- آپ کی اجازت سے۔
- ٹیسٹ سے پہلے اور بعد میں مشاورت کے ساتھ۔
- راز داری کے ساتھ۔

صرف آپ اور جن کو آپ چاہیں نتائج معلوم ہونا چاہئیں۔

ایک تجربہ کار ایچ آئی وی/ایڈز صلاح کار یہ فیصلہ کرنے میں آپ کی مدد کر سکتا ہے کہ آپ کو ایچ آئی وی ٹیسٹ کرانے کی ضرورت ہے کہ نہیں۔ اگر آپ کا ٹیسٹ پازیٹیو ہے۔ صلاح کار یہ فیصلہ کرنے میں آپ کی مدد کر سکتا ہے کہ یہ کس کو بتایا جائے اور آپ اپنی زندگی کی اس تبدیلی کا مقابلہ کس طرح کریں۔

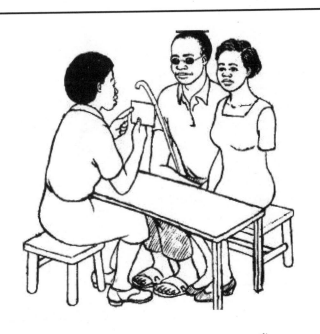

ایچ آئی وی سے متاثر افراد اور ان کے گھر والوں کے لئے مشاورت کا مطلب امید اور مایوسی کے مابین فرق ختم کرنا ہے۔ جیسا کہ کینیا کی ایچ آئی وی سے متاثر ایک عورت کہتی ہے "جب آپ کسی اچھے صلاح کار سے ملتے ہیں تو محسوس کرتے ہیں کہ آپ صحت مند ہو چکے ہیں"۔

ایچ آئی وی/ایڈز ایک لاعلاج عارضہ ہے۔ اس کا وجود تقریباً ہر جگہ ہے غیر محفوظ ماحول، صحت کی ناقص سہولیات اور عمومی عدم آگہی کے باعث یہ بڑی تیزی سے پھیل سکتا ہے۔ یہی وجہ ہے کہ عالمی ماہرین احتیاطی اقدام پر زور دے رہے ہیں۔ ایچ آئی وی/ایڈز معذور عورتوں کے لئے خطرہ ہے۔ دنیا بھر میں وہ صحت کی ناکافی سہولتیں رکھتی ہیں لہذا انہیں خصوصی احتیاط سے کام لینا چاہیئے۔ ایچ آئی وی ہونے کے بعد ایڈز کی علامات سامنے آنے تک متاثرہ فرد دوسرے صحت مند افراد کو بھی ایچ آئی وی میں مبتلا کر سکتا ہے حالانکہ وہ ظاہری طور پر صحت مند اور مستعد نظر آتا ہے

آپ کی رازداری کا تحفظ

کسی بھی عورت کو یہ فیصلہ خود کرنے کے قابل ہونا چاہئے کہ کس کو اس کے ایچ آئی وی ہونے کے بارے میں اور کس طرح بتایا جائے۔ عورت کے لئے ضروری ہے کہ وہ اپنے شوہر سے اس بارے میں بات کرے تا کہ وہ بھی ایچ آئی وی ٹیسٹ کرائے یا اپنی صحت کا تحفظ کرے۔ بہت سی عورتیں اپنے خاندان اور اپنی مدد کرنے والوں کو یہ بات بتاتی ہیں لیکن عورتیں عموماً اس بات سے خوفزدہ ہوتی ہیں کہ دوسروں کو یہ بات معلوم نہ ہوجائے۔

ایک معذور عورت کے لئے کسی صحت کارکن کے ساتھ رازدارانہ طور پر بات کرنا مشکل ہوسکتا ہے۔ یہ اس لئے ہوسکتا ہے کہ

- صحت کارکن نے کبھی یہ نہیں سیکھا ہو کہ معذوری کا شکار عورت کے ساتھ بھی دوسری عورتوں کی طرح احترام کے ساتھ پیش آنا چاہئے۔

- صحت کارکن معذور عورت کے گھر والوں اور احباب کو اس کے کچھ کہنے سے پہلے ہی اس کے صحت کے مسائل سے آگاہ نہ کر دے جن میں ایچ آئی وی یا کوئی جنسی بیماری بھی شامل ہے۔ یہ بات خصوصی طور پر اس صورت میں درست ہے جب معذور عورت ابلاغ میں دشواری رکھتی ہو۔

- عورت کے گھر والے اسے اپنے طور پر کسی صحت کارکن سے نہیں ملنے دیتے ہوں۔

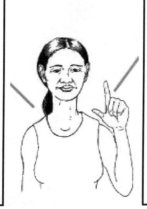

اگر میں اپنا ترجمان ساتھ لے جاتی ہوں تو مجھے وقت سے پہلے یاد دلانا پڑتا ہے کہ میں صحت کارکن سے نجی طور پر جو باتیں کروں گی وہ پرائیوٹ ہیں۔ میں یہ یقینی بنانے کی کوشش کرتی ہوں کہ میری ترجمان سمجھ لے کہ ٹیسٹ کے نتائج میرا نجی معاملہ ہیں۔ میں اس سے درخواست کرتی ہوں کہ وہ میری اجازت کے بغیر اس بارے میں کسی اور کو نہ بتائے، میرے گھر والوں کو بھی۔

کیونکہ میں بہری ہوں لہٰذا مجھے اکثر اپنی باتیں راز میں رکھنے میں دشواری ہوتی ہے۔ خاص طور پر اس وقت جب میں صحت کارکن سے بات کرنے کے لئے کسی اور کی مدد لیتی ہوں۔ وہ کلینک جہاں میں جاتی ہوں وہاں کی ترجمان جانتی ہے کہ میرے اور صحت کارکن کے درمیان جو باتیں ہوتی ہیں وہ رازدارانہ اور نجی ہوتی ہیں۔ وہ کسی کو بھی یہ نہیں بتائے گی، کسی دوسری صحت کارکن کو بھی نہیں کہ ہم نے کیا باتیں کیں۔

ایچ آئی وی/ایڈز کے مسائل صحت کے لئے مدد حاصل کرنا

ایک ایڈز زدہ فرد بڑی آسانی سے بہت سے مختلف مسائل صحت کا شکار ہوسکتا ہے۔ ایسے کچھ مسائل کے بارے میں عام معلومات درج ذیل ہیں لیکن مزید معلومات کے لئے کسی صحت کارکن سے بات کرنا 'ایچ آئی وی۔ ہیلتھ اینڈ یورکمیونٹی' جیسی کتاب پڑھنا بہترین ہے۔ ایچ آئی وی/ایڈز کے باعث پیش آنے والے انتہائی عام مسائل۔

بخار : بخار اکثر چڑھتا اور اترتا رہتا ہے۔ یہ جاننا مشکل ہے کہ بخار کا سبب ایسا کوئی انفیکشن ہے جس کا علاج ہوسکتا ہے۔ جیسے ٹی بی، پیٹرو کی بیماری (PID) یا ملیریا یا بخار کا سبب بذات خود ایچ آئی وی ہے۔ اگر بخار کسی انفیکشن کے باعث ہے تو یقینی بنائے کہ انفیکشن کا علاج ہو۔

اسہال: اسہال بھی اکثر ہوتے ہیں اور ان کا علاج مشکل ہوسکتا ہے۔ایڈز میں مبتلا فرد میں اسہال کا انتہائی عام سبب ایچ آئی وی کے باعث ہونے والے انفیکشنز یا کچھ ادویات کے ضمنی اثرات ہیں۔

جلد کے دھبے اور خارش: عموماً یہ جاننا دشوار ہوتا ہے کہ جلد کے دھبوں اور خارش کا سبب کیا ہے۔ایچ آئی وی /ایڈز سے تعلق رکھنے والے جلد کے کچھ مسائل مندرجہ ذیل کے باعث ہوسکتے ہیں۔

- ادویات کا الرجک ردِّ عمل

- منہ یا جلد پر بھورے یا اُودے دھبے، Kaposi's Sarcoma کہلانے والے خون کی رگوں یا لمف نوڈز کینسر کے باعث ہوتے ہیں۔

- ہرپیز زوسٹر (شنگلز)، عام طور پر ابتداء میں سرخ دانے چھالوں کے ساتھ نمودار ہوتے ہیں جن سے تکلیف ہوتی ہے۔یہ زیادہ تر چہرے، کمر اور چھاتی پر ہوتے ہیں۔

متلی اور قے: یہ انفیکشنز ، بعض ادویات، معدے اور آنتوں کی خراب یا ایچ آئی وی انفیکشن کے باعث ہوسکتا ہے۔

کھانسی: نمونیے یا ٹی بی جیسے پھیپھڑوں کے مسائل کی علامت ہوسکتی ہے۔ پھیپھڑے جب عفونت زدہ ہوں تو زیادہ مقدار میں بلغم بناتے ہیں جو کھانسی کا سبب بنتی ہے۔

ٹی بی: یہ پھیپھڑوں پر اثر انداز ہونے والے ایک جرثومے سے ہونے والا لاگین انفیکشن ہے۔ایڈز اور ٹی بی کی علامات ملتی جلتی ہیں مگر یہ دونوں مختلف بیماریاں ہیں۔ٹی بی میں مبتلا بیش تر عورتیں، مرد اور بچے ایڈز زدہ نہیں ہوتے ہیں لیکن ایڈز میں مبتلا کچھ لوگ جسم کی کمزور مدافعت کے باعث آسانی سے ٹی بی میں مبتلا ہو جاتے ہیں۔ایڈز کے باعث مرنے والے ہر تین میں ایک فرد کی موت کا سبب ٹی بی ہوتی ہے۔

منہ اور حلق کے مسائل: ان مسائل میں خراشیں، تڑخن، زخم، چھالے اور زبان پر سفید دھبے (تھرش دیکھئے صفحہ 260) شامل ہیں۔

وزن میں کمی اور ناقص غذائیت: ایڈز میں مبتلا فرد ڈائریا یا مسلسل بیمار رہنے سے ناقص غذائیت کا شکار ہوجاتا ہے جو جسم کو غذا میں موجود غذائیت جذب کرنے سے روکتی ہے۔بھوک کا احساس ختم ہونا اور منہ کے انفیکشنز بھی اس کے لئے کچھ کھانا مشکل بنا دیتے ہیں۔ایچ آئی وی میں مبتلا افراد میں وزن کی کمی اس قدر عام ہے کہ افریقہ کے کچھ علاقوں میں اسے ''دبلے پن کی بیماری'' کہتے ہیں۔

ایچ آئی وی/ایڈز کا علاج

نہ تو جدید ادویات نہ ہی صحت کے روایتی طریقوں میں ایڈز کا کوئی علاج ہے لیکن ایڈز میں مبتلا فرد کی مدد کے لئے ایسی بہت سی چیزیں ہیں جو کی جا سکتی ہیں۔ صاف پانی، اچھی غذا، صاف کپڑے، آرام اور نیند کے لئے صاف ستھری جگہ، دوستوں اور گھر والوں سے محبت آمیز تعلقات کسی بھی ایڈز زدہ فرد کو صحت مند رہنے میں ضروری مدد فراہم کر سکتے ہیں۔ ایڈز سے تعلق رکھنے والی بیماریوں میں وہی غذائیں اور ایسے فرد کے لئے موزوں ہیں جو کسی صحت مند فرد کے لئے موزوں اور اچھی ہیں (دیکھئے صفحہ 177 تا 178)

اگرچہ ایڈز کا کوئی علاج نہیں ہے، اے آر وی ادویات (antiretroviral medicines) ایڈز زدہ بیمار افراد کے مفاد میں کامیابی کے ساتھ استعمال کی جا رہی ہیں۔ اے آر وی ادویات مدافعتی نظام کو مضبوط بناتی ہیں لہٰذا ایچ آئی وی کا شکار افراد انفیکشنز سے لڑ کر صحت مند رہ سکتا ہے، لیکن ایچ آئی وی کا علاج نہیں ہوتا ہے۔ فرد کے جسم میں وائرسوں کی قلیل تعداد ہمیشہ رہتی ہے جب کسی شخص میں ایچ آئی وی ہو تو وہ کسی اور میں ایچ آئی وی منتقل کر سکتا ہے۔

معذوری والی عورتوں کے لئے صحت کی اچھی دیکھ بھال کا حصول آسان نہیں ہے اور جو ایچ آئی وی/ایڈز سے متاثر ہوں ان کے لئے بے حد مشکل ہو سکتا ہے۔ ممکن ہے صحت کا رکن ان کا ٹیسٹ یا علاج اس لئے نہیں کرنا چاہتے ہوں کہ ان کے خیال میں معذور عورتیں نہ تو جنسی ملاپ کر سکتی ہیں اور نہ ہی ایچ آئی وی/ایڈز کا شکار ہو سکتی ہیں یا یہ کہ وہ اس صورت میں بہت جلد مر سکتی ہیں۔ لیکن معذوری کا شکار عورتیں بھی دوسری عورتوں کی طرح ایچ آئی وی سے متاثر ہونے کا یکساں خطرہ رکھتی ہیں اور اگر انہیں علاج کی سہولت ملے تو وہ زیادہ عرصہ زندہ اور صحت مندرہ سکتی ہیں۔

اینٹی ریٹرو وائرل علاج (اے آر ٹی)

اے آر ٹی علاج کا مطلب میں کم از کم دو بار تین اینٹی ریٹرو وائرل دوائیں کھانا ہے۔ ایڈز زدہ فرد جب اے آر ٹی لینا شروع کرے تو یہ ادویات روزانہ پوری پابندی کے ساتھ لینا چاہئے۔ اے آر ٹی لینے والی عورت کے وزن میں اضافہ ہو گا اور وہ صحت مند نظر آئے گی اور خود کو صحت مند بھی محسوس کرے گی۔ لیکن اگر وہ اے آر ٹی روک دیتی ہے یا دوا کا ناغہ کرتی ہے یا غلط وقت پر کھاتی ہے تو اس کا ایچ آئی وی زیادہ طاقتور ہو سکتا ہے اور اسے دوبارہ بیمار ڈال سکتا ہے۔ (معذور عورتوں کے علاج یا ماں سے بچے میں ایچ آئی وی کی منتقلی سے بچاؤ کی مزید معلومات کے لئے دیکھئے صفحہ 385 سے 362)۔

اگرچہ اے آر ٹی ادویات مہنگی ہیں یہ بہت سے ممالک میں یہ سستی ہو رہی ہیں اور بڑے پیمانے پر دستیاب ہیں۔ سرکاری مراکز صحت اور دیگر پروگرام اے آر ٹی کم قیمت پر یا مفت فراہم کر سکتے ہیں۔

اس کے باوجود بہت سے مقامات پر ایچ آئی وی/ایڈز میں مبتلا بیشتر لوگوں کے لئے ادویات دستیاب نہیں ہیں۔ امیر ممالک کی بڑی ادویات ساز کمپنیوں کے اثر رسوخ نے دوسرے ممالک میں کم قیمت ادویات کی تیاری بند کرا دی ہے۔ اس سے ایچ آئی وی/ایڈز کے علاج کے لئے ضرورت مند لاکھوں عورتوں کی ادویات تک رسائی کی حق تلفی ہوتی ہے۔

ادویات کے ذریعے کچھ انفیکشنز سے بچاؤ

ایچ آئی وی/ایڈز میں مبتلا لوگوں کے لئے اینٹی بایوٹک دوا Cotrimoxazole کا باقاعدگی سے استعمال انہیں نمونیا، اسہال اور دیگر انفیکشنز سے تحفظ فراہم کرتا ہے۔ اگر آپ وزن میں کمی، ہونٹوں کے گرد زخموں یا تڑخن، خارشی دھبوں، شنگلز، منہ کے السر یا تواتر سے سردی کا شکار ہو رہی ہیں تو آپ یہ دوا لینا شروع کر دیں۔

علاج: روزانہ بہت سے پانی کے ساتھ 960 ملی گرام (دوہری قوت) کوٹری موکسازول کھائیں۔ اگر ممکن ہو تو یہ دوا روزانہ لیں خواہ آپ خود کو بیمار محسوس کریں یا نہ کریں۔

اہم بات ایڈز میں مبتلا افراد میں کوٹری موکسازول کے الرجک ردِعمل زیادہ عام ہیں۔ اگر جلد پر نئے دھبے پڑیں یا دوا کی الرجی کی دوسری علامات نمودار ہوں تو دوا لینا بند کر دیں۔

اہم بات کچھ عورتیں اینٹی بائیوٹک ادویات کھاتی ہیں تو وہ فرج کے یسٹ انفیکشنز کا شکار ہو جاتی ہیں۔ اس صورت میں دہی کھانا یا کھٹا دودھ پینا یا کچھ دہی یا سرکہ ملے پانی میں بیٹھنا مددگار ہو سکتا ہے۔ یسٹ انفیکشنز کے بارے میں مزید معلومات کے لئے دیکھئے صفہ 111 تا 113۔

کچھ ممالک میں ایچ آئی وی افراد کے لئے یہ بھی سفارش کی جاتی ہے کہ وہ ٹی بی سے تحفظ کی ادویات بھی لیں۔ اس بارے میں معلومات کے لئے کسی تجربے کار صحت کے رکن سے بات کیجئے۔

اچھی طرح کھانا

ایڈز جسم کی غذا ہضم کرنے کی صلاحیت پر اثر انداز ہوتا ہے اور یہ لوگوں میں بھوک محسوس کرنے کی صلاحیت ختم ہونے کا سبب بھی بنتا ہے لہذا وہ انتہائی دُبلے ہو جاتے ہیں۔ ایسا ادویات کے ضمنی اثرات، منہ اور حلق کے مسائل، ڈائریا اور چکنائی ہضم کرنے میں دشواری کے باعث بھی ہوتا ہے۔

اگر آپ ایچ آئی وی ہیں تو آپ کے لئے خصوصی طور پر اہم ہے کہ آپ اچھی طرح کھانا کھائیں تا کہ آپ کا وزن کم نہ ہو اور آپ کا جسم اور مدافعتی نظام جس حد تک ممکن ہو صحت مند رہے۔ اس کے لئے آپ مختلف یا متوازن غذائیں کھانے کی کوشش کریں (دیکھئے صفہ 86)۔ صاف پانی پئیں اور روزانہ ملٹی وٹامن لیں۔ اگر دستیاب ہو تو آپ وٹامن اے، سی اور ای کے سپلیمنٹس بھی لیں کیونکہ یہ آپ کے جسم میں ایچ آئی وی کی افزائش کی صلاحیت بھی کم کرتے ہیں۔

وٹامن اے پر مشتمل غذاؤں میں گاجر، آم، پپیتا، میٹھے آلو، دودھ، انڈے اور سبز پتوں والی ترکاریاں شامل ہیں۔

وٹامن سی والی غذاؤں میں لال اور ہری مرچیں، گہرے سبز رنگ کی ترکاریاں، سنگترہ، زرد اور سرخ پھل شامل ہیں۔

وٹامن ای والی غذاؤں میں انڈے، بادام، مٹکی، کھوپرے، مونگ پھلی، سورج مکھی کے بیج، گندم اور زیتون کا تیل شامل ہے۔

اگر آپ کی اشتہا ختم ہوگئی ہے تو آپ کے لئے صبح کے وقت زیادہ مقدار میں کھانا بہترین ہوسکتا ہے یا آپ دن میں چھ سے آٹھ بار بھی کھا سکتی ہیں۔ کھانے کے ساتھ ٹھنڈے مشروبات پینا، غذا کو نگلنا آسان بنا سکتا ہے۔

ایچ آئی وی / ایڈز کے ساتھ پُر امید انداز سے رہنا

مجھے ایڈز ہے لیکن میں اچھی غذا، صاف پانی اور ادویات حاصل کرسکتی ہوں۔ میں اچھی زندگی گزار رہی ہوں اور پوسٹ آفس میں ڈاک چھانٹنے کی اپنی ملازمت جاری رکھ سکتی ہوں۔

آپ صحت مند رہ سکتی ہیں اگر آپ

- صاف اور محفوظ پانی پئیں اور اپنا کھانا بھی اسی سے بنائیں۔
- بغیر دھلی کی سبزیوں سے گریز کریں۔ انہیں ہضم کرنا جسم کے لئے مشکل ہوتا ہے اور ان میں جراثیم بھی ہوسکتے ہیں۔
- مائعات خوب پئیں اور قلت آب کی علامات پر نظر رکھیں۔
- جب بھی آپ تھکن محسوس کریں، آرام کریں اور روزانہ کم از کم آٹھ گھنٹے سوئیں۔
- دوستوں اور گھر والوں کے ساتھ وقت گزاریں۔
- وہ کام کریں جن سے آپ لطف اندوز ہوں، اچھے احساسات بھی صحت مند رہنے کے لئے ضروری ہیں۔
- حد سے زیادہ فکرمندی سے بچنے کی کوشش کریں، دباؤ، کھچاؤ مدافعتی نظام کو نقصان پہنچا سکتا ہے۔
- اپنے روزمرہ کے کام کا انجام دے کر خود کو مستعد و فعال رکھنے کی کوشش کریں۔
- جس قدر ممکن ہو ورزش کریں (دیکھئے صفحہ 89 تا 95)۔
- تمباکو، الکحل اور دوسری منشیات سے گریز کریں۔
- ہاتھوں کو بار بار دھو کر انفیکشن سے تحفظ حاصل کریں۔
- نئے انفیکشنز اور غیر مطلوب حمل سے بچاؤ کے لئے محفوظ جنسی عمل طریقہ اپنائیں یہ مدافعتی نظام کو کمزور کر سکتے ہیں (دیکھئے صفحہ 180)۔
- طبی مسائل پر ابتداء میں توجہ دیں۔ ہر انفیکشن آپ کے مدافعتی نظام کو مزید کمزور کرتا ہے۔
- ڈائریا سے بچاؤ کے لئے اوری ری موکسازول لیں (دیکھئے صفحہ 339)۔
- اگر آپ کے علاقے میں ملیریا عام ہو تو مچھر دانی استعمال کریں۔

ان حالات کے خلاف جنگ کریں جو بیماری کے پھیلاؤ کا سبب ہیں، ان لوگوں کے خلاف نہیں جو اس سے متاثر ہیں۔ امتیازی رویہ دیکھ بھال کی راہ میں حائل رکاوٹ ہے۔ یہ لوگوں کو انفیکشن کے پھیلاؤ سے بچنے کے طریقے سیکھنے سے روک سکتا ہے۔

ایچ آئی وی/ایڈز اور رسوائی

اکثر معاشروں میں ایچ آئی وی یا ایڈز زدہ افراد کو حقارت سے دیکھا جاتا ہے۔ معاشرے میں کوئی بھی ان سے رابطہ نہیں رکھتا ہے۔ لوگ ایچ آئی وی/ایڈز میں مبتلا افراد کے گھر والوں کو سماج کے لئے توہین کا باعث سمجھتے ہیں۔

لاکھوں ایچ آئی وی پازیٹیو اپنی بیماری کو چھپائے ہوئے ہیں کیونکہ وہ خوفزدہ ہیں کہ ان کے دوست، گھر والے اور پڑوسی انہیں مسترد کر دیں گے حالانکہ عمومی رابط/میل جول سے ایچ آئی وی/ایڈز کسی میں منتقل نہیں ہوتا ہے۔

ایڈز میں مبتلا بہت سے لوگ اور ان کے گھر والے اس لئے صرف دوسروں سے مدد نہیں کرتے ہیں کہ لوگوں کے علم میں آنے کے بعد ان کے ردعمل سے ذلت اور شرمندگی محسوس کریں گے۔ یہ صورت کسی بھی ایڈز زدہ فرد کے لئے مدد حاصل کرنے اور علاج کو بے حد مشکل بناتی ہے حالانکہ ایسی ادویات بھی دستیاب ہیں جو ایڈز زدہ لوگوں کو طویل عرصہ صحت مند زندگی گزارنے کی سہولت فراہم کرتی ہیں۔

گھر میں انفیکشن سے بچاؤ

بہت سے لوگ سمجھتے ہیں کہ ایچ آئی وی آسانی سے پھیل سکتا ہے۔ یہ درست نہیں ہے اگر آپ رہنمائی کے مطابق عمل کریں تو کسی متاثرہ فرد سے اردگرد کے لوگوں کو ایچ آئی وی یا ہپپا ٹائٹس منتقل ہونے کا کوئی خطرہ نہیں ہوگا یا نہ ہی آپ ایچ آئی وی یا ہپپا ٹائٹس کا شکار ہو سکتی ہیں۔

- کوئی ایسی چیز استعمال مت کریں جس میں خون لگتا ہو، ان میں بلیڈ، سوئیاں، تیز دھار کوئی بھی آلہ جو کھال کاٹتا ہو اور ٹوتھ برش بھی شامل ہے۔ اگر آپ ایسی کوئی چیز مشترکہ طور پر استعمال کرتی ہوں تو استعمال سے پہلے اسے بیس منٹ پانی میں اُبالیں۔

- تمام زخموں کو صاف پٹی یا کپڑے سے ڈھک کر رکھیں۔ ایچ آئی وی یا ہپپا ٹائٹس ہو یا نہ ہو ہر ایک کو ایسا ہی کرنا چاہیئے۔

- استعمال شدہ پٹیوں کو جو دھوئی نہ جا سکتی ہوں جلا دیں یا زمین میں دفن کر دیں۔

- اپنے ہاتھوں پر جسمانی رطوبتیں نہ لگنے دیں۔ گندی پٹیوں، کپڑوں، خون، قے یا فضلہ کو ٹھکانے لگاتے ہوئے پلاسٹک یا کاغذ کے دستانے استعمال کریں۔

- گندی چادریں یا کپڑ چھونے کے بعد ہاتھوں کو صابن اور پانی سے اچھی طرح دھوئیں۔

- بستر اور کپڑوں کو صاف رکھیں۔ انہیں دھوتے وقت
  - ☆ دوسرے کپڑوں سے الگ رکھیں
  - ☆ ان کے غیر آلودہ حصے کو پکڑ کر جسمانی رطوبت والے حصوں پر پانی بہائیں
  - ☆ انہیں صابن آمیز پانی میں دھوکر خشک کریں۔ اگر ممکن ہو تو انہیں دھوپ میں سکھائیں اور عام کپڑوں کی طرح تہہ کرکے یا استری کرکے رکھیں

- جب بھی گندے کپڑے دھوئیں تو ضروری نہیں کہ دستانے یا پلاسٹک کی تھیلی ہاتھوں پر چڑھائیں لیکن ایسا کرنا بہتر ہے گا۔

ایچ آئی وی سے تحفظ کے دوسرے طریقے

- جنسی طور پر منتقل ہونے والے انفیکشنز کا علاج جلد از جلد کرائیں۔ کوئی بھی جنسی عارضہ ایچ آئی وی یا دوسرے جنسی امراض سے متاثر ہونا آسان بنا دیتا ہے۔

- اس وقت تک انجکشن نہ لگوائیں جب تک یہ اطمینان نہ ہو کہ سرنج اور سوئی کو محفوظ کر لیا گیا ہے۔ صحت کا رکن سرنج یا سوئی کو جراثیم سے پاک کئے بغیر استعمال نہ کریں۔

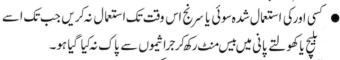

- کسی اور کی استعمال شدہ سوئی یا سرنج اس وقت تک استعمال نہ کریں جب تک اسے بلیچ یا کھولتے پانی میں بیس منٹ رکھ کر جراثیموں سے پاک نہ کیا گیا ہو۔

- یقینی بنائیے کہ ختنہ، کان چھدنے، ایکوپنکچر وغیرہ کے لئے استعمال ہونے والے آلات کو بیس منٹ اُبلتے ہوئے پانی میں رکھا گیا ہو۔

- جسمانی رطوبتوں مثلاً خون، قے، فضلہ، پیشاب وغیرہ کو محفوظ طریقے سے ٹھکانے لگائیں۔

- خون کے ایچ آئی وی اور ہیپاٹائٹس سے محفوظ ہونے کے یقین کے لئے بلڈ ٹیسٹ ہونا چاہئے۔ پھر بھی جب تک زندگی کو خطرہ لاحق نہ ہو خون کی منتقلی سے گریز کریں۔

محفوظ تر جنسی عمل

آؤ میرے ساتھ

پتہ نہیں اب تک یہ کہاں تھا؟

بیشتر صورتوں میں ایچ آئی وی اور دوسرے جنسی انفیکشنز جنسی ملاپ کے دوران ایک فرد سے دوسرے فرد میں منتقل ہوتے ہیں۔ محفوظ جنسی عمل کے بارے میں معلومات، کنڈوم کے حصول اور اپنے شوہر کے ساتھ اچھے ابلاغ کے ذریعے ہی آپ خود کو جنسی انفیکشنز سے بچا سکتی ہیں۔

لیکن جب کسی عورت کو جنسی معاملات میں خود فیصلے کرنے کی اجازت نہ ہو تو وہ اپنے شوہر سے کنڈوم استعمال کرنے کے لئے کہتے ہوئے جھجک سکتی ہے یا رد ہونے کا امکان رکھتی ہے یا وہ اپنے شوہر کے مطالبے پر ہر حال میں جنسی ملاپ کرنے پر مجبور ہو سکتی ہے اور ممکن ہے کہ اسے یہ معلوم نہ ہو کہ اس کے شوہر کے دوسری عورتوں کے ساتھ بھی جنسی تعلقات ہیں۔

ہر عورت کو جاننا چاہئے کہ وہ کس طرح جنسی ملاپ کو محفوظ تر بنا سکتی ہے؟

جنسی عمل کے محفوظ طریقے

محفوظ جنسی عمل کا مطلب جنسی ملاپ کے دوران جرثوموں کو ایک دوسرے میں منتقل ہونے سے روکنے کے لئے (کنڈوم جیسے) رکاوٹی طریقے استعمال کرنا یا اس انداز سے جنسی ملاپ کرنا ہے جو انفیکشنز ہونے کے امکانات کم تر کریں۔

مباشرت جنسی عمل کا ایک انتہائی عام طریقہ ہے لیکن جوڑے ایک دوسرے کو چھو کر اور گفتگو کر کے بھی لطف اندوز ہو سکتے ہیں اگر آپ کا شوہر کنڈوم کا استعمال پسند نہ کرتا ہو تو آپ دوسرے محفوظ طریقے اپنا سکتی ہیں اس طرح آپ کا شوہر بھی خوش رہے گا اور آپ بھی مختلف انفیکشنز سے محفوظ رہیں گی۔

انتہائی محفوظ

- انتہائی محفوظ طریقہ یہ ہے کہ جنسی عمل سے گریز کیا جائے۔ اس طرح آپ جنسی عوارض کا شکار نہیں ہوں گی تاہم ہمیشہ ایسا ہونا ممکن نہیں ہے۔

- ازدواجی زندگی میں باہمی ہم آہنگی اور ذہنی ربط بھی تحفظ کی علامت ہے کہ میاں بیوی ایک دوسرے سے اپنی جنسی زندگی میں تعاون اور ایک دوسرے کا خیال کرتے ہیں۔

- جنسی ملاپ کے لئے محفوظ طریقوں کا استعمال بھی اہم ہے لیکن اس کے لئے بھی میاں بیوی کے مابین ہم آہنگی ضروری ہے۔

محفوظ

- ہمیشہ لیٹکس کنڈوم استعمال کریں۔
- کوشش کریں کہ جسمانی رطوبتیں، آپ کے اندر منتقل نہ ہوں۔
- جنسی انفیکشنز کی معمولی علامات بھی نظر انداز مت کریں۔

محفوظ جنسی صحت

صحت مند زندگی کے لئے ضروری ہے کہ آپ کی جنسی صحت بھی محفوظ ہو۔ جنسی طور پر صحت مند رہنے کا انحصار خود آپ پر اور آپ کے شریک حیات کے طرز عمل اور تعاون پر ہے۔ بیشتر معاشروں میں عورت فیصلے کا اختیار نہیں رکھتی ہے اور اپنے شوہر کے احکامات کی پابند ہوتی ہے۔ شوہر کی بے راہ روی اس کے لئے مختلف سنگین خطرات کا سبب بن سکتی ہے۔ کوشش کریں کہ آپ کا شوہر بھی جنسی صحت کی اہمیت کو سمجھے اور اس کے لئے ہر امکانی تعاون کرے۔

جنسی ملاپ اور ماہواری

ماہواری کے دوران جنسی ملاپ بالکل بھی نہیں کرنا چاہئے۔ اس دوران کوشش کرنی چاہئے کہ خارج ہونے والا خون اور رطوبتیں کسی نئے خطرے کا سبب نہ بنیں۔ جسمانی رطوبتوں سے آلودہ پیڈز/ کپڑوں کو انتہائی احتیاط کے ساتھ تلف کیا جائے یہ رطوبتیں کسی کے لئے بھی خطرناک ثابت ہو سکتی ہیں۔

تبدیلی کے لئے کام

جنسی بیماریاں اور ایچ آئی وی/ ایڈز پورے سماج کی طرح معذور عورتوں کے لئے بھی سنگین خطرہ ہیں۔ بعض اوقات معذور افراد یہ سمجھتے ہیں کہ انہیں جنسی صحت کے لئے فکرمند ہونے کی ضرورت نہیں ہے لیکن جنسی صحت ان کے لئے ایک اہم مسئلہ ہے۔

جنسی صحت کے بارے میں معلومات اور جنسی انفیکشنز سے بچاؤ کے طریقے معذور عورتوں سمیت سب کے لئے دستیاب ہونی چاہئیں یہ معلومات بہری اور اندھی عورتوں کے لئے دستیاب ہونی چاہئیں۔

صحت دنیا کے ہر فرد کا حق ہے۔ معذور لڑکیاں اور عورتیں بھی صحت کی سہولت کے آسان حصول کا حق رکھتی ہیں۔ معذور افراد اپنی معذوری کے ساتھ غیر معذور افراد کو لاحق سارے خطرات بھی رکھتے ہیں جبکہ بیشتر صورتوں میں انہیں وہ سہولتیں بھی نہیں ملتیں ہیں جو ایک انسان کی حیثیت سے ان کا حق ہیں۔ معذور عورتیں اس اعتبار سے زیادہ محروم اور غیر محفوظ ہیں۔ انہیں اپنی بہتر زندگی اور صحت کے لئے زیادہ سے زیادہ باخبر رہنے کی ضرورت ہے جبکہ انہیں اپنے حقوق کے لئے جدوجہد بھی کرنی چاہئے۔

تبدیلی کے لئے اجتماعی طور پر کی جانے والی کوششیں کامیابی کے زیادہ امکان رکھتی ہیں۔ دنیا میں ایسے لوگ بھی موجود ہیں جو معذور افراد سے ہمدردی رکھتے ہیں اور ان کی مدد کے لئے آمادہ رہتے ہیں، ضرورت انہیں اپنے مسائل سے آگاہ کرنے کی ہے۔ معذور عورتیں ایسے افراد سے مل کر خود کو درپیش مسائل کے ایسے مؤثر حل یقینی بنا سکتی ہیں جو دوسروں کے لئے بھی مفید ثابت ہو سکتے ہیں۔

معذور عورتیں کیا کرسکتی ہیں؟

معذور افراد کو ایچ آئی وی/ایڈز اور جنسی صحت کے بارے میں آگاہ کریں

- معذور عورتوں کی دیکھ بھال کرنے والوں اور گھر والوں سے ملیں اور انہیں بتائیں کہ جنسی صحت کے لئے درست معلومات ہر ایک کے لئے کس قدر ضروری ہیں۔
- صحت کارکنوں اور دوسرے گروپوں کے ساتھ ایچ آئی وی/ایڈز اور جنسی صحت کی دیگر سہولتیں معذور افراد کے لئے قابل حصول بنانے کے لئے کام کریں۔
- اگر کوئی آپ کا جنسی طور پر استحصال کر رہا ہو تو کسی قابل بھروسہ فرد یا گھر والوں یا کسی صحت کارکن کو آگاہ کریں۔

گھرانے اور دیکھ بھال کرنے والے کیا کرسکتے ہیں؟

- معذور عورتوں کے لئے جنسی صحت، ایچ آئی وی/ایڈز اور جنسی انفیکشنز سے تحفظ کے طریقوں کے بارے میں معلومات یقینی بنائیں۔ انہیں یہ معلومات باوقار اور رازدارانہ طور پر فراہم کریں۔

یقینی بنائیں کہ کوئی بھی معذوری کا شکار عورتوں سے جنسی فائدہ نہ اٹھا سکے

- معذور بچوں کے والدین کو یہ سمجھائیں کہ جب ان کے بچے بڑے ہوں گے تو وہ غیر معذور افراد کی طرح جنسی تعلق کے بارے میں جاننا چاہیں گے۔

کمیونٹیز کیا کرسکتی ہیں؟

معاشرے میں ہر فرد کے لئے یہ جاننا ضروری ہے کہ ایچ آئی وی/ایڈز اور جنسی انفیکشنز کس طرح پھیلتے ہیں اور کس طرح ان سے بچا جاسکتا ہے۔ ان معلومات سے لوگ جان سکتے ہیں کہ یہ انفیکشنز کسی کو بھی ہوسکتے ہیں اور وہ اپنے آپ کو ان سے بچانے کے لئے اقدام کر سکتے ہیں۔ اور یہ معلومات لوگوں کو احساس دلا سکتی ہے کہ معذور عورتوں کا شکار لوگوں کو بھی معاشرے کے دوسرے افراد کی طرح صحت کی دیکھ بھال کی ضرورت ہوتی ہے۔

یہ انتہائی ضروری ہے کہ بیماری کے پھیلاؤ کا سبب بننے والے عوامل کے خلاف جدوجہد کی جائے، ان کے خلاف نہیں جوان کا شکار ہوئے ہوں۔ منصفانہ سماجی اور اقتصادی حالات کے لئے جدوجہد کے ذریعے ایچ آئی وی/ایڈز، جنسی طور پر منتقل ہونے والے انفیکشنز سے محفوظ رہا جاسکتا ہے۔ یہ جدوجہد ضروری ہے کہ معذور عورتوں سمیت تمام عورتوں کو فیصلہ سازی کا اختیار ملے، ان کی معاشی حالت بہتر ہو اور لوگوں کو اپنی ضروریات پوری کرنے کے لئے جسم فروشی کی ضرورت نہ پڑے۔

- معذور عورتوں سمیت تمام لوگوں کے لئے یقینی بنائیں کہ انہیں ایچ آئی وی/ایڈز اور دیگر جنسی انفیکشنز کے فروغ کو روکنے کے لئے کنڈوم سمیت جنسی صحت کی دیگر سہولتوں اور معلومات تک رسائی حاصل ہو۔
- یقینی بنائیں کہ ایچ آئی وی/ایڈز میں مبتلا لوگوں کے لئے ادویات صاف پانی اور پر غذائیت کھانا دستیاب ہو۔
- اپنی کمیونٹی میں لوگوں کو سکھائیں کہ وہ کس طرح معذوری رکھنے والی لڑکیوں اور عورتوں کو جنسی طور پر استحصال سے بچائیں اور انہیں یہ سمجھائیں کہ معذور عورتوں کے ساتھ جنسی ملاپ ایڈز کا علاج نہیں ہے۔

صحت کارکن کے لئے

صحت کی تعلیم میں معذوری والی عورتوں کو بھی شامل کیجئے اور ایسے مواقع تلاش کیجئے کہ گروپ کی صورت میں اکٹھی ہونے والی معذور عورتوں کو بھی صحت کے بارے میں معلومات فراہم کی جاسکیں۔ ہمیشہ معذور عورت کے وقار اور رازداری کے حق کا احترام کیجئے۔

کبھی بھی ایک عورت کے مسائل دوسروں کے سامنے بیان نہ کریں۔ اس کے گھر والوں کے سامنے بھی نہیں جب تک وہ عورت آپ کو اس کی اجازت نہ دے۔

- معذور عورتوں کو سمجھائیں کہ جنسی انفیکشنز اور ایچ آئی وی/ایڈز کس طرح ایک فرد سے دوسرے میں منتقل ہوتے ہیں اور کس طرح ان سے بچا جا سکتا ہے۔

- انہیں مردانہ اور زنانہ کنڈوم استعمال کرنے کی افادیت سمجھائیں۔

- جاننے کی کوشش کریں کہ معذوریوں کا شکار ایسی عورتوں کے امکانی مسائل کیا ہو سکتے ہیں جو جنسی امراض کے لئے مختلف ادویات لے رہی ہوں۔

- جب آپ صحت کے مسئلے سے دوچار کسی عورت سے ملیں تو جنسی زیادتی کی علامات کا جائزہ لیں۔

- یقینی بنائیں کہ معذور عورتوں کو ایچ آئی وی کے بارے میں مشاورت اور ٹیسٹ کی سہولت مل سکے۔

صحت کارکن والدین کو آگاہ کر سکتے ہیں کہ ایچ آئی وی/ایڈز سمیت جنسی امراض کے بارے میں جاننا معذور بچوں کو کس طرح آئندہ ان امکانی سنگین عارضوں سے تحفظ فراہم کر سکتا ہے۔

باب - 9

خاندانی منصوبہ بندی

عورتیں اس وقت صحت مندرہ سکتی ہیں جب وہ خود فیصلہ کرسکیں کہ وہ کب بچے پیدا کرنا چاہتی ہیں۔ یہ فیصلہ ہمیشہ ان کی مرضی سے ہونا چاہئے اور وہ عورتیں جو خاندانی منصوبہ بندی کے طریقے استعمال کرتی ہیں بہتر طور پر یہ انتخاب کرسکتی ہیں۔ آپ خاندانی منصوبہ بندی کو مندرجہ ذیل مقاصد کے لئے استعمال کرسکتی ہیں۔

ہر عورت صحت مند ہوگی اگر وہ فیصلہ کرسکے کہ کب جنسی ملاپ کیا جائے اور کب بچے پیدا کئے جائیں

- یہ فیصلہ کرنے کے لئے آپ کتنے بچے چاہتی ہیں اور یہ کہ یہ کب ہوں۔
- حمل سے اس وقت تک تحفظ جب تک آپ چاہیں۔
- حاملہ ہونے کے حوالے سے فکرمند نہ ہونے کے باعث آپ اور آپ کا شوہر جنسی عمل سے زیادہ لطف اندوز ہوسکیں۔

خاندانی منصوبہ بندی کے کچھ طریقے دوسرے فوائد بھی رکھتے ہیں مثلاً

- کنڈوم جنسی طور پر منتقل ہونے والے انفیکشنز بشمول ایچ آئی وی/ایڈز سے بچاتے ہیں۔
- ہارمونل طریقے (دیکھئے صفحہ 196) بے قاعدہ اخراج خون اور ماہواری کے دوران درد سے بچانے میں مدد فراہم کرسکتے ہیں۔

بدقسمتی سے دنیا بھر میں بہت سی عورتیں خاندانی منصوبہ بندی کے طریقوں تک رسائی نہیں رکھتی ہیں۔ اس کی بہت سی وجوہ ہوسکتی ہیں۔ کچھ لوگ سمجھتے ہیں کہ خاندانی منصوبہ بندی عورت کی صحت کے لئے خطرناک ہے یا یہ کہ خاندانی منصوبہ بندی کے طریقے استعمال کرنے والی عورت اپنے شوہر کی وفادار نہیں ہوگی یا وہ شادی سے پہلے مباشرت کرچکی ہوگی۔ بعض مذہبی عقائد بھی خاندانی منصوبہ بندی کو ممنوع قرار دیتے ہیں اور کچھ حکومتیں اس پر یقین نہیں رکھتی ہیں کہ عورتیں یہ فیصلہ کریں کہ یہ طریقے کب اور کس طرح استعمال کئے جائیں۔

معذور عورتوں کے لئے یہ زیادہ مشکل ہوسکتا ہے کہ وہ اس بارے میں معلومات حاصل کرسکیں یا خاندانی منصوبہ بندی کے طریقوں تک رسائی حاصل کرسکیں۔ بہت سے لوگ جن میں صحت کارکن بھی شامل ہیں یقین رکھتے ہیں کہ معذوری عورتیں نہ تو جنسی عمل کرسکتی ہیں اور نہ ہی حاملہ ہوسکتی ہیں اور انہیں اس حوالے سے معلومات یا مشورے نہیں دینے چاہئیں۔

اس باب میں خاندانی منصوبہ بندی کی مختلف اقسام کے بارے میں معلومات دی گئی ہیں اور یہ کہ آپ اپنے لئے موزوں ترین طریقہ کس طرح منتخب کریں۔

اس سے پہلے کہ آپ خاندانی منصوبہ بندی کا کوئی طریقہ استعمال کرنے کا فیصلہ کریں صفحہ 188 پر دیئے گئے چارٹ کا جائزہ لیں اور یہ دیکھیں کہ ہر طریقہ کس قدر بہتر طور پر حمل سے روکتا ہے۔ آپ کو مندرجہ ذیل باتوں پر بھی غور کرنا چاہئے۔

میں خاندانی منصوبہ بندی کے کس طریقہ کا انتخاب کروں

کنڈوم	پلز
اسپرمی سائیڈ	امپلانٹس
ڈایا فرام/ سرویکل کیپ	انجکشن
زنانہ کنڈوم	آئی یوڈی
دودھ پلانا	رطوبت کا طریقہ

- آپ کی کمیونٹی میں کون سے طریقے دستیاب ہیں؟

- کون سا طریقہ استعمال کرنا کتنا آسان ہے؟

- کیا آپ کو طریقہ استعمال کرنے سے کوئی نقصان پہنچ سکتا ہے؟

- کیا آپ کا شوہر خاندانی منصوبہ بندی کا طریقہ استعمال کرنے کے لئے آمادہ ہے؟

- کیا آپ کی معذوری استعمال کئے جانے والے طریقے پر اثر انداز ہوگی؟

جہاں خاندانی منصوبہ بندی کے بہت سے طریقے دستیاب ہیں، عورتیں استعمال کی سہولت، اخراجات، اپنے جسموں کی ساخت، کام کی نوعیت جو وہ کرتی ہیں، اپنی اور اپنے شوہروں کی ترجیح کے مطابق فیصلہ کرتی ہیں۔ اگر ان میں سے کچھ طریقے آپ کی کمیونٹی میں دستیاب نہیں ہیں تو آپ ان کے بارے میں مقامی صحت کارکنوں کے ذریعے جان سکتی ہیں اور ان کے ذریعے کوشش کرسکتی ہیں کہ وہ بھی دستیاب ہوں۔ آپ انہیں دوسری عورتوں کو سکھانے کے قابل بھی ہوسکتی ہیں۔

خاندانی منصوبہ بندی کے طریقے کس طرح کام کرتے ہیں

ہر مہینے کچھ دورانیہ ایسا ہوتا ہے جب ایک عورت بارور ہوتی ہے اور حاملہ ہوسکتی ہے۔ جب وہ بارور نہ ہوتو حاملہ بھی نہیں ہوگی۔ بیش تر عورتیں ہر ماہ ایک بیضہ پیدا کرتی ہیں۔ یہ بیضہ، بیضہ دانی سے خارج ہوتا ہے۔ بیضہ دانی سے خارج ہونے کے بعد تقریباً 24 گھنٹے کارآمد رہتا ہے (ایک دن اور ایک رات)۔

مرد کا مادہ حیات عورت کے جسم میں دو دن تک موثر رہ سکتا ہے۔ اگر عورت کا بیضہ اس وقت خارج ہو جب مرد کا مادہ حیات اس کے جسم میں ہو تو وہ حاملہ ہوسکتی ہے (حمل کے بارے میں مزید معلومات کے لئے دیکھئے صفحہ 77 تا 80)۔

خاندانی منصوبہ بندی کے طریقے عورت کی باروری کو تبدیل کرنے کے لئے کام کرتے ہیں اور مختلف طریقوں سے حمل سے بچاتے ہیں۔

- رکاوٹی طریقے۔ (مردوں کے لئے کنڈوم، عورتوں کے لئے کنڈوم، ڈایافرام، سرویکل کیپ) یہ طریقے مرد کے مادہ حیات کو عورت کے بیضے تک پہنچنے سے روکتے ہیں (دیکھئے صفحہ 189)

- رحم میں رکھے جانے والے آلات (آئی یو ڈی، آئی یو سی ڈی، آئی یو والیس، کاپرٹی، لوپ) مرد کے مادہ حیات کو عورت کا بیضہ بارور کرنے سے روکتے ہیں۔ (دیکھئے صفحہ 195)

- ہارمونل طریقے (گولیاں، انجکشن، امپلانٹس) عورت کی بیضہ دانی کو بیضہ خارج کرنے سے روکتے ہیں۔ کچھ طریقے اس کے رحم کے استر پر اثر انداز ہوتے ہیں لہذا وہاں بیضہ وہاں نشو ونما نہیں پا سکتا ہے (دیکھئے صفحہ 196)

- فطری طریقے عورت کو یہ جاننے میں مدد دیتے ہیں کہ وہ کب بارور ہوتی ہے (عورت کے ماہانہ چکر میں وہ وقت جب وہ حاملہ ہوسکتی ہے) لہذا وہ اس دوران جنسی عمل سے بچ سکتی ہے (دیکھئے صفحہ 200)

- مستقل طریقے (ناقابل تولید بنانا/ نس بندی) ایسے آپریشن ہیں جو ایک مرد کو مادہ حیات خارج کرنے سے یا عورت کو بیضہ خارج کرنے سے روکتے ہیں (دیکھئے صفحہ 203)

اگلے صفحے پر ایک چارٹ ہے جو بتاتا ہے کہ کون سا طریقہ حمل سے بچاؤ کے لئے کس طرح کام کرتا ہے اور جنسی طور پر منتقل ہونے والے امراض سے کس طرح تحفظ فراہم کرتا ہے۔ اور یہ کہ طریقہ کس صورت میں آپ کی معذوری پر اثر انداز یا اس سے متاثر ہوسکتا ہے۔ ہر طریقے کے ساتھ ستارے دیئے گئے ہیں جو بتاتے ہیں کہ یہ طریقہ کس قدر بہتر طور پر حمل روکتا ہے۔ چند طریقے کم ستارے رکھتے ہیں ہر چند کہ وہ خاصے موثر طریقے ہیں اس کی وجہ یہ ہے کہ اکثر لوگ یہ درست طور پر استعمال نہیں کرتے ہیں۔ درست استعمال کی صورت میں ہی طریقہ بہتر طور پر کام کرے گا۔

خاندانی منصوبہ بندی کا طریقہ	حمل سے تحفظ	ایس ٹی آئی سے تحفظ	دیگر مفید معلومات
مردوں کے لئے کنڈوم	★★★ بہت اچھا	اچھا	اسپرمی سائیڈ (تولیدی جرثومے ہلاک والے مادے)اور چکنائی کے ساتھ استعمال کیا جائے تو انتہائی موثر
عورتوں کے لئے کنڈوم	★★ اچھا	اچھا	اگر آپ ہاتھ کی محدود حرکت اور اپنی فرج تک رسائی نہیں رکھتی ہیں یا تو اپنی ٹانگوں کو بخوبی کھول نہیں سکتی ہیں یا آپ کے اوپری پیروں میں عضلاتی انجکشن ہوتی ہے تو موزوں نہیں ہوسکتا ہے۔
ڈایافرام یا سرویکل کیپ	★★ اچھا	کسی حد تک	اسپرمی سائیڈ کے ساتھ استعمال کے لیے زیادہ مؤثر ہے۔اگر آپ ہاتھ کی محدود حرکت اور اپنی فرج تک رسائی نہیں رکھتی ہیں، اپنی ٹانگوں کو بخوبی کھول نہیں سکتی ہیں یا آپ کے اوپری پیروں میں عضلاتی انجکشن ہوتی ہے تو موزوں نہیں ہوسکتا ہے۔
اسپرمی سائیڈ کے ساتھ اسفنج	★ کسی حد تک	کسی حد تک	اگر آپ ہاتھ کی محدود حرکت اور اپنی فرج تک رسائی رکھتی ہیں یا اپنی ٹانگوں کو بخوبی کھول نہیں سکتی ہیں یا آپ کے اوپری پیروں میں عضلاتی انجکشن ہوتی ہے تو موزوں نہیں ہوسکتا ہے۔
گھریلو ساختہ اسفنج	★ کسی حد تک	کسی حد تک	اگر آپ ہاتھ کی محدود حرکت اور اپنی فرج تک رسائی نہیں رکھتی ہیں یا اپنی ٹانگوں کو بخوبی کھول نہیں سکتی ہیں یا آپ کے اوپری پیروں میں عضلاتی انجکشن ہوتی ہے تو موزوں نہیں ہوسکتا ہے۔
اسپرمی سائیڈ	★ کسی حد تک	بالکل بھی نہیں	اگر آپ ہاتھ کی محدود حرکت اور اپنی فرج تک رسائی رکھتی ہیں یا تو اپنی ٹانگوں کو بخوبی کھول نہیں سکتی ہیں یا آپ کے اوپری پیروں میں عضلاتی انجکشن ہوتی ہے تو موزوں نہیں ہوسکتا ہے۔
ہارمونل طریقے (برتھ کنٹرول پلز، پیچ Patch ، انجکشن ،امپلانٹ)	★★★★ بہترین	بالکل بھی نہیں	کم ڈوز کی گولیاں ان عورتوں کے لئے درست ہیں جو مفلوج مگر باروزانہ ورزش کرتی ہیں۔ (مرگی والی عورتیں وہی گولیاں استعمال کریں جن میں صرف پروجیسٹین (Progestin) ہو)۔
آئی یوڈی، آئی یوایس	★★★★ بہترین	بالکل بھی نہیں	اگر آپ ہاتھ کی محدود حرکت رکھتی ہیں اور اپنی فرج تک رسائی نہیں رکھتی ہیں،تو اپنی ٹانگوں کو بخوبی کھول نہیں سکتی ہیں یا آپ کے اوپری پیروں میں عضلاتی انجکشن ہوتی ہے تو موزوں نہیں ہوسکتا ہے۔ یا آپ اپنی ماہواری برقرار رکھتے میں دشواری ہوں (یہ ہارمونز والی آئی یوایس کا ایک مسئلہ ہے)۔
باروری کی آگہی/علم	★★ اچھا	بالکل بھی نہیں	اگر آپ ہاتھ کی محدود حرکت رکھتی ہیں اور اپنی فرج تک رسائی نہیں رکھتی ہیں یا اپنی ٹانگوں کو بخوبی کھول نہیں سکتی ہیں یا آپ کے اوپری پیروں میں عضلاتی انجکشن ہوتی ہے تو موزوں نہیں ہوسکتا ہے۔
مباشرت کے بغیر جنسی تسکین	★ کسی حد تک	کسی حد تک	کیونکہ میاں بیوی اس طریقے پر کار بند نہیں رہ پاتے ہیں لہٰذا اکثر حمل ٹھہر جاتا ہے۔
دستبرداری	★ کسی حد تک	کسی حد تک	زیادہ موثر ہے اگر دوسرے طریقے جیسے اسپرمی سائیڈ یا ڈایافرام استعمال کیے جائیں۔
بریسٹ فیڈنگ (صرف ابتدائی چھ ماہ)	★★ اچھا	بالکل بھی نہیں	اس طریقے کو اپنانے کے لیے عورت کو چاہیے کہ وہ بچے کو صرف اپنا دودھ پلائے اور اس کی ماہواری شروع نہ ہوئی ہو۔
نس بندی	★★★★ بہترین	بالکل بھی نہیں	بعض اوقات یہ آپریشن غیر موثر بھی ہوسکتا ہے۔ مرد کی نس بندی کے بعد جوڑے کو تقریباً بارہ ہفتے کوئی اور طریقہ بھی استعمال کرنا چاہیے۔

خاندانی منصوبہ بندی کے رکاوٹی طریقے

خاندانی منصوبہ بندی کے رکاوٹی طریقوں میں مردوں کے لئے کنڈوم، عورتوں کے لئے کنڈوم، ڈایافرام، سروِیکل کیپ، اسفنج اور اسپرمی سائیڈ شامل ہیں۔

مردوں کے لئے کنڈوم (ربر، پروفائیلیٹک)

کنڈوم پتلے ربر کی تنگ سی تھیلی ہوتی ہے جو مرد جنسی عمل کے دوران اپنے عضوِتناسل پر پہن لیتا ہے۔ یہ مرد کے مادہ حیات کو جمع کرلیتی ہے لہٰذا یہ عورت کی فرج یا رحم میں نہیں جاتا ہے۔ کچھ مرد جنسی ملاپ کے لئے کنڈوم پہننا پسند نہیں کرتے ہیں، وہ کہتے ہیں کہ یہ ان کی جنسی لذت کم کر دیتا ہے۔ یہ رجحان برا ہے کیونکہ کنڈوم حمل اور جنسی طور پر منتقل ہونے والے انفیکشنز دونوں سے تحفظ فراہم کرتا ہے۔ چکنائی (Lubricant) کا استعمال عورت اور مرد دونوں کے لئے جنسی عمل کو بہتر بناتا ہے۔ یہ کنڈوم کو پھٹنے سے بھی بچاتی ہے۔ اس مقصد کے لئے پانی میں بنی چکنائی جیسے تھوک، کے وائی جیلی یا اسپرمی سائیڈ استعمال کریں۔ تیل، پیٹرولیم جیلی (ویسلین) اسکن لوشن یا مکھن وغیرہ استعمال نہ کریں، یہ کنڈوم کے پھٹنے یا پھٹنے کا سبب بن سکتے ہیں۔ کنڈوم کے اندرونی سرے پر چکنائی کا ایک قطرہ اسے عضوِتناسل پر زیادہ اطمینان بخش بناتا ہے۔ معمولی سی چکنائی کنڈوم پہننے کے بعد اس کی بیرونی سطح پر لگائی جاسکتی ہے۔ اس سے مباشرت کا عمل عورت کے لئے زیادہ آرام دہ بن سکتا ہے۔

انتہائی مؤثر کنڈوم Latex یا Polyurethane سے بنتے ہیں۔ Sheep skin سے نہیں۔ ہر بار جنسی عمل کے لئے نیا کنڈوم استعمال کیا جائے۔ صرف کنڈوم بھی استعمال کیا جا سکتا ہے جبکہ یہ خاندانی منصوبہ بندی کے دوسرے طریقوں کے ساتھ بھی استعمال ہوسکتا ہے۔ کنڈوم دوافروشوں، جنرل اسٹور، مخصوص مراکزِ صحت اور ایچ آئی وی/ ایڈز سے بچاؤ کے پروگراموں کے ذریعے بھی حاصل کئے جاسکتے ہیں۔

معذور عورتوں کے لئے حمل اور جنسی امراض سے بچاؤ کے وہی طریقے مؤثر ثابت ہو سکتے ہیں جو غیر معذور عورتوں کے لئے تجویز کئے جاتے ہیں۔ اکثر صورتوں میں معذور عورتوں کو زیادہ احتیاط کی ضرورت ہوتی ہے کیونکہ انہیں زندگی کے دوسرے معاملات کی طرح جنسی صحت اور دیکھ بھال میں بھی نظر انداز کیا جاتا ہے۔ معذوری عورت کے فطری جذبات اور ضروریات کو ختم نہیں کر سکتی ہے۔ یہ بات ہر معذور عورت کو اچھی طرح سمجھ لینا چاہئے۔

معذوری کا مطلب ہار مان لینا یا دستبردار ہونا ہرگز نہیں ہے۔ دنیا کی ہر معذور عورت اپنی معذوری کے باوجود ممکنہ حد تک بہتر اور فعال زندگی گزارنے کا حق رکھتی ہے۔ دنیا میں لاکھوں معذور عورتیں اپنی معذوریوں کے باوجود بھر پور زندگی بسر کر رہی ہیں ان میں ایسی عورتیں بھی شامل ہیں جو عمر کا ایک بڑا حصہ گزارنے کے بعد کسی حادثے یا بیماری کے بعد معذوری کا شکار ہوئی تھیں۔ یہ عورتیں اپنا گھر رکھتی ہیں، اپنے بچوں کی پرورش اور نگہداشت کرتی ہیں۔ ان میں سے بہت سی معذور عورتیں سماجی کاموں میں حصہ لیتی ہیں، دوسری معذور عورتوں کی مدد کرتی ہیں اور یہ فیصلے بھی خود کرتی ہیں۔ ایک معذور عورت کو فیصلے کرنے کا حق ہونا چاہئے۔ جنسی صحت کے لئے یہ بات خصوصی اہمیت رکھتی ہے لیکن بہت سے ممالک میں عورت اپنے اس حق سے محروم ہے۔ دستیاب حقائق اس بات کی حوصلہ افزائی کرتے ہیں کہ صحت کی اہمیت سے آگاہی اور ضروری معلومات رکھنے والی عورتیں اپنی روزمرہ زندگی کے علاوہ اپنی جنسی صحت، ازدواجی زندگی اور دیگر معاملات بھی درست رکھ سکتی ہیں۔ انہیں خاندانی منصوبہ بندی کے بارے میں معلومات حاصل کرنے کا حق ہے۔ اس میں خاندانی منصوبہ بندی کے طریقوں کے استعمال کی معلومات اور ان تک رسائی نمایاں اہمیت رکھتی ہے۔

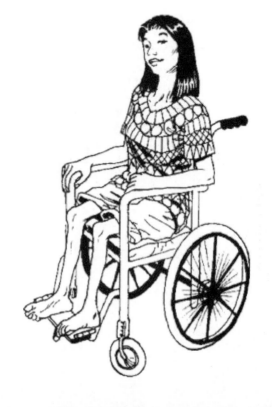

عورتوں کے لئے کنڈوم (زنانہ کنڈوم)

عورتوں کے کنڈوم بھی ایچ آئی وی اور دیگر ایس ٹی انفیکشنز ایک فرد سے دوسرے فرد میں منتقل ہونے سے بچاتے ہیں۔

آپ کسی مختلف طریقے کو ترجیح دے سکتی ہیں اگر

- اگر آپ ہاتھوں کی محدود حرکت رکھتی ہوں۔
- آپ کا ہاتھ فرج تک نہیں پہنچ سکتا ہو۔
- آپ کی ٹانگیں پوری طرح نہ کھل سکتی ہوں۔
- آپ اپنے اوپری پیروں میں عضلاتی اینٹھن کا عارضہ رکھتی ہوں۔

عورتوں کا کنڈوم فرج میں فٹ ہو جاتا ہے اور تناسلی اعضاء کے بیرونی لبوں (Vulva) کو ڈھانپ لیتا ہے۔ یہ حمل اور جنسی طور پر منتقل ہونے والے انفیکشنز اور ایچ آئی وی/ایڈز سے تحفظ فراہم کرتا ہے۔ بدقسمتی سے زنانہ کنڈوم مہنگے ہوتے ہیں اور ان کا حصول مردانہ کنڈوم سے زیادہ مشکل ہے۔ یہ اس وقت بہترین رزلٹ دیتے ہیں جب مباشرت کے وقت مرد اوپر اور عورت نیچے ہو۔

زنانہ کنڈوم کس طرح استعمال کیا جائے؟

1- کنڈوم کو پھاڑے بغیر احتیاط سے پیکٹ کھولیں

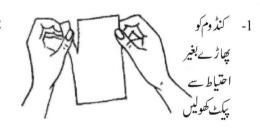

2- اندرونی چھلا (Ring) تلاش کریں جو کنڈوم کے بند سرے پر ہوتا ہے۔

بیرونی چھلّا

3- اندرونی چھلّے کو پکڑیں۔

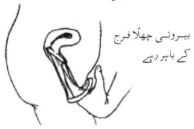

4- اندرونی چھلّے کو فرج کے اندر رکھیں انگلیوں کے ساتھ اندر دھکیلیں۔

بیرونی چھلّا فرج کے باہر رہے

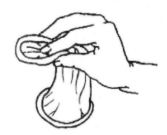

7- جنسی عمل کے فوراً بعد کھڑے ہونے سے پہلے زنانہ کنڈوم نکالیں۔ مرد کے مادہ حیات کو اندرونی پاؤچ میں جانے سے روکنے کے لئے اندرونی چھلّے کو پکڑ کر بل دے لیں۔ پاؤچ کو نرمی سے باہر نکال کر کنڈوم کو اس طرح ضائع کریں کہ یہ بچوں یا جانوروں کے ہاتھ نہ لگے۔ اسے کسی کوڑے دان میں ڈالیں یا پھر گڑھے میں دبا دیں۔

6- جنسی عمل کے دوران مرد کے عضو کو بیرونی چھلّے کے ذریعے فرج میں جانے دیں۔

اگر آپ زنانہ کنڈوم استعمال کرنا چاہتی ہیں لیکن آپ کی معذوری اسے مشکل بنا رہی ہو تو مختلف انداز آزمائیں یا اپنے شوہر سے مدد کی درخواست کریں۔

جب آپ جنسی ملاپ کریں تو ہر بار نیا زنانہ کنڈوم استعمال کریں لیکن نیا کنڈوم حاصل کرنا مشکل ہو یا فوری طور پر ممکن نہ ہو تو آپ زنانہ کنڈوم صاف کر کے اسے سات بار تک استعمال کر سکتی ہیں۔

زنانہ کنڈوم کس طرح صاف کیا جائے؟

جنسی ملاپ سے پہلے بڑے کپ میں (بیس حصے پانی میں ایک حصہ پاؤڈر یا مائع بلیچ ملائیں) بلیچ ایچ آئی وی کو ہلاک کرتا ہے۔

جنسی ملاپ کے بعد کنڈوم کو اپنی فرج سے نکالیں، خیال رکھیں کہ مرد کا مادہ حیات ذرا بھی نہ باہر ٹپکے۔ فوراً ہی بلیچ محلول کا نصف کنڈوم میں ڈالیں اور پھر بقیہ محلول کو بھی کنڈوم میں بھر لیں۔

کنڈوم کو صرف پانچ منٹ اسی طرح رکھیں۔ کنڈوم میں بلیچ محلول بھرنے سے پہلے اسے کسی اور طریقے سے صاف **کرنے کی کوشش مت کریں۔**

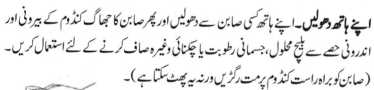

اپنے ہاتھ دھولیں۔ اپنے ہاتھ کسی صابن سے دھولیں اور پھر صابن کا جھاگ کنڈوم کے بیرونی اور اندرونی حصے سے بلیچ محلول، جسمانی رطوبت یا چکنائی وغیرہ صاف کرنے کے لئے استعمال کریں۔ (صابن کو براہ راست کنڈوم پر مت رگڑیں ورنہ یہ پھٹ سکتا ہے)۔

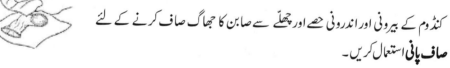

کنڈوم کے بیرونی اور اندرونی حصے اور چھلّے سے صابن کا جھاگ صاف کرنے کے لئے **صاف پانی** استعمال کریں۔

کنڈوم کے بیرونی اور اندرونی حصے کو کسی **صاف کپڑے سے پونچھ لیں** اور پھر اسے ہوا میں سوکھنے کے لئے رکھ دیں۔

کنڈوم کو روشنی میں رکھ کر اس کا جائزہ لیں اگر اس میں کوئی **چھوٹا سوراخ** بھی نظر آئے تو اسے پھینک دیں۔ اور نیا کنڈوم حاصل کریں۔ کنڈوم کے رنگ میں معمولی تبدیلی میں کوئی حرج نہیں اگر اس میں کوئی سوراخ نہ ہو تو اسے اگلے استعمال کے لئے صاف اور خشک جگہ محفوظ کر لیں۔

کنڈوم دوبارہ استعمال کرنے سے پہلے

کنڈوم کو پانی میں بنی چکنائی (Lubricant) سے چکنا کریں۔ زنانہ کنڈوم کے لئے آپ ویجیٹیبل آئل بھی استعمال کر سکتی ہیں۔ زنانہ کنڈوم (Latex) سے نہیں بنے ہیں لہٰذا ان پر تھوڑا سا تیل استعمال کیا جا سکتا ہے لیکن اس پر مونگ پھلی کا تیل کا ایسا لوشن نہ لگائیں جس میں بھیڑ کے اون کی چکنائی یا خوشبو شامل ہو۔ یہ جلد کے الرجک رد عمل کا سبب بن سکتے ہیں۔

ڈایافرام اور سرویکل کیپ

ڈایافرام اور سرویکل کیپ دونوں ہی نرم ربڑ کے بنے گہرے کپ ہوتے ہیں۔ یہ جنسی عمل کے دوران رحم میں پہنے جاتے ہیں۔ جنسی عمل کے بعد انہیں کم از کم چھ گھنٹوں کے لئے لگا رہنے دیا جائے۔ یہ 24 گھنٹے بھی لگے رہ سکتے ہیں لیکن اس سے زیادہ ہرگز نہیں۔

ڈایافرام اور کیپ دونوں ہی اچھے مانع حمل طریقے ہیں اگر انہیں ہر بار جنسی عمل کرتے ہوئے کسی مانع حمل کریم یا جیلی (اسپری سائیڈ) کے ساتھ استعمال کیا جائے۔ ڈایافرام اور سرویکل کیپ مختلف سائزوں میں ملتے ہیں۔ ایک تجربہ کار صحت کارکن آپ کے لئے درست سائز کا تعین کر سکتی ہے۔ ڈایافرام، کیپ سے بڑا ہوتا ہے اور کچھ چھوٹی عورتیں کہتی ہیں کہ کیپ ان کے لئے موزوں رہتا ہے۔ بچے کی پیدائش کے بعد اگر آپ کے وزن میں نمایاں کمی یا زیادتی ہوئی ہو تو ممکن ہے کہ آپ کو اپنے ڈایافرام کا سائز تبدیل کرنا پڑے۔

ڈایافرام اور کیپ عام طور پر سال بھر کار آمد رہتے ہیں۔ دونوں کو اس دوران با قاعدگی کے ساتھ روشنی میں چیک کیا جاتا رہے کہ ان میں سوراخ یا درخنے تو نہیں پڑ گئے ہیں۔ اگر ان میں معمولی سا سوراخ بھی ہو جائے تو انہیں فوراً بدل لیا جائے کیونکہ مرد کا مادہ حیات بہت ہی پتلا ہوتا ہے اور چھوٹے سے چھوٹے سوراخ سے بھی گز سکتا ہے۔ استعمال کے بعد انہیں گرم صابن آمیز پانی میں دھوئیں، صاف پانی میں نتھاریں اور خشک کر لیں۔ ڈایافرام یا سرویکل کیپ کو صاف ستھری اور خشک جگہ پر رکھیں۔

آپ کسی مختلف طریقے کو ترجیح دے سکتی ہیں اگر
• اگر آپ ہاتھوں کی محدود حرکت رکھتی ہوں۔
• آپ کی ٹانگیں پوری طرح نہ کھل سکتی ہوں۔
• آپ کا ہاتھ فرج تک نہیں پہنچ سکتا ہو۔
• آپ اپنے اوپری پیروں میں عضلاتی اینٹھن کا عارضہ رکھتی ہوں۔

یہ اشیاء ہر جگہ دستیاب نہیں ہیں لیکن اگر عورتوں کی زیادہ تعداد انہیں طلب کرے تو مقامی ادارے اور پروگرام ان کی دستیابی ممکن بنائیں گے۔

سرویکل کیپ

ڈایافرام

اسفنج

مانع حمل اسفنج

مانع حمل اسفنج نرم پلاسٹک سے بنا ہوتا ہے اور اس میں اسپری سائیڈ بھری ہوتی ہے۔ آپ جنسی عمل سے پہلے اسفنج کو اپنی فرج کے اندر رکھ سکتی ہیں۔ اسفنج رکھے جانے کے بعد آپ اس میں اسپری سائیڈ شامل کئے بغیر جتنی بار چاہیں جنسی عمل کر سکتی ہیں۔ یہ جنسی عمل کے بعد فرج میں چھ گھنٹے رہنا چاہیے۔ آپ اسے چوبیس گھنٹے بھی رکھ سکتی ہیں (مگر اس سے زیادہ نہیں)۔ بہت سے ممالک میں اسفنج دستیاب نہیں ہوتی ہے۔

آپ کسی مختلف طریقے کو ترجیح دے سکتی ہیں اگر
• آپ کی ٹانگیں پوری طرح نہ کھل سکتی ہوں۔
• اگر آپ ہاتھوں کی محدود حرکت رکھتی ہوں۔
• آپ کا ہاتھ فرج تک نہیں پہنچ سکتا ہو۔
• آپ اپنے اوپری پیروں میں عضلاتی اینٹھن کا عارضہ رکھتی ہوں۔
• آپ اپنے پیروں میں محسوسات نہ رکھتی ہوں۔

گھریلو ساختہ اسفنج

آپ سرکے یا لیموں کے رس میں بھیگی اسفنج بھی استعمال کرسکتی ہیں۔ یہ طریقہ مانع حمل اسفنج کی طرح مؤثر نہیں ہے لیکن کچھ حمل روک سکتا ہے۔ آپ یہ طریقہ اس وقت استعمال کرسکتی ہیں جب کوئی اور طریقہ ممکن نہ ہو۔

گھر میں اسفنج کس طرح تیار کریں

1- دو کھانے کے چمچے سرکہ ، ایک کپ   یا   ایک چمچہ لیموں کا عرق ایک کپ   یا   ایک چمچہ نمک چار چمچے اُبلے
اُبلے ہوئے صاف پانی میں ملائیں   اُبلے ہوئے صاف پانی میں ملائیں   ہوئے صاف پانی میں ملائیں

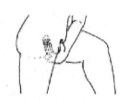

2- انڈے کے سائز کا اسفنج کا ٹکڑا اپانی میں اُبالنے کے بعد ان میں سے کسی محلول میں بھگولیں۔

3- اسفنج کو جنسی عمل سے تقریباً گھنٹہ بھر پہلے اپنی فرج میں اندر تک چڑھالیں۔

4- اسفنج کو جنسی عمل کے بعد چھ گھنٹے اندر ہی رہنے دیں۔ پھر اسے نکال لیں۔

اسفنج کو باہر نکالنا مشکل ہوسکتا ہے لیکن یہ فرج میں گم نہیں ہوسکتا ہے۔ اسے نکالنا آسان ہوسکتا ہے اگر آپ اکڑوں بیٹھ کر اس طرح زور لگائیں کہ جیسے فضلہ خارج کر رہی ہوں۔ اگر آپ کو اسفنج نکالنے میں مشکل ہو تو آپ اگلی بار اس کے ساتھ صاف ربن یا ڈوری باندھ سکتی ہیں۔

اسفنج کو صاف کیا، اُبالا اور کئی بار استعمال کیا جاسکتا ہے۔ اسے صاف اور خشک جگہ رکھیں۔ محلول وقت سے پہلے بنا کر بوتل میں رکھا جاسکتا ہے۔

اسفنج میں بھرا جانے والا اسپرمی سائیڈ یا محلول فرج کی اندرونی جِلد میں سوزش پیدا کر سکتا ہے جو عورت میں ایس ٹی آئی کی منتقلی آسان بنا سکتی ہے۔ اگر اس سے آپ کی فرج خشک، پُرسوزش اور تکلیف دہ ہو جاتی ہے تو انہیں استعمال کرنا بند کر دیں۔

اسپرمی سائیڈز

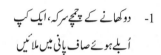

اسپرمی سائیڈز، فوم، گولیوں، کریم یا جیلی کی صورت میں ملتے ہیں، انہیں مباشرت سے پہلے فرج میں رکھا جاتا ہے۔ اسپرمی سائیڈ مرد کے جرثوموں کو رحم میں پہنچنے سے پہلے ہلاک کرتا ہے۔ یہ ایس ٹی آئی یا ایچ آئی وی/ایڈز سے تحفظ فراہم نہیں کرتا ہے۔ گولیاں فرج میں جنسی عمل سے دس تا پندرہ منٹ پہلے رکھی جاتی ہیں۔ فرج، جیلی یا کریم اگر مباشرت سے کچھ پہلے فرج میں لگائی جائے تو بہترین کام کرتی ہے۔ مباشرت کے بعد کم از کم چھ گھنٹے اسپرمی سائیڈ کو مت دھوئیں۔ کچھ اسپرمی سائیڈ فرج کے اندر جلن یا تکلیف کا باعث بن سکتے ہیں۔

فوم عام طور پر سوزش کا سب سے زیادہ معروف سبب مانا جاتا ہے۔ اگر آپ فوم سے حساسیت رکھتی ہیں تو اس کے بجائے مانع حمل جیلی یا کریم استعمال کریں۔

آئی یوڈی (انٹرایوٹیرائن ڈیوائسز ۔ آئی یوسی ڈی، کاپرٹی، لوپ)

انٹرایوٹیرائن ڈیوائس پلاسٹک یا پلاسٹک اور تانبے (کاپر) سے بنا چھوٹا سا آلہ ہے جس سے دو چھوٹی ڈوریاں منسلک ہوتی ہیں۔

<table>
<tr><td>

آپ دوسرے طریقے کو ترجیح دیں اگر

- اگر آپ بھاری ماہواری سے نہ نمٹ سکتی ہوں۔
- اپنی ٹانگوں کو چوڑا نہ کھول سکتی ہوں۔
- اپنے اوپری پیروں میں عضلاتی ایکٹھن رکھتی ہوں

</td></tr>
</table>

آئی یوڈی سے ایچ آئی وی/ ایڈز یا دیگر ایس ٹی آئی ز سے تحفظ نہیں ملتا ہے۔ ایک تجربہ کار صحت یا ڈ وائف رحم کے اندر آئی یوڈی اس طرح داخل کر سکتی ہے کہ اس کی ڈوریاں فرج میں لٹکی رہیں۔ آئی یوڈی مرد کی منی کو عورت کا بیضہ بارور کرنے سے روکتا ہے۔ آئی یوڈی کو محفوظ طور پر استعمال کرنے کے لئے آپ کو باقاعدگی سے اپنی فرج میں ڈوریوں کو چیک کرنے کے قابل ہونا ضروری ہے۔ ایسا کرنا ماہواری رکنے کے بعد فوراً بہترین ہے۔ اگر آپ خود ڈوریوں کو چیک نہیں کر سکتی ہیں تو اپنے شوہر یا کسی قابل بھروسہ فرد سے مدد کی درخواست کر سکتی ہیں۔

آئی یوڈی دس برس تک کے لئے رکھا جا سکتا ہے۔ آئی یوڈی استعمال کرنے والی عورتوں کو یہ اطمینان کرنے کے لئے وہ درست حال میں ہے باقاعدگی سے اپنے پیٹر و کا معائنہ کرانا چاہئے۔

عام ضمنی اثرات

آئی یوڈی رکھوانے کے بعد ممکن ہے پہلے ہفتے میں آپ کو معمولی سی بلیڈنگ ہو۔ کچھ عورتوں کو طویل، بھاری اور زیادہ تکلیف دہ ماہواری ہوتی ہے لیکن یہ عام طور پر ابتدائی ماہ کے بعد رک جاتی ہے۔ اگر آپ آئی یوڈی استعمال کرنا چاہتی ہیں تو کسی تجربہ کار صحت کارکن سے بات کریں تا کہ وہ دیکھ سکے کہ یہ طریقہ آپ کے لئے بہتر ہے گا۔

آئی یوایس (انٹرایوٹیرائن سسٹم): (ہارمونز کے ساتھ آئی یوڈی)

پروجسٹن ہارمون (Levonogestrel) پر مشتمل ایک قسم کی آئی یوڈی انٹرایوٹیرائن سسٹم (آئی یوایس) کہلاتا ہے۔ آئی یوایس ماہواری کے دوران خون کی مقدار کم کرتا ہے اور پانچ برس کے لئے حمل روکنے میں انتہائی مؤثر ہے۔ یہ دوسری آئی یوڈی سے زیادہ مہنگا ہے اور بہت سے ممالک میں دستیاب نہیں ہے۔ کسی صحت کارکن سے معلوم کیجئے کہ کیا یہ آپ کے ہاں دستیاب ہے۔

<table>
<tr><td>

تنبیہ! اگر آپ آئی یوڈی استعمال کر رہی ہیں اور اگر ان میں سے کوئی علامت ظاہر ہو تو فوری طور پر طبی امداد حاصل کیجئے۔

- تاخیر سے یا ماہواری نہ ہونا یا ماہواریوں کے درمیان غیر معمولی دھبے۔
- پیٹ کا درد جو ختم نہ ہوتا ہو یا جنسی عمل کے دوران درد۔
- انفیکشن کی علامات، غیر معمولی اخراج یا فرج سے ناگوار بد بو، بخار، کپکپاہٹ یا خود بیمار محسوس کرنا۔
- آئی یوڈی کی ڈوریوں کا چھوٹا یا بڑا ہونا یا گم ہو جانا یا آئی یوڈی کا رحم میں محسوس ہونا۔

</td></tr>
</table>

خاندانی منصوبہ بندی کے ہارمونل طریقے

ہارمونز وہ کیمیائی مادّے ہیں جو عام طور پر ایک عورت کا جسم بناتا ہے۔ ہارمونز عورت کے جسم کے کئی نظاموں پر اپنا اثر رکھتے ہیں اور انہیں با قاعدہ بناتے ہیں (دیکھئے صفحہ 72) ان میں عورت کی ماہواری اور حاملہ ہونے کی صلاحیت بھی شامل ہے۔ یہ عمل ایک معذور عورت میں بھی غیر معذور عورت سے مختلف نہیں ہے۔ خاندانی منصوبہ بندی کے ہارمونل طریقے آپ کے بیضہ دانیوں کو رحم میں بیضے خارج کرنے سے روکتے ہیں۔ **ہارمونل طریقے ایچ آئی وی/ ایڈز یا دیگر ایس ٹی آئی سے تحفظ نہیں دیتے ہیں۔**

ہارمونل طریقوں میں مندرجہ ذیل شامل ہیں۔

- روزانہ لی جانے والی گولیاں
- انجکشن جو ہر چند ماہ بعد لگائے جاتے ہیں۔
- امپلانٹس جو عورت کے بازو میں رکھے جاتے ہیں اور کئی برس باقی رہتے ہیں۔

زیادہ تر برتھ کنٹرول گولیاں اور انجکشن ایسے دو ہارمونز پر مشتمل ہوتے ہیں جو ان ہارمونز جیسے ہیں جو عورت کا جسم عام طور پر بناتا ہے۔ یہ ہارمون ایسٹروجن (ethinyl estradiol) اور پروجسٹرون (levonorgestrel) پر مشتمل ہوتے ہیں۔

نئے ہارمونل طریقے ابھی ایجاد ہونا باقی ہیں۔ کچھ نئے طریقے Contraceptive patch ایک چھلّا جو رحم کے منہ پر چڑھایا جاتا ہے اور ہارمونل آئی یو ڈی ہیں (دیکھئے صفحہ 195)۔

ضمنی اثرات

ہارمونل طریقوں کے بعض اوقات ضمنی اثرات سامنے آتے ہیں۔ یہ اثرات خطرناک نہیں لیکن یہ عموماً تکلیف دہ ہوتے ہیں۔ ہارمونل طریقے اپنانے کے بعد ایک عورت مندرجہ ذیل ذیلی اثرات کا شکار ہو سکتی ہے۔

ماہواری میں تبدیلیاں   وزن میں اضافہ   چھاتیوں کی سوجن   سر درد   جی متلانا

یہ اثرات عام طور پر چند ماہ بعد کم ہو جاتے ہیں اگر ایسا نہ ہو تو آپ خاندانی منصوبہ بندی کا کوئی دوسرا اور مختلف طریقہ اپنا سکتی ہیں۔

برتھ کنٹرول گولیاں (کھائی جانے والی مانع حمل گولیاں)

اگر آپ نے برتھ کنٹرول گولیاں کھانے کا فیصلہ کیا ہے تو یہ کم ڈوز کی ہونی چاہئیں۔ اس کا مطلب یہ ہے کہ یہ 35 مائیکروگرام یا کم ایسٹروجن اور ایک ملی گرام یا کم پروجسٹرون کی ہونا چاہئیں۔ 50 ملی گرام سے زائد ایسٹروجن کی گولیاں استعمال مت کریں۔ برتھ کنٹرول گولیوں کی مختلف برانڈ موجود ہیں (دیکھئے صفحہ 355 سے 356)۔

روزانہ یکساں وقت پر برتھ کنٹرول گولیاں لینا حمل سے بچنے کا انتہائی مؤثر طریقہ ہے۔ اگر یہ گولیاں نگلنا آپ کے لئے مشکل ہو تو انہیں پانی یا کسی مائع میں حل کرکے نلکی کے ذریعے پیا جاسکتا ہے۔

ان کے پیکٹ 21 یا 28 گولیوں پر مشتمل ہوتے ہیں۔ آپ پیکٹ کی پہلی گولی اپنی ماہواری کے پہلے دن کھائیں۔ اگر یہ ممکن نہ ہو تو پہلی گولی ماہواری شروع ہونے کے بعد ابتدائی سات دن میں کسی بھی وقت کھالیں۔ اگر آپ 21 گولیوں والا پیکٹ استعمال کر رہی ہیں تو روزانہ ایک گولی کے حساب سے 21 دن یہ گولیاں کھانے کے بعد نیا پیکٹ شروع کرنے سے پہلے سات دن انتظار کریں۔ عام طور پر آپ کی ماہواری اکیسویں دن کے بعد شروع ہوتی ہوگی لیکن اگر ایسا نہ ہو تو نیا پیکٹ سات دن میں شروع کر دیں۔ اگر 28 گولیوں والا پیکٹ استعمال کر رہی ہیں تو روزانہ ایک گولی لیں اور جیسے ہی گولیاں ختم ہوں نیا پیکٹ فوراً شروع کر دیں۔

اگر آپ گولی کھانے کے بعد تین گھنٹوں میں قے کر دیتی ہیں یا آپ کو ڈائریا ہو جاتا ہے تو کھائی جانے والی برتھ کنٹرول گولی آپ کے جسم میں اتنی دیر نہیں ٹھہرے گی کہ اپنا کام کر سکے۔ طبیعت بہتر ہونے اور اگلے سات دن تک روزانہ ایک گولی کھانے تک کنڈوم استعمال کریں یا جنسی عمل نہ کریں۔

ایسٹروجن اور پروجسٹرون دونوں پر مشتمل گولیاں دو ہفتے کے اندر حمل سے تحفظ فراہم کرنا شروع کرتی ہیں۔ اگر آپ نے اپنی ماہواری کے پہلے دن صرف پروجسٹن پر مشتمل گولی شروع کی ہے تو یہ ابتدائی چار ہفتے حمل نہیں روکے گی لہٰذا آپ کو حمل سے بچنے کے لئے کنڈوم یا کوئی اور طریقہ استعمال کرنا پڑے گا ورنہ آپ حاملہ ہوسکتی ہیں۔

تنبیہ ! اگر آپ مانع حمل گولیاں لے رہی ہیں اور ان میں سے کوئی بھی علامات آئے تو فوری طور پر طبی مدد حاصل کریں

- سینے میں درد یا سانس میں تنگی۔
- شدید سر درد
- بازوؤں یا پیروں میں بے حسی/سُن پن۔
- کسی ایک پیر میں شدید درد یا سوجن

ان علامات کا مطلب یہ ہوسکتا ہے کہ آپ کے جسم کے اندر کسی جگہ خون کا لوتھڑا موجود ہے جو آپ کے پھیپھڑوں، سینے، دماغ یا بازو یا پیر میں خون کی روانی متاثر کر رہا ہے۔

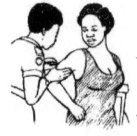

انجکشن کے ذریعے دیئے جانے والے ہارمونز

اس طریقے میں صحت کار کن عورت کو حمل سے بچاؤ کے لئے ہارمون کا انجکشن لگاتی ہے۔ ایک انجکشن ایک سے تین ماہ مؤثر رہتا ہے۔ بیش تر انجکشن صرف پروجسٹن پر مشتمل ہوتے ہیں۔ Depo Provera, Noristerat ان انجکشنوں کی سب سے زیادہ عام برانڈ ہیں۔ یہ انجکشن بریسٹ فیڈنگ کے دوران بھی محفوظ ہوتے ہیں اور ان عورتوں کے لئے بھی جو ایسٹروجن استعمال نہیں کرتیں (دیکھئے صفحہ 196)۔

انجکشن انتہائی مؤثر ہیں یہ طریقہ استعمال کرنے والی بہت ہی کم عورتیں حاملہ ہوتی ہیں۔ اس طریقہ کا ایک اور فائدہ یہ ہے کہ جنسی عمل سے پہلے آپ کو کچھ اور کرنے کی ضرورت نہیں پڑتی ہے اور آپ کی صحت کار کن کے علاوہ کوئی اور نہیں جانتا ہے کہ آپ خاندانی منصوبہ بندی کا کوئی طریقہ استعمال کر رہی ہیں۔ یہ طریقہ استعمال کرنے کے لئے آپ کو اگلے انجکشن کے لئے ہر ایک سے تین ماہ میں صحت کار کن سے ملنا ہوگا۔

پہلا انجکشن لگوانے کے بعد ابتدائی چند ماہ آپ کو معمول کے خلاف ماہواری یا شدید دھبوں کا سامنا ہو سکتا ہے۔ اس کے بعد ممکن ہے کہ آپ کو ماہواری بالکل بھی نہ ہو۔ یہ خطرناک بات نہیں ہے جب آپ انجکشن لگوانا بند کر دیں گی تو آپ کو حاملہ ہونے میں معمول سے زیادہ وقت (انتہائی بارہ ماہ سے زیادہ) نہیں لگے گا۔ اس لئے انجکشن اس صورت میں بہترین ہیں جب آپ اگلے برس کے دوران یا اس سے زیادہ عرصہ حاملہ نہ ہونا چاہتی ہوں۔

خاندانی منصوبہ بندی کے انجکشن لگوانے کے دوران ممکن ہے مرگی کی مریض عورتوں کو کم دورے پڑیں اور اگر آپ چھ ماہ یا اس سے زیادہ عرصہ انجکشن استعمال کریں تو اپنی ہڈیوں کو مضبوط رکھنے کے لئے ایسی غذائیں زیادہ مقدار میں کھائیں جن میں کیلشیئم موجود ہو (دیکھئے صفحہ 86)۔ طویل عرصہ انجکشن لگوانے سے آپ کی ہڈیاں کمزور ہو سکتی ہیں۔

امپلانٹس (Implants)

اس طریقے میں ایک تربیت یافتہ صحت کار کن عورت کے بازو میں کھال کے نیچے پروجسٹن کی چھوٹی سی نرم ٹیوب رکھتی ہے۔ امپلانٹ اپنی قسم کے مطابق آئندہ تین سے پانچ برس حمل سے تحفظ فراہم کرتا ہے۔ تین سے پانچ برس کے بعد امپلانٹ نکلوانا ضروری ہے اس کے بعد اگر آپ حاملہ نہیں ہونا چاہتی ہیں تو فوری طور پر آپ کو نئے امپلانٹ یا خاندانی منصوبہ بندی کا کوئی اور طریقہ استعمال کرنے کی ضرورت ہوگی۔ اگر آپ امپلانٹ غیر مؤثر ہونے کے وقت سے پہلے حاملہ ہونا چاہتی ہیں تو آپ کو کسی تجربہ کار صحت کار کن سے امپلانٹ نکلوانا ہوگا۔

امپلانٹس

اور یہ ایک تجربہ کار صحت کار کن کے ذریعے نکلوائے جا سکتے ہیں

امپلانٹس جِلد کے نیچے رکھے جاتے ہیں

آپ امپلانٹس خود نہیں نکال سکتی ہیں۔ یہ صرف تجربہ کار صحت کارکن کے ذریعے ہی نکالے جائیں۔ اگر آپ امپلانٹس استعمال کرنا چاہتی ہیں تو پہلے یہ یقینی بنائیں کہ آپ ایسی صحت کارکن سے رابطے میں رہیں جو انہیں نکالنا جانتی ہو۔

امپلانٹس کے بعد عورت کو حمل سے بچنے کے لئے مباشرت سے قبل کچھ بھی نہیں کرنا پڑتا ہے۔ امپلانٹس صرف پروجسٹن پر مشتمل ہوتے ہیں یہ ان عورتوں کے لئے محفوظ ہیں جو ایسٹروجن نہیں لیتی ہوں اور یہ بریسٹ فیڈنگ کے دوران بھی محفوظ طور پر استعمال کئے جاسکتے ہیں۔

ابتدائی مہینوں میں امپلانٹس بے قاعدہ اخراج خون (آپ کے ماہواری کے سلسلے کے وسط میں) کا سبب بن سکتے ہیں یا ممکن ہے کہ اخراج خون بالکل بھی نہ ہو۔ اس کا مطلب یہ نہیں ہے کہ آپ حاملہ ہوچکی ہیں یا کوئی اور گڑ بڑ ہے۔ جوں ہی آپ کا جسم زیادہ مقدار میں پروجسٹن کا عادی ہوتا ہے جنم لینے والی تبدیلیاں ختم ہوجاتی ہیں۔ اگر بے قاعدہ اخراج خون آپ کے لئے مسائل کا سبب بنے تو آپ کی معالج یا صحت کارکن آپ کو چند ماہ کے لئے کم ڈوز کی مشترکہ برتھ کنٹرول گولیاں بھی کھانے کے لئے دے سکتی ہے۔

تنبیہہ! امپلانٹس کے بعد ذیل علامات میں سے کوئی بھی علامت سامنے آنے پر طبی مدد حاصل کیجئے۔

- امپلانٹ کے قریب بازو میں درد
- امپلانٹس کے گرد مواد، سرخی یا خون کا اخراج
- امپلانٹس باہر نکل جائے

بریسٹ فیڈنگ

بچے کی پیدائش کے بعد ابتدائی چھ ماہ بچوں کو اپنا دودھ پلانے والی بیش تر عورتیں اپنی بیضہ دانیوں سے بیضہ خارج نہیں کرتی ہیں لہذا وہ جنسی عمل کے نتیجے میں حاملہ نہیں ہوسکتی ہیں۔

عورتیں عام طور پر حاملہ نہیں ہوتی ہیں اگر وہ بچے کو اپنا دودھ پلا رہی ہوں **اور**

1- بچے کی عمر چھ ماہ سے کم ہو **اور**

2- ولادت کے بعد عورت کو ماہواری نہ ہوئی ہو **اور**

3- عورت بچے کو صرف اپنا دودھ پلا رہی ہو۔

اگر آپ خاندانی منصوبہ بندی کا یہ طریقہ اپنانا چاہتی ہوں تو یاد رکھیں کہ اگر آپ بچے کو بیرونی دودھ، پانی یا دوسرے مشروبات دے رہی ہوں یا آپ بچے کو کپ سے دودھ پلانے کے لئے اپنا دودھ ہاتھ سے نکال رہی ہوں تو آپ با آسانی حاملہ ہوسکتی ہیں۔ آپ اس دوران اس صورت میں بھی حاملہ ہوسکتی ہیں اگر بچے کو دودھ پلانے کے درمیان وقفہ چھ گھنٹہ یا زیادہ ہو۔ چھ ماہ کے بعد بریسٹ فیڈنگ کے باوجود حاملہ ہونے کا امکان پہلے کی طرح ہوسکتا ہے۔ آپ اپنی ماہواری دوبارہ شروع ہونے سے دو ہفتے پہلے حاملہ ہوسکتی ہیں لہذا ماہواری شروع ہونے سے پہلے خاندانی منصوبہ بندی کا کوئی طریقہ استعمال کرنا شروع کر دیں۔

بریسٹ فیڈنگ سے ایچ آئی وی/ایڈز زیادہ تر جنسی انفیکشنز سے تحفظ نہیں ملتا ہے جبکہ ایچ آئی وی سے متاثر ہوا جاسکتا ہے، بریسٹ فیڈ بچے میں ایچ آئی وی منتقلی کا ذریعہ بن سکتی ہے۔ اگر آپ کے شوہر کو ایچ آئی وی/ایڈز ہونے کا کوئی امکان ہو تو آپ ہر بار جنسی عمل کے لئے کنڈوم استعمال کریں۔

فطری خاندانی منصوبہ بندی

فطری خاندانی منصوبہ بندی پر کچھ بھی خرچ نہیں ہوتا ہے اور نہ ہی اس کے ضمنی اثرات ہیں لیکن اُسے اپنانا دشوار ہوسکتا ہے۔ عورتوں کو عموماً یہ معلوم نہیں ہوتا ہے کہ وہ کب بارور ہیں اور اگر ان کی ماہواری میں ایک بار بے قاعدگی ہوتو وہ آسانی سے حاملہ ہوسکتی ہیں۔ یہ طریقے اس وقت بہترین رہتے ہیں جب آپ کی ماہواری کا سلسلہ انتہائی با قاعدہ ہو۔ اس کا مطلب یہ ہے کہ آپ کی ماہواری کا پہلا دن ہر ماہ یکساں وقفے کے بعد ہو۔ وقفہ کم از کم 26 دن ہو مگر 32 دن سے زیادہ نہیں ہو۔

ایک عورت مہینے میں ایک بار اپنی باروری کے زمانے میں جب بیضہ اس کی بیضہ دانی سے نکل کر اس کی بیض نالیوں اور رحم میں داخل ہوتا ہے، حاملہ ہوسکتی ہے۔ خاندانی منصوبہ بندی کا فطری طریقہ استعمال کرنے کے لئے آپ کو اپنے جسم کی علامات سمجھنا ہوں گی کہ آپ کب بارور ہوتی ہیں۔ باروری کے زمانے میں خاندانی منصوبہ بندی کا کوئی طریقہ استعمال کئے بغیر مباشرت نہ کریں اس عرصے میں آپ جنسی تسکین کے لئے دوسرے طریقے (مثلاً ایک دوسرے کو چھونا، پیار کرنا وغیرہ بھی اپنا سکتے ہیں یا پھر جنسی عمل کے لئے کنڈوم یا ڈایافرام استعمال کرکے حمل سے بچ سکتی ہیں۔

اب مجھ سے انتظار کی تاب نہیں ہے

فطری خاندانی منصوبہ بندی ان عورتوں کے لئے مناسب نہیں رہتی ہے جو جنسی عمل کے دوران خود کو کنٹرول نہیں رکھ پاتی ہیں۔ آپ کے باروری کے زمانے میں آپ کے شوہر کو کنڈوم یا ڈایافرام استعمال کرنے کے لئے رضامند ہونا چاہیے یا پھر اس دوران مباشرت نہ کی جائے۔ یہ طریقہ عام طور پر اس وقت کامیاب رہتا ہے جب جوڑے اسے استعمال کرنے کی ضروری تربیت رکھتے ہوں۔

اگر آپ نے حال ہی میں بچے کو جنم دیا ہو یا اسقاط حمل ہوا تو یہ طریقے اپنی ماہواری نظام کے کئی ماہ تک با قاعدہ ہونے تک استعمال نہ کریں۔ فطری خاندانی منصوبہ بندی کے بہت سے طریقے ہیں۔ اس کتاب میں ہم رطوبت کے طریقے اور دنوں کی گنتی کے طریقے پر بات کریں۔ یہ طریقے اس وقت مؤثر رہتے ہیں جب انہیں ملا کر استعمال کیا جائے۔ ایک طریقہ کسی اعتبار سے بہتر نہیں ہے۔

فطری خاندانی منصوبہ بندی ایچ آئی وی/ ایڈز وغیرہ ایس ٹی آئی ز سے تحفظ نہیں دیتی ہے۔

آپ کو کوئی اور طریقہ اپنانا چاہیے اگر

- آپ بازوؤں کی محدود حرکت رکھتی ہوں
- اپنی فرج تک ہاتھوں کی رسائی نہ رکھتی ہوں۔
- اپنی ٹانگوں کو مناسب حد سے چوڑا کھول سکتی ہوں۔
- آپ اپنے اوپری پیروں میں عضلاتی انجکشن کا عارضہ رکھتی ہوں۔
- اپنی انگلیوں میں محدود محسوسات رکھتی ہوں۔

رطوبت (Mucus) کا طریقہ

رطوبت کے طریقے میں یہ دیکھنے کے لئے کہ آپ بارور تو نہیں آپ کو روزانہ اپنی فرج کی رطوبت کا جائزہ لینا ہوگا۔ آپ کے باروری کے زمانے میں رطوبت کچے انڈے کی طرح کھینچنے والی (لیس دار) ہوتی ہے۔

رطوبت چیک کرنے کے لئے اپنی فرج کو انگلیوں، کاغذ یا کپڑے سے صاف کریں اور پھر فرج میں رطوبت ہوتو اسے اپنی انگلیوں سے محسوس کریں۔

شفاف، نم، پھسلنے والی رطوبت   سفید، خشک اور نہ کھینچنے والی رطوبت

باروری کے زمانے میں خارج   (یا کوئی رطوبت نہ ہو) دیگر دنوں میں

ہوتی ہے۔   خارج ہوتی ہے پہلے دو خشک

مباشرت مت کریں۔   دنوں کے بعد جنسی عمل/

مباشرت امکانی طور پر محفوظ رہے گا۔

آپ دو سے تین ماہ کی مشق کے بعد اپنی رطوبت کی ان تبدیلیوں
کے بارے میں آسانی سے جان اور سمجھ سکتی ہیں۔

رطوبت کا طریقہ کس طرح استعمال کریں؟

- روزانہ اپنی رطوبت ایک مقررہ وقت پر چیک کریں۔ یہ کام مباشرت سے پہلے کریں۔

- جب آپ رطوبت چکنی اور پھسلواں محسوس کریں اس دن مباشرت نہ کریں۔
  یا ان دنوں مباشرت کے لئے کنڈوم یا ڈایافرام استعمال کریں۔

- شفاف اور پھسلواں رطوبت محسوس کرنے والے دن کے بعد دو دن مباشرت نہ کریں۔

- کسی بھی وقت اپنی فرج میں پچکاری نہ ماریں یا اسے نہ دھوئیں۔ اس طرح رطوبت بھی صاف ہو جائے گی۔

اگر آپ فرج کے انفیکشن میں مبتلا ہیں یا آپ کو یقین نہیں ہے کہ یہ عرصہ باروری کا ہے یا نہیں تو خاندانی منصوبہ بندی کا
کوئی اور طریقہ استعمال کریں۔ رطوبت کا طریقہ اس وقت بہترین ہے جب یہ کسی اور طریقے کے ساتھ اپنایا
جائے۔

دنوں کی گنتی کا طریقہ

دنوں کی گنتی کے طریقے میں ایک عورت اپنی باروری کے زمانے میں مباشرت نہیں کرتی ہے۔ یہ وہی عورتیں استعمال کر سکتی ہیں
جن کی ماہواری با قاعدہ ہو۔ اس کا مطلب یہ ہے کہ آپ اپنی ایک ماہواری سے اگلی ماہواری کے درمیان دنوں کی تقریباً یکساں تعداد
رکھتی ہیں جو 26 دن سے کم اور 32 دن سے زیادہ نہیں ہوگی۔

اگر آپ کا ایک درمیانی وقفہ مختلف عرصے کا ہے تو آپ آسانی سے حاملہ ہو سکتی ہیں یہ عام سی بات ہے کہ اپنی بیماری یا حد
سے زیادہ دباؤ کے باعث کسی عورت کا یہ درمیانی وقفہ مختلف ہو جائے۔ ایسے زمانے میں آپ کے لئے بہتر ہے جب تک آپ
صحت مند نہ ہو جائیں اور آپ کی ماہواری کا سلسلہ سابقہ معمول کے مطابق نہ ہو خاندانی منصوبہ بندی کا کوئی اور مختلف طریقہ
اپنائیں۔

دنوں کی گنتی کا طریقہ کس طرح استعمال کریں

اس طریقے کے کارگر ہونے کے لئے ضروری ہے کہ آپ اپنی ماہواری کے سلسلے کے آٹھویں دن سے انیسویں دن تک مباشرت نہ کریں۔ اگر آپ اس دوران مباشرت کریں گی تو آپ کو اس کے لئے خاندانی منصوبہ بندی کا کوئی اور طریقہ استعمال کرنا ہوگا۔

میری ماہواری سات دن پہلے شروع ہوئی تھی لہذا آج ہم مباشرت نہیں کرسکتے ہیں اور اگلے گیارہ دن بھی نہیں۔

میں اپنی بہن سے بہتر جا رہی ہوں

آپ اپنی باروری کے دنوں کو یاد رکھنے کے لئے موتی، کوئی چارٹ یا کوئی اور طریقہ استعمال کرسکتی ہیں۔ دی گئی رہنمائی کے مطابق ایک ڈوری میں تین مختلف رنگوں کے 32 موتی پرولیں۔ ہر رنگ آپ کی ماہواری کے سلسلے کے مختلف حصے کی نشاندہی کرے گا۔

6 نیلے موتی ظاہر کرتے ہیں ان دنوں میں مباشرت عام طور پر حمل کا سبب نہیں بنتی ہے۔

12 سفید موتی باروری کا زمانہ ظاہر کرتے ہیں جب مباشرت حمل کا سبب بن سکتی ہے۔

سرخ موتی ماہواری کے پہلے دن کی نشاندہی کرتا ہے۔

13 نیلے موتی ظاہر کرتے ہیں کہ ان دنوں میں مباشرت عام طور پر حمل کا باعث نہیں بنتی ہے

اپنی ماہواری کے پہلے دن سرخ موتی پر کوئی چھلا یا ڈوری اس طرح چڑھائیں کہ اسے روزانہ اگلے موتی پر منتقل کرسکیں۔ ہر روز چھلا ایک موتی آگے بڑھاتی رہیں۔ جب یہ چھلا کسی سفید موتی پر آئے تو اس روز مباشرت کی صورت میں آپ حاملہ ہوسکتی ہیں۔ جب بھی آپ کی اگلی ماہواری شروع ہو تو چھلے کو دوبارہ سرخ موتی پر لے آئیں جو ابتداء ہے۔ اس قسم کا نیکلس بھی ملتا ہے جو Cycle Beads کہلاتا ہے۔

دنوں کی گنتی کرنے کے لئے آپ ایک ایک دائرے کے 32 حصے (آپ کی ماہواری کے سلسلے کے ہر دن کے لئے ایک حصہ) کرکے ذیل کے مطابق چارٹ بھی استعمال کرسکتی ہیں۔ چارٹ پر مرحلہ وار ہر حصے کو نشان زدہ کرنے سے آپ کو یاد رہے گا کہ کب آپ حاملہ ہوسکتی ہیں۔

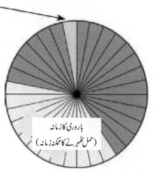

ماہواری کا پہلا دن

باروری کے اس زمانے میں مباشرت حمل کا سبب بن سکتی ہے

باروری کا زمانہ (حمل ٹھہرنے کا ممکنہ زمانہ)

مباشرت کے بغیر جنسی تسکین

مباشرت کے بغیر جنسی تسکین کے ایسے طریقے ہیں جو حمل کا باعث نہیں بنتے ہیں۔ بہت سے جوڑے اس طرح بھی لطف اندوز ہوتے ہیں۔ اس طرح ایچ آئی وی/ایڈز اور دیگر جنسی انفیکشنز منتقل ہونے کا امکان بہت ہی کم ہوتا ہے۔

مباشرت سے گریز حمل سے بچاؤ کا انتہائی یقینی طریقہ ہے۔ اس طرح ایچ آئی وی/ایڈز اور دیگر انفیکشنز کا خطرہ بھی نہ ہونے کے برابر رہ جاتا ہے۔ مباشرت نہ کرنا بہت سے جوڑوں کے لئے انتہائی دشوار ہوسکتا ہے۔ خصوصاً ان کے لئے جو اس کی مشق نہ رکھتے ہوں۔

عُزل (دستبرداری)

اس طریقے میں مرد انزال سے پہلے اپنا عضو باہر نکال لیتا ہے۔ اس سے اس کا مادّہ حیات عورت کی فرج میں داخل ہی نہیں ہوتا ہے۔ یہ طریقہ بھی ایک رکاوٹی طریقہ ہے لیکن اس میں مرد کا درست وقت پر دستبردار یا الگ ہونا لازمی ہے لیکن اکثر مرد ایسا نہیں کر پاتے ہیں یا ایسا نہیں چاہتے ہیں۔ اس صورت میں عورت حاملہ ہوسکتی ہے۔ یہ طریقہ اس وقت زیادہ مؤثر ہوتا ہے جب مرد جنسی عمل سے پہلے پیشاب کر لے اور اس کے ساتھ اسپرمی سائیڈ یا ڈایافرام کا جیسا کوئی طریقہ بھی استعمال کیا جائے۔

ناقابل تولید بنانا (نس بندی)

ایسے آپریشن ممکن ہیں کہ جن کے ذریعے عورت یا مرد کو اولاد پیدا کرنے کے لئے تقریباً ناقابل بنایا جاسکتا ہے۔ یہ آپریشن مستقل ہوتے ہیں لہٰذا ایسی عورتوں اور مردوں کے لئے اچھے ہیں جو یقینی طور پر مزید بچے نہیں چاہتے ہوں۔ صحت مرکز یا اسپتال میں تربیت یافتہ صحت کارکن یا ڈاکٹر یہ آپریشن کر سکتے ہیں۔

مردوں کی بہ نسبت عورتوں کے لئے یہ آپریشن نسبتاً زیادہ سنگین ہوتا ہے۔ مرد عموماً آپریشن کے ضمنی اثرات سے تیزی سے نجات پا لیتے ہیں۔ اگر ممکن ہو تو بہتر اور محفوظ یہ ہے کہ بیوی کے بجائے شوہر یہ آپریشن کرائے۔

عورتوں کے لئے آپریشن (Tubal ligation)

اس میں صحت کارکن ان نالیوں کو کاٹ دیتے یا باندھ دیتے ہیں جو بیضہ رحم میں لے جاتی ہیں۔ آپریشن میں 30 منٹ لگتے ہیں۔ اس سے عورت کی ماہواری پر کوئی فرق نہیں پڑتا ہے۔ آپریشن اس کے جنسی جذبے کو متاثر نہیں کرے گا اور وہ نارمل جنسی زندگی گزار سکے گی اور اس سے لطف اندوز ہو سکے گی۔

عورت کی نالیاں کاٹ دی گئی ہیں

اس جگہ سے

اور اس جگہ سے

آپریشن کے بعد اس بات کا معمولی سا امکان رہتا ہے کہ وہ حاملہ ہو جائے۔ اگر آپ کے ساتھ ایسا ہو تو صحت کارکن سے ملیں۔ اگر آپ کا حمل بیض نالیوں میں ہے تو یہ انتہائی خطرناک ہے (بیض نالی میں حمل کے لئے دیکھیے صفہ 220)۔

سمجھنے یا سیکھنے کے مسائل سے دوچار عورت کو ناقابل تولید (بانجھ) بنانا۔ سمجھنے اور سیکھنے کے مسائل سے دوچار بہت سی عورتیں شاندار ماں بنتی ہیں اور اپنے بچوں کی دیکھ بھال کرسکتی ہیں۔ تمام دوسری نئی ماؤں کی طرح انہیں بھی اپنے گھر والوں کی مدد کی ضرورت ہوتی ہے لیکن بعض اوقات گھر والے یا صحت کارکن یہ سمجھتے ہیں کہ سیکھنے اور سمجھنے کے مسائل سے دوچار عورتوں کو ماں بننے کی اجازت نہیں دی جانی چاہیئے۔ ممکن ہے کہ وہ فیصلہ کریں کہ ایسی عورت کو اس کی اجازت کے بغیر اور یہ بتائے بغیر کہ اس آپریشن کا مقصد کیا ہے بانجھ کرا دیا جائے۔ اگر وہ ایسا اس لئے بھی کریں کہ وہ اس کی صحت کی طرف سے فکرمندی اور اس کی بہتری کے خواہاں ہے تو جب بھی یہ اس کے انسانی حقوق کی پامالی اور انتہائی غلط ہے۔

اگر آپ ایک ایسی عورت ہیں جسے سمجھنے یا سیکھنے میں دشواری پیش آتی ہے تو آپ ناقابل تولید ہونے کا فیصلہ کرسکتی ہیں۔ لیکن یہ فیصلہ بہرصورت آپ کا ہونا چاہیئے۔ دوسرے لوگ تبادلہ خیال کے ذریعے یہ فیصلہ کرنے میں آپ کی مدد کرسکتے ہیں۔

کیا آپ جنسی عمل کے حوالے سے اچھے فیصلے کرسکتی ہیں؟ بعض اوقات عورت کو یہ سمجھنے میں دشواری ہوتی ہے کہ کب ایک مرد اسے جنسی عمل کے لئے استعمال کررہا ہے اور کب وہ اس کی دیکھ بھال کرتا ہے۔ عورت کو ناقابل تولید بنانا صرف حمل روکتا ہے۔ یہ آپ کو جنسی زیادتی کے جذباتی اور جسمانی نقصان سے تحفظ فراہم نہیں کرے گا۔ جنسی زیادتی کے ساتھ رہنا خود کو ناقابل تولید بنانے کی اچھی وجہ نہیں ہے۔ جنسی زیادتی کے خلاف آپ کیا کرسکتی ہیں؟ اس بارے میں معلومات کے لئے دیکھئے باب 14۔

کیا آپ خاندانی منصوبہ بندی کے طریقے استعمال کرنے کے بارے میں بہتر فیصلے کرسکتی ہیں۔ بعض اوقات ایک عورت بھول جاتی ہے کہ وہ اپنی ماہواری نظام کے کس مرحلے میں ہے (دیکھئے صفحہ 75) یا وہ برتھ کنٹرول گولی کھانا یا ڈایافرام استعمال کرنا بھول جاتی ہے۔ اگر وہ ناقابل تولید ہوتو اسے ان میں سے کسی بات کی پرواہ کرنے کی ضرورت نہیں ہوگی لیکن اگر آپ بعد میں کسی وقت ماں بننا چاہتی ہیں تو آپ کے لئے طویل المدتی طریقے جیسے امپلانٹس، انجکشن یا آئی یو ڈی بہتر انتخاب ہوسکتے ہیں۔

کیا آپ حمل کے دوران صحت مندرہ سکتی ہیں؟ جب ایک عورت حاملہ ہوتی ہے تو وہ جو کچھ کھاتی ہے یا پیتی ہے، اس کے نشوونما پاتے بچے پر بھی اس کا اثر پڑتا ہے۔ اچھی غذائیں کھانا، تمباکونوشی نہ کرنا، الکحل اور دیگر نشہ آور اشیاء سے گریز پیدائشی نقائص سمیت بہت سے مسائل سے تحفظ کے لئے ضروری ہے۔

کیا آپ اپنے بچے کی دیکھ بھال کرسکتی ہیں؟ بعض اوقات یہ یاد رکھنا مشکل ہوتا ہے کہ ماں ہونا بے پناہ توجہ، صبر اور کام کا دوسرا نام ہے۔ اس صورت میں بھی جب آپ تھکی ہوئی یا بیمار ہوں یا آپ کو دوسرے کام کرنا ہوں۔

کیا آپ محفوظ جنسی عمل کے حوالے سے بہتر فیصلے کرسکتی ہیں؟ ناقابل تولید بننا آپ کو ایچ آئی وی یا جنسی طور پر منتقل ہونے والے دیگر انفیکشنز میں مبتلا ہونے سے نہیں بچاتا ہے۔ اگر آپ نے یہ طریقہ اپنا رکھا ہو جب بھی آپ کو جنسی عمل کے محفوظ طریقوں پر عمل کی ضرورت ہوگی ان سوالات کا جواب دینا کسی کے لئے بھی مشکل ہوسکتا ہے اور آپ کو ان کے جوابات کے لئے بار بار تبادلہ خیال کی ضرورت ہوسکتی ہے۔ کیونکہ خود کو ناقابل تولید بنانا ایک اہم قدم ہے اس لئے ضروری ہے کہ آپ سمجھ لیں کہ اس آپریشن کا مطلب کیا ہے؟

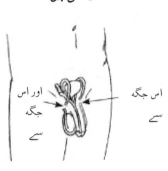

مرد کی نالیاں کاٹ
دی گئی ہیں

اس جگہ
سے

اور اس
جگہ
سے

ایک مرد کے لئے آپریشن (Vasectomy)

اس آپریشن میں مرد کے خصیوں سے اس کے عضوِتناسل میں مادّہ منویہ لانے والی نالیاں کاٹ دی جاتی ہیں۔اس آپریشن میں چندمنٹ لگتے ہیں اور اس سے مرد کی مردانہ صلاحیت،جنسی عمل سے لطف اندوز ہونے کی صلاحیت پر کوئی اثر نہیں پڑتا ہے۔آپریشن کے بعد بھی اسے انزال ہوتا ہے لیکن اس کے مادّے میں اسپرم نہیں ہوتے ہیں۔آپریشن کے بعد نالیوں میں تقریباً تین ماہ اسپرم رہتے ہیں لہٰذا ضروری ہے کہ اس دوران خاندانی منصوبہ بندی کا کوئی طریقہ استعمال کریں۔

خاندانی منصوبہ بندی کے ایمرجنسی طریقے (دی مارننگ آفٹر پلز)

ایمرجنسی خاندانی منصوبہ بندی غیر محفوظ جنسی عمل کے بعد حمل سے بچنے کا ایک طریقہ ہے۔اس طریقے میں برتھ کنٹرول کی ان گولیوں کی معمول سے زیادہ مقدار لی جاتی ہے جو کچھ عورتیں حمل سے تحفظ کے لئے روزانہ استعمال کرتی ہیں۔ ہائی ڈوز کی ایک یا دو گولیوں کی صورت میں "اسپیشل ایمرجنسی پلز" بھی دستیاب ہیں۔ دوا کتنی زیادہ مقدار میں کھائی جائے اس کا انحصار اس بات پر ہے کہ دوا کھاتے وقت آپ اپنی ماہواری نظام کے کس مرحلے میں ہیں۔ یہ طریقہ آپ کو بیضہ جاری ہونے سے بھی بچا سکتا ہے (دیکھئے صفحہ 75)۔

یہ طریقے اس وقت بہترین اثر دکھاتے ہیں جب آپ غیر محفوظ جنسی عمل کے بعد پانچ دن کے اندر جس حد تک جلد ممکن ہو منع حمل گولیاں کھائیں۔ آپ غیر محفوظ جنسی عمل کے بعد جس قدر جلد گولیاں کھائیں گی اسی مناسبت سے آپ کے حاملہ نہ ہونے کے امکانات بڑھ جاتے ہیں۔

ایمرجنسی خاندانی منصوبہ بندی، اسقاطِ حمل کی طرح نہیں ہے کیونکہ اگر آپ گولیاں کھانے سے قبل حاملہ ہو چکی ہیں تو نشوونما پاتے بچے کو کوئی نقصان نہیں پہنچے گا اور نہ ہی یہ ایسا طریقہ ہے جو آپ باقاعدہ خاندانی منصوبہ بندی کے لئے استعمال کرسکتی ہیں۔ اگر آپ نے جنسی عمل کیا ہے اور آپ حاملہ نہیں ہونا چاہتی ہیں تو صفحہ 188 پر دیئے گئے طریقوں میں سے کوئی ایک طریقہ استعمال کریں۔

اگر آپ گولیاں نہیں نگل سکتی ہیں یا پہلے ہی متلی اور قے کی شکایت رکھتی ہیں تو گولیاں فرج میں بھی رکھی جا سکتی ہیں جہاں یہ آپ کے جسم میں جذب ہو جائیں گی۔

ایک انٹر یوٹیرائن آلہ (آئی یو ڈی) غیر محفوظ جنسی عمل کے بعد پانچ دن تک رحم میں رکھا جائے تو یہ گولیوں کی بہ نسبت بہتر کام کرتا ہے لیکن یہ طریقہ وہ عورت استعمال نہ کرے جو اپنی باقاعدہ منصوبہ بندی کے لئے آئی یو ڈی استعمال کرنا چاہتی ہو۔

مقامی صحت کارکنوں سے تبادلہ خیال کیجیے کہ عورتوں کے لئے ایمرجنسی خاندانی منصوبہ بندی کتنی ضروری ہے۔ ان کے اور مقامی دوافروشوں کے ساتھ مل کر ہر عورت کے لئے جو ایسا چاہتی ہو ایمرجنسی خاندانی منصوبہ بندی کی دستیابی کے لئے کام کریں۔

فالج اور خاندانی منصوبہ بندی

اگر آپ کے نچلے دھڑ میں (پولیو یا ریڑھ کی ہڈی کی چوٹ) کے باعث محسوسات نہ ہوں اور آپ حاملہ نہ ہونا چاہتی ہوں تو کوئی طریقہ چننے میں آپ کی مدد کے لئے رہنما باتیں درج ذیل ہیں (ممکن ہے کہ ان میں سے کچھ طریقے ہر جگہ دستیاب نہ ہوں)۔

کم بچوں کا مطلب یہ ہوسکتا ہے کہ آپ کے پاس اپنے لئے اور موجود بچوں کے لئے زیادہ وقت ہے

رکاوٹی طریقے ۔ (مردوں کے لئے کنڈوم، عورتوں کے لئے کنڈوم، ڈایافرام، کیپ، اسفنج، اسپرمی سائیڈ) ایسے طریقوں کے لئے جو فرج میں داخل کئے جاتے ہوں تو آپ کو مدد کی ضرورت ہوسکتی ہے۔

ہارمونل طریقے (گولیاں، انجکشن، امپلانٹس، آئی یو ڈی معہ ہارمونز) اگر آپ متحرک رہتی ہیں

اپنی وہیل چیئر یا پہیہ گاڑی دھکیلنا، ورزش کرنا، جھاڑو وغیرہ دینے جیسے کام کرنا، باغبانی وغیرہ) تو آپ ایسے ہارمونل طریقے استعمال کرسکتی ہیں جن میں ایسٹروجن شامل ہو۔

آپ ایسے ہارمونل طریقے ہرگز استعمال نہ کریں جن میں ایسٹروجن شامل ہو اگر:

● سارا دن بیٹھی رہتی ہوں یا ورزش نہ کرتی ہوں۔

● کبھی آپ کے جسم کے اندر کہیں خون کا لوتھڑا اٹھا تھا۔

● فالج یا عارضہ قلب کی علامات رکھتی تھیں۔

● کسی قسم کا کینسر ہو۔

● 35 برس سے زائد عمر کی ہوں۔

● سگریٹ پیتی، تمباکو چباتی یا سونگھتی ہوں۔

اگر آپ بالغ ہونے کے بعد مفلوج ہوئی ہوں تو اپنے چھ ماہ تک ہارمونل طریقے استعمال کرنا شروع نہ کریں۔

انٹرایوٹیرائن ڈیوائس (آئی یو ڈی) ۔ آئی یو ڈی کے استعمال سے کچھ مسائل ہوسکتے ہیں۔ جیسے آئی یو ڈی باہر نکلنا یا انفیکشن۔

عام طور پر یہ ایسے درد کا سبب بنتا ہے جس سے عورت کو آگاہ ہوسکتی ہے کہ کوئی گڑ بڑ ہے۔ اگر آپ درد محسوس نہیں کر پاتی ہیں تو آپ کے لئے مناسب ترین بات یہی ہے کہ یہ طریقہ استعمال نہ کریں، اگر آپ یہ طریقہ استعمال کرنا چاہتی ہیں تو کسی صحت کار کن سے باقاعدہ معائنہ یقینی بنائیے۔

اسقاطِ حمل

جب حمل ختم کرنے کے لئے کچھ کیا جاتا ہے تو یہ اسقاطِ حمل کہلاتا ہے۔ حمل کا غیر طے شدہ گرنا حمل کا ضائع ہونا (miscarriage) کہلاتا ہے۔ حمل ضائع ہونے کے اسباب کے بارے میں مزید معلومات کے لئے دیکھئے صفہ 219۔

حمل ساقط کرانے کا فیصلہ کرنا دشوار ہو سکتا ہے۔ کچھ مذاہب کے مطابق اسقاطِ حمل ناجائز ہے اور کئی ممالک میں یہ قانونی اور محفوظ نہیں ہے۔ لیکن ایسی بہت سی وجوہ ہیں کہ ایک عورت کسی بھی حال میں اپنا حمل ساقط کرانے کی کوشش کرے۔ ایسا فیصلہ کرنے میں بیش تر عورتوں کو آبرو مندانہ مشاورت اور دوستانہ تائید مل سکتی ہے۔ ذیل میں چند مثالیں ہیں کہ کوئی عمل کیوں حمل ساقط کرا سکتی ہے۔

- اس کے پہلے ہی بچے ہوں جن کی اسے دیکھ بھال کرنی ہو۔
- حمل اس کی صحت اور زندگی کے لئے خطرناک ہو۔
- بچے کی پرورش کے لئے اس کو مدد حاصل نہ ہو۔
- وہ بچہ پیدا کرنا نہ چاہتی ہو۔
- وہ جبری جنسی عمل کے نتیجے میں حاملہ ہوئی ہو۔
- کوئی اسے اسقاط کے لئے مجبور کر رہا ہو۔

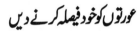

خواہ تم جو بھی فیصلہ کرو میں تمھاری دوست رہوں گی

عورتوں کو خود فیصلہ کرنے دیں

کچھ عورتیں انہیں مدد حاصل ہو یا نہ ہو بچہ پیدا کرنا چاہتی ہیں۔ وہ اس صورت میں بھی ولادت کا فیصلہ کر سکتی ہیں جب انہیں معلوم ہو کہ بچہ کسی سنگین مسئلے یا معذوری کا شکار ہوگا۔ بیش تر حاملہ عورتیں کہتی ہیں "مجھے یہ بچہ چاہئے۔" وہ جو بھی مشکلات درپیش ہوں ان سے نمٹنے کا عزم رکھتی ہیں۔

کچھ عورتوں کے لئے ان کی زندگی یا صحت کی صورتحال بچہ کا ہونا خراب انتخاب بنا دیتی ہے اور وہ

اسقاطِ حمل کا فیصلہ کرتی ہیں۔ وہ یہ فیصلہ اس لئے کر سکتی ہیں کہ وہ جانتی ہیں کہ انہیں بچی کی دیکھ بھال کے لئے مناسب مدد نہیں مل سکے گی یا بچہ کوئی معذوری یا صحت کا سنگین مسئلہ رکھتا ہوگا۔ یا جانتی ہوں کہ ان کے لئے معذوری یا معذوری کے بغیر کے ساتھ بچے کے معاملات سے نمٹنا بے حد مشکل ہوگا۔

بچہ پیدا کرنے کا فیصلہ ایک ذاتی فیصلہ ہے، تمام عورتوں کو یہ فیصلہ کرنے کا حق حاصل ہونا چاہئے۔ آپ کے عقائد جو بھی ہوں اگر وہ کوئی ایسا فیصلہ کرتی ہے جس سے آپ متفق نہ ہوں تو اس کے بارے میں کوئی اندازہ لگانے کی کوشش مت کریں۔ جذبے کے ساتھ اس کا خیال رکھیں اور اس کے ساتھ سلوک کریں وہ آپ اپنے ساتھ یا اپنی بیٹی کے ساتھ کرنا چاہتی ہوں

محفوظ اسقاط

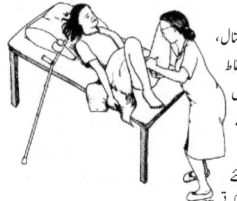

جہاں اسقاطِ حمل قانونی اور دستیاب ہے۔ ایک عورت اسپتال، مرکزِ صحت یا کلینک میں تجربہ کار صحت کارکن کے ذریعے محفوظ اسقاط کراسکتی ہے۔ عام طور پر یہ اس کے آئندہ حمل کو خطرے میں نہیں ڈالتا ہے۔ اسقاطِ حمل جب ابتدائی حمل میں کیا جائے تو سب سے محفوظ ہوتا ہے۔ تین قسم کے اسقاط ہیں جو محفوظ ہوسکتے ہیں۔

ویکیوم ایسپائریشن۔ اس طریقے میں رحم کو خالی کرنے کے لئے صحت کارکن ویکیوم مشین یا دستی ویکیوم ایسپائریشن (MVA) استعمال کرتی ہے۔ اگر ویکیوم ایسپائریشن درست طور پر کیا جائے تو اس سے کوئی نقصان نہیں پہنچتا ہے (دیکھئے 'اے بک فار مڈوائیوز'، باب 23)۔

ڈی اینڈ سی (ڈائی لیشن اور کیوریٹج)۔ صحت کارکن اس طریقے میں جراثیم سے پاک آلے کے ذریعے رحم کو کھرچ کر خالی کرتی ہے۔ تین سے زیادہ بار ڈی اینڈ سی (D&C) کرنے والی عورت کے رحم کے استر پر زخم کرسکتے ہیں جو آئندہ حمل کو مشکل بناسکتے ہیں۔

طبی اسقاط۔ حمل کے خاتمے کے لئے عورت دوائیں لیتی ہے جن سے رحم خالی ہوجاتا ہے۔ اس مقصد کے لئے مؤثر اور محفوظ دواؤں کے لئے کسی تجربے کار صحت کارکن سے بات کیجئے۔ (دیکھئے 'جہاں عورتوں کے لئے ڈاکٹر نہ ہو'، صفحہ 236-237)

غیر محفوظ اسقاط

جہاں اسقاط قانوناً ناجائز ہے۔ عورتوں کے حمل کو خود ختم کرنے کی کوشش ان کے لئے خطرناک ہوسکتی ہے یا وہ کسی ایسی عورت سے رابطہ کرتی ہیں جو غیر محفوظ انداز اور ماحول میں اسقاط کرتی ہے۔ غیر محفوظ اسقاط کہلاتا ہے۔ یہ بے تحاشا اخراجِ خون، سنگین انفیکشن یا بانجھ پن کا سبب بن سکتا ہے۔ یہ عورتوں کے لئے موت کا ایک بڑا سبب بھی بنتا ہے۔

ایسی غیر ضروری اموات کو روکا جا سکتا ہے اگر معذور عورتیں دوسری عورتوں اور مردوں کے ساتھ مل کر اپنی کمیونٹی کی تمام عورتوں کے لئے اسقاط کو محفوظ، قانونی اور قابلِ رسائی بنانے کے لئے کام کریں۔

اگر آپ نے کچھ عرصہ پہلے ہی محفوظ یا غیر محفوظ اسقاط کرایا ہو اور اس کے بعد آپ کسی انفیکشن یا زیادہ مقدار میں اخراجِ خون کا شکار ہوئی ہوں تو ممکن ہے کہ آپ کے رحم میں زخم ہوگئے ہوں جو آئندہ حمل یا ولادت میں مسائل پیدا کرسکتے ہیں۔ اگر آپ اس وقت حاملہ ہیں تو آپ کے لئے محفوظ ترین صورت یہی ہے کہ آپ بچے کو کسی اسپتال یا طبی مرکز میں جنم دیں اور اس بارے میں کسی صحت کارکن سے بات کریں۔

باب - 10

حمل

بچہ پیدا کرنے کا فیصلہ

حاملہ ہونا ذاتی فیصلہ ہے اور ہر عورت کو یہ حق حاصل ہونا چاہئے کہ وہ خود فیصلہ کرے کہ وہ کب ماں بننا چاہتی ہے۔ لیکن پوری دنیا میں عورتوں پر ان کے شریک حیات، گھر والے اور کمیونٹیز دباؤ ڈالتی ہیں کہ وہ جس قدر ممکن ہو بچے پیدا کریں۔

معذور عورتوں کے لئے اس سے برعکس رجحان پایا جاتا ہے۔ ان پر زور دیا جاتا ہے کہ وہ حاملہ نہ ہوں۔ بہت سی معذور عورتوں کو ان کی مرضی کے خلاف بانجھ کر دیا جاتا ہے، لہٰذا وہ حاملہ نہیں ہوسکتی ہیں۔ جو معذور عورتیں حاملہ ہو جاتی ہیں ان پر اسقاط غیر قانونی ہونے کے باوجود زور دیا جاتا ہے کہ وہ اپنا حمل ساقط کرالیں۔

لوگ غلط گمان رکھتے ہیں کہ معذور عورت ایک اچھی ماں نہیں ہوسکتی ہے یا اس کے بچے بھی معذور ہی ہوں گے۔

کوئی وجہ نہیں ہے کہ بیشتر معذور عورتیں محفوظ طور پر حاملہ نہ ہوں، صحت مند بچے کو جنم نہ دیں اور ایک اچھی ماں ثابت نہ ہوں (دیکھئے باب 12) تاہم چند احتیاطیں ہیں جو حمل کے دوران بعض قسم کی معذوریاں رکھنے والی عورتوں کو کرنی چاہئیں۔ اور کچھ عورتوں کو دوسری عورت سے زیادہ مدد کی ضرورت ہوتی ہے۔

اس باب میں دی گئی معلومات آپ کو یہ سمجھنے میں مدد دیں گی کہ حمل کے دوران کونسی تبدیلیاں ہوسکتی ہیں، یہ کس طرح مختلف معذوریوں پر اثر انداز ہوسکتی ہیں اور کس طرح محفوظ حمل اور ولادت کی منصوبہ بندی کی جائے۔

نومی کی کہانی :

کس طرح میں ماں بنی

جب میں جوان تھی اور میری سہیلیاں بچوں کے بارے میں گفتگو کرتی تھیں تو وہ سب مجھ سے کہا کرتی تھیں کہ اپنی معذوری کے باعث میں حاملہ نہیں ہوسکوں گی۔ وہ کہتی تھیں کہ میرے حاملہ ہونے کی صورت میں بچہ کی ولادت آپریشن کے ذریعے ہوگی اور وہ کسی نہ کسی معذوری کا شکار ہوسکتا ہے، میں حقیقتاً سمجھ نہیں پائی تھی کہ میری سہیلیوں کا مطلب کیا ہے کیونکہ میں سمجھتی تھی کہ میں بھی ان کی طرح ایک عورت ہوں۔ بس ان سے مختلف انداز سے چلتی تھی۔ میں نے ان باتوں پر یقین کرلیا جو انہوں نے کہی تھیں۔ پھر کسی ڈاکٹر نے اس کی تصدیق کے لئے میرا معائنہ نہیں کیا تھا میں اکثر بہت ہی غمزدہ رہتی تھی کیونکہ مجھے چھوٹے بچے بہت اچھے لگتے تھے۔ جب بھی میری کوئی سہیلی بچے کو جنم دیتی میں سوچتی تھی کہ کاش یہ بچہ میرا ہوتا۔

1987ء میں میں نے سنجیدگی سے سوچنا شروع کیا کہ مجھے لاحق خطرات کے باوجود اس کی کوشش کرنا چاہئے۔ میں نے اس پر غور کیا کہ کیوں نہیں اور پھر 27 دسمبر 1987ء کو میں حاملہ ہوگئی، جب مجھے معلوم ہوا کہ میں حاملہ ہوں تو میں بہت خوش تھی اور بے حد فکر مند بھی تھی کیونکہ پولیو زدہ تھی، میں نے یہ ثابت کرنے کے لئے کہ میں حاملہ ہوں ایک گائناِلوجسٹ سے رابطہ کیا اور مجھے یہ معلوم ہوا کہ میرے حمل اور ولادت کے دوران کچھ پیچیدگیاں بھی ہوسکتی ہیں۔

ڈاکٹر یہ سن کر حیران تھا کہ میں حاملہ ہوں۔ اس نے معائنے سے پہلے ہی مجھ سے کہا کہ میں چلنے کے انداز کے باعث حمل پورے دن بر قرار نہیں رکھ پاؤں گی۔ اس نے کہا کہ ابتدائی تین ماہ میں میرا حمل ضائع ہوسکتا۔ اس نے مجھے مشورہ دیا کہ میں تین ماہ انتظار نہ کروں اور اسی وقت اپنا حمل ساقط کرالوں۔ میں نے رضامندی ظاہر کردی اور 27 فروری 1988ء کی تاریخ لے لی۔ یہ بہت مہنگا کام تھا لیکن میں نے کسی نہ کسی طرح اتنی رقم جمع کرلی۔

میں نے اس وقت تک کسی کو نہیں بتایا تھا کہ میں حاملہ ہوں اور اس حوالے سے کتنی فکرمند ہوں۔ کینیا میں اسقاطِ حمل غیر قانونی ہے لہٰذا میں کسی کو بھی نہیں بتانا چاہتی تھی کہ میں اسقاط کرانے والی ہوں۔ پھر مجھے یہ اندازہ بھی نہیں تھا کہ اس پر میری سہیلیوں کا ردِعمل کیا ہوگا۔ کیا وہ مجھ پر ہنسیں گی یا پھر وہ مجھے مایوس کریں گی۔ میں نے اس لئے اس معاملے کو راز رکھا۔

میں نے بہت سی راتیں جاگتے ہوئے گزاریں ۔ میں دکھی اور خوفزدہ رہتی تھی، اول اس لئے کہ میں اپنی زندگی میں ماں نہ بننے کا تصور بھی نہیں کر سکتی تھی ۔ دوم اس لئے کہ اسقاطِ حمل خطرناک ثابت ہوتے تھے اور میں ایسی بہت سی نو جوان عورتوں کو جانتی تھی جو غیر محفوظ اسقاط کے باعث مر چکی تھیں ۔ سوم یہ کہ میں کرسچن تھی اور میرا یقین تھا کہ اسقاط گناہ ہے اور آخری بات یہ معاشرتی طور پر شادی کے بغیر حمل قابل قبول نہ تھا ۔ اس لئے آپ تصور کر سکتے ہیں کہ میں کس قدر مشکلات سے دوچار تھی ۔

بہرکیف وقت گزرتا رہا میں نے حوصلہ کیا اور خود کو اسقاط کے لئے تیار کر لیا ۔ طے شدہ دن میں اسپتال گئی اور ڈاکٹر کے کمرے کے باہر اپنے پکارے جانے کے انتظار میں بیٹھ گئی ۔ یہ میری زندگی کا انتہائی دشوار وقت تھا ۔ میں حوصلہ ہار چکی تھی، اور فکرمند تھی کہ میرے ساتھ کیا پیش آئے گا ۔ مجھے یقین تھا کہ اسقاط کے دوران مر جاؤں گی، میں نے دل ہی دل میں معافی مانگنا شروع کر دی ۔

اچانک مجھے یاد آیا کہ ڈاکٹر نے کہا تھا کہ میرا حمل بہرصورت 3 ماہ میں ضائع ہو جائے گا ۔ اس خیال نے میری سوچ بدل دی اور میں نے سوچا کہ مجھے اپنا حمل ضائع کرانے کی کوئی ضرورت نہیں ہے اور اگر اسقاط کے بجائے میرا حمل خود ہی ضائع ہو گیا تو اس میں کوئی بدنامی نہیں ہوگی ۔ میں گھر واپس آ گئی تاہم مجھے بالکل بھی یقین نہیں تھا کہ میں نے درست فیصلہ کیا ہے ۔

حمل کے ابتدائی چار ماہ بہت ہولناک تھے ۔ میرے وزن میں خاصی کمی ہوئی، میری بھوک ختم ہو گئی اور میں سارا وقت قے کرتی رہتی تھی، سب سے بڑھ کر یہ کہ میں خوف میں مبتلا رہتی تھی کہ کسی بھی وقت کوئی بھی بدترین واقعہ پیش آ سکتا ہے، جب میں نے پہلی بار بچے کی حرکت محسوس کی تو میں بہت خوفزدہ ہو گئی، میں سوچتی تھی کہ اب بچے کے باہر آنے کا وقت آ گیا ہے ۔

ہر چند کہ میں جانتی تھی کہ یہ ضروری ہے لیکن میں کچھ عرصے میڈیکل چیک اپ کے لئے جانے سے خوفزدہ رہی، لیکن پھر ایک دن میں نے قریبی ہیلتھ سینٹر جانے کا فیصلہ کر لیا، جہاں میں اس ڈاکٹر سے ملی جس نے میرا معائنہ کر کے مجھے یقین دلایا کہ میں حمل کے دن پورے کروں گی اور بچے کی ولادت بھی نارمل ہوگی ۔ اس نے مجھے مشورہ دیا کہ بچے کی ولادت اسپتال میں ہونی چاہیے ۔ میں نے خود پر اعتماد محسوس کیا اور تواتر سے میڈیکل چیک اپ کے لئے جانے لگی ۔ عملے نے مجھے بتایا کہ سب کچھ ٹھیک ہے ۔ نرسوں نے مجھے حمل، ولادت اور نومولود بچے کی دیکھ بھال پر کتابیں بھی دیں ۔ ان سے مجھے اچھی معلومات حاصل ہوئیں اور مجھے حمل جاری رکھنے کا حوصلہ بھی ملا ۔ میں اپنا بچہ چاہتی تھی کہ وہ کیسا ہوگا، کیا وہ معذور تو نہیں ہوگا اور سب سے بڑھ کر یہ مجھے وہ دوسری عورتوں کے بچوں کی طرح ماں کہہ کر بلائے گا ۔

سب حیران تھے کہ میرا حمل پورے عرصہ رہا اور میں نے 36 گھنٹے درد زہ سہنے کے بعد نارمل طریقے سے ایک صحت مند غیر معذور اور خوبصورت لڑکی کو جنم دیا ۔ اب میری بیٹی این 18 برس کی ہے ۔ وہ انتہائی صحت مند لڑکی ہے اور اس نے اپنے سیکنڈری اسکول میں بہترین کارکردگی دکھائی ہے ۔

حاملہ ہونے سے پہلے پوچھے جانے والے سوالات

ہر عورت کو یہ طے کرنے اور فیصلہ کرنے کی ضرورت ہے کہ اس کے کتنے بچے ہوں اور کب ہوں۔ ایک عورت کی عمر، صحت اور رہنے سہنے کی صورتحال اس کے ماں بننے کے فیصلوں پر اثر انداز ہو سکتی ہے۔

حاملہ ہونے سے پہلے ان سوالات پر غور معاون ثابت ہو سکتا ہے۔

- کیا آپ بچے چاہتی ہیں؟
- اگر آپ کے بچے ہیں تو کیا آپ مزید کسی بچے کی دیکھ بھال کر سکتی ہیں؟
- کیا گذشتہ حمل کے بعد آپ کا جسم صحت مند ہو چکا ہے؟
- کیا آپ خود بچے کی دیکھ بھال کر سکتی ہیں؟
- کیا بچے کی دیکھ بھال اور آپ کی مدد کے لئے کوئی فرد یا خاندان موجود ہے؟
- کیا کوئی آپ کو ماں بننے پر مجبور کر رہا ہے؟
- کیا حمل نے آپ کی معذوری پر کوئی اثر ڈالا تھا؟

کیا میرا بچہ کسی معذوری کے ساتھ پیدا ہوگا؟

بہت سی معذوریاں، ماں سے بچے میں منتقل نہیں ہوتی ہیں (وراثتی/خاندانی معذوریاں) لیکن کچھ معذوریاں ایسی ہیں جو منتقل ہو جاتی ہیں۔ بعض اوقات باپ کے ذریعے اور بعض اوقات ماں سے اور بعض اوقات دونوں سے۔ خاندان میں منتقل ہونے والی معذوریوں کے بارے میں مزید جاننے کے لئے دیکھئے صفحہ 14۔

اگر آپ سمجھتی ہیں کہ آپ کا بچہ ان میں سے کسی معذوری کے ساتھ جنم لے گا تو آپ کے لئے بہتر ہے کہ کسی پیچیدگی کی صورت سے بچنے کے لئے بچے کی ولادت کسی اسپتال میں ہونے کا انتظام کیجئے۔

بچہ لڑکا ہوگا یا لڑکی

یہ مرد کی منی (Sperm) ہوتی ہے جو لڑکا یا لڑکی کے ہونے کا تعین کرتی ہے، مرد کے مادہ منویہ کا نصف لڑکا تخلیق کرتا ہے اور بقیہ نصف لڑکی۔ صرف ایک جرثومہ عورت کے بیضے میں شامل ہوتا ہے اگر یہ نر جرثومہ ہے تو لڑکا ہوگا اور اگر وہ مادہ جرثومہ ہے تو لڑکی پیدا ہوگی۔ ایک معذور عورت یا غیر معذور عورت میں اس سے مختلف کوئی بات نہیں ہوتی ہے۔

سماج میں اکثر گھرانے لڑکوں کو ترجیح دیتے ہیں، اگر عورت بیٹا پیدا نہ کرے تو یہ اسے مورد الزام ٹھہراتے ہیں۔ یہ لڑکیوں کے لئے غیر منصفانہ ہے، جنہیں لڑکوں ہی کی طرح اہمیت دی جانی چاہئے اور یہ عورتوں کے ساتھ بھی ناانصافی ہے کہ یہ مرد ہوتا ہے جس کا مادہ منویہ بچے کی جنس کا تعین کرتا ہے۔

حمل اور ولادت کے لئے منصوبہ بندی

دنیا بھر میں بہت سی عورتیں مقامی مڈوائفز کی مدد سے گھروں میں بچوں کو جنم دیتی ہیں۔ یہ ولادتیں ماں اور بچے دونوں ہی کے لئے محفوظ اور صحت مند ہو سکتی ہیں خصوصاً جب مڈوائف تجربے کار ہو، ان مڈوائفز کے ذریعے حمل کے دوران معذوریاں رکھنے والی عورت کی دیکھ بھال اور وضح حمل عموماً محفوظ ہوتا ہے لیکن بعض اوقات مڈوائف کے ماہر ہونے کے باوجود عورتوں اور بچوں کو اسپتال میں دیکھ بھال کی ضرورت پڑ سکتی ہے۔

معذوری رکھنے والی کچھ عورتوں کو جو پیچیدگیوں کا زیادہ خطرہ رکھتی ہیں طبی نگہداشت کی ضرورت پڑ سکتی ہے جو صرف اسپتال میں ہی ملتی ہے۔ مثلاً اگر آپ

- **ایسی معذوری رکھتی ہوں جو آپ کی ٹانگوں کو کھلا رکھنے سے روکتی ہو۔** جیسے سیریبرل پالسی، گٹھیا، عضلات کا شدید جکڑاؤ وغیرہ۔ ولادت کے لئے آپ کو اپنی ٹانگیں دو سے تین گھنٹے کھلا رکھنے کی ضرورت ہوگی خواہ اپنے طور پر یا کسی کی مدد کے ساتھ یا پھر آپ کو بچے کی ولادت آپریشن سے کرانے کی ضرورت پڑ سکتی ہے۔

- **آپ چھوٹی جسامت کی (بونی) ہوں۔** اس صورت میں آپ کے پیٹرو کی ہڈیاں ممکن ہے اتنی چوڑی نہ ہوں کہ بچہ آپریشن کے بغیر محفوظ طور پر باہر آ سکے کیونکہ آپ کے جسم میں خون بھی کم ہوگا لہٰذا ممکن ہے آپ کو خون منتقل کرنے کی ضرورت پڑے، اس کا انحصار اس پر بھی ہے کہ بچے کی پیدائش کے دوران آپ کا کتنا خون ضائع ہوا۔

- **آپ ریڑھ کی ہڈی کی بلند چوٹ (T6 اور اس سے اوپر) رکھتی ہوں۔** اس صورت میں آپ ڈس ریفلیکسیا، خطرناک ہائی بلڈ پریشر کا خطرہ رکھتی ہیں (دیکھئے صفہ 117 تا 119)۔

جب آپ حاملہ ہونے کی کوشش کر رہی ہوں

خود کو اور اپنے بچے کو ممکنہ حد تک صحت مند رکھنے کے لئے باقاعدگی سے کھانا کھائیں اور مختلف صحت افزاء غذائیں کھانے کی کوشش کریں خصوصاً وہ غذائیں جو پیدائشی نقائص سے بچاؤ میں مدد دیتی ہیں (دیکھئے صفہ 86 اور 216)۔ صحت افزاء عادتیں جیسے اچھی غذائیں کھانا، تمباکو نوشی نہ کرنا، منشیات اور الکحل سے گریز، اہم ہیں کیونکہ بہت سے مسائل حمل میں بہت جلد شروع ہو جاتے ہیں، اس صورت میں بھی جب آپ کو اپنے حاملہ ہونے کا اندازہ بھی نہ ہو۔

ولادت کا منصوبہ بنائیں

ہر چند کہ اپنی ضرورت کے مطابق طبی علاج کا حصول معذور عورتوں کے لئے مشکل ہوسکتا ہے لیکن ہر حاملہ عورت کو ''برتھ پلان'' بنانا چاہیئے ۔ جیسے ہی آپ حاملہ ہوں، پری نیٹل (یا اینٹی نیٹل) معائنہ کرانا شروع کر دیں۔اگر ممکن ہو تو کوئی مڈوائف، ڈاکٹر یا صحت کارکن تلاش کریں جس پر آپ اعتماد کرسکیں۔ جب پہلی بار چیک اپ کے لئے جائیں تو اپنی کسی سہیلی یا گھر کے کسی فرد کو ساتھ لے جائیں۔ آپ امکانی مسائل پر بات کرسکتی ہیں کہ ایسی صورتوں میں کیا کیا جائے اور یہ کہ آپ کو کہاں سے بہترین مشورہ مل سکتا ہے۔

آپ کے برتھ پلان کے لئے مندرجہ ذیل معلومات مفید ہوسکتی ہیں۔

- آپ کے بچے کی پیدائش کے لئے محفوظ ترین جگہ کونسی ہے، گھر، میٹرنٹی ہوم یا اسپتال؟

- اگر آپ دوائیں باقاعدہ لیتی رہیں تو اس سے کوئی اثر آپ کے نشوونما پانے والے بچے پر پڑ سکتا ہے؟ ممکن ہے آپ کو دواؤں میں تبدیلی کی ضرورت ہو جو حمل کے دوران کے دوران زیادہ محفوظ ہوں، یہ بات خصوصی طور پر anti seizure ادویات کے لئے بھی درست ہے (دیکھئے صفحہ 213)۔

> بے فکر رہو، حمل کے دوران تمہیں جس چیز کی ضرورت ہوگی وہ میں فراہم کروں گی

- کیا حمل کے دوران آپ کی معذوری آپ کی صحت یا بچے کی نشوونما پر اثر انداز ہوگی؟

- کیا آپ کی معذوری زچگی یا ولادت کے وقت مسائل کا سبب بن سکتی ہے؟

- کیا پیچیدگیوں سے بچاؤ یا ان کا محفوظ تدارک ممکن ہے؟

- کیا آپ جانتی ہیں کہ حمل کے دوران کس طرح صحت مند رہا جاسکتا ہے؟

کس طرح جانیں کہ بچہ کب متوقع ہے

اپنی نارمل ماہواری شروع ہونے کی تاریخ میں نو ماہ اور سات دن جمع کریں آپ کا بچہ حاصل شدہ تاریخ سے دو ہفتے قبل یا بعد متوقع ہے۔

ایک عورت آخری ماہواری کے بعد دس چاند گزرنے کا حساب لگا کر جان سکتی ہے کہ اس کے ہاں بچے کی پیدائش کب ہوگی

حمل کے دوران صحت مند رہنا

اگر آپ حمل کے دوران اپنا بہتر طور پر خیال رکھ سکیں تو محفوظ طور حمل، ولادت اور بچے کے صحت مند ہونے کا زیادہ امکان ہے۔ اس لئے آپ

- جب آپ کے لئے ممکن ہو سوئیں اور آرام کریں۔
- پیدائش سے قبل پری نیٹل معائنے کے لئے جائیں۔
- اگر آپ نے کبھی ٹیٹنس امیونائزیشن نہیں کرائی تو جس قدر جلد ممکن ہو پہلی امیونائزیشن کرالیں اور دوسری امیونائزیشن اپنے حمل کی تکمیل سے کم از کم دو ہفتے پہلے کرالیں۔
- سکیڑنے کی ورزشیں کریں اگر آپ کے لئے ممکن ہو، اس طرح آپ کی فرج ولادت کے بعد طاقتور رہے گی (دیکھئے صفحہ 101)۔
- روزانہ کم از کم آٹھ گلاس پانی یا جوس پئیں اور مثانے اور گردوں کے انفیکشن سے بچاؤ کے لئے تواتر سے پیشاب کریں۔
- روزانہ ورزش کریں۔
- اگر آپ جنسی طور پر منتقل ہونے والے انفیکشنز یا دوسرے انفیکشنز میں مبتلا ہوں تو ان کا علاج کرائیں۔
- جدید ادویات یا جڑی بوٹیاں استعمال کرنے سے گریز کریں۔ یہ ادویات اس وقت تک استعمال نہ کریں جب تک کوئی صحت کارکن جو یہ جانتی ہو کہ آپ حاملہ ہیں، انہیں موزوں قرار نہ دے۔
- الکحل نہ پئیں، تمباکو کسی بھی انداز سے استعمال نہ کریں۔ یہ چیزیں آپ کے لئے نقصان دہ ہیں اور بچے کو بھی نقصان پہنچائیں گی۔
- پیسٹی سائیڈز، ہربی سائیڈز اور فیکٹری کیمیکلز سے دور رہیں۔
- ایسے بچے سے دور رہیں جس کے پورے جسم میں سرخ دانے ہوں۔ یہ جرمن خسرہ (ربیولا) کی وجہ سے بھی ہو سکتے ہیں۔
- اگر آپ فضلے کے اخراج کے لئے باؤل پروگرام پر عمل کرتی ہیں تو یہ عمل باقاعدگی سے کریں (دیکھئے صفحہ 107)۔

مختلف اقسام کی غذائیں کھائیں

اگر آپ حاملہ ہیں یا بچے کو دودھ پلاتی ہیں تو آپ کو معمول سے زیادہ کھانے کی ضرورت ہے۔ اضافی غذا آپ کو مناسب توانائی اور طاقت فراہم کرے گی اور اس سے آپ کے بچے کی افزائش میں مدد ملے گی جس قدر ممکن ہو مختلف اقسام کی غذائیں کھائیں۔ بنیادی غذائیں کاربوہائیڈریٹس، نشوونما والی غذائیں (پروٹین) گلوفوڈز (glow foods) (وٹامنز اور معدنیات) اور گوفوڈز (go foods) (چکنائی، تیل اور شکر) اور بہت سے مائعات استعمال کریں۔ (مزید معلومات کے لئے دیکھئے صفحہ 86)

اینیمیا (کمزور خون) سے بچاؤ

آپ کے لئے ضروری ہے کہ ایسی غذائیں کھائیں جن میں آئرن موجود ہوتا ہے کہ آپ کا خون طاقتور رہے۔ اگر ایک حاملہ عورت اینیمیا کا شکار ہو اور بچے کی ولادت کے دوران اس کا خون بھی خوب بہے تو وہ سخت بیمار پڑ سکتی ہے یا مر بھی سکتی ہے۔ اینیمیا کے بارے میں مزید معلومات کے لئے دیکھئے صفحہ 87 تا 88۔

فولک ایسڈ (فولیٹ)

فولک ایسڈ مناسب مقدار میں نہ لینے سے اینیمیا ہوسکتا ہے جبکہ اس باعث بچے میں سنگین پیدائشی نقائص جیسے ریڑھ کی ہڈی یا دماغ میں ابھار ہوسکتے ہیں۔ ان مسائل سے بچنے کے لئے ضروری ہے کہ آپ حاملہ ہونے سے پہلے اور حمل کے ابتدائی چند ماہ میں مناسب مقدار میں فولک ایسڈ لیں۔

ان غذاؤں میں فولک ایسڈ وافر مقدار میں ہوتا ہے۔

- گہرے سبز رنگ کے پتوں والی سبزیاں
- گوشت (خصوصاً جگر، گردے اور دیگر عضو کا گوشت)
- مٹر اور سیم
- سورج مکھی، کدو، لوکی
- غذائی اجناس (بھورے چاول، گندم)
- مچھلی
- انڈے
- مشرومز

کچھ عورتوں کو فولک ایسڈ کی گولیاں بھی لینی چاہئیں۔

فولک ایسڈ

روزانہ دن میں ایک بار منہ کے ذریعے 0.5 تا 0.8 ملی گرام (500 - 800 mcg) فولک ایسڈ لیں۔

Spina bifida والی عورتیں 800 mcg فولک ایسڈ روزانہ ایک بار رکھائیں۔

حمل کے دوران جنسی عمل

کچھ عورتیں جب حمل سے ہوں تو جنسی عمل پسند نہیں کرتیں۔ کچھ معمول سے زیادہ جنسی عمل چاہتی ہیں۔ دونوں ہی باتیں نارمل ہیں۔ عورت اور اس کے بچے کے لئے جنسی عمل ہونا یا نہ ہونا دونوں ہی باتیں ٹھیک ہیں۔ دوران حمل ماں کی کوکھ میں موجود بچے کے لئے عورت کا جنسی عمل خطرناک نہیں ہوتا ہے۔

بعض اوقات حمل میں جنسی عمل غیر اطمینان بخش ہوتا ہے، لہذا دوران حمل آپ جنسی عمل کے مختلف طریقے آزما سکتی ہیں۔ آپ اپنی معذوری کے اعتبار سے موزوں طریقہ جاننے کے لئے مختلف انداز استعمال کر کے فیصلہ کریں کہ کونسا طریقہ موزوں ترین ہے۔ اس دوران جنسی عمل کے علاوہ بھی ایک دوسرے کے قریب رہ کر مباشرت کے علاوہ دوسرے طریقوں مثلاً ہلکے پھلکے مساج، اپنے مستقبل کے خوابوں پر تبادلہ خیال سے بھی ایک دوسرے کو خوش رکھا جا سکتا ہے۔

محفوظ مباشرت

اگر آپ حاملہ ہیں اور مباشرت کرتی ہیں تو انفیکشن سے بچاؤ کے لئے ضروری ہے کہ آپ کے جسم میں جانے والی ہر چیز صاف ہو۔ اس میں عضو تناسل اور ہاتھ دونوں شامل ہیں۔ مباشرت کے دوران کنڈوم انفیکشنز، ایچ آئی وی/ایڈز اور دوسری بیماریوں سے بچاؤ کا بہترین ذریعہ ہیں، کنڈومز کے بارے میں مزید معلومات کے لئے دیکھئے صفحہ 189 تا 192۔

کنڈوم دوران حمل انفیکشن سے بچاؤ کا محفوظ ذریعہ ہے

مباشرت اور قبل از وقت زچگی

اگر آپ پہلے بھی ماں بن چکی ہیں اور زچگی وقت سے پہلے ہوئی ہو تو بہتر یہ ہے کہ چھٹے مہینے کے بعد مباشرت مت کریں۔

حمل کے نو مہینے

حمل عام طور پر نو ماہ کا ہوتا ہے اور اسے تین ماہ کے تین حصوں میں تقسیم کیا جاتا ہے ان تین ماہ میں عورت کا جسم مختلف تبدیلیوں سے دوچار رہتا ہے۔

پہلی سہ ماہی (ایک تا تین ماہ)

حمل ہونے کے بعد جب بچہ نشوونما پانے لگتا ہے، آپ کی چھاتیاں سوج جاتی ہیں اور ممکن ہے کہ دکھنے لگیں۔ ممکن ہے آپ معمول سے زیادہ تھکنے لگیں۔ اور آپ متلی محسوس کریں اور قے کریں۔ کچھ مقامات پر اسے (Morning sickness) کہا جاتا ہے۔

دوسری سہ ماہی (چار تا چھ ماہ)

بہت سی عورتیں حمل کے چوتھے، پانچویں اور چھٹے مہینے میں خوش رہتی ہیں۔ عام طور پر وہ معدے میں گڑبڑ محسوس نہیں کرتی ہیں، تھکن کا احساس بھی ختم ہو جاتا ہے۔ وہ عموماً زیادہ توانائی محسوس کرتی ہیں۔ یہ وقت ہے جب ان کا پیٹ بڑا ہو جاتا ہے اور بچہ حرکت شروع کرتا ہے۔ اس دوران آپ بچے کے دل کی دھڑکن محسوس کر سکتی ہیں۔

تیسری سہ ماہی (سات تا نو ماہ)

یہ آپ کے حمل کا سنسنی خیز زمانہ ہے۔ آپ روزانہ بچے کی حرکت محسوس کریں گی۔ اس کے ساتھ ہی آپ کا پیٹ بڑا ہوتا جائے گا۔ اس کا انحصار آپ کی معذوری پر ہے کہ آپ دن میں کس قدر مشکلات یا مسائل سے دوچار رہوں۔ اگر آپ نے ابتدائی چھ ماہ میں دشواریاں جھیلی ہیں تو یہ جاری بھی رہ سکتی ہیں اور سنگین تر بھی ہو سکتی ہیں۔

آخری ماہ میں ولادت سے تقریباً دو ہفتے پہلے۔ بچہ اکثر پیٹ کے نچلے حصے میں گر جاتا ہے۔ خاص طور پر پہلے بچے کی صورت میں ایسا ہوتا ہے اور اس صورت میں آپ کے لئے سانس لینا آسان ہو سکتا ہے۔

کیا متوقع ہوتا ہے

بچے کی حرکت

بیش تر عورتوں کے لئے پیٹ کے اندر بچے کی حرکت ان کے حمل کا سب سے سنسنی خیز حصہ ہوتا ہے، اور بیش تر عورتیں قطع نظر اپنی معذوری کے بچے کی حرکت کو محسوس کرسکتی ہیں تاہم ابتداء میں اس کا اندازہ لگانا مشکل ہوسکتا ہے۔ بہت سی عورتیں اسے لرزش جیسی حرکت کہتی ہیں تو دوسری اس کا گیس کے درد سے تقابل کرتی ہیں۔ کچھ عورتیں پیٹ کے اندر دباؤ سا محسوس کرتی ہیں اور حرکت کو محسوس کرنے کے لئے اپنے ہاتھ استعمال کرتی ہیں۔

جب چوتھے مہینے میں بچے حرکت کرنا شروع کرتا ہے حرکت انتہائی آہستہ ہوتی ہے اور ممکن ہے آپ اسے روزانہ محسوس نہ کریں لیکن پانچویں مہینے سے آپ یہ حرکات روزانہ محسوس کریں گی (دن بھر نہیں۔ ان حرکات کے دوران آرام کا دورانیہ بھی ہوگا) اگر آپ فکرمند ہیں کہ آپ نے کئی گھنٹے بچے کی حرکت محسوس نہیں کی تو آپ کچھ کھائی کرسی پر سکون جگہ تقریباً 30 منٹ لیٹی رہیں۔ اس دوران آپ کم از کم تین بار بچے کی حرکت محسوس کریں گی۔ اگر ایسا نہ ہو تو کسی مڈ وائف یا صحت کارکن سے رجوع کیجئے۔

بچے کے دل کی دھڑکن

یہ تقریباً پانچ ماہ بعد ممکن ہے اور جوں جوں دن گزرتے ہیں بچے کے دل کی دھڑکن سننا آسان ہو جاتا ہے۔ بچے کی دھڑکن بہت تیز اور مدھم ہوتی ہے۔ اسے سننا آسان نہیں حتیٰ کہ اسے اوسط درجے سے بہتر سماعتی ذریعے سے نہ سنا جائے۔ اس صورت میں بھی اسے محسوس کرنا یا سننا مشکل ہوسکتا ہے۔ مڈ وائف یا صحت کارکن بچے کے دل کی دھڑکن فیٹو اسکوپ (Fetoscope) کے ذریعے سن سکتی ہے یہ دھڑکن ماں کے لئے سننا مشکل ہے البتہ وہ اسٹیتھو اسکوپ کے ذریعے اسے آسانی سے سن سکتی ہے۔

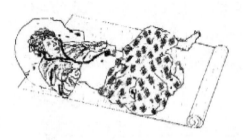

بچے کے دل کی دھڑکن تیز اور مدھم ہوتی ہے۔ یہ تکیئے کے نیچے رکھی گھڑی کی آواز کی طرح ہوتی ہے۔ بچے کی دھڑکن ایک صحت مند بالغ فرد کی دھڑکن سے دو گنی تیز ہوتی ہے۔ آپ اسے اسٹیتھو اسکوپ کے ذریعے سن سکتی ہیں۔

ایک کان لگا کر سننے والے فیٹو اسکوپ کے ذریعے بچے کے دل کی دھڑکن سنی جاسکتی ہے

فیٹو اسکوپ

حمل ضائع ہونا (Miscarriage)

معذور عورتیں غیر معذور عورتوں سے زیادہ حمل ضائع ہونے کے امکانات نہیں رکھتی ہیں۔ حمل ضائع ہونا کسی بھی ایسے فرد کے لئے صدمے کی بات ہے جو بچے کا خواہاں ہو لیکن یہ خاص طور پر معذور عورت کے لئے زیادہ سخت تجربہ ہے۔ بہت سے لوگ یہ نہ سوچتے ہوں کہ اسے حاملہ ہونا چاہیئے لیکن جب وہ حاملہ ہوتی ہے تو اسے اپنی کمیونٹی میں ناپسندیدگی کا سامنا کرنا پڑتا ہے، اگر اس کا حمل ضائع ہو جائے تو لوگ سمجھتے ہیں کہ اس کا سبب اس کی معذوری ہے۔ ممکن ہے وہ بھی یہی سوچے۔

حمل زیادہ تر ابتدائی تین ماہ میں ضائع ہوتا ہے۔ حمل ضائع ہونے کی بہت سی وجوہات ہو سکتی ہیں جیسے

حمل ضائع ہونا عام ہے اگر آپ کے ساتھ ایسا ہوا ہے تو اس کا یہ مطلب ہرگز نہیں ہے کہ آپ اگلی بار صحت مند طور پر حاملہ نہیں ہو سکتی ہیں

- غیر صحت مند بیضے یا مادہ منویہ
- رحم کی بناوٹ کا کوئی مسئلہ
- رحم میں بڑھوتری (رسولی)
- رحم یا فرج میں انفیکشن
- کوئی بیماری جیسے ملیریا
- سخت محنت کا کام یا کوئی حادثہ
- زہر
- ناقص غذائیت
- جذباتی تناؤ

اگر آپ کا حمل ضائع ہوا ہے تو ابتدائی چند دن اپنا بہتر خیال رکھیں۔ اس سے آپ کو انفیکشن سے تحفظ ملے گا اور آپ کا جسم تیزی سے صحت مند ہوگا۔ اس دوران کوشش کریں کہ

- خوب مائعات پئیں اور پُر غذائیت اشیاء کھائیں (دیکھئے صفحہ 86)
- آرام کریں
- سات دن مشقت والے کام نہ کریں۔
- پابندی سے غسل کریں لیکن خون کا اخراج رُکنے کے بعد چند دن تک اپنی فرج کو نہ دھوئیں یا پانی کے ٹب میں نہ بیٹھیں۔
- کم از کم دو ہفتے اور اخراج خون بند ہونے کے بعد چند دن تک اپنی فرج میں کوئی چیز نہ رکھیں اور نہ ہی مباشرت کریں۔
- دوبارہ حاملہ ہونے کے لئے تین بار ماہواری آنے دیں۔ اگر آپ انتظار کریں گی تو آئندہ حمل ضائع ہونے کے امکانات اسی قدر کم ہوں گے۔

حمل کے دوران ابتداء یا بعد میں حمل ضائع ہونا بے تحاشہ جذباتی، درد اور دکھ کا سبب بن سکتا ہے، یہ دکھ اس لئے بھی بڑھ جاتا ہے کہ آپ کے ارد گرد موجود لوگ یہ سمجھتے ہیں کہ یہ تکلیف دہ مرحلہ ختم ہونے کے بعد آپ ایک بار پھر نارمل ہو گئی ہیں۔ ممکن ہے کہ وہ یہ محسوس نہ کریں کہ آپ کس قدر دکھی ہیں۔

صدمہ برداشت کرنے کے ساتھ صدمہ کا اظہار بھی کریں۔ اپنے ان احباب کے ساتھ وقت گزاریں جو سمجھتے ہوں آپ کیا محسوس کر رہی ہوں گی، خوش رہنے کے لئے خود پر جبر نہ کریں۔ کچھ لوگ آپ سے کہہ سکتے ہیں آپ فوراً حاملہ ہونے کی کوشش کریں۔ دوبارہ حاملہ ہونے کے لئے وقت لیں۔

پیٹ کے نچلے حصے میں درد

ابتدائی تین ماہ میں مستقل طور پر شدید درد کا سبب یہ ہوسکتا ہے کہ حمل رحم کے باہر ٹیوب میں پروان چڑھ رہا ہو (Tubal or ectopic pregnancy) ٹیوبز کے کھنچنے سے درد ہوتا ہے۔اگر حمل حد سے زیادہ بڑھ جائے تو ٹیوب پھٹ جائے گی اور خون بہنے لگے گا، یہ نہایت خطرناک ہے۔اس صورت میں آپ کے پیٹ کے اندر خون بہے گا اور آپ مر سکتی ہیں۔

ٹیوب میں حمل کی علامات

ٹیوب میں پرورش پاتا حمل

- ماہواری کا نہ ہونا۔
- پیٹ کے نچلے حصے میں ایک طرف درد
- فرج سے معمولی اخراج خون
- چکر آنا،کمزوری یا بے ہوشی

اگر آپ میں ان میں سے کچھ علامات ہوں تو قریب ترین اسپتال جائیں۔

حمل کے دوران خون بہنا

اگر ابتدائی تین ماہ میں خون کے دھبے آئیں یا خون بہے تو فکر مند نہ ہوں۔ یہ غیر معمولی نہیں ہے۔ خصوصاً اس وقت جب درد یا مروڑ نہ ہو۔ لیکن اگر آپ مندرجہ ذیل صورتحال سے دوچار ہوں تو فوری طور پر طبی امداد کے لئے اسپتال جائیں۔

- حمل کے دوران کسی بھی وقت ماہواری کی طرح خون کا اخراج۔
- حمل کے دوران کسی بھی وقت درد کے ساتھ خون کا اخراج۔
- تین ماہ کے بعد درد کے بغیر اخراج خون۔

حمل کے دوران تکالیف

بہت سی عورتیں حمل کے دوران مختلف تبدیلیوں اور تکالیف سے دوچار ہوتی ہیں۔ کچھ معذور عورتوں کے لئے یہ تکالیف ان کی معذوری کو بدتر بناتی ہے اور کچھ کے لئے کم۔

حمل کی کچھ تکالیف جیسے تھکن یا کمر کا درد تمام عورتوں میں (بشمول معذور عورتیں) مشترک ہیں۔ ایک معذور عورت کے لئے کیا بات مختلف ہو سکتی ہے؟ یہ جاننا کہ اس کی تکلیف یا مسئلے کا سبب اس کا حمل ہے یا اس کی معذوری۔ یہ جاننے کے بعد ہی وہ مدد حاصل کرسکتی ہے اور سمجھ سکتی ہے کہ اسے کس وقت صحت کار کن سے رابطہ کرنا چاہئے۔

آپ اپنے جسم کو دوسروں سے بہتر جانتی ہیں لہذا جب آپ حاملہ ہوں تو غور کیجئے کہ کیا بات آپ کے جسم کے لئے نارمل ہے اور کیا نارمل نہیں ہے اس طرح آپ بتا سکتی ہیں کہ کیا چیز آپ کے حمل کے باعث ہے اور کیا معذوری کے سبب

اگر آپ اپنی معذوری کے باعث مثلاً ریڑھ کی ہڈی مجروح ہونے کے باعث یورین سسٹم کے انفیکشنز جیسے مختلف مسائل میں مبتلا ہوسکتی ہیں تو پھر آپ اپنے حمل کے دوران کئی اور مسائل کا شکار ہوسکتی ہیں۔ یہ مسائل اور تبدیلیاں آپ کے حمل کے دوران کسی بھی وقت سامنے آسکتی ہیں اس کا انحصار آپ کے جسم اور بچے کی نشوونما پر ہے۔

ہر عورت اپنی زندگی میں اپنے حمل سے مطابقت کے لئے مختلف اہتمام کرتی ہے۔

کچھ معذور عورتوں میں مبتلا عورتوں کے لئے چند تبدیلیاں درج ذیل ہیں جو سامنے آسکتی ہیں۔ ذیل میں تجاویز بھی ہیں کہ ان سے کس طرح نمٹا جائے۔

تھکن اور غنودگی محسوس کرنا

حمل کے ابتدائی تین یا چار ماہ میں بیش تر عورتیں تھکن اور غنودگی محسوس کرتی ہیں، دیگر امکانی اسباب کے متعلق مکمل معلومات کے لئے پڑھئے۔

- اینیمیا (جہاں عورتوں کے لئے ڈاکٹر نہ ہو، صفحہ 168)
- درست اقسام کا کھانا مناسب مقدار میں نہ کھانا (ناقص غذائیت)   (جہاں عورتوں کے لئے ڈاکٹر نہ ہو، صفحہ 168 تا 171)
- جذباتی مسائل (جہاں عورتوں کے لئے ڈاکٹر نہ ہو، باب 26)

نیند کے مسائل

حمل کے آخری چند ہفتوں میں بہت سی عورتوں کو رات کے وقت سونے میں مشکلات پیش آتی ہے۔ یہ اس لئے بھی ہوسکتا ہے کہ انہیں رات کے وقت پیشاب کے اخراج کی حاجت ہوتی ہے یا پیروں میں سوجن یا اکڑن ہوتی ہے۔ (دیکھئے صفحہ 222 اور 225) یا اس لئے کہ بچہ حرکت کرنا یا ضرب لگانا شروع کرتا ہے۔ سونے کا آرام دہ انداز ڈھونڈ نا مشکل ہوسکتا ہے۔ اگر ممکن ہو تو رات میں نیند کی کمی پوری کرنے کے لئے دن میں آرام کریں۔

آرام یا سونے کے لئے آرام دہ انداز اپنانا اہم ہے، انتہائی اہم بات یہ یاد رکھنا ہے کہ پشت کے بل سپاٹ انداز سے نہ سوئیں۔ اس سے آپ کے رحم کی خون کی رگیں دب جائیں گی اور دوران خون کے مسائل پیدا ہوں گے۔ یہ انداز کمر میں کھچاؤ اور سانس کے مسائل کے ساتھ ہاضمے کی خرابیوں کا سبب بھی بن سکتا ہے۔

کیا کرنا چاہئے

- سونے کی کوشش سے پہلے تھوڑا سا گرم دودھ یا سوپ پئیں۔
- قدرے بیٹھی ہوئی حالت میں سوئیں یا اپنے سر اور کندھوں کے نیچے کچھ رکھیں اور اپنے گھٹنوں کے نیچے کپڑا یا کوئی مناسب چیز گول تہہ کر کے رکھیں۔
- پہلو کے بل سوئیں۔ اگر ممکن ہو تو الٹے پہلو لیٹیں کہ یہ دوران خون کے لئے بہترین انداز ہے۔
- اپنے گھٹنوں اور کہنیوں کے درمیان کوئی آرام دہ شئے مثلاً گول تہہ کیا ہوا کپڑا یا اخبار رکھیں۔

غذائیت والی غذائیں کھائیں جن سے آپ مناسب مقدار میں پروٹین حاصل کرکے۔۔ اپنے کھانوں میں نمک کم مقدار میں استعمال کریں۔ (لیکن نمک استعمال ضرور کریں)

پیروں اور ٹانگوں کی سوجن

حمل کے دوران بہت سی عورتوں کے پیر اور ٹانگیں سوج جاتی ہیں، ایسا خصوصاً دوپہر کے وقت یا گرم موسم میں ہوتا ہے۔ پیروں کی سوجن عام طور پر خطرناک نہیں ہوتی ہے جب آپ صبح بیدار ہوں اور سوجن شدید ہو یا کسی بھی وقت آپ کے ہاتھوں اور چہرے پر سوجن نمودار ہوتو یہ pre-eclampia کی علامت ہوسکتی ہے (دیکھئے حمل کا ٹوکسیمیا صفحہ 232)

سوجے پیروں اور ٹانگوں کے لئے۔ دن میں دو یا تین بار، ہر بار تیس منٹ پہلو کے بل لیٹئے۔ اس کی کوئی اہمیت نہیں کہ کس پہلو لیٹا جائے۔ صرف پیروں کو اٹھا کر بیٹھنا کافی نہیں ہے۔ پہلو کے بل لیٹنا ہی بہترین ہے۔

pre eclampia سے بچاؤ کے لئے۔ پُر غذائیت خوراک لیں جس سے آپ کو مناسب مقدار میں پروٹین ملیں، خوب پانی پئیں اور کم مقدار میں نمک استعمال کریں۔ (لیکن نمک استعمال ضرور کریں)

نقل و حرکت اور توازن

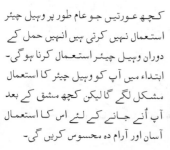

حمل کے دوران نو ماہ میں آپ کے جسم کی بناوٹ اس حد تک بدلے گی جس سے آپ کے چلنے پھرنے کے انداز پر اثر پڑ سکتا ہے۔ ایسا تقریباً تمام عورتوں کے ساتھ ہوتا ہے خواہ وہ معذور ہوں یا نہ ہوں۔ اس وجہ سے آپ اپنا توازن کھو سکتی ہیں اور آسانی سے گر سکتی ہیں۔ یا آپ کو اس کے باعث جھکنے اور اشیاء اٹھانے میں دشواری ہوسکتی ہے۔ اس کی وجہ سے معذوریاں رکھنے والی بہت سی عورتیں بچے کے پیدا ہونے تک چلنے پھرنے کے بعد کوئی چھڑی یا سہارا استعمال کرنا شروع کر دیتی ہیں۔

کٹی ہوئی ٹانگ

اگر آپ کی ٹانگ یا اس کا کوئی حصہ کٹا ہوا ہے اور آپ مصنوعی ٹانگ استعمال کرتی ہیں تو ممکن ہے آپ کو اپنا جسم بھاری ہونے یا کٹے ہوئے حصے کے اوپر کی جلد سوج جانے کی وجہ سے مصنوعی ٹانگ لگانے میں دشواری ہو۔ اگر ممکن ہوتو کسی ایسے فرد سے بات کریں جو آپ کی مصنوعی ٹانگ کو استعمال کے لئے موزوں بیسا کھیاں، واکر یا وہیل چیئر استعمال کرنے کی ضرورت پڑ سکتی ہے۔

کچھ عورتیں جو عام طور پر وہیل چیئر استعمال نہیں کرتی ہیں انہیں حمل کے دوران وہیل چیئر استعمال کرنا ہوگی۔ ابتداء میں آپ کو وہیل چیئر کا استعمال مشکل لگے گا لیکن کچھ مشق کے بعد آپ اُسے جانے کے لئے اس کا استعمال آسان اور آرام دہ محسوس کریں گی۔

کیا کرنا چاہئے؟

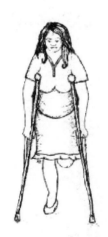

بانس یا لکڑی کے ٹکڑوں سے واکر بنائی جاسکتی ہے۔ ان کے سروں کو کسی مضبوط ڈوری یا ربڑ یا کار/سائیکل کی ٹیوب کی پٹیوں سے جوڑا جائے۔

دو پہیوں والا واکر، بغیر پہیے کے واکر کی بہ نسبت زیادہ آسان رہتا ہے۔ دو پہیوں کا واکر چار پہیوں والے واکر سے زیادہ بہتر رہتا ہے۔

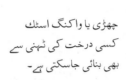

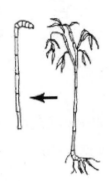

چھڑی یا واکنگ اسٹک کسی درخت کی ٹہنی سے بھی بنائی جاسکتی ہے۔

کہنی کی بیساکھی استعمال کے لئے بہترین ہے (دیکھئے صفحہ 94)۔ اگر آپ کو یہ نہ مل سکیں تو پورے سائز کی بیساکھیاں بھی درخت کی شاخوں سے بنائی جاسکتی ہیں۔

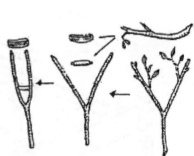

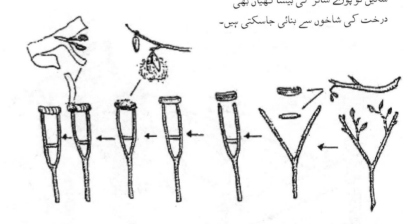

حمل کے آخری ہفتوں میں نقل و حرکت

حمل کے آخری ہفتوں میں کوئی معذوری نہ رکھنے والی عورت کو بھی توازن برقرار رکھنے میں دشواری ہوسکتی ہے۔ جسمانی معذوریاں جیسے نچلے دھڑ کا فالج یا عضلات پر محدود کنٹرول کی صورت میں ایسی عورتوں کے لئے مشکلات زیادہ شدید ہوتی ہیں۔ بڑھا ہوا پیٹ آپ کے روزمرہ کے کاموں مثلاً نہانا، کپڑے بدلنا کہیں آنے جانے پر اثر انداز ہوتا ہے۔

کیا کرنا چاہئے؟

لیٹی ہوئی حالت سے اٹھنا آسان ہوگا اگر آپ

اور پھر کھڑی ہوں .... پھر گھٹنوں کے بل اٹھیں .... پہلو میں مڑیں ....

اٹھنا اس صورت میں بھی آسان ہوگا اگر آپ
قریب موجود بھاری کرسی یا کوئی باکس سہارے
کے لئے استعمال کریں

حمل کے آخری دنوں میں جب چلنا پھرنا مشکل ہوجاتا ہے جسمانی معذوریوں والی بہت سی عورتوں کے لئے کچھ سادہ امدادی
اشیاء مددگار ہوسکتی ہیں۔

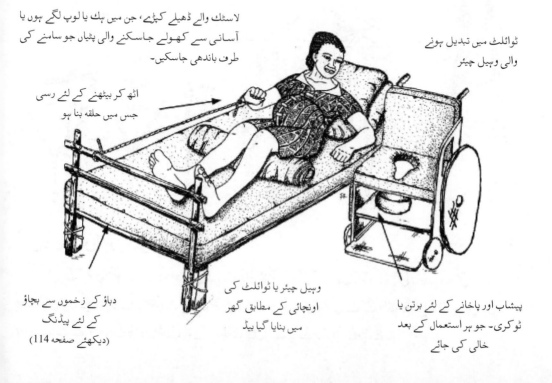

لاسٹک والے ڈھیلے کپڑے، جن میں ہک یا لوپ لگے ہوں یا
آسانی سے کھولے جاسکنے والی پٹیاں جو سامنے کی
طرف باندھی جاسکیں۔

ٹوائلٹ میں تبدیل ہونے
والی وہیل چیئر

اٹھ کر بیٹھنے کے لئے رسی
جس میں حلقہ بنا ہو

وہیل چیئر یا ٹوائلٹ کی
اونچائی کے مطابق گھر
میں بنایا گیا بیڈ

پیشاب اور پاخانے کے لئے برتن یا
ٹوکری- جو ہر استعمال کے بعد
خالی کی جائے

دباؤ کے زخموں سے بچاؤ
کے لئے پیڈنگ
(دیکھئے صفحہ 114)

عضلات میں اکڑاؤ یا کھچاؤ

رات کے وقت خاص طور پر ٹانگ کے نچلے حصے میں شدید تکلیف کے ساتھ عضلات میں کھچاؤ ممکن ہے۔ اگر آپ اکڑے عضلے کو چھوئیں تو وہ سخت ابھار کی طرح محسوس ہوسکتا ہے۔ ٹانگوں میں اکڑاؤ کا سبب یہ ہوسکتا ہے کہ آپ کی غذا میں کیلشیئم کی مناسب مقدار نہیں ہے۔

کیا کرنا چاہئیے

- پیروں کی انگلیوں کو نہیں اکڑائیں ...... اسٹریچنگ کرتے ہوئے بھی نہیں۔
- اسٹریچنگ کی ورزشیں با قاعدگی سے کیجیے (دیکھے صفحہ 90 تا 95)

- کیلشیئم والی غذائیں جیسے دودھ، چیز، دہی، سی سیم سیڈز اور سبز پتوں والی سبزیاں زیادہ مقدار میں لیں۔ کیلے بھی کھائیں۔
- پیروں کو معمولی سا موڑ کر رکھیں اور گھٹنوں کے درمیان نرم گول تہہ کیا ہوا کپڑا یا اخبار رکھتے ہوئے پہلو کے بل سوئیں۔
- بھاری چادر وغیرہ اوڑھ کرنہ تو لیٹیں یا سوئیں نہ ہی کوئی چیز سختی کے ساتھ اپنے بدن پر لپیٹیں۔

اگر آپ کے پیر یا ٹانگ میں اکڑن ہو۔

پھر اپنے پیر پر ضرب لگائیں ‌‌‌‌‌‌‌‌‌‌‌‌‌‌‌ نیچے مت کریں ‌‌‌‌‌‌‌‌‌‌‌‌‌ اپنے پیروں کی انگلیاں اوپر اٹھائیں

اپنی ٹانگ گرم پانی میں رکھیں یا اکڑے ہوئے حصے پر گرم پانی میں بھیگا کپڑا رکھیں۔ اس سے بھی آرام مل سکتا ہے۔

عضلاتی کھچاؤ / اینٹھن

یہ عضلات کی سختی یا کھچاؤ ہے جو کسی فرد کے لئے نقل و حرکت مشکل بناتا ہے۔ یہ زیادہ تر سیریبرل پالسی یا ریڑھ کی ہڈی کے زخم رکھنے والوں میں ہوتا ہے۔ ایسی عورتیں زچگی کے دوران عضلاتی کھچاؤ کا شکار ہوسکتی ہیں (دیکھیے صفحہ 243)

کیا کرنا چاہئیے

- سخت عضلات کو براہ راست کھینچیں یا دبائیں نہیں۔ یہ عمل کھچاؤ میں اضافہ کرے گا۔
- عضلات کے نرم پڑنے تک متاثرہ حصے کو نرمی سے گرفت کریں اور سپورٹ دیں۔

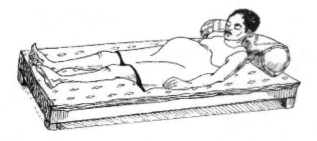

اگر عضلاتی کھچاؤ سے کمر یا پورا جسم متاثر ہے تو سر اور کندھوں کو تھوڑا اوپر رکھنے کے لئے ان کے نیچے کچھ رکھیں۔ اس سے پورے جسم کی سختی دور ہونے میں مدد ملے گی۔

- سخت عضلات پر گرم پانی استعمال کریں یا اگر ممکن ہو تو گرم پانی میں بیٹھیں یا لیٹیں۔

خیال رکھیں کہ جلد نہ جلے یا جسم کو غیر معمولی حرارت نہ ملے خصوصاً اس صورت میں جب آپ گرم یا ٹھنڈا محسوس نہ کر سکتی ہوں۔ حد سے زیادہ حرارت کوکھ میں موجود بچے کو نقصان پہنچا سکتی ہے۔

- دن میں دو سے تین بار آسان اسٹریچنگ ورزشیں کرنا عضلات کی سختی کم کرنے میں معاون ہو سکتا ہے۔
- وزن برداشت کرنے والی ورزشیں جیسے کھڑا ہونا بھی عضلات کو مضبوط بناتا ہے اور عضلات کی سختی کم کرتا ہے۔

صاف گرم پانی میں بیٹھنا عضلات کے کھچاؤ کو ختم کرنے میں معاون ہوگا۔

اہم بات ایک عام اصول کے مطابق اینٹھے ہوئے عضلات پر مساج مت کریں۔ کچھ ممالک میں لوگ بلکہ معالجین بھی اینٹھے ہوئے عضلات کو نرم کرنے کے لئے مساج یا مالش کرتے ہیں۔ ہر چند کہ مساج عضلاتی اکڑاؤ یا سخت عضلات کو جو دوسری وجوہات کے باعث ہوتا ہے آرام پہنچاتا ہے لیکن اینٹھن کی صورت میں عموماً مساج عضلات کی سختی بڑھاتا ہے۔

کمر کا درد

بیش تر حاملہ عورتیں خواہ وہ معذور ہوں یا غیر معذور خاص طور پر بچے کی پیدائش سے چند ہفتے قبل جب پیٹ بہت بڑا اور بھاری ہوتا ہے کمر کے درد میں مبتلا ہو جاتی ہیں۔ زیادہ تر اس لئے کہ پیٹ کے عضلات حمل کے دوران کھچ جاتے ہیں اور کمزور ہو جاتے ہیں اور کمر کے عضلات کو زیادہ مشقت جھیلنا پڑتی ہے۔

کچھ جسمانی معذور عورتوں کو کمر کا درد زیادہ شدید اور حمل کے دوران خاصی مدت پہلے ہونے لگتا ہے۔ حد تو یہ ہے کہ جسم کے نچلے حصے میں احساس نہ رکھنے والی عورتیں بھی، اکثر حمل کے دوران کمر کا درد محسوس کرتی ہیں۔

کیا کرنا چاہئیے؟

- کمر کے نچلے حصے اور پیٹ کے عضلات کو مضبوط رکھنے کے لئے حمل سے پہلے، حمل کے دوران اور حمل کے بعد مناسب ورزش کیجئے۔ کمر کے درد میں کمی اور کمر کو طاقتور رکھنے کے لئے پیرا کی ایک اچھا طریقہ ہے۔

- سیدھی پشت والی کرسی پر بیٹھئے۔

- کمر کا درد کم کرنے میں تکلیف دہ حصے کو آرام دینے کا طریقہ سیکھنا اور مساج کرنا مفید ہوسکتا ہے۔

  ☆ پیٹ کو سہارا دینے کے لئے کپڑے کی پٹی باندھنا مددگار ہوسکتا ہے کیونکہ اس سے آپ کی کمر کے عضلات پر زیادہ دباؤ نہیں پڑے گا۔

  ☆ اس طرح اپنے پیٹ کے گرد چار سے پانچ فٹ لمبی (ڈیڑھ میٹر) صاف اور پتلے کاٹن کی پٹی لپیٹیں۔

  ☆ پٹی اس قدر سخت نہ لپیٹیں کہ تکلیف محسوس کریں۔

  ☆ پٹی کو درست حالت میں رکھنے کے لئے سیفٹی پن استعمال کریں۔ یہ پٹی کے سروں پر لگائیں۔

سانس لینے میں دشواری

جوں جوں بچے کی افزائش ہوتی ہے یہ ماں کے پھیپھڑوں کو دباتا ہے اسے اپنے سینے میں سانس لینے کے لئے کم گنجائش ملتی ہے۔ یہ حمل کے دوران عام بات ہے لیکن جسمانی معذوریوں والی عورتیں جیسے چھوٹے قد والی یا سینے کے مفلوج عضلات والی عورتیں دوسری عورتوں کی بہ نسبت جلد سانس کی تنگی میں مبتلا ہوسکتی ہیں۔ بچہ ماں کے پھیپھڑوں سے آکسیجن حاصل کرتا ہے اس لئے حاملہ عورت کو اپنے پھیپھڑے صاف اور صحت مند رکھنا چاہئیے تاکہ نشو ونما پاتے بچے کو اس کی ضرورت کے مطابق آکسیجن مل سکے۔

کیا کرنا چاہئیے؟

- قدرے بیٹھی ہوئی حالت میں سوئیں۔ اگر آپ اپنے گھٹنوں کے نیچے کچھ رکھ لیں تو آپ زیادہ آرام دہ حالت میں رہیں گی۔

- زیادہ مقدار میں پانی پئیں۔ روزانہ کم از کم آٹھ گلاس پانی پئیں۔ اس سے آپ کے پھیپھڑوں کا بلغم نرم رہے تو آپ آسانی سے اسے باہر نکال سکیں گی۔ پھیپھڑوں میں بلغم انفیکشن کا سبب بن سکتا ہے۔

- با قاعدگی سے ورزش کریں۔

- اگر بلغم کے ساتھ پِس یا مواد بھی ہو تو صحت کارکن سے ملیں ممکن ہے آپ کو دوا کی ضرورت ہو۔ صحت کارکن فیصلہ کرسکتا ہے کہ حمل کے دوران آپ کے لئے کونسی دوا یا اینٹی بایوٹک موزوں اور محفوظ ہے گی۔

اہم بات اگر ماں کو سانس لینے میں دشواری ہوتی ہے اور وہ کمزور اور تھکی تھکی رہتی ہے یا وہ سارا وقت سانس کی تنگی محسوس کرتی ہے تو اسے صحت کارکن سے ملنا چاہئیے۔ ممکن ہے وہ دل کے کسی مسئلے کا شکار ہوا اور اسے طبی دیکھ بھال کی ضرورت ہو، یا اسے انیمیا ہو (دیکھئے صفحہ 87) ۔ وہ ناقص غذا، انفیکشن یا ڈپریشن کا ہوسکتا ہے (دیکھئے صفحہ 54) ۔

جوڑوں میں سوزش اور درد

ایک حاملہ عورت کا جسم بچے کی نشو ونما کی گنجائش پیدا کرنے اور ولادت کے لئے تیار ہونے کے لئے نرم اور ڈھیلا پڑتا ہے۔ بعض اوقات اس کے جوڑ بھی خصوصاً کولہے ڈھیلے اور غیر آرام دہ ہو جاتے ہیں۔ یہ عام طور پر حمل کے آخری چند ہفتوں کے دوران ہوتا ہے۔ یہ خطرناک نہیں ہے اور ولادت کے بعد سب کچھ بہتر ہو جاتا ہے۔

کیا کرنا چاہئیے؟

- تکلیف دہ جوڑوں کو آرام دیں۔ وقفہ وار حرکت میں رہیں تا کہ جوڑ سخت نہ ہوں لیکن آپ کا چلنا پھر ناسہولت کے ساتھ ہو۔

- درد میں کمی اور نقل وحرکت کو آسان بنانے کے لئے سرد اور گرم سنکائی کریں۔ عام طور پر سرد سنکائی گرم، پر سوزش جوڑوں پر موثر رہتی ہے جبکہ گرم سنکائی پر درد اور سخت جوڑوں کے لئے۔ یہ جاننے کے لئے تجربہ کریں کہ کون سی صورت آپ کے لئے بہتر ہے۔ اگر آپ گرم یا سرد محسوس نہیں کر پاتی ہیں تو خیال رکھیں کہ آپ جھلسیں یا حد سے زیادہ ٹھنڈک کا شکار نہ ہوں۔

سرد سنکائی کے لئے۔ برف کپڑے یا تولیئے میں لپیٹ کردس سے پندرہ منٹ سنکائی کریں۔

گرم سنکائی کے لئے۔ صاف اور گرم پانی میں بھیگا موٹا کپڑا (اضافی پانی نچوڑ دیں) استعمال کریں، اسے تکلیف دہ جوڑ کے گرد لپیٹ دیں۔ کپڑے کو باریک پلاسٹک سے ڈھانپ دیں اور کپڑے کی حرارت برقرار رکھنے کے لئے اس پر خشک کپڑا یا تولیہ لپیٹ دیں۔ جب بھیگا ہوا کپڑا ٹھنڈا ہونے لگے تو اسے دوبارہ گرم پانی میں رکھ کر نکالیں اور یہ عمل دوہرائیں۔ یا پھر ایک بوتل میں (مٹی، پلاسٹک یا شیشے کی) گرم پانی بھریں اور اسے کپڑے میں لپیٹ کر تکلیف دہ جوڑ پر رکھیں۔

حرارت سے تکلیف دہ اور سخت جوڑ کو آرام ملے گا

- درد کے لئے 500 ملی گرام کی پیراسیٹامول (acetaminophen) تین سے چار گھنٹے بعد کھائیں لیکن 24 گھنٹوں میں 8 سے زیادہ گولیاں (4000 ملی گرام) نہ کھائیں (دیکھئے صفحہ 350)۔

پیشاب خارج ہونا

بہت سی عورتوں کے لئے پیٹ بڑھنے کے ساتھ اپنا پیشاب خارج ہونے سے روکنا مشکل ہو جاتا ہے۔ حمل کے دوران عضلات پر محدود کنٹرول اور فالج یا نچلے جسم میں محسوس کرنے کی صلاحیت کھونے جیسی معذور عورتیں اکثر دوسری عورتوں کے مقابلے میں پیشاب کے بے قابو ہونے کے مسائل کا زیادہ شکار ہوتی ہیں۔

بچے کی نشو ونما کے ساتھ ماں کا پیٹ بھی بڑا ہو جاتا ہے، بچہ مثانے پر دباؤ بھی ڈال سکتا ہے، جس سے مثانے میں پیشاب کے لئے کم گنجائش بچتی ہے۔ اس صورت میں عورت کے کھانسنے یا چھینکنے کی صورت میں پیشاب خارج ہو سکتا ہے۔ بعض اوقات پیشاب اس قدر اچانک بہنے لگتا ہے کہ یہ کہنا مشکل ہوتا ہے کہ یہ پیشاب ہے یا 'bag of waters' پھٹ گیا۔ آپ اس کی بو سے کہہ سکتی ہیں کہ یہ پیشاب ہے یا نہیں۔ اگر ایسا ہو تو لیبر کی دوسری علامات کا جائزہ لیں اور مشورے کے لئے صحت کار کن یا مڈوائف سے رجوع کریں۔

اگر آپ عام طور پر پیشاب کے اخراج کے لئے ہر بار کیتھیٹر استعمال کرتی ہیں تو اگر اس سے کوئی مسئلہ پیدا نہ ہوتو اس کا استعمال جاری رکھیں۔ اگر معمول سے زیادہ وقت کیتھیٹر رکھنا زیادہ مشکل ہو تو پیشاب جذب کرنے کے لئے کپڑے کی دیز کی دو بارہ استعمال کریں۔ اپنے تناسلی اعضاء کے اطراف کی جلد کو سوزش یا انفیکشن سے بچنے کے لئے یہ پیڈ ز بار بار تبدیل کئے جائیں اور انہیں دو بارہ استعمال سے پہلے دھو کر خشک کیا جائے (دیکھئے صفحہ 111 تا 113)۔

کچھ عورتیں مستقل لگا رہنے والا کیتھیٹر (Foley catheter) استعمال کرنے لگتی ہیں لیکن اگر ممکن ہو تو یہ استعمال نہ کریں کیونکہ بچے کی پیدائش کے بعد اسے تبدیل کرنا مشکل ہوسکتا ہے۔ وہ عضلات جو آپ کے مثانے کو کنٹرول کرتے ہیں، بھول جائیں گے کہ پیشاب کو مثانے کے اندر کس طرح روکا جاتا ہے۔ اس کے ساتھ ہی سارا وقت کیتھیٹر لگا رہنا مثانے کے انفیکشن کا خطرہ بڑھا سکتا ہے۔ اگر آپ کو رات کے وقت پیشاب کے اخراج کا مسئلہ ہے تو پیڈ استعمال کریں یا پیشاب کے اخراج کے لئے قریب ہی کوئی برتن یا ٹوکری وغیرہ رکھیں۔ تبدیل کی جانے والی وہیل چیئر ایک اچھا حل ہوسکتی ہے (دیکھئے صفحہ 224)۔

میں بھینچنے کی ورزش کر رہی ہوں اور ایکتا کو اس کی خبر تک نہیں ہے

اگر آپ نچلے پیٹ کے عضلات استعمال کرنے کے قابل ہیں تو بعض اوقات بھینچنے والی ورزش مثانے کے اطراف کے عضلات کو مضبوط تر بنانے میں مفید ثابت ہوتی ہے۔ پیشاب کے مسائل کے بارے میں مزید معلومات کے لئے (دیکھئے صفحہ 105 تا 106)۔

قبض یا فضلے کے اخراج میں مشکل

بہت سی معذور عورتوں کو فضلہ کے اخراج میں دشواری ہوتی ہے۔ حمل اس نظام کو بہت آہستہ کرسکتا ہے اور اس سے فضلے کا اخراج زیادہ مشکل ہو جاتا ہے۔ قبض میں کمی اور اس سے بچاؤ کے لئے دیکھئے صفحہ 108۔

وہ عورتیں جو فضلے کے اخراج کے لئے باؤل پروگرام پر عمل کرتی ہیں (دیکھئے صفحہ 107) انہیں حمل کے دوران زیادہ بار فضلہ نکالنے کی ضرورت پڑسکتی ہے۔ سخت فضلہ نکالا نہ جائے تو ڈس ریفلیکسیا کا سبب بن سکتا ہے (دیکھئے صفحہ 117 تا 119) جو انتہائی خطرناک ہے۔

تنبیہہ حاملہ عورتیں قبض دور کرنے کے لئے سہل یا قبض کشا دوایات نہ لیں۔ یہ ادویات باؤل کو سخت کر کے یا اکڑا کر کام کرتی ہیں اور یہ لیبر وقت سے بہت پہلے شروع ہونے کا سبب بن سکتی ہے۔ بعض ادویات بچے کو نقصان پہنچا سکتی ہیں۔ حاملہ عورتوں کو اینما (Enema) کے ذریعے فضلے کا اخراج بھی نہیں کرنا چاہئے یہ بھی لیبر وقت سے بہت پہلے شروع ہونے کا سبب بن سکتا ہے۔

بواسیر (پھولی ہوئی رگیں)

بواسیر مقعد کے اندر اور اطراف میں سوجی ہوئی یا پھولی ہوئی رگیں ہوتی ہیں۔ یہ عموماً تکلیف دہ، پر سوزش یا خون کے اخراج کا ذریعہ بنتی ہیں اور انتہائی تکلیف کا سبب بن سکتی ہیں۔ بعض اوقات بواسیر بڑے خونی چھالوں کی طرح (blisters) نظر آتی ہے۔ قبض ہونے پر فضلے کے اخراج کے لئے زور لگانا انہیں بدتر کر سکتا ہے۔ معذور اور غیر معذور دونوں ہی عورتیں حمل کے دوران بواسیر میں مبتلا ہو جاتی ہیں۔ طویل عرصے بیٹھے رہنا تکلیف میں اضافہ کرتا ہے۔

کیا کرنا چاہئے؟

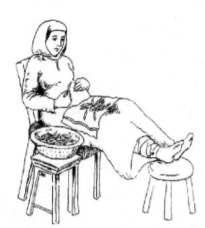

- پھولی ہوئی رگوں کو سکیڑنے کے لئے صاف کپڑے کو وچ ہیزل یا کاکٹس جیسے کسی پودے کے عرق میں بھگو کر تکلیف دہ حصے پر رکھیں۔
- دباؤ کم کرنے کے لئے بیٹھتے ہوئے نرم گدی استعمال کریں۔
- ہر گھنٹے میں ایک بار اٹھ کر چلیں۔
- اگر آپ سارا وقت لیٹی رہتی ہیں تو پہلو کے بل لیٹنے کی کوشش کریں اور تواتر کے ساتھ اپنی پوزیشن بدلنے کے لئے کسی کی مدد حاصل کریں۔
- اپنے پیروں اور ٹانگوں کو اونچار کھ کر بیٹھیں۔ اس دوران خون کے بہتر ہونے میں مدد ملے گی اور پھولی ہوئی رگیں زیادہ جلد درست ہو جائیں گی۔

مزید معلومات کے لئے دیکھئے''جہاں عورتوں کے لئے ڈاکٹر نہ ہو''، صفحہ 66۔

عام مسائل صحت

مثانے کے انفیکشنز

حمل کے دوران زیادہ تر عورتیں عام حالت کی بہ نسبت مثانے کے انفیکشن کے زیادہ امکانات رکھتی ہیں۔ رحم بڑھنے کے ساتھ، مثانے کو دباتا ہے اور مثانے میں موجود پیشاب مکمل طور پر خارج ہونے سے روکتا ہے۔ رہ جانے والے پیشاب میں جراثیم افزائش پا کر انفیکشن پیدا کر سکتے ہیں۔

عضلات کا محدود کنٹرول اور فالج یا نچلے جسم میں محسوس کرنے کی صلاحیت نہ رکھنے والی عورتیں دوسری عورتوں کی بہ نسبت پیشاب کے اخراج، مثانہ اور گردے کے انفیکشنز کے زیادہ امکانات رکھتی ہیں۔ مثانے کے مسائل ڈس ریفلیکسیا کا ایک عام سبب ہیں (دیکھئے صفحہ 117 تا 119)۔

اگر آپ عام طور پر پیشاب کے اخراج کے لے اپنے پیٹ کو دباتی ہیں تو حمل کے دوران بھی ایسا کرنا درست ہے۔ اس طرح آپ بچے کو کوئی نقصان نہیں پہنچائیں گی۔

وہ عورتیں جو "فکسڈ" کیتھیٹر استعمال کرتی ہیں، بعض اوقات ان کے پیشاب کا اخراج رُک سکتا ہے یہ اس وقت ہو سکتا ہے جب بچہ کیتھیٹر دباتا ہے۔

اگر آپ مثانے کے انفیکشن کا بروقت علاج کرسکیں تو آپ زیادہ سنگین مسائل مثلاً گردے کے انفیکشن قبل از وقت لیبر سے محفوظ رہ سکتی ہیں۔ انفیکشن کی علامات پر نظر رکھیں اور ضروری ہو تو صحت کارکن سے رجوع کریں۔

جب آپ حاملہ ہوں تو یورین انفیکشنز سے بچاؤ کے لئے

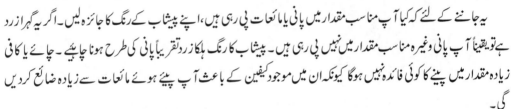

- خوب پانی یا پھلوں کا رس پئیں۔ کم از کم آٹھ گلاس روزانہ۔
- اپنے تناسلی اعضاء صاف رکھیں۔
- مباشرت کے بعد پیشاب کریں۔
- کیتھیٹر کو زیادہ بار صاف کریں (دیکھئے صفحہ 102 تا 104)۔

یہ جاننے کے لئے کہ کیا آپ مناسب مقدار میں پانی یا مائعات پی رہی ہیں، اپنے پیشاب کے رنگ کا جائزہ لیں۔ اگر یہ گہرا زرد ہے تو یقیناً آپ پانی وغیرہ مناسب مقدار میں نہیں پی رہی ہیں۔ پیشاب کا رنگ ہلکا زرد تقریباً پانی کی طرح ہونا چاہیے۔ چائے یا کافی زیادہ مقدار میں پینے کا کوئی فائدہ نہیں ہوگا کیونکہ ان میں موجود کیفین کے باعث آپ پئیے ہوئے مائعات سے زیادہ ضائع کردیں گی۔

مثانے اور گردے کے انفیکشنز کے لئے دیکھئے صفحہ 105 تا 106۔

دورے (کپکپی "convulsions"، فٹس، مرگی)

یہ کہنا مشکل ہے کہ مرگی کی مریضہ حمل کے دوران دوروں کا کس قدر زیادہ یا کم امکان رکھتی ہے۔ اگر آپ دوروں کا شکار ہوتی ہیں تو آپ بہتر جانتی ہوں گی، آپ کتنی بار ان کا شکار ہوتی ہیں اور یہ کس قدر سخت ہوتے ہیں۔ دوروں کی بعض ادویات خاص طور پر فینی ٹوئن (dephenylhydantoin, Dilantin) حاملہ عورتیں استعمال کریں تو پیدائشی نقائص کا خطرہ بڑھ سکتا ہے لیکن حمل کے دوران اینٹی سیزر (anti-seizure) ادویات استعمال کرنا بند مت کریں۔ یہ دوروں کو سنگین بنا سکتا ہے اور آپ مر بھی سکتی ہیں۔ کسی

تجربے کار صحت کارکن یا ڈاکٹر سے بات کریں جو مرگی کو سمجھتا ہو اور آپ کے لئے بہترین دوا تجویز کرسکتا ہو، فینوبار بی ٹون (phenobarbitone, Luminal) امکانی طور پر دوران حمل استعمال کے لئے محفوظ ترین دوا ہوتی ہے۔

حمل کا ٹوکسیمیا (Pre-Eclampsia)

حمل کے دوران ٹانگوں اور گھٹنوں کا کچھ سوج جانا نارمل بات ہے لیکن ہاتھوں اور چہرے کی سوجن بعض خصوصاً اس وقت جب آپ کے سر میں درد، نظر میں دھندلاہٹ یا پیٹ میں درد ہو، Pre-eclampsia کی علامت ہوسکتا ہے جسے حمل کا ٹوکسیمیا بھی کہتے ہیں۔ اچانک وزن میں اضافہ، ہائی بلڈ پریشر اور پیشاب میں پروٹین کی بڑی مقدار بھی ٹوکسیمیا کی علامات ہیں۔ یہ دورے یا فٹس کا سبب بن سکتا ہے اور آپ اور بچہ دونوں ہی مر سکتے ہیں۔ convulsions، مرگی کے دوروں سے مختلف ہوتی ہے (دیکھئے صفحہ 231)۔

اگر آپ کی والدہ یا بہنیں Toxemia کا شکار ہوچکی ہیں تو آپ بھی اس کا خطرہ رکھتی ہیں۔ یہ پہلے حملے یا نئے پارٹنر سے پہلے حمل کی صورت میں بھی ممکن ہے، ٹوکسیمیا ان عورتوں میں بھی عام ہے جو پہلے ہائی بلڈ پریشر، ذیابیطس، گردوں کے مسائل میں مبتلا عورتوں کو بھی ہوسکتا ہے۔

اگر آپ ٹوکسیمیا کی کوئی علامات رکھتی ہیں تو کسی مڈوائف یا صحت کارکن سے ملیں جو آپ میں ان علامات کا جائزہ لے کر تصدیق کرسکتی ہے۔

کیا کرنا چاہئے؟

- پرسکون رہ کر آرام کریں، اچھی صحت افزاء غذائیں، کم مقدار میں نمک کے ساتھ کھائیں خصوصاً وہ جن میں پروٹین کی مقدار خوب ہو، نمک آمیز غذائیں مت کھائیں۔

- **اگر آپ کی حالت فوری طور پر نہ سنبھلے یا آپ کو دیکھنے میں دشواری ہو یا چہرے پر سوجن بڑھ جائے یا آپ دورے کا شکار ہو جائیں تو فوری طور پر طبی مدد حاصل کریں۔ آپ کی زندگی خطرے میں ہے۔**

دباؤ کے زخم

زیادہ تر وقت بیٹھ کر یا لیٹ کر گزارنے والی عورتوں کو دباؤ کے زخم ہو سکتے ہیں۔ خاص طور پر ان عورتوں کو جوبہت تر وقت حرکت اور پوزیشن کی تبدیلی کے بغیر گزارتی ہوں۔ یہ ان مفلوج عورتوں کے لئے بھی درست ہے جو درد محسوس نہیں کر پاتی ہیں، جب آپ حاملہ ہوتی ہیں تو حمل آپ کے جسم کے ان حصوں پر زیادہ دباؤ ڈالتا ہے جہاں زیادہ تر زخم نمودار ہوتے ہیں۔

کیا کرنا چاہئے؟

اپنی پوزیشن معمول سے زیادہ بار بدلنے کی کوشش کریں۔ گھنٹہ بھر میں کم از کم ایک بار۔ زمانہ حمل سے قبل کی بہ نسبت اب دباؤ کے زخموں کا اندازہ لگانے کے لئے اپنی جلد کا معائنہ زیادہ توجہ سے کریں، دباؤ کے زخموں سے بچاؤ کے لئے معلومات کے لئے دیکھئے صفحہ 116۔

ایچ آئی وی/ایڈز اور حمل

اگرچہ ایچ آئی وی/ایڈز کا اب تک کوئی علاج دستیاب نہیں ہے مگر ایسی ادویات موجود ہیں جو ایچ آئی وی/ایڈز میں مبتلا افراد کی طویل عرصہ زندہ رہنے میں مدد کرتی ہیں۔ ایسی ادویات (ARVs) بھی موجود ہیں جو بچے کی ولادت کے وقت یا دودھ پلاتے وقت ایچ آئی وی کو حاملہ عورتوں سے اس کے بچے میں منتقل ہونے سے روکتی ہیں۔

اگر آپ ایچ آئی وی ہیں اور حاملہ بھی ہیں تو یہ آپ کے لئے ضروری ہے کہ آپ کو حمل کے دوران عمومی دیکھ بھال کے ساتھ ساتھ ایچ آئی وی کے علاج کی سہولت بھی ملے۔

ایچ آئی وی سے متاثر عورتیں حمل کے دوران زیادہ مسائل کا شکار ہو سکتی ہیں جیسے

- حمل ضائع ہونا
- بخار اور انفیکشنز
- فرج، منہ یا پیٹ کے یسٹ انفیکشنز
- جنسی طور پر منتقل ہونے والے امراض
- بچے کی پیدائش کے بعد مسائل جیسے جریان خون اور انفیکشن

یہ معلوم کرنے کی کوشش کریں کہ کیا آپ کے علاوہ آپ کے بچے کو ایچ آئی وی ہے۔ بچے کو بچانے کے لئے ابتدائی مرحلے میں دیکھ بھال کی سہولت مل سکتی ہے، اگر آپ کے علاقے میں طبی سامان سے آراستہ میڈیکل سینٹر موجود ہے تو آپ کے لئے بہتر ہوگا کہ آپ بچی کی ولادت وہاں ہونے دیں۔

تبدیلی کے لئے منصوبہ بندی

دوران حمل معذوریاں رکھنے والی عورتیں بھی دوسری حاملہ عورتوں کی طرح فکرمندی رکھتی ہیں۔ ہم صحت مند حمل اور صحت مند بچے کا جنم چاہتے ہیں

گھرانے اور دیکھ بھال کرنے والے کیا کر سکتے ہیں۔

- ہمیں مناسب غذا اور آرام کے مواقع دیں۔
- حمل کے بارے میں مثبت خیال رکھیں۔
- پری نیٹل دیکھ بھال کے لئے اور ہمارے ساتھ کلینک/اسپتال جائیں۔
- کسی بھی وقت مدد کے لئے دستیاب رہیں۔

مڈوائفز، ڈاکٹر اور دوسرے صحت کارکن کیا کر سکتے ہیں؟

صحت کارکن اسی وقت ہماری مدد کرنے کے قابل ہوں گے جب ہم حمل کی ابتداء ہی سے یا حمل ہونے سے پہلے سے ان کے پاس جائیں۔ کیونکہ بہت کم ڈاکٹر، نرسیں، مڈوائفز اور دیگر صحت کارکن معذور عورتوں کی دیکھ بھال کا تجربہ رکھتے ہیں، لہذا اہم انہیں یہ سکھا سکتے ہیں کہ کیا بات ہمارے لئے فطری ہے اور کس طرح ہماری معذوریاں حمل پر اثر انداز ہو سکتی ہیں یا نہیں ہو سکتی ہیں۔ صحت کارکن یہ بھی کر سکتے ہیں کہ

- وہ ایک معذور عورت کے ان امکانی مسائل کے بارے میں سیکھیں جو معذور عورت کو حمل کے زمانے میں پیش آ سکتے ہیں۔

- ان باتوں سے آگاہ رہیں جو ہم کر سکتے ہیں مثلاً یہ فرض مت کریں کہ ہم فطری طور پر بچے کو جنم نہیں دے سکتے ہیں۔ یاد رکھیں کہ اگر کوئی عورت معذور ہے تو اس کا یہ مطلب نہیں ہے کہ اس کا رحم بھی درست حالت میں نہیں ہے۔ اگر اس کا جسم اور ٹانگیں بھی مفلوج ہوں تو اس صورت میں بھی اس کا رحم بچے کو پروان چڑھانے اور دھکیلنے کے قابل ہوگا۔

- معذور عورتوں کی مشاورت کے لئے ایک گروپ تشکیل دیں جو انہیں زمانہ حمل میں کھانے پینے، ادویات اور طبی چیک اپ کے بارے میں رہنمائی فراہم کرے۔

- یقینی بنائیے کہ ہمیں بھی حمل کے دوران ضروری طبی دیکھ بھال دستیاب رہے۔

یوگنڈا میں معذور عورتوں کی صحت کے لئے بہتر رسائی

یوگنڈا میں ڈس ایبل ویمنز نیٹ ورک اور ریسورس آرگنائزیشن (DWNRO) ماہرین صحت کو معذور عورتوں کی ضروریات سے آگاہ کرنے کے لئے کام کرتی ہے۔ ان کے بنیادی مقاصد رسائی، دستیابی اور رویہ ہیں۔ مثلاً جب معذور حاملہ عورتوں سے اسپتال کا عملہ خراب سلوک کرتا ہے تو وہ اپنی خود اعتمادی گنوا دیتی ہیں اور پھر پری نیٹل معائنے کے لئے واپس نہیں آتی ہیں۔ وہ بعد میں حمل کے مسائل یا ولادت کے وقت مسائل کا شکار ہوسکتی ہیں جن سے معمول کے با قاعدہ چیک اپ کے ذریعے بچا جاسکتا ہے۔

آرگنائزیشن اسپتال کے وارڈوں اور سہولیات جیسے پوسٹ نیٹل دیکھ بھال، معائنہ ٹیبل سے مناسب ابلاغ کے فروغ کے لئے ریجنل ورکشاپس کا انعقاد کرتی ہے۔ کچھ اسپتالوں نے اپنے وارڈ زیادہ سہولت کے ساتھ قابل رسائی بنائے ہیں اور بہری عورتوں کے ایک گروپ نے مڈوائفز کے ایک گروپ کو علامتی یا اشاروں کی زبان سکھائی۔ DWNRO اب معذور عورتوں کو ان سہولیات کی دستیابی سے آگاہ کر رہی ہے تا کہ وہ ان سہولتوں کی فراہمی کا مطالبہ کر سکیں۔

میں نہیں سمجھتی کہ اس جیسی عورت بچہ پیدا کرسکتی ہے

سارا! اندر آجاؤ میں تمہاری مدد کے لئے تیار ہوں۔

باب - 11

زچگی اور ولادت

بچے کی ولادت ہر عورت کے لئے ایک مختلف تجربہ ہوتی ہے۔ بیش تر عورتوں کے لئے یہ بے تحاشہ قوت اور توجہ طلب مرحلہ ہوتا ہے اور بچے کی پیدائش کے بعد نڈھال محسوس کرنا کوئی غیر معمولی بات نہیں ہے۔ لیکن نومولود بچے کو گود میں لینے کی خوشی میں زیادہ تر عورتیں اس درد اور بے چینی کو بھلا دیتی ہیں جو انہیں برداشت کرنا پڑی ہو۔

زیادہ تر بچے کسی دشواری کے بغیر جنم لیتے ہیں لہٰذا ولادت کی اپنی فطری صلاحیت پر بھروسہ کریں۔ تاہم مسائل سامنے آسکتے ہیں اور اگر ایسا ہو تو آپ کو ماہرانہ دیکھ بھال کی ضرورت ہوگی۔ بچے کی ولادت سے پہلے اس فرد سے بات کرلیں جو آپ کے بچے کی ولادت کرائے گا تا کہ وہ آپ کی خصوصی ضروریات اور آپ کے اندیشوں کے بارے میں جان لے۔ اگر آپ کو روزمرہ زندگی میں ابلاغ کے لئے کسی کی مدد کی ضرورت پیش آتی ہے تو آپ کو زچگی اور ولادت کے دوران بھی اس مدد کی ضرورت ہوگی۔

حمل کی تمام تر تبدیلیوں کے بعد بھی آپ ہی اپنے جسم کو بہترین طور پر سمجھ سکتی ہیں۔ آپ یہ جان کر کہ ولادت کے دوران کیا پیش آئے گا، خود کو بہت سے مسائل سے بچا سکتی ہیں۔ درد زہ کے دوران سانس لینے کے طریقوں کی مشق (دیکھئے صفحہ 240) اور اپنے لئے آرام دہ انداز کے لئے مختلف پوزیشنوں کی آزمائش (دیکھئے صفحہ 240 اور 241) اسی سلسلے کی کڑی ہیں۔

کسی کو اپنے ہمراہ رکھیں

زچگی کے دوران تنہا ہونا کسی بھی عورت کے لئے مشکل ہوتا ہے۔ زچگی سے ولادت تک کسی ایسے فرد کو اپنے ہمراہ رکھنے کی کوشش کریں جسے آپ جانتی ہوں (شوہر، گھر کا کوئی فرد، سہیلی وغیرہ) اس کی موجودگی مندرجہ ذیل طور پر معاون ہوگی۔

- آپ کو یقین دلائے گی کہ آپ بہتر طرز عمل رکھتی ہیں۔
- سانس لینے میں آپ کی مدد کرے گی۔
- مختلف پوزیشنیں آزمانے میں مدد کرے گی تا کہ آپ سب سے اطمینان بخش پوزیشن کا تعین کر سکیں۔
- ڈاکٹر یا مڈوائف کو آپ کی فکرمندیوں یا مسائل سے آگاہ کرنے میں معاون ہوگی۔

درد زہ اور ولادت

ولادت کا مرحلہ اس وقت شروع ہوتا ہے جب رحم سکڑتا اور کھلتا ہے اس میں خاصا وقت لگ سکتا ہے۔ جب کوئی ماں اپنے پہلے بچے کو جنم دیتی ہے تو یہ وقت عام طور پر دس سے بیس گھنٹے یا اس سے بھی زیادہ ہوتا ہے بعد ازاں ولادتوں میں عموماً یہ سات سے دس گھنٹے ہوسکتا ہے جب رحم کا منہ پوری طرح کھل جاتا ہے تو بچے کو باہر دھکیلنے میں دو گھنٹے سے بھی کم وقت لگتا ہے۔ زچگی کا عمل ولادت کے بعد پلیسنیٹیا (Placentia) کے باہر آنے پر ختم ہوجاتا ہے۔

بچے کی ولادت کا یہ وقت ایک معذور عورت کے لئے بھی خواہ معذوری کوئی بھی ہو کسی اور عورت سے مختلف یا الگ نہیں جو عام طور پر تین سے چوبیس گھنٹے ہوسکتا ہے۔ جو بات مختلف ہے وہ یہ ہے کہ ایک معذور عورت بتائے کہ درد زہ شروع ہو چکا ہے اور وہ کون سی پوزیشن ہے جو ولادت کے دوران اس کے لئے ضروری اور بہتر ہوسکتی ہے۔

کس طرح جانیں کہ آپ زچگی کے مرحلے میں ہیں

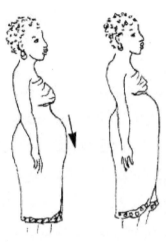

حمل سے دو ہفتے قبل بچہ اکثر و بیشتر پیٹ کے نچلے حصے کی جانب چلا جاتا ہے، خاص طور پر پہلا بچہ۔

زچگی کا مرحلہ عموماً اس وقت شروع ہوتا ہے جب آپ کو حاملہ ہوئے 8 ماہ سے زائد ہو چکے ہوں، بچہ آپ کے پیٹ کے نچلے حصے میں چلا جاتا ہے اور ممکن ہے اب آپ سانس لینے میں سہولت محسوس کریں۔

حمل کے آخری چند ہفتوں میں زیادہ تر عورتیں محسوس کرتی ہیں کہ ان کا رحم دن میں چند بار یا ہفتے میں چند بار سخت ہو رہا ہے۔ سختی کا یہ احساس، رحم کے سکڑاؤ کی مشق ہے حقیقی درد زہ نہیں۔ یہ عجیب سا محسوس ہوسکتا ہے اور چند منٹ برقرار رہتا ہے لیکن عام طور پر یہ نقصان نہیں پہنچاتا ہے اور نہ ہی با قاعدگی کے ساتھ ہوتا ہے۔

وہ عورتیں بھی جو مفلوج ہوں اور اپنے پیٹ کے عضلات میں حس نہ رکھتی ہوں عام طور پر بتا دیتی ہیں کہ بچہ کب باہر آنے کے لئے تیار ہے اگر مفلوج عورتوں کو حقیقی درد کا احساس نہ بھی ہوتو انہیں پیٹ اس حد تک محسوس ہوگا کہ وہ جان لیں کہ کوئی تبدیلی واقع ہو رہی ہے۔

سمجھنے اور سمجھانے کی دشواری رکھنے والی عورتوں کی زچگی کرانے والی مڈوائفوں اور صحت کارکنوں کے لئے

ولادت کا عمل آسان ہوگا اگر آپ بننے والی ماں بننے والی عورت کی مدد کے لئے دوران زچگی اور ولادت کسی فرد مثلاً اس کے شوہر، ماں، بہن، خالہ یا کسی اچھی سہیلی کو آمادہ کرلیں کہ وہ زچگی کے دوران آپ کی مدد کریں۔ یہ معاون صفحہ 235 پر دی گئی ہر بات میں مدد کرنے کے ساتھ مندرجہ ذیل معاونت کرسکتا ہے۔

- اس کا ہاتھ تھامے، اس سے پوچھے کہ وہ کیسا محسوس کر رہی ہے اور اسے یہ سمجھنے میں مدد دے کہ کیا ہو رہا ہے۔
- اسے یہ سمجھنے میں مدد دے کہ مڈوائف اس سے کیا چاہتی ہے۔
- حمل کے دوران اسے مختلف قسم کی تنفسی مشقیں کرائے تا کہ وہ زچگی کے دوران مناسب طور پر سانس لے سکے (دیکھئے صفحہ 240)۔

وضع حمل قریب ہونے کی علامات

مندرجہ ذیل تین علامات بتاتی ہیں کہ زچگی کا عمل شروع ہو رہا ہے یا جلد ہی شروع ہو جائے گا۔ ممکن ہے یہ سب نہ ہوں اور یہ کسی بھی ترتیب میں ہو سکتی ہیں۔ اگر ان میں سے کوئی بھی علامت سامنے آئے تو متعلقہ صحت کا رکن یا مڈوائف یا ڈاکٹر کو آگاہ کریں۔

1- فرج سے شفاف یا گلابی رنگ کا مادہ باہر آتا ہے۔ حمل کے دوران رحم کا منہ اس گاڑھے مادے سے بند ہوتا ہے۔ یہ بچے اور رحم کو انفیکشن سے بچاتا ہے۔ جب رحم کھلنا شروع ہوتا ہے تو یہ گاڑھا مادہ اور تھوڑا سا خون بھی خارج کرتا ہے۔

2- فرج سے صاف پانی بھی باہر نکلتا ہے۔ یہ اس تھیلی سے خارج ہوتا ہے جو رحم میں بچے کے گرد لپٹی ہوتی ہے اور اس کی حفاظت کرتی ہے۔ پانی کی یہ تھیلی درد زہ شروع ہونے سے پہلے یا زچگی کے دوران کسی بھی وقت پھٹ سکتی ہے۔

3- رحم سکڑنا شروع کر دیتا ہے اور پیٹ میں درد شروع ہو جاتا ہے۔ سکڑاؤ کے دوران رحم ابھرے گا اور سخت ہو جائے گا اور پھر نرم ہو جائے گا۔ ابتداء میں سکڑاؤ کا عمل دس سے بیس منٹ یا اس سے زیادہ وقت میں ہو سکتا ہے۔ جب یہ با قاعدہ صورت اختیار کرے (ہر بار یکساں وقت کے بعد) حقیقی درد زہ شروع ہو جاتا ہے۔

رحم کا سکڑاؤ عام طور پر تکلیف دہ ہوتا ہے لیکن اگر آپ اپنے پیٹ میں محسوس کرنے کی صلاحیت نہ رکھتی ہوں تو آپ تبدیلیوں کو اس طرح دیکھ یا محسوس کر سکتی ہیں۔

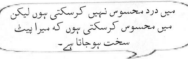

اور پھر دوبارہ نرم ہو جاتا ہے

میں درد محسوس نہیں کر سکتی ہوں لیکن میں محسوس کر سکتی ہوں کہ میرا پیٹ سخت ہو جاتا ہے۔

جب ان میں سے کوئی علامت سامنے آئے تو یہ ولادت کے لئے تیار ہونے کا وقت ہے۔ ذیل میں باتوں کی فہرست ہے جو آپ کر سکتی ہیں۔

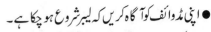

میں جاؤں گا اور دونا روزا سے کہوں گا کہ تمہیں درد شروع ہو چکے ہیں۔

- اپنی مڈوائف کو آگاہ کریں کہ لیبر شروع ہو چکا ہے۔
- خود کو صاف ستھرا کریں اور خاص طور پر اپنے تناسلی اعضا کو دھوئیں۔
- تھوڑی مقدار میں کھانا جاری رکھیں۔
- خوب مائعات (پانی، جوس وغیرہ) پئیں۔
- جب تک ممکن ہو آرام کریں۔
- اپنی نچلی پشت پر گرم سنکائی (گرم کپڑے یا تولیے سے) کریں یا اپنی معاون سے کہیں کہ وہ آپ کی کمر کا مساج کرے۔

مجروح ریڑھ کی ہڈی والی عورتوں کے لئے/ڈس ریفلیکسیا (Dysreflexia)

کے بارے میں اہم معلومات

اگر آپ ریڑھ کی ہڈی کی بلندی پر زخم (عام طور پر T6 یا اس سے اوپر) رکھتی ہیں تو آپ اچانک شدید ضرب لگاتے سر درد اور بے تحاشہ پسینے کے ساتھ خون کے دباؤ میں خطرناک اضافے کا شکار ہو سکتی ہیں (دیکھئے صفحہ 117 تا 119)۔ آپ زچگی کے دوران ان علامات کا شکار ہو سکتی ہیں

کیا کرنا چاہئے؟

صحت کارکن یا مڈوائف سے باقاعدگی سے ملیں تا کہ آپ کا بلڈ پریشر چیک کیا جاتا رہے۔ ابتدائی سات ماہ میں یہ ہفتے میں کم از کم ایک بار اور درد زہ شروع ہونے سے پہلے حمل کے آخری چند ہفتوں کے دوران روزانہ کیا جائے۔ اس سے یہ جاننے میں مدد ملے گی کہ آپ کا بلڈ پریشر بڑھنا تو شروع نہیں ہو رہا ہے۔

اپنے باؤل پروگرام (Bowel program) کا دھیان رکھیں۔ حمل کے دوران یہ خصوصی طور پر اہم ہے خوب پانی پئیں اور ایسی غذائیں کھائیں جو آپ کی باؤل موومنٹ کو آسان بنائیں آپ کے جسم میں قبض کی زیادتی ڈس ریفلیکسیا کا سبب بن سکتی ہے۔

یقینی بنائیں کہ آپ کا مثانہ خالی رہے۔ بھرا ہوا مثانہ ڈس ریفلیکسیا کا سبب بن سکتا ہے اگر آپ فولی کیتھیٹر استعمال کر رہی ہیں تو دھیان رکھیں کہ یہ مڑے یا خم نہ کھائے اور پیشاب کو خارج ہونے سے نہ روکے۔

بچے کی ممکنہ ولادت سے قبل آخری ہفتوں کے دوران روزانہ چند بار رحم کے سکڑاؤ کو دیکھنے یا محسوس کرنے کی کوشش کریں نیز زچگی کی دیگر علامات پر بھی نظر رکھیں۔ (دیکھئے صفحہ 236-237)

جیسے ہی آپ زچگی کی کوئی بھی علامت محسوس کریں فوری طور پر موضع حمل کے لئے اسپتال یا میٹرنٹی ہوم جائیں۔ آپ کو ریڑھ کی ہڈی میں Anesthesia لگانے کی ضرورت ہوگی یہ epidural کہلاتا ہے اور یہ آپ کو درد زہ کے باعث ہونے والے ڈس ریفلیکسیا سے محفوظ رکھے گا۔

حمل کے دوران زچگی اور ولادت کے مرحلے میں
مجروح ریڑھ کی ہڈی والی عورتوں کی دیکھ بھال کرنے والوں، مڈوائفز اور صحت کارکنوں کے لئے

ڈس ریفلیکسیا ایک میڈیکل ایمرجنسی ہے۔ مزید معلومات کے لئے دیکھئے صفحہ 117 تا 119 ہائی بلڈ پریشر دوروں کا سبب یا دماغ میں مہلک جریان خون کا سبب بن سکتا ہے۔ اس لئے ضروری ہے کہ ایسی معذور عورت کے حمل کے آخری دو ماہ کے دوران اس کا بلڈ پریشر روزانہ چیک کیا جائے۔ مجروح ریڑھ کی ہڈی والی عورت کے لئے انتہائی مناسب ہے کہ وہ درد زہ شروع ہونے سے قبل اسپتال کلینک پہنچ جائے تاکہ اسے زچگی اور ولادت کے دوران پیش آنے والے مسائل کے مطابق فوری طور پر طبی سہولت فراہم کی جا سکے۔

زچگی کے دوران ڈس ریفلیکسیا سے بچاؤ کے لئے ضروری ہے کہ حاملہ کو اس کی ریڑھ کی ہڈی میں anesthesia دیا جائے جو epidural کہلاتا ہے۔

- اسے تنہا نہ چھوڑا جائے۔

- یقینی بنایا جائے کہ وہ سپاٹ انداز سے لیٹی نہ رہے اور اس کا سر، کندھے اونچے اور گھٹنے مڑے ہوئے ہوں۔

- اس کا بلڈ پریشر تواتر سے ناپئے۔ کم از کم ہر دس منٹ بعد۔

- اگر اس کا فضلہ خارج کئے جانے کی ضرورت ہو۔ اگر آپ انگلی سے اس کا فضلہ نکال رہی ہوں تو انتہائی محتاط رہیں یا فضلہ خارج کرنے کے لئے اسے اینیما (enema) دیں۔ فضلہ ہٹانے سے ڈس ریفلیکسیا شروع ہو سکتا ہے۔ پہلے ریکٹم میں (دو فیصد سے چار فیصد) کی Lignocaine gel لگائیں۔

- اس کا مثانہ خالی رکھیں۔ اگر ضروری ہو تو پیشاب خارج کرنے کے لئے کیتھیٹر استعمال کریں (دیکھئے صفحہ 104-103)۔ کیتھیٹر لگاتے ہوئے پیشاب کے سوراخ میں Lignocaine gel لگائیں۔

زچگی کوآسان کس طرح بنائیں

پیشاب خارج کرنا یقینی بنائیں گھنٹے بھر میں ایک بار پیشاب کرنے کی کوشش کریں۔ اگر آپ کا مثانہ خالی ہوگا تو آپ زیادہ سکون محسوس کریں گی۔ اگر زچگی اور ولادت کے دوران پیشاب خارج کرنے کے لئے کیتھیٹر لگا رہے ہے تو ریڑھ کی ہڈی کا زخم رکھنے والی عورتیں ڈس ریفلیکسیا ہونے کے کمتر امکانات رکھتی ہیں۔

> زچگی کے دوران آپ کو بے تحاشہ پسینہ آسکتا ہے لہٰذا یہ بہت ضروری ہے کہ آپ جس حد تک ممکن ہو خوب پانی، جوس یا ہربل چائے پیتی رہیں تا کہ آپ قلت آب کا شکار نہ ہوں۔

اپنی پوزیشن کئی بار بدلیں (ہر گھنٹے میں کم از کم ایک بار) وقت سے پہلے ایک پوزیشن سے دوسری پوزیشن میں آنے کی مشق کریں تا کہ لیبر شروع ہونے کے بعد درد زہ کے دوران آسانی سے اپنی پوزیشن بدل سکیں اگر ضروری ہو تو کسی سے مدد طلب کریں۔ آپ جس قدر مطمئن اور پرسکون ہوں گی آپ کے مسلز بھی اتنے ہی پرسکون ہوں گے اور اس باعث ان میں اکڑاؤ یا کھنچاؤ کا امکان بھی کم ہوگا جب آپ اپنی پوزیشن بدلتی رہیں گی تو دباؤ کے زخموں کا امکان بھی اسی قدر کم رہے گا۔

اگر ممکن ہو تو آپ درد زہ کے دوران چہل قدمی کریں۔ چہل قدمی سے رحم کھلنے اور بچے کے آسانی سے باہر آنے میں مدد ملتی ہے۔

لیبر کے دوران سانس لینا

آپ جس انداز سے سانس لیتی ہیں وہ لیبر کی حالت میں گہرا اثر ڈالتا ہے۔ آپ حمل کے دوران سانس لینے کے مختلف انداز کی مشق کر سکتی ہیں تا کہ آپ لیبر شروع ہوتے ہی بہتر انداز سے سانس لے سکیں۔ مثال کے طور پر آہستہ اور نرمی سے سانس لینا۔ ناک کے ذریعے طویل مگر آہستہ سانس لیجیے سانس باہر نکالنے کے لئے ہونٹوں کو سکوڑ یئے اور آہستگی سے سانس خارج کیجیے۔

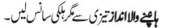

سی، سی
۔۔۔۔۔ سی، سی

- **Hee Breathing** آہستگی سے گہرا سانس لیں اور ہی ہی کی آوازیں نکالتے ہوئے تیزی سے سانس خارج کیجیے۔
- **ہانپنے والا انداز** تیزی سے مگر ہلکی سانس لیں۔
- **تیزی سے سانس لینا۔** تیزی سے سانس لیجیے اور خارج کیجیے

لیبر کے دوران آپ جو انداز بھی مناسب سمجھیں اس انداز سے سانس لے سکتی ہیں۔

ولادت کی پوزیشنز

زچگی اور ولادت کے دوران یہ پوزیشنز استعمال کی جاسکتی ہیں۔

اگر آپ پیروں یا بازوں کا کم یا بالکل بھی نہیں کنٹرول رکھتی ہیں تو آپ گود میں بیٹھ سکتی ہیں

یا آپ تکیے کے سہارے نیم دراز انداز میں رہ سکتی ہیں۔

یا آپ بازو اور پشت
والی ایسی کرسی
استعمال کرسکتی ہیں

اگر آپ بازو اور ہاتھ کا اچھا کنٹرول
رکھتی ہیں تو اس انداز کی ولادتی
کرسی (Birthing Chair) استعمال
کرسکتی ہیں(دوسری مثال کے لئے
دیکھئے صفحہ 242 فتوما کی کہانی)

یا کھڑی ہوسکتی ہیں.......

پیروں پر کچھ کنٹرول رکھنے والی
عورت کسی عورت کی مدد سے
اکڑوں بیٹھ سکتی ہے۔

آپ کو سہارا دینے والا فرد یا چیز مضبوط اور متوازن ہو۔ آپ کسی کرسی کی پشت پر بھی گرفت کرسکتی ہیں۔ جب ولادت کی رفتار کم ہو
یا ماں بچے کو دھکیلنے میں دشواری محسوس کررہی ہو، اکڑوں یا کھڑی پوزیشنز بچے کو نیچے لانے میں معاون ہوسکتی ہیں

اگر آپ کو پیروں اور بازوؤں پر کچھ کنٹرول ہے اور آپ ہاتھ اور گھٹنوں
کے بل جھکنے کا انداز آزمانا چاہتی ہیں تو یہ پوزیشن بعض اوقات
عضلاتی کھچاؤ سے بچاتی ہے یا اسے کنٹرول میں رکھتی ہے۔

یا پیر سیدھے ہوں.......

اگر آپ پیر کا کم یا بالکل بھی کنٹرول
نہیں رکھتی ہیں تو آپ پہلو کے بل اس
طرح لیٹ سکتی ہیں کہ کوئی آپ کے
اوپری پیر کو تھام لے اور آپ کے پیر
قدرے مُڑے ہوئے ہوں۔

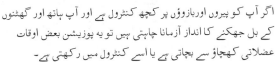

اگر آپ نابینا ہیں یا ناقص توازن رکھتی ہیں تو آپ فرش پر خود کو زیادہ محفوظ محسوس کرسکتی ہیں۔ زیادہ تر اسپتال اور مراکز صحت بچوں
کو جنم دینے والی عورتوں کے لئے خصوصی بیڈز رکھتے ہیں ان میں گھٹنوں کی سپورٹ کا انتظام ہوتا ہے اور یہ پیروں پر ناقص کنٹرول رکھنے
والی عورتوں کے لئے مفید ہوسکتا ہے۔

فتوما کا ولادتی اسٹول

فتوما اوچن یوگنڈا میں رہتی ہے وہ بچپن میں پولیو ہونے کے باعث دونوں پیروں سے مفلوج ہے۔ جب وہ حاملہ ہوئی تو دوسری بہت سی معذور عورتوں کی طرح اس سے بھی مقامی کلینک کی ڈاکٹروں نے کہا کہ اسے بچے کی ولادت آپریشن کے ذریعے کرانا ہوگی۔

فتوما نے آپریشن نہ کرانے اور بچے کی پیدائش نارمل طریقے سے کرانے کا فیصلہ کیا اس کی کمیونٹی میں بہت سی عورتیں بچوں کی ولادت اکڑوں بیٹھ کر کراتی تھیں لیکن اس کے پیر مفلوج تھے لہٰذا وہ جانتی تھی کہ اس کے لئے اکڑوں بیٹھنا ممکن نہیں ہوگا لیکن وہ جانتی تھی کہ اس کے بازو سارا وقت وہیل چیئر دھکیلنے کے باعث بہت مضبوط ہیں اس لئے اس نے ایک ایسا ولادتی اسٹول بنایا جو اسے اکڑوں پوزیشن میں رکھ سکے اس طرح اس کا بچہ فطری طریقے سے پیدا ہو سکتا تھا۔

ہر چند کہ فتوما مفلوج ہے اس کا رحم اب بھی اتنا مضبوط ہے کہ بچے کو باہر دھکیلنے کے لئے خود کو سکیڑ سکتا تھا اس اسٹول پر اس کے جسم کی پوزیشن نے بچے کو اس کے جسم سے فرج کے ذریعے باہر آنے میں اسی طرح مدد دی جس طرح اکڑوں بیٹھنا دوسری عورتوں کے لئے معاون ثابت ہوتا ہے۔

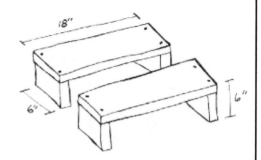

لکڑی سے بآسانی بنایا جانے والا ولادتی اسٹول

زچگی اور ولادت کے دوران عضلات کی اینٹھن اور اکڑاؤ (عضلات کا اچانک سخت ہوجانا)

سیریبرل پالسی، ریڑھ کی ہڈی کے زخم اور پولیو کے باعث مفلوج عورتوں کے عضلات زچگی اور ولادت کے دوران کسی وقت اکڑاؤ یا شدید اینٹھن کا شکار ہوسکتے ہیں۔ جسم کے کسی حصے میں سخت عضلات سر اور جسم کی پوزیشن سے متاثر ہوتے ہیں۔ اینٹھے ہوئے عضلات کو براہ راست کھینچنا یا دبانا انہیں اور سخت بنائے گا۔ زچگی کے دوران ایسی صورت پیش آنے پر اینٹھے ہوئے عضلات کو نرم بنانے کے لئے چند تجاویز درج ذیل ہیں۔

زچگی کے دوران دردزہ کے درمیانی وقفوں میں رینج آف موشن ورزشیں کیجئے (دیکھئے صفحہ 95) اگر ضروری ہو تو کسی کی مدد حاصل کریں۔ یہ ورزشیں مسلز کو گرم رکھیں گی اور اینٹھن اور اکڑاؤ سے تحفظ میں معاون ہوں گی۔

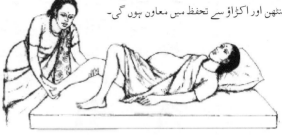

مسلز کو آرام دینے کے لئے صاف گرم پانی کے ٹب میں بیٹھنا مند فائدہ ہوسکتا ہے لیکن پانی کی تھیلی پھٹنے سے پہلے

صاف گرم پانی میں بھیگے کپڑے سے آرام مل سکتا ہے

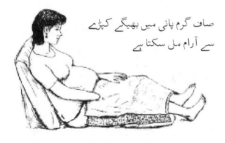

سر کے نیچے کوئی چیز رکھیں جو کندھوں کو بھی آگے کی طرف جھکائے، اس سے پورے جسم کی سختی میں کمی آئے گی۔

عورت کے پاؤں جانگھوں (جوڑوں) سے دور کرنے کی کوشش مت کریں اس طرح اس کے پیروں کو سختی سے کھینچنا ہوگا اس کے بجائے اس کے سر اور کندھوں کو اونچا کریں اور پیروں کو موڑیں اس کے پیر الگ الگ کرنے کے لئے پہلے اس کے گھٹنوں کو ملائیں اس سے اس کے پیر جکڑاؤ سے آزاد ہوں گے۔ اگر ایسا نہ ہو تو پیروں کو گھٹنوں کے اوپر سے گرفت کریں اس طرح اس کے پیر زیادہ آسانی سے کھل سکیں گے۔

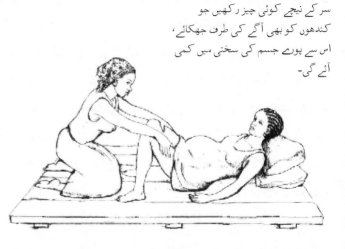

آپریشن کے ذریعے ولادت
(سیزیریں سیکشن ۔ سی سیکشن)

جب پیچیدگیوں کے باعث عورت کی زچگی اور بچے کی ولادت خطرناک ہو جاتی ہے تو ڈاکٹر فیصلہ کر سکتا ہے کہ بچے کو ماں کا پیٹ اور رحم کاٹ کر باہر نکال لیا جائے (رحم نکالا نہیں جاتا ہے) آپریشن رحم اور پیٹ پر زخم کا نشان چھوڑ دیتا ہے۔ ڈاکٹر رحم اور پیٹ کو سی دیتا ہے (رحم نکالا نہیں جاتا ہے) آپریشن رحم اور پیٹ پر زخم کا نشان چھوڑ دیتا ہے یہ آپریشن سیزرین سیکشن یا سی سیکشن کہلاتا ہے۔

جسمانی معذوریاں رکھنے والی زیادہ تر عورتیں ، خاص طور پر مفلوج عورتوں سے ڈاکٹر اور صحت کار کن کہتے ہیں کہ وہ بچے کی ولادت سی سیکشن کے ذریعے کرائیں یہ بات ہر معاملے میں صحیح نہیں ہے جسمانی معذوری کا شکار عورتوں یا وہ پیٹ میں محسوسات نہیں رکھتی ہیں ، تھوڑی سی مدد سے اپنے بچے کو فطری طریقے سے جنم دے سکتی ہیں قطع نظر اس کے کہ عورت کس قسم کی معذوری رکھتی ہے اس کے رحم کے عضلات اس حد تک سکڑاؤ کی صلاحیت رکھتے ہیں کہ بچے کو از خود باہر دھکیل سکیں۔

ان مخصوص معذوریوں کے لئے جو بچے کی ولادت کے دوران مسائل کا سبب بن سکتی ہیں (دیکھئے صفحہ 213)۔

بعض اوقات بعض صورتوں میں ضروری ہو جاتا ہے کہ عورت کا آپریشن کیا جائے مثلاً

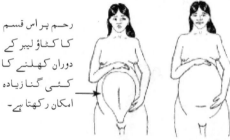

رحم پر اس قسم کا کٹاؤ لیبر کے دوران کھلنے کا کئی گنا زیادہ امکان رکھتا ہے۔

اور ممکن ہے کہ اس کے بیرونی سطح پر اس رحم کا کٹاؤ دوسرے انداز کا ہو

ایک کٹاؤ جو طرح نظر آ رہا ہو

- بچہ جسامت میں بڑا ہو یا ولادت کے لئے غیر موزوں پوزیشن میں ہو۔
- عورت پیٹرو کی خراب ساخت رکھتی ہو۔
- اس کی ریڑھ کی ہڈی میں خم ہو۔
- وہ اپنی ٹانگیں الگ رکھنے کے قابل نہ ہو۔
- اس کا پیٹرو چھوٹا اور بچہ بڑا ہو۔
- وہ لیبر کے لئے مناسب طاقت نہ رکھتی ہو۔

ہر چند کہ سی سیکشن بعض اوقات ضروری ہوتا ہے لیکن ممکن ہو تو اس سے بچنا سب سے بہتر ہے۔ یہ عموماً اس لئے کیا جاتا ہے کہ یہ ڈاکٹروں کے لئے آسان ہوتا ہے پھر یہ کہ یہ مہنگا ہوتا ہے اس میں امکان رہتا ہے کہ کوئی گڑ بڑ ہو جائے اور یہ کہ اس میں صحت یابی فطری ولادت کی بہ نسبت زیادہ عرصے میں ہوتی ہے۔

سی سیکشن کے ذریعے بچوں کو جنم دینے والی بہت سی عورتیں اگلی بار بچہ کو فطری طریقے سے جنم دے سکتی ہیں۔ یہ بات خصوصیت کے ساتھ درست ہے کہ سی سیکشن میں پیٹ کے نچلے حصے کو اوپر سے نیچے کی طرف کے بجائے افقی طور پر کاٹا جائے۔ اوپر سے نیچے کی طرف کا گھاؤ لیبر کے دوران کھلنے کا زیادہ امکان رکھتا ہے پھر یہ بھی ہے کہ افقی صورت میں اس بات کا بہت ہی کم امکان ہے کہ لیبر کے دوران رحم کا گھاؤ کھل جائے۔ اگر ایسا ہوتا ہے تو عورت کے اندرون حصے میں خون بہہ سکتا ہے اور وہ مر سکتی ہے۔ اگر چہ ایک عورت کو بچے کی ولادت کے لئے ممکن ہے دوسری بار آپریشن کی ضرورت نہ پڑے لیکن گزشتہ بار آپریشن کے ذریعے بچے کو جنم دینے والی عورت کے لئے سب سے محفوظ بات یہ ہے کہ اس کے بچے کی ولادت اسپتال میں ہو اور اگر یہ ممکن نہیں ہے تو وہ کسی اسپتال کے قریب دے اور بچے کی ولادت سے پہلے ہی زچگی کے دوران کسی بھی ہنگامی صورت میں اسے اسپتال لے جانے کا بند و بست ضرور کیا جائے۔

نسوانی تناسلی اعضا کا ٹنا (ایف جی سی، زنانہ ختنہ)

بہت سے ممالک میں بیشتر افریقہ کے، تاہم جنوبی ایشیاء، مشرق وسطیٰ اور دنیا کے دیگر حصوں میں لڑکیوں اور نوجوان عورتوں کے تناسلی اعضاء کو کاٹ چھانٹ دیا جاتا ہے۔ دوسری سماجی رسومات کی طرح عورتوں کا ختنہ بھی لڑکیوں کے جسموں میں تبدیلی کا ایک طریقہ ہے تا کہ وہ خوبصورت، قابل قبول اور صاف ستھری سمجھی جائیں لیکن یہ عمل ان لڑکیوں کی صحت اور زندگی پر خطرناک اور نقصان دہ اثرات مرتب کرتا ہے۔ اگلے برسوں میں زنانہ ختنہ پیشاب کے نظام میں انفیکشن، جذباتی نقصان، جنسی حساسیت یا کسی مرد سے مباشرت کی صلاحیت ختم ہونے اور طویل مشکل لیبر کا سبب بن سکتا ہے، جس کے باعث بچے یا ماں کی موت بھی واقع ہوسکتی ہے۔ اگر آپ بھی ایسی عورتوں میں شامل ہیں کہ آپ کے تناسلی اعضاء کاٹ کر انہیں جزوی طور پر سی دیا گیا تھا تو کسی تجربے کار مڈوائف یا صحت کار کن سے مل کر اسے اس بات سے آگاہ کریں۔ ولادت سے پہلے آپ کے تناسلی اعضاء کو کاٹ کر کھولنے کی ضرورت ہوگی۔

پاکستان میں نسوانی اعضاء کو کاٹنے کا عام رواج نہیں ہے۔ صرف ایک کمیونٹی میں اس کا رواج ہے اور یہ لوگ اب اس رواج کو چھوڑتے جا رہے ہیں۔

زچگی کے دوران خطرناک علامات

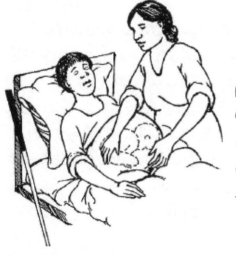

بیشتر عورتیں جن میں معذور عورتیں بھی شامل ہیں محفوظ طور پر بچوں کو جنم دیتی ہیں لیکن جب زچگی اور ولادت کے دوران کوئی خرابی ہو جاتی ہے تو ایسی عورت کے لئے ضروری ہے کہ اسے اپنی زندگی کے تحفظ کے لئے ضروری دیکھ بھال ملے (محفوظ زچگی اور ولادت، اور اس دوران ممکنہ مسائل کے بارے میں مزید معلومات کے لئے دیکھئے ایک بک فار مڈوائیوز) ذیل میں ایسی چند خطرناک علامات دی گئی ہیں جن میں فوری طور پر طبی مدد حاصل کرنی چاہیے۔

پانی بہہ جائے لیکن در دزہ 24 گھنٹوں میں شروع نہ ہو

کسی صحت مرکز یا اسپتال جائیں۔ جب پانی بہہ جائے تو اس بات کا انتہائی امکان ہوتا ہے کہ آپ یا آپ کا بچہ کسی سنگین انفیکشن میں مبتلا ہو جائے۔ آپ کو رگ کے ذریعے سیال یا ادویات کی ضرورت ہوسکتی ہے۔

بچہ ترچھا ہو

اسپتال جائیں۔ در دزہ شروع ہونے کے بعد بچے کی پوزیشن تبدیل کرنے کی کوشش مت کریں اس سے رحم پھٹ سکتا ہے یا پلیسینٹا رحم کی دیوار سے الگ ہوسکتا ہے۔ ترچھا بچہ بغیر آپریشن کے جنم نہیں لے سکتا ہے۔

بچے کی پیدائش سے پہلے خون کا اخراج

فوراً اسپتال جائیں۔ اگر بہنے والا خون چمکدار سرخ ہے تو اس کا مطلب یہ ہوسکتا ہے کہ پلیسینٹا (placenta) رحم کی دیوار سے الگ ہو رہا ہے یا رحم کے منہ کو ڈھانپ رہا ہے یہ انتہائی خطرناک ہے۔

بخار

بخار عام طور پر انفیکشن کی علامت ہوتا ہے۔ اگر بخار شدت کا نہیں ہے تو آپ کو صرف زیادہ مقدار میں مائعات کی ضرورت ہوسکتی ہے۔ خوب پانی، چائے یا جوس پیئں اور ہر چند گھنٹے بعد پیشاب کرنے کی کوشش کریں۔

اگر بخار بہت تیز ہے اور آپ کپکپاہٹ محسوس کر رہی ہیں تو **مرکز صحت یا اسپتال جائیں** آپ کو فوری طور پر اینٹی بایوٹک ادویات کی ضرورت ہے۔

حد درجہ طویل لیبر

مرکز صحت یا اسپتال جائیں۔ جب لیبر ایک دن اور ایک رات سے بڑھ جائے یا آپ دو گھنٹے سے زیادہ بچے کو دھکیلتی رہی ہیں تو آپ کو بچے کی ولادت کے لئے ادویات یا آپریشن کی ضرورت ہوسکتی ہے۔

سبز یا بھورا پانی

اگر لیبر کا ابتدائی مرحلہ ہے یا ماں نے بچے کو دھکیلنا شروع نہیں کیا تو بچے کے لئے بہتر یہی ہے کہ اس کی ولادت اسپتال میں ہو۔ جب پانی کی تھیلی پھٹ جائی ہے

زچگی کے دوران عورت دوبارہ سورج کو نکلتا ہوا نہ دیکھے۔
.... نائیجیرین کہاوت

۔ پانی کو شفاف یا قدرے گلابی ہونا چاہئے بھورے یا سبز پانی کا مطلب یہ ہے کہ بچے نے رحم میں پاخانہ خارج کر دیا ہے اور وہ مشکل میں ہوگا۔

اگر لیبر کا مرحلہ شروع ہوئے خاصا وقت گزر چکا ہے اور بچے کی ولادت ہونے والی ہے تو ماں سے کہیں کہ وہ جس قدر بچے کو دھکیل سکتی ہے دھکیلے اور بچے کو تیزی سے باہر آنے دیں۔ جیسے ہی بچے کا سر باہر آئے اور اپنی پہلی سانس لے ماں سے کہیں کہ وہ دھکیلنا بند کر دے اب بچے کے منہ اور ناک کو صاف کپڑے سے پونچھیں یا اس کے منہ اور ناک کا مواد، منہ سکشن بلب (Suction bulb) کے ذریعے کھینچیں جیسے ہی منہ اور ناک صاف ہو ماں بچے کے بقیہ جسم کو باہر نکالنے کے لئے دھکیل سکتی ہے۔

پری ایکلیمپسیا (ٹوکیسمیا آف پریگننسی)

پری ایکلیمپسیا دوروں اور حد تو یہ کہ موت کا سبب بن سکتا ہے اگر ماں میں مندرجہ پر خطر علامتوں میں سے کوئی نظر آئے تو اسے فوری طور پر اسپتال لے جائیں۔

- شدید سر درد
- اوورا ایکٹیو ریفلیکسز
- دھندلاہٹ یا دوہری بصارت
- ہائی بلڈ پریشر
- پسلیوں سے نیچے درمیانی حصے میں
- پیشاب میں پروٹین
- پیٹ کے اوپری حصے میں اچانک تیز اور شدید درد

اگر ماں پر دورہ پڑنا شروع ہو جائے اور آپ جانتی ہوں کہ ماں مرگی کی مریضہ نہیں ہے تو

- اس کے سر کی حفاظت کے لئے سر کے نیچے کوئی چیز رکھ دیں اور ممکن ہو تو اسے الٹے پہلو کے بل رکھیں یا جکڑنے یا گرفت میں رکھنے کی کوشش مت کریں۔

- اسے ٹھنڈک میں رکھیں۔

- **اسے قریب ترین اسپتال لے جائیں۔**

مرگی کی مریضہ عورت Toxemia کا شکار بھی ہو سکتی ہے
(مرگی کے بارے میں مزید معلومات کے لئے دیکھئے صفحہ 231)

بچے کی پیدائش کے بعد ابتدائی چند دنوں میں ماں کے لئے خطرے کی علامات

خون بہنا (اخراج خون)

بچے کو فوراً دودھ پلانا شروع کر دیں، اس سے جلد خون روکنے میں مدد ملے گی۔ بچے کی پیدائش کے بعد شروع ہونے والا اخراج خون ایک دن سے زیادہ ہو تو اس کا سبب عموماً پلیسینٹا کے ٹکڑے رحم میں رہ جانا ہوتا ہے **طبی مدد حاصل کریں۔**

حد سے زیادہ خون بہنے کی خطرناک علامات

- ولادت کے بعد پہلے دن دو سے زیادہ پیڈ زیا کپڑے کی دبیز تہہ کا خون سے بھر جانا

- پہلے دن پہلے گھنٹے میں ایک پیڈ یا کپڑے کی دبیز تہہ کا خون سے بھر جانا

- کم مقدار میں مگر مسلسل خون کا بہنا

کیا کرنا چاہئیے؟

1- اس کے رحم کے اوپری حصے کو رگڑیں تا کہ وہ انتہائی سخت ہو جائے اور خون بہنا رک جائے بچے کو اس کی چھاتیاں چوسنے دیں یا کوئی اس کے نپلز کو ہلائے۔

2- ضرورت کے مطابق ہر چھ گھنٹے میں اسے (0.2) Ergonovine ملی گرام) کھلائیں لیکن یہ دوا صرف 4 سے 7 دن دیں اس سے زیادہ نہیں۔

3- اگر خون بہنا بند نہ ہو تو طبی مدد حاصل کریں اسپتال لے جاتے ہوئے اس کے رحم کو رگڑنا جاری رکھیں۔

4- اگر انفیکشن کی علامات ہوں تو اسے رحم کے انفیکشن کے لئے صفحہ 248 کے مطابق اینٹی بایوٹک دوائیں دیں۔

رحم کا انفیکشن

رحم کا انفیکشن انتہائی خطرناک ہے، اس کا فوری علاج کیا جانا چاہیئے ورنہ عورت بانجھ ہوسکتی ہے یا مرسکتی ہے۔

رحم کے انفیکشن کی علامات

- بخار اور کپکپی
- پیٹ میں درد اور نرمی
- فرج سے ناگوار رطوبت کا اخراج

اگر ماں شکایت کرے کہ وہ خود کو بہتر محسوس نہیں کررہی ہے تو بار بار یک بینی سے اس میں انفیکشن کی علامات کا جائزہ لیں۔

رحم کے انفیکشن کے لئے ادویات

دوا	مقدار	کب تک/کتنی بار لی جائے
سپرو فلوکساسِن	500 ملی گرام	منہ کے ذریعے روزانہ دو بار
ڈوکسی سائیکلین	100 ملی گرام	منہ کے ذریعے دن میں دو بار
میٹرونیڈازول	500 ملی گرام	منہ کے ذریعے دن میں دو بار

بخار ختم ہونے کے بعد مزید دو دن تمام دوائیں جاری رکھیں لیکن ایک دن کے بعد مریضہ خود کو بہتر محسوس کرنا شروع نہ کرے تو اسے قریبی اسپتال لے جائیں۔ ممکن ہے اسے انجکشن کے ذریعے یا رگ کے ذریعے (آئی وی) ادویات کی ضرورت ہو۔

اہم بات اس سے مائعات خوب پینے کے لئے کہیں۔ میٹرونیڈازول لینے کے دوران الکحل استعمال مت کریں۔

زچہ کی دیکھ بھال

ماں کو بھی اسی طرح دیکھ بھال کی ضرورت ہوتی ہے جس طرح نومولود بچے کو۔ لوگ عموماً بچے کی دیکھ بھال میں اس قدر مصروف ہو جاتے ہیں کہ وہ اکثر ماں کو بھول جاتے ہیں۔ ان ضروری معلومات سے اپنے گھر والوں یا دیکھ بھال کرنے والوں بھی آگاہ کریں تا کہ وہ آپ کے لئے سہولتیں اور مدد حاصل کرسکیں۔

- **انفیکشن سے بچاؤ کے لئے۔** جب تک خون کا بہنا نہ رکے نہ تو مباشرت کریں اور نہ ہی فرج میں کوئی چیز رکھیں۔ غسل اپنے معمول کی طرح کریں لیکن ولادت کے بعد ایک ہفتے تک پانی میں نہ بیٹھیں اپنے جنسی اعضاء کو دھونا اور انتہائی صاف ستھرا رکھنا آپ کے لئے بہتر ہے۔

- **کم از کم چھ ہفتے تک آرام کریں۔**

- **معمول سے زیادہ غذا کھائیں۔** آپ کسی بھی قسم کی غذا کھا سکتی ہیں۔ مچھلی، گوشت، بیج، غذائی اجناس، سبزیاں اور پھل آپ کو جلد از جلد صحت یاب ہونے میں مدد دیں گے اور ان سے آپ کو اپنی اور اپنے بچے کی دیکھ بھال کے لئے ضروری توانائی ملے گی۔ ریشہ دار غذائیں کھانے سے قبض سے تحفظ ملے گا۔

- **مائعات خوب پئیں۔** اس سے قبض سے بچاؤ میں مدد ملے گی۔

- جس قدر زیادہ ممکن ہو متحرک رہیں اور چلیں پھریں۔

- اگر آپ کی چھاتیاں بہت زیادہ سوجی ہوئی، سخت یا تکلیف دہ ہوں، بچے کو جس قدر بار بار ممکن ہو رات اور دن میں دودھ پلائیں (ایک یا دو گھنٹے بعد، ہر چھاتی سے) ہر بار دودھ پلانے سے پہلے اپنی چھاتیوں پر پندرہ سے بیس منٹ گرم بھیگا ہوا کپڑا (گدی) رکھیں آپ درد کے لئے پیراسیٹامول لے سکتی ہیں (دیکھئے صفحہ 350)

- اگر آپ بچے کو دودھ نہیں پلانا چاہتی ہیں تو اپنی چھاتیوں سے دودھ نکالنے کی کوشش مت کریں اگر آپ ایسا کریں گی تو آپ کا جسم مزید دودھ بناتا رہے گا اس کے بجائے اپنے جسم کے گرد چھاتیوں پر نرمی سے کپڑے کی پٹی باندھیں اور ان پر ٹھنڈا نم کپڑا یا برف رکھیں۔ آپ درد میں آرام کے لئے پیراسیٹامول لے سکتی ہیں۔ (دیکھئے صفحہ 350)

- اگر آپ کے تناسلی اعضاء یا فرج میں خراشیں ہیں تو انفیکشن سے بچنے کے لئے اسے روزانہ مناسب صابن اور صاف پانی سے دھوئیں۔ خراشوں پر گرم نم کپڑا رکھیں اور شہد لگائیں تا کہ آپ کو آرام ملے اور وہ تیزی سے ٹھیک ہوں آپ معمولی مقدار میں نمک ملے صاف، گرم پانی کے ٹب میں بھی بیٹھ سکتی ہیں اگر خراشوں میں سوزش ہو تو ہر بار پیشاب کرتے ہوئے اپنے تناسلی اعضاء پر پانی بہائیں۔

- اگر آپ اپنے تناسلی اعضاء کی صحت کے لئے جڑی بوٹیاں یا پودے بطور دوا استعمال کرتی ہیں تو یقینی بنائیے کہ وہ انتہائی صاف ہوں (انہیں اُبالنا بہترین ہے) **اپنی فرج میں ایسی ادویات یا کوئی اور چیز مت رکھیں۔**

- دوبارہ جنسی عمل سے پہلے خاندانی منصوبہ بندی کا کوئی طریقہ استعمال کیجیے ورنہ آپ فوراً ہی دوبارہ حاملہ ہو سکتی ہیں آپ اپنی ماہواری دوبارہ شروع ہونے سے دو ہفتے پہلے حاملہ ہو سکتی ہیں اگر آپ اپنے بچے کو صرف اپنا دودھ پلا رہی ہیں تو اس سے عام طور پر تقریباً چھ ماہ دوبارہ حمل سے تحفظ مل سکتا ہے خاندانی منصوبہ بندی کے لئے دیکھئے باب 9۔

اگر آپ خود کو حد سے زیادہ پریشان یا افسردہ محسوس کرتی

زیادہ تر عورتیں بچے کی پیدائش کے بعد جذبات کے شدید جذبات محسوس کرتی ہیں۔ اگر آپ کے ساتھ بھی یہی صورتحال ہے تو ممکن ہے کہ آپ کے گھر والے یا صحت کا رکن یہ سمجھیں کہ اس کا سبب آپ کا معذور ہونا ہے خصوصاً اس صورت میں جب کہ آپ کے لئے خود اپنی روز مرہ دیکھ بھال انتہائی دشوار ہو اور آپ اپنے بچے کی دیکھ بھال کرنے کے قابل نہ ہوں۔ ممکن ہے وہ اس بات کو تسلیم نہ کریں کہ نئی ماں ابتدائی چند دن، چند ہفتے یا ممکن ہے مہینوں افسردگی یا فکر مندی میں مبتلا رہ سکتی ہے جب یہ جذبات انتہائی شدید ہوں اور آپ ٹھیک طرح سے سو کھانا نہ سکیں، خوب آہ و زاری کریں تو اسے ڈپریشن کہا جاتا ہے۔ گزشتہ بار ولادت کے بعد ایسے ہی جذبات کا شکار عورت ایک بار پھر یہی سیت یا سیت زدہ ہونے کے زیادہ امکانات رکھتی ہے۔ کسی ایسے فرد سے جس پر آپ اعتماد کرتی ہوں، اپنے جذبات کا اظہار کرکے آپ خود کو بہتر محسوس کرسکتی ہیں آپ کو خود اپنی اپنے گھر اور اپنے بچے کی دیکھ بھال کے لئے اضافی مدد کی ضرورت ہوسکتی ہے۔

اس صورت کے ازالے کے لئے جدید ادویات کے ساتھ کچھ روایتی طریقے بھی موجود ہیں۔ جدید ادویات مہنگی ہیں اور دوسرے مسائل پیدا کر سکتی ہیں لہٰذا انہیں شدید صورتوں میں ہی لیا جانا چاہیئے۔ اپنی مڈوائف یا صحت کار کن سے بات کریں۔ ذہنی صحت پر مزید معلومات کے لئے دیکھئے باب 3۔

نومولود کی دیکھ بھال

آپ کے بچے کے لئے بہترین غذا آپ کا دودھ ہے۔ اپنے بچے کو گرم اور صاف ستھرا رکھئے اور وہ جتنی بار چاہے اپنی چھاتیاں چوسنے دیں۔

عموماً پیدائش کے بعد ابتدائی چند ہفتے بچوں کی آنکھوں سے زردی مائل مواد خارج ہوتا ہے آپ اس کی آنکھیں اپنے دودھ یا اُبلے ہوئے ٹھنڈے پانی اور صاف کپڑے سے پونچھ سکتی ہیں۔ اگر بچے کی آنکھیں سرخ ہو جائیں یا سوجھ جائیں یا ان میں مواد زیادہ مقدار میں آئے تو بچے کو کسی صحت کار کن کو دکھائیں۔

ناف کی دیکھ بھال

بچے کی ناف کے بقیہ حصے کو صاف اور خشک رکھیں۔ اگر ممکن ہو تو اسے ہر بار ڈائپر (پوترا) تبدیل کرتے ہوئے صاف کپڑے اور الکحل سے صاف کریں۔ یہ پہلے ہفتے میں سیاہ پڑ کر جھڑ جائے گی اگر کھلیں اور گرد و غبار نہ ہو تو اسے ڈھانک کر رکھنے کی ضرورت نہیں ہے ڈھانکنے کی ضرورت ہو تو اسے صاف نرم کپڑے سے اس طرح ڈھکیں کہ اس طرح زور نہ پڑے۔

اگر آپ ناف کے اردگرد سرخی یا مواد محسوس کریں تو ممکن ہے کہ بچہ کسی انفیکشن میں مبتلا ہو۔ اسے فوری طور پر علاج کے لئے کسی صحت کار کن کو دکھائیں تشنج (Tetanus) کی دیگر علامات کے لئے جائزہ لیں جو عام طور پر بچے کی ناف کسی آلودہ چیز سے کاٹنے کی صورت میں ہوسکتا ہے۔

بچے کا تشنج

بچے کو تشنج ہوتو اسے فوری طور پر اسپتال لے جائیں۔ اگر اسپتال دو گھنٹے سے زیادہ مدت کی مسافت پر ہے تو گھر سے روانہ ہونے سے پہلے بچے کو Benzylpencillin کے 100,000 یونٹ کا انجکشن لگائیں۔

نومولود میں تشنج کی پُر خطر علامات

- بخار
- بچہ چھاتی نہ چوس سکے
- سارا وقت روتا رہے
- تیزی سے سانس لے
- بچے کا جسم سخت ہو جائے

تبدیلی کے لئے کام

گھرانے اور دیکھ بھال کرنے والے کیا کر سکتے ہیں؟

ہمارے بعد ہمارے گھر والے ہماری معذوریوں کو کسی اور سے بہتر سمجھتے ہیں۔ اس کا مطلب یہ ہے کہ وہ ہمیں زچگی اور ولادت دونوں کے دوران بہترین مدد فراہم کر سکتے ہیں وہ بچے کی ولادت کرانے والی مڈوائف یا صحت کارکن کو سمجھ سکتے ہیں کہ ہم معذوریاں رکھنے کے باوجود فطری طریقے سے بچے کو جنم دے سکتے ہیں اور اگر ہمیں ولادت کے لئے متبادل پوزیشن اختیار کرنے کی ضرورت ہوتو وہ ہمیں یہ بات سمجھانے میں بھی معاون ہو سکتے ہیں اور جب بچہ جنم لے تو وہ یقینی بنا سکتے ہیں کہ ہم بچے کو سنبھالیں خواہ اس کے لئے ہمیں کتنی ہی مدد کی ضرورت کیوں نہ ہو۔

> فکر مت کرو ڈاکٹر میری بیٹی معذور ضرور ہے لیکن وہ بہت طاقتور ہے اور اسے اپنے پہلے بچے کی پیدائش کے لئے آپریشن کی ضرورت نہیں ہوگی

مڈوائفز، ڈاکٹر اور دیگر صحت کارکن کیا کر سکتے ہیں؟

- یقینی بنائیں کہ کلینک یا اسپتال میں وہ کمرے یا جگہیں جہاں عورتیں بچوں کو جنم دیتی ہیں، ان تک پہنچنا ہمارے لئے آسان ہو۔ مثلاً یہ جگہ اوپری منزل پر ہو تو معذور عورتوں کی زچگی کے لئے گراؤنڈ فلور پر جگہ بنائی جائے۔
- یقینی بنائیں کہ تمام بیڈ اور معائنہ کی میزوں کی بلندی کم ہو اور ان میں پہیے نہ ہوں۔
- یقینی بنائیں کہ بہری یا نابینا عورت کا بچہ اس سے نزدیک تر رہے تاکہ ماں جو سن یا دیکھ نہ سکے تو پھر بھی وہ جان لے کہ اس کے بچے کو کسی چیز کی ضرورت ہے۔
- ماں اور بچے دونوں کو بعد از ان صحت کارکن سے دیکھ بھال کی سہولت ملتی رہے وہ ماں اور نومولود بچے کو پیدائش کے بعد دن میں کم از کم دو مرتبہ اور پھر آئندہ ہفتے میں کم از کم ایک بار ضرور دیکھے۔
- بچے کی ولادت کا اندراج کرانے کے لئے کمیونٹی کے مطابق قانونی ضروریات پوری کرنے میں ماں کی مدد کریں۔

انفیکشن سے بچاؤ

انفیکشن کے باعث اموات دنیا بھر میں اپنا وجود رکھتی ہیں مگر تیسری دنیا کے ممالک میں انفیکشن کے خطرات شدید ہیں نرسوں، مڈوائفز اور صحت کارکنوں کے لئے ایک رہنماء کتاب میڈیکل کے ہر طالب علم کے لئے یہ کتاب ضروری ہے

باب - 12

اپنے بچے کی دیکھ بھال

کوئی بھی ماں تنہا اپنے بچے کی پرورش نہیں کرسکتی ہے۔ نومولود بچے پر مستقل توجہ اور دیکھ بھال انتہائی تھکا دینے والا کام ہوتا ہے۔ تقریباً تمام مائیں اپنے بچوں کی دیکھ بھال کے لئے اپنے گھر والوں، احباب، پڑوسیوں، بچوں کی نگہداشت کرنے والوں اور ٹیچرز کی مدد پر انحصار کرتی ہیں۔ معذوریوں والی کچھ عورتیں بچے کی دیکھ بھال کے طریقے سے تیزی سے سیکھ لیتی ہیں لیکن اگر آپ اپنی معذوری کے باعث اپنے روزمرہ کے کاموں کے لئے دوسروں کی مدد لیتی ہیں تو یقیناً آپ کو اپنے بچے کی دیکھ بھال اور روزمرہ ضروریات کے لئے دوسروں کی مدد کی ضرورت ہوگی۔ نومولود بچوں کو بار بار دودھ پلانا اور کپڑے تبدیل کرنا ضروری ہے۔ اس لئے اگر آپ کو دوسروں کی

میری بچی جانتی ہے کہ میں اس کی ماں ہوں، وہ یہ بھی نہیں جانتی ہے کہ میں کوئی معذوری رکھتی ہوں

مدد کی ضرورت ہو تو اس میں جھجک محسوس نہ کریں تمام نئی مائیں نئی ضرورت پڑنے پر اپنے بچوں کے لئے دوسروں کی مدد حاصل کرتی ہیں۔

کوئی بات نہیں کہ آپ کو کتنی مدد کی ضرورت پڑسکتی ہے۔ آپ اس صورت میں بھی اپنے بچے کی ماں ہیں۔ بچے کی دیکھ بھال میں دوسروں کی مدد آپ کی حیثیت کم نہیں کرتی ہے۔ اس صورت میں بھی جب آپ جب بھی کسی اور فرد سے کسی بھی قسم کی مدد لیتی ہوں، آپ ہی اپنے بچے کی ضروریات، تحفظ اور بہتری کے لئے فیصلے کرنے والی واحد ہستی ہوتی ہیں، یہی وہ کام ہے جو ایک ماں کرتی ہے۔ اپنے بچے کو خود سے قریب رکھیں تا کہ وہ آپ کا چہرہ دیکھ سکے، آپ کی آواز سنے اور محسوس کرے، آپ کی بو سونگھے، اس طرح آپ کا بچہ جان لے گا کہ اس کی ماں کون ہے۔ یقیناً آپ۔

زندگی بھر کے لئے ایک تعلق بنانا

بچے اور ماں کے درمیان جو تعلق استوار ہوتا ہے وہ بچے کی جسمانی اور جذباتی نشوونما پر اثر انداز ہوتا ہے جیسے ہی یہ قریبی رشتہ تشکیل پاتا ہے بچہ اس میں اپنا تحفظ پانا سیکھ لیتا ہے اور اس سے آئندہ زندگی میں بچے کے لئے دوسروں کے ساتھ تعلق بنانا آسان رہے گا۔ اگرچہ گھر کے دوسرے افراد بچے کی دیکھ بھال میں آپ کی مدد کرسکتے ہیں لیکن یہ آپ کے لئے ضروری ہے کہ آپ کو ہی بنیادی دیکھ بھال کرنے والے کی حیثیت سے پہچانا جائے تاکہ آپ اپنے بچے کے ساتھ گہرا تعلق قائم کرسکیں۔

وہ عورتیں جو سیکھنے یا سمجھنے میں دشواری محسوس کرتی ہیں

بہت سی ایسی عورتیں جو سیکھنے یا سمجھنے میں دشواری محسوس کرتی تھیں، اچھی مائیں ہیں۔ آپ اپنے گھر والوں کے ساتھ تبادلہ خیال کرسکتی ہیں کہ آپ کو ایک اچھی ماں بننے کے لئے کن چیزوں کی ضرورت ہوگی۔

وہ چیزیں جن کے بارے میں سوچا جائے

چھوٹے بچوں کو دن اور رات دونوں وقت دودھ پلانے اور دیکھ بھال کی ضرورت ہوتی ہے لہٰذا آپ زیادہ دیر سو نہیں سکیں گی۔ کیونکہ آپ کا بچہ آپ کو رات میں کئی بار جگائے گا اور آپ دن میں خود کو انتہائی تھکا ہوا محسوس کریں گی لیکن دن میں بھی آپ کے بچے کو توجہ اور دیکھ بھال کی ضرورت ہوگی۔ کیا آپ مندرجہ ذیل کاموں کے لئے مدد مانگنے کے قابل ہیں؟

بچے بہت چھوٹے ہوتے ہیں، میں ڈرتی ہوں کہ میرا بچہ گر نہ جائے۔ ماما! کیا آپ میری مدد کریں گی؟

- بچے کو صاف ستھرا رکھنا

- یہ جاننا کہ بچے کو کب طبی دیکھ بھال کی ضرورت ہوتی ہے

- اگر ضروری ہو تو دودوا کی پیمائش

یہ یقینی بنانا کہ آپ کا بچہ مندرجہ ذیل سے محفوظ ہے

.... گرنے سے

.... جلنے سے

.... جانوروں سے

.... زہر سے

.... کوئی چیز نگلنے اور دم گھٹنے سے

.... ایسے حادثات سے جن میں ہڈی ٹوٹ سکتی ہو یا وہ زخمی ہوسکتا ہو

اگر آپ دودھ پلا سکتی ہیں تو پھر فارمولا دودھ کی کوئی ضرورت نہیں۔ لیکن آپ اپنا دودھ نہیں پلاتی ہیں تو پھر ضرورت ہوگی کہ بوتلیں صاف ستھری ہونا یقینی بنایا جائے اور متبادل دودھ درست طور پر تیار کیا جائے۔

بچے کو دودھ پلانا

ماں کا دودھ بہترین ہے

اگر ممکن ہو تو اپنے بچے کو خود دودھ پلائیں۔ بچے کی پیدائش کے بعد ابتدائی دو تین دن آپ کی چھاتیوں سے نکلنے والا زرد رنگ کا ابتدائی دودھ (Colostrum) بچے کے لئے بہترین غذا ہے۔ یہ بچے کے معدے کے لئے بہت اچھا ہے، اس میں بچے کی ضرورت کے مطابق ہر قسم کی غذائیت ہوتی ہے۔ یہ بچے کو بیماریوں سے تحفظ فراہم کرتا ہے۔ وہ بچے جنہیں اپنی ضرورت کے مطابق ماں کا دودھ ملتا ہے، **انہیں جڑی بوٹیوں، چائے یا میٹھے پانی کی کوئی ضرورت نہیں ہوتی ہے۔** اگر ممکن ہو تو ابتدائی چھ ماہ اپنے بچے کو سوائے اپنے دودھ کے کچھ اور نہ پلائیں۔ اگر آپ کے لئے بچے کو دودھ پلانا بے حد مشکل ہو تو اپنی چھاتیوں سے ہاتھوں کے ذریعے دودھ نکالیں (دیکھئے صفحہ 257 تا 258) تاکہ اسے دوسرے طریقے سے بچے کو پلا سکیں۔

بچے کو اپنا دودھ پلانا اس لئے اہم ہے کہ

- ماں کا دودھ بچے کی صحت مند اور مضبوط نشو و نما کے لئے مکمل غذا ہے۔
- بچے کو دودھ پلانا ولادت کے بعد رحم سے خون کا اخراج روکنے میں معاون ثابت ہوتا ہے۔
- ماں کا دودھ بچے میں ذیابیطس، کینسر جیسی بیماریوں اور ڈائریا اور نمونیا جیسے انفیکشنز کے خلاف مدافعت کی قوت پیدا کرتا ہے۔
- ماں کا دودھ پلانا، ماں اور بچے میں قربت اور تحفظ کا احساس پیدا کرتا ہے۔
- بچے کو دودھ پلانا کچھ عورتوں کو ابتدائی چھ ماہ حاملہ ہونے سے تحفظ فراہم کرتا ہے (دیکھئے صفحہ 199)۔
- ماں کا دودھ کسی قیمت کے بغیر ملتا ہے۔

معذوریوں والی بیشتر عورتیں اپنے بچوں کو دودھ پلا سکتی ہیں۔ کچھ معذور عورتوں کو ضرورت ہوتی ہے کہ انہیں بچے کو درست پوزیشن میں رکھنے کے لئے مدد دی جائے۔ کچھ معذوریاں عورت میں کمزوری اور تھکن کا احساس پیدا کرتی ہیں۔ یہ فیصلہ آپ کریں کہ آپ اپنے بچے کو دودھ پلا سکتی ہیں یا نہیں۔

دودھ کس طرح پلایا جائے

زیادہ تر بچے پیدائشی طور پر جانتے ہیں کہ دودھ کس طرح چوسا جائے۔ لیکن انہیں نپل منہ میں رکھنے کے لئے مدد کی ضرورت ہو سکتی ہے۔ بچے کے منہ میں چھاتی اس طرح ہونی چاہئے کہ نپل اس کے منہ میں اچھی طرح داخل ہو۔

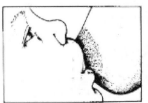

اس بچے کے منہ میں چھاتی کا مناسب حصہ نہیں ہے۔

اس بچے کے منہ میں چھاتی درست طور پر ہے۔

بچے کو کس طرح سنبھالیں

ابتداء میں دودھ پلانا تکلیف دہ ہوسکتا ہے لیکن اگر بچہ درست پوزیشن میں ہے تو آپ اس کے دودھ چوسنے کی عادی ہو جائیں گی اور درد ختم ہو جائے گا۔ اگر ایسا نہ ہو تو اپنی اور بچے کی پوزیشن تبدیل کر کے دیکھیں۔ یقینی بنائیں کہ بچے کے منہ میں چھاتی کا مناسب حصہ ہو۔ اگر آپ کے لئے دودھ پلانا تکلیف دہ ہو تو کسی صحت کارکن سے بات کریں۔ اس صورت میں کوئی اور مسئلہ بھی ہوسکتا ہے۔

کچھ عورتوں کے لئے دودھ پلانا اس صورت میں آسان ہوتا ہے کہ وہ کرسی پر یا بیڈ پر اس طرح بیٹھی ہوں کہ خود کو تھوڑا سا جھکا سکیں اور اپنے بازوؤں کو سہارا دے سکیں۔ اس طرح ان کے پیروں کو بھی آرام ملتا ہے۔ یقینی بنائیں کہ بچے کو بھی اچھی طرح سہارا مل رہا ہو۔

بیشتر عورتیں اپنے خاندان اور کمیونٹی میں دوسری عورتوں کو دیکھ کر اپنے بچوں کو دودھ پلانا سیکھ لیتی ہیں۔ اگر کمیونٹی میں کوئی اور عورت آپ کی طرح معذوری رکھتی ہے اور ماں بن چکی ہے تو اس سے مشورہ لیں۔

بہت سی معذور عورتیں آرام دہ پوزیشن جاننے پر اپنے بچوں کو دودھ پلا سکتی ہیں۔

اگر آپ اپنے بازوؤں اور اوپری جسم کو بخوبی استعمال کرسکتی ہیں تو آپ کسی الجھن کے بغیر اپنے بچے کو دودھ پلا سکتی ہیں۔ یقینی بنائیں کہ بچے کے سر کو اچھی طرح سہارا ملے اور یہ کہ آپ اس انداز سے بیٹھیں یا لیٹیں جو آپ کے لئے آرام دہ ہو۔

بہت سی عورتوں کے لئے آسان ہوتا ہے کہ وہ اپنے بچے کے ساتھ پہلو کے بل لیٹیں اور انہیں تکیے یا گول تہہ کئے ہوئے کپڑے سے سہارا دیں

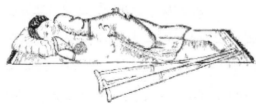

بچے کے نیچے تکیہ یا گول تہہ کیا ہوا کپڑا رکھیں

اگر آپ اپنے بازوؤں اور اوپری جسم کا محدود استعمال کر پاتی ہوں تو بچے کو دودھ پلانے کے لئے کم از کم ایک آرام دہ پوزیشن تلاش کریں۔ اگر ضروری ہو تو کسی سے مدد طلب کریں۔ ذیل میں چند تجاویز یہ ہیں کہ

اگر آپ اپنے بازو اور اوپری جسم استعمال نہ کر سکتی ہوں تو آپ گھر کے کسی فرد یا کسی سہیلی کی مدد سے اپنے بچے کو دودھ پلا سکتی ہیں۔ انہیں بتائیں کہ بچے کو کس انداز سے رکھا جائے کہ آپ اسے دودھ پلا سکیں۔ اگر ضروری ہو تو ان سے کہیں کہ وہ بچے کو اس پوزیشن میں تھامے رہیں خصوصاً بچے کے سر کو۔ اگرچہ آپ بچے کو اپنے بازوؤں میں تھامے ہوئے نہ بھی ہوں تو اب بھی بچہ آپ کا چہرہ دیکھ سکے اور آپ کے جسم کی گرمی اور مانوس بو محسوس کرسکے گا۔

بچے کے سر کو اپنے بازو کے خم میں یا اپنی چھاتی سے نزدیک تکئیے پر رکھیں

اگر آپ کے لئے اپنی چھاتی سنبھالنا مشکل ہے تو ایسی ''برا'' (بریزر) پہنیں جس میں نپل کے سائز کے مطابق سوراخ کیا گیا ہو۔ آپ ''نرسنگ برا'' بھی خرید سکتی ہیں جو اسی مقصد کے لئے بنائی جاتی ہے جس میں بریسٹ فیڈنگ کے لئے نپل کو کھولا جا سکتا ہے۔ یا آپ اپنے سینے کے گرد اس طرح کپڑا لپیٹ سکتی ہیں کہ اس میں نپل کے لئے سوراخ کیا گیا ہو۔ آپ اپنی چھاتیوں کے نیچے اپنے جسم پر کپڑا باندھ سکتی ہیں۔

اگر آپ دودھ نہ پلاسکتی ہوں

اگر آپ اپنے بچے کو دودھ پلانے کے قابل نہیں ہیں مگر اپنی چھاتیوں سے اپنے ہاتھ کے ذریعے دودھ نکال سکتی ہیں تو بوتل یا کپ کے ذریعے یہ دودھ بچے کو پلائیں۔ اگر آپ خود دودھ نہیں نکال سکتی ہوں تو کسی ایسے فرد سے مدد مانگیں جس پر آپ اعتماد کرتی ہوں۔

ہاتھ سے دودھ کس طرح نکالا جائے

1- ایک برتن اور اس کے ڈھکن کو صابن اور صاف پانی سے اچھی طرح دھوئیں اور سورج کی روشنی میں خشک کریں۔ اگر ممکن ہو استعمال سے پہلے برتن میں گرم پانی ڈال کر اسے کھنگال لیں۔ اس سے برتن میں موجود جراثیم ہلاک ہو جائیں گے اور نکالا ہوا دودھ بچے کے لئے محفوظ رہے گا۔

2- اپنے ہاتھ اچھی طرح دھوئیں۔

3- اپنی انگلیاں اور انگوٹھا اپنی چھاتی کے سیاہ حصے کے سرے (areola) پر رکھیں اور اسے سینے کی طرف دبائیں۔ انگلیوں کو نرمی سے ایک ساتھ دبائیں اور انہیں نپل کی طرف رول کریں، نپل کو نوچیں یا کھینچیں نہیں۔ دودھ نکالنے سے تکلیف نہیں ہونی چاہئے۔ انگلیوں کو پوری طرح سیاہ حصے کے گرد گھمائیں تا کہ دودھ پوری چھاتی سے باہر نکل سکے۔ ہر چھاتی سے اسی طرح دودھ نکال لیں۔

ابتداء میں زیادہ مقدار میں دودھ نہیں نکلے گا لیکن مشق ہونے پر زیادہ دودھ نکلے گا، اگر ممکن ہو تو ہر تین سے چار گھنٹوں میں یا 24 گھنٹوں میں کم سے کم 8 بار دودھ نکالیں تا کہ دودھ کی اچھی مقدار میں فراہمی یقینی رہے۔ اگر آپ پرسکون اور اطمینان بخش جگہ ہوں اور خود بھی مطمئن ہوں تو آپ زیادہ دودھ نکال سکتی ہیں۔ دودھ نکالتے ہوئے اپنے بچے کے بارے میں سوچنا اس عمل کو آسان بنا سکتا ہے۔ اگر دودھ نکالنے کی ابتداء مشکل ہو تو اپنی چھاتیوں کے گرد گرم، نم کپڑا یا تولیہ رکھ کر آزمائیں اور دودھ نکالنے سے پہلے چھاتیوں کا مساج کریں۔

آپ دودھ نکالنے کے لئے بریسٹ پمپ بھی استعمال کر سکتی ہیں۔ بعض مقامات پر کلینک اور طبی مراکز عارضی طور پر یا کرائے پر الیکٹرک پمپ بھی فراہم کرتے ہیں۔ سادہ دستی پمپ بھی کم قیمت پر مل سکتے ہیں۔

دودھ نکالنے کے لئے گرم بوتل استعمال کرنا

یہ طریقہ اس صورت میں بہترین ہے جب چھاتیاں دودھ سے خوب بھری ہوں یا بہت پُر درد ہوں،ایسا بچے کی پیدائش کے فوراً بعد یا پچے ہوئے نپل کے باعث یا بریسٹ انفیکشن کی وجہ سے بھی ہوسکتا ہے۔اگر آپ اپنی چھاتی اور بوتل گرفت نہ کرسکتی ہوں تو کسی قابل اعتماد فرد سے مدد کے لئے کہیں۔

3 سے 4 سینٹی
میٹر چوڑا منہ

1-   تین سے چار سینٹی میٹر چوڑے منہ کی ایک بڑی شیشے کی بوتل کو اچھی طرح دھولیں۔اس میں گرم پانی بھرلیں۔بوتل میں گرم پانی ایک دم نہ بھریں تا کہ وہ ٹوٹنے سے محفوظ رہے۔گرم پانی چندمنٹ بوتل میں رہنے دیں اور پھر پانی بوتل سے انڈیل دیں۔

2-   بوتل کے منہ اور ذیلی حصے کوصاف ٹھنڈے پانی کی مدد سے ٹھنڈا کرلیں تا کہ یہ حصہ چھاتی سے چھونے کے بعد آپ کی جلد کی نہ جلائے۔

3-   بوتل کے منہ کو نپل کے اوپر اچھی طرح جما دیں۔بوتل کو چندمنٹ اسی طرح تھامے رہیں۔جیسے ہی یہ ٹھنڈی ہوگی چھاتی سے دودھ نکل کراس میں جمع ہونے لگے گا۔

4-   جب دودھ نکلنے کی رفتار کم ہو جائے انگلیوں کی مدد سے چھاتی دبا کر بوتل الگ کریں۔

5-   یہی عمل دوسری چھاتی کے ساتھ دہرائیں۔

دودھ کس طرح رکھیں؟

اپنا دودھ کسی صاف اور بند برتن میں رکھیں۔ آپ دودھ اسی بوتل میں بھی رکھ سکتی ہیں جو دودھ نکالنے کے لئے استعمال کی گئی ہو۔

دودھ کا یہ برتن سورج کی روشنی سے دور کسی ٹھنڈی جگہ پر رکھیں۔ آپ اس برتن کو ٹھنڈے پانی میں رکھ کر دودھ کو ٹھنڈا رکھ سکتی ہیں یا اچھی طرح بند یہ برتن گیلی ریت میں بھی دبایا جاسکتا ہے یا اس کے گرد کپڑا لپیٹ کر سے سارا وقت گیلا رکھا جائے۔ ٹھنڈا بریسٹ ملک تقریباً 12 گھنٹے محفوظ رہ سکتا ہے۔

یہ دودھ ریفریجریٹر میں بھی رکھا جاسکتا ہے۔ ریفریجریٹر میں دودھ شیشے کے برتن میں دو سے تین دن رکھا جاسکتا ہے۔ دودھ میں موجود کریم الگ ہو جائے گی لہٰذا بچے کو دودھ پلانے سے پہلے برتن کو اچھی طرح ہلائیں تا کہ یہ کریم دودھ میں شامل ہو جائے پھر برتن کو گرم پانی میں رکھ کر ہلکا سا گرم کرلیا جائے۔ پلانے سے قبل دودھ کے چند قطرے اپنے بازو پر ٹپکا کر اطمینان کرلیں کہ یہ زیادہ گرم نہ ہو۔ دودھ گرم نہیں ہونا چاہیئے، یہ آپ کے بازو کے درجہ حرارت کے مطابق ہی محسوس ہونا چاہیئے۔

برتن کسی ٹھنڈی جگہ
مثلاً ٹھنڈے پانی سے بھرے
مٹی کے برتن میں بھی
رکھا جاسکتا ہے

تنبیہہ	دودھ ٹھنڈا نہ رکھا جائے تو خراب ہو جائے گا، اسے پھینک دیں، خراب دودھ بچے کو انتہائی بیمار کر سکتا ہے۔

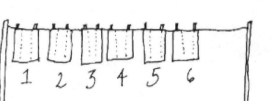

اگر بچے کا وزن بڑھ رہا ہو اور وہ دن میں کم از کم چھ بار پیشاب کر رہا ہو تو اسے مناسب مقدار میں دودھ مل رہا ہے۔

چھاتیوں کی دیکھ بھال

تکلیف دہ چھاتیاں

چھاتیوں میں درد تکلیف زدہ نپل یا پوری طرح بھری سخت چھاتیوں کے باعث ہو سکتا ہے۔ یہ درد عموماً ایک یا دو دن میں ختم ہو جاتا ہے۔ بچے کو دودھ پلانا جاری رکھنا اہم ہے خواہ چھاتیوں میں تکلیف ہی کیوں نہ ہو، بچے کو چھاتیاں چوسنے دیں۔

تکلیف دہ یا چٹختے ہوئے نپلز

نپل اس صورت میں تکلیف دہ یا چٹخ سکتے ہیں جب بچہ دودھ پیتے ہوئے نپل اور چھاتی کو اچھی طرح منہ میں لینے کے بجائے صرف نپل کو ہی چوسے۔

بچاؤ اور علاج

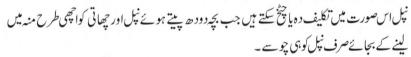

- بچہ جتنی دیر اور جتنی بار چاہے اسے دودھ پلائیں۔
- جب بچہ دودھ پینا بند کر دے تو دودھ کے چند قطرے نکال کر نپل پر ملیں۔
- اگر انفیکشن نہ ہو تو اپنی چھاتیوں پر صابن، کریم وغیرہ نہ استعمال کریں۔ آپ کا جسم خود ہی آئل بناتا ہے جو نپلز کو صاف اور نرم رکھتا ہے۔
- تنگ اور کھردرے کپڑے مت پہنیں۔
- بچے کے دودھ چوسنے پر درد حد سے زیادہ ہو تو اپنا دودھ ہاتھ سے نکال کر چمچے کے ذریعے بچے کو پلائیں۔
- نپل میں ہونے والی تڑخن دو دن میں صحیح ہو جاتی ہے۔
- اپنی چھاتیوں کو سخت اور دودھ سے لبریز مت ہونے دیں۔ اگر آپ کا دودھ بچے کی ضرورت سے زیادہ ہے تو بچے کو دودھ پلانے کے بعد اپنی چھاتیوں کو کپڑے یا تولیے سے ڈھانپ کر اپنی چھاتیوں کو ہاتھ سے خالی کریں (دیکھئے صفحہ 257 تا 258)۔ چند ہفتے کے بعد آپ کا جسم دودھ کی مناسب مقدار بنانے لگے گا اور پھر آپ کی چھاتیاں حد سے زیادہ نہیں بھریں گی۔

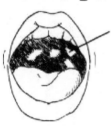

تھرش (Thrush)

بچوں کے حلق کا ایک مرض

تھرش (Thrush)

اگر بچہ دودھ چوستے ہوئے درست انداز سے ہو اور آپ کو اپنے نپلز میں درد محسوس ہو جو ہفتہ بھر سے زائد ہے تو اس کا سبب تھرش ہوسکتا ہے (یہ نپلز یا بچے کے منہ میں ہونے والا یسٹ انفیکشن ہے۔ اس میں آپ کے نپلز میں خارش بھی ہوسکتی ہے اور آپ چھاتی چبھتا ہوا پُرسوزش درد بھی محسوس کرسکتی ہیں۔ بچے کے منہ میں سفید دھبے یا سرخی ہوسکتی ہے۔ اگر بچے کے منہ میں تکلیف ہو تو وہ رو کر ہی اس کا اظہار کرے گا۔

تھرش نپلز میں تکلیف اور چبھن اور چھاتی کے انفیکشن کا سبب بن سکتا ہے۔ ماں اور بچے دونوں کا علاج ہونا چاہیئے۔

تھرش کا علاج کیسے کریں

0.25 فی صد کا محلول بنانے کے لئے Gentian violet اور پانی (دیکھئے صفحہ 344) ملائیے۔ اگر آپ کے پاس Gentian Violet کا ایک فیصد کا محلول ہے تو اس کا ایک چائے کا چمچہ، تین چمچے صاف پانی میں ملائیں۔

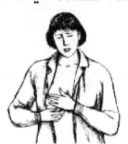

اپنے نپلز اور بچے کے منہ میں موجود سفید دھبوں پر یہ محلول کسی صاف کپڑے یا انگلیوں سے لگائیے۔ یہ عمل پانچ دن، روزانہ ایک بار کیجئے۔ Gentian Violet کپڑوں پر دھبے ڈالے اور بچے کے منہ اور آپ کے نپلز کو او دا کر دے گا۔ یہ عام سی بات ہے۔ آپ بچے کو دودھ پلاتی رہیں۔ اگر تھرش تین دن میں بہتر نہ ہوتو او بنایا ہوا محلول استعمال کرنا چھوڑ دیں اور طبی مشورہ حاصل کیجئے۔

بریسٹ انفیکشن (Mastitis)

تکلیف دہ چھاتیاں، زخم یا پھٹے ہوئے نپلز چھاتی کے اندرونی انفیکشن میں بدل سکتے ہیں۔

علامات:

● چھاتی کے کچھ حصے گرم، سرخ، سوجے ہوئے اور انتہائی تکلیف دہ ہوجاتے ہیں۔

● بخار یا سردی۔

● بغل کے لمف نوڈ عموماً درد کرتے ہیں اور سوج جاتے ہیں۔

چھاتی میں تکلیف دہ غدود (Absess) جو بعض اوقات پھٹ جاتے ہیں اور ان سے پیپ خارج ہونے لگتا ہے۔

علاج

- بچے کو دودھ پلانا جاری رکھیں۔ بچے کو متاثرہ چھاتی پہلے دیں یا متاثرہ چھاتی کا دودھ ہاتھ سے نکالیں انفیکشن بچے میں منتقل نہیں ہوگا۔

- آرام کریں اور بڑی مقدار میں مائعات پئیں۔

- دودھ پلانے سے پہلے ہر بار متاثرہ چھاتی پر پندرہ منٹ گرم سنکائی کریں۔ درد میں کمی کے لئے دودھ پلانے کے درمیانی وقفے میں ٹھنڈی سنکائی کریں۔

- بچے کو دودھ پلانے سے قبل تکلیف دہ چھاتی پر مساج کریں۔

- درد کے لئے پیراسیٹامول لیں (دیکھئے صفحہ 350)

- کوئی اینٹی بایوٹک استعمال کریں۔ Dicloxacillin استعمال کے لئے بہترین ہے (دیکھئے صفحہ 341) سات سے دس دن، روزانہ چار بار 500 ملی گرام دوا کھائیے۔ اگر یہ نہ ملے اور آپ پینسلین سے الرجک ہیں تو erythromycin (دیکھئے صفحہ 343) سات دن روزانہ دن میں چار مرتبہ 500 ملی گرام کھائیے۔

ایچ آئی وی/ایڈز اور بریسٹ فیڈنگ

ایچ آئی وی/ایڈز کے متعلق عمومی معلومات کے لئے پڑھئے صفحہ 169

ایچ آئی وی کچھ ماؤں سے ان کے بچوں کو بریسٹ فیڈنگ کے ذریعے منتقل ہو جاتا ہے لیکن ایچ آئی وی میں مبتلا کچھ عورتیں اپنے بچوں کو دودھ پلاتی ہیں لیکن انہیں ایچ آئی وی منتقل نہیں ہوتا۔ ایچ آئی وی ماں سے کچھ بچوں تک منتقل ہو جاتا ہے اور کچھ کو نہیں۔ کیوں؟ کوئی بھی اس کا سبب نہیں جانتا ہے۔ ایچ آئی وی امکانی طور پر بریسٹ فیڈنگ کے دوران زیادہ آسانی سے اس وقت منتقل ہوتا ہے جب

آپ جو بھی فیصلہ کریں مگر اپنے بچے کے ایچ آئی وی زدہ ہونے پر خود کو الزام مت دیں۔ اس وقت یہ جاننے کا کوئی طریقہ نہیں ہے کہ آپ اپنے بچے کو کس طرح تحفظ فراہم کریں۔

- ماں حال ہی میں ایچ آئی وی سے متاثر ہوئی ہو۔

- ماں ایڈز کی شدت میں مبتلا ہو۔

- ماں بریسٹ مِلک کے ساتھ فارمولا مِلک یا دیگر مائع اشیاء دیتی ہو۔

- ماں کے نپلز پھٹے ہوئے ہوں یا اسے بریسٹ انفیکشن ہو۔

- بچے کے منہ میں تھرش ہو۔

بیش تر ماؤں کے لئے، حتیٰ کہ ایچ آئی وی ماؤں کے لئے بھی بریسٹ فیڈنگ بچوں کو خوراک فراہم کرنے کا محفوظ ترین طریقہ ہے۔ ان علاقوں میں جہاں پانی محفوظ نہیں ہے بہت سے بچے ڈائریا کے باعث مر جاتے ہیں اور جب لوگ فارمولا مِلک کے اخراجات برداشت نہیں کر پاتے ہیں تو بچے ناقص غذائیت کے باعث مر جاتے ہیں۔

ایچ آئی وی ہونے پر بریسٹ فیڈنگ

ایک عورت جس کا علاج ایچ آئی وی کے لئے ادویات سے کیا گیا ہو بریسٹ فیڈنگ کے دوران اپنے بچے میں وائرس منتقلی کے امکانات بہت کم رکھتی ہے اور اگر ایسی عورت ART ادویات نہیں لے رہی ہو تو جب بھی بریسٹ فیڈنگ کو محفوظ عمل بنا سکتی ہے۔

● ابتدائی چھ ماہ صرف ماں کا دودھ ہی دیا جائے۔ فارمولا دودھ، چائے یا دیگر سیال اشیاء پر پلنے والے بچے ماں کا دودھ پینے والے بچوں کے مقابلے میں متاثر ہونے کا زیادہ امکان رکھتے ہیں۔ ایک چھوٹے بچے کے لئے دوسری غذائیں یا مائعات ہضم کرنا مشکل ہوتا ہے اور یہ چیزیں بچے کے معدے کے استر کو متاثر کر کتی ہیں۔ اس سے ایچ آئی وی زیادہ آسانی سے منتقل ہو سکتا ہے۔

● چھ ماہ بعد بریسٹ فیڈنگ روک دیں لیکن ایسا اچانک مت کریں۔ بچے کا دودھ رکنے کے عمل میں کئی دن لگتے ہیں (دیکھئے صفحہ 265)۔

● نپلز کو چٹخنے سے بچانے کے لئے دودھ پلاتے ہوئے بچے کو درست پوزیشن میں رکھیں۔

● تھرش، چٹخے ہوئے نپلز، بریسٹ انفیکشن کا علاج فوری طور پر کرائیں۔

● انفیکشن زدہ چھاتی سے بچے کو دودھ نہ پلائیں بلکہ اس کا دودھ نکال کر پھینک دیں اور اپنے بچے کو انفیکشن ٹھیک ہونے تک دوسری چھاتی سے دودھ پلائیں۔

● ماں کے دودھ میں ایچ آئی وی ختم کرنے کے لئے ماں کے دودھ کو ابالا جا سکتا ہے، پھر اسے ٹھنڈا کر کے کپ یا بوتل کے ذریعے بچے کو پلائیں۔ اس طرح کام تو بڑھتا ہے لیکن صاف پانی، ایندھن اور مدد موجود ہو تو یہ کیا جا سکتا ہے۔

ماں کا دودھ پاسچورائز کرنا

1- چھاتی کے دودھ کا برتن، پانی سے بھرے برتن میں رکھیں۔

2- اس برتن کے پانی کو اُبلنے دیں۔

3- برتن کو فوری طور پر آگ سے ہٹالیں۔

4- بچے کو دودھ پلانے سے پہلے دودھ کو ٹھنڈا ہونے دیں۔

دودھ کو پاسچورائزیشن کے بعد چند گھنٹے میں استعمال کرلیں، ماں کے دودھ کو اُبالا نہ جائے۔

دوسری اقسام کا دودھ استعمال کرنا

ماں کا دودھ بچے کے لئے بہترین ہے لیکن یہ ممکن نہ ہو تو فارمولا دودھ (مصنوعی دودھ)، ماں کے دودھ کا محفوظ متبادل ہو سکتا ہے۔ اگر آپ فارمولا دودھ نہیں پلا سکتیں تو شاید آپ کے خاندان یا سہیلیوں میں کوئی ایسی عورت ہو جو ایچ آئی وی/ایڈز نہ رکھتی ہو اور آپ کے بچے کو دودھ پلا سکے یا آپ اپنے بچے کو دوسرے جانوروں کا دودھ دے سکتی ہیں۔

بچے کو دیگر جانوروں کا دودھ دینا

گائے، بکری، اونٹ وغیرہ	بھیڑ یا بھینس کا دودھ
100 ملی لیٹر تازہ دودھ میں 50 ملی لیٹر پانی اور 10 گرام (2 چائے کے چمچے) شکر ملائیں۔	50 ملی لیٹر تازہ دودھ میں 50 ملی لیٹر صاف پانی اور 5 گرام شکر ملائیں۔

اس آمیزہ کو گرم کریں اور پھر ابال آتے ہی اسے آگ سے ہٹا دیں۔ اب اسے ٹھنڈا کریں اور بچے کو فوری طور پر پلا دیں۔

جانوروں کے دودھ میں وہ سارے وٹامنز نہیں ہوتے ہیں جو ایک افزائش پاتے ہوئے بچے کی ضرورت ہیں۔ اس لئے بچے کو مسلی ہوئی سبزیوں، پھلوں اور دیگر غذاؤں پر مشتمل اشیاء چھ ماہ کی عمر سے شروع کی جائیں۔

بچے کو کپ یا بوتل سے خوراک فراہم کرنا

اگر آپ بچے کو دودھ نہیں پلاتی ہیں تو آپ اپنے بچے کو کپ یا بوتل کی مدد سے چھاتی کا دودھ، جانوروں کا دودھ یا فارمولا دودھ پلا سکتی ہیں۔ اگر آپ کپ نہیں پکڑ نہیں سکتی تو پھر کسی سے مدد کے لئے کہیں۔

بچے کو کپ کے ذریعے خوراک فراہم کرنا

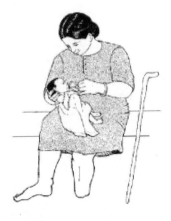

1. چھوٹا اور انتہائی صاف کپ لیں۔ اگر اسے ابالنا ممکن نہیں تو کپ کو صابن اور صابن پانی سے دھولیں۔

2. بچے کو آپ کے زانو پر ہونا چاہئے۔

3. کپ بچے کے ہونٹوں کے پاس اس طرح لائیں کہ دودھ بچے کے ہونٹوں کو چھوئے، کپ کو بچے کے نچلے ہونٹ پر ہلکا سا تھم رکھیئے جبکہ اوپری سطح کو بچے کے اوپر کے ہونٹ سے چھونے دیں۔

4. بچے کے منہ میں دودھ مت انڈیلیں۔ بچے کو کپ سے دودھ اپنے منہ میں جذب کرنے دیں۔

بچے کو بوتل سے دودھ پلانا

بچے کو بوتل سے دودھ پلانا اس وقت تک محفوظ نہیں جب تک آپ ان تمام سوالات کے جواب "ہاں" میں نہیں دے سکیں ۔

- کیا اس علاقے میں صاف پانی کی مستقل فراہمی کا کوئی ذریعہ ہے؟
- کیا وہاں پانی ابالنے کے لئے ایندھن کی مستقل فراہمی ہے؟
- کیا آپ یا آپ کے گھر والے بہت سی نئی بوتلیں اور نپلز خریدنے کی مالی حیثیت رکھتے ہیں؟
- کیا آپ یا آپ کے گھر والے انفنٹ فارمولا، ڈبہ بند دودھ یا جانوروں کا دودھ کم از کم چھ ماہ خریدنے کے لئے مناسب رقم رکھتے ہیں۔
- کیا آپ اور آپ کے گھر والے جانتے ہیں کہ بوتلوں اور نپلز کو کس طرح صاف کیا جائے اور دودھ کس طرح تیار کیا جائے۔

جب آپ فارمولا دودھ یا مویشی کا دودھ دیں تو ہر چیز نہایت ہی صاف ہونی چاہئے ۔ کپ، چمچہ، بوتل، ربڑ کے نپل اور دودھ رکھنے کا برتن ہر چیز مکمل طور پر صاف شدہ ہو اور استعمال سے قبل کم از کم ایک منٹ میں ابالی جائے ۔ تیار دودھ، ڈبے کا دودھ یا مویشی کا دودھ کمرے کے درجہ حرارت پر دو گھنٹے سے زیادہ نہ چھوڑا جائے ۔ فارمولا دودھ کو ریفریجریٹر میں بارہ گھنٹے تک رکھا جا سکتا ہے ۔

ڈکار لینے میں بچے کی مدد کرنا

دودھ پینے کے دوران کچھ بچے ہوا نگل لیتے ہیں جو انہیں پریشان کرتی ہے ۔ آپ اگر بچے کو اپنے کندھے یا سینے سے لگا کر اس کی پشت سہلائیں تو وہ ہوا خارج ہو سکتی ہے ۔ آپ بچے کو اپنی گود میں بٹھا یا لٹا کر بھی اس کی پیٹھ سہلا سکتی ہیں ۔

اگر آپ کا صرف ایک بازو ہے یا بازو میں محدود طاقت ہے تو بچے کو اپنے گھٹنے پر اس طرح بٹھائیں کہ بچے کا چہرہ آپ سے دور اور آپ کا صحت مند بازو بچے کے سینے کے گرد لپٹا ہو ۔ اب آگے اور پیچھے کی طرف جھکیں حتیٰ کہ بچہ ڈکار لے اور پُرسکون ہو جائے ۔

بچے کو غذا دینا

جب بچہ چھ ماہ کا ہو جائے تو آپ اسے اپنے دودھ کے ساتھ دوسری غذائیں بھی دے سکتی ہیں۔ ہمیشہ پہلے اپنا دودھ پلائیں پھر دوسری غذائیں دیں۔ اپنی معمول کی غذاؤں سے (دیکھئے صفحہ 87) ابتداء اچھی ہے لیکن بچے کے لئے ان کا اچھی طرح پکا ہوا اور مسلا ہوا ہونا ضروری ہے۔ ابتداء میں بچے کے لئے نگلنا آسان بنانے کے لئے ان میں ماں کا دودھ ملایا جا سکتا ہے۔

چند دن بعد (دیکھئے صفحہ 87) دیگر غذائیں شامل کیجئے لیکن ابتداء میں نئی غذا کی معمولی مقدار لیں اور اسے ایک بار دیں ورنہ ممکن ہے کہ بچے کو یہ ہضم کرنے میں دشواری ہو۔ زیادہ اہم بات ایسی غذاؤں کا اضافہ ہے جو اضافی توانائی دیں (جیسے تیل) اور جب ممکن ہو اضافی آئرن، (جیسے گہرے سبز رنگ کی سبزیاں) اپنے بچے کو صحت مند غذائیں کھلانے کے لئے مزید معلومات کے لئے دیکھئے ''جہاں عورتوں کے لئے ڈاکٹر نہ ہو'' صفحہ 106۔

یاد رکھیئے کہ چھوٹے بچے کا پیٹ بھی چھوٹا ہوتا ہے اور ایک بار میں زیادہ غذا نہیں کھا سکتا ہے لہٰذا اسے کئی بار کھلائیں اگر ممکن ہو تو دن بھر میں 5 سے 6 بار اور بنیادی غذا میں زیادہ توانائی والے اجزاء یا غذائیں ملائیں۔

اگر آپ اس کے کھانے کے وقت پر ہر چیز تیار رکھیں تو بچہ خوش خرم اور پرسکون رہے گا۔ اگر آپ یہ انتظار کریں کہ بچے کو بھوک لگے اور وہ روئے تو یہ آپ کے لئے بہت مشکل ہو گا کہ آپ خود کو غذا کی تیاری کے دوران مطمئن رکھ سکیں۔

اگر آپ بخوبی دیکھ نہیں سکتی ہوں

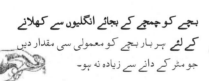

ہمیشہ یاد رکھیں کہ اپنے ہاتھ صابن اور پانی سے اچھی طرح دھولیں۔

بچے کو چمچے کے بجائے انگلیوں سے کھلانے

کے لئے ہر بار بچے کو معمولی سی مقدار دیں جو مٹر کے دانے سے زیادہ نہ ہو۔

بچے کو چمچے سے کھلانے کے لئے

1- کھانے کی تھوڑی سی مقدار چمچہ پر رکھنے کے لئے ایک ہاتھ استعمال کریں، چمچے کو اس طرح پکڑیں کہ چمچے میں موجود اضافی غذا اسی ہاتھ کی انگلی سے کم کرسکیں۔

2- اپنے دوسرے ہاتھ کا انگوٹھا بچے کی تھوڑی پر منہ کے عین نیچے رکھیں۔ انگوٹھے کو استعمال کرتے ہوئے غذا چمچے کے ذریعے بچے کے منہ میں ڈالیں۔

بچے کو کسی محفوظ جگہ آرام دہ پوزیشن میں بٹھائیں تاکہ وہ لڑھکیں یا گریں نہیں، کھانے کو کسی بڑے پیالے میں یا برتن میں اس طرح رکھیں کہ بچہ اس برتن کو لڑھکا نہ سکے۔

جب بچہ اتنا بڑا ہو جائے کہ وہ خود کھا سکے تو ابتداء میں وہ اچھا خاصا ہنگامہ مچا سکتا ہے۔ ممکن ہے کہ آپ کو گھر کے کسی فرد، دوست یا پڑوسی سے یہ جاننے کے لئے مدد لینی پڑے کہ کھانا کہاں کہاں گرا ہے تا کہ آپ اسے صاف کر کریں۔ صبر سے کام لینے کی کوشش کریں۔ بچہ جوں جوں بڑا ہو گا اسے خود پر زیادہ اعتماد ہو گا اور وہ کم سے کم کھانا خراب کرے گا۔

اگر آپ اوپری جسم کی محدود قوت اور ربط رکھتی ہیں تو آپ اپنے بچے کو اس کے پہلو میں بیٹھ کر کھانا کھلا سکتی ہیں۔ اس طرف آپ کو کھانا کھلانے کے لئے زیادہ آگے نہیں جھکنا پڑے گا لیکن اگر اسے خود کھانا نہیں کھلا سکتی ہیں تو آپ اس کے جس حد تک ممکن ہو قریب بیٹھ سکتی ہیں اور اس سے باتیں کر سکتی ہیں جبکہ کوئی اور اسے کھانا کھلائے۔ اس طرح وہ سوچے گا کہ یہ آپ ہی ہیں جو اسے بھوک لگنے پر کھانا دیتی ہیں۔

جب بچہ ایک برس یا اس سے بڑا ہو

جب آپ کا بچہ ایک برس کا ہو جائے تو وہ وہی غذا کھا سکتا ہے جو گھر کے دوسرے افراد کھاتے ہیں اس کے ساتھ اس کو معاون غذائیں بھی دیں جس سے اُسے بہترین توانائی پروٹین، وٹامنز، آئرن اور معدنیات ملیں تا کہ وہ صحت مند نشو و نما

پائے۔ یہ یقینی بنانے کے لئے کہ بچے کو مناسب مقدار کھانے کو ملے، اسے اس کی ڈش میں کھانا دیں اور اسے اپنا کھانا کھانے میں جو بھی وقت لگے وہ وقت دیں۔

بچے کو مطمئن کرنا

بچے کے لئے ضروری ہے کہ وہ اپنی ماں کے ساتھ خود محفوظ محسوس کرے۔ ماں کے لئے ضروری ہے کہ جب اس کا بچہ ناخوش ہو تو وہ اسے خوش اور مطمئن کرے۔ اگر آپ کا بچہ رونا شروع کر دے اور آپ اسے فوری طور پر اٹھا نہیں سکتی ہوں تو کوئی اور اسے آپ کے پاس لا سکتا ہے۔ پھر آپ کا بچہ آپ کا چہرہ دیکھ اور آپ کے محبت بھرے الفاظ سن سکتا ہے۔ بچے کے لئے یہ اہم ہے خواہ اسے آپ خود نہ اٹھا سکیں یا اسے اپنی گود میں نہ لے سکیں۔

اگر آپ اپنے بازو استعمال نہ کر پاتی ہوں یا اپنے بچے کو سنبھال نہ سکتی ہوں

دو طریقے ہیں جن سے آپ اپنے بچے کو مطمئن کر سکتی ہیں۔

یا کوئی آپ کے پیچھے بیٹھ کر بچے کو آپ کے سامنے اس طرح رکھے کہ آپ اس سے باتیں کر کے اسے مطمئن کر سکیں۔

کوئی بچہ کو آپ کے اس حد تک قریب لائے کہ وہ آپ کی آواز سن سکے اور اپنی ماں کی حیثیت سے آپ کی بو پہچان سکے۔

اگر آپ اچھی طرح سن نہ سکتی ہوں

ایک صحت مند بچہ جب بھوکا ہو یا بہتر محسوس نہ کر رہا ہو تو عام طور پر مخصوص آوازیں نکالتا ہے۔ اس لئے اگر آپ اچھی طرح سن نہ سکتی ہوں تو آپ کو جس حد تک ممکن ہو اپنے بچے کے قریب رہنے کی ضرورت ہے تا کہ آپ دیکھ سکیں کہ کب بچے کو آپ کی توجہ کی ضرورت ہے۔ رات کے وقت آپ اپنے بچے سے جس حد تک ممکن ہو قریب ہو کر سوئیں تا کہ آپ اس کی حرکت محسوس کر سکیں اور دن کے وقت بچے کو خود سے قریب رکھیں۔

یہ جاننے کے لئے کہ وہ کس طرح محسوس کرتا ہے، آپ کو دوسری عورتوں سے زیادہ اپنے بچے کا دھیان رکھنا ہوگا۔ وہ جلد ہی جان لے گا کہ آپ کس طرح سونگھتی اور محسوس کرتی ہیں اور آپ کی آواز کیسی ہے۔ یہ اسے آپ سے بہت ہی قریب اور محفوظ ہونے کا احساس دلائے گا۔

اگر آپ بول کر نہیں بلکہ علامتی یا اشاروں کی زبان میں بات کرتی ہیں تو آپ اپنے بچے سے علامتی زبان میں ہی بات کریں خواہ وہ سن بھی سکتا ہو۔ اس طرح آپ اور آپ کا بچہ ساری زندگی ابلاغ کر سکے گا۔ اس کے ساتھ ہی آپ کا بچہ گھر والوں اور دوستوں کے ساتھ بھی وقت گزارے جو بہرے نہیں ہیں تا کہ آپ کا بچہ بول بھی سکے۔

> میرا نواسا اس قدر اسمارٹ ہے کہ اس نے اپنی ماں سے علامتی زبان استعمال کرنا سیکھ لیا اور میں اسے سکھا رہی ہوں کہ کس طرح بولا جاتا ہے۔

اگر بچہ بے چین فطرت ہے

ابتدائی چند ماہ میں کچھ بچے بہت ہی پریشان کن ہو سکتے ہیں، خصوصاً شام کے وقت۔ ماں کا دودھ پینے والے بچوں میں یہ بہت کم ہوتا ہے لیکن ان میں بھی یہ ہو سکتا ہے۔ آپ اپنے بچے کو دودھ پلا کر، ڈکار دلا کر، گانا گا کر یا اس سے باتیں کر کے یا اسے گود میں لے کر ٹہل کر یا جھلا کر مطمئن کر سکتی ہیں۔ بچے حرکت کرنا پسند کرتے ہیں۔ بے چین اور مضطرب بچے اپنی ماں سے غیر مطمئن اور ناخوش ہو سکتے ہیں۔ بچے کا باپ یا گھر کا کوئی اور فرد یا دیکھ بھال کرنے والا بچے کو بہلانے میں مدد کر سکتا ہے اور اس طرح آپ کو آرام کے لئے زیادہ وقت مل سکتا ہے۔

چھوٹے بچے کے ساتھ سونا

چھوٹے بچوں کی بیش ترمائیں بہتر طور پر سو سکتی ہیں جب ان کا بچہ ان کے قریب ہو۔اس طرح بچے کے بھوک سے بیدار ہونے پر اسے دودھ پلانا بھی آسان رہتا ہے اور آپ بچے کو بغیر اٹھے سلاسکتی ہیں ۔ اگر آپ درست طور پر دیکھ یا سن نہیں سکتی ہیں تو اس صورت میں بھی آپ یہ جان لیں گی کہ آپ کے بچے کو کب دودھ پلانے یا کپڑا بدلنے کی ضرورت ہے۔

اگر آپ کو چلنے میں دشواری ہوتی ہو تو اپنے قریب نیپیز، ڈائپرز یا صاف کپڑے رکھیں تا کہ رات کے وقت آپ اٹھے بچے کی نپی وغیرہ بدل سکیں۔

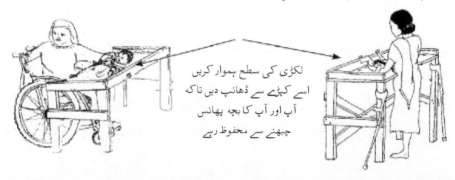

اگر آپ کی معذوری اس قسم کی ہے کہ آپ بچے کے اوپر لڑھک سکتی ہیں یا آپ کو سونے کے لئے بیٹھنے کی ضرورت ہوتی ہے تو آپ کو کچھ اور کرنے کی ضرورت ہوگی ۔ یہاں ایک مثال یہ ہے ۔

اگر آپ سمجھتی ہیں کہ آپ اپنے بچے کے اوپر لڑھک سکتی ہیں تو لکڑی کا ایک بیڈ بنائیں جس میں آپ کا بچہ آپ کے ساتھ سو سکے۔اس بیڈ کے ایک حصے کو قدرے کھلا رکھیں تا کہ آپ با آسانی بچے کی دیکھ بھال کر سکیں۔لکڑی کی سطح ہموار کریں یا اس پر کپڑا لپیٹیں تا کہ آپ یا بچے کو پھانس نہ چبھے۔

بچے کے کپڑے بدلنا

صحت مند بچہ متحرک ہوتا ہے، گھسیٹنا بہت جلد سیکھ لیتا ہے۔ جوں جوں وہ بڑا ہوتا ہے اس کے کپڑے بدلنا اتنا ہی مشکل ہو جاتا ہے۔ ایسے کپڑے استعمال کریں جو آسانی سے پہنائے اور اتارے جا سکتے ہوں۔نائیلون کی چپکنے والی پٹیاں بٹنوں کی بہ نسبت آسانی سے چپک اور کھل سکتی ہیں۔

اگر آپ جسمانی معذوری رکھتی ہیں

جسمانی معذوری رکھنے والی بہت سی عورتیں میز یا بیڈ پر اپنے بچے کے کپڑے آسانی اور محفوظ طریقے پر تبدیل کر سکتی ہیں، خصوصاً اس وقت جب وہ بیٹھ سکتی ہوں۔لیکن کچھ عورتیں یہ کرنے کے لئے ضروری توازن یا قوت نہیں رکھتی ہیں۔ ذیل میں لکڑی کی میزوں کی دو مثالیں دی گئی ہیں جن پر نہ صرف بچہ محفوظ رہے گا بلکہ اپنے آپ کو بھی مجروح نہیں کرے گا۔ آپ کے قد اور ضرورت کی مناسبت سے ایسی میزیں بنائی جا سکتی ہیں۔

لکڑی کی سطح ہموار کریں اسے کپڑے سے ڈھانپ دیں تا کہ آپ اور آپ کا بچہ پھانس چبھنے سے محفوظ رہے

بچے کی صفائی

جب آپ اپنے بچے کا جسم صاف کر رہی ہوں تو اس دوران بچے کے کھیلنے کے لئے کوئی چھوٹا سا کھلونا آپ کی مدد کرے گا کیونکہ اس طرح بچہ زیادہ حرکت نہیں کرے گا۔ بچے کے کھیلنے کے لئے دس مختلف کھلونے جمع کرنے کی کوشش کریں تا کہ وہ روزانہ ایک نئے کھلونے سے بہل سکے۔ گھر میں بھی ایسی بہت سی چیزیں بنائی جا سکتی ہیں جو کھلونوں کے بطور استعمال ہو سکیں یا مختلف چیزوں کو کھلونوں میں تبدیل کیا جا سکتا ہے مثلاً گھنٹی، کپڑے کی گڑیا، آئینہ، رنگین موتیوں والا بریسلٹ یا رنگین کاغذ۔ بچے کو ہر روز نئی چیز دیں، دس دن بعد پہلی چیز کا نمبر آئے گا جو بچے کو نئی محسوس ہوگی۔ سادہ کھلونوں کی چند مثالیں درج ذیل ہیں۔

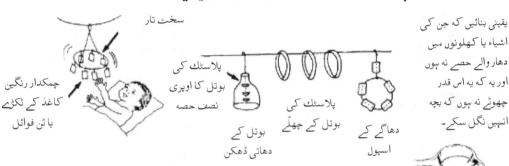

یقینی بنائیں کہ جن کی اشیاء یا کھلونوں میں دھار والے حصے نہ ہوں اور یہ کہ یہ اس قدر چھوٹے نہ ہوں کہ بچہ انہیں نگل سکے۔

سخت تار

چمکدار رنگین کاغذ کے ٹکڑے یا ٹن فوائل

پلاسٹک کی بوتل کا اوپری نصف حصہ

بوتل کے دھاتی ڈھکن

پلاسٹک کی بوتل کے چھلے

دھاگے کے اسپول

بچے کی صفائی یا کپڑے وغیرہ تبدیل کرنے کے بعد اور لیٹرین یا ٹوائلٹ استعمال کرنے میں بچے کی مدد کرنے کے بعد ہمیشہ اپنے ہاتھ اچھی طرح دھوئیں۔

اگر آپ اپنے ہاتھوں کا محدود استعمال کرتی ہوں

ہاتھوں کی محدود حرکت رکھنے والی بہت سی عورتیں بچے کے نچلے حصے اور تناسلی اعضاء کو صاف کر سکتی ہیں لیکن اکثر وہ نیپی یا ڈائپر نہیں پہنا سکتی ہیں خصوصاً جب انہیں لپیٹنے کے بعد سیفٹی پن استعمال کئے جاتے ہوں۔ ممکن ہے اس کام کے لئے آپ کو گھر والوں یا کسی مدد گار پر انحصار کرنا پڑے۔ اگر آپ خود بچے کو صاف نہیں کر سکتی ہیں تو ایسی جگہ یقینی بنائیے جہاں بچے کی صفائی اور نیپی وغیرہ تبدیل ہو سکے۔ یہ آپ سے قریب ہوتا کہ بچہ اس دوران آپ کی آواز سنے اور آپ کا چہرہ دیکھ سکے۔

اگر آپ کا ایک ہاتھ ہے یا ہاتھوں کا محدود استعمال ممکن ہے تو آپ اس وقت جب بچہ ایک ماہ کا ہو تو آپ اسے سکھا سکتی ہیں کہ وہ کس طرح نیپی بدلنے میں آپ کی مدد کرے۔

جب آپ اس کے کولہوں کے نیچے صاف کپڑا رکھیں اس کے کولہوں کو دو سے تین دفعہ اٹھائیں، ہر بار نیپی بدلتے ہوئے یہ عمل کریں وہ جلد ہی سیکھ جائے گا اور جب بھی آپ اس کے کولہوں کو چھوئیں گی تو وہ اپنے نچلے حصے کو خود اٹھانا شروع کر دے گا۔ یہ عمل آپ کے لئے اس کے نیچے کپڑا رکھنا آسان بنائے گا۔

بچے کا نیپی یا ڈائپر پہنوں کے بغیر چڈی (panties) پہنا کر بھی بدلا جا سکتا ہے۔ یقیناً چڈی بھی پیشاب میں بھیگ جائے گی لہٰذا اسے بھی ہر بار تبدیل کیا جائے۔ انہیں بھی اسی طرح دھویا اور خشک کیا جائے جس طرح ڈائپرز کو۔ ڈائپر کو باندھنے کے لئے چپکنے والی پلاسٹک کی پٹی بھی استعمال کی جا سکتی ہے۔

اگر آپ نابینا ہیں یا بخوبی دیکھ نہیں سکتی ہوں

یہ جاننا آپ کے لئے مشکل ہو سکتا ہے کہ کیا بچے کے جسم سے پاخانہ پوری طرح صاف کیا جا چکا ہے اس حوالے سے چند تجاویز یہ ہیں۔

بچے کو احتیاط سے میز کے سرے پر رکھیں اور صاف پانی اس کے نچلے دھڑ پر بہائیں۔

اگر پانی کافی مقدار میں دستیاب تو بچے کو ایک ہاتھ سے اچھی طرح سنبھال کر اس کے نچلے دھڑ سے پاخانہ صاف پانی سے بھری بالٹی یا ٹب میں دھوئیں۔

اگر پانی کافی مقدار میں نہ ہو تو بچے کے نچلے دھڑ کو پتلے اور نم کپڑے سے صاف کریں۔ موٹا کپڑا استعمال نہ کریں کیونکہ آپ محسوس نہیں کر پائیں گی کہ پاخانہ کہاں لگا ہے اس کے بعد کپڑے کو صابن اور صاف پانی سے اچھی دھوکر سورج کی روشنی میں خشک کر لیں۔

جب آپ بچے کو نہلائیں تو ٹب میں بچے کو بٹھانے یا لٹانے کے لئے تہہ کیا ہوا کپڑا رکھیں۔ اس سے بچہ پانی میں پھسلنے سے بھی محفوظ رہے گا

اگر آپ کے بچے کو قبض ہے تو تھوڑا سا کھانا پکانے کا تیل ریکٹم میں لگائیں یا آپ اپنی انگلی میں تیل لگا کر سخت فضلے کو توڑیں، بچے کو کاسٹر آئل یا وجیٹیبل آئل یا مسہل دوامت دیں۔

بچے کے ساتھ آنا جانا

اگر آپ اپنے بازو اور پیروں کو محدود طور پر استعمال کر پاتی ہیں تو بچے کو اٹھا کر کہیں آنا جانا آپ کے لئے مشکل ہوسکتا ہے۔ یہ ممکن ہے کہ آپ کے لئے توازن قائم رکھنا مشکل ہو اور آپ کی نچلی کمر مجروح ہو جائے۔ آپ کو اپنے تصور سے کام لینا ہوگا اور مختلف طریقے آزما کر وہ صورت معلوم کرنی ہوگی جو آپ کے لئے کارگر ہو۔ کچھ عورتوں کے لئے بچے کو پشت پر لادنا آسان ہوتا ہے تو دوسری عورتوں کے لئے بچے کو سامنے رکھنا آسان ہوتا ہے۔ آپ کا بچہ گزرتے وقت کے ساتھ وزنی اور متحرک ہوگا اور اس ماہ جو آپ کے لئے مناسب ہے وہ اگلے ماہ مؤثر نہ رہے۔

ابتداء میں آپ کا توازن بچے کے وزن کے باعث متاثر ہوگا لیکن اگر آپ شروع ہی میں خود کو اس کا عادی بنائیں تو جوں جوں بچہ بڑا اور بھاری ہوگا آپ کا توازن بھی اسی کے مطابق ہم آہنگ ہوتا چلا جائے گا۔

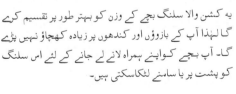

اگر آپ چلنے کے لئے بیساکھی یا چھڑی استعمال کرتی ہیں تو آپ کے لئے بچے کو بازوؤں میں لے جانا مشکل ہوسکتا ہے۔ لیکن آپ اپنے بچے کو اپنی پشت پر لاد کر اپنے ساتھ لے جاسکتی ہیں۔

اگر آپ اپنے بازو محدود طور پر استعمال کرتی ہوں

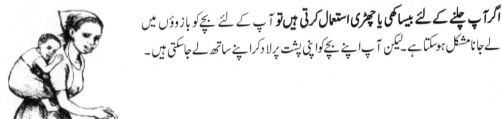

یہ کشن والا سلنگ بچے کے وزن کو بہتر طور پر تقسیم کرے گا لہٰذا آپ کے بازوؤں اور کندھوں پر زیادہ کھچاؤ نہیں پڑے گا۔ آپ بچے کو اپنے ہمراہ لانے لے جانے کے لئے اس سلنگ کو پشت پر یا سامنے لٹکاسکتی ہیں۔

دوسری معذور لڑکیوں اور عورتوں کی مدد کے لئے خیالات کا تبادلہ کیجئے کہ وہ کس طرح اپنے بچے کی دیکھ بھال کریں۔

اگر آپ وہیل چیئر یا پہیہ گاڑی استعمال کرتی ہیں

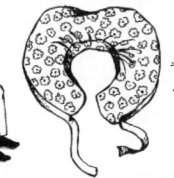

اگر آپ وہیل چیئر دھکیلنے کے لئے دونوں ہاتھ استعمال کرتی ہیں تو آپ کے لئے بچے کو اپنے بازوؤں میں یا گود میں رکھنا مشکل ہوسکتا ہے۔ لیکن اگر آپ اپنی گردن میں سلنگ ڈال لیں تو آپ اپنے بچے کو وہیل چیئر دھکیلتے ہوئے محفوظ طور پر سنبھال سکتی ہیں۔ سلنگ کو اپنی کمر کے گرد بھی ایک پٹی سے باندھیں تا کہ سلنگ بچے کو ادھر ادھر نہ ہونے دے۔

جب بچہ بڑا ہو تو آپ بچے کو اپنی گود میں بٹھانے کے لئے ڈوریاں (Harness) استعمال کرسکتی ہیں۔

اس طرح کا ہے بی کشن جو آپ اپنی کمر کے گرد باندھ سکیں بچے کو گود میں محفوظ طور پر رکھنے میں معاون ثابت ہوگا

اگر آپ علامتی زبان استعمال کرتی ہیں

اگر آپ اظہار کے لئے علامتی زبان استعمال کرتی ہیں تو آپ بچے کو بٹھانے کے لئے ایسی سلنگ استعمال کرسکتی ہیں جس سے آپ کے دونوں بازو بات کرنے کے لئے آزاد ہیں۔

اگر آپ پر دورے پڑتے ہیں تو آپ بہتر طور پر جانتی ہوں گی کہ یہ کتنی بار پڑتے ہیں اور کتنی شدت کے ہوتے ہیں۔ اگر آپ پر دورہ اس وقت پڑا ہو جب آپ اپنے چھوٹے سے بچے کو سنبھالے ہوئے ہوں تو وہ گر کر بری طرح زخمی بھی ہوسکتا ہے اور ہلاک بھی۔

اگر ممکن ہو تو اپنے ساتھ کسی ایسے فرد کو رکھیں جو دوروں سے محفوظ ہو۔ اگر آپ الگ رہتی ہیں اور بعض اوقات بچے کے ساتھ تنہا بھی رہتی ہیں تو کمرے یا گھر میں کوئی جگہ محفوظ بنائیں اور بچے کو سارا وقت اس جگہ رکھیں۔ بچے کو لے کر ادھر اُدھر مت گھومیں اور یقینی بنائیں کہ کرسی یا میز وغیرہ کے سرے دھار والے نہ ہوں۔ اس صورت میں اگر آپ پر دورہ پڑے جبکہ آپ بچے کے ساتھ تنہا ہوں تو بچہ دورہ ختم ہونے تک محفوظ رہے گا۔ اس طرح جب آپ بچے کو کچھ کھلائیں یا پلائیں، نہلائیں یا لباس تبدیل کریں تو بچے کا فرش پر رہنا محفوظ ہوگا۔

جب بچہ لڑھکنے یا چلنے کے قابل ہو جائے تو دروازے یا سیڑھیوں کے سامنے رکاوٹ رکھ دیں تا کہ بچہ آپ کو دورہ پڑنے اور آپ کی طبیعت سنبھلنے تک محفوظ رہے۔

بچے کوسنبھالنا

جب بچے لڑھکنا، گھسیٹنا اور چلنا سیکھ لیتے ہیں تو ماں یا باپ کے لئے انہیں سنبھالنا مشکل ہوسکتا ہے۔ نشوونما پاتے بچوں کے لئے چلنا اور دوڑنا وغیرہ صحت افزا بات ہے ابتداء میں چلنا سیکھتے ہوئے وہ بار بار گرتے ہیں۔ اس لئے اس بارے میں فکرمند نہ ہوں یہ نوعمر بچے کی نشوونما کا ایک عام حصہ ہے۔

اگر آپ تیزی سے حرکت نہ کرسکتی ہوں

چھوٹے بچے خاصی تیز رفتار سے حرکت کرسکتے ہیں اور ان کے گرکر زخمی ہونے کا امکان رہتا ہے اس لئے اگر آپ دوڑ کر انہیں پر خطر صورتحال میں جانے سے مثلاً کسی گاڑی کے سامنے دوڑنے، آگ کی طرف لپکنے سے نہ روکیں تو گر بڑ ہوسکتی ہے اگر آپ تیزی سے حرکت نہیں کرسکتی ہیں تو بچے کے بازو میں کوئی رسی باندھ لیں تا کہ آپ اسے تحفظ دینے کے لئے کھینچ سکیں۔ اگر آپ رسی پر گرفت نہ کرسکتی ہوں تو اس کا دوسرا سرا اپنی کلائی میں باندھ لیں۔

اگر آپ کو دیکھنے میں مشکل ہو یا آپ نا بینا ہوں

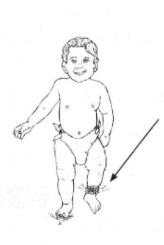

جب تک بچہ بہت چھوٹا ہے اس کے ساتھ ہی سوئیں۔ اس طرح آپ جان سکتی ہیں کہ اسے کس چیز کی ضرورت ہے یا وہ کہاں ہے۔

جیسے ہی بچہ بڑا ہوکر اپنے طور پر متحرک رہنا شروع کرتا ہے اس کی کلائی میں کوئی ایسی چیز باندھ دیں جو آواز کرتی ہو۔ اس طرح آپ اس چیز کی آواز سے جان سکتی ہیں کہ بچہ کہاں ہے۔

اس کے ساتھ ایک جگہ ایسی بنائیں جہاں بچہ خود کو زخمی کئے بغیر گھوم پھر یا کھیل سکے یقینی بنائیں کہ اس جگہ نوکیلے سرے یا کنارے نہ ہوں۔ کمروں کے داخلی دروازوں یا گھر سے باہر جانے والی سیڑھیوں کے سامنے رکاوٹ کھڑی کریں تا کہ بچہ اپنے طور پر محفوظ جگہ سے نہ نکل سکے۔

بچوں کی صحت کا تحفظ

بچوں میں بیماری اکثر بڑی تیزی سے شدت اختیار کرلیتی ہے۔ایسی بیماری جو کسی بالغ کو دنوں یا ہفتوں میں سخت نقصان پہنچاتی ہے یا ہلاک کرتی ہے بچے کو چند گھنٹوں میں ہلاک کر سکتی ہے۔ اس لئے ضروری ہے کہ بیماری کی ابتدائی علامات پر توجہ دیں اور صورتحال کے مطابق قدم اٹھائیں۔ بچوں میں ڈائریا (پتلے یا پانی جیسے پاخانے) عام ہے جو چھوٹے بچوں کے لئے بالغوں سے زیادہ خطرناک ہے۔ اگر آپ کے بچے کو ڈائریا ہو جائے تو فوری قدم اٹھائیں اور۔

- اسے دودھ پلانا جاری رکھیں
- غذا دینا جاری رکھیں
- اسے خوب مائعات پلائیں

ری ہائیڈریشن مشروب پانی کی کمی سے بچاتا ہے یا اس کمی کا علاج کرتا ہے خاص طور پر اس وقت جب بچے کو پانی جیسے پاخانے زیادہ تعداد میں ہوں۔

ری ہائیڈریشن مشروب بنانے کے دو طریقے

1- شکر اور نمک سے (شکر کے بجائے خام شکر یا راب بھی استعمال کی جاسکتی ہے (ایک لیٹر صاف پانی میں چائے کا نصف چھچی نمک ملائیں اطمینان کریں کہ اس کا ذائقہ آنسو سے کم نمکین ہو۔ اب اس میں آٹھ

چائے کے چمچے شکر ملائیں۔ انہیں اچھی طرح ملا کر بچے کو پلانا شروع کر دیں۔

2- پسے ہوئے سیریل اور نمک سے (پسے ہوئے چاول بہترین ہیں یا پھر باریک پسی ہوئی گندم کا آٹا یا ابلے اور پسے ہوئے آلو استعمال کریں۔

ایک لیٹر صاف **پانی** میں نصف چائے کا چمچہ نمک ملائیں، اطمینان کریں کہ اس کا ذائقہ آنسو سے کم نمکین ہو۔ پھر اس میں 8 چائے کے چمچے بھر کر (یا دو مٹھی) پسا ہوا سیریل ملائیں اور سیال دلیہ سا بنانے کے لئے پانچ سے سات منٹ اُبالیں۔ اسے ٹھنڈا کر کے بچے کو دینا شروع کر دیں۔

بچے کو یہ مشروب پلانے یا کھلانے سے پہلے ہر بار چکھیں کہ یہ خراب تو نہیں ہوا ہے سیریل مشروب گرم موسم میں چند گھنٹوں میں خراب ہو جاتا ہے۔ اگر دستیاب ہو تو اس میں نصف کپ پھلوں کا رس نارئیل کا پانی یا مسلا ہوا کیلا ملائیں اس سے بچے کو پوٹاشیم ملے گا جس سے بچے کو مزید کھانے یا پینے کی طلب ہو سکتی ہے۔

اہم بات اپنے علاقے کے مشروب استعمال کیجئے اور مقامی پیمائشوں کے مطابق اشیاء کی مقداریں استعمال کیجئے اگر آپ نو عمر بچوں کو دے رہی ہیں تو اس میں تھوڑا سا صاف پانی ملائیں تا کہ یہ سیال یا مائع جیسا بن جائے۔ بچے اور اپنی سہولت کے لئے کوئی آسان اور سادہ طریقہ تلاش کیجئے۔

بچوں کی صحت مند نشوونما اور بہت سی بیماریوں سے تحفظ کے تین اہم طریقے۔

- پُرغذائیت کھانا
- صفائی
- امیونائزیشن

پُرغذائیت کھانا

یہ اہم ہے کہ انتہائی پُرغذائیت کھانا کھانے والے بچوں کی نشوونما بہت اچھی ہوتی ہے اور وہ بیمار نہیں پڑتے ہیں۔ اس سے بڑھ کر یہ کہ تمام بچوں کو دن میں کئی بار مناسب مقدار میں کھانا ملنا چاہئے (دیکھئے 266-265)

صفائی

بچے زیادہ صحت مند رہ سکتے ہیں اگر ان کے گھر صاف ستھرے ہوں۔ اس سلسلے چند اہم باتیں مندرجہ ذیل ہیں۔

- بچوں کو صاف ستھرا رکھیں اور کئی بار ان کے کپڑے تبدیل کریں۔
- بچوں کو سکھائیں کہ وہ صبح بیدار ہونے کے بعد، پاخانہ کرنے کے بعد، کھانا کھانے یا کھانے کی چیزوں کو ہاتھ لگانے سے پہلے اپنے ہاتھ ضرور دھوئیں۔
- بچوں کو سکھائیں کہ وہ لیٹرین اور ٹوائلٹ کس طرح استعمال کریں۔
- جہاں ہک وارم Hook Worm موجود ہوں وہاں بچوں کو ننگے پاؤں نہ جانے دیں، انہیں سینڈل یا جوتے استعمال کرنے کا عادی بنائیں۔
- بچوں کو روزانہ دانت برش کرنا سکھائیں، انہیں بہت زیادہ میٹھی یا کاربونیٹڈ اشیاء نہ دیں۔
- ناخن ممکنہ حد تک چھوٹے کاٹیں۔
- ایسے بچے جو بیمار ہوں یا زخم رکھتے ہوں، اسکیبز، جوؤں یا رنگ وارم کا شکار ہوں انہیں دوسرے بچوں کے ساتھ سونے یا آپس میں کپڑے تولئے وغیرہ استعمال نہ کرنے دیں۔
- اسکیبی، رنگ وارم، آنتوں کے کیڑوں اور دوسرے انفیکشنز کا جو بچوں میں آسانی سے پھیل سکتے ہیں فوراً علاج کرائیں
- بچوں کو گندی چیزیں منہ میں نہ رکھنے دیں نہ ہی کتوں یا بلیوں یا دوسرے جانوروں کو ان کا چہرہ چاٹنے دیں۔
- مور یا کتے اور مرغیاں گھر کے باہر رکھیں۔
- پینے کے لئے صاف، اُبلا اور فلٹر کیا ہوا پانی استعمال کریں بچوں کے لئے بے حد ضروری ہے۔
- بچوں کو ملیریا سے بچانے کے لئے ہر ممکن قدم اٹھائیں۔ انہیں مچھر دانی میں سلائیں دیگر حشرات الارض سے بھی بچائیں۔

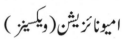

امیونائزیشن (ویکسینز)

ویکسینز بچوں کو بہت سی خطرناک بیماریوں سے یقینی تحفظ دیتی ہیں۔ اگر آپ کی کمیونٹی میں 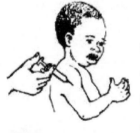 صحت ورکر امیونائزیشن کی سہولت فراہم نہیں کرتے ہیں تو اپنے بچوں کو امیونائزڈ کرانے کے لئے قریبی مرکزِ صحت لے جائیں۔ ان کے بیمار پڑنے کے بعد علاج سے بہتر ہے کہ انہیں ضروری ویکسین دلوائیں ویکسینز عموماً مفت دی جاتی ہیں (مختلف ممالک اس کے لئے مختلف شیڈول رکھتے ہیں) بچوں کے لئے انتہائی ضروری ویکسینز مندرجہ ذیل ہیں۔

نوٹ	کب دی جائے	امیونائزیشن
کچھ ممالک میں بچے کو 4 اور 6 برس کے درمیان ایک اضافی انجکشن لگایا جاتا ہے	دوسرے، چوتھے، چھٹے اور اٹھارہویں مہینے	**ڈی پی ٹی** ڈپتھیریا یا وہوپنگ کف (Pertursis) اور ٹیٹنس کے لئے
کچھ ممالک میں ابتدائی تین ڈوز ڈی پی ٹی انجکشن کے ساتھ دی جاتی ہیں اور چوتھی ڈوز بارہ سے پندرہ ماہ کی عمر میں اور پانچویں ڈوز 4 برس کی عمر میں دی جاتی ہے۔	کچھ ممالک میں ایک ڈوز پیدائش کے بعد اور تین مزید ڈوز ڈی پی ٹی کے انجکشن کے ساتھ دی جاتی ہیں۔	**پولیو** (شیر خوارگی کا فالج)
	پیدائش کے وقت یا اس کے بعد کسی وقت	**بی سی جی** (ٹی بی کے لئے)
بہت سے ممالک میں ایم ایم آر کہلانے والی تھری ان ویکسین (خسرہ، ممپس اور ریبولا جرمن خسرہ) 12 سے 15 ماہ کے درمیان دی جاتی ہے اور دوسرا انجکشن 4 سے 6 برس کے درمیان لگایا جاتا ہے۔	ایک انجکشن 9 ماہ سے پہلے اور اکثر دوسرا انجکشن پندرہ ماہ کی عمر میں یا اس کے بعد۔	**خسرہ**
کچھ ممالک میں یہ انجکشن ولادت پر، دوسرے مہینے اور چھٹے مہینے لگائے جاتے ہیں۔	تین انجکشن عام طور پر ڈی پی ٹی کے مطابق لگائے جاتے ہیں۔	**ہیپ بی** (ہپاٹائٹس بی)
	ڈی پی ٹی کے ساتھ تین انجکشن لگائے جاتے ہیں۔	**Hib** ہیموفولیسیس، انفلوئنزا ٹائپ ببنا جو نوعمر بچوں میں میگلٹس Mengtis اور نمونیا کا سبب بننے والا جرثومہ ہے۔
حاملہ عورتوں کو ہر حمل کے دوران امیونائزیشن کیا جائے تاکہ ان کے بچے ٹیٹنس سے تحفظ پائیں (دیکھئے صفہ 251)۔	ایک انجکشن ہر دس سال بعد بعض ممالک میں نو سے گیارہ برس کے درمیان آخری ڈی پی ٹی کے پانچ برس بعد اور پھر دس برس میں۔	(Tetanustoxid) **TT یا TD** ٹیٹنس کے لئے بالغ افراد اور 12 برس سے بڑے بچوں کے لئے

اپنے بچوں کو وقت پر امیونائز کرائیں
یقینی بنائیں کہ وہ ہر ویکسین کی سیریز مکمل کریں

باب-13

معذوری کے ساتھ عمر رسیدگی (بڑھاپا)

بڑھاپا وہ زمانہ ہوتا ہے جب آپ اپنے خاندان اور کمیونٹی میں زیادہ احترام پاتی ہیں تاہم یہ وہ زمانہ بھی ہوسکتا ہے جب آپ غربت، خراب برتاؤ اور صحت کے مسائل کا آسانی سے ہدف بن سکتی ہیں۔ آپ کوئی معذوری رکھتی ہوں یا نہ رکھتی ہوں جوں جوں آپ بوڑھی ہوں گی بہت سی تبدیلیوں سے گزریں گی۔ آپ کو اپنا جسم بوڑھا ہونے پر امکانی طور پر بہت سے کام کرنے کے انداز میں تبدیلیاں لانا ہوں گی۔ آپ کچھ کام نہیں کرسکیں گی کیونکہ انہیں کرنا آپ کے لئے مشکل ہوجائے گا۔ ممکن ہے آپ صحت کے ایسے مسائل یا معذوریوں سے دوچار ہوجائیں جو اس وقت نہیں تھیں جب آپ نسبتاً کم عمر تھیں۔ ممکن ہے کہ کچھ عورتیں عمر کے اس حصے میں چلنے یا آنے جانے کے لئے چھڑی یا وہیل چیئر استعمال کرنا شروع کردیں کیونکہ اب وہ چل نہ سکتی ہوں، کچھ نظر کے چشموں یا سماعتی آلے کا استعمال شروع کرسکتی ہیں۔

آپ بچپن ہی سے معذور ہوں یا بعد میں زندگی کے کسی حصے میں معذور ہوئی ہوں، اس باب کی معلومات آپ کو آگاہ کریں گی کہ کس طرح کچھ معذوریاں واقع ہوسکتی ہیں یا تبدیلیاں آ سکتی ہیں اور آپ کس طرح بڑھاپے میں اپنی صحت کا خیال رکھ سکتی ہیں۔

بڑھے ہونے پر مجھے معذور ہی کہہ کر پکارا گیا۔ اب مجھے چھڑی بردار بوڑھی عورت کی حیثیت سے دیکھا جاتا ہے۔

بڑھاپے میں صحت کے مسائل

عورتیں عموماً اس وقت تک اپنے بوڑھا ہونے کے بارے میں نہیں سوچتی ہیں جب تک ان کے بچے بڑے نہ ہوجائیں یا ان کے جسموں میں تبدیلیاں آنا شروع نہ ہوں۔ آپ یہ بھی نوٹ کرسکتی ہیں کہ آپ کا جسم بار بار تھک جاتا ہے، یہ کہ آپ پہلے کی طرح توانا اور مضبوط نہیں یا یہ کہ اب آپ کے لئے آنا جانا پہلے کی طرح آسان نہیں رہا ہے۔ بہتر یہ ہے کہ آپ ان تبدیلیوں کو سمجھ لیں جو عورتوں کے بوڑھا ہونے پر ہوتی ہیں۔ اگر آپ یہ جان لیں کہ آپ کے جسم میں آنے والی تبدیلیاں بڑھاپے کا ایک حصہ ہیں یا آپ کی معذوری کے باعث، تو ان سے نمٹنا اتنا ہی آسان ہو جائے گا۔ اپنے جسم کی دیکھ بھال کے لئے معلومات کے لئے دیکھئے صفحہ 85۔ معذور عورتیں بوڑھا ہونے پر درج ذیل مسائل کا شکار ہوسکتی ہیں۔

کمزور یا پُر درد عضلات اور جوڑ

اگر آپ کی معذوری کے باعث آپ کے جسم کا کوئی حصہ، دوسرے حصے کی طرح کام نہیں کرتا ہے تو یقیناً آپ نے گزرے وقت میں جسم کے اس حصے کی کمی کو پوری کرنے کے لئے بہتر طور پر کام کرنے والے حصے کو حد سے زیادہ استعمال کیا ہوگا۔ مثلأ

- اگر آپ کی ٹانگیں مفلوج ہیں تو آپ امکانی طور پر اپنی اچھی ٹانگ کسی ایسے فرد سے زیادہ استعمال کرتی ہوں گی جو اپنی دونوں ٹانگیں استعمال کرتا ہو اور آپ کے جوڑ اس کے زائد استعمال سے کمزور ہوگئے ہوں۔

- اگر آپ طویل عرصہ سے وہیل چیئر یا بیساکھیاں استعمال کرتی ہیں تو آپ کے ہاتھوں، بازوؤں اور کندھوں کے جوڑ مسلسل استعمال کے باعث تکلیف دہ اور خراب ہونا شروع ہوسکتے ہیں۔

اگر آپ چھوٹی جسامت کی ہیں تو آپ کے کندھوں، گھٹنوں اور کولہوں میں آپ کے ماضی کی حرکات و سکنات کے انداز کے باعث درد شروع ہوسکتا ہے۔ اگر آپ وہیل چیئر استعمال کر رہی ہوں یا زیادہ تر وقت بستر پر گزار رہی ہوں تو آپ کے لئے نہایت ہی ضروری ہے کہ آپ دباؤ کے زخموں سے بچنے کے لئے جس قدر ممکن ہو چلیں یا اپنی پوزیشن تبدیل کرتی رہیں (دیکھئے صفحہ 116)۔

وہیل چیئر استعمال کرنے والی عورتوں کے لئے

وہیل چیئر استعمال کرنے والی عورتیں عمر بڑھنے پر نسبتاً کم ورزش کرتی ہیں۔ دوسروں سے کہیں کہ وہ کھڑا ہونے یا اسٹینڈنگ فریم استعمال کرنے میں آپ کی مدد کریں تا کہ آپ اپنے پیروں کی ہڈیوں پر وزن ڈال سکیں۔ اس کے ساتھ آپ مختلف چیزیں اٹھا کر اپنے بازو بھی مضبوط بنائیں۔ مزید ورزشوں کے بارے میں جاننے کے لئے دیکھئے صفحہ 88 تا 95۔

پوسٹ پولیوسنڈروم (Post-Polio Syndrome)

اگر آپ پولیوزدہ ہیں تو پولیووائرس کے خاتمے کے برسوں بعد آپ شدید کمزوری، تھکن، درداور تنفّسی مشکلات میں مبتلا ہوسکتی ہیں۔اس صورت میں ضروری ہے کہ آپ ورزش کرتے ہوئے انتہائی محتاط رہیں۔عضلات کا حد سے زیادہ استعمال انہیں نقصان پہنچا سکتا ہے اور آپ کی کمزوری کو بدترین کرسکتا ہے۔اس کے بجائے آپ سادہ اسٹریچنگ کریں اور ایسی حرکات اپنائیں جو آپ کے جسم کواکڑنے سے بچائیں۔

چلنا اور توازن

اگر آپ مصنوعی ٹانگ استعمال کرتی ہیں تو اسے درست طور پر فٹ کئے جانے کی ضرورت پڑسکتی ہے کیونکہ ممکن ہے اب وہ درست طور پر فٹ نہ ہو۔خصوصاً اس صورت میں جب آپ اتنی ورزش نہ کرتی ہوں جو آپ کیا کرتی تھیں تو آپ کے عضلات کمزور اور نرم پڑسکتے ہیں۔

اگر آپ کسی سہارے کے بغیر چلتی ہیں تو اب آپ کو چھڑی، بیساکھیاں یا وہیل چیئر استعمال کرنے کی ضرورت پڑسکتی ہے۔ بہت سی عورتیں ایسی اشیاء استعمال کرنے کا فیصلہ تاخیر سے کرتی ہیں جوان کے لئے مفید نہیں ہوتا ہے۔چھڑی یا وہیل چیئر کا جلد استعمال آپ کو گرنے اور زخمی ہونے سے بچا سکتا ہے اور آسانی سے چلنے پھرنے میں آپ کا معاون ہوسکتا ہے۔اپنے اطراف میں آنا جانا آپ کے لئے بہتر ہے کہ اس طرح آپ اپنی کمیونٹی کے کاموں میں اسی مناسبت سے شریک ہوسکتی ہیں۔

آرتھرائٹس

آرتھرائٹس، جوڑوں کی تکلیف دہ سوجن اور جکڑن ہے۔ یہ بہت سے لوگوں کو متاثر کرتا ہے اوران کے لئے روزمرہ کے کاموں کو تکلیف دہ یا زیادہ مشکل بنا سکتا ہے۔اگر آرتھرائٹس ہاتھوں میں ہوتو یہ کچھ معذوریوں والے افراد کے خصوصی مسائل کا سبب بن سکتا ہے۔مثلاً

جذام اور آرتھرائٹس کے باعث ہاتھوں کی ساخت میں تبدیلی

- اگر آپ نابینا ہیں اور چھو کر دیکھنے یا پڑھنے کے لئے اپنے ہاتھ استعمال کرتی ہیں تو آپ اس صورت میں یہ کام اچھی طرح نہیں کرسکیں گی۔
- اگر آپ بہری ہیں تو ممکن ہے آپ علامتی زبان درست طور پر استعمال نہ کرسکیں۔
- اگر آپ جذام میں مبتلا ہیں اور آپ کے ہاتھ اس سے متاثر ہوچکے ہیں تو آرتھرائٹس آپ کے لئے ہاتھوں کا استعمال اور زیادہ مشکل بنا دے گا۔

جلد کے مسائل

جوں جوں آپ بوڑھی ہوتی ہیں آپ کی جلد بھی پتلی ہو جاتی ہے اور آپ کے لئے مشکل ہو سکتا ہے کہ آپ ان پر پڑنے والی خراشوں کو محسوس کر سکیں۔ یہ بیش تر عورتوں کے ساتھ ہوتا ہے۔

- اگر آپ زیادہ تر بیٹھی یا لیٹی رہتی ہیں تو پتلی جلد کا مطلب یہ ہے کہ آپ کو آسانی سے دباؤ کے زخم ہو سکتے ہیں (دیکھئے صفحہ 114)
- اگر آپ مصنوعی ٹانگ یا بازو استعمال کرتی ہیں تو یہ یقین کرنے کے لئے کہ یہ سرخ یا تکلیف دہ تو نہیں ہیں اپنی جلد کے ان حصوں کا ممکنہ حد تک زیادہ جائزہ لیتی رہیں جن سے مصنوعی اعضاء چھوتے ہیں۔
- اگر آپ جذام میں مبتلا ہیں تو آپ جلد کا روزانہ جائزہ لیں۔ پتلی جلد میں زخم اور انفیکشن آسانی سے ہو جاتے ہیں۔
- اگر آپ کی ریڑھ کی ہڈی مجروح ہے یا آپ فالج زدہ ہیں اور اپنی جلد میں محسوس کئے جانے کی صلاحیت نہیں رکھتی ہیں تو کسی سے کہیں کہ وہ دباؤ کے زخموں سے تحفظ کے لئے آپ کی جلد کا جائزہ لے، خصوصاً ان حصوں کا جو آپ دیکھ نہیں پاتی ہیں جیسے آپ کی کمر (دیکھئے صفحہ 117)۔

بصارت اور سماعت

بہت سے بوڑھے لوگ اس طرح نہیں دیکھ پاتے ہیں جس طرح وہ جوانی میں دیکھ لیتے تھے۔ اگر آپ بہری ہیں تو آپ کے لئے اس وقت کچھ سمجھنا مشکل ہو گا۔ جب کوئی آپ سے علامتی زبان میں بات کرے یا آپ لب ریڈنگ کا طریقہ استعمال کر رہی ہوں۔ اگر آپ جذام زدہ ہوں تو بڑھاپا آپ کی آنکھوں میں سوزش کا سبب بن سکتا ہے۔ جس کا علاج نہ کیا جائے تو آپ نابینا ہو سکتی ہیں۔ اگر آپ نابینا ہیں اور اس کے ساتھ سننے کی صلاحیت بھی گنوا رہی ہیں تو آپ کے لئے دوسروں سے ابلاغ اور گھومنا پھرنا زیادہ مشکل ہو جائے گا۔

اپنے گھر والوں سے کہیں کہ وہ ایسی تبدیلیاں ممکن بنائیں جو دیکھنے، سننے اور آسانی سے چلنے پھرنے میں آپ کی معاون ہوں۔ مثلاً اگر آپ درست طور پر دیکھ نہ سکتی ہوں تو وہ دیواروں پر سفید رنگ کر کے یا زیادہ روشنی والے بلب لگا کر گھر کے اندرونی حصوں کو روشن بنائیں۔ سیڑھیوں اور دروازوں کو مختلف رنگ دے کر واضح کریں تا کہ آپ انہیں بہتر طور پر دیکھ سکیں اور گریں یا ٹکرائیں نہیں۔

اگر آپ کے سننے کی صلاحیت کم ہو رہی ہو تو آپ لوگوں سے کہیں کہ وہ بات کرتے ہوئے آپ کے سامنے بیٹھیں اور واضح طور پر بات کریں مگر چلا کر بات نہ کریں۔ بہتر طور پر سننے کے لئے گفتگو کرتے ہوئے ریڈیو یا ٹی وی وغیرہ بند کر دیں۔

کمزور ہڈیاں (اوسٹیوپروسس)

آپ کی ماہواری بند ہونے کے بعد آپ کا جسم ایسٹروجن ہارمون کم بنانے لگتا ہے (دیکھئے صفحہ 72)۔ لہٰذا آپ کی ہڈیاں کمزور ہو سکتی ہیں۔ کمزور ہڈیاں آسانی سے ٹوٹ جاتی ہیں اور دیر سے جڑتی ہیں۔ اگر آپ کا توازن بڑھاپے کی وجہ سے متاثر ہے یا آپ کو مرگی کے دورے پڑتے ہیں یا آپ سیریبرل پالسی ہے تو آپ گرنے اور کمزور ہڈیوں کے ٹوٹنے کا زیادہ خطرہ رکھتی ہیں۔

آپ اپنی کمزور ہڈیوں کو اس طرح بچا سکتی ہیں کہ آپ

- کیلشیئم سے بھر پور غذائیں (دیکھئے صفحہ 86) ایسی غذاؤں کے ساتھ کھائیں جن میں وٹامن سی خوب ہو جیسے پھل اور زرد رنگ کی سبزیاں۔
- باقاعدگی سے ورزش کریں جو آپ کی ہڈیوں پر بوجھ ڈالیں (دیکھئے صفحہ 88 تا 90)۔

ذہنی الجھاؤ

کچھ بوڑھے یاد رکھنے یا توجہ مرکوز رکھنے میں دشواری محسوس کرتے ہیں ۔ بیشتر لوگوں کے لئے یہ کوئی سنگین مسئلہ نہیں ہے لیکن کچھ لوگ یاد داشت سے زیادہ سنگین مسائل الزہیمر ،ذہنی انحطاط ،عمر رسیدگی کے ذہنی خلل کا شکار اور آخر کا راس قدر الجھن زدہ ہو جاتے ہیں کہ وہ دوستوں اور افراد خانہ کو بھی نہیں پہچانتے ہیں یہ لوگ روزمرہ کی چیزیں جن سے یہ اچھی طرح واقف ہوتے ہیں استعمال کرتے ہوئے انتہائی خوفزدہ اور الجھے الجھے سے رہتے ہیں ۔

ڈاؤن سنڈروم کا شکار بوڑھا فرد زیادہ آسانی سے الجھن زدہ ہو جاتا ہے اور ممکن ہے اسے مرگی کے دورے پڑنے لگیں ۔

نئے طریقے تلاش کرنا

بڑھاپے کے ساتھ ہونے والی تبدیلیوں کا مطلب یہ ہے کہ آپ کو روزمرہ کے کام کرنے اور دوسروں کی مدد حاصل کرنے کے لئے نئے طریقے ڈھونڈنے کی ضرورت ہوگی ۔ اور ممکن ہے آپ کوئی امدادی اشیاء جیسے آلہ سماعت ،چھڑی یا وہیل چیئر استعمال کرنے کی ضرورت ہو ۔ آپ جوں ہی محسوس کریں کہ آپ کے جسم نے تبدیل ہونا شروع کر دیا ہے ،آپ اپنے کاموں کے کرنے کے لئے نئے طریقے تلاش کرنا شروع کریں ۔ یہ جان لینا کہ کیا ہو سکتا ہے ،اپنی جسمانی دیکھ بھال میں آپ کی مدد کر سکتا ہے اور آپ بوڑھا ہونے کے ساتھ جس قدر ممکن ہو صحت مندرہ سکتی ہیں ۔

مدد یا معاونت حاصل کرنا

اگر آپ کے لئے مختلف کام جیسے کھانا ،غسل کرنا ،لباس بدلنا یا لیٹی ہوئی حالت سے اٹھنا مشکل ہو رہا ہے تو اپنے دوستوں اور افراد خانہ کی دیکھ بھال کرنے والوں اور جن پر آپ بھروسہ کر سکتی ہوں انہیں آگاہ کریں کہ وہ کس طرح آپ کی مدد کر سکتے ہیں ۔ آپ اپنے ساتھ کسی رشتہ دار یا دوست کے رہنے کا بندوبست بھی کر سکتی ہیں ۔ آپ کی مدد کے عوض وہ رہنے کے لئے جگہ بھی حاصل کرے گا ۔

اگر آپ چیزیں اور کام بھولنے لگی ہیں تو ان کاموں کی فہرست بنانا مددگار ہوگا یا پھر ہر روز اپنے گھر والوں سے ان کاموں کے بارے میں تبادلہ خیال کریں جو آپ اس روز کرنا چاہتی ہوں تا کہ وہ آپ کو ان کے بارے میں یاد دلا سکیں ۔

> اینٹار میرے لئے پڑھنے کا بہت بہت شکریہ، میرے ہاتھ کی انگلیاں اب اس قدر مڑ چکی ہیں کہ میرے بریل کے ذریعے کچھ پڑھنا بہت مشکل ہو گیا ہے ۔

ڈپریشن
(انتہائی افسردگی یا کوئی بھی احساس نہ ہونا)

کچھ لوگ بوڑھا ہونے پر اداسی اور ڈپریشن محسوس کرنا شروع کر دیتے ہیں۔ عموماً یہ تنہائی، صحت میں تبدیلیوں یا وہ کچھ کر سکنے کے قابل نہ ہونے کے باعث ہوتا ہے جو وہ عموماً کر لیتے تھے۔ کچھ معذور عورتیں جو لوگوں کی بے توجہی اور تحقیر کا شکار ہوں بڑھاپے میں خود کو زیادہ تنہا اور ڈپریشن زدہ محسوس کر سکتی ہیں۔

ڈپریشن کی چند علامات مندرجہ ذیل ہیں۔

- زیادہ تر اداسی محسوس کرنا
- سونے میں دشواری یا حد سے زیادہ سونا
- واضح طور پر سوچنے میں دشواری
- خوشیوں والی سرگرمیوں، کھانے پینے یا جنسی عمل میں دلچسپی کا فقدان
- سر درد یا پیٹ کے مسائل جیسے جسمانی عوارض جو بیماری کی وجہ سے نہیں ہوں۔
- گفتگو اور حرکات میں سست روی
- روزمرہ کے کاموں کے لئے توانائی کا فقدان
- موت یا خودکشی کے بارے میں سوچنا

ڈپریشن سے تحفظ کے لئے کیا کیا جائے؟

جس حد تک ممکن ہو متحرک رہا جائے، ورزش کی جائے اور اچھی طرح کھایا جائے۔ اس سے بڑھ کر یہ کوشش کی جائے کہ زیادہ وقت تنہا نہ گزرے۔ اپنی کمیونٹی کے چھوٹے بچوں کی دیکھ بھال میں مدد کریں۔ تبادلہ خیال اور وقت گزارنے کے لئے دوسری بوڑھی عورتوں سے ملیں۔ اگر آپ اکثر اداس رہتی ہوں یا آپ کو نیند نہیں آتی ہو تو اپنے گھر کے کسی ایسے فرد سے بات کیجئے جس پر آپ کو اعتماد ہو یا کسی صحت کارکن سے ملیے۔ دماغی صحت کے بارے میں مزید معلومات کے لئے دیکھئے باب 3۔

جب ماہواری بند ہو جائے (سن یاس)

عام طور پر ماہواری بتدریج ایک سے دو برس میں عموماً 45 سے 55 برس کے درمیان بند ہوتی ہے۔ یہ اس لئے ہوتا ہے کہ آپ کی بیضہ دانی بیضے بنانے بند کر دیتی ہے اور آپ کا جسم ایسٹروجن اور پروجسٹرون ہارمونز کم بناتا ہے۔ ڈاؤن سنڈروم کا شکار عورتوں کی ماہواری دوسری عورتوں کی یہ نسبت جلد بند ہو جاتی ہے۔

علامات

- آپ کی ماہواری کا سلسلہ تبدیل ہو جاتا ہے۔ یہ بھی ممکن ہے کچھ عرصہ آپ کی ماہواری معمول کی یہ نسبت زیادہ بار آئے یا ممکن ہے آپ کی ماہواری چند ماہ رک جائے اور پھر شروع ہو جائے۔
- بعض اوقات آپ اچانک ہی بہت زیادہ حدت محسوس کریں یا آپ کو پسینہ آئے۔(hot flashes)
- آپ کی فرج کم گیلی اور چھوٹی ہو جائے۔
- آپ کے احساسات آسانی سے تبدیل ہونے لگیں۔

یہ علامات اس وقت ختم ہو جاتی ہیں جب آپ کا جسم کم مقدار میں ایسٹروجن کا عادی ہو جاتا ہے۔

اگر آپ ماہواری کے خاتمے (سنِ یاس) کے دوران بے اطمینانی محسوس کریں

اگر آپ بے اطمینانی یا بے چینی محسوس کریں تو مندرجہ ذیل تدابیر آزمائیں۔

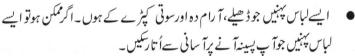

- ایسے لباس پہنیں جو ڈھیلے، آرام دہ اور سوتی کپڑے کے ہوں۔ اگر ممکن ہوتو ایسے لباس پہنیں جو آپ پسینہ آنے پر آسانی سے اُتار سکیں۔

  - گرم اور مصالحے دار غذائیں یا مشروبات سے گریز کریں۔ یہ hot flashes کا سبب بن سکتے ہیں۔

  - روزانہ ورزش کریں

  - کافی، چائے یا سافٹ ڈرنک (پاپ، سوڈا، کولا) وغیرہ زیادہ نہ پئیں۔ ان میں کیفین ہوتی ہے جس سے آپ خود کو نروس محسوس کرسکتی ہیں اور آپ کی نیند متاثر ہوسکتی ہے۔

- اگر آپ الکحل استعمال کرتی ہیں تو کم مقدار میں پئیں۔ الکحل خون کا اخراج اور ہاٹ فلیشز بڑھا سکتی ہے۔

- تمبا کو پینا، سونگھنا یا کھانا بند کردیں یہ خلاف معمول اخراج خون کا سبب بن سکتی ہے اور کمزور ہڈیوں کو زیادہ خستہ کرسکتی ہے۔

- اپنے گھر والوں کو بتائیں کہ آپ کے احساسات آسانی سے تبدیل ہوجاتے ہیں۔ سنِ یاس سے گزرنے والی دوسری عورتوں سے یہ تبادلہ خیال بھی معاون ہوسکتا ہے کہ اس دوران آپ کیا محسوس کرتی ہیں۔

- اپنی کمیونٹی میں استعمال ہونے والی روایتی دواؤں کے بارے میں معلوم کیجئے۔ عموماً سنِ یاس سے گزرنے والی عورتیں وہ طریقے جانتی ہیں جنہیں اپنا کر آپ بہتر محسوس کرسکتی ہیں۔ ماضی میں ڈاکٹرس سنِ یاس کے دوران آنے والی تبدیلیوں اور بے اطمینانی کے لئے ہارمون ری پلیسمنٹ تھراپی (HRT) تجویز کرتے تھے، بدقسمتی سے یہ طریقہ عورتوں کے لئے چھاتی کے کینسر، عارضہ قلب، خون میں لوتھڑے بننے اور فالج کے خطرات میں اضافے کا سبب ثابت ہوا ہے لہٰذا اس طریقہ کا استعمال نہ کرنا بہتر ہے۔

بڑھاپا زندگی کا ایک لازمی حصہ ہے۔ عمر کے اس حصے میں ہر فرد اپنی معلومات، بہتر حکمت عملی اور منصوبہ بندی سے خود کو بڑھاپے کے مختلف مسائل سے بچا سکتا ہے۔ معذور عورتوں کے لئے یہ مسائل ان کے امکانی حل سمجھنا ضروری ہیں۔

ماہواری کے خاتمے کے بعد جنسی تعلقات

کچھ عورتوں کے لئے سن یاس کا مطلب جنسی تقاضوں سے نجات ہے جبکہ دوسری عورتیں اس لئے جنسی عمل میں زیادہ دلچسپی لیتی ہیں کہ اب وہ غیر مطلوب حمل کا کوئی خوف نہیں رکھتی ہیں ۔ تاہم تمام عورتوں کو محبت اور چاہت کے اظہار کی ضرورت ہوتی ہے ۔

جوں جوں آپ بوڑھی ہوتی ہیں آپ کے جسم میں آنے والی تبدیلیاں آپ کے جنسی تعلقات پر اثر انداز ہو سکتی ہیں ۔ یہ ہو سکتا ہے کہ آپ جنسی عمل کے دوران ہیجان میں آنے میں زیادہ وقت لیں (ایسا مردوں کے ساتھ بھی ہوتا ہے) اور کیونکہ آپ کا جسم اب زیادہ مقدار میں ایسٹروجن جن ہارمون نہیں بناتا ہے اس لئے آپ کی فرج زیادہ خشک ہو سکتی ہے ۔ یہ صورت فطری مباشرت کو تکلیف دہ بنا سکتی ہے یا آپ فرج یا یورین سسٹم کے انفیکشنز کا زیادہ آسانی سے شکار ہو سکتی ہیں ۔

ایسی کوئی وجہ نہیں کہ آپ جب تک زندہ ہیں جنسی عمل سے لطف اندوز نہ ہوں

اس کے ساتھ ہی آپ کی فرج کی اندرونی جلد بھی پتلی ہو جائے گی ، اس لئے مباشرت سے پہلے زیادہ وقت صرف کرنا یقینی بنائیں تا کہ آپ کی فرج اپنے طور پر قدرتی نمی بنا سکے ۔ آپ اس مقصد کے لئے تھوک ، وجیٹیبل آئل (کورن آئل ، زیتون کا تیل) یا پانی سے بنی چکنائی استعمال کر سکتی ہیں ۔

اہم باتیں :

- اگر آپ کنڈوم استعمال کر رہی ہوں تو تیل استعمال نہ کریں ۔ تیل کنڈوم کو کمزور کر سکتا ہے جس سے وہ پھٹ سکتا ہے ۔

- فرج میں نمی بڑھانے کے لئے خوشبو والی پیٹرولیم جیلی یا تیل استعمال نہ کریں اس سے سوزش پیدا ہو سکتی ہے ۔

- اپنی فرج کو خشک کرنے کے لئے کوئی چیز استعمال نہ کریں اس سے بھی فرج میں سوزش ہو سکتی ہے جس سے آپ ایچ آئی وی یا دوسرے انفیکشنز میں آسانی سے مبتلا ہو سکتی ہیں دیکھئے صفحہ 169 ۔

- پیشاب کے مسائل سے بچنے کے لئے مباشرت سے پہلے پانی یا جوس وغیرہ پئیں تا کہ آپ اس کے فوراً بعد جس قدر جلد ممکن ہو پیشاب کر سکیں ۔ اس طرح جرثوموں کے بہہ جانے سے انفیکشنز سے تحفظ ملے گا جو دوسری صورت میں یورین ٹیوب (Urethra) کے ذریعے مثانے میں داخل ہو سکتے ہیں ۔

اپنے آپ کو حمل اور جنسی طور پر منتقل ہونے والے انفیکشنز (STIs) سے بچائیے

ماہواری بند ہونے کے بعد سال بھر کے دوران آپ اب بھی حاملہ ہوسکتی ہیں غیر مطلوب حمل سے بچنے کے لئے سال بھر (بارہ مہینے) ماہواری نہ ہونے تک خاندانی منصوبہ بندی کا کوئی طریقہ اپنائیے (دیکھئے صفحہ 188) اگر آپ خاندانی منصوبہ بندی کے لئے ہارمونل طریقہ (گولیاں، انجکشن، امپلانٹ) استعمال کررہی ہیں تو پچاس برس کی عمر کے لگ بھگ اسے بند کر دیں۔ ہارمونز کے بغیر خاندانی منصوبہ بندی کا کوئی اور طریقہ استعمال کریں حتیٰ کہ آخری ماہواری کو بارہ ماہ ہوجائیں۔ جب تک اس بات کا یقین نہ ہوکہ آپ اور آپ کا شریک حیات ایچ آئی وی/ ایڈز سمیت کسی ایس ٹی آئی میں مبتلا نہیں ہے مباشرت کے وقت کنڈوم استعمال کیجئے۔ اس صورت میں بھی جب آپ حاملہ نہیں ہوسکتی ہوں۔

فعال زندگی گزارنا

جس قدر ممکن ہو اپنی زندگی فعال گزارنے کی کوشش کریں۔ معاونت کسی فرد کی ہو یا کسی مددگار آلے کی آپ اپنے خاندان اور کمیونٹی میں ایک بھرپور اور فعال زندگی گزار سکتی ہیں اور اپنی خوشی کے لئے زیادہ کام کر سکتی ہیں۔ خود کو ذہنی اور جسمانی طور پر فعال رکھنے کے لئے کمیونٹی میں گھوم پھریں، دوسرے لوگوں سے ملیں

آپ مطالعے یا دوسرے لوگوں کے ساتھ کھیلوں میں حصہ لے کر اپنے دماغ کو فعال رکھ سکتی ہیں کارڈ گیمز، لفظوں پر مشتمل گیمز (اسکریبل) ، شطرنج یا اپنی کمیونٹی میں مقبول دوسرے گیمز آپ کو اپنے طور پر لطف اندوز ہونے اور دوسرے لوگوں کے ساتھ رہنے اور بات کرنے کے مواقع فراہم کرتے ہیں۔ بچوں کو پڑھنا سکھائیں، انہیں اپنی کمیونٹی کے پس منظر سے آگاہ کریں یا اسکول کے کام میں ان کی مدد کریں آپ ایک عمر کی دانش اور تجربہ رکھتی ہیں۔ اہل خانہ، دیکھ بھال کرنے والوں اور کمیونٹی کے دوسرے بوڑھوں، معذور اور بوڑھی عورتوں کے ساتھ مل کر کام کرنا انتہائی بھرپور ثابت ہوسکتا ہے۔

نئی باتیں سیکھنے کا وقت

یوگنڈا کے پالیسا ضلع کی امیلڈا کی عمر 67 برس ہے حال ہی میں اس نے یونیورسل پرائمری ایجوکیشن پروگرام میں داخلہ لیا ہے اور اس طرح وہ ایک بار پھر اسکول پہنچ گئی ہے۔ اب وہ انگریزی لکھ اور بول سکتی ہے اسکول کے دوسرے طلبہ اسے پسند کرتے ہیں اور اُسے دادی ماں کہہ کر پکارتے ہیں۔

تبدیلی کے لئے کام

ہر چند کہ عمر بڑھنے کے ساتھ آپ کی معذوری سنگین ہوسکتی ہے لیکن آپ ایسی سرگرمیاں تلاش کریں جن میں آپ بہتر طور پر شامل ہوسکیں۔ آپ اپنے تجربات کی بدولت کے ساتھ معذورعورتوں کی حالت زار بہتر بنانے کے لئے بہت سی چیزیں کرسکتی ہیں۔

بوڑھی معذور عورتیں کیا کرسکتی ہیں؟

- بہت سی حکومتیں بوڑھے معذور افراد کو ماہانہ پنشن، رہنے اور صحت کی دیکھ بھال کی سہولت فراہم کرتی ہیں اگر آپ کی حکومت ایسا نہیں کرتی ہے تو آپ دوسری معذورعورتوں اور غیر معذور ماؤں، بہنوں، بیٹیوں، بیٹوں اور پڑوسیوں کے ساتھ مل کر تبدیلیوں کے لئے کام کرسکتی ہیں۔ اس قسم کی تبدیلی میں وقت لگتا ہے۔

- ہم معذورعورتوں کے گروپ بنا سکتے ہیں جو کمتر اخراجات کے لئے مل کر رہیں اور ایک دوسرے کی مدد بھی کریں مثلاً ایک نا

مل جل کر کام کرتے ہوئے ہم اپنی کمیونٹی کے لیڈروں سے ملاقات کرکے ان سے بوڑھی معذورعورتوں کے لئے مختلف خدمات اور سہولیات جیسے کم لاگت کی رہائش کے لئے درخواست کرسکتے ہیں۔

ہم صحت کارکنوں سے معذور بوڑھی عورتوں کی صحت کی ضروریات کے حوالے سے تبادلہ خیال کرسکتے ہیں اور یہ کہ کس طرح مراکز صحت کو قابل رسائی بنایا جاسکتا ہے۔

بینا عورت ایک بہری عورت کے لئے کان (سننے کا ذریعہ) بن سکتی ہے اور ایک بہری عورت نابینا عورت کے لئے آنکھیں (دیکھنے کا ذریعہ) بن سکتی ہے۔

- ہم زندگی کا وسیع تجربہ رکھتے ہیں اور اپنی معلومات میں نوجوان معذورعورتوں اور لڑکیوں کو شامل کر سکتے ہیں۔ ہم انہیں درپیش مسائل کے حل کے لئے مدد فراہم کر سکتے ہیں کیونکہ ہم نے بھی وہی مسائل اس وقت جھیلے تھے جب ہم نوجوان تھے۔

باب -14

زیادتی، تشدد اور ذاتی تحفظ

زیادتی کسی بھی عورت کے ساتھ ہوسکتی ہے۔ دنیا بھر میں بہت سی عورتوں کے ساتھ اجنبی یا پہچانے جانے والے لوگ انتہائی بُرا اسلوک کرتے ہیں، وہ مار پیٹ، عصمت دری، جنسی حملوں یا دیگر طریقوں سے زیادتی کا نشانہ بنتی ہیں یا ہلاک تک کردی جاتی ہیں۔ بہت سی صورتوں میں کوئی بھی ان پر ہونے والی زیادتی سے آگاہ نہیں ہوتا ہے کیونکہ وہ اس بارے میں بات کرتے ہوئے شرمندگی یا خوف محسوس کرتی ہیں۔ وہ سمجھتی ہیں کہ کہ کوئی بھی ان کی پرواہ نہیں کرے گا یا وہ خوفزدہ ہوتی ہیں کہ خود انہیں ہی زیادتی کے باوجود مورد الزام ٹھہرایا جائے گا۔

بہت سی عورتوں کے ساتھ ناروا اسلوک اس لئے کیا جاتا ہے کہ وہ زیادتی کرنے والے فرد کے مقابلے میں کمتر حیثیت اور اختیار رکھتی ہیں یا اس لئے کہ وہ تنہا، کمزور اور آسان ہدف ہوتی ہیں۔ غیر معذور عورتوں کی بہ نسبت معذور عورتیں اور لڑکیاں زیادتی، ظلم اور جنسی حملوں کے زیادہ امکانات رکھتی ہیں۔ انہیں بے بس اور کمزور اور نسبتاً کم اہم سمجھا جاتا ہے۔ ایک عورت کی معذوری اسے کسی بھی لحاظ سے تشدد، زیادتی یا نظر انداز کئے جانے کے قابل نہیں بناتی ہے۔ معذور عورتیں بھی ان لوگوں کے ساتھ محفوظ طور پر زندگی بسر کرنے کا حق رکھتی ہیں جو ان کی دیکھ بھال کریں اور ان کے ساتھ اچھی طرح پیش آئیں۔

ایک معذور عورت یا لڑکی کے ساتھ کوئی بھی مرد یا عورت زیادتی کرسکتا ہے۔ وہ اپنے گھرانے کے کسی فرد، اپنے شوہر یا سہیلی یا کسی اور فرد یا دیکھ بھال کرنے والے کی زیادتی کا نشانہ بن سکتی ہیں۔ اس کے ساتھ کوئی پڑوسی، خاندانی دوست یا آجر یا ساتھ کام کرنے والا یا کوئی اجنبی بھی زیادتی کرسکتا ہے۔

اگر زیادتی کرنے والا ایسا کوئی فرد ہے جسے عورت جانتی ہے تو ممکن ہے وہ یہ محسوس کرے کہ وہ اس کے خلاف کسی اور سے فریاد نہیں کرسکتی ہے، خصوصاً اس صورت میں جب وہ اپنی گزر بسر یا روزمرہ معاملات کے لئے اس پر انحصار کرتی ہو۔ لیکن جب ایک عورت کسی زیادتی پر خاموشی اختیار کرتی ہے تو اور زیادہ الگ تھلگ اور خطرے کا آسان ہدف بن جاتی ہے۔ کسی ایسے فرد سے رابطہ کرنا جس پر وہ بھروسہ کرسکتی ہو تشدد کی مزاحمت میں اس کا معاون بن سکتا ہے اور اسے مدد مل سکتی ہے۔

کسی بھی عورت کے ساتھ زیادتی یا اس پر تشدد کسی بھی حال میں درست نہیں ہوتا ہے۔ کیونکہ معذور عورتوں کو عموماً کم اہمیت دی جاتی ہے، انہیں بعض اوقات تحفظ دیئے جانے کے قابل نہیں سمجھا جاتا ہے، اس سے کچھ لوگ یہ سمجھتے اور سیکھتے ہیں کہ معذور عورتوں سے زیادتی کوئی مسئلہ نہیں ہے اور ان کے حقوق کا احترام کرنے کی کوئی ضرورت نہیں ہے۔

نقصان دہ سوچ

حقیقت یہ ہے کہ ایک معذور لڑکی بھی دوسری لڑکیوں کی طرح دیکھ بھال کا حق رکھتی ہے۔ زیادتی کا شکار ہونے والا کوئی بھی فرد "خوش قسمت" نہیں ہے۔

> معذور لڑکیاں خوش قسمت ہیں کہ کوئی ان کی دیکھ بھال کرنے والا ہو- خواہ انہیں کچھ زیادتیاں بھی جھیلنا پڑیں

نقصان دہ سوچ

حقیقت یہ ہے کہ کسی معذور عورت کے ساتھ زیادتی درست نہیں ہے۔ زیادتی کسی کے ساتھ بھی نہیں ہونی چاہئے۔ خصوصاً اس عورت کے ساتھ جو آموزشی مشکلات سے دو چار ہو۔ خواہ وہ نظر انداز کیا جانا ہو یا جذباتی ضرر ہو، دیکھ بھال نہ کرنا یا انہیں اسکول جانے سے روکنا، ان سے مشاورت کے بغیر فیصلے کرنا، جسمانی زیادتی یا جنسی تشدد زیادتی کی ہر صورت ختم ہونی چاہئے۔

> اگر آپ کسی معذور کے ساتھ زیادتی کرو تو اس سے کوئی فرق نہیں پڑتا ہے، معذور عورتیں تو احمق ہوتی ہیں اور کوئی ان کی بات سننے کی زحمت نہیں کرتا ہے

معذور عورتوں کے لئے سماج کے تصورات یا عمومی سوچ میں مثبت تبدیلی ضروری ہے۔ معذوری کسی بھی فرد کو کمتر نہیں بناتی ہے۔ معذور افراد کو بھی احترام اور تحفظ ملنا چاہئے۔ کسی بھی معذور کے ساتھ زیادتی نہیں ہو، اس مقصد کے لئے بہت کچھ کرنے کی ضرورت ہے۔

زیادتی کی مختلف اقسام

اکثر لوگ جب زیادتی کے بارے میں سوچتے ہیں تو وہ عام طور پر تصور کرتے ہیں کہ کسی پر پُرتشدد حملہ کیا گیا ہوگا۔ جس کے نتیجے میں وہ زخمی، زدوکوب، زنا بالجبر، جنسی حملے کا نشانہ بنا ہو یا ہلاک ہو گیا ہو۔ تاہم معذور عورتیں جسمانی تشدد کا آسان ہدف ہیں لیکن وہ زیادتیوں کی دوسری اقسام کی زد میں بھی آتی ہیں۔

مثلاً وہ عورتیں جو اپنی روزمرہ دیکھ بھال کے لئے دوسروں کی مدد پر انحصار کرتی ہیں، ان کی تذلیل کی جاسکتی ہے، انہیں غذا، پانی اور دواؤں سے محروم رکھا جاسکتا ہے، انہیں اتنی مدت کے لئے الگ تھلگ چھوڑا جاسکتا ہے کہ وہ غلاظت میں لتھڑ جائیں یا ان کی ضرورت کے مطابق دیکھ بھال نہ کی جائے۔ کچھ لوگ دیکھ بھال کے بدلے میں عورت کو جنسی تبادلے کے لئے مجبور کرسکتے ہیں۔ معذوری رکھنے والی کچھ لڑکیوں اور عورتوں کو دوسرے لوگوں سے ملنے یا گھر سے باہر جانے کی اجازت بہت ہی کم ملتی ہے۔ بہت سی عورتیں تنہا چھوڑ دی جاتی ہیں یا انہیں دوسرے طریقوں سے زیادتی کا نشانہ بنایا جاتا ہے۔

ایک معذور عورت کسی ایسے فرد کی زیادتی کا شکار بھی ہوسکتی ہے جو خود بھی معذور ہو۔ اگر کوئی معذور فرد اپنی معذوری کے باعث خود کو بے بس محسوس کرتا ہے اور غصہ کرتا ہے تو وہ اپنی طاقت یا برتری کی تسکین کے لئے اپنی شریک حیات کو زدوکوب کرکے اپنے غصے کا اظہار کرسکتا ہے۔

میں سمجھتی ہوں کہ ایلن کے ساتھ کانگ بُرا سلوک کرتا ہے

لیکن کانگ تو معذور ہے وہ کسی کو بھی مجروح نہیں کرسکتا ہے

جسمانی تشدد۔ دیگر اقسام کی زیادتیوں اور تشدد کی طرح ذہنی صحت کے مسائل پیدا کرسکتا ہے۔ وہ عورتیں جن پر اکثر تشدد یا زیادتی کی جاتی ہو خوفزدہ، دکھی اور بعض اوقات ڈپریشن میں مبتلا ہوجاتی ہیں جب کسی عورت کے ساتھ زیادتی ہوتی ہے تو اس کے جسم کو بھی صحت یابی کی ضرورت ہوتی ہے اور اس کے ذہن اور روح کو بھی سکون درکار رہتا ہے۔

جذباتی توہین/ زیادتی

جذباتی توہین یا زیادتی اس وقت ہوتی ہے جب کوئی کسی عورت کی توہین کرتا ہے یا اسے خوفزدہ کرتا ہے یا اسے تنہا چھوڑ دیتا ہے یا اس سے ایسا سلوک کرتا ہے کہ جیسے اس کی کوئی اہمیت ہی نہ ہو، کچھ لوگ معذور عورتوں سے یہ کہہ کر زیادتی کے مرتکب ہوتے ہیں کہ بہتر یہی ہے کہ وہ مرجائیں یا یہ کہ وہ تو ایک بوجھ ہیں اور انہیں زندہ رہنے کا کوئی حق نہیں ہے۔

جذباتی زیادتی اس صورت میں ہوسکتی ہے جب کوئی شخص

● کسی بہری عورت کے بارے میں خراب لہجے میں ایسے لوگوں سے بات کرے جو سن سکتے ہوں۔

● یا وہ مختلف کوتاہیوں پر اسے الزام دے یا اس پر چیخے یا چلّائے۔

جذباتی زیادتی عورت کو کمزور بناتی ہے

جذباتی توہین/زیادتی ایک عورت سے اس کا اعتماد چھین لیتی ہے۔ایک ایسی عورت جسے الگ تھلگ چھوڑ دیا گیا ہو،محسوس کرسکتی ہے کہ وہ سماج میں اپنی اہمیت اور مقام گنوا چکی ہے۔یہ صورت اسے غمزدہ اور کمزور بنا سکتی ہے۔

اگر ایک عورت کی مسلسل توہین کی جائے اور اُسے غلط ناموں سے پکارا جائے تو وہ خود کو احمق اور دکھی محسوس کرنے لگے گی اور اگر کوئی اس کے احباب یا پڑوسیوں کے سامنے اس کا مذاق اُڑاتا ہے تو وہ شرمندگی محسوس کرسکتی ہے اور باہر نکلنا کم ہی پسند کرے گی۔کچھ عرصے بعد وہ یہ سمجھنے لگے گی کہ وہ ڈھنگ کا کوئی بھی کام کرنے کے لائق نہیں ہے۔ایک ایسی عورت جو اکثر جذباتی توہین کا نشانہ بنے،دماغی صحت کے مسائل یا ڈپریشن کا شکار ہو جاتی ہے،دماغی صحت پر مزید معلومات کے لئے دیکھئے باب۔3۔

الگ تھلگ چھوڑ دینا

بعض اوقات لوگ معذور فرد کو الگ تھلگ چھوڑ دیتے ہیں یا اس کی دیکھ بھال کرنے سے انکار کردیتے ہیں۔ اگر کوئی گھرانا معذور بچے کی موجودگی پر شرمندہ ہو یا یہ سمجھتا ہو کہ وہ اس کی ضروری دیکھ بھال نہیں کر سکے گا تو وہ اسے گھر سے نکال دیتا ہے۔ ایک عورت سے اس کا شوہر یا گھرانا اس لئے قطع تعلق کرلیتا ہے کہ وہ اس کے جسم میں آنے والی تبدیلی کو قبول نہیں کرتا ہے۔

بہت سی معذور عورتیں جنہیں گھر سے نکال دیا جاتا ہے اپنے رشتہ داروں کے پاس آ جاتی ہیں جو ممکن ہے اسے ناپسند کریں۔ جب کرنے کے لئے بہت سا کام ہو یا وہ گھرانہ غریب ہو تو معذور عورتوں کو بوجھ سمجھا جاتا ہے۔ بعض اوقات رشتہ دار معذور عورت کو اس کی بدقسمتی کا ذمہ دار ٹھہراتے ہیں خصوصاً اس صورت میں جب اس کے بچے بھی ہوں۔

> ہم اتنے غریب ہیں کہ اپنے گھرانے کے لئے بمشکل کھانے کا بندوبست کرپاتے ہیں اور اب ہمیں اسے بھی کھلانا پڑے گا۔ اب ہمیں زیادہ بھوک برداشت کرنی ہوگی اور اس کی ساری ذمہ داری اس پر ہے

> میں یہاں نہیں آتی اگر میرے شوہر نے مجھے نہیں چھوڑا ہوتا۔

تنہائی

کسی معذور عورت کو الگ تھلگ کسی کمرے میں بند کرنا زیادتی کی بدترین صورت ہے۔

جب کوئی کمیونٹی معذور افراد کا احترام نہیں کرتی ہے یا انہیں خود سے خارج کردیتی ہے تو کچھ لوگ اپنے گھر میں موجود معذور عورت یا لڑکی پر شرمندگی محسوس کرتے ہیں۔ ممکن ہے وہ دوسرے لوگوں کو اپنے گھرانوں میں موجود معذور عورتوں یا لڑکیوں کی موجودگی سے لاعلم رکھنا چاہتے ہیں یا یہ ظاہر کرتے ہوں کہ ان کا کوئی وجود ہی نہیں ہے۔ اکثر معذور عورتوں اور لڑکیوں کو تعلیم حاصل کرنے یا کمیونٹی کی تقریبات یا مذہبی رسومات میں حصہ لینے کی اجازت نہیں دی جاتی ہے۔

> میری خواہش ہے کہ میں باہر جاؤں، میں یہاں سارا وقت تنہا رہتی ہوں اور انتہائی دکھ محسوس کرتی ہوں

بعض کمیونٹیز میں معذور عورتیں الگ تھلگ رہتی ہیں اس لئے دوسرے لوگ یہ خوف رکھتے ہیں کہ معذور افراد کی موجودگی انہیں بھی معذور بنا دے گی۔ اور کچھ یہ یقین رکھتے ہیں کہ اگر کوئی معذور عورت کسی حاملہ عورت کو چھوئے تو اس کا بچہ بھی معذور پیدا ہوگا۔ ان میں سے کوئی بھی بات درست نہیں ہے۔ آپ کسی فرد سے معذوری میں مبتلا نہیں ہو سکتے ہیں۔

غفلت یا نظر انداز کرنا

جب کوئی معذور کی دیکھ بھال کرنے والا اسے نظر انداز کرے یا اس کی مدد نہ کرے تو یہ غفلت کہلائے گی۔ مثال کے طور پر اگر کوئی

- اسے کھانا نہ کھلائے یا درست طور پر کھانا نہ کھلائے۔
- جب اسے ضرورت ہو تو دوا دینے میں مدد سے انکار کرے۔
- ٹوائلٹ میں اس کی مدد نہ کرے۔

غفلت کی دوسری مثالیں یہ ہیں۔

- معذور عورت کو طویل عرصے بستر پر رہنے دینا۔
- لباس بدلنے یا نہانے یا دھونے میں اس کی مدد نہ کرنا۔
- دباؤ کے زخموں سے تحفظ کے لئے نقل و حرکت یا پوزیشن بدلنے میں اس کی مدد نہ کرنا۔
- گندی چادر یا گیلے کپڑے نہ بدلنا۔

لوگ معذور عورتوں کو گھر میں چھوڑ کر یا انہیں اچھی تعلیم، مناسب غذا یا لباس نہ دے کر بھی غفلت برتتے ہیں۔

غفلت ایک عورت کو مجروح کرنے کے ساتھ اسے اداس اور خوفزدہ کر سکتی ہے۔ اگر وہ زیادہ عرصے بستر پر رہے یا خود کو حرکت نہ دے سکے تو دباؤ کے زخموں کا شکار ہو سکتی ہے۔ دباؤ کے زخموں سے تحفظ کی مزید معلومات حاصل کرنے کے لئے دیکھئے صفحہ 116

جسمانی زیادتی/تشدد

جسمانی زیادتی میں زدوکوب کرنا، گرانا، تھپڑ یا گھونسہ مارنا یا ضرب لگانا شامل ہے۔ بعض اوقات لوگ معذور عورت کے بچوں کو بھی ہراساں کرتے ہیں۔ اس صورت میں وہ زیادہ خوفزدہ ہوسکتی ہے کیونکہ وہ محسوس کرتی ہے کہ وہ اپنے بچوں کو تشدد سے نہیں بچاسکتی ہے۔

مار پیٹ یا حملے کے علاوہ ایک معذور عورت کے ساتھ اس صورت میں جسمانی زیادتی ہوتی ہے جب

معذور عورتوں کے ساتھ کلینک، اسکولوں، گھروں، کام کرنے کی جگہوں یا پبلک ٹرانسپورٹ وغیرہ میں مختلف انداز سے نامناسب سلوک کے ذریعے جسمانی زیادتی کی جاسکتی ہے۔

● کوئی جان بوجھ کر اس کے استعمال میں آنے والے آلات جیسے اس کا سماعتی آلہ یا بیساکھیاں توڑ دے۔

● کوئی جان بوجھ کر اس کے چلنے پھرنے کو مشکل بنانے کے لئے کسی نابینا عورت کے فرنیچر کی جگہ بدل دے۔

● اسے کسی پبلک اسپتال یا کلینک میں معائنے کے دوران اجنبی لوگوں کے سامنے برہنہ ہونے پر مجبور کیا جائے۔

صحت کارکن معذور عورت کے جسم کو جس حد تک ممکن ہو اس کے اپنے کپڑوں یا چادر سے چھپا کر اس قسم کی صورتحال سے محفوظ کر سکتے ہیں یا اگر دستیاب ہو تو اس کا معائنہ کسی کمرے میں کر سکتے ہیں۔

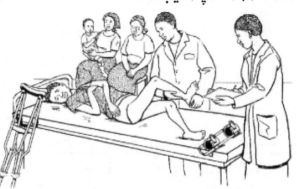

زیادتی سے تحفظ

معذور عورتوں کے لئے زیادتی سے بچاؤ کا ایک طریقہ یہ ہے کہ وہ کمیونٹی میں ممکنہ حد تک زیادہ شریک ہوں۔ دوسروں سے ملنا اور بات کرنا مدد حاصل کرنے میں معاون ثابت ہوسکتا ہے۔

- **ایک سے زیادہ فرد سے رابطہ۔** اگر ابتداء میں دوسرے لوگوں کے لئے آپ کی گفتگو سمجھنا مشکل ہو تو مشق کے بعد وہ آپ کی باتیں بہتر طور پر سمجھ سکیں گے۔ سادہ تصویر کشی بھی آپ کی مدد کرسکتی ہے۔

- **زیادتی کے بارے میں دوسری عورتوں سے جن پر آپ بھروسہ کرتی ہوں بات کیجیے۔** کیا پیش آیا یا یہ بیان کرنا مشکل ہوسکتا ہے، ممکن ہے آپ شرمندہ ہوں یا خوفزدہ ہوں کہ آپ کے ساتھ زیادتی کرنے والا اس سے آگاہ ہوجائے گا۔ آپ اس خیال سے بھی فکرمند ہوسکتی ہیں کہ کوئی آپ کی بات پر اعتبار نہیں کرے گا۔ بعض اوقات آپ گفتگو کرنے کے بعد بے کیفی محسوس کرسکتی ہیں۔ خصوصاً اس وقت جب وہ شخص جس سے آپ بات کررہی ہوں آپ کی بات پر توجہ سے نہ سن رہا ہو۔ لیکن کسی سے بات کرنا عام طور پر مدد حاصل کرنے کا بہترین طریقہ ہے۔ سپورٹ گروپس کے بارے میں مزید معلومات کے لئے دیکھئے صفحہ 65۔

میں خود کو پرسکون رکھنے کی کوشش کرتی ہوں اور براہ راست اس شخص سے بات کرتی ہوں جو میرے ساتھ اچھی طرح پیش نہ آرہا ہو۔ میں یہ واضح کردیتی ہوں کہ مجھے وہ انداز پسند نہیں ہے

جب مجھے معاونت اور مدد کی ضرورت ہوتی ہے میں قابل اعتماد بزرگ عورتوں سے بات کرتی ہوں

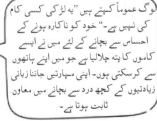

لوگ عموماً کہتے ہیں "یہ لڑکی کسی کام کی نہیں ہے۔" خود کو ناکارہ ہونے کے احساس سے بچانے کے لئے میں نے ایسے کاموں کا پتہ چلا لیا ہے میں اپنے ہاتھوں سے کرسکتی ہوں۔ اپنی مہارتیں جاننا زبانی زیادتیوں کے کچھ درد سے بچانے میں معاون ثابت ہوتا ہے۔

جب میں اسکول میں پڑھتی تھی تو ایک لڑکے نے مجھے دھکا دینے کے بعد میری بیساکھیاں چھین لی تھیں، میں نے اس کی شکایت اپنی ٹیچر سے کی تو انھوں نے اسے سزا دی۔ انھوں نے مجھے یہ بھی سمجھایا کہ میں ہمیشہ دوسرے بچوں کے ساتھ رہا کروں

میں خود کو تنہا محسوس کرتی تھی کہ صرف میرا شوہر ہی علامتی زبان جانتا تھا۔ کمیونٹی میں دوسرے لوگوں کے علامتی زبان سیکھنے، نئے دوست بنانے میں میری مدد کی اور میں دوسرے لوگوں کے ساتھ مل کر زیادہ کام کرنے لگی۔ اب مجھے دوسری عورتوں کی مدد حاصل ہے اور میں ضرورت پڑنے پر ان کی مدد حاصل کرسکتی ہوں۔

مختلف صلاحیتیں رکھنے والی عورتیں اگر مل کر کوئی ایسا کام کرتی ہوں جس میں وہ ماہر ہوں تو ایک دوسرے سے اپنی مہارتوں کا تبادلہ کرسکتی ہیں۔

- **اپنا کام دوسری عورتوں کے ساتھ کریں۔** جہاں دوسرے لوگ دیکھ سکتے ہوں وہاں امکان ہے تو آپ کو ہراساں یا خراب برتاؤ کم کریں۔ اگر آپ کسی پبلک مقام پر زخمی ہوں تو دوسری عورتیں خوفزدہ ہوسکتی ہیں یا آپ کی مدد کے لئے کچھ کرتے ہوئے جھجک سکتی ہیں۔ لیکن بعد میں آپ انہی عورتوں سے مدد بھی حاصل کرسکتی ہیں۔

- اگر آپ کے ساتھ یا آپ کی کسی جاننے والی عورت کے ساتھ زیادتی ہوئی ہو تو کسی کمیونٹی گروپ کی دوسری عورتوں سے مدد حاصل کرنے کی کوشش کریں۔

- اگر آپ جسمانی طور پر مجروح ہوئی ہو تو کسی صحت کارکن سے بات کیجیے۔

زیادتی کرنے والے عموماً عورت کی عزت نفس کو مجروح کرتے ہیں، اسے یہ یقین دلاتے ہیں کہ وہ (اور اس کے بچے) اس کے بغیر زندہ نہیں رہ پائیں گے۔ یاد رکھئے کہ آپ زیادتی کرنیوالے کے بغیر بھی رہ سکتی ہیں۔

تشدد کے خلاف اپنے دفاع کے طریقے جاننے کے لئے دیکھئے صفحہ 308۔

میرا شوہر ہمیشہ یہ کہتا ہے کہ اس نے مجھ سے شادی کرکے مجھ پر احسان کیا ہے۔ کیونکہ میرے گھر والے غریب تھے لہٰذا وہ مجھے الزام دیتا ہے کہ میں نے اس کی دولت کی وجہ سے اس سے شادی کی ہے۔ وہ مجھ سے کہتا ہے میں کوئی اور شریک حیات پانے کے قابل نہیں ہوں اور وہ میرے ساتھ زیادتی کرتا ہے۔ میں نے خاتون وکیل سے مشورہ کرنے اور اپنے شوہر کے رشتہ داروں سے تبادلہ خیال کرنے کے بعد اپنے شوہر کو چھوڑنے کا فیصلہ کرلیا۔

ایک معذور بچی کو جنم دینے کے بعد میری کمیونٹی کی ایک عورت کو اپنے شوہر کے جسمانی تشدد کا سامنا کرنا پڑا، آخر کار اس نے اپنے شوہر سے علیحدگی حاصل کرکے اپنے والدین کے گھر جانے کا فیصلہ کرلیا۔

میری دوست کے ساتھ زیادتی کا سلسلہ اس کے معذور ہونے کے بعد شروع ہوا۔ اس کے شوہر نے اس کی مدد کرنے سے انکار کردیا اور دوسری عورتوں سے تعلقات قائم کرلئے۔ میری دوست اپنے گھر والوں کے پاس لوٹ گئی جہاں اسے پیار اور اچھی دیکھ بھال ملی۔ اس کی حالت بہتر ہونے لگی اور وہ زیادہ صحت مند اور خوش و خرم رہنے لگی

تشدد پسند ساتھی چھوڑنے کے بعد عورتوں کی مدد

بہت سے مقامات پر عورتیں تشدد کے خلاف تحفظ کے قوانین اور عدالتوں کی تشکیل اور تشدد سے تحفظ کے لئے عورتوں کے منظّم ہونے کے باعث وہ محفوظ ہیں۔ دوسرے بہت سے مقامات پر وہ لوگ جو قانون کا نفاذ کرتے ہیں خصوصاً پولیس، قانون داں اور جج صاحبان وغیرہ کسی عورت کی مدد کے لئے قابل بھروسہ نہیں ہیں لیکن تمام کمیونٹیز میں عورتوں کو بہترین تحفظ اس وقت ملتا ہے جب وہ مل کر کام کرتی ہیں۔ ذیل میں اس کی ایک مثال ہے۔

اگر آپ کے شوہر نے آپ کو زدوکوب کیا.......

اور آپ نے اسے چھوڑنے کا فیصلہ کرلیا ہے.......

تو اس بارے میں کسی ایسے فرد سے بات کیجئے جس پر آپ بھروسہ کرتی ہوں (کوئی پڑوسی، دوست یا رشتہ دار)

پھر اپنے مسائل پر گفتگو کے لئے کمیونٹی کی دوسری عورتوں کو اکھٹا کرنے کی کوشش کیجئے۔

پھر آپ سب مل کر اس مرد سے بات کر سکتی ہیں۔ اگر وہ اپنے طور طریقے تبدیل کرنے پر آمادہ نہ ہو تو سب مل کر کمیونٹی کے لئے کسی بزرگ سے رجوع کریں.......

.......یا پولیس میں اس کے خلاف رپورٹ درج کرائیں۔ اگر آپ گروپ کی صورت میں ہوں تو امکان ہے کہ وہ آپ کی شکایت پر سنجیدگی سے کارروائی کرے گی۔

جنسی زیادتی

لڑکیاں خاص طور پر جنسی زیادتی کا خطرہ رکھتی ہیں کیونکہ وہ چھوٹی، کمزور اور اپنی کمیونٹی میں موجود جنس کے حوالے سے سماجی ضابطوں اور طریقہ کار سے کم آگاہ ہوتی ہیں۔ لڑکیوں کے ساتھ باپ یا ماں یا چچا یا ماموں یا دوسرے رشتہ دار یا بھائی یا دوسرے افراد زیادتی کر سکتے ہیں۔ اگر کوئی لڑکی کی خاندان میں کسی کو اپنے ساتھ ہونے والی زیادتی سے آگاہ کرتی ہے تو خاندان عموماً زیادتی کرنے والے کو تحفظ دیتے ہیں اور لڑکی کو اس کا ذمہ دار ٹھہراتے ہیں لیکن جس کے ساتھ زیادتی ہوئی ہو خاص طور پر کسی بچے یا معذور عورت کو اس کا الزام دینا درست نہیں ہے۔

معذور لڑکیاں اور عورتیں زیادتی کا زیادہ خطرہ رکھتی ہیں خاص طور پر جب وہ اپنی معذوریوں کے باعث کمزور ہوں یا ابلاغ میں دشواری رکھتی ہوں یا انہیں سماج میں مکمل طور پر قبول نہ کیا جاتا ہو اور ان کے ساتھ جو بھی پیش آئے سماج اس کی پروا نہیں کرے۔

ایک معذور عورت کے ساتھ اس کا شوہر، اس کے گھرانے کے دوسرے افراد اس کی دیکھ بھال کرنے والا یا کوئی اجنبی جنسی زیادتی کر سکتا ہے۔ زیادہ تر عورتیں کسی جاننے والے مرد کی جنسی زیادتی (عصمت دری) کا نشانہ بنتی ہیں۔ کیونکہ معذور عورت کے گھر والے عموماً اسے سماجی ماحول میں رہنے کی اجازت نہیں دیتے ہیں جہاں وہ دوست بنا سکتی ہے اور مردوں اور عورتوں کے مابین جنسی تعلق کے بارے میں جان سکتی ہے لہذا ممکن ہے کہ وہ یہ سمجھے کہ اس کے لئے زیادتی برداشت کرنے کے علاوہ اور کوئی راہ نہیں ہے۔ وہ یہ بھی سوچ سکتی ہے کہ زیادتی کرنے والے کے علاوہ کوئی اور اس میں دلچسپی نہیں لے گا۔

عام طور پر مختلف انداز سے جنسی زیادتی کے واقعات ہوتے ہیں لیکن بہت ہی کم لوگ انہیں جنسی حملہ یا عصمت دری سمجھتے ہیں۔ جنسی زیادتی کا مطلب عورت سے کوئی بھی ایسا جنسی رابطہ ہے جو وہ نہیں چاہتی ہو۔ ایک لڑکی یا عورت جنسی زیادتی کا شکار اس وقت ہوتی ہے جب

- اس کے بالجبر مباشرت کی گئی ہو یا اس کی مرضی کے بغیر جنسی عمل کیا گیا ہو۔
- اس کی چھاتیوں یا جنسی اعضاء یا جسم کے دوسرے حصوں کو اس کی اجازت کے بغیر چھوا گیا ہو۔
- اسے اپنی ملازمت جاری رکھنے کے لئے یا امتحان میں پاس ہونے کے لئے کسی کے ساتھ مباشرت کے لئے مجبور کیا گیا ہو۔
- دیکھ بھال کے بدلے میں مباشرت کے لئے مجبور کیا گیا ہو۔
- کیونکہ وہ اپنی کفالت کا کوئی اور ذریعہ نہیں رکھتی ہے لہذا اسے رقم یا غذا کے عوض مباشرت قبول کرنی پڑی ہو۔
- رقم، غذا یا دیکھ بھال کے بدلے عریاں تصاویر بنوانی پڑی ہوں۔
- اسے دوسرے افراد کے مابین مباشرت کی باتیں دیکھنی یا سننی پڑی ہوں۔
- اس نے جنسی موضوع پر باتیں سنی یا کی ہوں یا اسے مضطرب کرنے والی جنسی زبان میں گفتگو یا لطیفے سننے پر مجبور کیا گیا ہو۔
- یا پورنوگرافی دکھائی گئی ہو۔

جنسی زیادتی کسی بھی لڑکی یا عورت کے ساتھ ہو سکتی ہے۔ یہ اس کا قصور نہیں ہوتا ہے۔

سیکھنے اور سمجھنے میں دشواریوں سے دوچار عورت سے جنسی زیادتی کے بارے میں گفتگو

تمام بچوں کو سکھایا جاتا ہے کہ وہ بڑوں کا کہنا مانیں۔ جب وہ بڑے ہوتے ہیں تو وہ سیکھتے ہیں کہ کب کہنا نہ مانا جائے۔ لیکن وہ عورتیں جو آموزشی مشکلات رکھتی ہیں عموماً انہیں دوسرے لوگوں پر اعتماد کرنا، خاموش رہ کر خود کو اچھا ثابت کرنا، حجت نہ کرنا اور ان سے جو کہا جائے وہ کرنا سکھایا جاتا ہے۔ یہ صورت انہیں ان لوگوں کے لئے آسان ہدف بنا دیتی ہے جو ان سے ہر قسم کا ناجائز فائدہ اٹھا سکتے ہیں۔

ایسی لڑکیوں اور عورتوں سے جنسی زیادتی کے بارے میں گفتگو کریں جو آموزشی مشکلات رکھتی ہیں اور انہیں یہ سمجھائیں کہ وہ زیادتی سے محفوظ رہنے کا حق رکھتی ہیں۔ انہیں یقینی طور پر آگاہ کریں کہ وہ چھوئے جانے یا زیادتی کی صورت میں کسی ایسے فرد سے بات کر سکتی ہیں جس پر انہیں بھروسہ ہو اور یہ کہ ان کی بات پر یقین کیا جائے گا اور وہ محفوظ رہیں گی۔

آموزشی مشکلات رکھنے والی عورتوں کو یہ بھی سکھائیں کہ وہ کس طرح اپنا دفاع کر سکتی ہیں۔

جنسی حملہ اور زنا

جنسی حملے سے مراد عورت کو غیر مطلوب جنسی رابطہ پر مجبور کرنا ہے۔ زنا جنسی حملے کی سب سے پُرتشدد صورت ہے۔ زنا اس صورت میں ہوتا ہے جب کوئی مرد عورت کی رضامندی کے بغیر اپنا تناسلی عضو، انگلی یا کوئی چیز عورت کی فرج، مقعد یا منہ میں ڈالے۔

لڑکیوں پر جنسی حملہ اور زنا

جنسی حملہ یا زنا کسی کے لئے بھی انتہائی نقصان دہ ہوتا ہے لیکن اس کے اثرات خاص طور پر سنگین اور طویل المدتی لڑکیوں پر ہو سکتے ہیں۔ کیونکہ لڑکیاں جنسی طور پر پختہ نہیں ہوتیں اور ممکن ہے کہ وہ یہ نہ سمجھتی ہوں کہ ان کے ساتھ جو ہوا اسے کس طرح درست طور پر بیان کیا جائے۔ اکثر وہ کسی کے ان باتوں پر یقین کرنے تک رسائی حاصل کرنے میں دشوار وقت گزارتی ہیں۔ بعض کمیونیٹیز میں زنا کا شکار ہونے والی لڑکیوں کو شادی کے قابل نہیں سمجھا جاتا ہے۔

ان جگہوں پر جہاں فوجی تسلط ہو بعض اوقات لڑکیوں کو جبری طور پر فوجیوں اور مسلح گروپوں کی جنسی دل جوئی کے لئے مجبور کیا جاتا ہے۔ یہ لڑکیاں عموماً جسمانی اور جذباتی طور پر نا کارہ ہو جاتی ہیں۔

بعض مقامات پر لوگ یقین رکھتے ہیں کنواری لڑکی کے ساتھ مباشرت ایڈز کا علاج ہوتی ہے اور اس یقین کی وجہ سے بہت ہی کم عمر لڑکیوں اور بچیوں کے ساتھ زنا کیا جاتا ہے۔ کیونکہ ان کے جسم چھوٹے ہوتے ہیں لہٰذا ان کے جنسی اعضاء کو سخت نقصان پہنچتا ہے اور یہ آسانی سے ایڈز اور دیگر جنسی طور پر منتقل ہونے والے انفیکشنز میں مبتلا ہو جاتی ہیں۔

عصمت دری یا جنسی حملہ ہونے کی صورت میں مدد حاصل کرنا

عصمت دری کے معاملے میں ہر عورت کا تجربہ مختلف ہوتا ہے تاہم مدد حاصل کرنے سے قبل خود پر قابو پانے میں مدد کے لئے کچھ چیزیں آپ کی معاون ثابت ہوسکتی ہیں۔ سب سے پہلے خود سے یہ سوالات کریں۔

- آپ مدد کے لئے کس سے کہہ سکتی ہیں؟
- کیا آپ عصمت دری کے بارے میں پولیس کو بتانا چاہتی ہیں؟
- آپ طبی دیکھ بھال کے لئے کہاں جاسکتی ہیں؟
- کیا آپ جرم کے مرتکب کو سزا دلانا چاہتی ہیں؟

عصمت دری کا نشانہ بننے والی معذور عورت کو بھی دوسری عورتوں کی طرح مدد کی ضرورت ہوتی ہے۔ کسی ایسے فرد کو اس بارے میں آگاہ کرنا اہم ہے جو آپ کے ساتھ کسی صحت کارکن کے پاس جا سکے اور اگر آپ اس سے آگاہ کرنا چاہتی ہیں تو اس ضمن میں فیصلہ کرنے میں آپ کی مدد کرے۔ ممکن ہے آپ طویل عرصہ خود کو دکھی، مجروح، خوفزدہ یا برہم محسوس کریں، لہٰذا آپ کو کسی ایسے فرد سے تبادلہ خیال کی ضرورت ہوگی جو آپ کے جذبات سمجھ سکے۔ کسی ایسے فرد کا انتخاب کریں جو آپ کا خیال کرتا ہو، طاقتور اور قابل بھروسہ ہو اور جس پر آپ کو یقین ہو کہ وہ دوسروں سے اس کا ذکر نہیں کرے گا۔ آپ کے گھر والے یا مددگار اپنے ذہنی انتشار کے باعث آپ کو مطلوبہ مدد فراہم کرنے سے قاصر ہوسکتے ہیں۔

عصمت دری کا داغ

کچھ جگہوں پر اس عورت کو جس کی عصمت دری ہوئی ہو، اس کے خاندان یا پوری کمیونٹی کے لئے شرمندگی یا تذلیل کا سبب قرار دیا جاتا ہے۔ اس کو غیر منصفانہ طور پر اپنے اوپر ہونے والے حملے پر نہ صرف اپنے اوپر ہونے والے حملے بلکہ پوری کمیونٹی کی اخلاقی پامالی کا ذمہ دار ٹھہرایا جاتا ہے۔ یہ رسوائی یا داغ کہلاتا ہے۔ اس کے باعث عصمت دری کا نشانہ بننے والی عورت خود پر بیتنے والے سانحے کے بارے میں بتاتے ہوئے خوفزدہ ہوسکتی ہے۔ وہ اس بات سے خوفزدہ ہوسکتی ہے کہ اگر لوگوں کو اس بارے میں پتہ چلے گا تو کمیونٹی اس کے ساتھ مختلف انداز سے پیش آئے گی یا عورت کے گھر والے اسے اپنے گھر انے کی تحقیر کے باعث دوسروں کے علم میں لانا نہیں چاہتے ہیں۔ یہ رسوائی ایک معذور عورت کے لئے بدترین ہوسکتی ہے کیونکہ اس سے اس رسوائی میں اضافہ ہوتا ہے جو لوگوں کے خیال میں پہلے ہی معذور افراد یا خاندان میں معذور افراد کے ہونے کے باعث ہوتی ہے۔

ایسی عورت جس کی عزت لٹی ہو اسے اس پر موردِ الزام نہیں ٹھہرانا چاہئے۔ ایسی عورت کو اپنے گھر والوں اور کمیونٹی کی مدد کی ضرورت ہوتی ہے۔ بدنامی ایک عورت کی بحالی اور مستقبل میں جنسی حملوں کو روکنے کی راہ میں ایک رکاوٹ ہے۔

تم اطمینان سے بات کرو میں تمہاری باتیں ضرور سنوں گی

اور .... اور .... اور پھر ....

اگر آپ مختلف انداز سے باتیں کرتی ہوں تو آپ کے لئے یہ بیان کرنا مشکل ہوسکتا ہے کہ کیا ہوا؟ اگر آپ شرمندہ یا خوفزدہ ہوں تو آپ کے لئے کچھ بولنا مشکل ہوسکتا ہے بعض اوقات جو پیش آیا ہو اس کی وضاحت کے لئے تصویر کشی معاون ہوتی ہے۔ صحت کارکن اور احباب معاملات کو آسان بناسکتے ہیں۔

اگر آپ بہری ہوں

وہ عورتیں جو بہری ہوں یا بولنے میں دشواری رکھتی ہوں انہیں عصمت دری یا زیادتی کی صورت میں مدد حاصل کرنے میں مشکل پیش آسکتی ہے۔ ہر چند کہ وہ خود پر حملہ کرنے والے کے بارے میں یا اس کا حلیہ بتاسکتی ہو، کسی کے بھی علامتی زبان نہ سمجھنے کی صورت میں اسے ہونے والا واقعہ بیان کرنے میں مشکل پیش آئے گی۔

کیونکہ میرا شوہر مجھے مارتا ہے لہٰذا میں پولیس کے پاس گئی، وہ میری علامتی زبان نہیں سمجھتے تھے اور تحمل سے کام نہیں لے رہے تھے۔ میری سوکن نے بھی میرے شوہر کا دفاع کیا اور کسی نے بھی مجھ پر یقین نہیں کیا۔

اگر آپ کی کسی شناسا کے ساتھ زیادتی (عصمت دری) ہوئی ہو

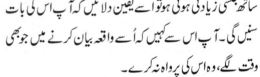

اگر آپ کسی ایسی عورت سے بات کر رہی ہوں جس کے ساتھ جنسی زیادتی ہوئی ہو تو اسے یقین دلائیں کہ آپ اس کی بات سنیں گی۔ آپ اس سے کہیں کہ اُسے واقعہ بیان کرنے میں جو بھی وقت لگے، وہ اس کی پرواہ نہ کرے۔

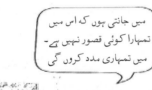

میں جانتی ہوں کہ اس میں تمہارا کوئی قصور نہیں ہے۔ میں تمہاری مدد کروں گی

- اسے یقین دلائیں کہ یہ اس کا قصور نہیں ہے۔

- اس کی معاون بنیں، اس کے جذبات سے آگاہ ہوں، اسے یہ فیصلہ کرنے میں مدد دیں کہ اسے کیا ضرورت ہے؟ اور اسے یقین دلائیں کہ وہ اپنی زندگی اس کے ساتھ بھی گزار سکتی ہے۔

- پرائیوسی اور تحفظ کے لئے اس کی خواہشات کا احترام کریں۔ جب تک وہ نہ چاہے کسی سے اس کے بارے میں کچھ نہ کہیں۔

- اس کے ساتھ صحت کارکن کے پاس، پولیس میں رپورٹ درج کرانے یا کسی تربیت یافتہ معاون کے پاس جائیں، کسی وکیل سے ملنے میں اس کی مدد کریں اور عدالت میں جائیں۔

- اگر آپ اس کی عزت لوٹنے والے کو جانتی ہیں تو اسے تحفظ فراہم کریں نہ کریں اگر ممکن ہو تو دوسری عورتوں کو اس مرد کے بارے میں آگاہ کریں۔ وہ کمیونٹی میں موجود ہر عورت کے لئے ایک خطرہ ہے۔

اگر آپ پولیس سے رجوع کرتی ہیں

بیشتر مقامات پر عصمت دری یا زنا ایک جرم ہے لیکن یہ ثابت کرنے میں طویل وقت لگ سکتا ہے اور انتہائی مشکلات پیش آ سکتی ہیں کہ آپ کے ساتھ زنا کیا گیا ہے۔ پولیس سے رجوع کرنے کا فیصلہ اچھی طرح سوچ سمجھ کر کیجیے۔ کیا پولیس اس سے قبل آپ کی کیمونٹی میں عصمت دری کا شکار دوسری عورتوں کی مدد کر چکی ہے؟ اگر آپ چاہتی ہیں کہ آپ کا معاملہ راز رہے تو کیا پولیس اسے دوسروں سے علم میں لانے سے باز رہے گی؟

پولیس کے پاس تنہا نہ جائیں۔ کچھ کیمونٹیز میں پولیس کے پاس تنہا جانے والی عورت ایک بار پھر پولیس کے ذریعے عصمت دری کا خطرہ رکھتی ہے۔ یقینی بنائیں آپ کے ہمراہ کوئی اور بھی جائے۔

اگر آپ پولیس میں عصمت دری کی رپورٹ درج کرانا چاہتی ہیں تو آپ ممکنہ حد تک جلدی ایسا کریں۔ جانے سے پہلے نہ تو اپنا بدن دھوئیں، غسل نہ کریں اور کپڑے بھی نہ بدلیں۔ اس سے یہ ثابت کرنے میں مدد مل سکتی ہے کہ آپ کی عصمت دری کی گئی ہے۔ پولیس ممکن ہے اپنے لئے کام کرنے والے کسی ڈاکٹر سے آپ کے طبی معائنے کے لئے کہے۔ یہ معائنہ بھی یہ ثابت کرنے میں معاون ہو سکتا ہے کہ آپ کی عصمت دری ہوئی ہے۔

اگر مجرم گرفتار ہو تو آپ کو اسے پولیس کے سامنے یا عدالت میں جج کے سامنے شناخت کرنا ہوگا۔ عصمت دری کے معاملے میں عدالت جانا آسان نہیں ہوتا ہے یہ بیان کرنا کہ کیا ہوا تھا، آپ کو ایک بار پھر عصمت دری ہونے کے احساسات سے دو چار کر سکتا ہے۔ کوئی بھی یہ نہیں سمجھے گا۔ کچھ آپ کو ہی مورد الزام ٹھہرانے کی کوشش کر سکتے ہیں یا یہ کہہ سکتے ہیں کہ آپ جھوٹ بول رہی ہیں۔

اور کچھ لوگ صرف آپ کی معذوری کی وجہ سے آپ کی بات نہیں سنیں گے۔ ممکن ہے کہ وہ سمجھتے ہوں کہ ایک معذور عورت سچی یا باوثوق گواہ نہیں ہو سکتی ہے تاہم کچھ معذور عورتوں نے عدالت میں کامیابی حاصل کی ہے خصوصاً ان عورتوں نے جنہیں اپنی کیمونٹیز کی تائید حاصل تھی جب آپ عدالت جانے کا فیصلہ کریں تو اپنے کسی قابل اعتماد فرد کو ضرور لے جائیں۔

عصمت دری کے باعث صحت کے مسائل

عصمت دری کے بعد سب سے بہتر یہ ہے کہ کسی صحت کارکن سے ملا جائے، اس صورت میں جب بھی آپ مجروح نہ ہوئی ہوں۔ صحت کارکن کو بتائیں کہ آپ کی عصمت دری کی گئی ہے۔ وہ عصمت دری کے باعث ہونے والے صحت کے عمومی مسائل کے تدارک اور علاج میں آپ کی مدد کر سکتی ہے۔

حمل

اگر فوری قدم اٹھائیں اور ایمر جنسی فیملی پلاننگ کے طریقے استعمال کریں تو آپ حمل سے بچ سکتی ہیں۔ صحت کارکن سے اس بارے میں بات کیجیے۔ عصمت دری کے فوراً بعد جس قدر جلد ممکن ہو ایمر جنسی فیملی پلاننگ کیجیے پانچ دن (120 گھنٹوں) کے بعد نہیں۔ دیکھیے صفحہ 357۔

اگر ایمر جنسی طریقہ استعمال کرنے کے بعد بھی اگلی بار ماہواری اپنے وقت پر نہ آئے تو فوری طور پر یہ اطمینان کرنے کے لئے کہ آپ حاملہ تو نہیں اپنا معائنہ کرائیں۔ اگر آپ سمجھتی ہیں کہ آپ حاملہ ہیں تو صحت کارکن سے رجوع کیجیے۔ کچھ ممالک میں عصمت دری کا شکار ہونے والی لڑکی یا عورت کا اسقاط حمل قانونی طور پر جائز ہے۔

جنسی طور پر منتقل ہونے والے انفیکشنز (ایس ٹی آئی)اور ایچ آئی وی/ ایڈز

ممکن ہے آپ کی عصمت دری کرنے والا فرد جنسی طور پر منتقل ہونے والے انفیکشنز یا ایچ آئی وی/ ایڈز میں مبتلا ہوا اورآپ میں منتقل کرسکتا ہو۔ایک صحت کارکن آپ کو جنسی طور پر منتقل ہونے والے امراض جیسے گنور یا، سفلس اور کلے مائیڈیا وغیرہ سے تحفظ کی ادویات دے سکتا ہے۔ بجائے اس کے کہ آپ انفیکشن کی علامات کا انتظار کریں ان سے تحفظ بہتر ہے۔

آپ دو سے چار ہفتوں میں ایچ آئی وی ٹیسٹ کی کوشش بھی کرسکتی ہیں (دیکھئے صفحہ 172)۔ جب تک منفی نتیجہ سامنے نہ آئے اپنے پارٹنر کو ممکنہ انفیکشن سے محفوظ رکھنے کے لئے کنڈوم استعمال کیجئے۔ اگر آپ کسی ایسے علاقے میں رہتی ہیں جہاں بہت سے لوگ ایچ آئی وی/ ایڈز میں مبتلا ہیں تو آپ صحت کارکن سے انفیکشن میں مبتلا ہونے کا خطرہ کم کرنے کے لئے ادویات لینے کے بارے میں بات کرسکتی ہیں۔

خراشیں اور زخم

عصمت دری سے تناسلی اعضاء پر خراشوں یا زخم کے باعث انہیں نقصان پہنچ سکتا ہے۔ یہ درد کا سبب ہوسکتا ہے لیکن یہ درد وقت کے ساتھ ہی ختم ہوگا۔اگر خون خوب بہا ہوتو صحت کارکن سے رجوع کیجئے جو زخموں کو سی سکے اور آپ کو انفیکشن سے تحفظ کی دوا دے سکے۔چھوٹے زخموں اور خراشوں کے لئے:

● اُبال کر ٹھنڈا کئے ہوئے پانی میں اپنے تناسلی اعضاء کو دن میں تین بار رکھئے۔ پانی اُبالتے ہوئے اس میں بابونہ (Chamomile) کے پھول ڈالنے سے زخموں کے بھرنے میں مدد ملتی ہے یا آپ خراشوں اور زخموں پر گھیکوار کے پتوں کا عرق بھی لگا سکتی ہیں۔

● پیشاب کرتے ہوئے اپنے تناسلی اعضاء پر پانی بہائیے تا کہ جلن محسوس نہ ہو۔خوب پانی بہنے سے پیشاب کے دوران سوزش میں کمی آتی ہے۔

● انفیکشنز کی مندرجہ علامات پر نظر رکھئے۔ حدت، زرد مائع (پس) ، ناگوار بو اور درد جو بڑھتا جائے۔

مثانے اور گردے کے انفیکشنز

پُرتشدد جنسی عمل کے بعد عورتوں میں مثانے یا گردے کا انفیکشن عام ہے۔ اگر آپ پیشاب کرتے ہوئے درد محسوس کریں یا پیشاب میں خون آئے تو صحت کارکن سے ملیئے۔ آپ کو دواؤں کی ضرورت ہوسکتی ہے، خوب پانی پئیں (دن بھر میں کم از کم آٹھ گلاس) دیکھئے صفحہ (106-105)۔

صحت کارکن کے لئے

اگر آپ کسی ایسی عورت سے ملیں جو عصمت دری یا زیادتی کا شکار ہوئی ہو تو اس کے ساتھ ہمدردی اور معاملہ فہمی کے ساتھ پیش آئیں۔ اسے یہ بتانے پر مائل کریں کہ اس کے ساتھ کیا پیش آیا۔ اس کی باتیں توجہ سے سنیں اور اسے احساس دلائیں کہ آپ اس پر یقین کر رہی ہیں۔ اسے الزام مت دیں۔ ممکن ہے وہ چھوا جانا یا دیکھا جانا ناگوار محسوس کرے لہٰذا اسے چھونے سے پہلے بتائیں کہ آپ کس طرح اس کا معائنہ کریں گی۔ اس کے آمادہ ہونے کا انتظار کریں۔ یاد رکھیں کہ عصمت دری اور تشدد کے حوالے سے اس کے احساسات طویل مدت حتیٰ کہ برسوں قائم رہ سکتے ہیں۔

اس کے مسائل صحت کا علاج کریں۔ اسے ایس ٹی آئی اور حمل سے بچاؤ اور ایچ آئی وی/ایڈز کا خطرہ کم کرنے کی ادویات دیں۔ اگر وہ عصمت دری کے باعث حاملہ ہو تو اسے یہ فیصلہ کرنے میں مدد دیں کہ اسے کیا کرنا چاہئے۔

لکھئے کہ اس کی عصمت دری کس نے کی اور اس کے ساتھ کیا پیش آیا۔ اگر آپ کا کلینک ریکارڈ محفوظ نہیں رکھتا ہے تو آپ یہ ریکارڈ کسی محفوظ جگہ رکھیں۔ اس کے جسم کے سامنے اور پشت کے حصوں کا خاکہ بنا کر نشان لگائیے کہ اسے کہاں کہاں زخم آئے ہیں۔ آپ نے جو کچھ لکھا وہ اسے دکھائیں یا بتائیں اور وضاحت کریں کہ اگر وہ پولیس میں اس کی رپورٹ درج کرائے یا عصمت دری کرنے والے کے خلاف قانونی کارروائی کرے گی تو یہ ریکارڈ اس کی تائید میں استعمال ہو سکتا ہے۔

اس کے جذباتی اور ذہنی مسائل کا علاج کریں۔ اس سے پوچھیں کہ کیا اس نے اس بارے میں کسی سے بات کی ہے۔

عزت نفس اور زندگی پر اپنا اختیار بحال کرنے میں اس کی مدد کیجئے۔

خود فیصلے کرنے میں اس کی مدد کیجئے۔ اگر وہ پولیس میں رپورٹ درج کرانا چاہتی ہے تو قانونی خدمات کے حصول میں اس کی مدد کیجئے۔ عصمت دری کا نشانہ بننے والی عورتوں کے لئے کمیونٹی میں موجود دیگر سہولیات کے حصول میں اس کی معاون بنیں۔

اگر اس کا شریک حیات یا گھر والے اس بارے میں نہیں جانتے ہیں تو انہیں آگاہ کرنے میں اپنی مدد کی پیشکش کیجئے۔

آپ از سر نو بحالی تک اس کی مدد کے طریقوں کے جاننے میں ان کی مدد کر سکتی ہیں۔ یاد رکھیے کہ عام طور پر گھر والوں کو بھی عصمت دری کے حوالے سے اپنے احساسات پر قابو پانے کے لئے مدد کی ضرورت ہوتی ہے۔

اگر آپ صحت کارکن ہیں تو عصمت دری یا زیادتی کا شکار ہونے والی معذور لڑکی یا عورت کے معائنے سے قبل اس کی اجازت حاصل کیجئے اس سے یہ سمجھنے میں مدد ملے گی کہ وہ خود کو چھوئے جانے پر اپنا اختیار رکھتی ہے

عصمت دری کے بعد ازدواجی تعلق

عصمت دری کے بعد آپ نارمل جنسی تعلقات قائم کرسکتی ہیں بس آپ کو انتظار کرنا ہوگا کہ آپ کے تناسلی اعضاء کے زخم اور خراشیں وغیرہ بھر جائیں۔ بہت سی عورتیں مباشرت کے دوران اپنی عصمت دری کے حوالے سے سوچنے لگتی ہیں اگر آپ کے ساتھ یہ ہو تو اپنے شوہر سے اس بارے میں بات کیجیے کہ کیوں جنسی عمل آپ کو حسب سابق محسوس نہیں ہوتا ہے، کیوں یہ آپ کو خوفزدہ کردیتا ہے اور کیوں آپ کو مہلت درکار ہے۔ آپ اپنے شوہر سے درخواست کریں کہ وہ آپ کے تناسلی اعضاء کو چھوئے بغیر آپ کو نرمی سے چھوکر اور لپٹا کر اس خوف پر قابو پانے میں آپ کی مدد کرے۔

جوں ہی آپ خود کو مطمئن اور محفوظ محسوس کریں گی تو آپ خود کو دوبارہ مباشرت کے قابل محسوس کرسکتی ہیں۔ لیکن اس میں وقت لگے گا اور آپ دونوں کو صبر و تحمل سے کام لینا ہوگا۔

ایسی عورت جس کی عصمت دری ہوئی ہو، اس کا شوہر مہربانی اور معاملہ فہمی سے اس کی نمایاں مدد کرسکتا ہے۔ بعض اوقات ایسی عورت کا شوہر اسے چھوڑ بھی سکتا ہے۔ اگر وہ یقین رکھتا ہو یا اس کی کمیونٹی کا بھی یہی خیال ہو کہ عورت کی عصمت دری نے اس کی بے عزتی کی ہے تو وہ شرمندہ اور برہم ہوسکتا ہے۔ اگر وہ کمیونٹی میں کسی فرد سے اپنے احساسات بیان کرسکتی ہو تو اس سے اسے مدد مل سکتی ہے۔

عصمت دری کے بعد آپ کیا محسوس کرتی ہیں؟

جسم صحت مند ہونے کے بعد عصمت دری طویل عرصہ تک آپ کو پریشان کرسکتی ہے۔ چند عام ردعمل مندرجہ ذیل ہیں۔

ایسی عورت کے لئے جس کی عصمت دری ہوئی ہو یہ اہم ہے کہ وہ کسی سے تبادلہ خیال کرے یا ایسا کام کرے جو بہتر سوچنے میں اس کی مدد کرے۔ ہر عورت اپنی بحالی کے طریقے خود تلاش کرتی ہے۔ کچھ عورتیں عبادت کرتی ہیں۔ دوسری عورتیں زنا کار کو سزا دینے کی کوشش کرتی ہیں یا دوسری عورتوں کو اس سے بچانے کے لئے کام کرتی ہیں۔ آپ جو بھی کریں، خود بھی صبر سے کام لیں اور دوسروں کو بھی اس کی تاکید کریں۔ مزید معلومات کے لئے دیکھئے باب 3، ذہنی صحت۔

اداروں میں زیادتیاں

بعض اوقات جب لوگ معذور عورت کی دیکھ بھال میں دشواری پاتے ہیں تو وہ اُسے کسی ادارے یا اقامتی مرکز میں رکھنے کا فیصلہ کرتے ہیں۔ وہ سمجھتے ہیں کہ ادارہ ان کی بیٹی، بہن یا ماں کی دیکھ بھال ان سے بہتر طور پر کر سکے گا۔ ایسے اداروں یا مراکز میں رہنے والے بہت سے لوگ خود کو ایک کمیونٹی کی طرح محسوس کرتے ہیں۔ ہر چند کہ اقامتی مراکز، اسپتالوں یا محتاج گھروں میں رہنے والے معذور افراد اچھی دیکھ بھال کے باعث بہتر طور پر زندگی گزار سکتے ہیں لیکن وہ ان جگہوں پر زیادتی کا شکار بھی ہو سکتے ہیں۔

کیونکہ اداروں میں رہنے والے لوگ عموماً الگ تھلگ، تنہا اور بے اختیار ہوتے ہیں لہٰذا وہ زیادہ زیادتی کا نشانہ بننے کا زیادہ امکان رکھتے ہیں۔ یہ لوگ زیادہ تر وقت اپنے گھر والوں سے دور رہ کر گزارتے ہیں یا اپنے گھر میں دیکھ بھال کرنے والے نہیں رکھتے ہیں۔

اداروں میں رہنے والے لوگ عموماً اپنی زندگی میں بہت معمولی اختیار رکھتے ہیں۔ عام طور پر ان سے کہا جاتا ہے کہ وہ کیا کریں؟ وہ خود کوئی فیصلہ نہیں کر سکتے ہیں۔ آموزشی مشکلات رکھنے والی عورتیں خاص طور پر ان اداروں میں الگ تھلگ ہو سکتی ہیں کیونکہ وہ سمجھنے یا اپنی بات سمجھانے سے قاصر رہتی ہیں۔

ان اداروں میں رہنے والے لوگوں کے دیگر مسائل کا تعلق اداروں کے انتظامی ڈھنگ سے ہوتا ہے۔ بہت سے اداروں میں مقیم افراد کی تعداد زیادہ ہوتی ہے جبکہ سرمایہ ناکافی ہے۔ عموماً ان اداروں میں کام کرنے والے کام کی زیادتی، فرسٹریشن اور عدم برداشت کا شکار ہوتے ہیں۔ بعض اوقات ان اداروں میں کام کرنے والوں کو حد سے زیادہ اختیارات حاصل ہوتے ہیں، یہ ضابطے بنانے، دیکھ بھال کرنے اور حالات قابو میں رکھنے کے ذمے دار سمجھے جاتے ہیں۔

اس سے قبل بیان شدہ زیادتیوں کے علاوہ معذور عورتیں ان اداروں میں زیادتیوں اور تشدد کی دوسری قسموں کا شکار بھی ہو سکتی ہیں۔

- کام کرنے والوں، دیکھ بھال کرنے والے اور دیگر مقیم افراد کے ساتھ جبری جنسی تعلق
- مار پیٹ اور پُر تشدد ضرب
- کام یا خوشی کے لئے کوئی سرگرمی نہ ہونا، سارا وقت بیزار اور بے کیف گزارنا
- جبری بانجھ پن یا اسقاط
- کمرے میں تنہا بند کیا جانا
- سزا کے بطور سرد پانی سے غسل
- جبری طور پر ادویات دیا جانا (مسکن ادویات)
- دوسرے لوگوں کے ساتھ زیادتی ہوتے ہوئے دیکھنا
- باندھ کر رکھا جانا یا جکڑ کر رکھنا

اداروں میں کام کرنے والے افراد

معذورعورتوں کی دیکھ بھال کے اداروں میں کام کرنے والے بہت سے لوگ نیک نیت ہوتے ہیں لیکن کچھ کام کرنے والے معذوروں کے ساتھ بُرا سلوک بھی کر سکتے ہیں۔ یہ لوگ دوسرے افراد پر اختیار رکھنا پسند کرتے ہیں۔ ان اداروں میں کام کرنے والے دوسرے افراد کا ممکن ہے کہ عورتوں کے ساتھ ہونے والے سلوک پر ذہنی طور پر پریشان ہو سکتے ہیں اور مختلف طور پر کام کرنا چاہتے ہوں یہ دیکھ بھال کرنے والے عام طور پر کم معاوضے پر طویل دورانیے کام کرتے ہیں، ان سے توقع کی جاتی ہے کہ یہ وہی کریں جو ان سے کہا جائے اور یہ ان حالات کار کو بدلنے کا اختیار بہت ہی کم رکھتے ہیں۔

بعض اوقات دیکھ بھال کرنے والے کچھ بھی نہیں کہتے ہیں کیونکہ وہ نہیں جانتے ہیں کہ کس سے

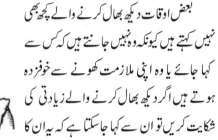

میں مریضوں کے ساتھ زیادہ سے زیادہ وقت گزارنا چاہتی ہوں لیکن اس صورت میں اپنا کام پورا نہیں کرسکوں گی اور اپنی ملازمت سے محروم ہوسکتی ہوں

کہا جائے یا وہ اپنی ملازمت کھونے سے خوفزدہ ہوتے ہیں اگر دیکھ بھال کرنے والے زیادتی کی شکایت کریں تو ان سے کہا جاسکتا ہے کہ یہ ان کا مسئلہ نہیں ہے۔ یا انہیں خوفزدہ کیا جاسکتا ہے یا ان کا مذاق اُڑایا جاسکتا ہے۔ اکثر اوقات دیکھ بھال کرنے والے خراب دیکھ بھال کو معمول کا حصہ سمجھ لیتے ہیں۔ ممکن ہے کہ ادارے کے نگراں ہونے والی زیادتی سے باخبر نہ ہوں یا اگر جانتے ہوں تو وہ ظاہر کرتے ہوں کہ کچھ بھی نہیں ہوا یا وہ کہتے ہوں کہ معذور افراد پر تشدد کوئی اہمیت نہیں رکھتا ہے۔

اداروں کی خراب صورتحال اور معذور افراد کے ساتھ زیادتیاں پوری کمیونٹی کا مسئلہ ہیں۔ لوگوں کی اچھی دیکھ بھال اور زیادتیوں سے تحفظ کے لئے اچھے وسائل کی ضرورت ہوتی ہے۔

اداروں کو تبدیل کرنے کے لئے کام کرنا

اگر آپ کسی ایسے فرد کو جانتی ہوں جسے کسی ادارے میں بھیجا ہوا ہو اور آپ سمجھتی ہوں کہ اس کے ساتھ اچھا برتاؤ نہیں ہو رہا ہے تو تبدیلی کے لئے چند آئیڈیے درج ذیل ہیں۔

● والدین یا گھرانوں کا گروپ بنائیں اور ادارے کے نگراں افراد سے بات کیجئے۔ اگر آپ اکیلے جانے کے بجائے گروپ کی صورت میں جائیں تو آپ کی باتوں پر توجہ دیے جانے کے زیادہ امکان ہیں۔

● کمیونٹی میں بامقصد بیرونی سرگرمیوں اور میل جول کے مواقع پیش کر کے اداروں اور ان میں مقیم افراد کے ساتھ کمیونٹی کا عمل دخل یقینی بنائیں۔

● اداروں میں مقیم افراد سے ملاقات کے لئے کمیونٹی میں مہم چلائیں جس سے اداروں میں مقیم افراد کو ملاقات کے لئے آنے والے فرد کے ساتھ باہر جانے، ملاقات کے لئے آنے والوں کے ساتھ الگ تھلگ رہ کر وقت گزارنے کی سہولت ملے۔

● کمیونٹی پروگراموں اور گھر میں رہتے ہوئے خدمات کی فراہمی کے لئے جدوجہد کیجئے تا کہ معذور افراد کو اداروں میں بھیجنے کی ضرورت ہی نہ رہے۔

آپ تشدد سے بچنے کے لئے کیا کرسکتی ہیں

ذاتی تحفظ/ دفاع

معذور ہونے کا یہ مطلب نہیں ہے کہ آپ یہ قبول کرلیں کہ آپ کمزور ہیں لہٰذا ہمیشہ دوسروں پر انحصار کریں۔ آپ خود کو ممکنہ زیادتی، تشدد یا جنسی حملے سے بچانے کے لئے اپنے تحفظ کے طریقے سیکھ سکتی ہیں۔

کبھی یہ مت سمجھیں کہ آپ تنہا ہیں۔ آپ عورتوں اور مردوں کی ایک عالمگیر تحریک کا حصہ ہیں جو عورتوں کے خلاف تشدد کے خاتمے کے لئے کام کررہی ہے۔

اس کی ابتداء ان لوگوں سے بچ کر کرسکتی ہیں جو آپ کی مرضی یا ضرورت کے خلاف مدد کرنا چاہتے ہیں۔ یہ لوگوں کو اپنی مضبوطی، اپنے لئے بات کرنے کی صلاحیت اور خود فیصلہ کرنے کی خوبی سے آگاہ کرنے کا ایک طریقہ ہے۔ ہر چند کہ مدد کی پیشکش کرنے والا فرد آپ کو نقصان نہ پہنچانا چاہتا ہو، اسے منع کرتے ہوئے خوفزدہ نہ ہوں خواہ وہ برہم ہی کیوں نہ ہو۔ اگر قریب میں دوسرے لوگ موجود ہوں تو بلند آواز میں بات کیجئے تاکہ وہ بھی سن سکیں۔ مستقل مزاجی سے کام لیجئے لیکن ایسے لوگوں سے اکھڑ پن نہ آئیں جو حقیقتاً مگر غیر مطلوبہ مدد کر رہے ہوں۔

جب مرد سمجھتے ہوں کہ وہ آپ کو بلاروک ٹوک چھو سکتے ہیں تو یہ سمجھ سکتے ہیں کہ وہ آپ سے اضافی فوائد بھی حاصل کرسکتے ہیں۔ اگر کوئی آپ کو آپ کی اجازت کے بغیر چھوئے تو اس سے درج ذیل تین باتیں کہہ سکتی ہیں۔

1- تم مجھے کیوں چھور ہے ہو۔
2- میں یہ پسند نہیں کرتی ہوں۔
3- اپنے ہاتھ مجھ سے دور ہٹاؤ۔

اگر کوئی آپ کے بازو پکڑتا ہے یا آپ کی وہیل چیئر دھکیلنا شروع کر دیتا ہے تو ٹھوس اور بلند لہجے میں کہیں:

"تم میرا بازو جکڑ رہے ہو۔"
"میرا بازو مت پکڑو۔"

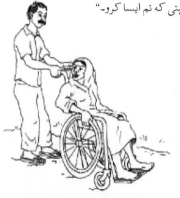

"تم میری وہیل چیئر کیوں دھکیل رہے ہو؟"
ہٹ جاؤ "میری وہیل چیئر مت دھکیلو۔"
"میں نہیں چاہتی کہ تم ایسا کرو۔"

اگر کوئی آپ کی طرف آ رہا ہو اور آپ سمجھتی ہوں کہ وہ آپ کو نقصان پہنچا سکتا ہے تو یہ آزمائیں

"رک جاؤ، میرے قریب مت آؤ۔"

اکثر یہ تنبیہ ایسے شخص کو دور بھگانے کے لئے کافی ہوتی ہے۔ وہ سوچے گا کہ اگر اس نے کوئی فائدہ اٹھانے کی کوشش کی تو آپ اس کے لئے مشکل پیدا کر سکتی ہیں۔ اگر وہ آپ کے نزدیک بڑھنا جاری رکھے تو مدد کے لئے چلّا ئیے۔

حملہ آور اکثر ایسے لوگوں کا انتخاب کرتے ہیں جو آسانی سے ان کا شکار بن سکیں اور ایک معذور عورت خواہ وہ جو بھی معذوری رکھتی ہو، آسانی سے شکار ہو سکتی ہے خصوصاً اس صورت میں جب یہ محسوس ہوتا ہو کہ وہ گمشدہ ہے یا یہ نہیں جانتی ہے کہ وہ کہاں ہے؟ اس لئے بھر پور اعتماد کے ساتھ ردِعمل بھی اسی طرح اہم ہے جس طرح یہ جاننا کہ عملی طور پر کیا کیا جائے۔ جب ایک عورت یقینی انداز میں نقل و حرکت کرتی ہے، بات کرتی ہے یا کوئی قدم اٹھاتی ہے تو وہ ظاہر کرتی ہے کہ جیسے وہ جہاں بھی ہے اپنے ہونے کا جواز رکھتی ہے۔ وہ پُر اعتماد انداز سے ہر کام کرتی ہے اور ایک بھر پور اور مضبوط شخصیت نظر آتی ہے۔ عزتِ نفس کے موضوع پر معلومات کے لئے دیکھئے صفحہ 62 تا 65۔

جب آپ پر حملہ ہو تو کیا کریں

اگر ایک عورت خود کو نقصان پہنچانے والے کے خلاف مزاحمت کرتی ہے تو وہ عموماً خود کو آبرو ریزی سے بچا سکتی ہے۔ کچھ لوگ سمجھتے ہیں کہ آبرو ریزی کے خلاف مزاحمت حملہ آور کو برہم کر دے گی لیکن حملہ آور تو پہلے ہی خطرناک ہوتا ہے۔

یہ جاننا ناممکن ہے کہ اگر کوئی آپ کی آبرو ریزی کی کوشش کرے تو آپ کس طرح ردِعمل کا اظہار کریں گی۔ کچھ عورتیں غصے سے بھر جاتی ہیں اور ایسی قوت محسوس کرتی ہیں جس سے وہ واقف نہیں ہوتی ہیں۔ کچھ محسوس کرتی ہیں کہ وہ حرکت نہیں کر سکتی ہیں۔ لیکن یاد رکھیئے اگر آپ کی آبرو ریزی ہوئی ہے تو اس لئے نہیں کہ آپ اپنا تحفظ کرنے میں ناکام رہیں۔ آبرو ریزی کبھی بھی آپ کا قصور نہیں ہوتی ہے۔

اگر کوئی آپ پر حملہ کرتا ہے یا آپ کی عزت لوٹنے کی کوشش کرتا ہے تو آپ اس سے بچنے کے لئے جو کچھ کر سکتی ہیں ضرور کریں۔

- کچھ ایسا کریں جو اس کے کراہیت کا باعث بنے جیسے رال ٹپکانا یا اس پر تھوکنا وغیرہ۔
- اس کے جسم کے نازک حصوں جیسے آنکھوں، ناک، یا خصیوں کو نوچ کر یا ضرب لگا کر مجروح کریں۔

شور مچائیں، چلائیں یا جس قدر زور سے
ممکن ہو "مدد، مدد" پکاریں

نہیں!

اس کی آنکھوں میں مرچیں، مٹی ڈالیں۔ اس سے وہ کچھ دیر
دیکھ نہ پائے گا اور اسے شدید تکلیف بھی ہوگی۔ آپ اس دوران
دور بھاگ سکتی ہیں۔

اگر آپ اپنا توازن کھودیتی ہے تو بہتر ہے کہ اپنا دفاع
کرنے سے گھٹنوں کے بل بیٹھ جائیں اور اس کے خلاف
مزاحمت کریں

جب آپ پر حملہ کرنے والا جھکے اس کی ناک
یا آنکھوں پر ضرب لگائیں۔ آپ اس کی ناک پر
اپنا سر بھی زور سے ٹکراسکتی ہیں

ایسی عورتیں جو بیساکھیاں استعمال کرتی ہیں یا جن کی ٹانگیں کمزور ہیں یا وہ اپنے پیروں پر کھڑی نہیں ہو پاتی ہیں ان کے لئے بیٹھنا یا گھٹنوں کے بل ٹکنا
محفوظ دفاعی پوزیشن ہے۔ جیسے ہی آپ اپنے لئے مناسب انداز اختیار کریں اپنی بیساکھی یا چھڑی سے وار کریں۔

- اپنی وہیل چیئر جس قدر رشدت اور تیزی سے ممکن ہو اس سے ٹکرائیں۔

اگر آپ چھڑی یا لاٹھی استعمال کرتی ہیں تو یہ چھن جانے کے بعد آپ اپنا توازن کھوسکتی ہیں۔ اگر آپ سمجھتی ہیں کہ آپ پر حملہ ہونے والا
ہے تو اپنی چھڑی کے موٹے سرے کا رخ اس طرح اس کی طرف کریں کہ چھڑی کا طویل حصہ دوسری طرف رہے۔ اسے آپ جتنی قوت سے ممکن
ہو حملہ آور پر گھونپیں۔ اپنی چھڑی کو بیس بال یا کرکٹ بلے کی طرح مت گھمائیں۔ اس صورت میں چھڑی کو پکڑنا یا چھین لینا آسان ہوتا ہے۔

اگر آپ نابینا ہیں

نابینا عورتیں حملہ ہونے کی صورت میں اپنی چھڑی وغیرہ سے محروم ہوسکتی ہیں لیکن آپ حملہ آور کے جسم کو اپنی مدد کے لئے استعمال
کرسکتی ہیں۔ کوشش کریں کہ اس کے کندھوں کے اوپر اس حصے کو پالیں جہاں گردن شروع ہوتی ہے۔ تیزی کے ساتھ یہ حصہ پانا سب

چھڑی کو گھمانے کے
بجائے اس کو کسی کے
جسم میں گھونپنا بہتر
رہتا ہے۔

اگر آپ بیساکھیاں
استعمال کرتی ہیں تو
انہیں حملہ آور کو مارنے
کے لئے ہتھیار کی طرح
استعمال کریں

سے زیادہ آسان ہے اور اس طرح آپ اس کے جسم کے بقیہ حصے کی پوزیشن کے بارے میں بہتر طور پر اندازہ لگا سکتی ہیں۔اب آپ اس کے جسم کے نازک حصوں پر ضرب لگا سکتی ہیں۔

اپنی کسی دوست سے کہیں کہ وہ کندھے کا اوپر مطلوبہ حصہ تلاش کرنے اور پھر جسم کے نازک حصوں کا اندازہ لگانے میں آپ کی مدد کرے۔ آپ کی دوست، پھینکی جانے والی چھڑی تلاش کرنے کی مشق میں بھی آپ کی مدد کرسکتی ہے۔

اپنا گھٹنا اٹھا کر تیزی اور جس قدر ممکن ہو زور سے حملہ آور کے خصیوں پر ماریں

اپنی مہارتوں کی مشق کریں

ذاتی دفاع یا تحفظ مشق کرنے سے آپ خود کو زیادہ محفوظ اور پُر اعتماد محسوس کرکتی ہیں خواہ آپ پر کبھی حملہ نہ ہوا ہو۔عملی مشق بھرپور اور یقینی اثر ڈالتی ہے۔اپنے بچاؤ کے مختلف طریقوں کے بارے میں سوچیے اور دوسری عورتوں کے ساتھ ان کی مشق کیجیے۔ آپ ایک گروپ بھی بنا سکتی ہیں تا کہ آپ مل کر یہ سیکھیں۔ ذاتی دفاع کی کلاسوں میں عورتیں ان کے لئے جس قدر ممکن ہو شدت سے ضرب لگانے کی مشق کرسکتی ہیں۔ اس عمل کے لئے وہ ڈمی یا پیڈز وغیرہ سے محفوظ کی ہوئی کسی اور عورت کو ضرب لگاتی ہیں۔ یہ مشق ان عورتوں کے لئے بے حد مفید ہے۔

> یاد رکھیے اگر آپ خود اپنا دفاع نہیں کرپائیں تو یہ آپ کی غلطی نہیں کہ آپ پر حملہ کیا گیا یا آبروریزی ہوئی

گھرانے اور دیکھ بھال کرنے والے زیادتی سے تحفظ دلا سکتے ہیں

معذوریوں کے ساتھ جوان ہونے والی لڑکیاں اور عورتوں کو ان کے گھر کے افراد، مددگار اور صحت کار کرکن بار بار چھوتے ہیں۔ اکثر یہ ان کی اجازت کے بغیر ہوتا ہے۔ ایک عورت خواہ وہ معذور ہو یا نہ ہو، یہ کہنے کا حق رکھتی ہے کہ اُسے کون چھو سکتا ہے۔

گھر والے اور دیکھ بھال کرنے والے ایک لڑکی کو اچھے لمس اور بُرے لمس کا فرق سکھا کر اسے جنسی زیادتی سے تحفظ حاصل کرنے میں مدد دے سکتے ہیں۔ اسے چھونے سے پہلے اس کی اجازت طلب کریں۔ اگر اسے ذاتی دیکھ بھال یا روزمرہ کاموں کے لئے مدد کی ضرورت ہو تو اسے ہمیشہ یہ کہنے کا موقع دیں کہ وہ کیا چاہتی ہے جو آپ اس کے لئے کریں۔ اسے یہ بتانے دیں کہ اسے کس طرح چھوایا اس کے جسم کو کس طرح حرکت دی جائے جو اس کے لئے اطمینان بخش ہو۔ معذور لڑکیوں کو ناپسندیدہ یا ناگوار لمس پر ''نہیں'' کہنا سکھا سکیں۔

معذور لڑکیوں سے عصمت دری اور جنسی زیادتی کے حوالے سے بات کریں اور یقینی بنائیں کہ وہ اپنا دفاع خود کرسکیں۔

جب آپ کسی معذور بچی کی پرورش محبت اور احترام کے ساتھ کرتے ہیں تو اس میں ایک پُر اعتماد، پُریقین شخصیت پروان چڑھتی ہے اور دوسرے لوگ اس کے ساتھ خراب سلوک کم ہی کرسکیں گے۔

کمیونٹیز تشدد اور زیادتی سے بچا سکتی ہیں

جب کمیونٹی سمجھتی ہے کہ زیادتی افسوسناک حرکت ہے تو ایسا بہت ہی کم ہوتا ہے کہ کسی عورت کے ساتھ زیادتی ہو۔ جب معذور عورتیں کمیونٹی کا اہم حصہ ہوتی ہیں تو بہت ہی کم معذور عورتوں کے ساتھ زیادتی ہوتی ہے لیکن جن کمیونٹیز میں یہ سمجھا جاتا ہے کہ عورتوں کی کوئی اہمیت نہیں، معذور عورتوں کی بڑی تعداد کے ساتھ زیادتی ہوتی ہے۔

ایسی عورتیں جن کے ساتھ زیادتی کی گئی ہوخصوصاً معذور عورتوں کو مدد فراہم کریں۔ عورتوں کی مدد کرنے کے رپ کرائسز سنٹر، ایمرجنسی ہومز، شیلٹر اور دیگر پروگرام معذور عورتوں کے خصوصی پروگرام شامل کئے جا سکتے ہیں۔ یقینی بنائیں کہ ان کے لئے عمارتیں قابل رسائی ہوں اور نابینا، بہری اور آموزشی دشواریاں رکھنے والی عورتوں کے لئے مناسب معلومات دستیاب ہوں۔

مراکز صحت، اسکول، مشاورتی مراکز، عبادت گاہیں اور کمیونٹی کے بزرگ زیادتی کا شکار ہونے والوں کی ذہنی صحت کی دیکھ بھال کے لئے مدد کر سکتے ہیں۔ مشاورت سے زیادتی کا شکار ہونے والے افراد کے اعتماد، عزت نفس اور بہتری کے احساس کی بحالی ممکن ہو سکتی ہے۔

زیادتی کیوں غلط ہے، اس موضوع پر تبادلہ خیال میں مردوں کو شامل کیجئے اور یقینی بنائیے کہ پولیس اور کمیونٹی کے دوسرے حکام یہ سمجھیں کہ معذور عورتوں کے ساتھ زیادتی درست نہیں ہے۔ کمیونٹی سروسز کے تمام اداروں جیسے پولیس تھانوں، کلینک اور اسپتالوں میں ایسے افراد کو بھرتی کریں جو مقامی علامتی زبان جانتے ہوں۔

ایسے قوانین کے بارے میں جانیں جو زیادتی کا شکار عورت کو تحفظ فراہم کرتے ہیں، ان کے بارے میں دوسروں کو بتائیں۔ عورت کے خلاف تشدد اور زیادتیوں کے خلاف تبادلہ خیال اور احتجاج کے لئے معذور عورتوں سمیت عورتوں کے عوامی اجتماع کا اہتمام کیجے۔ جب معذور عورتیں، صحت کارکن اور کمیونٹی کے دوسرے لوگ ان حوالوں سے تبادلہ خیال کریں گے تو تمام عورتوں کے لئے زیادتی سے محفوظ رہنا آسان ہو سکتا ہے۔

زیادتی محض خاندانی معاملہ نہیں، زیادتی ایک سماجی مسئلہ صحت ہے

اگر کوئی عورت کہتی ہے کہ اس کے ساتھ زیادتی ہوئی ہے تو اس پر یقین کیا جانا چاہئیے۔ قطع نظر اس کے کہ وہ کون ہے۔ اس کی مدد کی جائے۔ اور اگر کوئی معذور عورت زیادتی کا شکار ہو تو یہ زیادہ بُری بات ہے کیونکہ وہ کمزور پوزیشن میں ہوتی ہے

پاکستان میں عورتوں اور لڑکیوں پر تشدد کا ایک عام واقعہ ہے۔ انہیں جسمانی تشدد کا نشانہ بنایا جاتا ہے۔ وہ جنسی ہوس کا شکار ہوتی ہیں اور انہیں قتل بھی کر دیا جاتا ہے۔ عزت کے نام پر عورتوں اور لڑکیوں کا قتل عام ہے اور تقریباً ہر صوبے میں ہر سال سینکڑوں کی تعداد میں عورتیں اس کا نشانہ بنتی ہیں۔

تشدد کا شکار ہونے والی عورتوں کے لئے انصاف کا حصول تقریباً ناممکن ہے۔ اس سلسلے میں حکومتی اقدامات تسلی بخش نہیں ہیں یہ بھی دیکھنے میں آیا ہے کہ تشدد اور جنسی زیادتی کا شکار ہونے والی عورتوں کو پولیس اسٹیشنوں پر دوبارہ تشدد اور جنسی ہوس کا نشانہ بنایا گیا ہے۔ عورتوں کو ہر صورت میں اس صورت کا سامنا ہونے کے بعد احتجاج کرنا چاہئے۔ پاکستان میں عورت فاؤنڈیشن، شرکت گاہ، ویمنز فرنٹ اور وار اگینسٹ ریپ (WAR) جیسے ادارے عورتوں کی مدد کے لئے تیار رہتے ہیں۔ تشدد کا شکار ہونے والی عورتوں اور ان کے رشتہ داروں کو ایسی صورتحال میں ان اداروں سے ضرور رابطہ کرنا چاہیے تاکہ اس واقعہ کی مناسب رپورٹ لکھوائی جا سکے اور ان مظلوم عورتوں کو انصاف مل سکے۔

باب -15

دیکھ بھال کرنے والوں کی تائید و حمایت

ہر فرد کو کسی نہ کسی وقت مدد کی ضرورت پڑسکتی ہے۔ یہ بہت ہی کم ہوتا ہے کہ ہم پورا دن کسی کی مدد حاصل کئے بغیر یا گھر والوں،
پڑوسیوں اور حتیٰ کہ اجنبیوں کی مدد کے بغیر گزار سکتی ہوں۔ ایک دوسرے کی مدد کرنا انسانی فطرت ہے۔

ایک معذور عورت کو اس کے روزمرہ کے کاموں میں مدد کی ضرورت ہوتی ہے۔ اگر وہ اپنی ضرورت کے مطابق مدد حاصل کرلیتی
ہے تو وہ زیادہ صحت مند اور خوشگوار زندگی گزار سکتی ہے اور اپنے خاندان اور کمیونٹی کے لئے خدمات انجام دے سکتی ہے۔

معذور عورتوں کی دیکھ بھال کا کام با معاوضہ ہوسکتا ہے۔ اگر معذور عورت کو بہت زیادہ دیکھ بھال کی ضرورت ہوتو یہ نہایت ہی
بوجھل اور پریشان کن کام بھی ہوسکتا ہے۔ یہ باب خاص طور پر ان گھرانوں اور لوگوں کے لئے ہے جو معذور عورتوں کی دیکھ بھال کرتے
ہیں اور یہ معذور عورتوں کے لئے بھی، ان لوگوں کی ضرورت یات سمجھنے میں معاون ثابت ہوگا جو ان کی (معذور عورتوں کی) دیکھ بھال اور
نگہداشت کرتے ہیں۔

رومیلا نے مجھے بتایا ہے کہ اسے کس قسم کی مدد کی ضرورت ہے۔ وہ ایسے معاملات طے کرنے میں خودمختار ہے

مل جل کر کام کرنا

سب سے بڑھ کر بات یہ ہے کہ ہر وہ فرد جو کسی معذور عورت کی مدد کرتا ہو یہ یاد رکھے کہ وہ عورت بچہ نہیں ایک مکمل عورت ہے۔ اگر وہ بتا سکتی ہے تو اسے بتانے دیں کہ اسے کس قسم کی مدد درکار ہے اور پھر باہمی تبادلہ سے آپ فیصلہ کر سکتے ہیں کہ کون سا طریقہ بہترین رہے گا۔

جہاں تک ممکن ہو ایک معذور عورت کو اپنی دیکھ بھال اور زندگی کا خود ذمہ دار ہونا چاہئے۔ دیکھ بھال کریں، معذور عورت کی حوصلہ افزائی کریں کہ وہ خود کو ٹیم کا کپتان سمجھے اس طرح وہ اپنی ضرورت کے مطابق مطلوبہ مدد حاصل کر سکتی ہے، وہ معاونت نہیں جس کو وہ غیر مناسب، غیر مددگار اور احترام سے محروم سمجھتی ہے۔

جہاں تک ممکن ہو معذور افراد سے معلوم کیجئے کہ وہ کیا چاہتے ہیں۔ ان سے پوچھئے کہ وہ کن کاموں (ذمہ داریوں) میں ہاتھ بٹا سکتے ہیں اور دیکھ بھال کرنے والوں کو کیا کرنے کی ضرورت ہے اور وہ کیا نہ کریں۔ ممکن ہے وہ دوسروں کی مدد مانگنے کے بجائے کام خود کرنا پسند کرتی ہوں۔ اگر دیکھ بھال کرنے والا اور معذور عورت کھل کر تبادلہ خیال کریں تو اچھی دیکھ بھال کرنا نہایت ہی آسان ہوگی۔ اور اگر یہ ممکن نہ ہو تو وہ خود کو اس کی جگہ رکھ کر سوچیں کہ اس صورت میں وہ کیا جذبات رکھتیں۔

اگر وہ بہری ہے تو اس سے ابلاغ کے لئے علامتی زبان استعمال کریں، اس سے مطابقت کے لئے جس قدر جلد ممکن ہو اشاروں کی زبان سیکھیں۔

اگر وہ نابینا ہے تو اسے بتانے دیں کہ وہ اپنے لئے کیا مدد چاہتی ہے۔ اسے چلانے کے لئے اس کا بازو یا ہاتھ نہ تھامیں، اسے اپنا ہاتھ تھامنے دیں اگر وہ چلنے پھرنے کے لئے چھڑی یا عصا استعمال کرتی ہے تو یقینی بنائیے کہ چھڑی ہمیشہ اس کے نزدیک اور قابل رسائی رہے۔

معذور عورتیں کیا کر سکتی ہیں؟

اگر آپ معذور ہیں اور اپنے روزمرہ کے کاموں یعنی غسل کرنے، لباس پہننے، کھانے یا اٹھنے بیٹھنے میں مدد کی ضرورت پڑتی ہے تو اس فرد سے بات کریں جو آپ کی معاونت کرنے والا ہو۔ یقینی بنائیں کہ اسے یہ معلوم ہو کہ آپ کو کس قدر مدد کی ضرورت ہے یا نہیں ہے۔

تحمل سے کام لیں: ایک فرد کو یہ جاننے میں کچھ وقت لگے گا کہ کس قسم کے کاموں میں آپ کو مدد کی ضرورت ہے۔

دیکھ بھال کرنے والوں کو بھی احترام کی ضرورت ہوتی ہے

بیشتر معاونت اور دیکھ بھال کرنے والے محنت سے کام کرتے ہیں۔ انہیں روزانہ کچھ وقت کے لئے اور ہفتے میں کم از کم ایک دن آرام کی ضرورت ہوتی ہے اگر آپ کی دیکھ بھال کرنے والا چاق و چوبند ہے تو وہ آپ کو بہترین مدد فراہم کرے گا۔ فیصلہ مل کر کیجئے۔

اگرچہ آپ بہتر جانتی ہیں کہ آپ کو کس قسم کی معاونت کی ضرورت ہے لیکن اپنے نگراں یا دیکھ بھال کرنے والے کے آئیڈیے بھی سننے۔ بعض اوقات یہ آئیڈیے بہت ثابت ہو سکتے ہیں۔

دوسری معذور عورتوں سے ملاقات کیجئے۔ ذاتی دیکھ بھال کرنے والے کی مدد، بہتر طور پر کرنے کے بارے میں خیالات کا تبادلہ کے لئے دوسری معذور عورتوں سے ملاقات کیجئے۔

کمیونٹی دیکھ بھال کرنے والوں کو اہمیت دے

معذوریاں رکھنے والی اور دیکھ بھال کرنے والی دونوں عورتیں، خواہ وہ ایک گھر سے تعلق رکھتی ہوں یا باوضہ مددگار ہوں، وہ مرد ہوں عورتیں یا بچے، ہمارے سماج کے قیمتی اور اہم افراد ہیں۔ انہیں بھی ان گھرانوں اور کمیونٹیز میں حقیقی اور پرجوش تعلقات کی ضرورت ہوتی ہے جہاں ہم ایک ساتھ رہتے، کام کرتے اور اپنے دکھ سکھ بانٹتے ہیں۔ لیکن دوسرے گھریلو کاموں کی طرح معذور عورتوں پر بیٹھا عورتوں کی معاونت کرنے والی عورتوں کی عزت کی شاذ و نادر ہی جاتی ہے یا انہیں اہم سمجھا جاتا ہے۔ بعض اوقات ایک مددگار محسوس کرتا ہے کہ ایک معذور عورت بذات خود مدد کو اپنا حق سمجھتی ہے۔

دیکھ بھال کرنے والی عورتیں

زیادہ تر گھر کی عورتیں اور لڑکیاں ہی گھر کے ان افراد کی دیکھ بھال کرتی ہیں جو بیمار ہوں یا کوئی معذوری رکھتے ہوں اور یہ کام وہ اپنے معمول کے دیگر سارے کاموں کے ساتھ ہی کرتی ہیں۔ بہت سی عورتوں کے لئے ان کے معمول کا کام صبح ہونے سے قبل شروع ہوتا ہے اور رات دیر تک مکمل نہیں ہوتا ہے۔ جب کوئی عورت اس کے ساتھ کسی کی دیکھ بھال بھی کر رہی ہوتو اسے اپنے معمول سے زیادہ کام کرنا پڑتا ہے۔

دیکھ بھال کرنے والے بچے

یہ بھول جانا آسان ہے کہ بچے خصوصاً بیٹیاں جو اپنی ماؤں کی مدد کرتی ہیں، خود اپنی ضرورتیں بھی رکھتی ہیں۔ بچوں کو سیکھنے اور کھیلنے کے لئے دوسرے بچوں کے ساتھ وقت گزارنے کی ضرورت ہوتی ہے۔ معذوریاں رکھنے والی مائیں اپنے بچوں کی مدد کے لئے یقینی بنا سکتی ہیں کہ وہ ہمیشہ ان پر انحصار کرنے کے بجائے دوسرے بالغ افراد کا تعاون حاصل کرنے کی کوشش بھی کریں۔ ماں اپنی ضرورت کے مطابق مدد کے لئے ہر ایک سے کہہ سکتی ہے اور گھر کے افراد اس کی مدد کے لئے ایک ٹیم بن کر کام کر سکتے ہیں۔

دیکھ بھال کرنے والے مرد

بعض خاندان میں کوئی مرد یا لڑکا ہوتا ہے جو بیوی، بہن یا ماں کی دیکھ بھال کرتا ہے اگر ایسا ہوتو اسے خاندان کی کسی دوسری عورت اور اس عورت کی مدد کی ضرورت ہوگی جس کی وہ دیکھ بھال کر رہا ہو، تا کہ وہ سمجھ سکے کہ معذور عورت کے لئے زندگی کیوں کسی مرد سے مختلف ہوتی ہے۔ مردوں اور عورتوں کے جسموں کے مابین فرق اہم ہے لیکن مرد اور عورت جس طرح گھرانے میں پروان چڑھتے ہیں اور ان کے ساتھ جو سلوک کیا جاتا ہے وہ زیادہ اہم ہے۔

بامعاوضہ مددگار

بعض اوقات معذور عورت ذاتی مددگار کو اس کی خدمات کا معاوضہ ادا کر سکتی ہے، بعض کمیونٹیز میں حکومت، معذور افراد کو اپنی روزمرہ کی دیکھ بھال کے لئے کسی کی خدمات حاصل کرنے کے لئے ادائیگی کرتی ہیں یا افراد خانہ اور دوست اس ضمن میں ادائیگی کرتے ہیں، بعض اوقات ایک معذور عورت اپنی مدد کرنے والے کو کھانا اور رہنے کی جگہ فراہم کرتی ہے۔

ہر چند کہ وہ کام جو ایک مددگار روزانہ کی صحت وصفائی کے ضمن میں انجام دیتا ہے اور جس میں پیشاب اور پاخانے کے اخراج کی دیکھ بھال بھی شامل ہے، ایک فرد کی صحت کے لئے نہایت ہی اہم ہوتا ہے لیکن نہایت ہی کم حیثیت کا روزگار سمجھا جاتا ہے اور اس کے بدلے میں بہت معمولی ادائیگی کی جاتی ہے۔ بہت سے ذاتی دیکھ بھال کرنے والے بتاتے ہیں کہ اکثر خاندان کے دوسرے افراد ان پر حکم چلانا چاہتے ہیں اور غیر معقول مطالبے کرتے ہیں اور بغیر کسی تو جہیہ کے انہیں ملازمت سے نکال دیتے ہیں اور اگر معذور فرد تنہا رہتا ہو تو وہ سمجھ ہی نہیں پاتا کہ اس کی دیکھ بھال کرنے والے کے ساتھ کس قدر بدترین سلوک کیا گیا ہے۔

اگر کرسٹائن یہ مان لے کہ دن بھر میں مجھے بھی اپنے لئے کچھ وقت اور ہفتے بھر بعد ایک چھٹی کی ضرورت ہے تو مجھے کسی اور کی ہمدردی کی ضرورت نہیں۔

بامعاوضہ دیکھ بھال کرنے والوں کو دوسرے محنت کشوں کی طرح منصفانہ معاوضوں، وقت پر چھٹی، تعطیلات اور بیماری کے باعث چھٹی کی ضرورت ہوتی ہے۔ آرگنائزیشنز اور کمیونٹی گروپ جو ذاتی معاونت کے لئے تربیت اور روزگار فراہم کرتے ہیں یہ کر سکتے ہیں کہ

- شرائط کار کے تعین میں مدد کریں۔
- تنازعات سے تحفظ اور کمی کے طریقوں کے بارے میں سکھائیں۔
- معذور عورتوں کی جذباتی ضروریات کے مطابق عمل کے لئے مشاورتی مہارتوں کی تربیت فراہم کریں۔
- وزن اٹھانے، ورزش میں مدد کرنے اور انفیکشن سے بچاؤ کی مہارتیں سکھائیں۔

کمیونٹی کے صحت کارکن، دیکھ بھال کرنے والوں کے لئے وقت نکالیں

گھانا میں بوڑھے افراد کی مدد کرنے والے کمیونٹی ورکرز کے ایک گروپ نے مل کر اس بارے میں تبادلہ خیال کیا کہ بوڑھوں کے لئے کس طرح مختلف امور آسان بنائے جائیں۔ انہوں نے یہ گروپ اس لئے شروع کیا تھا کہ جب انہوں نے گھانا کے مختلف گاؤں میں سفر کیا تو مشاہدہ کیا کہ ہر جگہ بوڑھے لوگوں کی معاونت کرنے والے موجود ہیں۔ انہوں نے اندازہ لگایا کہ بوڑھے افراد کی مدد کے ساتھ ان لوگوں کو بھی مدد کی ضرورت ہے جو روزمرہ کے کاموں میں بوڑھوں کی مدد کرتے ہیں۔ اب جب بھی کوئی کمیونٹی لیڈر کسی بوڑھے شخص سے ملنے جاتا ہے تو وہ اس کی دیکھ بھال کرنے والے سے بھی تبادلہ خیال کرتا ہے۔ وہ ان کے خیالات اور مسائل سنتا ہے۔ اگر ضرورت ہو تو کمیونٹی لیڈران کی مدد کرتا ہے اور یقینی بناتا ہے کہ دیکھ بھال کرنے والے کو کچھ دیر آرام کا وقفہ ضرور ملے۔

دیکھ بھال کرنے والوں کو بھی آپ کے خیالات / جذبات سمجھنے کے لئے مدد کی ضرورت ہوتی ہے

اگر میں لی ہنگ کے بارے میں اور زیادہ جان لوں تو میں اس کی مدد بہتر طور پر کرسکتی ہوں، ہم میں سے کوئی بھی نہیں جانتا ہے کہ اسے کس چیز کی ضرورت ہے۔ یہ لاعلمی مجھے جھنجھلاہٹ میں مبتلا کرتی ہے اور میں خود کو اس کے لئے معاون محسوس نہیں کرتی ہوں۔

دیکھ بھال کرنے والے، معذور فرد کی بہتر خدمات کے لئے مصروف عمل رہتے ہیں۔ وہ اکثر سوچتے ہیں کہ وہ فرد کس طرح سوچتا ہوگا اگر آپ دیکھ بھال کرنے والی ہیں تو آپ کے لئے بھی یہ اہم ہے کہ آپ اپنے جذبات کا جائزہ لینے کے لئے کچھ وقت وقف کریں۔ اگر آپ کسی معذور عورت کی معاونت کر کے خوش بھی ہیں تو اس صورت میں بعض اوقات آپ تھک جائیں گی یا خود کو دباؤ یا فرسٹریشن زدہ یا اپ سیٹ محسوس کریں گی۔ دیکھ بھال کرنے والے کی حیثیت سے آپ کو مختلف انداز میں ایک نرس، ایک کونسلر، ایک ڈرائیور، ایک باورچی، ایک اکاؤنٹنٹ اور ایک ہاؤس کیپر کی حیثیت سے کام کرنا پڑ سکتا ہے اور اس کے ساتھ اس دوران آپ کو ایک ایسے فرد کی دیکھ بھال بھی کرنی ہو جو نہایت ہی بیمار یا ڈپریس ہوتو آپ انتہائی دباؤ محسوس کرسکتی ہیں۔

معذور عورتوں کی معاونت کرنے والے افراد خانہ اور دیگر لوگوں کے لئے یہ فطری ہے کہ وہ خود کو دوسروں کے غلط برتاؤ کے باعث شرمندہ اور غلطی کا مرتکب، تھکا، برہم اور اپ سیٹ محسوس کریں۔ وہ شخص بھی جو کسی قریبی رشتہ دار یا شریک حیات کی دیکھ بھال کر رہا ہو غیر اطمینان بخش جذبات اور خیالات کا شکار ہوسکتا ہے۔

اگر آپ یہ اندازہ نہیں لگا سکتی ہیں کہ کیا بات آپ کو برہم، مایوس اور بے چارگی کا شکار کرتی ہے تو آپ اپنے جذبات کی تہہ میں موجود اسباب کے مطابق تبدیلی کے طریقے ڈھونڈ سکتی ہیں۔

وہ ٹھیک ہوجائے گی، لیکن اس میں طویل وقت لگے گا بارودی سرنگ نے اسے شدید نقصان پہنچایا ہے لہذا اسے دوبارہ بہتر ہونے تک ہماری ڈھیر ساری مدد کی ضرورت ہے۔

میں خوفزدہ ہوں ڈیڈی کیا ماں ٹھیک ہوجائے گی

اچانک معذوری پورے خاندان کو متاثر کرتی ہے

جب آپ کا کوئی قریبی فرد کسی حادثے یا بیماری کی وجہ سے اچانک معذور ہوجائے، اس سے گھر کے سارے افراد متاثر ہوتے ہیں۔ اس کی زندگی میں آنے والی غیر متوقع تبدیلی خوفزدہ کرنے والی ہوسکتی ہے۔ ایسا ہونا معذور افراد اور گھر کے دیگر ارکین کے لئے انتہائی دکھاور غصے کا سبب ہوسکتا ہے۔

اپنے جذبات کو سمجھنا آپ کو ایسے اقدام پر مائل کر سکتا ہے جو آپ کی اور اس معذور عورت کی زندگی کو بہتر بنا سکتے ہیں جس کی آپ دیکھ بھال کرتی ہوں۔

فرسٹریشن زدہ، بے یارومددگار یا تنہا ہونے کا احساس، آپ کو اپنی کمیونٹی میں، ایسے دوسرے افراد تک پہنچنے میں مدد دے سکتا ہے جو ایسی عورتوں کے ساتھ رہتے ہیں جو معذور ہوں یا جو بذاتِ خود معذور ہوں۔ یہ لوگ آپ کے لئے مددگار ہو سکتے ہیں۔

> مجھے اپنی بیوی سے پیار ہے اور میں اس کی مدد کرنا چاہتا ہوں، لیکن اس کے ساتھ پیش آنے والے حادثے کے بعد میری زندگی انتہائی مشکل بن گئی ہے، مجھے اس کے لئے ہر کام کرنا پڑتا ہے جو بعض اوقات مجھے اس پر برہم کر دیتا ہے۔

> برہمی کا احساس، آپ کو حکومت سے رجوع کرنے کے لئے، دیکھ بھال کرنے والوں اور دوسرے افراد کا ایک گروپ، منظم کرنے کی توانائی دے سکتا ہے جو معذوریاں رکھنے والی عورتوں کے لئے اچھی ہیلتھ کیئر کے حصول ٹرانسپورٹ اور عوامی عمارتوں تک رسائی کی کوشش کرے۔

> میرے گاؤں کے لوگ معذوریاں رکھنے والی عورتوں کے ساتھ اچھا سلوک نہیں کرتے ہیں۔ میری خواہش ہے کہ میں یہ صورتحال بدل سکوں اور سیتا کے لئے زندگی بہتر بنا سکوں۔ یہ صورت مجھے غمگین اور بے یارومددگار ہونے کا احساس دلاتی ہے۔

اپنے جذبات کا اظہار کرنے کے صحت مند طریقے

جب لوگ ایک ساتھ بہت سا وقت گزارتے ہوں مگر ایک دوسرے سے اپنے احساسات کے بارے میں گفتگو نہیں کرتے ہوں تو وہ ایک دوسرے سے خفا اور برہم ہو سکتے ہیں۔ اگر گفتگو آپ کے احساسات کی وجوہات تبدیل کرنے کا طریقہ نہیں ڈھونڈ سکتی ہے تو پھر بھی اس سے آپ کو ان احساسات پر ردِعمل کی تبدیلی میں مدد مل سکتی ہے۔

لوگ اپنے احساسات کا اظہار مختلف انداز سے کرتے ہیں، اظہار کے صحت مند اور غیر صحت مند طریقے موجود ہیں جیسے

یہ جذبات کے اظہار کا ایک خطرناک اور غیر صحت مند انداز ہے۔

> تم ناشکری عورت ہو میں نہیں جانتا ہوں کہ میں کیوں تمہاری مدد کے لئے زحمت اٹھاتا ہوں۔

یہ جذبات کے اظہار کا صحت مند اور محفوظ طریقہ ہے۔

> جب تم مجھ سے کہتی ہو کہ وہ تمام چیزیں جو میں کر رہا ہوں غلط ہیں اور تم کبھی بھی اس چیز کا ذکر نہیں کرتی ہو جو میں درست کر رہا ہوں تو مجھے یہ سوچ کر غصہ آجاتا ہے کیونکہ تم میری مدد کا اعتراف نہیں کرتی ہو۔

نائجیریا کی ایک ماں بتاتی ہے کہ
وہ کس طرح اپنی معذور بیٹی کا خیال رکھتی ہے

میری بیٹی پولیوزدہ ہے۔ جب وہ چھوٹی سی تھی تو ہم اس کے لئے وہیل چیئرنہیں خرید سکتے تھے یا اس کی دیکھ بھال میں مدد کے لئے کسی کو ملازم نہیں رکھ سکتے تھے۔ میں اسے اپنی پیٹھ پر لاد کر ہر روز اسکول لے جاتی تھی حتٰی کہ وہ بارہویں جماعت میں پہنچ گئی۔ اسے لے جانا میرے لئے بہت مشکل ہوگیا تھا کہ وہ بہت بڑی اور بھاری ہوگئی تھی لہٰذا میں بہت تھک جاتی تھی۔ جب وہ یونیورسٹی جانے لگی تو اسے وہیل چیئر مل گئی اور اب اس کے پاس اپنی کار بھی ہے اس لئے اب میں چند برس پہلے کے مقابلے میں زیادہ آرام کر سکتی ہوں۔ لیکن اب اسے جذباتی مدد کی ضرورت پڑتی ہے کیونکہ وہ بعض اوقات اپنے رویے میں غیر یقینی ہو جاتی ہے میں سمجھتی ہوں کہ وہ عمر کے جس دور سے گزر رہی ہے یہ اس کا ایک حصہ ہے لہٰذا میں اس کی مدد کرنے کی کوشش کرتی ہوں لیکن اگر میں اس کو جذباتی طور پر زیادہ مستحکم کرنے کے لئے کسی کی خدمات حاصل کر سکتی تو میں خود اپنی دیکھ بھال بہتر طور پر کر سکتی تھی اور اس طرح ہر وقت نڈھال نہیں ہوتی۔

اپنی دیکھ بھال کرنا

کچھ دیکھ بھال کرنے والے خود کو ان لوگوں کی ضروریات پوری کرنے کے لئے مکمل طور پر وقف کر دیتے ہیں۔ وہ دوسروں کی مدد کرنے میں اس طرح منہمک ہو جاتے ہیں کہ خود اپنا خیال رکھنا بھول جاتے ہیں بعض اوقات وہ اپنی بہتری اور زندگی کی خوشیاں بھی قربان کر دیتے ہیں کچھ عرصے بعد وہ مدد کرنے والے جو خود اپنے بارے میں نہیں سوچتے فرسٹریشن زدہ ہونا شروع کر دیتے ہیں اور ان پر برہم ہونے لگتے ہیں جن کی وہ مدد کر رہے ہوتے ہیں۔ یہ صورت دونوں ہی کے جذبات مجروح کر سکتی ہے اگر آپ اپنا خیال نہیں رکھیں گی تو آپ میں دوسروں کی مدد کے لئے اتنائی نہیں رہے گی۔ کسی کی اچھی دیکھ بھال کے لئے آپ کو بھی مناسب نیند اور آرام، اپنی جسمانی ضروریات کا خیال رکھنے اور زندگی کے دوسرے رشتوں سے تعلق برقرار رکھنے کی ضرورت ہے۔

جسم اور دماغ کو پرسکون رکھنے اور اپنی باطنی قوت کو پروان چڑھانے کے لئے یوگا، مراقبہ، تائی چی اور دوسری مشقیں یا روایتی طریقوں پر عمل کریں ان طریقوں پر باقاعدگی سے عمل اس تناؤ سے نمٹنے میں معاون ہو سکتا ہے جو کسی کی دیکھ بھال کے نتیجے میں جنم لیتا ہے۔

اپنی صحت کا خیال رکھیں

- اچھی غذا کھائیں تا کہ آپ کا جسم مضبوط رہے۔
- اتنی نیند لیں جو آپ کی توانائی دن بھر برقرار رکھے۔
- کسی کی دیکھ بھال کرنے سے پہلے جتنی ورزش کرتی تھیں اس سے زیادہ ورزش کریں۔
- مساج آپ کے جسم کو پرسکون رکھنے میں معاون ثابت ہو سکتا ہے اس سے دباؤ اور انتشار دور ہونے میں بھی مدد ملے گی۔

کچھ وقت نکالیں اور کچھ ایسا کریں جس سے آپ لطف اندوز ہوسکیں۔ یہ معذور عورت اور اس کی دیکھ بھال کرنے والے کے لئے ضروری ہے کہ دونوں کے دوست اور دلچسپیاں ایک دوسرے مختلف ہوں۔ بھرپور اور مطمئن زندگی کے لئے ضروری ہے کہ آپ میں سے ہر ایک اپنا کچھ وقت دوسرے لوگوں کے ساتھ گزارے۔

محتاط رہیں کہ خود کو مجروح نہ کرلیں

دیکھ بھال کے کام میں عموماً جسمانی مشقت بھی شامل ہوتی ہے جیسا کہ اس فرد کو اٹھانا جس کی آپ دیکھ بھال کر رہی ہوں یہ کام آپ کی کمر کو مجروح کرسکتا ہے کسی کو اٹھانے یا کچھ محفوظ طور پر اٹھانے کے لئے۔

- اپنی کمر کے عضلات نہیں بلکہ اپنے پیروں کے عضلات استعمال کیجئے جب آپ زمین پر رکھی کوئی بھاری چیز اٹھائیں تو اسے جھک کر اٹھانے کے بجائے گھٹنوں کے بل بیٹھیں۔

- جب آپ اپنی ٹانگیں سیدھی کریں تو اپنی پشت، کندھے اور گردن کو جس حد تک ممکن ہو سیدھا رکھیں۔

- اس فرد کو اٹھانے کے لئے جس کی آپ دیکھ بھال کر رہی ہیں کسی سے مدد کے لئے کہیں۔ یہ کام بذات خود ممکن ہے آپ کو مشکل محسوس نہ ہو لیکن اس طرح اپنی کمر مجروح کرلیں تو بعد میں ممکن ہے کہ آپ کوئی مدد کرنے کے قابل نہ رہیں۔

دوسروں سے مدد کے لئے درخواست کرنا

دیکھ بھال کرنے والے اور فرد سارا وقت الگ تھلگ ہوسکتے ہیں۔ جب کوئی معذور فرد سارا وقت ایک ہی مددگار پر انحصار کرتا ہے تو دوسرے لوگ سمجھ سکتے ہیں کہ "ماہر مددگار" وہی فرد ہے جو مدد کرنے کا درست طریقہ جانتا ہے لیکن کسی ایک فرد کو معذور عورت کا واحد مددگار نہیں ہونا چاہئے خاندان کے دوسرے افراد، دوست اور پڑوسی بھی مختلف طریقوں مثلاً کھانا پکانا، مارکیٹ جانا، صفائی یا صرف ملاقات کرکے مدد کرسکتے ہیں اس سے بھی آپ کو آرام کرنے اور توانائی بحال کرنے کا موقع مل سکتا ہے۔

مدد کرنا اور مدد حاصل کرنا

کوئنگ چِنگ اپنے خاندانی پولٹری بزنس کا سارا حساب کتاب رکھتی ہے آپ اسے ہماری ریذیڈنٹ منیجر کہہ سکتے ہیں۔

ایسے طریقے ڈھونڈیں جن سے ایک معذور عورت اپنے گھرانے کے معمول کے مطابق ایک حصہ بن کر کام کرسکے۔ اس صورت میں وہ مدد حاصل کرنے کے بجائے مدد دینے والی بن سکتی ہے۔ اچھی اور حقیقت پسندانہ توقعات رکھیں ایک عورت سے جس حد تک وہ بہتر ہوسکتی ہے، ہونے کی توقع رکھیں نئی چیزیں آزمانے اور نئی مہارتیں پروان چڑھانے کے لئے اس کی حوصلہ افزائی کریں۔

دیکھ بھال کرنے والوں کا گروپ بنا ئیے

دیکھ بھال کرنے والوں کی صحت بہتر بنانے کا ایک اہم طریقہ یہ ہے کہ وہ ایک دوسرے سے تبادلہ خیال کریں۔ دیکھ بھال کرنے والوں اور معذور افراد کو ایسے افراد کی مدد کی ضرورت ہے جو ملتے جلتے تجربات اور جذبات رکھتے ہوں۔ دوسروں سے اپنی ضروریات اور جذبات کے حوالے سے تبادلہ خیال کو الگ تھلگ ہونے کے احساس سے بچا سکتا ہے۔ آپ دوسرے دیکھ بھال کرنے والوں سے تبادلہ خیال کر کے یہ کہ کس طرح چیزوں کو آسان بنایا جا سکتا ہے اور ان تمام معذور لوگوں کے لئے تعاون اور مدد کا سلسلہ قائم کیا جا سکتا ہے۔

اگر ایسا کوئی گروپ موجود نہیں ہے لیکن آپ کمیونٹی میں موجود دوسرے دیکھ بھال کرنے والوں سے واقف ہیں تو یہ آپ پر منحصر ہے کہ ایسے گروپ کی ابتداء کریں۔ بہت سے طاقت ورترین اور انتہائی فعال گروپوں کی ابتداء محض ایک فرد کی سوچ سے ہوئی ہے۔ مل جل کر کام کرنے والا ایک گروپ آسانی سے مسائل حل کر سکتا ہے اور انفرادی طور پر کام کرنے والوں سے زیادہ کام کر سکتا ہے۔

مجھے ہر وقت دیکھ بھال کی ضرورت ہوتی ہے۔ میری دیکھ بھال کرنے والوں اور دوستوں کا ایک گروپ ہے جس نے "مائی کی دیکھ بھال" نیٹ ورک بنا لیا ہے یہ ایک دوسرے سے تبادلہ خیال کرتے ہیں ایک دوسرے کی مدد کرتے ہیں یہ سب مجھے پسند کرتے ہیں اور میں بھی انھیں پسند کرتی ہوں یہ ایک شاندار سلسلہ ہے۔

گروپ شروع کرنا

دو یا زائد دیکھ بھال کرنے والے تلاش کریں جو سپورٹ گروپ بنانا چاہتے ہوں۔ اگر آپ ایسے گروپوں سے واقف نہیں جہاں کوئی معذور فرد موجود ہو تو ایک صحت کارکن قرب و جوار میں ایسے گھرانوں سے واقف ہو سکتا ہے۔ طے کیجئے کہ کب اور کہاں ملاقات کی جائے۔ اس سے ایسی جگہ منتخب کرنے میں مدد ملے گی جہاں ہر ایک کو اطمینان سے بات کر سکے گا یہ کسی صحت مرکز یا کمیونٹی سینٹر کا کمرہ یا کوئی عبادت گاہ بھی ہو سکتی ہے۔ پہلی ملاقات میں اس پر بات کیجئے کہ آپ کیوں مل رہے ہیں اور کیا کرنے کی توقع رکھتے ہیں۔ امکانی طور پر ابتدائی چند ملاقاتوں میں ایک فرد لیڈر رہو گا لیکن اہم بات یہ ہے کہ کوئی ایک فرد گروپ کے لئے فیصلے نہ کرے ہر ایک کو بات کرنے کا موقع ملنا چاہئے۔ گفتگو کو اس ملاقات کی بنیادی وجہ تک محدود رکھنے کی کوشش کریں۔ ابتدائی چند ملاقاتوں کے بعد گروپ کی قیادت باری باری تبدیل کیجئے ہر ملاقات میں مختلف فرد کی قیادت شر ملیے اراکین کو گفتگو میں شامل ہونے پر مائل کرے گی۔

مل جل کر ہم اپنے بچوں اور اپنی مدد کر سکتے ہیں

بنگلور بھارت کی ایک آبادی میں معذور بچوں اور بالغوں کی مدد کے لئے بہت سے خاندانوں نے ایک سپورٹ گروپ تشکیل دیا ہے۔ یہ ہفتے میں ایک بار معذور افراد کے لئے کمیونٹی میں سہولتوں کی بہتری اور ضروریات کے لئے ملاقات کرتے ہیں۔ اس گروپ کے پاس معذور بچوں کو اسکول لے جانے اور واپس لانے کے لئے کئی آٹو رکشہ بھی ہیں۔

گروپ میں ایک دوسرے کی مدد کرنا سیکھیں

اکثر لوگ جو دوسروں کی مدد کرتے ہیں اس قدر مصروف ہوتے ہیں کہ خود اپنے بارے میں سوچنے کے لئے فرصت نہیں رکھتے ہیں یا وہ سمجھتے ہیں کہ انہیں ذہنی طور پر پریشان ہونے کا کوئی حق نہیں یا یہ کہ معذوری رکھنے والی عورت ہی اپ سیٹ ہو سکتی ہے حتیٰ کہ جب لوگ ایک دوسرے کو اچھی طرح جان لیتے ہیں تو جب بھی انہیں احساسات، تجربات اور دیکھ بھال کرنے والے کی حیثیت سے درپیش چیلنجوں کے بارے میں اطمینان سے بات کرنے میں وقت لگتا ہے۔ دوسرے لوگوں کی بہ نسبت گروپ میں رہ کر کچھ لوگوں کے لئے بات کرنا آسان ہوتا ہے لیکن صرف بات کرنا ہی وہ طریقہ نہیں ہے جس سے لوگ اپنے خیالات اور احساسات کا اظہار کر سکیں۔ اس کے لئے مختلف سرگرمیاں اپنائیں جیسے گانے گانا، نظمیں لکھنا یا ایک دوسرے کی شرکت کے لئے راہ ہموار کرنے کے لئے کہانیاں سنانا وغیرہ۔ کچھ لوگ خاکہ کشی کے ذریعے یا تصویر بنا کر بہتر طور پر اپنے جذبات کا اظہار کر سکتے ہیں۔

ہم اپنی مشکلات سے تمام ہمسائیوں کو کیوں آگاہ کریں۔

مگر ہمارے دوستوں سے زیادہ بہتر اور کون ہماری مدد کر سکتا ہے؟ ہم سب کسی معذور فرد کی دیکھ بھال کرتے ہیں لہٰذا ہم سب ہی کچھ یکساں مسائل رکھتے ہوں گے۔

گروپ میں شامل افراد کے اطمینان اور ایک دوسرے پر اعتماد بحال کرنے کے لئے چند تجاویز مندرجہ ذیل ہیں۔

سنیں کہ دوسرے کیا کہتے ہیں: سوچئے کہ آپ اپنی بات سنے جانے کے لئے دوسروں سے کیا چاہتے ہیں اور پھر خود بھی اسی طرح دوسروں کو سننے کی کوشش کیجئے دوسرے لوگوں سے یہ کہنے کی کوشش مت کریں کہ کیا کریں؟

آپ دوسروں کو یہ سمجھنے میں مدد دے سکتے ہیں وہ کس طرح محسوس کر رہے ہیں اور انہیں اپنے تجربات میں شامل کیجئے لیکن ہر فرد معذور افراد کی مدد کرنے کے بہترین طریقے کے حوالے سے خود اپنے طور پر فیصلے کرے۔ سپورٹ گروپ ایک ایسی جگہ بھی ہوسکتا ہے جہاں دیکھ بھال کرنے والے اپنے غصے اور فرسٹریشن کا اظہار کر سکتے ہیں۔ تجربات اور خیالات میں شرکت کر کے آپ ان احساسات کی وجہ جان کر تبدیلی کے طریقے ڈھونڈنے میں ایک دوسرے کی مدد کر سکتے ہیں۔

باہمی مشاورت اور تبادلہ خیال کے ذریعے وہ باتیں بھی سامنے آ سکتی ہیں جو ایک مددگار کے لئے پریشانی یا فکرمندی کا سبب ہوں۔ سپورٹ گروپ اسی لئے اہم سمجھے جاتے ہیں کہ ملتے جلتے حالات سے دو چار افراد نہ صرف اپنے دبے جذبات کا اظہار کر کے ذہنی طور پر پُرسکون ہو سکتے ہیں بلکہ وہ سمجھ سکتے ہیں کہ آئندہ وہ کس طرح اپنے مسائل سے نمٹ سکتے ہیں۔ ہمارے معاشرے میں ایسے معاون گروپوں کی کمی ہے۔ ہمیں اس پر توجہ دینی چاہیے۔

عمل کے لئے منصوبہ بندی

مل جل کر کام کرنے والا گروپ بہت سے مسائل حل کرنے کے لئے عملی اقدامات کرسکتا ہے۔ عملی اقدام کے لئے چند تجاویز مندرجہ ذیل ہیں۔

1- کسی ایسے مسئلے کا انتخاب کیجئے جسے گروپ کے بیشتر لوگ اہم سمجھتے ہوں۔ ہر چند کہ بہت سی تبدیلیوں کی ضرورت ہوسکتی ہے آپ کا گروپ ایک وقت میں ایک مسئلہ پر سوچ بچار کرے تو یہ زیادہ موثر ہوسکتا ہے۔ سب سے پہلے وہ مسئلہ چنیں جسے آپ کا گروپ فوری طور پر حل کرنے کا موقع رکھتا ہو۔ آپ پیچیدہ مسائل کے حل کے لئے بھی کام کرسکتے ہیں۔

2- فیصلہ کریں کہ آپ مسئلہ کس طرح حل کرنا چاہتے ہیں۔ مسئلہ کن طریقوں سے حل ہوسکتا ہے ان کی فہرست بنائیے اور ان میں وہ بہترین حل منتخب کیجئے جو آپ کے گروپ کی قوت اور وسائل کے استعمال سے ممکن ہو۔

3- ایک منصوبہ بنائیں کہ کام کرنے کے لئے گروپ کے اراکین کو کیا کیا کرنے کی ضرورت ہوگی ہر ایک کے لئے تکمیل کی کوئی تاریخ طے کرنے کی کوشش کریں۔

4- جب آپ دوبارہ ملیں تو تبادلہ خیال کریں کہ کام کیسا جا رہا ہے۔ اگر مشکلات سامنے آئی ہوں تو اپنے منصوبے میں رد و بدل کریں۔

صحت کارکنوں کے لئے

صحت کارکن بھی دیکھ بھال کرنے والوں کی مدد کر سکتے ہیں۔ جب وہ معذوری والی کسی عورت کا علاج کریں تو اس سے اس کی دیکھ بھال کرنے والے کے ساتھ رابط کے حوالے سے تبادلہ خیال کریں اگر اس کی دیکھ بھال کرنے والا/والی اس کے ساتھ آتی ہے تو دونوں سے ان کے مابین تعلقات اور ایک دوسرے کی توقعات کے بارے میں بات کریں۔

دیکھ بھال کرنے والے کو مائل کریں کہ وہ اپنے احساسات کے حوالے سے بات کرے اس کی باتیں سنیں اور اسے بولنے دیں اسے اس کے خیالات کے انتشار اور الجھاؤ پر برا بھلا نہ کہیں۔ کسی فرد کی معاونت کرنا

مشکل کام ہے۔ دیکھ بھال کرنے والے کو احساس دلائیں کہ کبھی کبھار فرسٹریشن کا شکار ہونا، افسردگی یا غصہ محسوس کرنا فطری بات ہے۔

دیکھ بھال کرنے والے سے اس کی اپنی ضروریات پوچھیں اسے معذور فرد کی مدد کرنے کے ساتھ ساتھ خود اپنی دیکھ بھال کے لئے راغب کریں۔

کوئی ایسا فرد تلاش کرنے کی کوشش کریں جو دیکھ بھال کرنے والے کو وقفہ دے سکے۔ ہر فرد کو اپنے لئے بھی وقت کی ضرورت ہوتی ہے۔ جائزہ لیں کہ کیا اس خاندان یا کمیونٹی میں کوئی ایسا فرد ہے جو کچھ مدت کے لئے مدد کر سکے۔

اگر ضروری ہو تو وہ اپنی کمیونٹی کے دیکھ بھال کرنے والوں کو صحت کی دیکھ بھال کے طریقے اور ایسی مشاورتی مہارتیں سکھائیں جن کی انہیں معذور عورت کی بہتر دیکھ بھال میں ضرورت پڑ سکتی ہے۔

اہم بات : فرسٹریشن محسوس کرنے اور جس فرد کی مدد کی جا رہی ہو اس کے جذبات مجروح کرنے میں واضح فرق ہے۔ بعض اوقات دیکھ بھال کرنے والے اس حد تک تناؤ اور غصے میں ہوتے ہیں کہ وہ اس فرد کے لئے خطرناک ہو سکتے ہیں جس کی وہ دیکھ بھال کر رہے ہوں کسی بھی معذور عورت کا معائنہ کرتے ہوئے ناروا اسلوک کی علامات کے لئے الرٹ رہیں اور یہ اطمینان کرنے کے لئے معذور عورت سے تنہائی میں بات کرنے کی کوشش کریں کہ کہیں وہ کسی بھی طرح اپنی دیکھ بھال کرنے والے کے ذریعے زیادتی کا نشانہ تو نہیں بن رہی ہے۔ معذور عورتوں سے زیادتی اور تشدد کے بارے میں مزید معلومات کے لئے باب 14 دیکھئے۔

دیکھ بھال کرنے والی عورت سے یہ جائزہ لینے کے لئے گفتگو کیجئے کہ اسے کوئی ایسا مسئلہ تو درپیش نہیں جس میں آپ اس کی مدد کر سکتی ہوں۔

میں نے ان معذور عورتوں سے بہت کچھ سیکھا ہے جن کی میں نے مدد کی ہے۔ میں ان کے عزم سے بہت متاثر ہوں۔ انہوں نے مجھے سکھایا ہے کہ ہم کس طرح ان رکاوٹوں پر قابو پا سکتے ہیں کہ ہم اپنی بھرپور اور مکمل زندگی میں سامنا کرتے ہیں۔

خصوصی صفحات

ان صفحات سے کس طرح فائدہ اٹھائیں

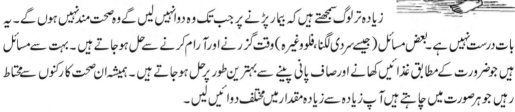

اس حصے میں ان جدید ادویات کے بارے میں بتایا گیا ہے جن کے استعمال کے بارے میں اس کتاب میں کہا گیا ہے۔ اگر آپ مقامی یا روایتی ادویات استعمال کرنا چاہتی ہیں تو اس بارے میں مقامی روایتی معالج سے معلوم کیجئے کہ کیا وہ آپ کو درپیش طبی مسئلے کا کوئی حل یا علاج جانتا ہے۔ روایتی علاج ہر جگہ بدل جاتا ہے لہٰذا ضروری نہیں کہ کوئی دوا کسی اور جگہ بھی دستیاب ہو یا وہاں بھی کارگر ہو۔

دواؤں کو محفوظ طور پر کس طرح استعمال کیا جائے؟

دوائیں صرف اس وقت استعمال کی جائیں جب ان کی ضرورت ہو

زیادہ تر لوگ سمجھتے ہیں کہ بیمار پڑنے پر جب تک وہ دوا نہیں لیں گے وہ صحت مند نہیں ہوں گے۔ یہ بات درست نہیں ہے۔ بعض مسائل (جیسے سردی لگنا، فلو وغیرہ) وقت گزرنے اور آرام کرنے سے حل ہو جاتے ہیں۔ بہت سے مسائل ہیں جو ضرورت کے مطابق غذائیں کھانے اور صاف پانی پینے سے بہترین طور پر حل ہو جاتے ہیں۔ ہمیشہ ان صحت کارکنوں سے محتاط رہیں جو ہر صورت میں چاہتے ہیں آپ زیادہ سے زیادہ مقدار میں مختلف دوائیں لیں۔

دواؤں کی پوری مقدار لیں

آپ اگر کچھ دوا کھانے کے بعد ہی خود کو بہتر محسوس کریں تو جب بھی تجویز شدہ مدت تک بتائی گئی مقدار میں دوا کھاتی رہیں۔ بعض اوقات دوا کم مقدار میں کھانے سے بیماری لوٹ آتی ہے۔ اس سے دوا کی مزاحمت بھی پیدا ہو سکتی ہے جس کا مطلب یہ ہے کہ اب وہی دوا اس بیماری کے خلاف مفید ثابت نہیں ہو گی۔

زیادہ مقدار میں دوا نہ لیں

تجویز شدہ مقدار سے زیادہ دوا کھانے سے جلد صحت یابی ممکن نہیں۔ اس طرح آپ اور زیادہ بیمار بھی ہو سکتی ہیں۔

مسائل/خرابیوں کی علامات سمجھیں اور ان پر نظر رکھیں

بعض ادویات نقصان دہ ضمنی اثرات رکھتی ہیں یا الرجک ردِعمل کا سبب بن سکتی ہیں جو آپ کے لئے انتہائی مہلک ثابت ہو سکتا ہے (دیکھئے صفحہ 329)

دوا کے بارے میں جس قدر ممکن ہو معلومات حاصل کریں

جن دواؤں کی آپ کو ضرورت ہو سکتی ہے یا جو دوا آپ استعمال کر رہی ہوں، اس کے بارے میں صحت کارکن یا دوا فروش سے معلومات حاصل کریں۔ ان صفحات پر اور ہیسپرین فاؤنڈیشن کی دوسری کتابوں میں بھی آپ کو یہ معلومات مل سکتی ہیں۔

(Books: Where Women have no Doctor . Where There is No Doctor)

کھائی جانے والی دوائیں عموماً انجکشنوں سے زیادہ محفوظ ہوتی ہیں

اس کتاب میں بیشتر کھائی جانے والی ادویات تجویز کی گئی ہیں۔ صرف ان انجکشن کے ذریعے لی جانے والی دواؤں کے بارے میں معلومات دی گئی ہیں جو کھائی نہیں جاسکتی ہیں۔ اگر آپ کو انجکشن کی ضرورت ہے تو صحت کارکن سے ملیں۔ انجکشن محفوظ طور پر کس طرح لگایا جائے اس بارے میں معلومات کے لئے اُردو کتاب ''جہاں عورتوں کے لئے ڈاکٹر نہ ہو'' کے صفحات 472 تا 474 دیکھئے۔

ملی جلی دواؤں کے مقابلے میں ایک دوا زیادہ محفوظ اور سستی ہوتی ہے

تاہم کچھ ادویات خصوصاً ایچ آئی وی/ ایڈز کے لئے لی جانے والی ادویات ملی جلی صورت میں بہتر اور مفید ہوتی ہیں۔

اہم بات

- اگر ممکن ہو تو دوا اس وقت کھائیں جب آپ کھڑی ہوئی یا بیٹھی ہوئی ہوں۔ اس کے ساتھ آپ ہر بار ایک گلاس کوئی مناسب مشروب بھی پی سکتی ہیں۔

- اگر دوا کھانے کے بعد اُلٹی یا تقے ہو جائے اور اس میں دوا نظر آرہی ہو تو آپ کو دوبارہ دوا کھانے کی ضرورت ہوگی۔

- اگر آپ کو برتھ کنٹرول کی گولی کھانے کے بعد تین گھنٹے میں تقے ہو جائے تو حمل سے بچنے کے لئے ایک گولی اور کھائیں۔

معذوری کی دواؤں کے ساتھ دوسری دوائیں کھانا

اگر آپ اپنی معذوری کے باعث کوئی دوا باقاعدگی کے ساتھ کھا رہی ہیں تو ممکن ہے کہ یہ دوا اس کتاب میں دی گئی کچھ ادویات کے ساتھ مؤثر نہ رہے۔ کچھ دوسری دوائیں بھی آپ کی معذوری والی دوا کو کم مؤثر بنا سکتی ہیں یا دوسری دواؤں پر اثر انداز ہوسکتی ہیں۔ مثلاً اگر آپ مرگی کے باعث فینی ٹوائن (Phenytoin) لے رہی ہیں تو اس صورت میں آپ برتھ کنٹرول کے لئے ایسی گولیاں مت کھائیں جن میں ایسٹروجن (Estrogen) اور پروجسٹین (Progestin) دونوں موجود ہوں۔ اس سے آپ کے دورے شدید ہو سکتے ہیں۔ اپنی معمول کی دوا کے نئی دوا پر ممکنہ اثرات اور اس کے بجائے کوئی اور دوا (متبادل دوا) کے بارے میں جاننے کے لئے کسی تجربہ کار صحت کارکن سے معلوم کیجئے۔ خوش قسمتی سے اس کتاب میں دی گئی ہر دوا باہمی رد عمل یا منفی اثر نہیں رکھتی ہے۔ چند نئی ادویات جو اس کا امکان رکھتی ہیں۔ ان کے بارے میں ''دیگر ادویات سے باہمی رد عمل'' کے تحت معلومات دی گئی ہیں اور اس کے لئے مندرجہ علامت استعمال کی گئی ہے۔

الرجی

کچھ لوگ مختلف دواؤں سے الرجی کا شکار ہو جاتے ہیں ۔ جب کسی ایسے فرد کو دوا دی جاتی ہے تو اس کا جسم ایک خاص ردِعمل کا اظہار کرتا ہے ۔ یہ ردِعمل (جیسے جلد کا سرخ ہونا، جلد یا آنکھوں میں خارش، ہونٹوں یا چہرے پر سوجن، خرخراہٹ وغیرہ) تکلیف دہ یا اس کی زندگی کے لئے انتہائی سخت اور مہلک ہو سکتا ہے (جیسے زرد پڑ جانا، کپکپاہٹ یا پسینہ بہنا، کمزوری یا تیز نبض یا دل کی دھڑکن، سانس لینے میں دشواری، خون کا کم دباؤ یا حواس گم ہو جانا) اگر کوئی فرد ڈر جِک شاک کا شکار ہو اسے فوری طور پر طبی مدد کی ضرورت ہوتی ہے ۔ اسے ایپائن فرائن (epine phrine) دیں (دیکھئے صفحہ 342)

وہ دوا مت لیں جو آپ کے لئے الرجی کا سبب بنتی ہو ۔ اسی سلسلے کی دوسری دوائیں بھی مت لیں ۔ (اینٹی بائیوٹک ادویات کے بارے میں معلومات کے لئے دیکھئے صفحہ 330 سے 331)۔

دوا کا نام

دوا کا نام ←

جنرک نام ←

دواؤں کے عام طور پر دو نام ہوتے ہیں ۔ جنرک (یا سائنٹیفک) نام پوری دنیا میں ایک ہوتا ہے ۔ کچھ کمپنیاں جو دوا بناتی ہیں، اپنی ہر دوا کو کاروباری نام (برانڈ نیم) دیتی ہیں ۔ دو مختلف کمپنیوں کی تیار شدہ ایک ہی دوا کے دو مختلف کاروباری نام ہوں گے ۔ اس کتاب میں جنرک نام استعمال کئے گئے ہیں ۔ اگر جنرک نام یکساں ہیں تو آپ ایک دوا کے بجائے دوسری دوا استعمال کر سکتی ہیں خواہ اس کا کاروباری نام جو بھی ہو ۔ بعض برانڈ ادویات دوسروں کی بہ نسبت کم قیمت ہوتی ہیں ۔

دوا کتنی مقدار میں لینی ہے

زیادہ تر گولیوں، کیپسولز، جسم میں داخل کی جانے اور انجکشن کے ذریعے دی جانے والی ادویات کی پیمائش گرام (g)، ملی گرام (mg) مائیکروگرام (mcg) اور یونٹس (U) میں کی جاتی ہے ۔

ایک گرام = 1000 ملی گرام   (ایک ہزار ملی گرام، ایک گرام کے برابر ہوتے ہیں)

0.001 گرام = ایک ملی گرام   (ایک ملی گرام، ایک گرام کا ہزارواں حصہ ہوتا ہے)

بعض ادویات جیسے برتھ کنٹرول کی گولیوں کا وزن مائیکروگرام (mcg/ucg) میں ہوتا ہے

$$1 Ucg = 1 mcg = 1/1000 mg = 0.001 \ mg$$

(اس کا مطلب یہ ہے کہ ایک ملی گرام میں ایک ہزار مائیکروگرام ہوتے ہیں)۔

انجکشن کے ذریعے دی جانے والی ادویات کی پیمائش یونٹس (U) یا انٹرنیشنل یونٹس (IU) میں ہو سکتی ہے ۔

دواؤں کی نوعیت

دوائیں مختلف صورتوں میں ہوتی ہیں اس کتاب میں یہ بتانے کے لئے کون سی دوا کس طرح استعمال کی جائے مندرجہ ذیل تصاویر استعمال کی گئی ہیں۔

| جب یہ تصویر ہو تو سیرپ (مائع دوا) استعمال کریں | جب یہ تصویر ہو تو قطرے استعمال کریں | جب یہ تصویر ہو تو آئنٹ یا کریم استعمال کریں | جب یہ تصویر ہو تو گولیاں، پلر، کپسول لیں یا داخل کریں | جب یہ تصویر ہو تو انجکشن کے ذریعے |

انجکشن کے خطرات سے بچنے کے لئے منہ کے ذریعے دوا جانا دیا جانا بہترین ہے لیکن ایمرجنسی میں دوا انجکشن کے ذریعے دی جانی بہتر ہو سکتی ہے کیونکہ انجکشن دوا کو زیادہ تیزی سے کارگر بنا سکتا ہے۔

 یہ تصویر تنبیہہ کے لفظ کے ساتھ اس وقت نظر آئے گی جب حاملہ عورت یا بچوں کو اپنا دودھ پلانے والی عورتوں کو خصوصی احتیاط کی ضرورت ہو۔

دواؤں کی اقسام

اس کتاب میں درج دواؤں کی مختلف اقسام ہیں۔ ایک گروپ کی ادویات اینٹی بایوٹکس کے بارے میں ایک گروپ کے بطور وضاحت کی ضرورت ہے۔

اینٹی بایوٹکس (ادویات)

اینٹی بایوٹکس ادویات بیکٹیریا کے باعث ہونے والے انفیکشنز کے لئے استعمال کی جاتی ہیں۔ **اینٹی بایوٹکس ادویات وائرسوں کے ذریعے ہونے والے امراض جیسے سردی، ہپاٹائٹس یا ایچ آئی وی/ ایڈز جیسی بیماریوں کا علاج نہیں کرتی ہیں۔** ایک ہی جیسی اینٹی بایوٹکس کو ایک ہی گروپ میں شمار کیا جاتا ہے۔ اس لئے اگر اس گروپ کی کوئی دوا نہ ملے تو آپ اسی گروپ کی دوسری دوا استعمال کرسکتی ہیں۔ اگر آپ پر کسی اینٹی بایوٹک دوا سے الرجک ردعمل ہوتا ہے تو آپ کو اسی گروپ کی دوسری ادویات سے بھی الرجی ہوسکتی ہے لہذا اس گروپ کی دوسری ادویات بھی مت لیں۔

ذیل میں اینٹی بایوٹک ادویات اور ان کے گروپ دیے گئے ہیں۔

پنسلین: ایموکسی سیلین، ایمپی سیلین، بینزا تھائن پنسلین، بینزائل پنسلین، ڈائی کلوکساسلین، پروکین پنسلین اور دیگر

میکرولائیڈز: ازیتھرومائسین، اریتھرومائسین اور دیگر

ٹیٹراسائیکلین: ڈوکسی سائیکلین، ٹیٹراسائیکلین

سلفا (سلفونامائیڈز): سلفامیتھوکسازول، (کوٹری موکسازول کا جزو) اور دیگر

امینوگلائیکو سائیڈز: جینٹا مائی سین، اسٹریپٹو مائی سین اور دیگر

سیفلو اسپورنز: کیفکسائم، سیفالیکسن اور دیگر

اینٹی بایوٹک ادویات حد سے زیادہ استعمال کی جاتی ہیں لیکن اینٹی بایوٹک ادویات صرف اسی وقت استعمال کی جائیں جب ضروری ہو، انہیں احتیاط کے ساتھ استعمال کیا جائے (دیکھئے صفحہ 327)۔

مسائل

یہ اس کتاب میں بیان کردہ صحت کے ان مسائل کی فہرست ہے جن کا علاج دواؤں سے ہوسکتا ہے۔ سیدھے ہاتھ پر پہلے کالم میں مسائل، درمیان میں صفحہ نمبر جن میں موجود متعلقہ مسئلہ جس کے لئے دوا لی جاسکتی ہے اور اُلٹے ہاتھ پر اس کے علاج کی استعمال کی جاسکنے والی دواؤں کے نام ہیں۔ صفحہ 333 سے ادویاتی چارٹ شروع ہو رہا ہے۔

اگر آپ یہ فیصلہ نہیں کرسکیں کہ کون سی دوا لی جائے یا وہ دوا جو آپ لے رہی ہیں، مؤثر محسوس نہ ہو رہی ہو تو کسی تجربے کار صحت کارکن سے رابطہ کیجئے۔ یہ آپ کے لئے بہتر اور دستیاب دوسری دوا کے یقین میں آپ کی مدد کرسکتا ہے۔

ایڈرینالین یا ایپی نیفرائن
Adrenaline or Epinephrine

ایڈرینالین اور ایپی نیفرائن ایک ہی دوا کے دو نام ہیں۔ یہ مختلف الرجک ردِ عمل یا الرجک شاک اور دمے کے شدید حملوں میں استعمال ہوتی ہے۔ دیکھئے ایپی نیفرائن صفحہ 342

اسیٹامینوفین یا پیراسیٹامول
acetaminophen or Paracetamol

(اے پی اے پی، پیناڈول، ٹیمپر اٹامیلینول اور دیگر)
APAP, Penadol, Tempra, Tylenol, other

اسیٹامینوفین اور پیراسیٹامول ایک ہی دوا کے دو نام ہیں جو درد میں آرام اور بخار میں کمی کے لئے استعمال کی جاتی ہے۔ دیکھئے پیراسیٹامول صفحہ 350۔

ایسی کلوور
(Acyclovir)

زووائریکس (Zovirax)

ایسی کلوور وائرسوں کو ہلاک کرتی ہے اور ہرپیز کے علاج کے لئے استعمال کی جاتی ہے جو تناسلی اعضاء، مقعد اور منہ میں تکلیف دہ چھالوں کا سبب بنتا ہے۔

اہم بات: ایسی کلوور ہرپیز دوبارہ ہونے سے نہیں روکتی ہے لیکن یہ ہرپیز کی تکلیف کم کرتی ہے اور اسے پھیلنے سے روکتی ہے۔

ضمنی اثرات: بعض اوقات سر درد، غنودگی متلی اور قے کا سبب بنتی ہے۔

عام طور پر ملتی ہیں: 200، 400 اور 800 ملی گرام کی گولیاں
5 فیصد آئنٹمنٹ

آئنٹمنٹ کی بہ نسبت گولیاں کئی گنا زیادہ مؤثر اور عام طور پر کم قیمت ہوتی ہیں۔ گولیاں زیادہ پانی کے ساتھ لیں۔

استعمال کس طرح کریں: تناسلی ہرپیز (دیکھئے صفحہ 165)۔
200 ملی گرام، منہ کے ذریعے، دن میں پانچ بار، سات سے دس دن لیجئے۔ ہاتھ فوراً ہی دھولیں۔

 تنبیہہ: اگر گردے کا عارضہ ہو تو یہ دوا مت کھائیں

ایمپی سیلین
(Ampicillin)

(ایمسل، ایپی سین، اومنی پین، پین بریٹین، پولی سیلین)
(Amcil, Amicin, Omnipen, Penbritin, Polycillin)

ایپی سیلین، پنسلین گروپ کی اینٹی بایوٹک ہے جو کئی اقسام کے انفیکشنز کے لئے استعمال ہوتی ہے۔ دوائی مزاحمت کی شدت کے باعث یہ اب پہلے کے مقابلے میں کم مفید ہے۔

اہم بات: ایپی سیلین، کھانے سے قبل لیں، اگر آپ تین دن میں بہتری محسوس نہ کریں تو طبی مدد حاصل کریں ممکن ہے آپ کو دوسری دواؤں کی ضرورت ہو۔

ضمنی اثرات: معدے میں گڑبڑ، دست اور جلد پر سرخ چکتے۔ عورتوں میں یسٹ انفیکشن اور بچوں کے پوتڑوں کی جگہ ددورے۔

عام طور پر ملتی ہیں: 250اور500 ملی گرام کی گولیاں یا کیپسولز۔

کس طرح استعمال کریں:

حمل کے دوران رحم کے انفیکشن کے لئے دن میں چار مرتبہ 500 ملی گرام کھائیں جب تک آپ کو طبی امداد نہ مل سکے۔

لیبر کے دوران انفیکشن کے لئے دن میں چار بار، ہر بار 2 گرام منہ کے ذریعے، سات سے دس دن تک۔

ولادت کے بعد انفیکشن کے لئے (دیکھئے صفحہ 248) دن میں چار بار، منہ کے ذریعے، جب تک بخار 48 گھنٹے تک نہ ہو۔

دوسری ادویات جو فائدہ مند ہو سکتی ہیں:

حمل کے دوران رحم کے انفیکشن کے لئے، میٹرونیڈازول
ولادت کے بعد رحم کا انفیکشن، اموکسی سیلین، سپروفلوکساسین، ڈوکسی سائیکلین، میٹرونیڈازول

 تنبیہہ: اگر آپ پنسلین گروپ کی دواؤں سے الرجی ہوں تو ایپی سیلین استعمال مت کریں۔

اموکسی سیلین
(Amoxicillin)

(اموکسی فار، اموکسل، ہیموکس، میگاموکس، سوموکسل)
(Amoxifar, Amoxil, Himox, Megamox, Sumoxil)

اموکسی سیلین، پنسلین گروپ کی اینٹی بایوٹک ہے جو کئی اقسام کے انفیکشنز کے علاج کے لئے استعمال کرتی ہے۔ ادویاتی مزاحمت کے باعث ماضی کے مقابلے میں اب یہ مفید ہے۔

اہم بات: یہ دوا کھانے کے ساتھ لیں۔ اگر آپ تین دن میں خود بہتر محسوس کرنا شروع نہ کریں تو طبی مدد حاصل کریں۔ ممکن ہے آپ کو مختلف دوا کی ضرورت ہو۔

ضمنی اثرات: ڈائریا، سرخ چکتے، متلی، قے۔ عورتوں میں ممکن یسٹ انفیکشن کا سبب بنے یا بچے کے پوتڑوں کی جگہ ددورے ہو جائیں۔

عام طور پر ملتی ہیں: 250اور500 ملی گرام کی گولیاں

کس طرح استعمال کریں:

مثانے کے انفیکشن کے لئے (دیکھئے صفحہ 105) 500 ملی گرام کی گولی دن میں تین بار سات دن تک کھائیں۔ کلے مائیڈیا کے لئے 500 ملی گرام کی گولی دن میں تین بار سات دن تک کھائیں۔

وجائنل اخراج کے لئے

ادویاتی اشتراک کے لئے دیکھئے صفحہ 162۔

گردے کے انفیکشن کے لئے (دیکھئے صفحہ 106) 500 ملی گرام کی گولی دن میں تین بار سات دن تک کھائیں۔

بچے کی پیدائش کے بعد رحم کے انفیکشن کے لئے۔ ایک گرام دن میں تین بار، دس دن تک کھائیں (دوسری ادویات بھی استعمال کریں۔ دیکھئے صفحہ 248)

دوسری ادویات جو مؤثر ہو سکتی ہیں:

مثانے یا گردے کے انفیکشن کے لئے سیفٹرائم، سپروفلوکساسین، کوٹری سوکسازول، نائٹروفیورانٹوان، نورفلوکساسین،

بچے کی پیدائش کے بعد رحم کے انفیکشن کے لئے۔ ایپی سیلین، سپروفلوکساسین، ڈوکسی سائیکلین، میٹرونیڈازول

 تنبیہہ: اگر پنسلین گروپ کی دواؤں سے الرجی ہو تو استعمال مت کریں۔

تنبیہ

اسپرین
(Aspirin)

(ایسٹل سیلسلک ایسڈ،اےایس اے)
(Acetylsilicylic acid, ASA)

اسپرین، سوجن، دردبشمول جوڑوں کے درداور بخار میں استعمال ہوتی ہے۔

اہم بات: اسپرین،غذاؤں یا دودھ یا زیادہ مقدار میں پانی کے ساتھ لیں۔اسپرین آرتھرائٹس میں جوڑوں کے درد میں آرام کے لئے بھی استعمال کی جاسکتی ہے۔

ضمنی اثرات: پیٹ کی گڑبڑ، پیٹ کے درد یا جریان خون کا سبب بن سکتی ہے۔

اسپرین کے حد سے زیادہ استعمال کی علامات: کانوں میں جھنجھناہٹ، سر درد،غنودگی،ابہام اور تیز تنفس۔

عام طور پر ملتی ہیں:300اور600 ملی گرام کی گولیاں۔

کس طرح استعمال کریں:

درد یا سوجن اور بخار کے لئے 300 سے 600 ملی گرام منہ کے ذریعے ضرورت کے مطابق مگر دن میں چھ بار سے زیادہ مت کھائیں۔

دوسری ادویات جو فائدہ مند ہوسکتی ہیں:

درداور بخار کے لئے پیراسیٹامول

درد، بخاراور سوجن کے لئے آئبوپروفین

دوسری ادویات کے ساتھ ردِعمل:

والپورک ایسڈ کے ساتھ والپروٹیک ایسڈ کا ارتکاز بڑھا سکتی ہے۔فینی ٹوئن کے ساتھ فینی ٹوئن کا ارتکاز بڑھا سکتی ہے۔

تنبیہ: حمل کے آخری تین ماہ میں عورتیں اسپرین نہ لیں۔ پیٹ کے السریا جریان خون کے مسائل رکھنے والوں کو اسپرین نہیں لینی چاہیے۔سرجری سے پہلے استعمال مت کریں۔ بریسٹ فیڈنگ کے پہلے ہفتے میں استعمال نہ کریں۔اگر کانوں میں جھنجھناہٹ شروع ہوجائے (جو پوائزننگ کی ابتدائی علامت ہے) تو جھنجھناہٹ ختم ہونے تک اسپرین مت لیں۔اس کے بعد دوبارہ اسپرین کم مقدار میں لیں۔

ازیتھرو مائی سین
(Azithromycin)

(زیتھرومیکس)
(Zithromax)

ازیتھرو مائی سین، میکرولائیڈ گروپ کی اینٹی بایوٹک ہے۔ یہ جنسی طور پر منتقل ہونے والے انفیکشنز کے لئے استعمال کی جاتی ہے۔ یہ مہنگی دوا ہے اور بعض اوقات نہیں ملتی ہے۔لیکن جب دوسری اینٹی بایوٹکس ناکام ہو جائیں تو یہ مؤثر ثابت ہوتی ہے۔

اہم بات: کھانے سے کم از کم ایک گھنٹے پہلے یا دو گھنٹے بعد استعمال کریں۔ ازیتھرو مائی سین جنسی امراض کا شاندار علاج ہے جو رطوبت کے اخراج یا تناسلی زخموں کا سبب بنتے ہیں جب مریض میں دوسری دواؤں کے خلاف مزاحمت پائی جاتی ہو تو ازیتھرو مائی سین علاج کے لئے بہترین ہے۔

ضمنی اثرات: ڈائریا،متلی، قے ، پیٹ کا درد ۔

عام طور پر ملتے ہیں:250 ملی گرام کے کیپسول ۔

کس طرح استعمال کریں:

کلے مائیڈیا، شنکرائیڈ یا پیپرو کے عارضے کے لئے ایک گرام منہ کے ذریعے صرف ایک بار لیں۔

دوسری ادویات جو فائدہ مند ہوسکتی ہیں

شنکرائیڈ کے لئے سپروفلوکساسین،ازیتھرو مائی سین ۔

پیپرو کے عارضے کے لئے دیکھئے صفحہ 162۔

تنبیہ: اگر آپ ازیتھرو مائی سین گروپ کی ادویات سے الرجک ہوں تو یہ دوا مت کھائیں۔

بینزائل پینسلین
(Benzyl penicillin)

(سیلینکس ۔ ہائی ۔ ڈو۔ پین، پینسلین جی، پوٹاشیم یا سوڈیم)

(Celinex, Hi-Do-Pen, Penicillin G Potassiam or Sodium)

بینزائل پینسلین، پینسلین گروپ کی اینٹی بایوٹک ہے جومختلف شدید انفیکشنز کے علاج کے لئے استعمال ہوتی ہے۔

اہم بات : الرجک ردِعمل کے تدارک کے لئے تیار رہیے (دیکھئے صفحہ 329)۔

ضمنی اثرات : عورتوں میں یسٹ انفیکشنز یا بچے میں پوٹروں کے ددوڑوں کا سبب بن سکتی ہے۔

عام طور پر ملتی ہے: ایک یا 5 ملین یونٹس کے انجکشنوں کے لئے پاؤڈری کی شکل میں۔

کس طرح استعمال کریں:

نومولود بچوں میں تشنج کے لئے ، صرف ایک بار عضلے میں 100,000 یونٹس / فی کلوگرام وزن اور پھر طبی مدد حاصل کی جائے۔

تنبیہہ : الرجک ردِعمل یا شاک کی علامات پر نظر رکھی جائے۔ ایسے لوگوں کو نہ دی جائے جو پینسلین گروپ سے الرجی ہوں۔

بینزاتھائن پینسلین
(Bezathine Penicillin)

(ہائی سیلین ایل اے، پینا ڈورایل اے، پیر ماپین)

(Bicillin, L-A, Penadur L-A, Permapen)

بینزاتھائن پینسلین سفلس ، تناسلی اعضاء کے السر اور دوسرے انفیکشنز کے علاج کے لئے استعمال ہونے والی پینسلین گروپ کی اینٹی بایوٹک ہے۔

اہم بات : یہ ہمیشہ انجکشن کے ذریعے بڑے عضلے (سل) میں لگائی جاتی ہے۔

ضمنی اثرات : کچھ لوگوں پر سوزش اُبھار یا ددوڑے ہوجاتے ہیں۔

بہت ہی کم : خطرناک ردِعمل (الرجک شاک) پینسلین لگانے کے فوراً بعد مریض زرد پڑ جاتا ہے (ٹھنڈے پسینے، کمزوری، نبض یا دل کی تیز دھڑکن، سانس لینے والی دشواری اور حواس گم ہونا) اس کی علامات ہیں۔ فوری طور پر اپنی نیفرائن انجکشن کے ذریعے دی جائے (دیکھئے صفحہ 342)

عام طور پر ملتے ہیں: 1.2 یا 2.4 ملین یونٹس کے انجکشنوں میں ملانے کے لئے پاؤڈر پانچ ایم ایل کے وائل میں۔

کس طرح استعمال کریں:

سفلس کے لئے، اگر زخم ہو تو صرف ایک بار 2.4 ملین یونٹس بڑے عضلے (جیسے ران) میں انجکشن کے ذریعے۔ اگر بلڈ ٹیسٹ ہو چکا ہے یا زخم پہلے ہی غائب ہو چکے ہیں تو پھر تین ہفتے تک ہر ہفتے میں ایک بار انجکشن لگا ئیں۔

دوسری ادویات جو فائدہ مند ہو سکتی ہیں:

سفلس کے لئے، ڈوکسی سائیکلین ، ٹیٹرا سائیکلین ، اریتھرومائی سین۔

تنبیہہ : اپی نیفرائن، پینسلین لگاتے ہوئے اپنے پاس رکھیں، الرجک ردِعمل یا شاک کی علامات پر نظر رکھیں، جو تیس منٹ میں سامنے آ سکتی ہیں۔

سیفکزائم
(Cefixime)

سپریکس (Suprax)

سیفکزائم سیفالوپورین گروپ کا اینٹی بایوٹک ہے جو گونوریا، پیشاب کی بیماری اور گردے کے انفیکشن سمیت کئی اور انفیکشنز کے علاج کے لئے استعمال ہوتی ہے۔

اہم بات: الرجک ردِعمل پر نگاہ رکھیں۔

ضمنی اثرات: پیٹ کی گڑبڑ، ڈائریا، سردرد۔

عام طور پر ملتی ہیں:200یا400 ملی گرام کی گولیوں یا5 ملی لٹر میں 100 ملی گرام مائع کی صورت میں۔

کس طرح استعمال کریں:

گونوریا کے لئے،400 ملی گرام، منہ کے ذریعے صرف ایک بار
فرج سے رطوبت کے اخراج یا پیشرو کے عارضے کے لئے دیکھئے صفحہ 162-
گردے کے انفیکشن کے لئے500 ملی گرام منہ کے ذریعے دن میں دوبار، دس دن تک۔

دوسری ادویات جو فائدہ مند ہو سکتی ہیں:

گونوریا کے لئے، سپروفلوکساسین ڈوکسی سائیکلین، نورفلوکساسین

تنبیہہ: جگر کا عارضہ رکھنے والے سیفکزائم لیتے ہوئے محتاط رہیں۔ اگر آپ سیفالوپورین گروپ سے الرجی ہیں تو اسے استعمال نہ کریں۔

سپروفلوکساسین
(Ciprofloxacin)

(سلوکسان، سپرو، سپروبے)
(Ciloxan, Cipro, Ciprobay)

سپروفلوکساسین، کیونولون گروپ کی ایک طاقتور اینٹی بایوٹک ہے جو جلد اور گردے کے انفیکشنز اور گونوریا، شنکرائڈ اور پیشرو کے انفیکشنز کے علاج کے لئے استعمال کی جاتی ہے۔

اہم بات: خوب پانی پئیں، سپروفلوکساسین لیتے ہوئے آپ کچھ کھا سکتی ہیں صرف ڈیری مصنوعات نہ کھائیں۔

ضمنی اثرات: متلی، ڈائریا، قے اور سردرد۔

عام طور پر ملتی ہیں:250, 500یا750 ملی گرام کی گولیاں۔

کس طرح استعمال کریں:

گونوریا کے لئے 500 ملی گرام، منہ کے ذریعے صرف ایک بار
(فرج سے رطوبت کے اخراج کے علاج کے لئے دوسری ادویات لینے کے لئے) دیکھئے صفحہ 162-
شنکرائڈ کے لئے 500 ملی گرام، منہ کے ذریعے، دن میں دوبار، تین دن تک۔
پیشرو کے انفیکشنز کے لئے دیکھئے صفحہ 162-
بچے کی ولادت کے بعد انفیکشن کے لئے، 500 ملی گرام منہ کے ذریعے دن میں دوبار۔
گردے کے انفیکشن کے لئے 500 ملی گرام منہ کے ذریعے دن میں دوبار دس دن تک۔

دوسری ادویات جو فائدہ مند ہو سکتی ہیں:

گونوریا کے لئے، سیفکزائم
شنکرائڈ کے لئے، ازیتھرومائسین، اریتھرومائسین
گردے کے انفیکشن کے لئے، سیفکزائم، کوٹری موکسازول

تنبیہہ: یہ دوا کیفین (کافی، چاکلیٹ، کولامشروب) کے ساتھ مل کر ردِعمل کا سبب بن سکتی ہے۔ ڈیری مصنوعات کے ساتھ مت لیں اگر حاملہ ہوں یا بچے کو دودھ پلا رہی ہوں یا عمر سولہ برس سے کم ہو تو استعمال مت کریں۔

کلوٹرائی میزول
(Clotrimazole)

(کینسٹین، گائنی لوٹری من، مائی سیلیکس)

(Canesten, Gyne-lotrimin, Mycelex)

کلوٹرائی میزول یِست اور فرج، ، منہ اور جلد کے دیگر فنجائی انفیکشنز کے علاج کی اینٹی فنجائی دوا ہے۔

اہم بات : فرج کے انفیکشن کے لئے کلوٹرائی میزول استعمال کرنے کے بعد تین دن مباشرت مت کریں۔ یہ کنڈوم یا ڈایا فرام کو کمزور کرسکتی ہے۔ احتیاط کریں کہ یہ آنکھوں میں نہ لگے، اگر یہ جلن یا تکلیف کا سبب بنے تو استعمال بند کردیں۔

عام طور پر ملتے ہے: ایک فیصد، دو فیصد اور دس فیصد کی کریم، سو، دوسو اور پانچ سو ملی گرام کی بتیاں، 10 ملی گرام کے لوزینجر۔

ضمنی اثرات: سوزش، پیٹ کی گڑ بڑ (لوزینجر کے استعمال پر)

کس طرح استعمال کریں:
فرج کے یِست انفیکشنز کے لئے: 100 ملی گرام کی بتی یا ایک فیصد کریم، سات راتوں تک ہر رات ایک بتی یا 5 گرام کریم فرج میں اندر تک رکھیں۔

200 ملی گرام کی بتی یا 2 فیصد کریم: تین راتوں تک ہر رات ایک بتی یا 5 گرام کریم فرج میں رکھیں۔

منہ کے یِست انفیکشنز (تھرش) کے لئے: ایک لوزینجی دن میں پانچ مرتبہ چودہ دن تک لوزینجی کو چوسیں، چبائیں یا نگلیں مت۔ **جلد کے انفیکشنز کے لئے:** متاثرہ حصے پر کریم نرمی سے دن میں دو مرتبہ، دو سے آٹھ ہفتے تک لگا ئیں۔

دوسری ادویات جو فائدہ مند ہوسکتی ہیں:

حبیثیان وائلٹ، نیسٹاٹن، مییکونیزول

کلینڈا مائی سین
(Clindamycin)

سیلوسین، ڈیلاسین

(Celocin, Dalasin)

کلینڈا مائی سین، لنکوسامائیڈ گروپ کی اینٹی بایوٹک ہے جو فرج، پیٹرو اور جلد کے انفیکشنز کے علاج کے لئے استعمال ہوتی ہے۔

اہم بات : اریتھرو مائی سین یا کلورامفینیکول کے ساتھ اس دوا کا استعمال دونوں ادویات کے اثر کو کم کرسکتا ہے۔ اگر کریم استعمال کرتے ہوئے آپ کی ماہواری آرہی ہو تو ٹیمپٹون (Tempton) مت استعمال کریں کیونکہ یہ دوا یہ جذب کرے گا۔

ضمنی اثرات : متلی، تے اور ڈائریا دوا استعمال کرنے کے ابتدائی چند ہفتوں میں ہوسکتے ہیں۔ اگر کلینڈا مائی سین کے استعمال سے جلد پر دردوڑے پڑیں تو اس کا استعمال بند کردیں اور اپنے معالج/صحت کارکن سے ملیں۔

عام طور پر ملتے ہیں: 300, 150, 75, 25 گرام کے کپسولز، 2 فیصد کریم۔

کس طرح استعمال کریں:
بیکٹر یائی وجائی نوسز کے لئے: 300 ملی گرام منہ کے ذریعے، دن میں دو بار، سات دن تک **یا** سات راتوں تک ہر رات سوتے وقت فرج میں 5 گرام کریم رکھیں۔

ٹرائیکومو ناس کے لئے: 300 ملی گرام منہ کے ذریعے، دن میں دو بار، سات دن تک ۔

فرج سے رطوبت کے اخراج یا پیٹرو کی بیماریوں کے علاج کے لئے دوسری ادویات ملانے کے لئے دیکھیں صفحہ 162۔

دوسری ادویات جو فائدہ مند ہوسکتی ہیں:
بیکٹر یائی وجائی نوسز کے لئے، میٹرونیڈازول

تنبیہہ : تین دن سے زیادہ دوا کا استعمال تھرش اور یِست انفیکشن کا سبب بن سکتا ہے اور گردے یا جگر کے مسائل والے افراد کو نقصان پہنچا سکتا ہے۔ فرج کی کریم استعمال کے بعد کنڈوم کو ناکارہ کرسکتی ہے۔ اگر آپ بریسٹ فیڈنگ کر رہی ہیں تو اس دوا سے آپ کے بچے کو دست آ سکتے ہیں۔ اس صورت میں دوا استعمال کرنا بند کردیں۔

کوٹرائی موکسازول ....تسلسل

ایڈز زدہ افراد، خونی اسہال کے لئے، 480 ملی گرام کی دو گولیاں، منہ کے ذریعے دن میں دوبار، دس دن تک

ایڈز زدہ افراد، نمونیے کے لئے، 480 ملی گرام کی دو گولیاں، منہ کے ذریعے، دن میں تین بار، ایک دن تک

دوسری دوائیں جو فائدہ مند ہوسکتی ہیں:

مثانے اور گردے کے انفیکشنز کے لئے، سیفلگزائم، سپروفلوکساسن، نائٹروفیوران ٹوئن

ایڈز زدہ افراد میں ڈائریا کے لئے، نورفلوکساسن، میٹرونیڈازول

دوسری ادویات کے ساتھ ردعمل:

فینی ٹوان کے ساتھ: فینی ٹوان کی سطح بڑھ سکتی ہے جس سے جسم کی حرکات پر (ataxia) یا آنکھوں کی حرکات (Nystagmus) میں دشواری ہوسکتی ہے۔

ڈیپزون کے ساتھ: ٹرائم تھوپرم کی سطح بڑھ سکتی ہے اور خون کی کمی کا خطرہ بڑھ سکتا ہے۔

تنبیہہ: حمل کے آخری تین ماہ میں یا اگر آپ سلفا گروپ سے الرجی ہیں تو کوٹرائی موکسازول مت استعمال کریں۔

کوٹرائی موکسازول
(Cotrimoxazole)

(ٹرائم تھوپرم+سلفامیتھوکسازول)

(ازوگانتانول، بیکٹریم، کوپٹن، گنتانول، پولوگریم، سپٹرا، سلفاٹرم، ٹی ایم پی/ایس ایم ایکس، ٹرم پیکس اور دیگر)

(Trimethoprim + Sulfamethoxazole)
(Azogantanol, Bactrim, Coptin, Gantanol, Pologrim, Septra, Sulfa trim, TMP/SMX, Trim Pex others

کوٹرائی موکسازول دو اینٹی بایوٹکس سے مل کر بنتی ہے (ایک کا تعلق سلفا گروپ سے ہے جو مثانے، گردے کے انفیکشنز، گونر یا اور شنکرائڈ کے باعث فرج سے رطوبت کے اخراج کے علاج کے لئے استعمال ہوتی ہے۔ یہ ایچ آئی وی افراد کے ڈائریا اور نمونیا اور دیگر انفیکشنز بھی تحفظ میں معاون ہے۔

اہم بات: دوا کھاتے ہوئے پانی زیادہ مقدار میں پئیں۔

ضمنی اثرات: اگر یہ خارش یا دوددوڑوں جیسے الرجک ردعمل کا سبب بنے تو دوا کھانا بند کردیں۔ یہ متلی، قے کا سبب بھی بن سکتی ہے۔ دوا حد سے زیادہ لینے کی علامات میں متلی، ڈائریا، الجھن اور پسینے کا اخراج شامل ہے۔

عام طور پر ملتی ہے: 120 ملی گرام (20 ملی گرام ٹرائم تھوپرم+ 100 ملی گرام سلفامیتھوکسازول)۔ 480 ملی گرام (80 ملی گرام ٹرائم تھوپرم + 400 ملی گرام سلفامیتھوکسازول) جسے (Single strength) کہا جاتا ہے اور 960 ملی گرام (160 ملی گرام ٹرائم تھوپرم + 800 ملی گرام سلفامیتھوکسازول) جسے Double Strength کہا جاتا ہے، گولیوں کی صورت میں 420 ملی گرام مائع کی صورت (40 ملی گرام ٹرائم تھوپرم +200 ملی گرام سلفامیتھوکسازول) فی پانچ ملی لٹر)

کس طرح استعمال کریں:

مثانے کے انفیکشنز کے لئے، 480 ملی گرام کی دو گولیاں، منہ کے ذریعے، دن میں دوبار تین دن تک

گردے کے انفیکشنز کے لئے، 480 ملی گرام کی دو گولیاں، منہ کے ذریعے، دن میں دوبار، دس دن تک

ایچ آئی وی افراد، نمونیے اور ڈائریا سے تحفظ کے لئے، 480 ملی گرام کی دو گولیاں روزانہ

ڈائی کلوکساسیلین
(Dicloxacillin)

ڈائی کلوکساسیلین، پینسلین گروپ کی اینٹی بایوٹک ہے جو چھاتی اور جلد کے انفیکشنز کے علاج کے لئے استعمال ہوتی ہے۔

اہم بات: الرجک ردعمل کے علاج کے لئے تیار رہیں۔

ضمنی اثرات: متلی، قے، ڈائریا، عورتوں میں یِیسٹ انفیکشنز یا بچوں کے نچلے دھڑ پر دردوڑے۔

عام طور پر ملتے ہیں: 125 ملی گرام، 250 اور 500 ملی گرام کے کیپسول، 625 ملی گرام فی 5 ملی لٹر سیال

کس طرح استعمال کریں:

چھاتی یا جلد کے انفیکشنز کے لئے، 500 ملی گرام، منہ کے ذریعے، دن میں چار بار، 7 سے 10 دن تک

دوسری ادویات جو فائدہ مند ہو سکتی ہیں:
سیفالیکسن، اریتھرومائی سین، پینسلین

✋ **تنبیہہ:** اگر آپ پینسلین گروپ کی ادویات سے الرجی ہیں تو ڈائی کلوکساسیلین مت استعمال کریں۔

ڈائزیپام
(Diazepam)

(انکسیوئل، کامپوز، ویلیم)
(Anxionil, Calmpose, Valium)

ڈائزیپام سکون بخش دوا ہے جو دوروں اور جھٹکوں کے علاج اور تحفظ کے لئے دی جاتی ہے۔ یہ انزائٹی سے نجات دیتی ہے اور نیندلاتی ہے۔

اہم بات: ڈائزیپام نشہ آور دوا ہے جو عادت بن سکتی ہے۔ اسے دوسری نشہ آور اشیاء خصوصاً الکحل کے ساتھ مت لیں ورنہ آپ کو حد سے زیادہ نیند آ سکتی ہے۔

ضمنی اثرات: (حد سے زیادہ مقدار لینے کی علامات) نیند، توازن کھونا، ذہنی الجھن

عام طور پر ملتی ہیں: 5 اور 10 ملی گرام کی گولیاں، 5 ملی گرام فی ملی لٹر یا 10 ملی گرام فی 2 ملی لٹر کے انجکشن کے لئے سیال حالت میں۔

کس طرح استعمال کریں:

حمل کے دوران دوروں کے لئے، 20 ملی گرام سوئی کے بغیر سرنج سے 20 ملی گرام انجکشن کے ذریعے دی جانے والی ڈائزیپام مقعد میں داخل کریں۔ دس منٹ بعد ضروری ہو تو یہی عمل دوہرائیں۔ دوروں کے بعد 15 ملی گرام استعمال کریں اگر انجکشن والی دوا نہ ہو تو گولیاں پیس کر پانی میں ملا لیں۔

✋ **تنبیہہ:** حمل کے دوران ڈائزیپام کا بار بار یا زیادہ مقدار میں استعمال بچے میں پیدائشی نقص کا سبب بن سکتا ہے۔ دوا دودھ کے ذریعے بھی بچے میں منتقل ہوتی ہے لہذا اپنے بچے کو دودھ پلانے والی مائیں سوائے ایمرجنسی کے ڈائزیپام استعمال نہ کریں۔

ڈوکسی سائیکلین
(Doxycycline)

(بائیوکولائن، ڈوریکس، مونو ڈوکس، وائبرامائی سن، وائبرا۔ گولیاں)
(Biocolyn, Doryx, Monodox, Vibramycin,
Vibra-Tabs)

ڈوکسی سائیکلین، ٹیٹرا سائیکلین گروپ کی اینٹی بایوٹک ہے اور جنسی انفیکشنز ، پیٹ و اور جلد کے انفیکشنز کے علاوہ متعدد انفیکشنز میں استعمال ہوتی ہے۔

اہم بات : ڈوکسی سائیکلین دودھ یا ڈیری مصنوعات یا antaacids کے ساتھ نہ لیں۔ لیٹنے سے فوراً پہلے مت لیں۔ بیٹھی ہوئی حالت میں دوا خوب پانی کے ساتھ کھائیں تا کہ دوا کے باعث ممکن گڑبڑ سے بچ سکیں۔

ضمنی اثرات: ڈائریا یا پیٹ کی گڑبڑ ، کچھ لوگ دوا لینے کے بعد دھوپ میں رہیں تو دوڑے دوڑے پڑسکتے ہیں۔ عورتوں میں ایسٹ انفیکشن یا بچوں میں پوٹری کی جگہ دوڑے ہوسکتے ہیں۔

عام طور پر ملتی ہیں: 50اور 100 ملی گرام کی گولیاں

ڈوکسی سائیکلین ......تسلسل

کس طرح استعمال کریں:

کلے مائیڈیا کے لئے، 100 ملی گرام، منہ کے ذریعے، دن میں دو مرتبہ، سات دن تک۔

سفلس کے لئے جب تاسلی زخم نظر آ رہا ہو، 100 ملی گرام، منہ کے ذریعے، دن میں دو مرتبہ، 14 دن تک۔

بچے کی ولادت کے بعد انفیکشنز کے لئے، 100 ملی گرام، منہ کے ذریعے، مسلسل دو دن بخار نہ ہونے تک۔

وجائنل ڈسچارج یا پیٹرو کی بیماری کی ادویاتی اشتراک کے لئے دیکھئے صفہ 162۔

دباؤ کے زخموں یا دوسرے جلدی انفیکشن کے لئے، 100 ملی گرام، منہ کے ذریعے، دن میں دوبار، 14 دن تک۔

دوسری ادویات جو فائدہ مند ہوسکتی ہیں:

سفلس کے لئے، بینزاتھائن پینسیلین، اریتھرومائی سن، ٹیٹرا سائیکلین۔

گونوریا کے لئے، سیفکزائم، سپروفلوکساسن۔

جلد کے انفیکشنز کے لئے، ڈائی کلوکساسیلین، اریتھرومائی سن، پینسیلین، ٹیٹرا سائیکلین۔

ولادت کے بعد انفیکشنز کے لئے، ایمپی سیلین، سپروفلوکساسن، میٹرونیڈازول۔

تنبیہہ: حاملہ اور دودھ پلانے والی مائیں ڈوکسی سائیکلین مت لیں۔ دھوپ میں رکھی یا استعمال کی مدت ختم ہونے والی ڈوکسی سائیکلین استعمال مت کریں۔

ارگومیٹرائن
(Erogmetrine maleate)
(Methylergonovine maleate)

(انورہیک، ارگونوائن، ارگوٹریٹ، میتھرگین، میتھائل ارگونوائن)
(Anurrage, Ergonovine, Ergotrate,
Methergine, Methylergonovine)

ارگومیٹرائن رحم اور اس کی رگوں میں سکڑاؤ پیدا کرتا ہے اور بچے کی پیدائش کے بعد زیادہ مقدار میں اخراج خون روکنے کے لئے استعمال کی جاتی ہے۔ ارگومیٹرائن اور میتھائل ارگونوائن ایک ہی دوا ہیں۔ یہ دوا دینے کے بعد طبی امداد حاصل کیجئے۔

ضمنی اثرات: متلی، قے، غنودگی اور پسینہ بہنا۔

عام طور پر ملتی ہے: 0.2 ملی گرام کی گولیوں میں۔

کس طرح استعمال کریں:

ولادت کے بعد بھاری اخراج خون کے لئے، پلیسینٹا باہر آنے کے بعد 0.2 ملی گرام کی ایک گولی کی ضرورت کے مطابق ہر چھ سے بارہ گھنٹے بعد کھلائیں۔

تنبیہہ: لیبر کی رفتار بڑھانے یا اسقاط کے لئے ارگومیٹرائن استعمال مت کریں، بچے اور پلیسینٹا کے باہر آنے سے پہلے یہ دوا مت دیں۔

ایپی نیفرائن یا ایڈرینالین
(Epinephrine or adrenaline)

(ایڈرینالین) ...... (Adrenalin)

ایپی نیفرائن اور ایڈرینالین ایک ہی دوا کے دو نام ہیں۔ یہ الرجک ردِعمل یا الرجک شاک (مثلاً پینسلین کے شاک) کے علاج کے لئے استعمال کی جاتی ہے۔

اہم بات: دوا انجکٹ کرنے سے پہلے مریض کی نبض پر ہاتھ رکھئے۔ تین ڈوز سے زیادہ ہرگز نہ دیں۔ اگر نبض پہلے انجکشن کے بعد فی منٹ تیں بار سے زیادہ ہو جائے تو دوسری ڈوز مت دیں۔

ضمنی اثرات: خوف، بے چینی، نروس ہونا، ٹینشن، سردرد، غنودگی، دھڑکن میں اضافہ۔

زیادہ مقدار لینے کی علامات: ہائی بلڈ پریشر، دل کی تیز دھڑکن، اسٹروک۔

عام طور پر ملتی ہے: انجکشن کے لئے ایک ملٹی لٹر میں ایک گرام کے ایمپول میں۔

کس طرح استعمال کریں:

درمیانہ شدت کے الرجک ردِعمل یا الرجک شاک کے لئے آدھا ملی گرام (آدھا ملی لٹر) اوپری بازو کی کھال میں (عضلے میں نہیں) انجکشن کے ذریعے داخل کریں۔ اگر ضروری ہوتو بیس سے تیس منٹ بعد دوسری ڈوز اور اس کے بیس سے تیس منٹ بعد تیسری ڈوز دی جا سکتی ہے۔

تنبیہہ: تجویز کردہ مقدار سے زیادہ دوا مت دیں۔ کولہے میں انجکشن لگانے سے گریز کریں۔ اس کے بجائے اوپری بازو کے پچھلے حصے میں انجکشن لگائیں۔

اریتھرومائی سن
(Erythromycin)

(ای ای ایس، ای مائی سن، ایری میکس، ای تھرل، ایلوسون، ایلوٹائسن)
(E.E.S, E-mycin, Ery-max, Ethril, Ilosone, Ilotycin)

اریتھرومائی سن، میکرولائیڈز گروپ کی اینٹی بایوٹک ہے جوبعض جنسی امراض اور جلد کے انفیکشنز سمیت بہت سی اقسام کے انفیکشنز کے علاج کے لئے استعمال کی جاتی ہے۔ یہ حمل کے دوران بھی محفوظ طور پر استعمال کی جاتی ہے اور دنیا بھر میں دستیاب ہے۔

اہم بات: اریتھرومائی سن کھانے سے ایک گھنٹہ پہلے یا بعد میں لی جائے تو بہترین اثر دکھاتی ہے۔ اگر یہ پیٹ میں حد سے زیادہ گڑ بڑ کرے تو اس کے ساتھ کچھ غذا کھالیں۔

گولیوں کو تو ڑئیے نہیں، اس پر عموماً ایک تہہ چڑھائی جاتی ہے جو اسے اپنا کام شروع کرنے سے پہلے، معدے کی رطوبتوں سے تحفظ دینے کے لئے ہوتی ہے۔

ضمنی اثرات: پیٹ میں گڑ بڑ یا متلی، قے اور ڈائریا۔

عام طور پر ملتے ہیں: 250,200 یا 500 ملی گرام کے کپسول یا گولیاں، ایک فیصد کا مرہم، ہر پانچ ملی لٹرمحلول کے لئے 125 ملی گرام کا پاؤڈر۔

کس طرح استعمال کریں:

کلے مائیڈیا کے لئے، 500 ملی گرام، منہ کے ذریعے، دن میں چار بار، سات دن تک۔

فرج سے اخراج یا پیٹرو کی بیماریوں کے علاج کے لئے دیکھئے صفحہ 162۔

شنکرائڈ کے لئے، 500 ملی گرام، منہ کے ذریعے، دن میں چار بار، سات دن تک۔

سفلس کے لئے، 500 ملی گرام، منہ کے ذریعے، دن میں چار بار، پندرہ دن تک

چھاتی کے انفیکشن کے لئے: 500 ملی گرام، منہ کے ذریعے، دن میں چار بار، سات دن تک۔

اریتھرومائی سن......تسلسل

دباؤ کے زخموں یا جلد کے انفیکشنز کے لئے، 250 ملی گرام، منہ کے ذریعے، دن میں چار بار، سات دن تک۔

آنکھوں کے انفیکشن (Conjunctivitis) کے لئے، تھوڑا مرہم نچلی پلک کے اندر دن میں چار بار، دو سے تین دن تک۔

نومولود کی آنکھوں کی دیکھ بھال کے لئے، ایک فیصد کا مرہم ہر آنکھ میں ولادت کے بعد دو گھنٹے کے اندر ڈالیں۔

دوسری ادویات جو فائدہ مند ہو سکتی ہیں:

کلے مائیڈیا کے لئے، اموکسی سیلین، ٹیٹرا سائیکلین، از یتھرومائی سن، ڈوکسی سائیکلین۔

شنکرائڈ کے لئے، از یتھرومائی سن، سپروفلوکساسن۔

سفلس کے لئے، بینزا تھائن پینسلین، ڈوکسی سائیکلین، ٹیٹرا سائیکلین۔

چھاتی کے انفیکشن کے لئے، ڈائی کلوکساسیلین۔

جلد کے انفیکشن کے لئے، سیفلیکس، ڈائی کلوکساسیلین، ڈوکسی سائیکلین، پینسلین، ٹیٹرا سائیکلین۔

بچے کی آنکھوں کے لئے، ٹیٹرا سائیکلین آئی آئنمنٹ

تنبیہہ: اگر میکرولائیڈز گروپ کی ادویات سے الرجی ہیں تو اریتھرومائی سن مت استعمال کریں۔

حینٹیان وائلٹ (Gentian Violet)

(کرسٹل وائلٹ، میتھائل روزانیلینیم کلورائڈ)
(Crystal Violet, Methylrosanilinium Chloride)

جینٹیان وائلٹ جلد، منہ اور فرج کے انفیکشنز کے لئے معاون جراثیم کش دوا ہے۔

اہم بات: اسے بچے کے منہ میں لگانے کے بعد اس کے چہرے کا رخ زمین کی طرف کر دیں تا کہ اس کے حلق میں دوا کی زیادہ مقدار نہ جائے۔ جینٹیان وائلٹ آپ کی جلد اور کپڑوں پر اودے داغ ڈال دے گی۔

ضمنی اثرات: طویل عرصہ تک استعمال سے خارش ہوسکتی ہے، کسی زخم یا پھٹی ہوئی جلد پر لگانے سے زخم درست ہونے کے بعد اس حصہ کی جلد اودی ہوسکتی ہے۔

عام طور پر ملتی ہے: 0.5 فیصد، ایک فیصد اور 2 فیصد مائع حالت میں، 0.5 فیصد میں ٹنکچر، ایک چائے کے چمچے کے برابر دوا کے کرسٹل نصف لٹر پانی میں 2 فیصد کا محلول بناتے ہیں۔

کس طرح استعمال کریں:

فرج کے یسٹ انفیکشن کے لئے صاف روئی ایک فیصد کے محلول میں بھگو کر رات میں فرج کے اندر کی طرف رکھیں۔ یہ عمل تین رات کریں۔ ہر صبح روئی نکالنا یاد رکھیں۔

منہ میں یسٹ انفیکشن (تھرش) کے لئے، ایک فیصد محلول سے دن میں دو بار ایک سے دو منٹ غرارے کریں۔ دوا کو نگلیں نہیں۔

جلد کے انفیکشنز کے لئے، پہلے متاثرہ حصہ کو صابن اور پانی سے دھو کر خشک کر لیں پھر جلد، منہ یا Vulva پر دن میں تین بار لگائیں۔ یہ عمل پانچ دن تک کریں۔

ایڈز زدہ افراد میں جلد کے انفیکشنز کے لئے، جلد کو صابن اور پانی سے دھو کر خشک کریں پھر تکلیف ختم ہونے تک دن میں دوبار اس جگہ یہ دوا لگا ئیں۔

جینٹیان وائلٹ ...... تسلسل

دوسری ادویات جو فائدہ مند ہوسکتی ہیں:

جلد کے انفیکشنز کے لئے، اینٹی بایوٹک آئنٹمنٹ، آیوڈین۔

منہ میں تھرش کے لئے، لیموں (بچوں کے لئے نہیں) نسٹاٹن۔

فرج کے یسٹ انفیکشن کے لئے، نسٹاٹن، میکونوزول، کلوٹرائی میزول۔

تنبیہہ: فرج کے انفیکشن میں جینٹیان وائلٹ کے استعمال کے دوران مباشرت مت کریں تا کہ آپ کے شوہر میں انفیکشن منتقل نہ ہو۔ اگر اس دوا سے تکلیف ہو تو دوا کا استعمال بند کر دیں، دوا کو آنکھوں میں نہ لگنے دیں۔

آئبوپروفین
(Ibuprofen)

ایکٹی بروفین،ایڈول،جین پرل،موٹرین،نیوپرن،ریفین اور دیگر
(Actiprofen, Advil, Genpril, Motrin, Nuprin, Refen, others)

آئبوپروفین، درد، سوجن اور بخار کے خلاف کام کرتی ہے۔ یہ ماہواری کے دوران، آرتھرائٹس کے درد اور ایڈز میں آرام کے لئے انتہائی مفید ہے۔

اہم بات: اگر غذا خصوصاً ڈیری مصنوعات کے ساتھ لی جائے تو پیٹ میں کم گڑبڑ کا سبب بنتی ہے۔

ضمنی اثرات: پیٹ میں گڑبڑ یا درد، کانوں میں جھنجھناہٹ، قبض۔

عام طور پر ملتی ہیں: 200 ملی گرام یا زائد مقدار کی گولیاں، 5 ملی لیٹر میں 100 ملی گرام کا سیال

کس طرح استعمال کریں:
200 سے 400 ملی گرام دن میں 4 سے 6 بار لیں۔ ایک دن میں 2400 ملی گرام سے زیادہ نہ لیں۔

دوسری ادویات جو فائدہ مند ہو سکتی ہیں:
درد،سوجن اور بخار کے لئے اسپرین
درد اور بخار کے لئے ایسٹا مائنوفین

دوسری ادویات کے ساتھ رد عمل:
فینی ٹوئن کے ساتھ جگر کے مسائل کا سبب بن سکتی ہے۔

تنبیہ: سرجری سے ایک ہفتہ قبل یا بعد لینے سے گریز کریں۔ حمل کے ابتدائی تین ماہ میں لینے سے گریز کریں۔

ہائیڈروکارٹی زون یا کارٹی سول
(Hydrocortisone or Cortisol)

(ایکزا کورٹ، ہائیکوٹل، سولیوکورٹیف وغیرہ)
(Eczacort, Hycocotil, Solu-Cortef   others)

ہائیڈروکارٹی زون، جلد کی سوجن اور کھجلی دور کرنے والے اسکن کریم ہے جو جلد کی خارش اور دودڑے وغیرہ دور کرنے کے لئے استعمال ہوتی ہے۔ یہ بواسیر کے علاج کے لئے بھی مفید ہے۔

اہم بات: کریم لگا کر پٹی مت باندھیں، حاملہ اور دودھ پلانے والی عورتیں بھی کریم استعمال کر سکتی ہیں البتہ گولیاں احتیاط کے ساتھ کھائیں۔

ضمنی اثرات: کریم جلد کو پتلا اور خراش زدہ کر سکتی ہے اگر دس دن سے زیادہ لگائی جائے۔

عام طور پر ملتی ہیں: ایک فیصد کی کریم یا مرہم کی صورت میں۔

کس طرح استعمال کریں:
سرخی، خارش، یا بواسیر کے لئے، کریم کو براہ راست اس جگہ دن میں تین سے چار بار لگائیں۔

میبینڈازول
(Mebendazole)

ورموکس.........(Vermox)

میبینڈازول مختلف ورم انفیکشنز (Work infection) میں بشمول ہک ورم استعمال کی جاتی ہے۔

ضمنی اثرات : پیٹ کا درد یا ڈائریا ہوسکتا ہے البتہ ضمنی اثرات عام نہیں ہیں۔

عام طور پر ملتی ہیں: 100 ملی گرام کی گولیاں۔

کس طرح استعمال کریں:

ہک ورم کے لئے ، 100 ملی گرام، منہ کے ذریعے دن میں دو مرتبہ، 3 دن تک

دوسری ادویات جو فائدہ مند ہوسکتی ہیں:

البینڈازول

تنبیہہ : اگر آپ حاملہ ہیں تو استعمال مت کریں، دو برس سے کم بچوں کو مت دیں۔

لڈوکین
(Lidocaine)

(ٹوپیکین، زائلوکین)
(Topicaine, Xylocaine)

لڈوکین جیل جلد کو تحفظ فراہم کرنے اور جلد کی معمولی خارش، زخموں، جلنے اور حشرات الارض کے کاٹنے پر استعمال کی جاتی ہے۔

ضمنی اثرات : جلد کی رنگت میں تبدیلی (جو عام طور پر اصل رنگت پر آ جاتی ہے) جلد پر آبلے

عام طور پر ملتی ہے: 2 فیصد تا 4 فیصد آئنمنٹ

کس طرح استعمال کریں:

ڈس ریفلیکسیا سے تحفظ میں مدد کے لئے، سخت فضلہ ہاتھ سے نکالنے سے پہلے دوا کی تھوڑی سی مقدار مقعد میں لگائیں یا پیشاب کے سوراخ میں کیتھیٹر چڑھانے سے پہلے لگا ئیں۔

دائیں کالم

میکونازول
(Miconazole)

(ڈیکٹران، فنگٹو پک، میکاٹن، مونسٹاٹ)
(Daktarin, Fungtopic, Micatin, Monistat)

میکونازول ایک اینٹی فنگس دوا ہے جو ییسٹ اور دیگر فنجائی انفیکشنز کے علاج کے لئے استعمال ہوتی ہے۔

اہم بات : میکونازول استعمال کرتے ہوئے ابتدائی تین چار دن اپنے شوہر کو انفیکشن سے بچانے کے لئے جنسی ملاپ مت کریں۔ دوا کو آنکھوں سے دور رکھیں۔

ضمنی اثرات : سوزش اور بے چینی۔

عام طور پر ملتی ہیں: 2 فیصد کریم، 100 اور 200 ملی گرام کی بتیاں (Inserts)

کس طرح استعمال کریں:
فرج کے ییسٹ انفیکشنز کے لئے، 5 گرام کریم سات راتوں تک ہر رات فرج میں اندر تک لگا ئیں۔
100 گرام کی بتی، سات راتوں تک ایک بتی فرج میں ہر رات اندر تک چڑھائیں۔

دوسری ادویات جو فائدہ مند ہو سکتی ہیں:
تمام قسم کے ییسٹ انفیکشنز کے لئے، جینٹیان وائلٹ، نیسٹاٹن، کلوٹری میزول۔

تنبیہہ : حمل کے ابتدائی تین ماہ میں میکونازول مت استعمال کریں۔

بائیں کالم

میٹرونیڈازول
(Metronidazole)

(فلیجل، میتھو پروٹوسٹیٹ، میٹرو، میٹروزائن، سیٹرک)
(Flagyl, Methoprotostat, Metro, Metroxyn, Satric)

میٹرونیڈازول پیٹ کی بیماریوں، فرج اور جلد کے انفیکشنز اور امیبائی اسہال میں استعمال ہوتی ہے۔

اہم بات : آپ کے ساتھ آپ کے شوہر کا بھی علاج ہوگا۔ میٹرونیڈازول لینے کے دوران الکحل نہ پی جائے ورنہ آپ انتہائی متلاہٹ محسوس کریں گی۔

ضمنی اثرات : منہ کے مزے میں تبدیلی، گہرے رنگ کا پیشاب، متلی، پیٹ کی گڑبڑ، سردرد

عام طور پر ملتی ہیں: 250, 400 اور 500 ملی گرام کی گولیاں، 375 ملی گرام اور 500 ملی گرام کی بتیاں۔

کس طرح استعمال کریں:
بچے کی پیدائش کے بعد رحم کے انفیکشن کے لئے، 500 ملی گرام کی گولی، منہ کے ذریعے، دن میں 2 بار، دو دن بخار کے بغیر گزارنے تک۔

بیکٹریائی وجائی نوسز یا ٹرائیکونو ماس کے لئے، 2 گرام صرف ایک بار کھائیں اگر آپ حاملہ نہیں ہیں۔

یا ٹرائیکونو ماس کے لئے، اگر حاملہ ہیں تو 400 سے 500 ملی گرام منہ کے ذریعے روزانہ دن میں دو بار، سات دن تک۔

فرج سے رطوبت کے اخراج یا پیٹ کی بیماری کے لئے، دیکھئے صفحہ 162

ایڈز زدہ افراد میں خونی اسہال کے لئے (بخار کے ساتھ یا بخار کے بغیر) 500 ملی گرام منہ کے ذریعے، دن میں تین بار، سات دن تک۔

دوسری ادویات جو فائدہ مند ہو سکتی ہیں:
ایڈز زدہ افراد میں اسہال کے لئے، کوٹری میکسازول، نورفلوکساسن

تنبیہہ : اگر آپ بے حسی (سن پن) محسوس کریں تو دوا کھانا بند کردیں۔ حمل کے ابتدائی تین ماہ میں میٹرونیڈازول مت استعمال کریں۔ اگر ضروری ہو، حمل کے دوران دوا کی ایک بڑی مقدار مت لیں۔ اگر آپ بچے کو دودھ پلا رہی ہوں تو بڑی مقدار لینا محفوظ ہے۔ جگر کے مسائل جیسے یرقان میں مبتلا افراد یہ دوا استعمال نہ کریں۔

نیفیڈیپائن
(Nifedipine)

(اڈالٹ، نیلاپائن، نیفی کارڈ، نیفیڈ، پروکارڈیا)
(Adalat, Nelapine, Nifecard, Nifed, Procordia)

نیفیڈ یپائن ہائی بلڈ پریشر کو تیزی سے کم کرنے کے لئے استعمال کی جاتی ہے۔

اہم بات : نیفیڈ یپائن کے ساتھ گریپ فروٹ کا جوس مت پئیں۔ دوا بہتر طور پر اثر نہیں دکھائے گی۔

ضمنی اثرات: سردرد، چکر، دوردے، چہرے میں جلن۔

عام طور پر ملتی ہیں: 10 ملی گرام کے کیپسول اور گولیاں۔

کس طرح استعمال کریں:

صرف ڈس ریفلیکسیا کے باعث بلڈ پریشر اچانک زیادہ ہونے پر:

کیپسول: کیپسول دانتوں سے کچل کر نگل لیں یا کیپسول میں سوراخ کر کے دوا زبان کے نیچے رکھیں۔

گولیاں: گولی پیس کر صاف پانی میں ملا کر نرم سا آمیزہ بنائیں اور زبان کے نیچے رکھیں یا گولی چبا کر تھوڑے سے پانی سے نگل لیں۔

دوسری ادویات کے ساتھ ردعمل:

کاربامیزیپائن، فینوباربی ٹل یا فینی ٹوئن کے ساتھ، جب نیفیڈ یپائن لی جا رہی ہو تو یہ دوائیں ممکن ہے مؤثر ثابت نہ ہوں۔

تنبیہہ : ریڑھ کی ہڈی کے مجروح افراد میں ڈس ریفلیکسیا کے لئے نیفیڈ یپائن ہرگز استعمال نہ کریں (دیکھئے صفہ 117)۔ یہ دوا بڑھتے ہوئے ہائی بلڈ پریشر، عام ہائی بلڈ پریشر یا دل کے مسائل کے لئے استعمال نہ کی جائے۔

نائٹروفیورانٹوان
(Nitrofurantoin)

(فیوراڈانٹین، میکروبڈ، میکروڈانٹن)
(Furadantin, Macrobid, Macrodantin)

نائٹروفیورانٹوان مثانے کے انفیکشنز کے علاج کے لئے استعمال کی جانے والی اینٹی بایوٹک ہے۔

ضمنی اثرات: متلی یا قے، سردرد، ریاح کا اخراج، ان پر قابو کے لئے دوا غذا یا دودھ کے ساتھ لیں۔

عام طور پر ملتی ہیں: 25، 50 یا 100 ملی گرام کی گولیاں۔ 25mg/5ml کا سسپینشن

کس طرح استعمال کریں:

مثانے کے انفیکشنز کے لئے، 100 ملی گرام دن میں دو بار، تین دن تک۔

دوسری ادویات جو فائدہ مند ہو سکتی ہیں:

مثانے کے انفیکشنز کے لئے، اموکسی سیلین، کوٹری موکسازول فار فلوکساسن۔

تنبیہہ : گردے کے عوارض میں مبتلا یہ دوا نہ لیں، حمل کے آخری ماہ میں عورتیں یہ دوا ہرگز نہ کھائیں۔

نارفلوکساسن
(Norfloxacin)

(لیکسی نار، نارُوکسن، یوریٹراسن)
(Lexinor, Noroxin, Uritracin)

نارفلوکساسن کوئنولون گروپ کی اینٹی بایوٹک ہے جو گونوریا مثانے اور گردے کے انفیکشنز اور اسہال کے شدید مریضوں کے علاج کے لئے استعمال کی جاتی ہے۔

ضمنی اثرات: یہ سر کے ہلکے پن کا سبب بن سکتی ہے اور کیفین کے اثرات بڑھا سکتی ہے۔ ان اثرات کو محدود کردینے کے لئے یہ دوا کھانے سے آدھ گھنٹہ پہلے یا کھانے کے دو گھنٹے بعد کھائیں۔

عام طور پر ملتی ہیں: 400 ملی گرام کی گولیاں۔

کس طرح استعمال کریں:

مثانے کے انفیکشنز کے لئے، 400 ملی گرام، منہ کے ذریعے دن میں دوبار، تین دن تک

گردے کے انفیکشنز کے لئے، 400 ملی گرام، منہ کے ذریعے دن میں دوبار، دس دن تک

گونوریا کے لئے، 800 ملی گرام منہ کے ذریعے صرف ایک بار

ایڈز زدہ افراد کے ڈائریا کے لئے، 400 ملی گرام صرف ایک بار

دوسری ادویات جو فائدہ مند ہوسکتی ہیں:

مثانے کے انفیکشنز کے لئے، اموکسی سیلین، سیفکزائم، سپرو فلوکساسن، کوٹری موکسازول

گونوریا کے لئے، سیفکزائم، سپرو فلوکساسن، ڈوکسی سائیکلین

ایڈز زدہ افراد کے ڈائریا کے لئے، میٹرونیڈازول، کوٹری موکسازول

تنبیہہ : زیادہ پانی کے ساتھ نگلیں، انٹاسڈز (antacids) یا ایسے وٹامنز کے ساتھ مت کھائیں جن میں آئرن یا زنک ہو۔ اگر نارفلوکساسن یا کوئنولون گروپ سے الرجک ردِعمل ہوتو استعمال مت کریں، حاملہ اور دودھ پلانے والی عورتیں یا سولہ برس سے کم عمر افراد نارفلوکساسن استعمال مت کریں۔

نیسٹاٹن
(Nystatin)

(ڈرموڈیکس، مائی کوسٹائن، نلسٹیٹ، نائسٹیٹ)
(Dermodex, Mycostatin, Nilstat, Nystat)

نیسٹاٹن ایک اینٹی فنگس دوا ہے جو منہ، فرج یا جِلد کے ییسٹ انفیکشنز کے علاج کے لئے استعمال کی جاتی ہے۔

اہم بات: نیسٹاٹن صرف Candida ییسٹ انفیکشنز کے خلاف کام کرتی ہے جبکہ میکونازول دیگر فنجائی انفیکشنز کے خلاف بھی کام کرتی ہے۔ کلوٹری میزول نسبتاً کم مہنگی اور استعمال کرنے میں آسان ہوسکتی ہے۔

عام طور پر ملتی ہیں: 100,000 یونٹ کی بتیاں، منہ کے لئے 200,000 یونٹس کے لوزنجر، 100,000 یونٹ فی گرام کی کریم، 100,000 یونٹ فی ملی لیٹر کا سیال۔

کس طرح استعمال کریں:

منہ یا حلق کے انفیکشنز کے لئے، دن میں تین سے چار بار ایک ملی لیٹر لیکوئڈ منہ میں رکھیں، ایک منٹ اسے منہ کے دونوں حصوں میں گھمائیں اور نگل لیں۔ پانچ دن تک یہ عمل کریں۔

جِلد کے انفیکشنز کے لئے، متاثرہ حصے کو خشک رکھیں اور اس پر انفیکشن ختم ہونے تک دن میں تین بار کریم لگائیں۔

فرج کے انفیکشنز کے لئے، دن میں دو بار فرج کے اندر کریم لگائیں، یہ عمل دس سے چودہ دن تک کریں۔ چودہ راتوں تک ہر رات سونے سے قبل فرج میں 100,000 یونٹ کی بتی اندر تک چڑھائیں۔

دوسری ادویات جو فائدہ مند ہوسکتی ہیں:

میکونازول، کلوٹری میزول، سرکہ یا جینئیان وائلٹ

تنبیہہ: اگر نیسٹاٹن آپ کے لئے تکلیف کا سبب بنے تو اس کا استعمال روک دیں۔ نیسٹاٹن کے استعمال کے دوران جنسی ملاپ مت کریں تاکہ آپ کے شوہر میں انفیکشن منتقل نہ ہو۔

پیراسیٹامول یا اسیٹامینوفین
(Paracetamol or Acetaminophen)

(اے پی اے پی، پیناڈول، ٹیمپرا، ٹائلیٹول اور دیگر)
(APAP, Panadol, Tempra Tylenol o and others)

پیراسیٹامول یا اسیٹامینوفین ایک ہی دوا کے دو نام ہیں جو درد میں آرام اور بخار کی کمی کے لئے استعمال کی جاتی ہے۔ یہ ایک محفوظ ترین درد کش دوا ہے۔ یہ پیٹ میں گڑ بڑ کا سبب نہیں بنتی ہے لہٰذا یہ اسپرین یا ابیوپروفین کی جگہ استعمال کی جاسکتی ہے۔ یہ حاملہ عورتوں کو استعمال کرائی جاسکتی ہے اور کم ڈوز میں بچوں کو بھی۔

اہم بات: پیراسیٹامول بیماری کا علاج نہیں کرتی ہے یہ صرف درد یا بخار کم کرتی ہے، درد یا بخار کا اصل سبب معلوم کرکے اس کا علاج ضروری ہے۔

حد سے زیادہ ڈوز کی علامات: متلی، قے، پیٹ میں درد۔

عام طور پر ملتی ہیں: 325, 100 اور 500 ملی گرام کی گولیاں۔ 120 اور 160 ملی لیٹر فی لیٹر مائع حالت میں، 325, 300, 120, 80 یا 80 ملی گرام فی 0.8 ملی لیٹر کے قطرے۔

کس طرح استعمال کریں:
درد اور بخار میں کمی کے لئے، 500 سے 1000 ملی گرام، منہ کے ذریعے، ضرورت کے مطابق ہر چار سے چھ گھنٹے میں۔

دوسری ادویات جو فائدہ مند ہوسکتی ہیں:
اسپرین ابیوپروفین مؤثر ہوسکتی ہے لیکن انہیں حمل کے دوران استعمال مت کریں۔

تنبیہہ: جگر یا گردہ کی خرابی میں استعمال مت کریں۔ اگر الکحل استعمال کرنے کے دوران یا بعد میں با قاعدگی سے استعمال کی جائے تو نقصان کا سبب بن سکتی ہے۔

پینسلین
(Penicillin)

(بیٹاپین وی کے، پین وی کے، فینوکسی میتھائل پینسلین)
Bentapen VK, Pen Vee K, Phenoxymethyl Penicillin)

پینسلین ایک اینٹی بایوٹک دوا ہے جو منہ، دانتوں، جلد، رحم کے انفیکشنز کے علاوہ متعدد انفیکشنز کے لئے استعمال کی جاتی ہے۔ بدقسمتی سے گزرے برسوں پینسلین کے خلاف خاص مزاحمت پروان چڑھ چکی ہے لہٰذا یہ اب ماضی کے مقابلے میں کم استعمال کی جا رہی ہے۔

اہم بات: الرجک ردعمل اور الرجک شاک کی علامت پر نظر رکھیں۔

عام طور پر ملتی ہیں: 250 اور 500 ملی گرام کی گولیاں۔ 125 یا 250 ملی گرام فی 5 ملی لیٹر کا سیال۔

کس طرح استعمال کریں:
ولادت کے بعد رحم کے انفیکشنز کے لئے، 250 ملی گرام (جو 400,000 یونٹس کے برابر ہے) منہ کے ذریعے، دن میں چار بار، دس دن تک۔

جلد کے انفیکشنز یا زخموں کے لئے، 250 ملی گرام، منہ کے ذریعے دن میں چار بار، دس دن تک۔

دوسری ادویات جو فائدہ مند ہوسکتی ہیں:
ولادت کے بعد رحم کے انفیکشنز میں، اموکسی سیلین، ایمپی سلین، سپرو فلوکساسن، ڈوکسی سائیکلین، میٹرونیڈازول
جلد کے انفیکشنز میں، ڈائی کلوکساسن، ڈوکسی سائیکلین، اریتھرومائی سن، ٹیٹرا سائیکلکلین

تنبیہہ: اگر آپ پینسلین گروپ کی کسی بھی دوا سے الرجی ہیں تو یہ دوا مت لیں۔

پروبینیسڈ
(Probenicid)

(بینی مڈ، پروبیلان)
(Benemid, Probalan)

پروبینیسڈ کچھ پنسیلین گروپ کی اینٹی بایوٹک ادویات کی پنسیلین کو جسم کے اندر دیر تک رکھنے کے لئے استعمال کی جاتی ہے۔ یہ علاج کو زیادہ مؤثر بناتی ہے۔

اہم بات: دو برس سے کم عمر بچوں کو پروبینیسڈ مت دیں۔

ضمنی اثرات: بعض اوقات سر درد، متلی اور قے

عام طور پر ملتی ہیں: 500 ملی گرام کی گولیاں

کس طرح استعمال کریں:
500 ملی گرام یا ایک گرام پنسیلین گروپ کی دوا کھاتے وقت

تنبیہہ: حمل کے دوران اور رضاعت میں یا اگر معدے کے السر میں مبتلا ہوں تو احتیاط کے ساتھ استعمال کریں۔

پوڈوفائلین
(Podophyllin)

(کونڈائی لوکس، پوڈوکان۔25، پوڈوفائلوم ریسن)
(Condylox, Podocon-25, Podophylom resin)

پوڈوفائلین ایک مائع ہے جو تناسلی اعضاء کے مسّوں کو ختم کرنے کے لئے ان پر براہ راست لگایا جاسکتا ہے۔

اہم بات: پوڈوفائلین صحت مند جلد کے لئے نہایت ہی پُرسوزش ہے۔ ایک وقت میں بہت کم (آدھا ملی لٹر یا کم) مقدار میں استعمال کریں۔ استعمال سے پہلے مسّے کے اِرد گرد کی جلد کو پیٹرولیم جیلی وغیرہ لگا کر محفوظ کر لیں۔

ضمنی اثرات: جلد کے لئے انتہائی پُرسوزش ہو سکتی ہے۔

عام طور پر ملتی ہیں: 10 فیصد یا 25 فیصد میں مائع حالت میں

کس طرح استعمال کریں:
روٹی یا ٹوتھ پک یا نوک کو صاف بنا کر صاف کپڑے پر تھوڑی سی دوا لگائیں اور مسّوں پر لگائیں۔ چار گھنٹے کے بعد اسے صابن اور صاف پانی سے احتیاط کے ساتھ دھولیں، ہفتے میں ایک بار، چار ہفتے یہ عمل دوہرائیں۔

دوسری ادویات جو فائدہ مند ہو سکتی ہیں:
تناسلی مسّوں کے لئے، ٹرائی کلورو ایسٹک یا بائی کلورو ایسٹک ایسڈ

تنبیہہ: جن مہاسوں سے خون بہہ رہا ہو ان پر، پیدائشی نشانوں، مسّوں، بال والے مہاسوں یا منہ کے اندر مت لگائیں۔ اگر جلد میں سخت سوزش ہو تو دوبارہ دوا مت لگائیں۔ اگر آپ حاملہ ہوں یا دودھ پلا رہی ہوں تو استعمال مت کریں۔

ٹیٹنس ٹوکسائڈ
(Tetanis Toxid)

ٹیٹا ویکس .... (Tetavax)

ٹیٹنس ٹوکسائڈ، ٹیٹنس سے بچاؤ کے لئے دی جانے والی دوا ہے۔ حمل کے بعد یا حمل کے دوران یا حمل ضائع ہونے کے بعد دی جاسکتی ہے۔ اگر عورت کو حمل کے دوران دو انجکشن (یا اب تک بہتر ثابت ہوا کہ تین انجکشن) لگائے جائیں تو اس سے اس کے نومولود بچے کو بھی اس مہلک انفیکشن سے تحفظ ملے گا۔

اہم بات: یہ ٹیٹنس امیونائزیشن کے لئے ہے، اس کی ابتداء بچپن سے ہوتی ہے۔

ضمنی اثرات: درد، سرخی، حرارت، ہلکی سوجن

عام طور پر ملتی ہے: 4، 5، 10 یونٹس فی 0.5 ملی لٹر کے انجکشن کے لئے سیال حالت میں۔

کس طرح استعمال کریں:

عمر بھر ٹیٹنس سے تحفظ کے لئے، آپ کو بچاؤ کے پانچ انجکشن لگوانا چاہئیں، بعد ازاں ہر دس سال کے بعد ایک انجکشن۔

ہر امیونائزیشن کے لئے، اوپری بازو کے عضلے میں 0.5 ملی لٹر کا ایک انجکشن۔

پروکین پینسلین
(Procaine Penicillin)

(بینزائل پینسلین، پروکین پیسلین سی آر، کرسٹیسیلین، ڈیوراسیلین اے ایس، پینڈ اڈیور، فزیپین اے ایس، وائی سیلین)
(Benzyl Penicillin, Procaine, Bicillin C-R, Crysticillin, Duracillin AS, Penadur, Pfizepen AS, Wycillin)

پروکین پینسلین رحم کے اور دیگر انفیکشنز مثلاً انفیکشنز زدہ دباؤ کے زخم، گونوریا وغیرہ کے لئے استعمال ہونے والی اینٹی بایوٹک ان لوگوں کے لئے ہے جو پینسلین مزاحمت نہیں رکھتے ہیں۔

اہم بات: جب پروبینسڈ کے ساتھ لیں (دیکھئے صفہ 351 تو خون میں پینسلین کی مقدار بڑھتی ہے اور دیر تک رہتی ہے جس سے علاج زیادہ مؤثر ہوتا ہے۔

ضمنی اثرات: عورتوں میں یسٹ انفیکشن اور چھوٹے بچوں میں پوٹڑوں کی جگہ دو دڑے

عام طور پر ملتے ہیں: انجکشن کے لئے 300,000 اور 400,000 اور 600,000 یونٹس کے وائل پاؤڈر ایک گرام انجکشن میں ملانے کے لئے = ایک ملین یونٹس۔

کس طرح استعمال کریں:

حمل کے دوران بخار کے لئے، عضلے میں 1.2 ملین یونٹس انجکشن کے ذریعے ہر بارہ گھنٹے بعد تا ہم عورت کو طبی علاج کے لئے لے جائیں۔

دوسری ادویات جو فائدہ مند ہوسکتی ہیں:

حمل کے دوران بخار کے لئے، ایمپی سلین، میٹرونیڈازول

تنبیہ: پروکین پینسلین دمہ کے مریضوں میں دمہ کے حملے کا سبب بن سکتی ہے۔ یہ دوا ٹیٹرا سائیکلین کے ساتھ کبھی استعمال مت کریں۔ اگر آپ پینسلین گروپ کی دواؤں سے الرجی ہوں تو پروکین پینسلین مت استعمال کریں، یہ دوا رگ میں نہ لگائیں۔

ٹیٹراسائیکلین
(Tetracycline)

(ایکرومائیسن، سائیمسن، ٹیرامائی سن، تھیراسن، یونی مائی سن)
(Achromycin, Sumycin, Teramycin,
Theracin, Unimycin)

ٹیٹراسائیکلین، ٹیٹراسائیکلین گروپ کی اینٹی بایوٹک ہے۔ یہ کلے مائیڈیا، سفلس، پیٹرو کی پُرسوزش بیماری، گردے اور مثانے کے انفیکشنز سمیت کئی طرح کے انفیکشنز کے علاج کے لئے استعمال ہوتی ہے۔ ڈوکسی سائیکلین بھی ان تمام انفیکشنز کے لئے کارگر ہے جس کے اخراجات کم اور استعمال آسان ہے (دیکھئے صفحہ 341)۔

اہم بات: ٹیٹراسائیکلین عام سردی یا جنسی انفیکشنز سے تحفظ فراہم نہیں کرتی ہے۔

ضمنی اثرات: اگر آپ دن میں بیشتر وقت دھوپ میں گزارتی ہیں تو یہ جلد کی تکلیف کا سبب بن سکتی ہے۔ یہ ڈائریا یا پیٹ کی گڑبڑ کا سبب بھی ہوسکتی ہے۔

عام طور پر ملتی ہے: 100، 250یا500 ملی گرام کے کیپسول
ایک فیصد کا مرہم۔

ٹیٹراسائیکلین ...... تسلسل

کس طرح استعمال کریں:

کلے مائیڈیا کے لئے، 500 ملی گرام، منہ کے ذریعے دن میں چار بار، سات دن تک۔

سفلس کے لئے، 500 ملی گرام، منہ کے ذریعے، دن میں چار بار، 14 دن تک۔

پیٹرو کی بیماری کے لئے، 500 ملی گرام، منہ کے ذریعے، دن میں چار بار، دو دن تک بخار نہ ہونے تک (پیٹرو کی بیماریوں اور فرج سے رطوبت کے اخراج کے لئے دوسری ادویات کے لئے دیکھئے صفحہ 162)

جلد کے انفیکشنز کے لئے، 250 ملی گرام منہ کے ذریعے، دن میں چار بار، 14 دن تک۔

نومولود کی آنکھوں کے تحفظ کے لئے، تھوڑا سا مرہم (آئنٹمنٹ) ولادت کے بعد صرف ایک بار ہر آنکھ میں لگائیں۔

دوسری ادویات جو فائدہ مند ہوسکتی ہیں:

کلے مائیڈیا کے لئے، ازیتھرومائی سن، ڈوکسی سائیکلین، اریتھرومائی سن۔

سفلس کے لئے، بینزاتھائن پینسلین، ڈوکسی سائیکلین، اریتھرومائی سن۔

پیٹرو کی بیماریوں کے لئے، دیکھئے صفحہ 162۔

نومولود کی آنکھوں کے لئے، اریتھرومائی سن آئنٹمنٹ

تنبیہ: ڈیری مصنوعات یا انٹاسڈ (antacid) کھانے کے بعد ایک گھنٹہ تک دوا مت کھائیں۔ دھوپ میں رکھی یا استعمال کی تاریخ ختم ہونے والی ٹیٹرا سائیکلین مت کھائیں۔ حاملہ یا دودھ پلانے والی عورتیں ٹیٹراسائیکلین استعمال مت کریں۔

ٹرائی کلوروایسٹک ایسڈ (TCA)
بائی کلوروایسٹک ایسڈ (BCA)
(Trichloroacetic acid (TCA
Bicholoroacetic acid (BCA)

ٹرائی کلوروایسٹک ایسڈ یا بائی کلوروایسٹک ایسڈ تناسلی مہاسوں کے علاج کے لئے استعمال کیا جاسکتا ہے۔

اہم بات: پہلے مسّوں کے اردگرد کی جلد پر پیٹرولیم جیلی لگا کر اُسے محفوظ کرلیں پھر مسّوں پر ٹرائی کلوروایسٹک ایسڈ لگائیں۔ یہ پندرہ سے تیس منٹ تکلیف دے گا۔ اگر یہ صحت مند جلد پر لگ جائے تو اس حصے کو صابن اور پانی سے دھولیں۔ آپ اس حصے پر بے بی پاؤڈر یا کھانے کا سوڈا بھی لگا سکتی ہیں۔

ضمنی اثرات: ٹرائی کلوروایسٹک ایسڈ عام جلد پر لگ کر اُسے خراب یا مجروح کرسکتا ہے۔

عام طور پر ملتا ہے: 10 فیصد اور 35 فیصد کی درمیانی طاقت میں سیال۔

کس طرح استعمال کریں:

صرف مسّوں یا مہاسوں پر ہفتے میں ایک بار، ضرورت کے مطابق ایک سے تین ہفتے لگائیں

دوسری ادویات جو فائدہ مند ہو سکتی ہیں:

پوڈوفائلکن

تنبیہہ: انتہائی احتیاط کے ساتھ استعمال کریں، یہ عام جلد کو اس حد تک جلا سکتا ہے کہ اس پر زخم ہو جائیں۔

کھانے والی مانع حمل (حمل روک) ادویات (پلز، برتھ کنٹرول پلز)

کھانے والی مانع حمل ادویات کے دوسری ادویات کے ساتھ ردِعمل: کچھ ادویات، ملی جلی مانع حمل ادویات (وہ جن میں ایسٹروجن اور پروجیسٹن دونوں شامل ہوں) کا اثر کم کرتی ہیں یا بالکل ختم کر دیتی ہیں۔ اگر آپ با قاعدگی سے مندرجہ ذیل دوائیں کھا رہی ہوں تو مشترکہ ہارمونز والی مانع حمل ادویات مت کھائیں۔

- کاربامیزے پیائن (ٹیگر یٹول)
- فینی ٹوئن (ڈائی فینائل ہائی ڈانٹوئن، ڈیلانٹن)
- فینوباربیٹل (فینوباربی ٹون، لیمینال)
- والپرائٹک ایسڈ (ڈیپاکینی)

کھانے والی مانع حمل ادویات (اور خاندانی منصوبہ بندی کے دوسرے ہارمونل طریقے) آپ کی معذوری پر کیا اثر ڈال سکتی ہیں، اس بارے میں معلومات کے لئے دیکھئے صفحہ 196۔ برتھ کنٹرول پلز ہر ہارمون کی مختلف قوت میں دستیاب ہیں اور مختلف تجارتی ناموں سے فروخت ہوتی ہیں۔ ہم نے نیچے دیئے دیئے چارٹ میں چند برانڈز کا اندراج کیا ہے۔

عام طور پر وہ برانڈز سب سے محفوظ ہوتی ہیں اور بیش تر عورتوں کے لئے مؤثر رہتی ہیں جن میں دونوں ہارمونز کی مقدار کم ہو۔ کم ڈوز والی یہ ادویات گروپ نمبر 1, 2 اور 3 میں ملتی ہیں۔

گروپ 1 - ٹرائی فیسک پلز

ان میں ایسٹروجن اور پروجیسٹن دونوں کی کمتر مقدار شامل ہوتی ہے جو مہینہ کے مختلف دنوں میں تبدیل ہوتی رہتی ہے۔ کیونکہ ان میں ہارمونز کی مقدار تبدیل ہوتی ہے لہٰذا یہ پلز ترتیب سے کھانا ضروری ہے۔

برانڈ کا نام لوگے نون (Logynon)	ٹرائی سائیکلین (Tricyclen)	ٹرائی نووم (Trinovum)
سائنوفیز (Synophase)	ٹرائنوڈائیول(Trinodiol)	ٹرائی کیولر (Triquilar)
ٹرائی فیسیل (Triphasil)		

گروپ 2- کم ڈوز والی پلز

ان میں ایسٹروجن کی مقدار کم ہوتی ہے (35 مائیکروگرام ایسٹروجن، ایتھنائل ایسٹراڈیئول یا 50 مائیکروگرام ایسٹروجن، میسٹرانول) اور پروجیسٹن بھی شامل ہوتا ہے جس کی مقدار پورے مہینے یکساں رہتی ہے۔

برانڈ کا نام بریویکون 1+35 (Brevicon 1+35)	نوری نائل 50 1+ (Norinyl, 1+35, 1+50)
اووس مین 135 (Ovysmen 1/35)	نوریمین (Norimin)
نوری ڈے 1+50 (Noriday 1+50)	اورتھو نوویم 1/50,1/35 (Ortho-Novum 1/35, 1/50)
نیووکون (Neocon)	پرلے (Perle)

گروپ 3- کم ڈوز والی پلز

ان پلز میں پروجیسٹن زیادہ مقدار میں اور ایسٹروجن کم مقدار میں(30 یا 35 مائیکروگرام،ایسٹروجن،ایتھنائل ایسٹراڈیئول) ہوتا ہے۔

برانڈ کا نام لو فیمینل (Lo-Feminal)	لواورل (Lo ovral)	مائیکرو ولر (Microvlar)
مائیکرواوگینون 30 (Microgynon30)	نورڈیٹ (Nordette)	

اثر انگیزی یقینی بنانے اور خون کے دھبوں (ماہواری کے علاوہ عام دنوں میں ہلکا ہلکا خون آنے) کو کم سے کم کرنے کے لئے پلز ہر روز ایک مقررہ وقت پر ہی لیں، خاص طور پر وہ پلز جن میں ہارمونز کی مقدار کم ہوتی ہے۔ اگر خون کے دھبے تین سے چار ماہ کے بعد بھی جاری رہیں تو گروپ 3 کی کوئی اور برانڈ آزمائیں۔ اگر اب بھی تین ماہ بعد خون کے دھبے جاری رہیں تو پھر گروپ 4 کی کوئی دوا استعمال کریں۔

لیکن اگر کسی عورت کو کئی ماہ سے ماہواری نہ آ رہی ہو یا وہ بہت ہی معمولی ماہواری کے باعث ہی پریشان ہو تو وہ گروپ 4 کی زیادہ ایسٹروجن والی پلز کا انتخاب کر سکتی ہے۔

ایسی عورت جسے ماہواری زیادہ مقدار میں آ رہی ہو یا جس کی چھاتیاں ماہواری شروع ہونے سے قبل تکلیف دہ ہو جاتی ہوں، اس کے لئے کم مقدار میں ایسٹروجن والی ایسی پلز بہتر ہو سکتی ہیں جن میں پروجسٹن زیادہ مقدار میں شامل ہو۔ ایسی پلز گروپ 3 میں شامل ہیں۔

ایسی عورتیں جنہیں گروپ 3 کی کوئی دوا استعمال کرتے ہوئے خون کے دھبے آنا جاری رہیں یا انہیں ماہواری نہ ہو یا جو کسی اور قسم کی پلز استعمال کرنے سے پہلے حاملہ ہوگئی ہوں، اپنی پلز کے بجائے ایسٹروجن کی قدرے زیادہ مقدار والی پلز استعمال کر سکتی ہیں۔ زیادہ مقدار والی یہ پلز گروپ چار میں شامل ہیں۔

گروپ 4 - زیادہ مقدار والی پلز

ان پلز میں ایسٹروجن کی زیادہ مقدار (50 مائیکروگرام ایسٹروجن، ایتھنائل ایسٹراڈائیول ہوتی ہے اور زیادہ تر میں پروجسٹن کی زیادہ ہوتی ہے)

برانڈ کا نام	یوگی نون (Eugynon)	نیوگینون (Neogynon)	اورل (Ovral)
	فیمینل (Feminal)	نورڈیئول (Nordiol)	پرانموالر (Prinovlar)

وہ عورتیں جو دودھ پلا رہی ہوں یا جو سر درد یا معمولی ہائی بلڈ پریشر کے باعث با قاعدگی سے پلز نہ کھاتی ہوں مگر صرف پروجسٹن والی پلز کھانا چاہتی ہوں وہ گروپ 5 کی پلز کھا سکتی ہیں جنہیں منی پلز بھی کہا جاتا ہے۔

گروپ 5 - پروجسٹن والی منی پلز

منی پلز کہلانے والی ان پلز میں صرف پروجسٹن ہوتا ہے۔

برانڈ کا نام	فیمیولین (Femulen)
	مائیکرونور (Micronor)
	مائیکرونوویم (Micronovum)
	نور کیو ڈی (Nor-QD)

مائیکرولیوٹ (Microlut)
مائیکرواوول (Microoval)
نیوگیسٹ (Neogest)
نیوگیسٹون (Neogeston)
اوور ریٹ (Overette)

یہ برانڈ زیر ایمرجینسی خاندانی منصوبہ بندی کے لئے بھی استعمال کی جا سکتی ہیں (دیکھئے اگلا صفحہ 357)

صرف پروجسٹن پر مشتمل پلز ہر روز مقررہ وقت پر کھانی چاہئیں، ماہواری کے دوران بھی، کیونکہ ماہواری عموماً بے قاعدہ ہوتی ہے۔

ایمرجنسی خاندانی منصوبہ بندی (ایمرجنسی میں استعمال کی جانے والی مانع حمل ادویات)

اگر آپ عام طور پر خاندانی منصوبہ بندی کی پلز استعمال نہیں کرتی ہیں تو جب بھی ایمرجنسی خاندانی منصوبہ بندی کی پلز کھانا درست ہے (دیکھئے صفہ 355)۔ کیونکہ آپ یہ پلز مختصر عرصے کے لئے استعمال کرتی ہیں لہذا یہ آپ کی معذوری کی دواؤں پر اثر انداز یا ان سے متاثر نہیں ہوں گی۔

آپ غیر محفوظ جنسی عمل کے بعد حمل سے بچنے کے لئے پانچ دن کے اندر ایمرجنسی پلز یا عام برتھ کنٹرول پلز استعمال کرسکتی ہیں۔ آپ کی ضرورت کے مطابق پلز کی تعداد کا انحصار ان میں موجود ایسٹروجن یا پروجیسٹن کی مقدار پر ہوگا۔ اس چارٹ میں ہر قسم کی پلز کی چند عام برانڈز کو شامل کیا گیا ہے۔ ایمرجنسی میں استعمال سے پہلے پلز میں موجود ہارمونز کی مقدار چیک کریں۔

ایمرجنسی منصوبہ بندی کے لئے گولیاں کس طرح لیں

دوسری ڈوز بارہ گھنٹے کے بعد	پہلی ڈوز	ایمرجنسی خاندانی منصوبہ بندی کے لئے پلز
مزید دو پلز کھائیں	دو پلز کھائیں	50 مائیکروگرام ایتھنائل ایسٹرا ڈی یول پر مشتمل زیادہ ڈوز والی پلز (اوول، اوجسٹریل)
مزید چار پلز کھائیں	4 پلز کھائیں	30 یا 35 مائیکروگرام ایتھنائل ایسٹرا ڈی یول پر مشتمل پلز (لوفیمینل، لو اورل مائیکرو ویجیسون، نورڈیٹ)
مزید 5 پلز کھائیں	5 پلز کھائیں	20 مائیکروگرام ایتھنائل ایسٹرا ڈی یول پر مشتمل کم ڈوز والی پلز (ایلیسی، لیسینا، لیوٹیرا)
مزید 2 پلز کھائیں	2 پلز کھائیں	اسپیشل ایمرجنسی پلز جو ایتھنائل ایسٹرا ڈی یول اور لیوونورجسٹریل پر مشتمل ہوتی ہیں (ٹیرا اگینون)
40 پلز صرف ایک بار کھائیں یا پھر 20 پلز ایک ساتھ ایک بار کھانے کے بارہ گھنٹے بعد 20 پلز ایک ساتھ کھائیں۔		75 مائیکروگرام لیوونورجسٹریل پروجسٹن والی پلز (اوورریٹ، نیوجیسٹ)
50 پلز صرف ایک بار کھائیں یا 25 پلز کی ایک ساتھ کھانے کے بارہ گھنٹے بعد 25 پلز کی دوسری ڈوز کھائیں۔		20 مائیکروگرام لیوونورجسٹریل پروجسٹن والی منی پلز مائیکرولیوٹ، مائیکرو واول، نورجسٹون
	صرف ایک بار ایک گولی کھائیں	اسپیشل ایمرجنسی پلز جو 1500 مائیکروگرام لیوونورجسٹریل پر مشتمل ہوتی ہیں (پوسٹینور 1)
	ایک بار صرف 2 پلز کھائیں	اسپیشل ایمرجنسی پلز جو 750 مائیکروگرام لیوونورجسٹریل پر مشتمل ہوتی ہیں۔ پوسٹینور-2، پلان بی، پلز 72، پوسٹ ڈے

ایمرجنسی کی صورت میں 28 پلز والے پیک کی پہلی ایکس پلز استعمال کریں۔ اس میں آخری سات پلز استعمال مت کریں کیونکہ ان میں کوئی بھی ہارمون نہیں ہوتا ہے۔

ایمرجنسی خاندانی منصوبہ بندی کی ملے جلے ہارمونز والی پلز کے مقابلے میں صرف پروجیسٹن پر مشتمل پلز اور اسپیشل ایمرجنسی پلز کے کچھ ضمنی اثرات (سردرد، متلی وغیرہ) ہوتے ہیں۔

ایڈز کے لئے ادویات، اینٹی ریٹروواٸرل تھراپی (ART)

ایڈز کے مریضوں کو دی جانے والی ادویات اینٹی ریٹروواٸرل ادویات (ARVs) کہلاتی ہیں۔ یہ دواٸیں ایڈز زدہ مریض کو زیادہ عرصہ اور صحت مند زندگی برقرار رکھنے میں مدد دے سکتی ہیں۔ ان دواؤں میں کم از کم تین دواٸیں ایک ساتھ دی جاتی ہیں اور اسے اینٹی ریٹروواٸرل تھراپی یا اے آر ٹی کہا جاتا ہے۔ اس حصے میں اے آر ٹی کے تحت دی جانے والی چند دواؤں کے گروپوں کے بارے میں معلومات دی گئی ہیں۔ آپ اپنی معذوری کی باقاعدہ دوا اور ایڈز کے لئے دی جانے والی دواؤں کے باہمی اثرار دعمل کے بارے میں جاننے کے لئے صفحہ 361 تا 362 پر دیئے گئے چارٹ دیکھئے۔

ایچ آٸی وی کیئر اور اے آرٹی پروگرام

ایچ آٸی وی/ایڈز ایک پیچیدہ بیماری ہے جو انسانی جسم کے ہر حصے پر اثر انداز ہوتی ہے جیسے ہی آپ کا ایچ آٸی وی ٹیسٹ پازیٹیو آٸے تو فوری طور پر کوٸی ایچ آٸی وی کیئر گروپ تلاش کیجئے جس کے تربیت یافتہ صحت کارکن باقاعدگی سے آپ کا معاٸنہ کریں اور آپ کو صحت مند رکھنے میں ضروری مدد فراہم کریں۔ ایچ آٸی وی کیئر پروگرام بیماریوں سے بچاؤ اور علاج کے لئے دواٸیں مشورے اور دوسری مدد فراہم کر سکتے ہیں۔ یہ اے آر ٹی شروع کرنے اور ضمنی اثرات سے نمٹنے میں آپ کی مدد کر سکتے ہیں۔ اگر دواٸیں مؤثر نہ ہو رہی ہوں تو ان میں تبدیلی کر سکتے ہیں۔ اے آر ٹی پروگرام کے تحت دواٸیں حاصل کرنا کسی اور ذریعے کی بہ نسبت زیادہ معتبر اور کم قیمت ہوتا ہے۔

اے آر ٹی کی ضرورت کب ہوتی ہے؟

ایسے ایچ آٸی وی افراد جن کا مدافعتی نظام صحت مند ہو، انہیں اے آر ٹی کی ضرورت نہیں ہوتی ہے۔ ان ایچ آٸی وی پازیٹیو افراد کو جن میں ایڈز کی علامات سامنے آ رہی ہوں یا جن کا مدافعتی نظام مؤثر نہ رہا ہو، اے آر ٹی کی ضرورت ہوتی ہے۔ بلڈ ٹیسٹ CD-4 (Count) سے مدافعتی نظام کی حالت معلوم ہو سکتی ہے۔ اگر یہ ٹیسٹ ہو سکے اور پتہ چلے کہ آپ کا سی ڈی فور کاؤنٹ 200 سے کم ہے تو آپ اور آپ کا صحت کارکن فیصلہ کر سکتے ہیں کہ اے آر ٹی کب شروع ہونی چاہیئے۔

اے آر ٹی شروع کرنے سے پہلے اپنے صحت کارکن سے تبادلہ خیال ضروری اور اہم ہے۔

- کیا آپ پہلے کبھی اے آر وی لے چکی ہی ہیں؟ یہ بات اب لی جانے والی ادویات پر اثر انداز ہو سکتی ہے۔

- کیا آپ ٹی بی، شدید انفیکشنز یا بخار میں یا کسی اور بیماری میں مبتلا ہیں۔ پہلے اس کا علاج کرنا ہوگا۔

- اے آر ٹی کے فواٸد، خطرے اور ممکنہ ضمنی اثرات کیا ہیں؟ صحت کارکن کے علاوہ کسی ایسے فرد سے بھی معلوم کرنے کی کوشش کریں جو اے آر وی ادویات استعمال کر رہا ہو۔

- کیا آپ ہر روز درست وقت پر دواٸیں کھانے کے لئے تیار ہیں؟ یہ اے آر ٹی کے لئے نہایت اہم ہے۔

- کیا آپ کو کسی قابل بھروسہ فرد یا ایچ آٸی وی/ایڈز سپورٹ گروپ کی مدد حاصل ہے جو آپ کو معلومات اور مدد فراہم کر سکے۔

آپ کو یقینی طور پر معلوم ہونا چاہئے کہ اے آر وی ادویات لینے کے بعد کسی بھی مسئلے کی صورت میں ضمنی اثرات سامنے آنے پر یا دوسرے مسائل صحت کے لئے مدد کہاں سے ملے گی۔

اپنے طور پر خود اے آرٹی شروع مت کریں۔ یہ آپ کے لئے نقصان دہ بھی ہوسکتا ہے اور شدید ضمنی اثرات مرتب کر سکتا ہے۔

اپنی اے آر وی ادویات میں کسی اور کو شریک مت کریں خواہ وہ آپ کا شوہر ہو یا بچہ۔ ڈوز سے کم مقدار میں دوائیں لگانا، دواؤں کو غیر مؤثر بنانے کے علاوہ آپ اور دوا میں شریک فرد کے لئے نقصان دہ ہوسکتا ہے۔

اے آر وی ادویات کسی ایسے فرد یا جگہ سے مت خریدیں جس کا منظور شدہ ایچ آئی وی یا کیئر یا اے آرٹی پروگرام سے تعلق نہ ہو۔

اے آر وی ادویات کا اشتراک (دواؤں کو ملانا)

اے آر وی ادویات اسی صورت میں مؤثر ہوتی ہیں جب انہیں کم از کم تین دواؤں کے مخصوص گروپوں میں دیا جائے۔ ہم نے ایسے چار عام اشتراک صفحہ 360 پر دیئے ہیں اور ہر دوا کے بارے میں مزید معلومات فراہم کی ہیں۔ ایچ آئی وی کے بارے میں جس قدر زیادہ معلوم ہوگا کہ اسے کس طرح کم رفتار کیا یا روکا جا سکتا ہے، دواؤں کی تقسیم بھی تبدیل ہوگی۔ اپنے صحت کارکن سے معلوم کیجئے کہ کون سی ادویات دستیاب ہیں اور آپ کے لئے بہتر طور پر کام کر سکتی ہیں۔ اگلے صفحے (360) پر موجود باکس میں دی گئی تقسیم استعمال میں آسان ترین ہے یہ دوائیں غذا یا غذا کے بغیر کھائی جا سکتی ہیں۔ یہ نسبتاً سستی اور آسانی سے مل جاتی ہیں کچھ دوائیں (تینوں ادویات یا بعض اوقات تین میں سے دو) صرف ایک پلز کی صورت بھی دستیاب ہیں جسے فکسڈ ڈوز کمبی نیشن کہا جاتا ہے۔

اے آرٹی کس طرح لیں

- آپ جو بھی کمبی نیشن لیں، تینوں دوائیں روزانہ ایک طے شدہ وقت پر کھائیں۔
- اگر دوائیں دن میں دو بار کھانی ہوں تو ان کے درمیان 12 گھنٹے کا وقفہ ہونا چاہئے مثلاً آپ پہلی ڈوز صبح چھ بجے کھائیں تو دوسری ڈوز شام چھ بجے ہی کھائی جانی چاہئے۔ اگر ان کے درمیان وقفہ بارہ گھنٹے سے زیادہ ہوتو آپ کے جسم میں دوا کی مقدار اس حد تک کم ہوسکتی ہے جو چند گھنٹوں کے لئے آپ کے اندر دوا کی مزاحمت پیدا کر سکتی ہے۔
- اگر آپ کسی وقت دوا کھانا بھول جائیں تو پانچ گھنٹوں کے اندر دوا کھانے کی کوشش کریں۔ اگر وقفہ پانچ گھنٹوں سے زیادہ ہو جائے تو اگلی ڈوز کے وقت تک انتظار کریں۔
- اے آرٹی کی تقسیم کے مطابق دواؤں میں سے کوئی بھی دوا کھانا بند مت کریں۔ جب تک آپ کا صحت کارکن آپ کا معائنہ کرنے کے بعد یہ نتیجہ اخذ نہ کرے کہ آپ کی دواؤں کو باری باری یا پھر ایک ساتھ روکنا ہے۔

اے آر ٹی کے ضمنی اثرات

اے آر ٹی نے بہت سے لوگوں کو طویل عرصہ اور صحت مند زندگی گزارنے میں مدد دی ہے لیکن متعدد دوسری دواؤں کی طرح اے آر وی ادویات بھی ضمنی اثرات رکھتی ہیں۔ بہت سے لوگوں کے جسم دوا کے عادی ہو جاتے ہیں اور ضمنی اثرات کم تر ہوکر مکمل طور پر ختم بھی ہو سکتے ہیں۔ اے آر ٹی کے چند عام اثرات میں ڈائریا، تھکن، سر درد اور پیٹ کے مسائل مثلاً متلی، قے، پیٹ کا درد یا کھانے میں عدم رغبت وغیرہ شامل ہیں۔ اگر آپ اے آر وی ادویات کھا رہیں ہوں تو ضمنی اثرات سامنے آنے پر بھی اس وقت تک دوائیں کھاتی رہیں جب تک آپ کا صحت کارکن دوا تبدیل کرنے یا بند کرنے کا نہ کہے۔

بعض ضمنی اثرات سامنے آنے کا مطلب یہ ہے کہ دوا تبدیل کرنے کی ضرورت ہے۔ شدید ضمنی اثرات میں ہاتھوں اور پیروں میں جھجھناہٹ یا جلن کا احساس، بخار، دودڑے، زردا نکھیں، سانس میں دشواری کے ساتھ تھکن، خون کی کمی اور خون کے دوسرے مسائل اور جگر کے مسائل شامل ہیں۔ آپ ایسی علامات سامنے آنے پر فوری طور پر صحت کارکن سے رابطہ کریں۔

بالغ اور نو بالغ افراد کے لئے آر وی تقسیم (یہ بچوں کے لئے نہیں ہے)
d4T (stavudine), 30 or 40 mg +3TC (lamivudine), 150 mg + NVP (nevirapine), 200 mg   اور   یہ تینوں ادویات ایک ساتھ ایک گولی کی صورت میں ملتی ہیں جو Triomune کہلاتی ہے۔
d4T (stavudine), 30 or 40 mg +3TC (lamivudine), 150 mg +EVF (efavirenz), 600 mg   اور   دن میں ایک بار۔
AZT (ZDV, zidovudine), 300 mg +3TC (lamivudine), 150 mg +NVP (nevirapine), 200 mg   اور   AZT اور 3TC ایک گولی میں مشترکہ طور پر ملتی ہیں جو Combvir کہلاتی ہے۔
AZT (ZDV, zidovudine), 300 mg +3TC (lamivudine), 150 mg + EVF (efavirenz),600 mg   AZT اور 3TC ایک گولی میں مشترکہ طور پر ملتی ہیں جو Combvir کہلاتی ہے، دن میں ایک بار

اہم بات

کچھ اے آر ٹی ادویات دوسری ادویات کے مقابلے میں زیادہ سخت ضمنی اثرات کا سبب بنتی ہیں، ان میں سے ایک اسٹاویوڈائن (d4T) ہے۔ **عالمی ادارہ صحت کے مطابق d4T پر مبنی علاج رفتہ رفتہ ترک کیا جائے۔** بہت سے ممالک میں ایڈز زدہ لوگ اسٹاروویوڈائن زیادہ عرصہ نہیں کھاتے ہیں۔ دوسرے ممالک میں یہ ممکن نہیں ہوتا ہے کہ وہاں کوئی اور دوا دستیاب نہیں ہوتی ہے یا دوسری دوائیں مقابلتاً زیادہ مہنگی ہوتی ہیں۔ اگر آپ ایڈز زدہ ہیں اور اے آر ٹی ادویات کھانا چاہتی ہیں تو کسی تجربہ کار ایچ آئی وی/ ایڈز صحت کارکن سے ملیں اور معلوم کریں کہ کیا دوسری دوائیں آپ کے ہاں دستیاب ہیں۔

ایچ آئی وی/ایڈز کے لئے ادویات

ایفاوائرنز
(Efavirenz)

ای ایف وی،ای ایف زیڈ،سسٹیوا
(EFV, EFZ, Sustiva)

ایفاوائرنز ایڈز کے علاج میں دوسری اے آر وی ادویات کے ساتھ ملا کردی جانے والی ایک اے آر وی دوا ہے۔

ضمنی اثرات: ای ایف وی چکر، ابہام، موڈ میں تبدیلی اور عجیب وغریب خوابوں کا سبب بن سکتی ہے لیکن یہ علامات عام طور پر چند ہفتے بعد ختم ہوجاتی ہیں۔ اگر ختم نہ ہوں تو اپنی صحت کارکن سے بات کریں۔

عام طور پر ملتے ہیں: 200، 100، 50 ملی گرام کے کپسول، 600 ملی گرام کی گولیاں، 150mg/5ml کا پینے کا محلول۔

کس طرح استعمال کریں:

600 ملی گرام، منہ کے ذریعے، دن میں ایک بار

دوسری ادویات کے ساتھ ردِعمل:

ریفامپسن کے ساتھ، ایفاوائرنز کی اثر انگیزی کم ہوجاتی ہے۔ آپ کو ای ایف یو کی زیادہ مقدار لینے کی ضرورت ہوسکتی ہے (مثلاً 600 ملی گرام کے بجائے 800 ملی گرام)

تنبیہہ: حمل کے ابتدائی تین ماہ میں ایف یو وی مت لیں یہ پیدائشی نقائص کا سبب بن سکتی ہے۔ ای ایف یو لینے والی عورتوں کو خاندانی منصوبہ بندی کا کوئی طریقہ اپنانا چاہیے۔ اس دوران حاملہ ہونے والی عورتیں کوئی حمل روک طریقہ استعمال کریں۔

لیمی ویوڈائن
(Lamivudine)

3TC، ایپی وائر......(3TC, Epivir)

لیمی ویوڈائن، ایڈز کی دوسری اے آر وی ادویات کے ساتھ ملا کردی جانے والی ایک اے آر وی ہے۔ اس کے ضمنی اثرات بہت ہی کم ہیں۔

عام طور پر ملتی ہیں: 150 ملی گرام کی گولیاں، 50mg/5ml کا پینے کا محلول۔

کس طرح استعمال کریں:

150 ملی گرام، منہ کے ذریعے، دن میں دو مرتبہ۔

نی وائراپائن
(Nevirapine)

این وی پی، وائرامیون......(NVP, Viramune)

نی وائراپائن، ایڈز کی دوسری اے آر وی ادویات کے ساتھ ملا کر دی جانے والی ایک اے آر وی ہے۔ این وی پی حمل اور ولادت کے دوران ماں سے بچے کو ایچ آئی وی منتقلی سے تحفظ کے لئے بھی استعمال ہوتی ہے۔

ضمنی اثرات: زرد آنکھوں، جلد کی سرخی، بخار، سانس میں دشواری کے ساتھ تھکن اور بھوک کے احساس میں کمی کی صورت میں فوری طور پر طبی مدد حاصل کریں۔

عام طور پر ملتی ہیں: 200 ملی گرام کی گولیاں، 50mg / 5ml کا اورل سسپنشن۔

کس طرح استعمال کریں:

ضمنی اثرات کے امکانات کو کم ترکرنے کے لئے این وی پی کی نصف ڈوز ابتدائی دو ہفتے (دن میں ایک بار 200 ملی گرام) کھائیں۔ دو ہفتے بعد این وی پی دن میں دو بار ہر بار 200 ملی گرام لیں۔

دوسری دواؤں کے ساتھ ردِعمل:

ریفامپسین، این وی پی کی اثر انگیزی کم کرسکتی ہے۔

اسٹاویوڈائن
(Stavudine)

d4T، زیریٹ
(d4T, Zerit)

اسٹاویوڈائن ایڈز کے لئے دی جانے والی دوسری اے آر وی کے ساتھ ملا کردی جانے والی اے آر وی ہے۔

ضمنی اثرات : بازوؤں یا پیروں میں جھجھناہٹ، بے حسی/سُن پن یا جلن کا احساس، متلی، قے یا پیٹ کے شدید درد میں، سانس میں دشواری کے ساتھ تھکن اور جسمانی چربی میں تبدیلیوں کی صورت میں فوری طبی مدد حاصل کریں۔

عام طور پر ملتے ہیں: 40, 30, 20, 15 ملی گرام کے کیپسول، 5mg / 5ml کا پینے کا محلول کے لئے پاؤڈر

کس طرح استعمال کریں:

اگر آپ کا وزن 60 کلوگرام سے زیادہ ہے تو دن میں دوبار 40 ملی گرام دوا کھائیں۔

اگر آپ کا وزن 60 کلوگرام سے کم ہے تو 30 ملی گرام دوا دن میں دوبار کھائیں

تنبیہہ: اگر وزن 60 کلوگرام سے کم ہے تو دوا کی کم مقدار استعمال کی جائے۔ اگر دوسری دوائیں دستیاب ہوں تو حاملہ عورتیں اور بھاری بھرکم عورتیں اسٹاویوڈائن مت لیں کیونکہ یہ سخت ضمنی اثرات مرتب کرسکتی ہے۔ عالمی ادارہ صحت نے یہ دوا بتدریج ترک کرنے کی سفارش کی ہے۔

زیڈوویوڈائن
(Zidovudine)

(AZT، ZDV، ازیڈوتھائمیڈائن، ریٹروروائر)
(AZT, ZDV, Azidothymidine Retrovir)

زیڈوویوڈائن ایڈز کے علاج کے لئے دی جانے والی اے آر وی ادویات کے ساتھ دی جانے والی ایک اے آر وی دوا ہے۔ یہ حمل اور ولادت کے دوران ماں سے بچے کو ایچ آئی وی کی منتقلی سے تحفظ کے لئے بھی استعمال کی جاتی ہے۔

ضمنی اثرات : تھکن اور سانس کی تنگی، زرد جلد یا خون کی کمی کی دوسری علامات پر فوری طبی مدد حاصل کریں۔

عام طور پر ملتے ہیں: 100 یا 200 ملی گرام کے کیپسول، 300 ملی گرام کی گولیاں۔ 50mg / 5 ml کا پینے کا محلول یا سیرپ، 20 ملی لٹر کے وائل میں 10mg / ml کا آئی وی انفیوژن انجکشن محلول۔

کس طرح استعمال کریں:

300 ملی گرام ،دن میں دوبار

دوسری ادویات کے ساتھ رد عمل:

ڈاپوزون کے ساتھ خون کی کمی کا سبب بن سکتی ہے۔

والپروئیک ایسڈ کے ساتھ زیڈوویوڈائن کی سطح بڑھ سکتی ہے اور متلی، قے، تھکن/کمزوری کا سبب بن سکتی ہے۔

ریفامپسین کے ساتھ، زیڈوویوڈائن کا اثر کم ہوسکتا ہے۔

اپنے آلات کی دیکھ بھال

اس باب میں بتایا گیا ہے کہ مختلف معذوریاں رکھنے والی عورتیں کس طرح اپنے آلات کی دیکھ بھال کرسکتی ہیں تا کہ یہ بخوبی کام کریں اور ممکنہ حد تک طویل مدت کارآمد رہیں۔

سماعتی آلات

سماعتی آلات مہنگے ہوتے ہیں لیکن صرف آلہ خرید لینا ہی کافی نہیں۔ اگر آلے کی ساخت کان کی بناوٹ کے عین مطابق نہ ہوتو یہ درست طور پر کام نہیں کرے گا۔ اگر آلے کا کان سے جڑنے والا حصہ (Ear mold) چبھنے لگے یا سکڑ جائے تو اسے تبدیل کرنے کی ضرورت ہوگی۔ یہ حصہ عام طور پر دو سال سے زیادہ کارآمد نہیں رہتا ہے۔ اس کے ساتھ سماعتی آلات میں بیٹری کی ضرورت ہوتی ہے۔ دو انتہائی عام سماعتی آلات درج ذیل ہیں:

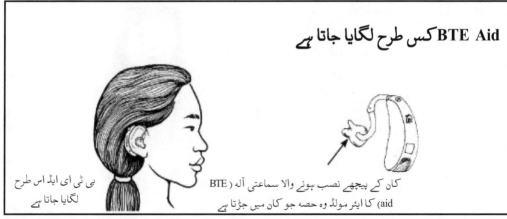

BTE Aid کس طرح لگایا جاتا ہے

بی ٹی ای ایڈ اس طرح لگایا جاتا ہے

کان کے پیچھے نصب ہونے والا سماعتی آلہ (BTE aid) کا ایئر مولڈ وہ حصہ جو کان میں جڑتا ہے

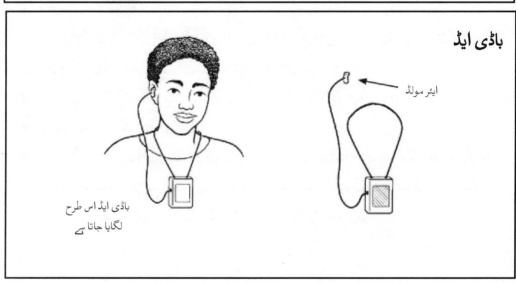

باڈی ایڈ

ایئر مولڈ

باڈی ایڈ اس طرح لگایا جاتا ہے

ان دونوں سماعتی آلات کے لئے ہر ہفتے یا دو سے تین ماہ میں نئی بیٹریوں کی ضرورت ہوتی ہے۔ آپ کو کتنے عرصے میں یا کتنی بار بیٹریوں کی ضرورت ہوسکتی ہے اس کا انحصار آلے کے روزانہ استعمال کی مدت، آلے کی قسم اور استعمال ہونے والی بیٹریوں کی نوعیت پر ہے۔

بوٹسوانہ (جنوبی افریقہ) کی ایک آرگنائزیشن نے ایسا سماعتی آلہ بنایا ہے جس کی بیٹریاں ری چارج کر کے بار بار استعمال کی جاسکتی ہیں۔ (مزید معلومات کے لئے دیکھئے صفحہ 377)۔

نیا ایئر مولڈ اور بیٹریاں حاصل کرنا انتہائی مہنگا ہوسکتا ہے

بعض ممالک سماعتی آلہ اور بیٹریاں مفت فراہم کرتے ہیں اگر آپ کے ملک میں یہ سہولت ہو تو اپنے ملک کی وزارت صحت سے رابطہ کیجئے۔

> سماعتی آلے اور بیٹریاں ہر اس شخص کو ملنی چاہئیں جسے ان کی ضرورت ہو، صرف ان لوگوں کو نہیں جو ان کے اخراجات برداشت کرسکتے ہیں۔

سماعتی آلے کی دیکھ بھال کرنا

خواہ آپ کسی بھی قسم کا سماعتی آلہ استعمال کرتی ہوں۔ اگر اس کی با قاعدہ دیکھ بھال کی جائے تو یہ طویل مدت چلے گا اور بہتر آواز فراہم کرے گا۔ سماعتی آلے کی دیکھ بھال کے لئے چند گر کی باتیں درج ذیل ہیں۔

- جب آپ اپنا سماعتی آلہ استعمال نہ کر رہی ہوں اسے برقی آلات مثلاً ریفریجریٹر یا ٹی وی سے دور رکھیں۔
- اسے بہت گرم یا سرد جگہ پر نہ رکھیں۔
- اسے خشک رکھیں۔ پسینہ یا نمی اسے نقصان پہنچائے گی۔ دن کے وقت اسے وقفہ وار کان سے الگ کر کے پسینہ یا نمی صاف کریں۔ نہانے یا پیراکی کرنے سے پہلے یا برسات میں اسے اتار دیں رات کے وقت اسے نمی جذب کرنے والے مادے (Silica gel) میں رکھیں سماعتی آلے پر پرفیوم یا کسی اور قسم کا اسپرے نہ کریں۔

صفائی

- سماعتی آلے کے ایئرمولڈ کوصاف کرنے کے لئے نرم اور خشک کپڑا استعمال کریں۔کلینگ لوشن استعمال مت کریں۔

- ویکس کے لئے ایئرمولڈ کا باقاعدگی سے جائزہ لیں۔

- ایئرمولڈ کو گرم پانی سے صاف کریں اور استعمال کرنے سے پہلے اچھی طرح خشک کر لیں۔

بیٹری

- بیٹری دیر تک کارآمد رکھنے کے لئے سماعتی آلہ استعمال نہ ہونے کے دوران بند رکھیں۔

- بیٹری صاف رکھیں اور جب آلہ زیادہ عرصہ استعمال میں نہ رہے تو اس سے بیٹری نکال دیں مثلاً رات کے وقت سوتے ہوئے۔

- بیٹری باقاعدگی سے تبدیل کریں۔ بیٹری تبدیل کرنے کے وقت جاننے کے لئے آلے کی آواز حد تک پوری کھول دیں اگر آلے سے سیٹی جیسی آواز آئیں تو بیٹری ٹھیک ہے اگر نہیں تو نئی بیٹری کی ضرورت ہے۔ اگر ضروری ہو تو کسی سے مدد کی درخواست کریں اگر بیٹری معمول سے ہٹ کر اپنی قوت کھو رہی ہو تو یہ سماعتی آلے کی کسی خرابی کی علامت ہو سکتی ہے۔

- بیٹریاں ٹھنڈی اور خشک جگہ رکھیں۔ ریفریجریٹر میں رکھی بیٹری کو استعمال کرنے سے پہلے کچھ دیر کمرے کے درجہ حرارت پر رکھیں۔

- سماعتی آلے کو مناسب وقفے سے کسی ماہر سے (کلینک یا سماعتی آلات کے ادارے سے) چیک کرانے کی کوشش کریں۔

- سماعتی آلات کو مرمت کی ضرورت ہو سکتی ہے عام طور پر یہ کام بڑے شہروں میں ممکن ہو سکتا ہے لیکن اکثر ملک میں ڈیف آرگنائزیشنوں نے بہرے افراد کو اس سلسلے میں تربیت دینی شروع کر دی ہے تا کہ وہ کانوں کی ساخت کے مطابق ایئر مولڈ بنا سکیں اور سماعتی آلات کی مرمت کر سکیں۔

چھڑی کی مناسب ترین لمبائی یہ ہے کہ چھڑی زمین سے آپ کی کمر اور کندھوں کے درمیان نصف حصے تک پہنچے

چلنے پھرنے کے لئے چھڑی استعمال کریں

اگر آپ نابینا ہیں یا دیکھنے میں یاد کھنے میں دشواری محسوس کرتی ہیں تو آنے جانے کے لئے چھڑی کا استعمال آپ کو اعتماد دیتا ہے۔ خاص طور پر ان مقامات پر جو آپ کے لئے مانوس نہ ہوں چھڑی جتنی لمبی ہوگی آپ اتنا ہی تیز چل سکیں گی کیونکہ چھڑی راستے کے بڑے حصے کو محسوس کرنے میں آپ کی مدد کرے گی۔

چھڑی مضبوط لکڑی کی ہو لیکن یہ بھاری یا موٹی نہ ہوتا کہ اسے دن بھر ہاتھ میں رکھنا آسان رہے۔ چھڑی کا اوپری حصہ جسے آپ تھامیں، نسبتاً موٹا ہو یہ کھایا ہوا یا سیدھا ہو سکتا ہے آپ اس میں فیتہ بھی لگا سکتی ہیں جو آپ کی کلائی کے گرد مگر ڈھیلا رہے اس سے چھڑی گرنے یا کھونے سے بچاؤ ہوگا۔

ہر روز گھر سے روانہ ہونے سے پہلے چھڑی کا جائزہ لیں کہ وہ کہیں سے ٹوٹی یا چٹخی ہوئی تو نہیں ہے اگر ضروری ہو تو کسی سے مدد کے لئے کہیں۔

اپنی وہیل چیئر کی دیکھ بھال

اگر آپ اپنی وہیل چیئر کی دیکھ بھال کریں تو یہ زیادہ مدت کار آمد رہے گی اور آپ کے لئے اطمینان بخش سواری ثابت ہوگی۔ وہیل چیئر کی دیکھ بھال کے لئے آپ کو چند بنیادی اوزاروں کی ضرورت ہوگی (دیکھئے نیچے دیا گیا باکس) یہ صفحات (366 تا 368) وہیل چیئر کی دیکھ بھال میں آپ کی عمومی مدد اور رہنمائی کریں گے۔ آپ کی وہیل چیئر کو دیکھ بھال اور مرمت کی کتنی ضرورت ہے اس کا انحصار ان سڑکوں اور راستوں کی حالت یا ساخت پر ہے جو آپ استعمال کرتی ہیں۔

اگر آپ کے پاس مندرجہ ذیل چیزیں ہیں تو ان سے بھی مدد ملے گی۔

وہیل چیئر کی دیکھ بھال کے لئے ضروری اوزار

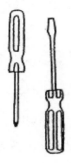

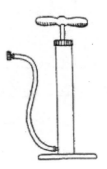

اسکرو ڈرائیور   ایڈجسٹ ایبل رینج   ہوا بھرنے کا پمپ

دھاتی فائل

اسپوک رینج

- میڈیم وہیل مشین آئل

- نرم موم۔ جوکاروں کو چمکانے کے لئے استعمال ہوتی ہے۔ یا Jojoba یا Shea tree سے بنائی گئی موم یا بھیڑ کی اون کا بنا lanolin یا کوکا بٹر۔ موم بتی کا موم کام نہیں کرے گا کیونکہ یہ حد سے زیادہ سوکھا ہوا ہوتا ہے۔

روزانہ کامعمول

- کپڑے سے وہیل چیئر کے فریم کو روزانہ صاف کریں۔
- وہیل چیئر کے اگلے اور پچھلے پہیوں کے بیرنگز میں پانی اور گرد داخل نہ ہونے دیں (وہیل چیئر کو ایسی جگہ نہ رکھیں جہاں پر بھیگنے کا امکان رکھتی ہو)۔
- وہیل چیئر کے ٹائر چیک کریں۔اگر ضروری ہو تو ان میں پمپ سے ہوا بھریں دبانے پر یہ ٹائر سخت ہونے چاہئیں۔

اپنی وہیل چیئر کا خیال رکھیں
یہ بھی آپ کا خیال رکھے گی

هر هفتے

- ٹائروں کی ممکنہ خرابی (ادھڑنے یا پنکچر) کا اندازہ لگانے کے لئے انہیں چیک کریں اگر آپ کے ٹائروں میں کوئی کیل یا سخت چیز پیوست ہو تو اسے اس وقت تک مت نکالیں جب تک ٹیوب کی درستگی کا بندوبست نہ ہو۔اگر ضروری ہو تو ٹائر تبدیل کریں اگر آپ کی وہیل چیئر میں سائیکل کے ٹائر استعمال کئے گئے ہوں تو ان کی مرمت یا تبدیلی بائیسکل شاپ پر ہو سکے گی۔ٹائر میں موجود ٹیوب میں پنکچر لگانا بہت مشکل نہیں ہے کسی جاننے والے سے یہ ہوتا دیکھنے کی درخواست کیجیے۔
- پچھلے پہیوں کا جائزہ لیں کہ یہ سہولت سے گھومتے ہوں۔اگر یہ ڈگمگاتے ہوئے یا پُرشور ہوں تو ممکن ہے کہ ان کے بال بیرنگ خراب ہو چکے ہوں اور انہیں تبدیل کرنے کی ضرورت ہو وہیل چیئر کے کسی بھی حصے کو جو ڈھیلا ہو کسنے کی ضرورت ہے (ممکن ہے بڑے اوزاروں کے لئے آپ کو کسی مکینک سے رابطے کی ضرورت پڑے) نٹ کسنے کے بعد انہیں معمولی سا ڈھیلا کرنے کی ضرورت ہوگی تا کہ وہیل چیئر آزادی سے گھوم سکے۔
- اگر سامنے کا حصہ (Front forks) آسانی سے اطراف میں نہیں گھومتا ہے تو یقینی بنائیے کہ وہ پائیدان سے نہ ٹکرائے۔ بڑے پچھلے پہیوں کے اسپوک لہراتے ہوں تو ڈھیلے اسپوک کس لیں، ٹوٹے اسپوک تبدیل کر لیں۔
- اگر پورا عقبی حصہ (Fork) ڈگمگاتا ہو تو اوپر کے نٹ کو کس لیں اور پھر اسے معمولی سا ڈھیلا کریں تا کہ یہ آسانی سے مڑ سکے۔
- اگر سامنے کا پہیہ آسانی سے نہیں گھومتا ہے تو اسے نئے بیرنگ کی ضرورت ہو سکتی ہے۔ بہتر ہے کہ اسے ورکشاپ میں چیک کرا لیں۔
- استعمال شدہ کپڑے سے تمام وہیل ایکسل ہاؤسنگز کو اچھی طرح صاف کریں اور ان پر تیل کے چند قطرے ٹپکائیں۔

ماہانہ دیکھ بھال

وہیل چیئر کے فریم کو صاف کرنے، اسے کھولنے اور بند کرنے میں سہولت کے لئے اس پر موم ملیں۔

- پہیوں کے وسط میں نصب بال بیرنگ فیکٹری سے ڈھکی حالت میں ہو سکتے ہیں۔ اگر نہیں ہیں تو انہیں باہر نکال کر کسی محلول سے صاف کریں، پھر انہیں خشک کرکے تازہ گریس لگائیں اور دوبارہ لگا دیں اور اگر بیرنگ میں پانی چلا جائے تو یہ زنگ کا سبب بن سکتا ہے اور پھر وہیل چیئر ہموار انداز نہیں چلے گی۔

- بازو، پیر اور ہاتھ رکھنے کی جگہوں پر کھردرے دھبوں یا دھار والی سطحوں کا جائزہ لیں اور انہیں ہموار کر لیں۔

- اپنی وہیل چیئر کے اسکرو اور بولٹ چیک کریں اور ضروری ہو تو کس لیں۔ سیٹ اور پشت کے، ہینڈ رِم، ایکس براس، پائیدانوں، فرنٹ ایکسلز، ریئر ایکسلز بریک، سامنے کے چھوٹے پہیے اور گھومنے والے حصے (Pivot) کے اسکرو چیک کیجئے۔

- وہیل چیئر کے فریم پر ٹوٹ پھوٹ اور ڈینٹ چیک کیجئے کچھ اقسام کی ٹوٹ پھوٹ ویلڈنگ سے درست ہو سکتی ہیں۔

ہر چار سے چھ ماہ میں

- ایکس براس کے وسطی اور نچلے حصے میں میڈیم ویہیٹ مشین آئل ڈالئے۔ چیئر کے گھومنے والے حصوں میں آئل ڈالئے۔

- سیٹ کا کپڑا (یا کور) چیک کیجئے اگر یہ پھٹ گیا یا خراب ہو گیا ہو تو اسے تبدیل کیجئے۔ بہت ضروری ہے کہ پھٹی ہوئی یا بوسیدہ سیٹ تبدیل کی جائے کیونکہ یہ دباؤ کے زخموں کا ذریعہ بن سکتی ہے۔

اہم بات سامنے کے ٹائر خواہ ٹھوس ربر کے ہوں یا ہوا والے، انہیں وہیل چیئر ورکشاپ میں تبدیل کرانے کی ضرورت ہو سکتی ہے۔ اگر وہیل چیئر ٹوٹ پھوٹ جاتی ہے تو اس کی ورکشاپ میں مرمت کرانے کی ضرورت ہوگی اگر آپ کے علاقے میں وہیل چیئر ورکشاپ نہیں ہے تو اس کی مرمت سائیکل مرمت کرنے والی دکان یا میٹل ورک ورکشاپ میں ہو سکتی ہے۔

صحت کے لئے علامتی زبان

دنیا بھر میں بہرے افراد نے خود اپنی علامتی زبانیں ایجاد کرلی ہیں۔ جسے یہ لوگ تقریباً عام لوگوں کی طرح تیزی سے اور بخوبی ایک دوسرے سے ابلاغ کے لئے استعمال کرتے ہیں۔ ذیل میں مختلف ممالک میں استعمال ہونے والی علامتیں یا اشارے دکھائے گئے ہیں۔ مختلف مقامات پر ایک جیسے حروف کے لئے علامات مختلف ہوسکتی ہیں ان علامات کو بطور گائیڈ استعمال کیجے۔ اگر آپ صحت کا رکن، گھر کا فرد یا دیکھ بھال کرنے والے ہیں تو اپنی کمیونٹی کی بہری عورتوں سے دوسری علامات سکھانے کے لئے کہیں جو وہ استعمال کرتی ہوں۔ علامتی زبان آپ کے کمیونٹی میں موجود بہری خواتین کے لئے صحت کی اچھی دیکھ بھال کو یقینی بنانے میں معاون ہوسکتی ہے۔

وہ علامات جو صحت کے کارکنوں کے لئے جاننا ضروری ہیں

ذیل میں وہ چند علامات دی گئی ہیں جو بہری خواتین کی بہتر دیکھ بھال کے لئے جاننا ضروری ہیں یہ علامات مختلف ممالک کی ہیں۔ ہم نے انہیں آپ کو اپنے آئیڈئے سوچنے میں مدد دینے کے لئے شامل کیا ہے۔ بعض علامات اشاروں پر مبنی ہیں اور بعض کے لئے انگلیوں کی مدد سے حروف تہجی بنائے جاتے ہیں۔

وہ علامات جو جسم کے مختلف حصوں کی نشاندہی کرتی ہیں

درد

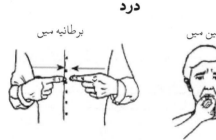

برطانیہ میں   چین میں

درد کی نشاندہی کے لئے جسم کے اس حصے کی نشاندہی کی جاسکتی ہے جہاں درد ہو۔ مثلاً پیٹ میں درد کی نشاندہی کے لئے درد کی علامت بنا کر پیٹ کی طرف اشارہ کیا جائے۔

علامات جن میں اشارے استعمال ہوتے ہیں

ذیل میں مراکز صحت میں استعمال ہونے والی دوسری علامات کی مثالیں ہیں ان علامات میں زیادہ تر اشارے استعمال ہوتے ہیں۔

خاندانی منصوبہ بندی		ایڈز	انفیکشن		ادویات	
ویتنام میں		کینیا میں	چین میں	کینیا میں	برطانیہ میں	ویتنام میں	امریکہ میں	کینیا میں

سیکس

کینیا میں   ویتنام میں

کینیا میں

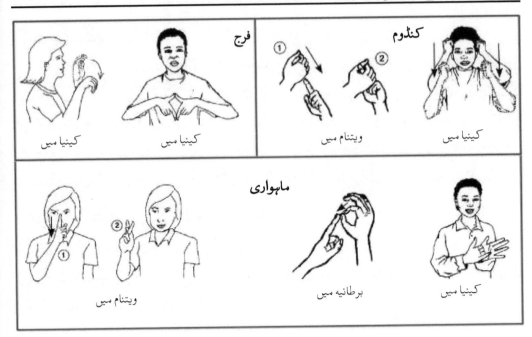

فرج

کنڈوم

کینیا میں   کینیا میں

ویتنام میں   کینیا میں

ماہواری

ویتنام میں

برطانیہ میں   کینیا میں

وہ علامات جن میں انگلیاں استعمال ہوتی ہیں

بہت سے الفاظ انگلیوں سے بنے حروف تہجی سے سمجھے جا سکتے ہیں مثال کے طور پر

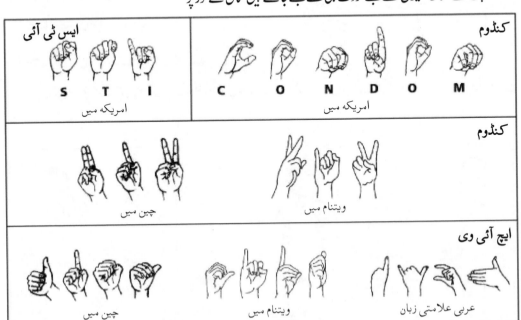

ایس ٹی آئی   کنڈوم

S T I

امریکہ میں   C O N D O M

امریکہ میں

کنڈوم

چین میں   ویتنام میں

ایچ آئی وی

چین میں   ویتنام میں   عربی علامتی زبان

دیگر مفید الفاظ جن کی علامات سیکھنا چاہئیں

ذیل میں دوسرے ایسے الفاظ دیئے گئے ہیں جن کی مقامی علامتی زبان سیکھنا ایک صحت کارکن کے لئے کار آمد ہے گا۔

پیشاب کرنا	بے حسی (محسوس نہ ہونا)	کپکپاہٹ/سردی لگنا	اسقاطِ حمل
پیڑو کا	بخار	صفائی/دھونا	زیادتی
مردانہ عضو تناسل	بڑھوتری (جلد پر یا جسم کے اندر)	اینٹھن/عضلات کا سکڑنا	مقعد
حاملہ	کھجلی/خارش	کٹاؤ/زخم/خراشیں	ولادت
زنا/عصمت دری/آبروریزی	زچگی	ڈپریشن	اخراج خون
نیند کے مسائل	رسولی	ڈائریا	دھندلی بصارت
پسینہ آنا	ادویات	ڈسچارج	چھاتی
سوجن	اسقاط	چکرانا	چھاتی سے دودھ پلانا
نرمی/نرماہٹ	پیپ/مواد	منشیات	تنفسی مسائل
قے/الٹی	متلاہٹ	خشک	سوزش (جلنا)
کمزوری	حواس باختگی	معائنہ	رنگ میں تبدیلیاں
رحم	فضلہ خارج کرنا	ورزش	سکڑاؤ خصوصاً رحم کا سکڑنا
		بے ہوش ہونا	کھانسی

نمبروں کی علامات

نمبروں کو یہ بتانے کے لئے استعمال کیا جا سکتا ہے کہ ایک فرد کب سے مسئلے میں مبتلا ہے (کتنی بار، کتنے دنوں، ہفتوں یا مہینوں سے) ایک سے پانچ تک کے نمبر درد کی مقدار یا شدت کے لئے استعمال کئے جا سکتے ہیں ایک کا مطلب کم درد اور پانچ کا مطلب انتہائی شدت کا درد ہے۔

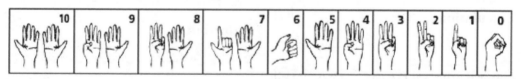

برطانوی علامتی زبان سے مثالیں

علامتی حروف تہجی

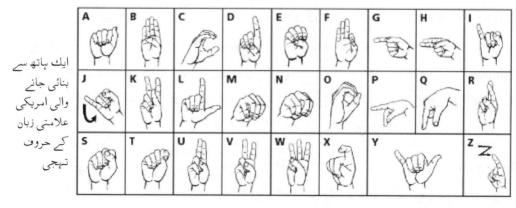

ایک ہاتھ سے بنائی جانے والی امریکی علامتی زبان کے حروف تہجی

فہرست مشکل الفاظ

ذیل میں ایسے الفاظ دیئے گئے ہیں جن کا مفہوم سمجھنا مشکل ہوسکتا ہے۔ یہ مفہوم سمجھ کر ہی اس کتاب کو بہتر طور پر سمجھا جاسکتا ہے۔ کچھ الفاظ یا اصطلاحات کی تشریح بھی کی گئی ہے۔ یہ سب اُردو حروف تہجی کے لحاظ سے ہے۔

آ	ا	ب	پ	ت	ٹ	ث	ج	چ	ح	خ	د	ڈ
ذ	ر	ڑ	ز	ژ	س	ش	ص	ض	ط	ظ	ع	غ
ف	ق	ک	گ	ل	م	ن	و	ہ	ے			

❖ ا ۔ آ ❖

Nerves — اعصاب

وہ مہین ریشے جن کے ذریعے پیغامات جسم کے ایک حصے سے دوسرے حصے تک پہنچتے ہیں۔ کچھ اعصاب کا تعلق محسوس کرنے سے اور کچھ کا تعلق ہماری جسمانی نقل و حرکت سے ہے۔

Infection — انفیکشن

بیکٹریا، وائرس اور دیگر جرثوموں کے باعث پیدا ہونے والی خرابی یا بیماری۔ انفیکشن جسم کے کسی حصے یا پورے جسم پر اثر انداز ہوسکتا ہے۔

Injection — انجکشن

دوا یا دیگر مائعات کا سرنج اور سوئی استعمال کرکے جسم کے اندر داخل کیا جانا۔

Immunization — امیونائزیشن

مخصوص بیماریوں سے تحفظ کے لئے ایک خاص شیڈول کے تحت دوا یا ویکسین بذریعہ انجکشن یا منہ سے دینا۔

Spasticity — اینٹھن/اکڑاؤ

عضلات کا بے قابو انداز سے سخت ہونا یا کھچ جانا۔ سیریبرل پالسی، ریڑھ کی ہڈی کے زخم یا دماغی ضرر کی صورت میں ایسا عموماً ہوتا ہے۔

X-ray — ایکس رے

جسم کے اندرونی حصوں مثلاً ہڈیوں، پھیپھڑوں وغیرہ کے عکس حاصل کرنے کے لئے استعمال ہونے والی شعاعیں۔ عام طور پر ایسے عکس بھی ایکس رے کہلاتے ہیں۔

Antiretrovirals — اینٹی ریٹروو وائرلز

وہ ادویات جو ایڈز کے مریض کو صحت مند رکھنے کے لئے دی جاتی ہیں لیکن یہ ایڈز کا علاج نہیں ہیں۔

Alergy — الرجی

کچھ چیزیں سانس کے ذریعے، غذا میں مل کر انجکشن یا دوا کی صورت میں جسم میں جانے یا جسم پر لگنے کے باعث جسمانی ردِعمل مثلاً سوزش، جلن، دھبے یا چھوٹے چھوٹے اُبھار یا سانس لینے میں دشواری کا عارضہ۔ یہ کیفیات الرجک ردِعمل کہلاتا ہے اگر ردِعمل انتہائی شدید ہو تو الرجک شاک کہلائے گا۔

Ejaculation — انزال

مرد کی منی خارج ہونا۔

Gang Rape — اجتماعی آبرو ریزی/عصمت دری

کسی عورت کے ساتھ ایک سے زائد افراد کا زنا کرنا۔

Infant Formula — انفنٹ فارمولا

ماں کے دودھ کی جگہ دیا جانے والا مصنوعی دودھ۔

Enema — اینیما

قبض دور کرنے یا فضلہ خارج کرنے کے لئے مقعد میں آبی محلول چڑھانا یا جسم میں سیال کی مقدار میں اضافہ کرنا۔

Estrogen — ایسٹروجن

عورتوں کے جسم میں پیدا ہونے والا ایک ہارمون۔

لئے بطور دوا استعمال کئے جاسکتے ہیں۔

Progestrone — پروجسٹرون

ایک نسوانی ہارمون۔

Progesten — پروجسٹن

قدرتی طور پر عورت کے جسم میں بننے والے پروجسٹرون سے مشابہ مصنوعی طور پر بنایا جانے والا ہارمون، پروجسٹن خاندانی منصوبہ بندی کے کچھ ہارمونی طریقوں میں استعمال کیا جاتا ہے۔

Pus — پس۔ پیپ

جراثیم سے بھری سفید یا زرد رطوبت جو انفیکشن زدہ کھلے زخم میں پائی جاتی ہے۔

Douche — پچکاری مارنا

فرج کو دھونے کے لئے پچکاری استعمال کرنا۔

Paralegia — پیرالیجیا

ریڑھ کی ہڈی کے مجروح ہونے یا بیماری کے باعث دونوں پیروں کے عضلات میں حرکت ختم ہونا یا فالج۔

───❀───ت───❀───

Convulsion — تشنج

بے اختیار دورہ، جسم یا اس کے کسی حصے میں جھٹکے

Genitals — تناسلی اعضاء

مرد یا عورت کے اندرونی و بیرونی جنسی اعضاء۔

Reproductive Health — تولیدی صحت

افزائش نسل کے لئے مطلوب عورت کی صحت۔

Thermometer — تھرمامیٹر

جسم کا درجہ حرارت ناپنے کا آلہ۔

───❀───ٹ───❀───

Trichomonas — ٹرائی کوموناس

جنسی ملاپ کے ذریعے منتقل ہونے والا جنسی انفیکشن۔

Operation — آپریشن

علاج یا کسی خرابی کو دور کرنے کے لئے جسم کے کسی حصے کو کاٹا جانا۔ آپریشن کے ذریعے بچے کی ولادت بھی ہوتی ہے جس میں سرجن عورت کے پیٹ اور رحم کو کاٹتا ہے اور ولادت ہونے کے بعد کٹے ہوئے حصوں کو سی دیا جاتا ہے۔

───❀───ب───❀───

Hemorroids — بواسیر

مقعد کے اندر یا ارد گرد تکلیف دہ اُبھار جو ایک قسم کی سوجی ہوئی وریدیں ہوتی ہیں جو جلن، خارش یا زخم کا سبب بن جاتی ہیں۔

Biopsy — بائیوپسی

بیماری کی تصدیق یا تردید کے لئے جسم کے کسی حصے کا ٹشو یا رطوبت نکالنا۔

Brand Name — برانڈ نیم

دوا کو دوا یا دوا ساز کمپنی کا نام۔

Inserts — بتیاں

دوا کی وہ صورت جو زخم میں یا شگاف میں اندر داخل کی جا سکے۔

───❀───پ───❀───

Abdomen — پیٹ

جسم کا وہ حصہ جس میں معدہ، آنتیں، جگر اور تولیدی اعضاء ہوتے ہیں۔

Birth Defects — پیدائشی نقائص

وہ جسمانی یا ذہنی خرابیاں جو بچے میں پیدائشی طور پر ہوتی ہیں۔

Birth Spacing — پیدائش میں وقفہ کرنا

بچوں کی پیدائش کے درمیان وقفے کے لئے مانع حمل طریقے استعمال کرنا۔

Placenta — پلیسینٹا

حمل کے دوران عورت کے رحم میں بچے کی نشوونما کے لئے ضروری اشیاء فراہم کرنے والا اسفنجی عضو جس سے بچہ اپنی نال کے ذریعے جُڑا ہوتا ہے۔ بچے کی پیدائش کے بعد پلیسینٹا بھی رحم سے باہر نکل جاتا ہے۔

Plant Medicines — پلانٹ میڈیسین

کسی پودے کے پھول، پتّے، جڑ اور دیگر حصے جو کسی بیماری کے علاج کے

ٹ

Tissue ٹشو

جسم کے عضلات چکنے حصے اور اعضاء بنانے والا میٹریل

Toxemia ٹوکسیمیا

حمل کے دوران ایک خطرناک صورت جو دوروں کا سبب بن سکتی ہے (اس میں خون زہریلا ہوجاتا ہے)۔

ڈ

Diaphragm ڈایافرام

خاندانی منصوبہ بندی کے لئے استعمال ہونے والی نرم اور پیالہ نما شے جو جنسی ملاپ سے پہلے رحم کے منہ پر چڑھالی جاتی ہے۔

Side Effect ذیلی/ضمنی اثرات

دوا کے استعمال کے بعد ایسے اثرات جن کا تعلق بیماری کے علاج سے نہ ہو۔

ج

Generic Name جزک نام

کسی دوا میں موجود بنیادی جزو کا نام۔ دوا کا اصل نام

Lever جگر

سیدھے حصے کی نچلی پسلیوں کے نیچے موجود ایک اہم عضو جو خون صاف کرنے اور زہریلے اجزاء سے نجات میں مدد دینے کے علاوہ اور کئی اہم کام انجام دیتا ہے۔

ر

Uterus رحم

بچہ دانی، عورت کے جسم کا وہ حصہ جہاں بچہ نشوونما پاتا ہے۔

Access رسائی

سہولتوں کا حصول ممکن ہونا۔

Tumor رسولی

جسم کے اندر ٹشوز کی غیر معمولی نشوونما۔

Intravenous (IV) رگ کے ذریعے

رگ کے ذریعے جسم میں داخل کی جانے والی دوائیں/انجکشن۔

چ

Chart چارٹ

کسی چیز کے بارے میں معلومات کے لئے خاکہ/فائل۔ طبی اصطلاح کے مطابق جس میں بیماری اور علاج کے بارے میں معلومات کا اندراج کیا جائے۔

س

Liquid سیال/مائع

پانی جیسی حالت میں

خ

Cell خلیہ

کسی جاندار کا سب سے چھوٹا حصہ

Testicles خصیے

مرد کے تناسلی اعضاء کا ایک حصہ جو مرد کا تولیدی مادہ (منی) بناتے ہیں۔

ش

Acute شدید

انتہائی درجے کا، شدت کا (درد یا انفیکشن وغیرہ)۔

د

Chronic دائمی

طویل عرصہ رہنے والی خرابی یا بیماری۔

Temperature درجہ حرارت

کسی فرد کے جسم کی حرارت کی مقدار یا درجہ۔

ص

Health Worker صحت کارکن

وہ فرد جو صحت کی معلومات اور سہولتیں فراہم کرے۔

م

Blood Transfusion منتقلی خون

ایک فرد کا خون دوسرے فرد کے جسم میں منتقل کرنے کا عمل۔

By Mouth منہ کے ذریعے

مراد دوا کھائی جائے۔

Rectal Exam مقعد کا معائنہ

بیماری کی تشخیص کے لئے مقعد کا جائزہ لیا جانا۔

Warts مسّے / مہاسے

ایک خاص قسم کے اُبھار جو نقصان دہ بھی ہوسکتے ہیں خصوصاً تناسلی مہاسے / مسّے۔

Ointment مرہم

دوا کی وہ قسم جو ٹھوس اور مائع کے درمیان ہوتی ہے۔ کسی ٹھوس چیز کا مائع میں گاڑھا آمیزہ۔

Resistance مزاحمت

کسی کے سامنے رکاوٹ بننا، دوا کے اثرات روکنا۔

ہ

Hormones ہارمونز

وہ کیمیائی مادے جو جسم بناتا ہے۔

Hepatites ہپاٹائٹس

جگر کی سنگین بیماری جو وائرس بیکٹریا، الکحل یا کیمیائی زہر کے باعث ہوتی ہے۔ ہپاٹائٹس کی کچھ اقسام جنسی طور پر بھی منتقل ہوتی ہیں۔

ی

Jaundice یرقان

وہ بیماری جس میں جلد اور آنکھوں کا رنگ زرد ہوجاتا ہے یہ ہپاٹائٹس کی علامت بھی ہوتی ہے۔

Gender Role صنفی کردار

کمیونٹی کے مطابق ایک مرد یا عورت کا اس صنف کے لحاظ سے طے شدہ کردار۔

Shock صدمہ

قلت آب، بے تحاشہ جریان خون، زخموں یا سنگین بیماری کے باعث ہونے والی صورت۔

ض

Side Effect ضمنی اثرات

دوا یا ہارمون کے غیر مطلوبہ اثرات جو اس دوا کا ایک حصہ ہوتے ہیں۔

ع

Penis عضو تناسل

مرد کا جنسی عضو۔

Muscle عضلہ

پر گوشت حصہ، پٹھا۔

ف

Paralysis فالج

جسم یا اس کے کچھ حصوں میں حرکت کی صلاحیت ختم ہوجانا۔

ک

Cancer کینسر

ایک خطرناک بیماری جس میں خلیے ابنارمل انداز سے پروان چڑھتے ہیں۔ کینسر جسم کے کسی بھی حصے کو متاثر کر سکتا ہے۔

Scar کٹاؤ / زخم

جسم کی کھال کا پھٹ جانا۔

گ

Home Remedies گھریلو علاج

علاج کے گھریلو یا روایتی طریقے۔

To learn more

Here is a small selection of organizations, printed materials and internet resources that can provide useful information for women with disabilities, organizations for and by people with disabilities and information about the rights of persons with disabilities. We have tried to cover as many of the topics in this book as possible, and to include groups working in all areas of the world. Many of the printed materials are easy to adapt and often include other helpful resource lists.

Organizations

Many of the organizations listed here do most of their work with disability communities around the world. Contact them to see if they have projects or programs in your country. Some organizations provide general health information.

AIFO (Associazione Italiana Amici di Raoul Follereau)
Via Borselli 4-6, 40135 Bologna, Italy
Tel: (39-051)439-3211; Fax: (39-051)434-046
Email: info@aifo.it
Web: www.aifo.it

AIFO, an Italian NGO, provides support to projects in the global South. At present AIFO is involved in over 180 projects in 57 countries, supporting projects on leprosy and primary health care; rehabilitation of disabled persons; and support for vulnerable children.

ABILIS
Aleksanterinkatu 48 A, 00100 Helsinki, Finland
Tel: (358-9)612-40333
Email: abilis@abilis.fi
Web: www.abilis.fi

ABILIS is part of the international Independent Living and Disability Rights Movement. They give grants for projects run by disabled people in developing countries that focus on human rights and women with disabilities.

Action on Disability and Development (ADD)
Vallis House 57 Vallis Road
Frome Somerset, BA11 3EG, UK
Tel: (44-137)347-3064; Fax: (44-137)345-2075
Email: add@add.org.uk
Web: www.add.org.uk

ADD supports active networks of disabled people in several countries. Many of ADD's staff have disabilities themselves.

Arab Resource Collective
Lebanon: PO Box 13-5916, Beirut, Lebanon
Visiting Address: 5th Floor, Dkeik Building Hanra, Beirut, Lebanon
Tel: (9611)742-075; Fax: (9611)742-077
Email: arcleb@mawared.org
Cyprus: PO Box 27380, Nicosia1644, Cyprus
Tel: (357)2277-6741; Fax: (357)2276-6790
Email: arccyp@spidernet.com.cy
Web: www.mawared.org/english/

Written and audio-visual materials on general health care, community development, and skills training for grass-roots organizations in the Middle East.

CBM (Christoffel-Blindenmission/ Christian Blind Mission)
Nibelungenstraße 124
64625 Bensheim, Germany
Tel: (49-6251)131-392; Fax: (49-6251)131-338
Web: www.cbm.org

CBM supports community-based rehabilitation programs run by local partner organizations in developing countries.

Danish Council of Organizations
of Disabled Persons (DSI)
Kløverprisvej 10B
DK-2650 Hvidovre, Denmark
Tel: (45)367-517-77; Fax: (45)367-514-03
Email: dsi@handicap.dk
Web: www.handicap.dk/english

DSI supports organizations in the global South to improve their organizational structures, educate representatives of disability organizations and develop negotiation structures between authorities and disability organizations.

Disability and Development Partners
(formerly Jaipur Limb Campaign)
404 Camden Road, London N7 0SJ, England
Tel/Fax: (44-207)700-7298
Email: info2006@
disabilityanddevelopmentpartners.org
Web: www.disabilityanddevelopmentpartners.org

Disability and Development Partners (DDP)

works with local partner organizations in developing countries to bring social and economic benefits to people with disabilities—especially to those who have lost limbs or the use of limbs through war, accidents, or preventable diseases.

Disability India Network
Disability India Network
c/o Society For Child Development
Cottage 15, Oberoi Apts, 2 Sham Nath Marg
Delhi 110 054, India
Email: vijay@disabilityindia.org
Web: www.disabilityindia.org

Disability India Network has a website with comprehensive information related to disability, and works for the empowerment of persons with disabilities and equal access to healthcare, education and employment.

Disabled Peoples' International (DPI)
902-388 Portage Avenue, Winnipeg
Manitoba, R3C 0C8 Canada
Tel: (1-204)287-8010; Fax: (1-204)783-6270
Email: info@dpi.org
Web: www.dpi.org

DPI promotes the human and economic rights, and social integration of people with disabilities. Information on issues including women's health care, human rights, independent living, and social justice. Special focus on grass-roots development. Has local offices in over 110 countries.

Disabled People of South Africa (DPSA)
Office of the Secretary General
P.O. Box 3467, Cape Town, 8000, South Africa
or
Room 705, 7th Floor, Dumbarton House
1 Church Street, Cape Town, South Africa
Tel: (0-21)422-0357; Fax: (0-21)422-0389
Email: info@dpsa.org.za
Web: www.dpsa.org.za

DPSA is a cross-disability umbrella body of disability organizations in South Africa. Programs include the Disabled Women's Development Programme, which promotes disabled women's participation through leadership and self-help skills development.

Global Fund for Women
1375 Sutter Street, Suite 400
San Francisco, CA 94109, USA
Tel: (1-415)202-7640; Fax: (1-415)202-8604
Email: gfw@globalfundforwomen.org
Web: www.globalfundforwomen.org

Gives small grants to community-based women's groups, especially those working on controversial issues and in difficult conditions. Areas of special interest are human rights, communications technology, and economic independence.

Godisa
P.O. Box 142
Otse, Botswana
Tel: (267)533-7634; Fax: (267)533-7646
Email: mwb@info.bw
Web: www.godisa.org

The only hearing aid maker in Africa, Godisa distributes low-cost, durable, solar rechargeable hearing aids for people in developing countries.

Handicap International
Waterman House101-107, Chertsey Road
Woking, Surrey GU21 5BW, England
Tel: (44-0870)774-3737
Email: hi-uk@hi-uk.org
Web: www.handicap-international.org.uk

*Handicap International works in over 50
countries to help people disabled by natural
disasters, violence, and armed conflict, including
land-mine victims.*

Healthlink Worldwide (formerly AHRTAG)
56-64 Leonard Street, London, EC2A 4JX, UK
Tel: (44-207)549-0240; Fax: (44-207)549-0241
Email: info@healthlink.org.uk
Web: www.healthlink.org.uk

*Healthlink Worldwide works to strengthen the
provision, use and impact of health information,
to increase social inclusion of people working on
issues such as disability and HIV/AIDS. Their
partners publish Disability Dialogue (earlier
called CBR News).Together with the Overseas
Development Group, they publish a series of
disability research papers as part of the
Disability Knowledge and Research Programme.*

Helen Keller International
352 Park Avenue South, 12th Floor
New York, NY 10010, USA
Tel from US (toll free): (1-877) 535-5374
Tel: (1-212) 532-0544; Fax: (1-212) 532-601
Email: info@hki.org
Web: www.hki.org

*Has programs in 25 countries. Material on
blindness from lack of vitamin A. Information on
blindness prevention and visual chart.*

Inclusion International
Inclusion International administrative office
c/o The Rix Centre, University of East London
Docklands Campus, London E16 2RD, England
Tel: (44-208)223-7709; Fax: (44-208)223-7411
Email: info@inclusion-international.org
Web: www.inclusion-international.org

*Inclusion International (II) is a global federation
of family-based organizations advocating for the
human rights of people with intellectual
disabilities. II represents 200 member
federations in 115 countries.*

**International Disability
and Development Consortium (IDDC)**
IDDC Administrator
c/o Handicap International, Waterman House
101-107 Chertsey Road, Woking
Surrey GU21 5BW5, England
Tel: (44-0870)774-3737
Fax: (44-0870)774-3738
Email: administrator@iddc.org.uk
Web: www.iddc.org.uk

*IDDC is an international network on disability
and development. Members in over 100
countries implement programs, and fund,
disseminate information and advocate human
rights of people with disabilities.*

International Federation
of Anti-Leprosy Associations (ILEP)
The Teaching and Learning Materials
Coordinator
234 Blythe Road, London, W14 OHJ, England
Email: ilep@ilep.org.uk
Web: www.ilep.org.uk

*ILEP coordinates support to leprosy programs.
Their website includes a large list of funders as
well as international partners.*

International HIV/AIDS Alliance
Queensberry House, 104-106 Queens Road
Brighton BN1 3XF, England
Tel: (44-0127)371-8900
Fax: (44-0-127)371-8901
Web: www. aidsalliance.org

*The International HIV/AIDS Alliance works to
reduce the spread of HIV and meet the
challenges of AIDS in developing countries. It
channels funds and provides technical and
organizational support to community-based
groups. It also produces and distributes useful
information on HIV/AIDS.*

**Latin American and Caribbean Women's
Health Network**
Simón Bolivar 3798, Ñuñoa, 6850892
Casilla 50610, Santiago 1, Santiago, Chile
Tel: (56-2)223-7077; Fax: (56-2)223-1066
Email: secretaria@reddesalud.org
Web:www.reddesalud.org

*Promotes women's health and women's human
rights through cultural, political and social
transformation.*

Leonard Cheshire International (LCI)

30 Millbank
London, SW1P 4QD, England
Tel: (44-207)802-8224
Fax: (44-207)802-8275
Email: m.ekanger@london.leonard-cheshire.org.uk
Web: www.lcint.org

The work of LCI includes projects in education, employment, economic empowerment, rehabilitation and day care services, short and longer-term residential care, and community programs. Their website lists resources and training materials.

Mobility International USA (MIUSA)

132 E. Broadway, Suite 343
Eugene, Oregon 97401, USA
Tel: (1-541)343-1284; Fax: (1-541)343-6812
Email: info@miusa.org
Web: www.miusa.org

MIUSA works to empower people with disabilities through international exchange and development to achieve human rights. MIUSA's website has a comprehensive listing of international disability organizations.

Mobility International India

1st and 1st A Cross, 2nd Phase, JP Nagar
Bangalore 560 078, India
Tel: (91-80)2649-2222; Fax: (91-80)2649-4444
Email: E-mail@mobility-india.org
Web: www.mobility-india.org

Promotes community-based rehabilitation and mobility for persons with disabilities, especially in rural areas. Trains women with disabilities to make artificial limbs and provide rehabilitation services in their communities.

Motivation

Brockley Academy, Brockley Lane, Backwell
Bristol, BS48 4AQ, England
Tel: (44-0127)546-4012
Fax: (44-0127)546-4019
Email: info@motivation.org.uk
Web: www.motivation.org.uk

Motivation designs and provides low-cost mobility products (wheelchairs, tricycles and artificial limbs). It also trains people to meet the needs of their local disabled communities.

Musasa Project

Physical address: 64 Selous Ave
Cnr 7th Street, Harare, Zimbabwe
Postal address: PO Box A712, Avondale
Harare, Zimbabwe
Tel: (263-04)725-881, (263-04)734-381
Fax: (263-04)794-983
Email: musasaproj@africaonline.co.zw

Provides information and support to abused women and education programs on domestic violence and rape.

National Union of Women with Disabilities of Uganda (NUWODU)

Plot No. 62 Ntinda Road, Ntinda
P.O. Box 24891, Kampala, Uganda
Tel: (256)41-285240; Mobile: (256)77-475186
Fax: (256)41-540178
Email: nuwodu@infocom.co.ug
Web: www.disability.dk/site/countryindex.
php?section_id=28

An umbrella promoting equal rights and opportunities for women and girls with disabilities. Provides information on grassroots groups of women with disabilities and assists development of newly formed groups.

Teaching Aids at Low Cost (TALC)

P.O. Box 49, St. Albans,
Herts AL 1 5TX, England
Tel: (44-0172)785-3869
Fax: (44-0172)784-6852
Email: info@talcuk.org
Web: www.talcuk.org

Low-cost books, slides, and accessories in English, French, Spanish, and Portuguese on health care and development for use in poor communities. Free booklist and CD libraries.

The Southern Africa Federation of the Disabled (SAFOD)

PO Box 2247, 19 Lobengula Street
Bulawayo, Zimbabwe
Tel: (263-9)69356; Fax: (263-9)74398
Email: safod@netconnect.co.zw
Web: www.safod.org

Human rights organization in Zimbabwe by disabled people for disabled people.

Swedish Organizations of Disabled Persons International Aid Association (SHIA)
Liljeholmstorg 7A, 11763 Stockholm, Sweden
Tel: (46-8)462-3360; Fax: (46-8)714-5922
Email: shia@shia.se
Web: www.shia.se

SHIA aims to strengthen the efforts of persons with disabilities to achieve equality and participation through development co-operation and partnerships between persons with disabilities in Sweden and elsewhere.

Voluntary Health Association of India (VHAI)
B-40, Qutab Institutional Area, South of IIT
New Delhi 110016, India
Tel: (91-112)651-8071-72
Email: vhai@vsnl.com
Web: www.vhai.org

Health education materials in English and local Indian languages. Also publishes Health for the Millions, a journal on low-cost health care.

Whirlwind Wheelchair International (WWI)
San Francisco State University
1600 Holloway Avenue - SCI. 251
San Francisco, California 94132, USA
Tel: (1-415)338-6277; Fax: (1-415)338-1290
Email: info@whirlwindwheelchair.org
Web: www.whirlwindwheelchair.org

WWI developed the Whirlwind wheelchair, a lightweight, low-cost, sturdy wheelchair designed for rough urban and rural conditions in developing countries. WWI works with an international network of wheelchair workshops to continually update its designs.

World Blind Union
Enrique Pérez, Secretary General
C/Almansa 66, 28039 Madrid, Spain
Tel: (34)914-365-366; Fax: (34)915-894-749
Email: umc@once.es
Web: www.worldblindunion.org

The World Blind Union (WBU) unites people with visual disabilities in member organizations around the world. They work to promote equal opportunities for the blind; to raise the status of blind women, DeafBlind persons, blind people with multiple disabilities; and to prevent blindness worldwide.

World Federation of the Deaf (WFD)
Postal Address: P.O. Box 65, FIN-00401
Helsinki, Finland
Physical address: Light House, Ilkantie 4 Haaga, Helsinki, Finland
Fax: (358-9)580-3572
Email: info@wfdeaf.org
Web: www.wfdeaf.org

WFD is an organization of national associations of deaf people. They work to improve the status of national sign languages and education for deaf people, improve access to information and services, improving human rights for deaf people, and promote the establishment of deaf organizations.

World Health Organisation (WHO)
Rehabilitation Section
20 Avenue Appia
1211 Geneva 27 Switzerland
Tel: (41-22)791-2111; Fax: (41-22)791-3111
Email: info@who.int
Web: www.who.int

WHO provides advice and financial support to health-related programs.

Books and other printed materials

A Woman's Guide to Coping with Disability
Esther Boylan (ed.)
Resources for Rehabilitation
33 Bedford Street, Suite 19A
Lexington, MA 02420, USA

Written by women with disabilities; personal accounts and experiences cover the stigma of disability, its challenges, and prevention. Includes general information and further resources on specific disabilities: arthritis, multiple sclerosis, osteoporosis, and spinal cord injury.

AgeWays and Aging and Development
HealthAge International
67-74 Saffron Hill, London EC1R OBE, England
Tel: (44-207)278-777
Fax: (44-207)713-7993, (44-171)447-203
Email: helpage@gn.apc.org
Web: www.helpage.org

These free journals are printed 2 times a year. HelpAge has a network of over 50 organizations worldwide, working to achieve a lasting improvement in the quality of life for older persons.

Breast Awareness:

Taking Care of Our Breasts (booklet)
Women's Health Information & Support Centre
120 Bold Street, Liverpool, L1 4JA. England
Tel: (44-0151)707-1826;Fax: (44-0151)709-2566
Email: women@whisc.org.uk
Web: www.whisc.org.uk

Easy-to-read, heavily-illustrated booklet on how to give a breast exam. Written by and for women with learning troubles, but useful for all.

Women's Health Information & Support Centre has also published:

Having a Mammogram: X-ray of the Breasts, Going to the Breast Assessment Clinic at the Breast Unit, and Having a Smear Test.

Building an Inclusive Development Community: A Manual on Including People with Disabilities in International Development Programs
Karen Heinicke-Motsch and Susan Sygall (eds)
Mobility International USA
P.O. Box 10767, Eugene, Oregon 97440, USA
Web: www.miusa.org

Strategies for including persons with disabilities in a wide range of development programs, including health care.

Also available from Mobility International USA:

Loud, Proud and Passionate:
Including Women with Disabilities in
International Development Programs
C. Lewis and S. Sygall (eds.)

A great resource for starting a group to empower women with disabilities and train them in leadership skills. Includes profiles of individual women as well as groups, their strategies, successes and failures.

Couples with Intellectual Disabilities
Talk about Living and Loving
Karin Schwier Melberg, Woodbine House
5615 Fishers Lane, Rockville, MD 20852, USA
Tel: (1-301)897-3570
Tel in the US (toll-free): (800)843-7323
Fax: (1-301)897-5838
Web: www.woodbinehouse.com

Fifteen interviews with couples with disabilities reveal their personal stories on finding love, companionship, and happiness.

Disability Dialogue

56-64 Leonard Street, London, EC2A 4JX, England

Tel: (44-207)549-0240; Fax: (44-207)549-0241

Email: info@healthlink.org.uk

Web: www.healthlink.org.uk

Disability Dialogue (formerly CBR News) newsletter exchanges information between disabled people and development, health and rehabilitation workers, and aims to promote disability equality, and good policy and practice. Practical information about community approaches and appropriate equipment, published in various language editions.

Bangla:

Social Assistance for the Physically Vulnerable

(SARPV)

3/8, Block-F, Lalmatia, Dhaka 1207, Bangladesh

Email: shaque@bd.drik.net

Web: www.sarpv.org

English for Africa and Portuguese (planned):

Southern Africa Federation of

the Disabled (SAFOD)

PO Box 2247, 19 Lobengula Street

Bulawayo, Zimbabwe

Tel: (263-9)69356; Fax: (263-9)74398

Email: safod@netconnect.co.zw

Web: www.safod.org

English for India and Hindi (Akshamata Samvad):

Amar Jyoti Rehabilitation and Research Centre

Kakardooma, Vikas Marg, Delhi 110 092, India

Email: amarjoti@del2.vsnl.net.in

French:

Mauritium, 5 Avenue Buswell

Quatre Bonnes, Mauritius

Gujarati:

Centre for Health Education Training and Nutrition Awareness (CHETNA)

Lilavatiben Lalbhai's Bungalow

Civil Camp Road, Shahibaugh, Ahmedabad

Gujarat 380 004, India

Tel: (91-079)286-696; Fax: (91-079)286-6513

Email: chetna@icenet.net

Web: www.chetnaindia.org

Nepali (Apangata Kurakani):

Partners for Rehabilitation

International Nepal Fellowship

PO Box 28, Pokhara, Nepal

Email: kurakani@bigfoot.com

Tamil (Oonam Seithi):

Rural Unit for Health and Social Affairs

(RUHSA)

Christian Medical College and Hospital RUHSA Campus Post Office, Vellore District Tamil Nadu 632 209, India

Email: abel-rajaratnam @hotmail.com

*Audio-cassette (International edition) Action on Disability and Development (ADD), UKdistributed by **Healthlink Worldwide** at the address above*

English braille

and *Indian language audio-cassette:*

Blind People's Association of India

Dr Vikram Sarabhai Rd Vastrapur, Ahmedabad 380 015, India

Tel: (91-079)2630-5082; Fax: (91-079)2630-0106

Email: blinab@sancharnet.in

Disability, Inclusion and Development:

Key Information Resources (directory)

Source, c/o Healthlink Worldwide

50-64 Leonard Street, London EC2A 4JX, England

Tel: (44-207)549-0240; Fax: (44-207)549-0241

Web: www.asksource.info/res_library/disability.htm

Directory of over 300 information resources with abstracts and details of distributors and linked to the full resource online. For people with limited access to the Internet, the directory is available in print with a CD-ROM, which holds the fully searchable and browseable text of many resources.

Our Bodies Ourselves

Boston Women's Health Book Collective, OBOS

34 Plympton Street, Boston, MA 02118

Tel: (1-617)451-3666; Fax: (1-617)451-3664

Email: office@bwhbc.org

Web: www.ourbodiesourselves.org

Includes a chapter on Sex and Disabilities, including the effects of the disability on sexuality, and helpfulhints and special implications.

Multiplying Choices:

Improving Access to Reproductive Health

Services for Women with Disabilities

Barbara Waxman Fiduccia
Berkeley Planning Associates
Oakland, CA 94610, USA
Tel: (1-510)465-7884; Fax: (1-510)465-7885
Email: info@bpacal.com
Web: www.berkeleypolicyassociates.com

Excellent book compiled by disabled women describes their concerns regarding health services, offers suggestions to meet the reproductive health needs of women with disabilities. Has information on the Americans with Disabilities Act, and offers lists of books and articles about sexual and reproductive health for women with disabilities.

Reproductive Issues

for Persons with Physical Disabilities

F. Haseltine, S. Cole, and D. Gray (eds.)
Paul H. Brooks Publishing Co.
P.O. Box 10624
Baltimore, Maryland, 21285-0624, USA
Tel: (1-410)337-9580; Fax: (1-410)337-8539
Email: custserv@brookespublishing.com
Web: www.brookespublishing.com

Written for professionals, caregivers, researchers and individuals with disabilities, it provides research findings and personal stories exposing myths surrounding disability and reproductive issues.

South African Women's Health Book

Order through: Heinemann Publishers
Grayston Office Park, 128 Peter Road
PO Box 781940 Sandown
Johannesburg, South Africa 2146
Tel: (27-11)322-8660
Fax: (27-11)322-8715 or -8716
Email: cust_services@heinemann.co.za

Comprehensive information on women's health in South Africa including stories from women in local communities. Includes chapters on gender, culture, healthy living, violence, work, sexuality, and reproductive health.

The Disabled Woman's Guide

to Pregnancy and Birth

Judith Rogers, Demos Medical Publishing
386 Park Avenue South, Suite 301
New York, NY 10016, USA
Tel: (1-212)683-0072
Tel from US (toll free): (1-800)532-8663
Fax: (1-212) 683-0118
Email: orderdept@demosmedpub.com
Web: www.demosmedpub.com/default.
aspx?page_id=0

*Pregnancy and childbirth information for women with disabilities. Revised and re-titled version of **Mother To Be: A Guide to Pregnancy and Birth for Women with Disabilities**. Based on the experiences of 90 women with disabilities who chose to have children.*

The Kenyan's Deaf Peer Education Manual

Sahaya International, USA
c/o Koen Van Rompay
2949 Portage Bay Avenue, Apartment #195
Davis, CA 95616, USA
Tel:(1-530)756-9074
Email: kkvanrompay@ucdavis.edu
Web: www.sahaya.org

A training manual with activities for basic understanding of sexual health, HIV and AIDS.

Welner's Guide to the Care

of Women with Disabilities

Sandra L. Welner and F. Haseltine
Lippincott Williams & Wilkins
530 Walnut Street, Philadelphia, PA 19106, USA

Guide for caregivers focusing on management of pain, osteoporosis, anesthesia, incontinence, infertility, depression and psychotropics, substance abuse, and hormonal management. Also addresses more fundamental issues such as nutrition, sexuality, exercise, physical examination, etc.

Internet-based resources

The World-Wide Web is an incredible resource for those who have access to it. Besides the very useful websites listed for the organizations and print materials on the previous pages, here we list just a few of the websites we use most frequently when we look for disability and health information.

Contact
www. wcc-coe.org/wcc/news/contact.html

*Contact, published in French, Spanish, English and Portuguese by **World Council** of Churches, focuses on issues of health and healing.*

Disability Knowledge and Research
www.disabilitykar.net/learningpublication/
references.html

*The Disability Knowledge and Research program, an effort of **Healthlink Worldwide** and the **Overseas Development Group** of East Anglia University, aims for better health and quality of life for people in developing countries. This website presents their papers, research, etc., including:*

Mainstreaming disability in development:
lessons from gender mainstreaming
Carol Miller and Bill Albert (March 2005)

Explains mainstreaming, as a strategy of feminist advocacy in the context of development, and draws comparisons with the history of the disability movement. Includes recommendations for mainstreaming disability, and compares these with 'good practices' in gender mainstreaming.

Mainstreaming disability in development:
India country report
Philippa Thomas (June 2005)
Examines disability in relation to poverty and social exclusion. The report also highlights ways to integrate disability into development.

Disability, poverty and
the 'new' development agenda

David Seddon and Rebecca Yeo (July 2005) Looks at changes that have taken place in recent years around poverty, disability and the relationship between the two.

Disability World
www.disabilityworld.org

This is a great site that is an internationally focused news web magazine covering disability issues. The site is useful in sorting out relevant information and is in both English and Spanish. Internet users with slower connections can load the site as text only.

Source International Information
Support Centre
www.asksource.info

Source, is an international information support centre designed to strengthen the management, use and impact of information on health and disability. Source now has over 25,000 information resources on a range of subjects including HIV and AIDS, disability and inclusion, mother and child health, information and communication technology and participatory communication. Resources are available online, by CD-ROM and as printed materials.

Table Manners and Beyond:
The Gynecological Exam for Women
with Developmental Disabilities
and Other Functional Limitations.
www.bhawd.org/sitefiles/TblMrs/cover.html

An excellent guide, edited by Katherine M Simpson and Kathleen Lankasky, for both women with disabilities and health care workers of how to do pelvic and breast exams for women with a wide range of disabilities.

World Enable
www.worldenable.net/women/

A project of several disability organizations with a great list of resources and a page specifically on women with disabilities.

معذور افراد کی تعلیم و تربیت اور بحالی کے پاکستانی ادارے

• گورنمنٹ اسپیشل سینٹر
رانجھن روڈ نزد کہوٹ، راولپنڈی
فون نمبر 051-3313468

⊙

• گورنمنٹ ڈیف اینڈ ڈی میکیٹو ہیئرنگ
پرائمری اسکول سوان کیمپ
جی ٹی روڈ، راولپنڈی
فون نمبر 051-4840588

⊙

• گورنمنٹ انسٹیٹیوٹ فار بلائنڈ (مڈل گرلز)
شمس آباد، راولپنڈی
فون نمبر 051-4840588

⊙

• گورنمنٹ قندیل انسٹیٹیوٹ فار بلائنڈ (پی آئی)
کوہاٹی بازار، راولپنڈی
فون نمبر 051-550465

⊙

• گورنمنٹ اسپیشل ایجوکیشن سینٹر
ہاؤس 5N، آل رضا مواہرا شاہ ولی شاہ
نزد دھبیان اسٹاپ ٹیکسلا، راولپنڈی
فون نمبر 051-4543153

⊙

• گورنمنٹ اسپیشل ایجوکیشن سینٹر
تھانہ روڈ کلا رسیداں، راولپنڈی
فون نمبر 0345-5297554

⊙

• گورنمنٹ اسپیشل ایجوکیشن سینٹر فار مینٹلی
ریٹارڈ چلڈرن راولپنڈی، ہاؤس نمبر 61
اسٹریٹ نمبر 15 گلزاری ہاؤسنگ کالونی
0333-5154433 راولپنڈی

⊙

• امید اسپیشل اکیڈمی
B-2/9 ایریا نمبر 2 بابر مارکیٹ لانڈھی
بالمقابل سندھ گورنمنٹ ڈسپنسری لانڈھی
فون نمبر 5022693

⊙

• انٹرنیشنل اسکول آف اسٹڈیز
پلاٹ SNPA-20A بلاک 7/8
کے ایم سی ایچ ایس، نزد ڈبل پارک،
فون نمبر 021-4388965

⊙

• آٹزم انسٹی ٹیوٹ
C-91 بلاک 2، کے اے ایچ ایس،
کراچی، فون نمبر 021-4544134

⊙

• مہران اسپیشل ایجوکیشن سینٹر
ہاؤس نمبر B-11، بلاک 19،
سیٹلائٹ ٹاؤن، میر پور خاص

⊙

• گورنمنٹ آف پاکستان
شاہ لطیف اسپیشل ایجوکیشن سینٹر فار مینٹلی
ریٹارڈ چلڈرن
ہاؤس نمبر A-38، جی او آر کالونی، حیدرآباد
فون نمبر 022-9200915

⊙

• سچل سرمت اسپیشل ایجوکیشن
قاضی احمد روڈ بالمقابل سکندر دھیراج فروٹ
فارم، نوابشاہ فون نمبر 0241-9370149

⊙

• اسکول فار مینٹلی ریٹارڈ چلڈرن
جی او آلہ کالونی، حیدرآباد
فون نمبر 022-9201301

⊙

• ACELP
197/8 رفیقی شہید روڈ
بالمقابل سندھ میڈیکل کالج کراچی
فون نمبر 021-5662458

⊙

• ال امیدری ہیبلیٹیشن ایسوسی ایشن (Aura)
پلاٹ St-2: بلاک 3
کے ڈی اے اسکیم 36
گلستان جوہر کراچی۔
فون نمبر 021-4617496
فیکس 021-4014574

⊙

• دارالخضود
247۔ اسٹاف لین، فاطمہ جناح روڈ کراچی
فون نمبر 021-5660570

⊙

• مشعل اسپیشل ایجوکیشن سینٹر
عقب رشین مارکیٹ اسٹیل ٹاؤن
بن قاسم کراچی
فون نمبر 021-4592227

⊙

• بحریہ کالج کارساز
پی این ایس اسپیشل چلڈرن اسکول
D-37/2 حبیب رحمت اللہ روڈ پی این ایس
کارساز روڈ کراچی، فون نمبر 56633175

⊙

• اُجالا
61۔ خیابان ہلال فینر فائیوڈی ایچ اے کراچی
فون نمبر 5846075-5876075

⊙

گورنمنٹ اسپیشل ایجوکیشن سینٹر فارسلولرز چلڈرن، راولپنڈی
فون نمبر 051-4840588

⊙

گورنمنٹ ریڈی میڈ گارمنٹس سینٹر
راولپنڈی سیٹلائٹ ٹاؤن، راولپنڈی
فون نمبر 051-9290573

⊙

آرمی اسپیشل ایجوکیشن
93۔مرگلہ روڈ نزد جی ایچ کیو آڈیٹوریم
راولپنڈی کینٹ
فون نمبر 56132763

⊙

برائٹ ہورایزونز اسکول فار اسپیشل چلڈرن
اے پی ایس اینڈ سی برال کالونی، منگلا کینٹ
فون نمبر 544639355

⊙

سی ایف ایس پی سینٹر فار دا میفغی ہینڈی کیپ
چلڈرن، اسلام آباد
فون نمبر 2103163

⊙

چنیلی انسٹیٹیوٹ فار دا میفغی ریٹارڈ چلڈرن
B-20 سیٹلائٹ ٹاؤن، راولپنڈی
فون نمبر 4842757

⊙

گورنمنٹ انسٹیٹیوٹ فار میفغی ریٹارڈ اینڈ
فزیکلی ہینڈی کیپ چلڈرن
S-30 غازی گوٹھ ٹاؤن شپ، مانسہرہ
فون نمبر 987303060

⊙

گورنمنٹ آف پاکستان اسپیشل ایجوکیشن سینٹر
فار وی ایچ سی چارسدہ
فون نمبر 921510978

⊙

گورنمنٹ اسکول فار ڈیف چلڈرن
تیمر گرہ ودیر (لوئر)
فون نمبر 945823757

⊙

گورنمنٹ اسپیشل ایجوکیشن سینٹر
نزد بوائز ڈگری کالج
پنڈی روڈ، تلہ گنگ، ڈسٹرکٹ ساہیوال

⊙

گورنمنٹ اسپیشل ایجوکیشن سینٹر
نزد ریلوے اسٹیشن بلال مسجد روڈ، فتح جنگ

⊙

گورنمنٹ اسپیشل ایجوکیشن سینٹر
نزد دال بنی فلنگ اسٹیشن چنگی نمبر 1
پنڈی گھیب

⊙

حسن اکیڈمی اسپیشل ایجوکیشن
علی آباد، مہر آباد، پشاور روڈ
راولپنڈی کینٹ
فون نمبر 5460644

⊙

انسٹیٹیوٹ فار مینٹلی ریٹارڈ اینڈ
فزیکلی ہینڈی کیپ چلڈرن، خیلہ بٹ
ٹاؤن شپ، سیکٹر 2 # مرینا چوک ہری پور
فون نمبر 995619486

⊙

انسٹیٹیوٹ فار دی ہیبلیٹیشن آف
ڈس ایبلڈ پرسنز
جنرل اسپتال نوشہرہ کینٹ
فون نمبر 9220025

⊙

کنگسٹن اسکول فار ہینڈی کیپ چلڈرن
کہیل نزد اقبال خان کا گھر ایبٹ آباد
فون نمبر 992340699

⊙

مشعل انسٹیٹیوٹ فار اسپیشل ایجوکیشن
سوبراسٹی تربیلا، تحصیل غازی ہری پور
فون نمبر 995660464

⊙

نفیس اسکول فار اسپیشل ایجوکیشن
دی مال، واہ کینٹ

⊙

پی اے سی اسپیشل ایجوکیشن اسکول
پی اے سی کامرہ ڈسٹرکٹ اٹک، کامرہ اٹک
فون نمبر 22102445

⊙

پی اے ایف اسکول فار اسپیشل ایجوکیشن
معرفت پی اے ایف کالج آف ایجوکیشن
فار ویمنز، چکلالہ، راولپنڈی
فون نمبر 50503383

⊙

پی ٹی اے اسپیشل ایجوکیشن سینٹر فار مینٹلی ہینڈی
کپڈ چلڈرن، شیخ ناصر ہاؤس، توصیف آباد
ڈیرہ بان روڈ، ڈیرہ اسماعیل خان
فون نمبر 9280150

⊙

پی ٹی اے اسپیشل ایجوکیشن کمپلیکس فار مینٹلی
ہینڈی کپڈ چلڈرن
شیخ مالٹون ٹاؤن سیکٹر ایم فیئر II مردان
فون نمبر 931781011

⊙

اسپیشل ویلفیئر آرگنائزیشن پروجیکٹ سینٹر
فار ہینڈی کپڈ
محلہ پادان ویلج تحصیل تخت بھائی مردان
فون نمبر 536088

⊙

سینٹ تھامس کمیونٹی ہیلتھ نیٹ ورک
حکیم لقمان روڈ سیکٹر G-7/2،اسلام آباد
فون نمبر 2890476
⊙

والینٹری سوشل ویلفیئر ایجنسی فار
فزیکلی ڈس ایبلڈ،ڈیرہ اسماعیل خان
فون نمبر 961715900
⊙

علامہ اقبال اسپیشل ایجوکیشن سینٹر
اسپیشل ایجوکیشن کمپلیکس
45-B-II،جوہر ٹاؤن،لاہور
فون نمبر 5868140-5116585
⊙

چمن سینٹر فار مینٹلی ریٹارڈ چلڈرن
بلاک 3،سیکٹر D-1،نزد عمر چوک ٹاؤن شپ
لاہور،فون نمبر 843578-5121531
⊙

رائزنگ سن اسکول ڈیفنس
A-11،ڈیفنس ہاؤسنگ اتھارٹی لاہور،کینٹ
فون نمبر 5734133
⊙

چائلڈ ویلفیئر سینٹر فار اسپیشل نیڈز
پنجاب یونیورسٹی ریذیڈنشل ایریا
نزد گیٹ نمبر 1،لاہور
فون نمبر 92031158
⊙

دورتھیا سینٹر فار اسپیشل چلڈرن
G-7،مین جی ٹی روڈ
نزد فرخ آباد،گھاٹی شادمان لاہور
فون نمبر 0303-6406608
⊙

انسٹیٹیوٹ آف چائلڈ ہیلتھ
دی چلڈرن اسپتال
فیروز پور روڈ،لاہور
فون نمبر 9230901-9-ext 2210
⊙

ال شاہی،اسپیشل ایجوکیشن سینٹر
اسٹریٹ نمبر 2،بلاک V
طارق بن زیاد کالونی،ساہیوال
فون نمبر 0441-65201
⊙

کہکشاں اسپیشل ایجوکیشن سینٹر
ضلع کونسل اسٹاف کالونی
کچہری روڈ،جھنگ صدر
فون نمبر 0471-626491-622668
⊙

آرمی اسپیشل ایجوکیشن اسکول
در بارمحل،نزد ون یونٹ چوک،بھاولپور
فون نمبر 0621-8914126
⊙

شاہین اسپیشل ایجوکیشن سینٹر
پی اے ایف ڈبلیو اے۔
ایئر فورس بیس،سرگودھا
فون نمبر 0451-720050-710100
⊙

شہاب ٹریننگ اسکول فار مینٹلی ریٹارڈ چلڈرن
بلاک x،منڈی چوک،نیو ملتان
فون نمبر 061-9220173
⊙

بہبود دین اسپیشل ایجوکیشن سینٹر
1640،گلیکسی ٹاؤن
نیل کوٹ،بوسن روڈ،ملتان
فون نمبر 061-851356
⊙

خواجہ فرید اسپیشل ایجوکیشن سینٹر
46-A،سیٹلائٹ ٹاؤن،بھاولپور
فون نمبر 0621-80830-80885
⊙

اسپیشل ایجوکیشن سینٹر فار ایم آر سی
گل روڈ،گوجرانوالہ
فون نمبر 0431-254168
⊙

شاداب ٹریننگ اسکول فار مینٹلی ریٹارڈ چلڈرن
بلاک Y،سیٹلائٹ ٹاؤن
ڈی ای او آفس،سرگودھا
فون نمبر 0451-215772
⊙

نیو ہورائزن اسپیشل ایجوکیشن سینٹر
فار مینٹلی ریٹارڈ چلڈرن،سیالکوٹ
فون نمبر 0432-5552011
⊙

دی لائٹ ری ہیبلیٹیشن سینٹر
فار مینٹلی ہینڈی کیپ چلڈرن
170،مریدیا روڈ،ماڈل ٹاؤن سیالکوٹ
فون نمبر 0432-550843-589210
⊙

اے کے انسٹیٹیوٹ آف اسپیشل ایجوکیشن
ڈی آئی ہاؤسنگ اسکیم،کوٹلی آزاد کشمیر
فون نمبر 005866044912
⊙

تعمیر ویلفیئر آرگنائزیشن
پی-138،اسٹریٹ-2
مین کالونی پی سی 2،فیصل آباد
فون نمبر 38090
⊙

پاکستان نیشنل فورم آن ویمنز ہیلتھ کی شائع کردہ کتابیں

قیمت

آسان مڈوائفری	اردو/سندھی	542 صفحات	450/- روپے
درسی کتاب برائے مڈوائف	اردو/سندھی	670 صفحات	550/- روپے
جہاں عورتوں کے لئے ڈاکٹر نہ ہو	اردو/سندھی	500 صفحات	400/- روپے
حمل اور زچگی کی پیچیدگیوں کا علاج		420 صفحات	300/- روپے
مڈوائفری باتصویر		120 صفحات	175/- روپے
زچگی کی ہنگامی دیکھ بھال		72 صفحات	80/- روپے
انفیکشن سے بچاؤ		90 صفحات	100/- روپے
نرسنگ اور مڈوائفری تدریس کے طریقے		120 صفحات	200/- روپے
نرسنگ کی تدریس کے عملی طریقے		152 صفحات	200/- روپے
رہنماء کتاب برائے میڈیکل ٹیکنیشنز		156 صفحات	300/- روپے

خریداری کے لئے رجوع کریں صبح 11 سے 7 بجے

فلیٹ نمبر 12 ۔پہلی منزل،عبدالرسول بلڈنگ،ڈاکٹر بلموریا اسٹریٹ، آئی آئی چندریگر روڈ کراچی
موبائل: 0345-2500409